EUROPEAN MINERALOGICAL UNION
NOTES IN MINERALOGY

Commissioning Editor: R. Oberti

Volume 18

MINERAL FIBRES: CRYSTAL CHEMISTRY, CHEMICAL-PHYSICAL PROPERTIES, BIOLOGICAL INTERACTION AND TOXICITY

UNIVERSITY TEXTBOOK

Edited by

A. F. GUALTIERI

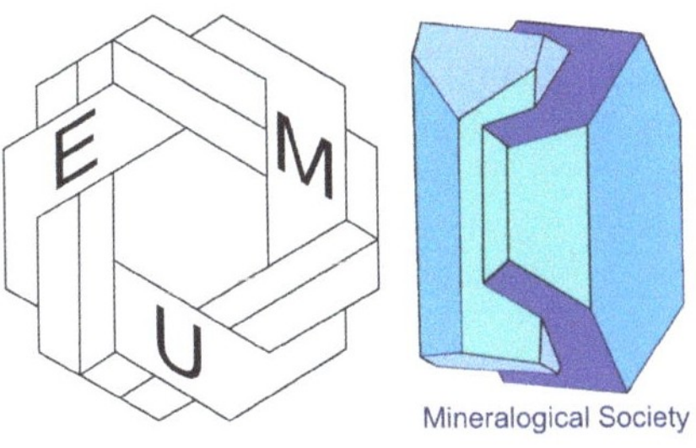

Published by the European Mineralogical Union and the
Mineralogical Society of Great Britain & Ireland, London, 2017

The publication of this textbook is supported by the European Mineralogical Union

EMU Notes in Mineralogy
A series published under the auspices of the European Mineralogical Union (EMU).

Initiator of the EMU Schools and the EMU Notes in Mineralogy
Giovanni Ferraris, Torino, President of the EMU 1992–1996

Commissioning Editor: R. Oberti, Pavia (previous editors: Giovanni Ferraris, Torino, Tamás G. Weiszburg and Gábor Papp, Budapest)

Editor of this Volume
Alessandro F. Gualtieri, Università di Modena e Reggio Emilia, Modena, Italy

Managing Editor and Indexer: Kevin Murphy, London

Front cover design: Michel H. Guay

On the front cover: Original SEM images of mineral fibres in different environments. A crocidolite fibre bundle inside the lung tissues of Sprague-Dawley rats, coated with ferruginous asbestos bodies (left); a curious association of chrysotile fibres and a plant trichome inside a cement-asbestos slate (centre); a bundle of chrysotile fibres inside MeT5A (SV40-immortalized pleural mesothelial cells) culture (right). The images were kindly assembled by Nicola Bursi Gandolfi.

ISSN: 1417 2917
ISBN: 978-0903056-65-6

Published by the European Mineralogical Union and the Mineralogical Society of Great Britain & Ireland (12, Baylis Mews, Amyand Park Road, Twickenham TW1 3HQ, UK)
Printed by Lightning Source, USA

The EMU Notes in Mineralogy Series

Published volumes

Volume	*Year*	*Editors*	*Title*
1	1997	S. Merlino	*Modular aspects of minerals*
2	2000	D.J. Vaughan R. Wogelius	*Environmental mineralogy*
3	2001	C.A. Geiger	*Solid solutions in silicate and oxide systems*
4	2002	C. Gramaccioli	*Energy modelling in minerals*
5	2003	D.A. Carswell R. Compagnoni	*Ultrahigh pressure metamorphism*
6	2004	A. Beran E. Libowitzky	*Spectroscopic methods in mineralogy*
7	2005	R. Miletich	*Mineral behaviour at extreme conditions*
8	2010	F.E. Brenker G. Jordan	*Nanoscopic approaches in earth and planetary sciences*
9	2010	G. Christidis	*Advances in the characterization of industrial minerals*
10	2010	M. Prieto	*Ion-partitioning in ambient-temperature aqueous systems*
11	2011	M.F. Brigatti A. Mottana	*Layered mineral structures and their application in advanced technologies*
12	2012	J. Dubessy M.-C. Caumon F. Rull	*Applications of Raman Spectroscopy to earth sciences and cultural heritage*
13	2013	D.J. Vaughan R.A. Wogelius	*Environmental mineralogy II*
14	2013	F. Nieto K.J.T. Livi	*Minerals at the nanoscale*
15	2015	M.R. Lee H. Leroux	*Planetary mineralogy*
16	2017	W. Heinrich R. Abart	*Mineral reaction kinetics: microstructures, textures, chemical and isotopic signatures*
17	2017	I.A.M. Ahmed K.A. Hudson-Edwards	*Redox reactive minerals: properties, reactions and applications in clean technologies*
18	2017	A.F. Gualtieri	*Mineral fibres: crystal chemistry, chemical-physical properties, biological interaction and toxicity*

Copies of the EMU Notes (volumes 1–7) are distributed in Europe by the larger member societies of the European Mineralogical Union:

Società Italiana di Mineralogia e Petrologia:
www.socminpet.it

Mineralogical Society of Great Britain & Ireland:
www.minersoc.org

Société Française de Minéralogie et de Cristallographie:
www.sfmc-fr.org

and by the Secretary of the EMU, Prof. Juraj Majzlan:
juraj.majzlan@uni-jena.de

in America by the Mineralogical Society of America:
www.minsocam.org

Institutional orders as well as individual requests from outside Europe and America should be sent to the Secretary of EMU, Prof. Juraj Majzlan:
juraj.majzlan@uni-jena.de

For volumes 8 onwards, the Mineralogical Society of Great Britain acts as co-publisher and copies may be ordered from www.minersoc.org

or from:

Mineralogical Society
12 Baylis Mews,
Amyand Park Road,
Twickenham TW1 3HQ
UK

E-mail: admin@minersoc.org
Tel. + 44 (0)20 8891 6600
Fax: + 44 (0)20 8891 6599

Contents

Preface

Asbestos is probably one of the most studied substances ever. Asbestos is synonymous with argument and controversy: it is magic but feared, essential but dreaded, a strategic natural raw material but a source of concern and hazard; it is banned but still used safely, and so the list goes on. Asbestos-related diseases are certainly of significant concern in terms of occupational and public health. Asbestos World Health Organisation officials estimate that 125,000,000 people worldwide are exposed annually to asbestos in occupational settings, and >100,000 people die annually of diseases associated with asbestos exposure. Use of asbestos has been banned in most developed countries, but chrysotile asbestos is still used in many developing countries.

This book presents the state of the art in the vast multidisciplinary research field of asbestos and of mineral fibres in general. The protagonists of the book are the mineral fibres with their immense complexity and poorly understood biochemical interactions. The approach of the chemist/mineralogist/crystallographer puts the fibre in focus whereas the approach of the biochemist/toxicologist/doctor assumes the perspective of the organism interacting with the fibre. The perspectives of both the 'invader' and the 'invaded' must be considered together to establish a conclusive model to explain the toxicity of mineral fibres. In fact, this sharing of different perspectives and working in a multidisciplinary way is the key to understanding the mechanism of asbestos-induced carcinogenesis.

With this in mind, the state of the art in the field of mineral fibres is illustrated and discussed here, with a multidisciplinary approach taking into account all the different scientific strands (biology, chemistry, epidemiology, mineralogy, physics, toxicology *etc.*). The different views have been considered in an attempt to assemble the pieces of the jigsaw and to present the reader with an up-to-date and complete picture.

Some of the results presented here arise from a long-term Italian Research Project of National Interest (PRIN), in operation since 2011. The project is aimed at the characterization of mineral fibres and at understanding the nature of their biological interaction mechanisms, with special attention to chrysotile, amphibole asbestos and fibrous erionite. A general model to classify mineral fibres on the basis of their toxicity would be of paramount importance to predict *a priori* their potential toxicity, to identify their occurrences, and eventually to plan a risk-assessment analysis for the population. Because we live in a fibre-rich world, it is very likely that mineral fibres other than asbestos, erionite and fluoro-edenite will soon be recognized as potential carcinogens, and mineral fibres of little or no commercial/economic value may become sources of concern in the future. Having a predictive universal model would be a powerful tool in risk assessment and prevention, especially for reducing exposures to naturally occurring asbestos (NOA), now a source of considerable concern worldwide.

The first part of the volume deals with the characterization of mineral fibres: the crystal structure of mineral fibres is described in Chapter 2. Chapter 3 describes the crystal habit of mineral fibres and the problems related to the definition of 'mineral fibre'. Chapter 4 reports the bulk spectroscopy of mineral fibres and Chapter 5 explores the application of vibrational spectroscopies (FTIR, Raman) to the analysis of asbestos minerals. Chapter 6 explores the surface and bulk properties of mineral fibres relevant to toxicity. Chapter 7 enters the world of the thermal stability of mineral fibres. The middle part of the volume deals with the *in vitro* (Chapter 8) and *in vivo* biological activity of mineral fibres (Chapter 9); Chapter 10 describes their dissolution and biodurability. The core of the volume dealing with epidemiological, biochemical and medical studies on mineral fibres includes Chapters 11, 12, 13 and 14, with focus on the differential pathological response and pleural transport of mineral fibres, their biological activities, and lung epithelial cell toxicity and pulmonary fibrosis. Chapter 11 is about the epidemiological approaches to the health effects of mineral fibres and the development of knowledge and current practice, Chapter 12 deals with the 'Differential pathological response and pleural transport of mineral fibres'. 'Biological activities of asbestos and other mineral fibres' is the subject in Chapter 13 while Chapter 14 gives an insight into mineral fibre-induced lung epithelial cell toxicity and pulmonary fibrosis. More general issues concerning the global asbestos problem and perspectives to establish a general model to predict mineral-fibre toxicity are described in the final Chapter, no. 15.

Alessandro F. Gualtieri
Modena, April 2017

Disclaimer

Introduction

ALESSANDRO F. GUALTIERI

Dipartimento di Scienze Chimiche e Geologiche, Università di Modena e Reggio Emilia,Via Campi 103, I-41125, Modena, Italy
e-mail: alessandro.gualtieri@unimore.it

This introductory chapter presents the state of the art in the multidisciplinary research field of asbestos and mineral fibres in general. The book describes the world of mineral fibres with its huge complexity and poorly understood, detrimental bio-chemical interaction with the human body. The approach of the chemist/mineralogist/ crystallographer adopts the perspective of the fibre itself (the invader of the body) whereas the approach of the bio-chemist/toxicologist/physician adopts the perspective of the organism (the invaded) interacting with the fibre. Both perspectives must be considered in synergy in an attempt to outline a conclusive model explaining the toxicity of mineral fibres, and provide a robust scientific basis that can be used by political and social partners to resolve finally the global issue of asbestos.

1. Introduction

When it comes to asbestos, everybody knows more or less what it is and (competently or not) has an opinion about it. As a matter of fact, if one searches for the word 'asbestos' on the World Wide Web 40,600,000 hits arise (Google.com, January, 2016). Despite its universal popularity and apparent simplicity, this general, elusive, threatening, misused, magic term actually hides an incredible complexity. Hence, it is not incorrect to assert that when it comes to asbestos, nothing is really simple, not even its definition and terminology.

What we know for sure is that the term asbestos, or something like it, comes from Greek and refers to an outstanding fibre known and used for millennia (Skinner *et al.*, 1988). The various terms used in ancient times, including asbestos, asbestus, asbestinon, asbest, asbeste, asbeston, abeston, amiantos, amiantus, amianthus, amiant and amiante, can be traced to the writings of the ancient Greek philosophers and their use of two words – αμιαντος and ασβεστος. The Greek word αμιαντος (translated as *amiantos*), when used as a noun, is synonymous with the English word asbestos, and when used as an adjective means pure or undefiled (Ross and Nolan, 2003). The Greek word ἄσβεστος (properly transliterated as *asvestos* and not *asbestos*), when used as a noun means lime, quicklime or unslaked lime, and when used as an adjective, means inextinguishable, unquenchable (Ross and Nolan, 2003).

There is evidence that the first time amphibole asbestos was used was to manufacture ceramics in Finland during the Stone Age, 7000–10,000 years ago, and, more

DOI: 10.1180/EMU-notes.18.1

commonly, during the so-called Early Metal Age. Ceramic types such as Sär 1, Kierikki and Pöljä (in the collection of The Regional Museum of Lapland, Rovaniemi, Finland) belong to that period. Figure 1 shows an anthophyllite-rich ceramic fragment (labelled 111 kt, and found during a field search in 1989–1991) consisting of two wall-pieces (of a vessel for holding liquid or food) of so-called Kjelmöy-type ceramic from northern Finland (Sierijärvi, Rovaniemi) from the period 900–800 B.C. to 0–400 A.D. In this type of ceramic, anthophyllite asbestos is always mixed with the clay to form thin, strong vessels which could withstand heat.

Chrysotile asbestos was discovered and utilized for the first time in Cyprus, perhaps as long as 5000 years ago (Dilek and Newcomb, 2003). In fact, the Greek physician Pedanius Discorides of Cilicia (40–90 A.D.) reported an *aminatos lithos* that occurs in Cyprus and resembles fissile alum that can be woven and is not consumed by fire (Ross and Nolan, 2003).

Pliny the Elder (23–79 A.D., *Natural History*, Book 19), apparently misunderstanding the use of this word by the early Greek philosophers, replaced the Greek noun for quicklime (ασβεστος) with the dubious Greek word ασβεστινον (G.E.L., 1940, p. 255), which he interpreted to be a non-combustible material. Pliny then transliterated ασβεστινον into the Latin noun *asbestinon*, alluding to an incombustible linen, cleansed by fire, and used as shrouds for royalty during cremation (Rackman, 1961). Over the years, asbestos continued to attract the attention of kings, alchemists and magicians from western Europe to China. In the 16th century, Georg Agricola

Figure 1. Two wall-finds of the Kjelmöy-type anthophyllite-rich ceramics (labeled 111 kt from Sierijärvi, Rovaniemi, northern Finland) found during a field campaign in 1989–1991 by Dr Hannu Kotivuori, archaeologist and Director of the Regional Museum of Lapland, Rovaniemi (Finland).

(1490–1555) provided a critical boost to the scientific understanding of this substance (Alleman and Mossman, 1997).

The modern term asbestos was conceived in the 18th century in Germany to indicate mineral species that occur as bundles of flexible fibres and that can be separated into thin, durable threads. The start of the modern asbestos industrial age began in the second half of the 19th century with the exploitation of chrysotile deposits in northern Italy and in Canada. Chrysotile asbestos mining in Quebec (Canada) dates back to 1878 and by 1885 seven mines were active in the Thetford area. By 1890, the modern asbestos industry was full-blown, with hundreds of mining applications being made (Alleman and Mossman, 1997) and the opening of large-scale asbestos industries in Scotland, Germany and England for the manufacture of asbestos-containing products. Note that asbestos was also discovered and exploited in other parts of the world. Chrysotile deposits were found in the Ural Mountains near the city of Ekaterinburg in early 18th century, although extensive exploitation of the Ural deposits began when the huge Bazhenovskoye deposit of chrysotile asbestos was discovered near Asbest City in 1884. Since 1918, the mines and mills have been operated by the Uralasbest Company (Shcherbakov *et al.*, 2001); this is the biggest mining company exploiting asbestos today. Crocidolite asbestos was discovered in 1812 in the Profret area, Northern Cape Province, South Africa, but it was not until 1893 that asbestos production began near the town of Koegas and in 1926 near the Pomfret area (Beukes and Dreyer, 1986). Amosite (A = asbestos M = mines O = of S = South Africa) asbestos was also discovered near the town of Penge, Transvaal Province, in 1907, and commercial production began in 1916 (Bowles, 1955). Since then, asbestos has rapidly become an every-day commodity the world over. It is not surprising that the contribution of asbestos to humanity was celebrated in the New York World Fair in 1939 (Alleman and Mossman, 1997).

Nowadays the term asbestos is both commercial and regulatory and may be a source of confusion. In an attempt to rationalize the various existing definitions, a few basic classifications are required: the term 'fibrous' refers to a (mineral) phase the crystal habit of which consists of fibres, whether the phase contains separable fibres or not (Hawthorne *et al.*, 2007); the term 'asbestiform' refers to a (mineral) phase the crystal habit of which consists of a strong, flexible structure made up of an ensemble of single, very narrow, elementary fibres (Veblen and Wylie, 1993). Such small segments (see Fig. 2 showing fibres with a fibrous habit and fibres with a fibrous-asbestiform habit) are called 'fibrils' (Skinner *et al.*, 1988). The length of a single fibril typically ranges from a few microns to decimetres, as does the length of a fibre, with a diameter typically <0.5 μm. Even these definitions display areas of uncertainty. For example, it may be argued that fibres or fibrous particles should be called acicular instead, to indicate a needle-like morphology (Skinner *et al.*, 1988). Of some help is the definition of aspect ratio, the relationship of the fibre length to its diameter.

'Asbestos' has been defined as:

(1) a collective generic term encompassing the (fibrous) asbestiform varieties of various minerals, used to identify well known silicate

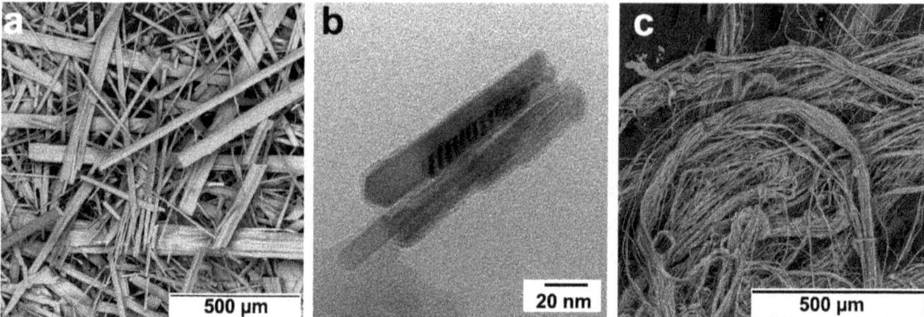

Figure 2. Examples of fibres with a fibrous crystal habit: (a) tremolite fibres observed with SEM; (b) curious crocidolite fibrils 'with the tie' (due to an interference effect in the TEM imaging); and fibres with a fibrous-asbestiform crystal habit (c) chrysotile observed with SEM.

minerals which break down into thin and flexible fibres when crushed. Some of these minerals were of industrial and economic importance and have been used extensively. The term asbestos has no definitive mineralogical significance but is applied to several minerals, which, under certain circumstances crystallize with an asbestiform habit (Case *et al.*, 2011);

(2) an industrial/commercial term indicating a manufactured product obtained by mining and processing primarily fibrous-asbestiform minerals (Campbell *et al.*, 1977);

(3) both a commercial and regulatory definition applied to a group of minerals that has grown with a specific crystal habit (fibrous-asbestiform) and which exhibits unique physical-chemical and technological properties such as flexibility, large surface area and heat resistance (Williams *et al.*, 2013);

(4) a detailed mineralogical term coined in 1982: a term applied to six minerals exploited commercially for their desirable physical properties, which are in part derived from their (fibrous) asbestiform habit. The six minerals are the serpentine mineral chrysotile (also known as white asbestos) and the amphibole minerals amosite (fibrous-asbestiform variety of grunerite, also known as brown asbestos), crocidolite (fibrous-asbestiform variety of riebeckite, commercially known as blue asbestos), as well as anthophyllite, tremolite and actinolite asbestos (Alleman and Mossman, 1997; Gualtieri, 2012). Individual mineral particles, however processed and regardless of their mineral name, are not demonstrated to be asbestos if the length-to-width ratio is <5:1 (Ross *et al.*, 1984);

(5) an early regulatory term indicating the six minerals described above, originally nominated to carry the asbestos label (U.S. Department of Labor, 1975);

(6) a detailed regulatory term which specifies physical parameters, such as length and diameter. The definition used by the Occupational Safety and Health Administration (OSHA), National Institute for Occupational Safety and Health (NIOSH) and World Health Organization (WHO) for a regulated form of asbestos is limited to those structures longer than 5 μm and with a defined aspect ratio of 3:1 (Case *et al.*, 2011).

The inadequate and incomplete definition of asbestos has resulted, as noted by an IARC (International Agency for Research on Cancer) Consensus panel, in "taxonomic confusion and lack of standardized operating definitions for fibres" (Kane *et al.*, 1996). The regulatory definition given by NIOSH has recently been integrated and published in a Bulletin entitled *Asbestos Fibers and Other Elongate Mineral Particles: State of the Science and Roadmap for Research* (NIOSH, 2011; Williams *et al.*, 2013). In that contribution it is made clear that all particles, including amphibole minerals in a non-asbestiform habit (*e.g.* acicular or prismatic amphibole particles) may be counted as fibres (and misclassified as asbestos) when their length is >5 μm and aspect ratio is ⩾3:1 when viewed using optical microscopy. NIOSH actually clarified the term 'fibres' by replacing it with the term 'elongated mineral particles', a term intended to include both asbestiform and non-asbestiform mineralogical habits that meet the specified dimensional criteria, according to the established protocol (NIOSH, 2011). The morphological aspects of mineral fibres are treated in Chapter 3.

2. Mineralogy of asbestos

As we have seen in the first paragraph, both mineralogical and regulatory definitions point to the division of asbestos minerals into two major families of silicates: serpentine and amphibole asbestos. The fibrous-asbestiform variety of serpentine is called chrysotile. Chrysotile and amphibole asbestos are both silicates sharing a fibrous-asbestiform crystal habit but holding very different structural units at a molecular scale (Whittaker, 1956; Yada, 1971; Bailey, 1988; Devouard and Barronet, 1995). The way Nature assembled the different molecular units to realize similar outputs at macroscopic scale is very elegant (Gualtieri, 2012). Chrysotile is a layer silicate with ideal formula $Mg_3(OH)_4Si_2O_5$, composed of Si-centred tetrahedral (T) sheets in a pseudo-hexagonal network joined to Mg-centred octahedral (O) sheets in units with a 1:1 (TO) ratio (Fig. 3a: chrysotile from the molecule – micro-, to the fibre – macro). Because the TO unit is polar and a misfit exists between the smaller parameters of the T sheet and the larger ones of the O sheet (Bailey, 1988), a differential stress occurs between the two sides of the layer. The stress is relieved by rolling the TO layer around the fibre axis. Thus, the structure of chrysotile fibrils is composed of layers curved concentrically or spirally, usually around the crystallographic *a* axis (clinochrysotile and orthochrysotile), and seldom around the crystallographic *b* axis (parachrysotile), into a tubular structure (Yada, 1971). In this way, a layer silicate with the layers rolled up assumes a fibrous crystal habit. Because the layers cannot, energetically, withstand too-tight a curvature, the rolls possess hollow cores with a diameter of ~5–8 nm

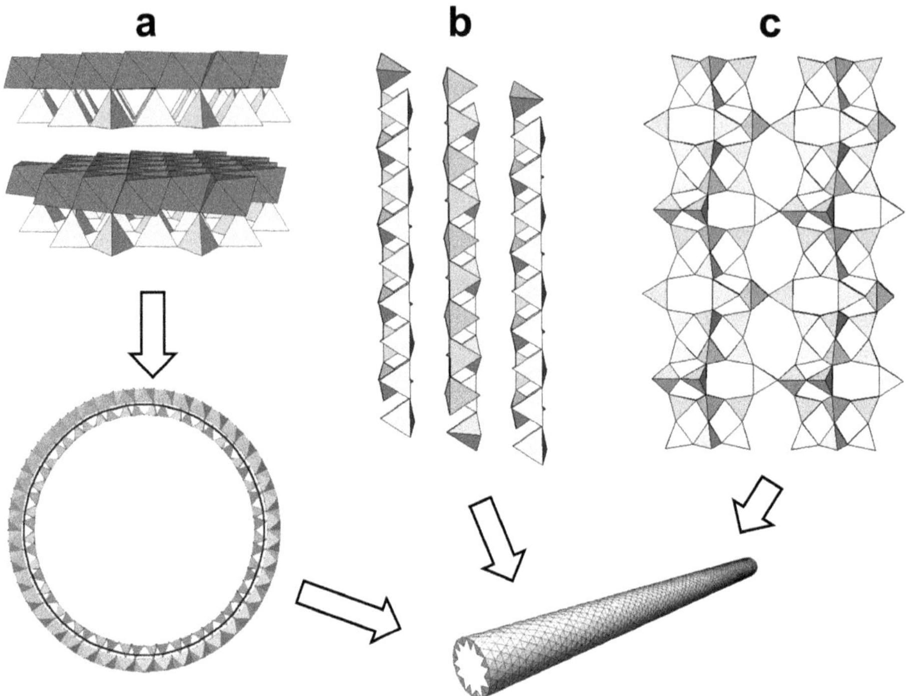

Figure 3. From the microscopic to the macroscopic worlds: the molecular structure of the three mineral fibres and their macroscopic appearance: (a) chrysotile; (b) amphibole; (c) zeolite erionite.

(Cressey and Whittaker, 1993). The peculiar curved structure of chrysotile, showing remarkable distortions of the unit cell with respect to the conventional crystal structures, prompted the development of a new theory specially formulated for cylindrical lattices (Jagodzinski and Bagchi, 1953; Whittaker, 1955; Devouard and Barronet, 1995).

Amphiboles are double-chain silicates with a Si(Al):O ratio of 4:1. The oxygen atoms of the chains coordinate not only Si(Al) but a variety of other cation sites, yielding the following simplified general formula (Veblen, 1981; Hawthorne *et al.*, 2012): $AB_2C_5T_8O_{22}W_2$ where A (in the [12] cavity with a complex nomenclature used to describe the positional disorder of the cations) = □, Na, K, Pb, Li; B (in an [8]-coordinated site) = Na, Ca, Mn^{2+}, Fe^{2+}, Mg; C (in octahedrally coordinated sites) = Mg, Fe^{2+}, Mn^{2+}, Al, Fe^{3+}, Mn^{3+}, Ti^{4+}, Li; T (in the tetrahedrally coordinated sites within the silicate chain) = Si, Al, Be; and W = (OH), F, Cl, O^{2-}. Because of the presence of strong bonds, amphiboles are normally elongated along the *c* crystallographic axis. Hence, the fibrous crystal habit is due to the mono-dimensional character of their structural units (chains) (Fig. 3b: amphiboles from the molecule – micro- to the fibre – macro).

The structure details of both asbestos serpentine and amphiboles will be described in Chapter 2 of this book.

The definitions given in the first paragraph pointed out that a huge number of minerals may occur as elongated particles in the natural environment (Skinner *et al.*, 1988). There is a growing consensus that the regulatory definition of 'asbestos' should be expanded to include other amphibole minerals with asbestiform habit that have been shown to cause disease or which are potentially toxic to humans: winchite asbestos [ideally $(CaNa)Mg_4(Al,Fe^{3+})Si_8O_{22}(OH)_2$] and richterite asbestos [ideally $(Na)CaNaMg_5Si_8O_{22}(OH)_2$] are typically mentioned, as are other minerals with asbestiform crystal habits (NIOSH, 2011). An excellent case relates to the amphibole fluoro-edenite, ideally $NaCa_2Mg_5(Si_7Al)O_{22}F_2$, which has caused many cases of pleural mesothelioma in the town of Biancavilla (Sicily, Italy), on the south-west slopes of the Etna volcano, from 1980 to 1997 (Soffritti *et al.*, 2004). An example of an asbestiform mineral which showed cytotoxic and oxidative properties similar to those exerted by crocidolite asbestos is balangeroite [$(Mg,Fe^{2+},Fe^{3+},Mn^{2+})_{42}Si_{16}O_{54}(OH)_{36}$] (Gazzano *et al.*, 2005; Groppo *et al.*, 2005). This mineral is associated with chrysotile in the Italian Balangero mine (Belluso and Ferraris, 1991) and is believed to have had carcinogenic effects on chrysotile miners and millers (Case *et al.*, 2011). Without doubt, the best known example concerns erionite, a common, natural, fibrous zeolite. Erionite belongs to the so-called ABC-6 family (Gottardi and Galli, 1985) and usually occurs in volcanic ash altered by weathering processes. It is hexagonal, space group $P6_3/mmc$, topological code [ERI] (Gualtieri *et al.*, 1998; Baerlocher *et al.*, 2007). The framework consists of columns of double six-rings (D6R) alternating with cancrinite cages, and columns of erionite cages (Fig. 3c: erionite from the molecule – micro- to the fibre – macro). A mean chemical formula $K_2(Na,Ca_{0.5})_8[Al_{10}Si_{26}O_{72}]\cdot30H_2O$ has been proposed for erionite, but owing to a relevant chemical variability, three different species, erionite-K, erionite-Ca and erionite-Na are recognized (Ballirano and Cametti, 2015). The erionite structure is described in detail in Chapter 2. Fibrous erionite was found to be the cause of an unprecedented epidemic of malignant mesothelioma (MM) among three villages, Karain, Tuzkoy and Sarihidir, in Cappadocia, Turkey (Baris *et al.*, 1978; Carbone *et al.*, 2007).

All of these examples demonstrate that we live in a fibre-rich world, with the tip of the iceberg being the most widespread species chrysotile, crocidolite and amosite, which rapidly became strategic raw materials for various industrial applications, but of public health concern later. As explained below, the focus of this book will be not just on the notorious fibre species but also on the less common or rare species that were discovered to be hazardous to the environment and to humans.

3. Uses of commercial asbestos species

The six asbestos minerals included in definitions 2 to 5 above, all exhibit outstanding properties that have been exploited to create asbestos-containing materials (ACMs) used in a huge number (>3000) of practical and industrial applications. The major chemical-physical and technological properties of commercial chrysotile, amosite and crocidolite asbestos minerals are resistant to abrasion, heat (non-flammable even at very high temperatures) and chemicals, and are flexible, resilient, have low sound-transmission coefficients, large surface areas, extremely high tensile strengths and low thermal conductivity (Gualtieri, 2012).

Chrysotile is by far the most used asbestos fibre. The asbestos-cement industry is the largest user of chrysotile fibres (~85% of all applications). For the reasons illustrated in the next paragraph, chrysotile is the only asbestos species used. It is estimated that >95% of commercially developed asbestos ore deposits were chrysotile asbestos (Ross *et al.*, 2008).

ACMs can be divided into friable and compact materials. Friable asbestos designates any ACM that can be crumbled or pulverized easily when dry, with loose asbestos fibres that can be scratched effortlessly by hand. This ACM typology represents 70–95 wt.% asbestos fibres. Some major examples of applications of friable asbestos in building materials are: artificial ashes, cavities in partitions in floors and ceilings, corrugated paper, fire-door gaskets in furnaces, fireproofing spray, insulating boards/panels, acoustic panels and finishes, fireboards, insulation systems and covering of ventilation and air-conditioning systems, refrigerators/freezers, laggings and soundproofing or decorative spray coatings for ceilings (Gualtieri, 2012). Figure 4a is an example of application of friable asbestos used to paint a cement-asbestos chimney. Compact asbestos is a composite material with asbestos fibres (generally from 4 to 15 wt.%)

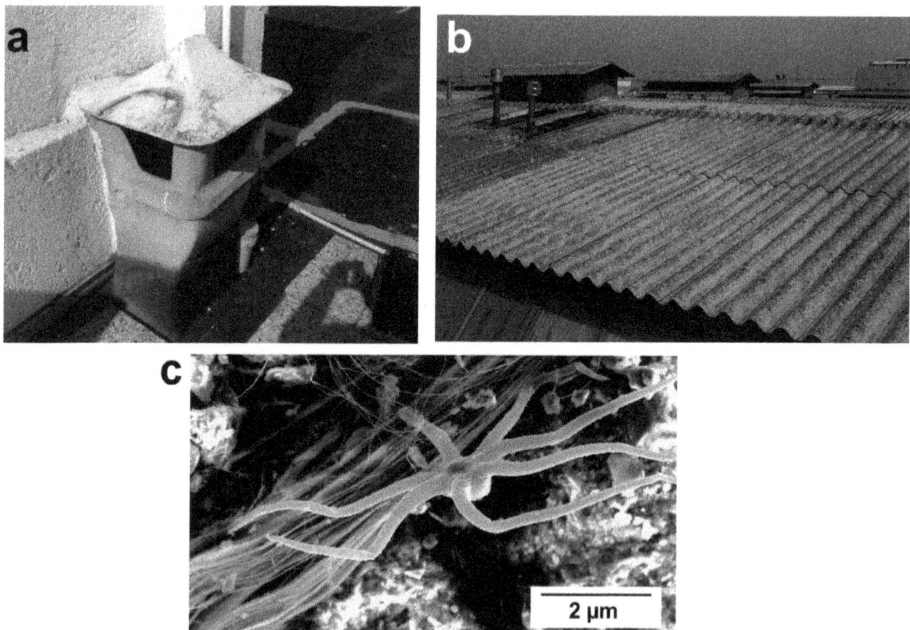

Figure 4. (a) Friable asbestos used to paint a cement-asbestos chimney; (b) a corrugated cement-asbestos roof in an industrial site; and (c) a SEM image of a cement-asbestos matrix showing the peculiar association of a chrysotile fibre bundle and a leaf trichome.

embedded in a cement or polymeric matrix. This ACM typology is not prone to release fibres unless it is sawn or scratched by mechanical tools. Examples of applications of compact asbestos in building materials are: masonry filler, mortars, mastics, caulk, putties, ceiling tiles, asbestos-cement products with 4–15 wt.% chrysotile asbestos and/or 0–6 wt.% amphibole asbestos such as planar or corrugated roofing (example in Fig. 4b and microscopic example in Fig. 4c, which curiously displays an association of a chrysotile fibre bundle and a leaf trichome within a cement matrix), pipes, chimney flues and tops, floor tiles, textiles and composites (linoleum, vinyl asbestos, asphalt, insulating seals), manufactured mixing of Portland cement with chrysotile asbestos (generally 12–50 wt.%), and water tanks (Gualtieri, 2012).

In countries where chrysotile asbestos is still used in a 'safe way' (which means that it can be handled using individual protective devices), new applications for the material are discovered every day: catalytic magnetic composites based on layered vermiculite and chrysotile with carbon nanostructures have been prepared in Brazil (Teixeira *et al.*, 2012); InSb nanowires in the channels of chrysotile to be used for non-linear optics applications have been realized in Russia (Uryupin *et al.*, 2014); chrysotile nanotubes for Pb(II) removal from aqueous solutions have been developed in China (Cheng *et al.*, 2015).

4. Toxicity of asbestos and current global situation

The distinctive fibrous-asbestiform crystal habit and surface activity, responsible for the outstanding technological properties of asbestos minerals, seem, also, to be the cause of their potential toxicity. The first evidence that asbestos exposure may cause lung diseases was discovered during the period of peak use of asbestos minerals during the Industrial Age (end of the 19th century). The history of the epidemiological reports of asbestos-related diseases was described well by Skinner *et al.* (1988). Since the early 1980s, although there is significant epidemiological evidence that amphibole asbestos crocidolite and tremolite are apparently more dangerous than chrysotile asbestos (Hodgson and Darnton, 2000), all six asbestos minerals (chrysotile, actinolite, amosite, anthophyllite, crocidolite and tremolite) have been assumed to be toxic to human health and therefore regulated.

The toxicological research became relevant to the investigation of possible hazards from other fibres and led to the so called "fibre paradigm" (Miller *et al.*, 1999). The toxicity/carcinogenicity of asbestos is known to be related to the following parameters: (1) a fibre length $>\sim 8$ μm, less than which the fibre can, in principle, be removed by lung macrophages (white blood cells which are an integral part of the human immune system – their job is to locate and engulf foreign particles); (2) a fibre diameter of $<\sim 3$ μm, allowing the fibre to be inhaled into the gas-exchanging part of the lung; (3) fibre insolubility in the acidic environment (pH = 4–4.5) developed during intracellular phagocytosis; and (4) a threshold dose to the target organ. It appears that the basic mechanism leading to lung cancer is 'frustrated phagocytosis', where the macrophage is injured (inflammatory burst) in the attempt to engulf fibres longer than its diameter, with release of cytokines, mitogens and oxidants extracellularly which initiate the process of fibrosis and carcinogenesis (Seaton *et al.*, 2010). Biological mechanisms involving mineral fibres are discussed in Chapters 8, 9, 12, 13 and 14 of this volume.

The results of all the epidemiological, *in vitro* and *in vivo* cohort studies indicate that inhalation of all regulated asbestos minerals may cause the following lung diseases (Skinner *et al.*, 1988; Dilek and Newcomb, 2003): (1) asbestosis: a non-malignant diffuse interstitial fibrosis of the lung tissue. High asbestos exposure leads to scarring of the lung, causing it to become stiff, resulting in a restriction in pulmonary function and reduction in the lung's ability to exchange CO_2 for O_2. It is a typical lung disease developed in an industrial environment (Filkenstein, 1983) with a latency period of ~10–20 years; (2) lung cancer or carcinoma: which includes squamous carcinoma, small- or oat-cell carcinoma and adeno-carcinoma. Lung cancer is also a typical lung disease developed in industrial environments with a latency period of ~15–20 years; (3) MM: a cancer of the pleura, pericardium and peritoneal membranes which surround the lung, heart and abdominal cavities, respectively. The MM may develop as a consequence of both working and environmental exposure with a latency period of ~20–40 years; and (4) pleural plaques: localized scars consisting of collagen deposits, sometimes calcified, normally found in the parietal pleura.

Asbestos was declared a proven human carcinogen by the US Environmental Protection Agency (EPA), the International Agency for Research on Cancer (IARC) of the World Health Organization (WHO) and the National Toxicology Program more than 20 years ago (Nicholson, 1986; IARC 2012).

Since the late 1980s, in response to the demands of large sections of the scientific community, asbestos victims' associations, environment protection groups and many political parties, many countries worldwide began to ban or regulate the use of ACMs. For example, in 1989 the United States EPA issued the *Asbestos Ban and Phase Out Rule*.

Over the past two decades, asbestos minerals have been the subject of intensive multidisciplinary investigations because the mechanisms by which they induce cyto- and geno-toxic damage remain poorly understood. Unfortunately, the cause-effect relationship between exposure to the fibres and the onset of mesothelioma and other lung diseases remains ambiguous. The difficulties arise mainly from the fact that mineral fibres display great variability in their chemistry, molecular arrangement, size and diameter, surface activity (Pollastri *et al.*, 2014), ability to generate reactive oxygen species and biopersistence (Donaldson *et al.*, 2010; Pollastri *et al.*, 2014) so that a general model explaining the toxicity has, thus far, proven elusive.

Because of such uncertainties in our scientific knowledge, the global scientific and political community is divided into two groups: one group assumes that all regulated asbestos fibres are indistinctly classified as potentially toxic substances; the other promotes the 'safe use' of chrysotile. The result is that the use of regulated asbestos minerals is restricted or banned today in only 55 of 195 countries (28%) in the world whereas 'safe use' of chrysotile is permitted in the remaining countries (72%). This apparent contradiction is referred to as the 'asbestos global issue'. The pro-chrysotile side claims that only amphibole asbestos minerals should be considered as highly toxic whereas the toxicity potential of chrysotile asbestos should be assumed to be very limited (the so called 'amphibole hypothesis'). Such an hypothesis is based on the assumption that chrysotile asbestos has little potential for causing mesothelioma (*e.g.* Liddell *et al.*, 1997; McDonald *et al.*, 1997; Camus, 2001) and that lung diseases are, in fact, due to amphibole minerals (especially tremolite) which probably contaminate chrysotile asbestos. The 'amphibole hypothesis' is supported by knowledge of the behaviour of chrysotile and amphibole in the lungs. Chrysotile dissolves reasonably quickly (because of its low biodurability) in the intracellular macrophage environment, during phagocytosis, whereas amphiboles are much more durable (high biodurability) and remain in the lungs for a very long time (Hume and Rimstidt, 1992; Bernstein *et al.*, 2008; Oze and Solt, 2010). Although the position is obviously not agreed universally (Stayner *et al.*, 1996; Hogson and Darnton, 2000; Berman and Crump, 2008), there are few studies in which dose-response relationships have been estimated separately for cancer risk and exposure to different fibre types in the same exposed population and this leaves the issue open to debate. Wagner (1997) claimed that no MMs have been reported to have occurred in workers exposed to chrysotile, unless the exposure has been intense and for >20 years.

Hence, the current global situation in the asbestos market assumes that chrysotile is a natural fibre that can be used in a 'safe way' for industrial purposes. The 2014 (chrysotile) asbestos trade data (source United States Geological Service and IBAS) show that the top five asbestos producers (t/year) in decreasing order are: Russia (1,100,000), China (400,000), Brazil (284,000), Kazakhstan (240,000) and India (270) while the top five asbestos users (t/year) in decreasing order are: Russia (608,000), China (507,000), India (379,000), Brazil (154,000) and Kazakhstan (68,000).

Besides the uncertainties in current scientific models, it is clear that solving the global asbestos issue will not be an easy task, due to the political and economic pressure in favour of chrysotile from the so-called BRIC (Brazil, Russia, India, China) countries. In those countries, chrysotile asbestos is a primary strategic raw material the economic importance of which appears to prevail over health concerns. In this scenario, it is not surprising that chrysotile asbestos has not been included to date (May 1st, 2017) in the list of dangerous products covered by the Rotterdam convention. Consequently, ACMs are freely distributed worldwide without warning labels. This situation generates enormous problems and international legal actions, and is certainly an example of a failure of globalization. The lack of warning labelling for ACMs means that any purchaser would be unaware of the contents of the product. Hence, critical situations may occur: consider the case of an ACM produced in a country such as China, where chrysotile asbestos is used legally, and exported to a country such as Italy where the import, export and use of ACMs has been banned since 1992.

5. Final introductory remark

Some of the results presented in this volume are the outcome of a long-term Italian Research Project of National Interest (PRIN) conducted since 2011. Supported by a significant systematic chemical-physical, mineralogical and structural characterization of the fibres, the main goal of this multidisciplinary project was to contribute to the understanding of the nature of the biological interaction mechanisms of chrysotile, amphiboles and erionite and compare them systematically: the aim being to draw up a convincing rank of toxicity of mineral fibres. A general model to classify mineral fibres on the basis of their toxicity is of paramount importance to predict *a priori* the toxicity potential of each mineral fibre, identify its occurrences and eventually plan a risk assessment analysis for the population. The cases of the carcinogenic erionite in Turkey and fluoro-edenite in Italy are striking examples of mineral fibres whose toxicity was assessed *ex post* at the expense of human lives. In an environment rich in various kinds of fibres it is likely that other cases will materialize sooner or later. Having a universal predictive model, could, potentially, be a powerful tool of risk assessment and prevention, especially for reducing exposures to naturally occurring asbestos (NOA), which is now sources of considerable concern worldwide. NOA in fact occurs in the natural environment (rocks and soils) as a result of geological processes. Weathering and human activity may disturb NOA-bearing rock or soil and release mineral fibres into the air, which pose a greater potential for human exposure by inhalation. For this reason, mineral fibres of little or no commercial/economic value may come into prominence and become causes of concern. Consider for example the naturally occurring zeolite, laumontite [$CaAl_2Si_4O_{12}\cdot 4H_2O$], and the sorosilicate pumpellyite-(Al) [$Ca_2(Al,Fe^{2+},Mg)Al_2[(OH,O)_2SiO_4Si_2O_7]\cdot H_2O$], which are common in many different rock types. Their habit can be fibrous (*e.g.* Fig. 5a,b) and such fibres can be classified as regulated. Ferric or ferrous iron can be present as a surface impurity (laumontite) or in the crystal lattice (pumpellyite). How are we to classify these mineral fibres in terms of toxicity? Should populations living near green stones containing such fibres be made aware of their presence? The present volume should be a strong and comprehensive base of knowledge for any scientist who wants to work in the field of mineral fibres and be a valid resource in terms of building a model capable of answering the questions raised above.

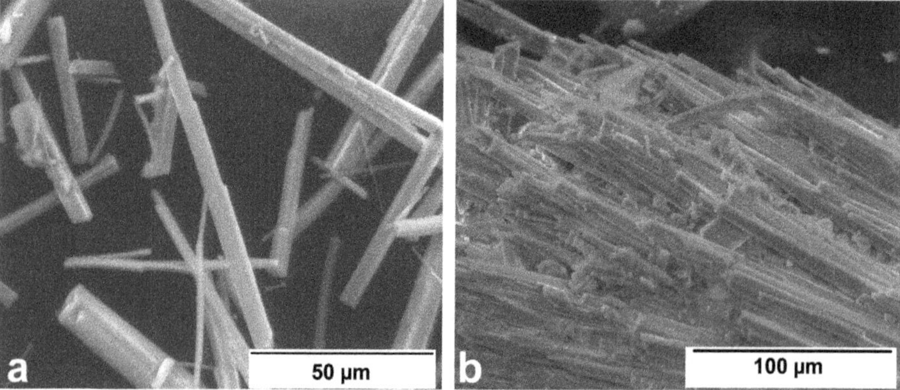

Figure 5. SEM images of (a) laumontite fibres and (b) pumpellyite fibres.

References

Alleman, J.E. and Mossman, B.T. (1997) Asbestos revisited. *Scientific American*, **277**, 70–75.

Baerlocher, Ch., McCusker, L.B. and Olson, D.H. (2007) *Atlas of Zeolite Framework Types. Sixth Revised Edition.* Elsevier, Amsterdam, 327 pp.

Bailey, S.W., editor (1988) *Hydrous Phyllosilicates (Exclusive of Micas).* Reviews in Mineralogy and Geochemistry **19**. Mineralogical Society of America, Chantilly, Virginia, USA, 725 pp.

Ballirano, P. and Cametti, G. (2015) Crystal chemical and structural modifications of erionite fibers leached with simulated lung fluids. *American Mineralogist*, **100**, 1003–1012.

Baris, Y.I.A., Sahin, A., Ozesmi, M., Kerse, I., Ozen, E., Kolacan, B., Altinors, M. and Goktepeli A. (1978) An outbreak of pleural mesothelioma and chronic fibrosing pleurisy in the village of Karain/Urgup in Anatolia. *Thorax*, **33**, 181–192.

Belluso E. and Ferraris, G. (1991) New data on balangeroite and carlosturanite from alpine serpentinites. *European Journal of Mineralogy*, **3**, 559–566.

Berman, D.W. and Crump, K.S. (2008) A meta-analysis of asbestos-related cancer risk that addresses fiber size and mineral type. *Critical Reviews in Toxicology*, **38**, 49–73.

Bernstein, D.M., Donaldson, K., Decker, U., Gaering, S., Kunzendorf, P., Chevalier, J. and Holm, E. (2008) A biopersistence study following exposure to chrysotile asbestos alone or in combination with fine particles. *Inhalation Toxicology*, **20**, 1009–1028.

Beukes, N.J. and Dreyer, C.J.B. (1986) Amosite asbestos deposits of the Penge Area. Pp. 901–910 in: *Mineral Deposits of Southern Africa, Vols I and II* (C.R. Anhaeusser and S. Maske, editors). Geological Society of South Africa. Johannesburg.

Bowles, O. (1955) *The Asbestos Industry.* U.S. Bureau of Mines, Bulletin, **552**. Washington, D.C., 122 pp.

Campbell, W.J., Blake, R.L., Brown, L.L., Cather, E.E. and Sjoberg, J.J. (1977) Selected silicate minerals and their asbestiform varieties: Mineralogical definitions and identification-characterization. *Bureau of Mines Information Circular IC-8751*, Bureau of Mines.

Camus, M. (2001) A ban on asbestos must be based on a comparative risk assessment. *Journal of the Association of Medicine Canadian*, **164**, 491–494.

Carbone, M., Emri, S., Umran Dogan, A., Steele, I., Tuncer, M., Pass, H.I. and Baris, Y.I. (2007) A mesothelioma epidemic in Cappadocia: scientific developments and unexpected social outcomes. *Nature Reviews*, **7**, 147–154.

Case, B.W., Abraham, J.L., Meeker, G., Pooley, F.D. and Pinkerton, K.E. (2011) Applying definitions of "asbestos" to environmental and "low dose" exposure levels and health effects, particularly malignant mesothelioma. *Journal of Toxicology and Environmental Health, Part B*, **14**, 3–39.

Cheng, L., Ren, X., Wei, X. and Sun, X. (2015) Facile synthesis and characterization of chrysotile nanotubes and their application for lead(II) removal from aqueous solution. *Separation Science and Technology*, **50**, 700–709.

Cressey, B.A. and Whittaker, E.J.W. (1993) Five-fold symmetry in chrysotile asbestos, revealed by transmission electron microscopy. *Mineralogical Magazine*, **57**, 729–732.

Devouard, B. and Baronnet, A. (1995) Axial diffraction of curved lattices; geometrical and numerical modeling; application to chrysotile. *European Journal of Mineral*ogy, **7**, 835–846.

Dilek, Y. and Newcomb, S. (2003) *Ophiolite Concept and the Evolution of Geological Thought.* Special Paper 373, The Geological Society of America, Boulder, Colorado (USA), GSA Books Science Editor.

Donaldson, K., Murphy, F.A., Duffin, R. and Poland, C.A. (2010) Asbestos, carbon nanotubes and the pleural mesothelium: a review of the hypotheis regarding the role of long fibre retention in the parietal pleura, inflammation and mesothelioma. *Particle and Fibre Toxicology*, **7(5)**, 1 17.

Filkenstein, M.M. (1983) Mortality among long-term employees of an Ontario asbestos-cement factory. *British Journal of Industrial Medicine*, **40**,138–144.

Gazzano, E., Riganti, C., Tomatis, M., Turci, F., Bosia, A., Fubini, B. and Ghigo, D. (2005) Different cellular responses evoked by natural and stoichiometric synthetic chrysotile asbestos. *Journal of Toxicology and Environmental Health*, **68**, 41–49.

G.E.L. (1940) *A Greek-English Lexicon, 9th edition.* Oxford University Press, Oxford, UK, 2111 pp.

Gottardi, G. and Galli, E. (1985) *Natural Zeolites*. Springer-Verlag, Heidelberg, Geermany, 409 pp.
Groppo, C., Tomatis, M., Turci, F., Gazzano, E., Ghigo, D., Compagnoni, R. and Fubini, B. (2005) Potential toxicity of non regulated asbestiform minerals: Balangeroite from the western Alps. Part 1: identification and characterization. *Journal of Toxicology and Environmental Health, Part A*, **68**, 1–19.
Gualtieri, A.F., Artioli, G., Passaglia, E., Bigi, S., Viani, A. and Hanson, J.C. (1998) Crystal structure-crystal chemistry relationships in the zeolites erionite and offretite. *American Mineralogist*, **83**, 590–606.
Gualtieri, A.F. (2012) Mineral fibre-based building materials and their health hazards. Pp. 166–195 in: *Toxicity of Building Materials* (F. Pacheco-Torgal, S. Jalali and A. Fucic, editors). Woodhead Publishing, Cambridge, UK.
Hawthorne, F.C., Oberti, R., Della Ventura, G. and Mottana, A. (editors) (2007) *Amphiboles: Crystal Chemistry, Occurrence, and Health Issues*. Reviews in Mineralogy and Geochemistry, **67**. Mineralogical Society of America and Geochemical Society. Chantilly, Virginia, USA. 545 pp.
Hawthorne, F.C., Oberti, R., Harlow, G.E., Maresch, W.V., Martin, R.F., Schumacher, J.C. and Welch, M.D. (2012) Nomenclature of the amphibole supergroup. *American Mineralogist*, **97**, 2031–2048.
Hodgson, J.T. and Darnton, A. (2000) The quantitative risks of mesothelioma and lung cancer in relation to asbestos exposure. *Annals of Occupational Hygiene*, **44**, 565–601.
Hume, L.A. and Rimstidt, J.D. (1992) The biodurability of chrysotile asbestos. *American Mineralogist*, **77**, 1125–1128.
IARC (2012) Metals, arsenic, dusts and fibres. *Monographs on the Evaluation of Carcinogenic Risks to Humans*, **100C**, 527 pp, Lyon, France.
Jagodzinski, H. and Bagchi, S.N. (1953) Die gerollte Struktur des Chrysotils. *Neues Jahrbuch für Mineralogie, Monatshefte*, 97–100.
Kane, A.B., Boffetta, P., Saracci, R. and Wilbourn, J.D. (1996) Mechanisms of fibre carcinogenesis. *IARC Scientific Publication* 140. Lyon: International Agency for Research on Cancer, WHO.
Liddell, F.D.K., McDonald, A.D. and McDonald, J.C. (1997) The 1891–1920 birth cohort of Quebec chrysotile miners and millers: development from 1904 and mortality to 1992. *Annals of Occupational Hygiene*, **41**, 13–36.
McDonald, A.D., Case, B.W., Churg, A., Dufresne, A., Gibbs, G.W., Sébastien, P. and McDonald, J.C. (1997) Mesothelioma in Quebec chrysotile miners and millers: epidemiology and aetiology. *Annals of Occupational Hygiene*, **41**, 707–719.
Miller, B.G., Searl, A., Davis, J.M.G., Donaldson, K., Cullen, R.T., Bolton, R.E., Buchanan, D. and Soutar, C.A. (1999). Influence of fibre length, dissolution and biopersistence on the production of mesothelioma in the rat peritoneal cavity. *Annals of Occupational Hygiene*, **43**, 155–166.
Nicholson, W.J. (1986) 'Airborne asbestos health assessment update'. *Report EPA-600/8-84-003F*. Washington, D.C: Office of Health and Environmental Assessment, US Environmental Protection Agency, 200 pp.
NIOSH (2011) *Asbestos fibers and other elongate mineral particles: State of the science and roadmap for research*. DHHS Publication No. 2011-159. National Institute for Occupational Safety and Health (NIOSH).
Oze, C. and Solt, K.L. (2010) Biodurability of chrysotile and tremolite asbestos in simulated lung and gastric fluids. *American Mineralogist*, **95**, 825–831.
Pollastri, S., Gualtieri, A.F., Lassinantti Gualtieri, M., Hanuskova, M., Cavallo, A. and Gaudino, G. (2014) The zeta potential of mineral fibres. *Journal of Hazardous Materials*, **276**, 469–479.
Rackham, H. (1961) *Pliny Natural History with an English Translation in Ten Volumes*. Harvard University Press, Cambridge, Massachusetts, USA, 544 pp. (Books 17–19).
Ross, M., Kuntze, R.A. and Clifton, R.A. (1984) A definition for asbestos. Pp. 139–147 in: *Definitions for Asbestos and other Health-related Silicates* (B. Levadie, editor). ASTM STP 834. Philadelphia: American Society for Testing and Materials.
Ross, M. and Nolan, R.P. (2003) History of asbestos discovery and use and asbestos-related disease in context with the occurrence of asbestos within ophiolite complexes. *Geological Society of America Special paper*, **373**, 447–470.
Ross, M., Langer, A.M., Nord, G.L., Nolan, R.P., Lee, R.J., Van Orden, D. and Addison, J. (2008) The mineral nature of asbestos. *Regulatory Toxicology and Pharmacology*, **52(S1)**, S26–S30.
Seaton, A., Tran, L., Aitken, R., Donaldson, K. (2010) Nanoparticles, human health hazard and regulation.

Journal Royal Society Interface, **7**, S119–S129.

Shcherbakov, S.V., Kashansky, S., Domnin, S.G., Koggan, F.M., Kozlov, V., Kochelayev, V.A. and Nolan, R.P. (2001) The health effects of mining and milling chrysotile: The Russian experience. Pp. 187–198 in: *The Health Effects of Chrysotile Asbestos: Contribution of Science to Risk Management Decisions* (R.P. Nolan, A.M. Langer, M. Ross, F.J. Wicks and R.F. Martin, editors). *The Canadian Mineralogist Special Publication* **5**. Ottawa, Canada.

Skinner, H.C.W., Ross, M. and Frondel, C. (1988) *Asbestos and Other Fiberous Materials*. Oxford University Press, New York, 204 pp.

Soffritti, M., Minardi, F., Bua, L., Degli Esposti, D. and Belpoggi, F. (2004) First experimental evidence of peritoneal and pleural mesotheliomas induced by fluoro-edenite fibres present in Etnean volcanic material from Biancavilla (Sicily, Italy). *European Journal of Oncology*, **9**, 169–175.

Stayner, L.T., Dankovic, D.A. and Lemen, R.A. (1996) Occupational exposure to chrysotile asbestos and cancer risk: a review of the amphibole hypothesis. *American Journal of Public Health*, **86**, 179–186.

Teixeira, A.P.C., Purceno, A.D., Barrosa, A.S., Lemos, B.R.S., Ardisson, J.D., Macedo, W.A.A., Nassor, E.C.O., Amorim, C.C., Moura, F.C.C., Hernández-Terrones, M.G., Portela, F.M. and Lago, R.M. (2012) Amphiphilic magnetic composites based on layered vermiculite and fibrous chrysotile with carbon nanostructures: Application in catalysis. *Catalysis Today*, **190**, 133–143.

Uryupin, O.N., Kartenko, N.F. and Tabachkova, N.Yu. (2014) On the structure of InSb nanowires in the channels of chrysotile asbestos. *Semiconductors*, **48**, 974–978.

United States Department of Labor (1975) Occupational exposure to asbestos. *Federal Regulation*, **40**, 47652–47665.

Veblen, D.R. (1981) Non-classical pyriboles and polysomatic reactions in biopyriboles. Pp. 189–236 in: *Amphiboles and Other Hydrous Pyriboles: Mineralogy* (D.R. Veblen and P.H. Ribbe, editors). Reviews in Mineralogy, **9A**. Mineralogical Society of America, Chantilly, Virginia, USA.

Veblen, D.R. and Wylie, A.G. (1993) Mineralogy of amphiboles and 1:1 layer silicates. Pp. 61–138 in: *Health Effects of Mineral Dusts* (G.D. Guthrie, Jr. and B.T. Mossman, editors). Reviews in Mineralogy, 28, Mineralogical Society of America, Chantilly, Virginia, USA.

Wagner, J.C. (1997) Asbestos-related cancer and the amphibole hypothesis. *American Journal of Public Health*, **87**, 687–688.

Whittaker, E.J.W. (1955) A classification of cylindrical lattices. *Acta Crystallographica*, **8**, 571–574.

Whittaker, E.J.W. (1956) The structure of chrysotile. II. Clino-chrysotile. *Acta Crystallographica*, **9**, 855–862.

Williams, C., Dell, L., Adams, R., Rose, T. and Van Orden, D. (2013) State-of-the-science assessment of non-asbestos amphibole exposure: Is there a cancer risk? *Environmental Geochemical Health*, **35**, 357–377.

Yada, K. (1971) Study of microstructure of chrysotile asbestos by high-resolution electron microscopy. *Acta Crystallographica A*, **27**, 659–664.

EMU Notes in Mineralogy, Vol. 18 (2017), Chapter 2, 17–64

The crystal structure of mineral fibres

PAOLO BALLIRANO[1,2], ANDREA BLOISE[3], ALESSANDRO F. GUALTIERI[4], MARCO LEZZERINI[5], ALESSANDRO PACELLA[1], NATALE PERCHIAZZI[5], MERAL DOGAN[6,7] and A. UMRAN DOGAN[7,8]

[1]*Dipartimento di Scienze della Terra, Sapienza Università di Roma, Piazzale Aldo Moro 5, IT-00185, Roma, Italy, e-mail: paolo.ballirano@uniroma1.it*
[2]*Laboratorio Rettorale Fibre e Particolato Inorganico, Sapienza Università di Roma, Piazzale Aldo Moro 5, IT-00185, Roma, Italy*
[3]*Dipartimento di Biologia, Ecologia e Scienze della Terra, Università della Calabria, Via P. Bucci, IT-87036, Arcavacata di Rende (Cs), Italy*
[4]*Dipartimento di Scienze Chimiche e Geologiche, Università di Modena e Reggio Emilia, Via Campi 103, IT-41125, Modena, Italy*
[5]*Dipartimento di Scienze della Terra, Università di Pisa, Via S. Maria 53, IT-56126, Pisa, Italy*
[6]*Geological Engineering Department, Hacettepe University, TR-06800 Beytepe, Ankara, Turkey*
[7]*Center for Global and Regional Environmental Research, The University of Iowa, Iowa City, US-52242, Iowa, USA*
[8]*Chemical and Biochemical Engineering, The University of Iowa, Iowa City, US-52242, Iowa, USA*

This chapter deals with the crystal structure of regulated and unregulated mineral fibres. The aim is to provide readers, both specialists and researchers broadly interested in environmental problems, with up-to-date information on a topic that is expanding daily. The chapter describes specifically the structure of the fibrous modification whenever available and outlines possible differences from the corresponding prismatic variety. Details of the experimental techniques used for structure determination/refinement are reported also, if appropriate, to outline the experimental difficulties faced due to the small dimensions, sensitivity and chemical complexity of mineral fibres.

1. Introduction

Because of its links with several lung diseases, asbestos has been studied extensively, stimulated by public concern, and aimed at guiding government regulations. Simultaneously, much effort has been devoted to identifying the mechanism/s by which such agents lead to severe damage to human health. A series of monographs published by the International Agency for Research on Cancer (IARC) has evaluated

DOI: 10.1180/EMU-notes.18.2

and classified (as of October 26, 2015: http://monographs.iarc.fr/ENG/Classification) 985 agents (natural and man-made) on the basis of their degree of carcinogenicity for both humans and animals, leading to a final overall evaluation of carcinogenicity to humans (Table 1). A few minerals have been evaluated so far and are listed in Table 2.

Table 1. Classification of agents by IARC (2011).

Degree of carcinogenicity in humans	Overall evaluation of carcinogenicity to humans
Sufficient evidence: *A causal relationship has been established between exposure to the agent and human cancer.*	**Group 1**: The agent is carcinogenic to humans. *This category is used when there is sufficient evidence of carcinogenicity in humans.*
Limited evidence: *A positive association has been observed for which a causal interpretation is considered to be credible.*	**Group 2A**: The agent is probably carcinogenic to humans. *This category is used when there is limited evidence of carcinogenicity in humans and sufficient evidence in experimental animals.* **Group 2B**: The agent is possibly carcinogenic to humans. *This category is used when there is limited evidence of carcinogenicity in humans and less than sufficient evidence in experimental animals.*
Inadequate evidence: *The available studies are of insufficient quality, consistency or statistical power to permit a conclusion.*	**Group 3**: The agent is not classifiable as to its carcinogenicity to humans. *This category is used when the evidence of carcinogenicity is inadequate in humans and inadequate or limited in experimental animals.*
Evidence suggesting lack of carcinogenicity: *There are several adequate studies covering the full range of levels of exposure that humans are known to encounter, which are mutually consistent in not showing a positive association.*	**Group 4**: The agent is probably not carcinogenic to humans. *This category is used when there is evidence suggesting lack of carcinogenicity in both humans and experimental animals.*

Table 2. Evaluation of carcinogenicity of breathable minerals as in IARC Monographs 1–114.

Mineral	Crystal habit	Group
Asbestos	Fibrous	1
Erionite	Fibrous	1
Fluoro-edenite	Fibrous	1
Palygorskite/attapulgite	Fibrous (fibre length >5 μm)	2B
	Fibrous (fibre length <5 μm)	3
Sepiolite	Fibrous	3
Silica dust	Various	1
Talc*	Platy	3
Wollastonite	Fibrous	3
Zeolites other than erionite (clinoptilolite, mordenite, phillipsite)	Fibrous	3

* For talc containing asbestiform fibres, see Asbestos.

With the exception of crystalline silica all of them occur commonly with a fibrous crystal habit. Therefore, mineral fibres play a significant role in particle toxicology, which is a branch of toxicology dealing with solid particles entering the human body through inhalation. As can be seen from Table 2, a few mineral species other than asbestos have been listed as having significant carcinogenicity and others possessing "...inadequate evidence of carcinogenicity in humans and inadequate or limited in experimental animals". In addition, different research groups are currently testing several other fibrous minerals showing possible dangerous properties. This topic is of particular relevance as we are all in daily contact with mineral fibres, several of them possessing chemical, physical and morphological features similar to those of species classified unambiguously as having deleterious effects on human health.

The present chapter will describe the crystal structure of the mineral fibres that are considered to have or potentially have carcinogenic effects.

2. Regulated and unregulated amphibole group minerals (actinolite, amosite, anthophyllite, crocidolite, tremolite, fluoro-edenite, winchite/fibrous riebeckite)

The crystal chemistry, classification and structural features of amphiboles have been discussed in great detail in two excellent volumes of the *Reviews in Mineralogy and Geochemistry* series (formerly *Reviews in Mineralogy*) (vol. **9A**: Veblen and Ribbe, 1981; vol. **67**: Hawthorne *et al.*, 2007). Therefore, we will focus on the structural features and crystal chemistry of fibrous amphiboles, both regulated (actinolite, amosite, anthophyllite, crocidolite, tremolite) and unregulated (fluoro-edenite, winchite). Note that common features have been observed by several researchers involved in the crystal chemical and structural characterization of amphibole fibres, *i.e.* the relevant chemical variability of fibres from the same locality often leading to the classification as different amphibole species from the same hand specimen. This has been pointed out clearly at Libby, Montana, USA (Meeker *et al.*, 2003; Gunter *et al.*, 2003) and Biancavilla, Italy (Gianfagna *et al.*, 2007; Mazziotti *et al.*, 2009; Andreozzi *et al.*, 2009) two classic localities where amphiboles have been related to abnormally high numbers of asbestos-related diseases. At present, it is unclear if this extended compositional variability is amplified partly by the experimental difficulties encountered whenever analysing fibres of very small dimension. Hawthorne and Oberti (2007) have stressed appropriately the fact, apparently obvious, that the availability of detailed chemical data is the crucial point in the crystal chemical study of amphiboles. This is due to the occurrence of mixed valence cations (Fe, Mn among the more relevant) and that of light elements (H, Li) that renders the simple task of determining a reliable crystal chemical formula of amphiboles extremely tricky. In the case of fibrous amphiboles the small dimensions of the samples amplifies this effort by impeding the use of conventional analytical procedures. Besides, very important parameters are represented by the Fe^{2+}/Fe^{3+} ratio and the corresponding partition among the various cation sites, which have been obtained consistently by combining X-ray Powder Diffraction (XRPD), Mössbauer and Fourier transform-infrared (FTIR) spectroscopy. The relevance of those parameters is fundamental in order to model properly the chemical and biological reactivity of fibrous amphiboles that has been related unequivocally to the effect of surface iron ions acting as catalytic sites generating free radicals and reactive oxygen species (ROS) (Fubini and Otero Aréan, 1999).

Because the general structural features of amphiboles have been described extensively by Hawthorne and Oberti (2007) they will not be discussed in detail in the present chapter.

The chemical formula of amphiboles can be written as $A\ B_2\ C_5\ T_8\ O_{22}\ W_2$
where
A = Na, K, □ (vacancy), Ca, Li;
B = Na, Li, Ca, Mn^{2+}, Fe^{2+}, Mg;
C = Mg, Fe^{2+}, Mn^{2+}, Al, Fe^{3+}, Mn^{3+}, Ti^{4+}, Li;
T = Si, Al, Ti^{4+};
W = (OH), F, Cl, O^{2-}

Amphiboles crystallize in six different space groups (three monoclinic: *C*2/*m*, *P*2$_1$/*m*, *P*2/*a*; two orthorhombic: *Pnma*, *Pnmn*; one triclinic: $C\bar{1}$). However, other than anthophyllite (orthorhombic, *Pnma*), the amphiboles of environmental interest described in the present chapter crystallize in the monoclinic system, space group *C*2/*m* (Table 3). A sketch of the monoclinic *C*2/*m* amphibole structure, indicating the position of the cation sites, is shown in Fig. 1a. The double chain of tetrahedra contains two *T*(1) and *T*(2) sites that regularly alternate along the single chain that is fastened to an adjacent one *via* bridging between two *T*(1) tetrahedra. The *M*(1), *M*(2) and *M*(3)

Table 3. List of fibrous amphiboles of environmental interest.

Industrial/ commercial term	Mineralogical name	Regulated	Space group	Idealized chemical formula
Actinolite	Actinolite	Y	*C2/m*	$\square Ca_2Fe_5^{2+}[Si_8O_{22}](OH)_2$
Amosite	Grunerite	Y	*C2/m*	$\square Fe_2^{2+}Fe_5^{2+}[Si_8O_{22}](OH)_2$
Anthophyllite	Anthophyllite	Y	*Pnma*	$\square Mg_2Mg_5[Si_8O_{22}](OH)_2$
Crocidolite	Riebeckite	Y	*C2/m*	$Na_2(Fe_3^{2+}Fe_2^{3+})[Si_8O_{22}](OH)_2$
Tremolite	Tremolite	Y	*C2/m*	$\square Ca_2Mg_5[Si_8O_{22}](OH)_2$
Fluoro-edenite*	Fluoro-edenite	N	*C2/m*	$NaCa_2Mg_5[Si_7AlO_{22}]F_2$
Winchite*	Winchite	N	*C2/m*	$\square CaNa(Mg_4Al)[Si_8O_{22}](OH)_2$

* Does not apply, as they are not used for industrial purposes.

sites are conventionally grouped together as they host the so-called *C* cations. They are octahedrally coordinated but in the case of *M*(1) and *M*(3) the coordination comprises two and four O(3) anions, respectively, which, according to chemistry, may be (OH), F, Cl and O^{2-}. The *C* cations identify a strip of edge-sharing octahedra that is linked to the double chain of tetrahedra along the *a* and *b* axes. On the other hand the *M*(4), hosting *B* cations, is located at the rim of the strip of octahedra and is bonded to eight oxygen atoms in a square antiprism coordination. Finally, the *A* site occurs within a cavity located between the back-to-back double chain of tetrahedra being displaced with respect to the centre towards sites at 2 [*A*(2)] or *m* [*A*(*m*)] symmetry. The structure of orthorhombic *Pnma* amphiboles (Fig. 1b) is characterized by a doubling of the *a* parameter with respect to the monoclinic *C2/m* unit cell. The axial doubling results in the occurrence of four *T*1A, *T*1B, *T*2A and *T*2B tetrahedral sites that are arranged in such a way to build up two symmetry-independent double-chains. *M* cation sites, labelled *M*1, *M*2, *M*3 and *M*4, show comparable neighbouring bonding with respect to the corresponding ones in monoclinic *C2/m* amphiboles. Conversely, the *A* site is located at exactly the centre of the cavity originating between the back-to-back double-chain of tetrahedra.

Natural samples commonly have compositions that differ slightly in their major elements. A considerable amount of the trace metals that represent lattice substitution in certain crystallographic sites (specifically *M*1, *M*2, *M*3 and *M*4) was detected in asbestos amphiboles (Table 7) (Holmes *et al.*, 1971; Morgan and Cralley 1973; Bowes and Farrow 1997; Upreti *et al.*, 1984; Bloise *et al.*, 2016a). Amphiboles have a significant capability for hosting many geochemical trace elements (Scambelluri *et al.*, 1997; Tiepolo *et al.*, 2007). In fact, at present, because of both their common occurrence in a variety of igneous and metamorphic rock types and their ability to host trace elements, trace-element contents of amphiboles have been determined mainly for petrological purposes (*e.g.* Vils *et al.*, 2008). For example, knowledge of the

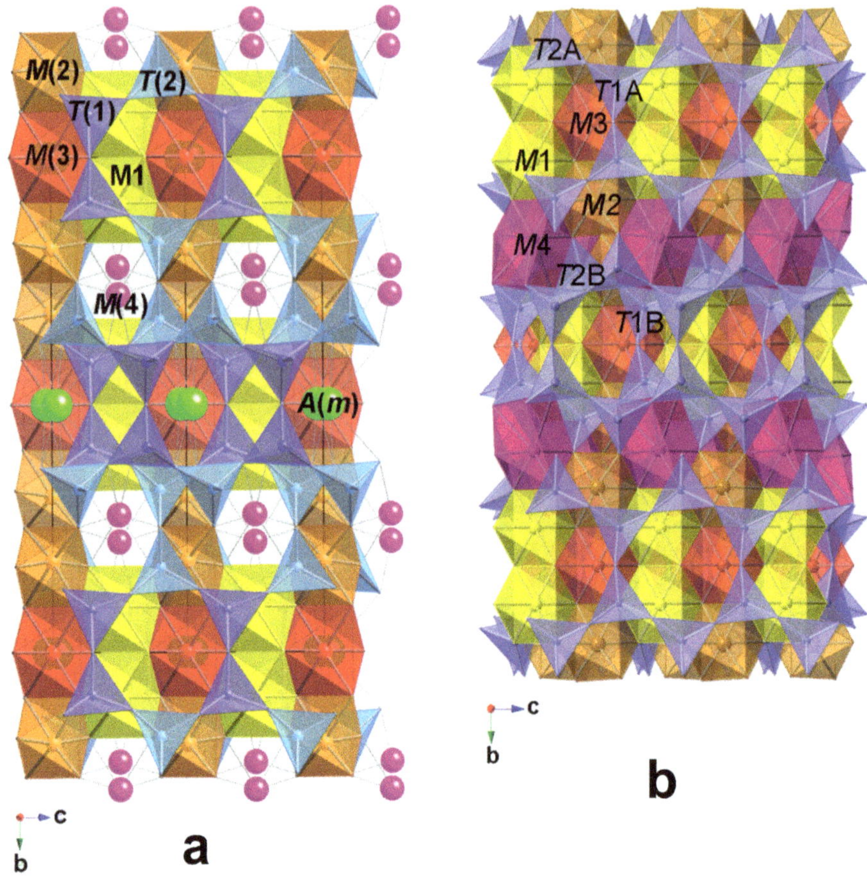

Figure 1. The structure of: (a) monoclinic *C2/m* amphiboles projected onto (001). *M*(1) polyhedra are yellow, *M*(2) orange, *M*(3) red. *M*(4) sites are pink, the *A*(*m*) sites are green. *T*(1) tetrahedra are mauve and *T*(2) are cyan; (b) orthorhombic *Pnma* amphiboles projected onto (001). *M*1 polyhedra are yellow, *M*2 orange, *M*3 red, *M*4 pink. All *T* tetrahedra are mauve.

partitioning behaviour of trace elements between amphiboles and/or serpentines and melt is important for understanding lithospheric processes (Veblen and Ribbe, 1981). The finding of relatively large concentrations of trace metals in asbestos samples is also consistent with the data in the literature showing large trace-metal contents (*e.g.* Mn, Cr, Co, Ni) in soils deriving from serpentinite (Lyon *et al.*, 1968; 1970; Smith *et al.*, 2007; Bloise *et al.*, 2016b). However, the role of trace elements in influencing asbestos-related health effects was rarely recognized in previous papers. In general, there are few contributions (Holmes *et al.*, 1971; Morgan and Cralley, 1973; Upreti *et al.*, 1984; Bowes and Farrow, 1997) that take into account the rare earth elements (*REE*) in this kind of chemical content. In relation to the water content (hydroxyl group) in amphiboles, extensive experiments on the dehydration of fibrous amphiboles were conducted with the principal aim of gaining knowledge of their crystal chemistry. However, after an initially intense experimental phase, few works addressed these topics in more recent years. Only recently has this issue received renewed attention in view of the possibility of fibrous amphiboles recycling *via* crystal chemical transformation induced by thermal treatment (see Bloise *et al.*, 2017, this volume). The application of thermogravimetry offers a number of possibilities for the investigation of asbestos amphiboles. The investigation may focus on the determination of the structural water content or the temperature at which structural breakdown of the fibrous amphiboles occurs (Bloise *et al.*, 2017, this volume). Note that the presence of other phases (*e.g.* carbonates) leads to variation in the mass loss (*i.e.* H_2O, CO_2) of the samples. It is also interesting to note that the thermal transformation sequence of materials, composed of a variety of phases, is very different from the transformation sequence of pure fibrous amphiboles. Moreover, the results of dehydroxylation for amphiboles often indicate an anomalously high water content. Hodgson *et al.* (1965a) attributed this excess to water molecules held tightly among the fibres. The hydroxyl groups of the amphiboles were driven out of the structure only at high temperatures (Bloise *et al.*, 2017, this volume) when – measured by thermogravimetric analyses – the values of water were between 1.37% for amosite and 2.62% for anthophyllite (Bush and Schumacher, 1982).

2.1. Fibrous actinolite

Fibrous actinolite from the Ligurian Alps, Italy, has been described from the crystal-chemical and structural point of view by Vignaroli *et al.* (2014) in order to define the asbestos hazard in ophiolitic rock mélanges. The samples analysed were both actinolite and tremolite. Fe^{2+}/Fe^{3+} site partitioning was determined indirectly from calculation of the $\langle M(1,2,3){-}O\rangle$ bond distance from the site population arising from the assignment to the *C* group site, as indicated by Hawthorne (1981). Despite different FeO_{total} contents, ranging from 1.9 to 12.3 wt.%, an Fe^{2+} $M(2) > M(1) > M(3)$ site-preference has been observed consistently in actinolite. An interesting result of the TEM dimensional analysis of the various fibres was that according to the criterion chosen to classify the fibres by an aspect ratio *vs.* width diagram a very different asbestos hazard scenario would be hypothesized. Moreover, it was observed that the site scattering (s.s.) of the *C* group sites calculated from electron probe micro-analysis (EPMA) data and that from structure refinement differs. In particular s.s. from site partitioning was found to be consistently higher than that calculated from Rietveld refinements; with increasing deviation the s.s. increased. This phenomenon was modelled by the following linear fit (recalculated): s.s. from site partitioning = $-16(7) + 1.25(11) \times$ s.s. from refinement ($R^2 = 0.989$) considering only high-quality Rietveld refinements from transmission mode experiments on capillaries (Pacella *et al.*, 2008; Ballirano *et al.*, 2008; Andreozzi *et al.*, 2009; Vignaroli *et al.*, 2014) (Fig. 2). As observed by Gianfagna *et al.* (2007) the mismatch is taken up mainly by excess s.s. at $M(2)$. No straightforward justification for this systematic error is available at present except for the hypothesis that such an effect is generated by an imperfect absorption correction procedure.

2.2. Amosite

Although grunerite structure models are reported in the literature, there are difficulties in obtaining a structure model of its fibrous variety, amosite, from diffraction data, probably because of texture and disorder effects. In fact, a fibrous specimen of amosite studied by electron diffraction and high-resolution electron microscopy showed twinning on (100) leading to stacking disorder, and an explanation is provided of previously supposed anomalies in their diffraction patterns (Chisholm, 1973; Hutchinson *et al.*, 1975). Sections perpendicular to [001] of amosite examined by high-resolution TEM revealed two kinds of dislocation with about equal frequency. One lies on [001] and has a Burgers vector **a**. The other is on [001] and has a Burgers vector 1/2**a**+1/2**b** (Whittaker *et al.*, 1981).

We report in this chapter the results of a structure refinement of amosite using synchrotron powder diffraction data. The sample is the UICC standard amosite sample with chemical formula determined by EPMA, partition determined from Mössbauer data (Pollastri *et al.*, 2015) and H_2O from TG data (Bloise *et al.*, 2016c):

$$Ca_{0.02}Na_{0.01}Fe^{2+}_{2.00}(Fe^{2+}_{3.00}\ Fe^{3+}_{0.46}Mg_{1.47}Mn_{0.06})_{4.99}(Si_{7.86}\,{}^{IV}Al_{0.01})_{7.87}O_{21.96}(OH)_{2.04}$$

The sample is contaminated with very minor impurities of calcite, hematite and quartz. Details of the data collection, refinement strategy and results of the structure refinement are reported by Pollastri *et al.* (2017). The calculated structure model (space group *C*2/*m*) is very similar to that of grunerite (Finger, 1969) and in perfect agreement

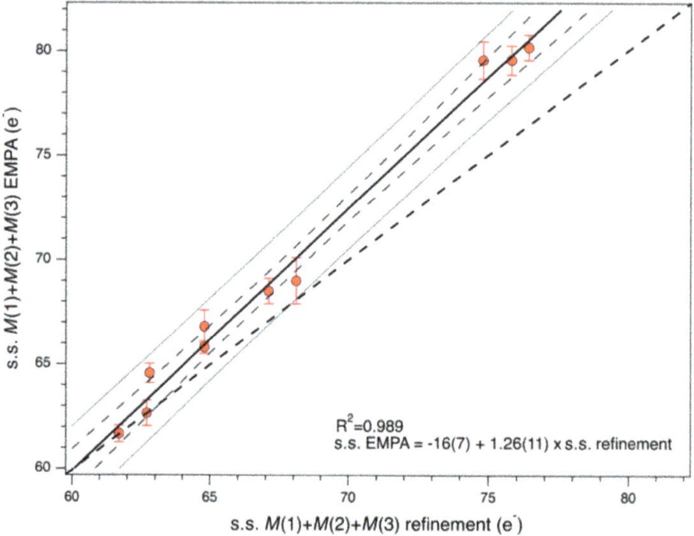

Figure 2. Comparison between the s.s. of the C cations as calculated from EMPA and Rietveld refinements. Modified from Vignaroli *et al.* (2014). See original reference for samples details. Regression, confidence (95%) and prediction intervals are reported as full black, dotted and short-dash grey lines, respectively. As a visual guide a short-dash black line as been drawn indicating ideal behaviour.

with the chemical data from the EPMA analysis. The Rietveld agreement factors are: R_{wp} = 8.23%, R_p = 5.00%, and χ^2 = 5.44. The refined unit-cell parameters are: *a* = 9.5484(2) Å, *b* = 18.3395(4) Å, *c* = 5.3346(1) Å, and β = 101.825(2)°. The refined cell parameters and volume are smaller than those of the grunerite model determined by Finger (1969) because of the smaller iron content of our sample (the Fe/(Fe+Mg) ratio is 0.88 for the Finger sample and 0.78 for our sample). The values of the <*T*–O> distances for both <*T*(1)–O> and <*T*(2)–O> are consistent with the literature values for virtually Al-free *C*2/*m* amphiboles (Hawthorne and Oberti, 2007). The geometry of the tetrahedra in amosite is very similar to that of Finger's grunerite with <*T*(1)–O> = 1.628 Å (*vs*. 1.627 Å) and <*T*(2)–O> = 1.633 Å (*vs*. 1.622 Å). Polyhedron distortion, Δ, shows very small values for both *T*(1) and *T*(2) tetrahedra (Table 4).

The results of the structure refinement indicate that Mg is disordered over the *C* sites with a preference for the *M*(2) site whereas Mg is not found at the *B* site *M*(4). This cation distribution scheme is in perfect agreement with Finger's grunerite model where 15% Mg is found at the *M*(1) site, 22.7% Mg is found at the *M*(2) site, 11.2% Mg is found at the *M*(3) site and virtually no Mg at the *M*(4) site (Finger, 1969). Our cation distribution scheme is also in agreement with the single-crystal X-ray structure refinements of natural members of the cummingtonite-grunerite series (Ghose, 1961; Fischer, 1966; Finger, 1969; Ghose and Ganguly, 1982) showing that Fe^{2+} strongly prefers the *M*(4) site and Mg prefers *M*(2). The cation distribution scheme of amosite is also in agreement with the grunerite model of Ghose and Hellner (1959).

The octahedrally coordinated cations at the *M*(1), *M*(2) and *M*(3) sites display a fairly regular environment. The mean <*M*(1,2,3)–O> is 2.084 Å and considering an empirical ionic radius of 0.78 Å for iron and 0.72 Å for magnesium in an octahedrally coordinated environment, our data are in agreement with the estimated dependence of the <<*M*(1,2,3)–O>> distance as a function of the mean aggregate radius of the *M*(1,2,3) cations in *C*2/*m* amphiboles (Hawthorne and Oberti, 2007).

The *M*(4) site, which occurs at the junction of the strip of octahedra and the double-chain of tetrahedra in all amphibole structure types, is occupied by Fe^{2+} in only our sample. The *B* cation at the *M*(4) site bonds to oxygen atoms of both the strip of octahedra and the double chain of tetrahedra, and is the basic link between these two parts of the structure. As a result, this site and its constituent cations have a major influence on the symmetry, crystal chemistry and chemical composition of the amphiboles (Hawthorne and Oberti, 2007). The distance <*M*(4)–O> = 2.223 Å of amosite is comparable to that of Finger's grunerite (2.293 Å); although the latter includes unrealistic values such as the distance *M*(4)–O(4) of 1.988(4) Å. It is confirmed that <*M*(4)–O> increases slightly in the cummingtonite–grunerite series with macroscopic Fe/(Fe+Mg), due primarily to large changes in the longer *M*(4)–O(6) distance. It is also confirmed that the *M*(4)–O(4) bond distance actually decreases with increasing Fe/(Fe+Mg), which is consistent with the inference that this bond has considerable covalent character (Ghose and Ganguly, 1982). The octahedral environment of Fe^{2+} at the *M*(4) site is highly distorted as witnessed by the enormous value of the polyhedron distortion (Δ = 97.263). This value is even greater for Finger's grunerite (211.294).

Table 4. Structural data of magnesium- and fluoro-riebeckite samples.

Sample	Reference	*a* (Å)	*b* (Å)	*c* (Å)	β (°)	*V* (Å^3)
Magnesio-riebeckite	Whittaker (1949)	9.72*	17.95	5.31	103.9*	899
Magnesio-riebeckite	Hawthorne *et al.* (2008a)	9.700(3)	17.917(6)	5.300(2)	103.70(3)	894.91
Magnesio-riebeckite	Hawthorne *et al.* (2008a)	9.703(2)	17.917(3)	5.300(1)	103.74(2)	895.03
Magnesio-riebeckite	Hawthorne *et al.* (2008a)	9.708(4)	17.909(8)	5.300(4)	103.78(4)	894.94
Magnesio-riebeckite	Hawthorne *et al.* (2008a)	9.773(2)	17.912(3)	5.292(1)	103.87(2)	899.37
Fluoro-riebeckite	Hawthorne (1978)	9.811(3)	18.013(5)	5.326(2)	103.68(1)	914.54

* Recalculated from original data.

Sample	Reference	<*M*(1)–O>^{VI} (Å)	<*M*(2)–O>^{VI} (Å)	<*M*(3)–O>^{VI} (Å)	<*M*(4)–O>^{VIII} (Å)	<*A*–O>^{X} (Å)
Magnesio-riebeckite	Whittaker (1949)	2.11	2.06	2.11	2.54	–
Magnesio-riebeckite	Hawthorne *et al.* (2008a)	2.074	2.034	2.081	2.526	2.925
Magnesio-riebeckite	Hawthorne *et al.* (2008a)	2.074	2.034	2.079	2.526	2.924
Magnesio-riebeckite	Hawthorne *et al.* (2008a)	2.071	2.036	2.079	2.527	2.921
Magnesio-riebeckite	Hawthorne *et al.* (2008a)	2.071	2.043	2.082	2.537	2.919*
Fluoro-riebeckite	Hawthorne (1978)	2.114	2.03	2.113	2.550	2.944

As far as the trace-metals content (Cr, Mn, Co, Ni, Cu and Zn) is concerned, in standard UICC amosite analysed by different authors (Holmes *et al.*, 1971; Morgan and Cralley, 1973; Bowes and Farrow, 1997; Bloise *et al.*, 2016a) the Mn content has often been detected in large amounts (>1000 ppm). This is in accordance with the results of a previous study (Oberti *et al.*, 2007), reporting that, in magnesium-iron-manganese-lithium amphiboles, Mn may substitute for Fe or Mg in all the *M*(1), *M*(2), *M*(3) and *M*(4) sites (Oberti *et al.*, 2007). Also, the other trace metals (*i.e.* Co, Ni, Cu, Zn) substitute for Mg and Fe^{2+} in the cummingtonite–grunerite series (Oberti *et al.*, 2007). Kohyama *et al.* (1996) found a large Mn content (1.8 wt.% as MnO) in the standard UICC amosite even though, in a sample of amosite from Chamoli (India) analysed by XRF, the Mn content was only 200 ppm (Upreti *et al.*, 1984). These differences in Mn content of Indian samples as compared to UICC samples were attributed to the geochemical processes involved in the formation of silicate ores in different places (Upreti *et al.*, 1984). However, Morgan and Cralley (1973) reported that the levels of trace metals excluding Mn (*i.e.* Cr, Ni and Co) in UICC amosite from South Africa were generally one or two orders of magnitude lower than in standard UICC chrysotile (Table 7). For *REE*, Bloise *et al.* (2016a), who investigated standard UICC amosite from Penge (South Africa) using inductively coupled plasma-mass spectrometry (ICP-MS), reported that they were present in small amounts compared to the other asbestos samples (*i.e.* crocidolite, tremolite, anthophyllite and chrysotile).

With respect to water content (hydroxyl group) in amosite, a huge effort was made by Hodgson *et al.* (1965b) to understand the thermal behaviour of amosite from Penge (South Africa). Those authors reported that when amosite was heated, part of the hydroxyl water was lost as hydrogen, and the amount of water produced was between 1.40 and 1.85%. Bush and Schumacher (1982) showed a TG (thermogravimetry) curve in which the water (hydroxyl group) loss by amosite from South Africa was 1.37%. Bloise *et al.* (2016c) showed a TG curve between 110 and 690°C, in which a mass loss of 1.94% was due to dehydroxylation of the UICC amosite (South Africa) (see Bloise *et al.*, 2017, this volume).

2.3. Fibrous anthophyllite

The literature reports many structure models of non-fibrous orthorhombic anthophyllites (Warren and Modell, 1930; Lindemann, 1964; Finger, 1970; Walitzi *et al.*, 1989; Evans *et al.*, 2001; Schindler *et al.*, 2008; Hawthorne *et al.*, 2008b). Like amosite, difficulties in obtaining a structure model of these mineral fibres are plausibly due to texture and disorder effects. In fact, Veblen (1980) showed that a natural anthophyllite specimen consists primarily of small amphibole crystallites that are greatly elongated in the *c* direction. Adjacent crystallites are rotated with respect to each other, but are not related by a twinning operation. Anthophyllite possesses pervasive chain-width disorder. Central screw dislocations are not present in the crystallites, indicating that they did not grow by a spiral growth mechanism around screw dislocations.

In this section, we report the results of a structure refinement of fibrous anthophyllite. The sample is the UICC standard sample from Paakkila (Finland) with chemical formula determined by EPMA, partition determined from Mössbauer data (Pollastri *et al.*, 2015) and H_2O from TG data (Bloise *et al.*, 2016c):

$$Ca_{0.04}Mg_2(Mg_{3.83}Fe^{2+}_{1.14}Mn_{0.05})_{5.02}(Si^{IV}_{7.86}Al_{0.02})_{7.88}O_{21.62}(OH)_{2.38}$$

The fibres are contaminated with biotite, clinochlore/vermiculite and talc. Details of the data collection are reported by Pollastri *et al.* (2017). Refinements were accomplished using the *GSAS* package (Larson and Von Dreele, 1994) and its graphical interface *EXPGUI* (Toby, 2001). The Rietveld agreement factors of the refinement conducted in space group *Pnma* are: R_{wp} = 5.930%, R_p = 4.180%, and χ^2 = 8.530. The refined unit-cell parameters are: a = 18.5770(8) Å, b = 18.035(2) Å, and c = 5.27285(9) Å. The results of the structure refinement are reported extensively in Pollastri *et al.* (2017). The orthorhombic anthophyllites by Walitzi *et al.* (1989), Schindler *et al.* (2008) and Hawthorne *et al.* (2008b) share many crystal-chemical features with our sample: they are Al-poor (Al <0.5 a.p.f.u.); they display no Fe^{3+}; Fe^{2+} is in the range 1.14–1.66 a.p.f.u. (our sample has the lowest Fe^{2+} content); they possess comparable unit-cell dimensions, with 18.544 $\leqslant a$ (Å) $\leqslant$ 18.577, 18.026 $\leqslant b$ (Å) $\leqslant$ 18.353 and 5.273 $\leqslant c$ (Å) $\leqslant$ 5.290. The values of the mean <*T*–O> distance of 1.611 Å for our fibrous anthophyllite (with Al = 0.01 a.p.f.u.) is in very good agreement with the grand <*T*–O> distance of 1.621(1) Å for virtually $^{[4]}$Al-free *Pnma* amphiboles (Schindler *et al.*, 2008) and with the anthophyllite of Walitzi *et al.* (1989) having Al = 0.5 a.p.f.u. and <*T*–O> = 1.625 Å. Polyhedron distortion Δ shows very small values for all the tetrahedra.

The results of the structure refinement indicate that only Mg is found at the *M*1, *M*2 and *M*3 sites with Fe^{2+} at the *M*4 together with Mn, and minor Ca at the *A* site. This result is virtually identical to that found by Walitzi *et al.* (1989) for their orthorhombic anthophyllite. Schindler *et al.* (2008) also assigned most of Fe^{2+} (1.02–1.39 a.p.f.u.) to *M*4 with minor populations at the *M*1 (0.17–0.32 a.p.f.u.), *M*2 (0.04–0.17 a.p.f.u.), and *M*3 (0.05–0.14 a.p.f.u.) sites in orthorhombic anthophyllites. The octahedrally coordinated Mg at the *M*1, *M*2, and *M*3 sites displays a regular geometry as the polyhedron distortion Δ shows small values for all the octahedra. The mean <*M*1,2,3–O> is 2.078 Å and considering an empirical ionic radius of 0.72 Å for magnesium in an octahedrally coordinated environment, our data are in agreement with the estimated dependence of the <<*M*1,2,3–O>> distance as a function of the mean aggregate radius of the *M*(1,2,3) cations in figure 3 of Schindler *et al.* (2008). Site *M*4 shows some differences with respect to the literature data. The distance <*M*4–O> = 2.244 Å for a 6-fold coordination is comparable to the value (2.261 Å) reported by Walitzi *et al.* (1989) for their anthophyllite with *M*4 in an octahedrally coordinated environment. Such distances are shorter with respect to the values reported by Schindler *et al.* (2008) for orthorhombic anthophyllite with *M*4 in a 7-fold coordinated environment: <$^{[7]}$*M*4–O> = 2.349(2) Å for sample *A*(5) and 2.351(3) Å for sample *A*(18) (see table 5a,b in Schindler *et al.*, 2008). The octahedral environment at the *M*4 site is highly distorted as evidenced by the enormous value of the polyhedron distortion (Δ = 135.1).

Freeman (1966) suggested that, in anthophyllite, the weight loss is greater than that required for 2(OH) per formula unit, and water is lost in two steps. This two-stage dehydroxylation behaviour was noted previously by Thilo and Rogge (1939). The first stage was attributed to loss of hydroxyls bonded to Si while the second stage was attributed to hydroxyls bonded to *M*1 and *M*3 cations. Bush and Schumacher (1982) showed a TG curve in which the water (hydroxyl group) loss of anthophyllite from Paakkila (Finland) was 2.62%. However, analysing the same sample, Bloise *et al.* (2016c) showed that the mass loss was slightly lower, 2.30%. It is worth mentioning that standard UICC anthophyllite asbestos from Paakkila (Finland) has the largest amounts of trace metals (Mn, Cr, Co, Ni, Cu and Zn) compared to the other asbestos samples (Morgan and Cralley, 1973; Bloise *et al.*, 2016a). In particular, the amount of Zn in standard UICC anthophyllite asbestos is generally several orders of magnitude greater than the concentrations found in other asbestos (Bowes and Farrow, 1997; Bloise *et al.*, 2016a).

2.4. Crocidolite

Crocidolite represents the first example of a crystal-structure refinement of a fibrous amphibole (Whittaker, 1949). As a matter of fact, it is the first detailed structural description of an amphibole following the pioneering work of Warren (1930a,b). According to its formula $(Na_{1.38}K_{0.13}Ca_{0.17}Mg_{0.25})(Mg_{2.81}Fe^{3+}_{1.66}Fe^{2+}_{0.48}Al_{0.05})[Si_{7.94}Al_{0.06}O_{22}](OH)_2$, it may be classified as a magnesio-riebeckite, different from the South African samples that are commonly classified as riebeckite (Hodgson, 1979; Miyano and Beukes, 1997; Pacella *et al.*, 2014) owing to a larger iron content. Despite being the fibrous amphibole most used by industry (~50% of production and consumption: Virta, 2005), it represents the only structural data available for fibrous riebeckite. Whittaker (1949) indicated that the fibrous amphibole contains fibres approximating to single crystals. The refined structure resembles closely that reported by Hawthorne (1978) for fluoro-riebeckite and by Hawthorne *et al.* (2008) for several magnesio-riebeckite samples (Table 4). The main difference is related to the inability to locate electron density at the *A* site from the old photographic data. Single-crystal analysis points consistently to an almost complete segregation of Fe^{3+} at *M*(2) (Hawthorne *et al.*, 2008) in the case of prismatic samples. As for trace elements, standard UICC crocidolite from Penge (South Africa) contains the smallest amount of trace metals compared to the other UICC asbestos samples (Holmes *et al.*, 1971; Morgan and Cralley, 1973; Bowes and Farrow, 1997; Bloise *et al.*, 2016a). For instance, Cr was concentrated mainly in standard anthophyllite (Finland) and decreased in other asbestos amphiboles, *i.e.* UICC crocidolite that has the lowest Cr content (up to 20 ppm) (Holmes *et al.*, 1971; Bloise *et al.*, 2016a). However, Bloise *et al.* (2016a) suggested that the decreasing concentration of Cr from magnesium (anthophyllite) to sodium amphibole (crocidolite) was controlled solely by its geochemical availability. In fact, an experimental synthesis has shown that Cr allocated mainly in the *M*(2) site can be a major component in calcic, sodic-calcic and sodic amphiboles (Oberti *et al.*, 2007) On the other hand, standard UICC crocidolite, analysed by ICP-MS, has the largest Be content (0.55 ppm) compared to the other asbestos samples (Bloise *et al.*, 2016a). It is worth remembering that several studies claim that Be increases both asbestos carcinogenesis (*e.g.* Dixon *et al.*, 1970) and mortality from lung cancer (Mancuso, 1970). In crocidolite, Be^{2+} replace Si^{4+} at the *T* site and the electroneutrality required is obtained by the occurrence of Pb in the *A* site (Oberti *et al.*, 2007). UICC crocidolite has a large Pb content (4.99 ppm) (Bloise *et al.*, 2016a).

With respect to water content (hydroxyl group) in crocidolite, Hodgson *et al.* (1965a) studied the thermal decomposition of fibrous crocidolite from Koegas (South Africa) concluding that the main mass loss of 1.92% (range 500–700°C) was related to crocidolite dehydroxylation. Some years later, with the aim of investigating the thermal behaviour of different types of asbestos, Bush and Schumacher (1982) showed a TG curve in which the water (hydroxyl group) loss of crocidolite from South Africa (in the range 700–900°C) was 1.63%. Recently, using very sensitive DSC/DTG (differential scanning calorimetry and derivative thermogravimetry), Bloise *et al.* (2016c) showed a full thermal behaviour of crocidolite from Koegas (South Africa) highlighting the downward peaks which related to the dehydrogenation process (Bloise *et al.*, 2017, this volume). Those authors concluded that standard UICC crocidolite dehydroxylation took place within the 350–700°C range, and the TG plot showed a mass loss of 2.18% in this range. This finding was in agreement with the work of Fujishige *et al.* (2007) who showed that weight loss of crocidolite was not observed above 700°C, and that the weight loss related to breakdown of crocidolite was ~2.00%.

2.5. Fibrous tremolite

In recent years several investigations have been performed on samples of fibrous tremolite from different localities in Italy. This is due to its relatively widespread diffusion in rocks that are currently subjected to excavation for numerous construction projects (railway lines, motorways, *etc.*). A nearly end-member of tremolite from the Susa Valley, Italy, was analysed with particular regard to the oxidation state and cation-site distribution of Fe (Ballirano *et al.*, 2008). The results were obtained by combining FTIR and Mössbauer spectroscopy, EPMA and Rietveld refinements on laboratory X-ray powder diffraction (XRPD) data. The sample, characterized by FeO_{total} of 1.17 wt.%, was found to contain 84% of Fe^{2+} located primarily at *M*(1) (0.05 a.p.f.u.) and *M*(2) (0.04 a.p.f.u.) and subordinately at *M*(3) (0.02 a.p.f.u.). All Fe^{3+} (0.02 a.p.f.u.) was assigned to *M*(2). The same approach was applied to a sample from Ala di Stura, Lanzo Valley, Italy (Pacella *et al.*, 2008). The sample was characterized by a greater FeO_{total} content of 2.42 wt.% and by a correspondingly larger cell volume (909.07(2) rather than 907.37(1) $Å^3$). The same Fe^{2+}-Fe^{3+} site partition scheme was observed in both fibrous tremolite samples albeit in this case a preferential partition of Fe^{2+} at *M*(2) was observed (*M*(1) = 0.09 a.p.f.u.; *M*(2) = 0.12 a.p.f.u.; *M*(3) = 0.05 a.p.f.u.). Similarly to the sample from the Susa Valley, Fe^{2+} is dominant (90%) with respect to Fe^{3+}.

Four samples of tremolite asbestos from Italy and the USA have been characterized by a multi-analytical approach to correlate crystal chemistry and chemical reactivity (Pacella *et al.*, 2010). Samples analysed from Italy were from: (1) Castelluccio Superiore, Potenza, Basilicata; (2) Mt Rufeno, Acquapendente, Latium (previously described by Burragato *et al.*, 2001); and (3) San Mango, Catanzaro, Calabria. A further sample from Montgomery County, Maryland, USA, was also characterized. The bulk Fe^{3+}/Fe_{total} ratio, determined by Mössbauer spectroscopy, highlighted very different values ranging from 6 to 24%. For all samples Fe^{2+} was found to be distributed among the *M*(1,2,3) sites whereas Fe^{3+} was segregated at *M*(2). Notably, the $HO^{\cdot}$ radical production was correlated positively to Fe occupancy at the *M*(1)+ *M*(2) sites. It was pointed out that the refined cell volume was larger (908–909 Å^3) than that of prismatic tremolite samples (906–907 Å^3: Yang and Evans, 1996; Gottschalk *et al.*, 1999; Verkouteren and Wylie, 2000), the only exception being the sample from Mt Rufeno. This expansion was related to the occurrence of Fe^{2+}. In the case of the sample from Mt Rufeno the occurrence of Al^{3+} at *C* was assumed partly to counterbalance the expansive effect of Fe^{2+}. The authors reported an excellent agreement between s.s. refined and calculated from chemical data, the largest difference being <4%.

A very interesting exception to the general trend is represented by the sample pertaining to the actinolite-tremolite amphibole suite from the Ligurian Alps described by Vignaroli *et al.* (2014). In fact, one (TB) sample, containing minor antigorite and traces of quartz, has an anomalously small Fe^{2+}/Fe^{3+} ratio of 0.48, different from both other associated actinolite samples (Fe^{2+}/Fe^{3+} ~0.9) and reference fibrous tremolite (Table 5). Fe^{3+} was located at *M*(1)+*M*(2) whereas Fe^{2+} was allocated to the three *M* sites. The indirect Fe^{2+}/Fe^{3+} partition was performed by calculating the <*M*(1),(2),(3)–O> bond distance from the site population, devised from the conventional assignment to the *C* group sites (Hawthorne, 1981) and using the cation ionic radii from Shannon (1976) and a ^{III}O radius of 1.36 Å, and subsequent iterative optimization of the corresponding <*M*(1),(2),(3)–O> bond distance from the Rietveld refinements. In a recent work, Bloise *et al.* (2016a) showed that trace metals such as Mn (>1000 ppm), Co (27 ppm) and Ni (473 ppm) in tremolite asbestos from Val d'Ala (Italy) were present in large amounts. This finding is in line with the results of Upreti *et al.* (1984) who analysed tremolite asbestos from Bhilwara (India) by X-ray fluorescence, and concluded that it contained a considerable amount of Mn, Co and Ni. It is likely that both Ni and Co in tremolite occupy the specific crystallographic *M*(1) and *M*(3) sites as in synthetic calcic amphibole (*i.e.* pargasite) (Della Ventura *et al.*, 1997). It is also worth mentioning that among the asbestos trace elements known as being dangerous for human health, Ni is of greatest concern (Nackerdien *et al.*, 1991). Tremolite asbestos from Val d'Ala (Italy) has the greatest *REE* concentration with respect to other asbestos (Bloise *et al.*, 2016a). Light rare earth elements (LREE: Sc, La, Ce, Pr, Nd, Sm, Eu and Gd), which have ionic radii larger than heavy rare earth elements (*HREE*: Y, Tb, Dy, Ho, Er, Tm, Yb and Lu), preferentially occupy the *M*(4) site, while *HREE* are preferentially sited closer to the strip of octahedra at the *M*(4') site (Bottazzi *et al.*, 1999).

As far as water content (hydroxyl group) in tremolite is concerned, one of the first works on tremolite thermal behaviour was reported by Posnjak and Bowen (1931) who – through static weight-loss experiments – showed that, at 900°C, 2.22% of water (hydroxyl group) content in tremolite was expelled. However, Schaller, (1916), who

Table 5. Fe speciation and site partition of selected fibrous tremolite samples.

Sample	Reference	Fe^{2+}/Fe_{tot}	Fe_{tot} (a.p.f.u.)	*M*(1)	*M*(2)	*M*(3)
Ala di Stura, Italy	Pacella *et al.* (2008)	0.883	0.281	$Fe^{2+}_{0.09}$	$Fe^{2+}_{0.12}+Fe^{3+}_{0.03}$	$Fe^{2+}_{0.05}$
Mt. Rufeno, Italy	Pacella *et al.* (2010)	0.758	0.264	$Fe^{2+}_{0.08}$	$Fe^{2+}_{0.08}+Fe^{3+}_{0.06}$	$Fe^{2+}_{0.04}$
Castelluccio, Italy	Pacella *et al.* (2010)	0.790	0.238	$Fe^{2+}_{0.08}$	$Fe^{2+}_{0.07}+Fe^{3+}_{0.05}$	$Fe^{2+}_{0.04}$
San Mango, Italy	Pacella *et al.* (2010)	0.938	0.341	$Fe^{2+}_{0.13}$	$Fe^{2+}_{0.13}+Fe^{3+}_{0.02}$	$Fe^{2+}_{0.04}$
Montgomery Co., Md, USA	Pacella *et al.* (2010)	0.841	0.521	$Fe^{2+}_{0.18}$	$Fe^{2+}_{0.17}+Fe^{3+}_{0.08}$	$Fe^{2+}_{0.09}$
Ligurian Alps, Italy	Vignaroli *et al.* (2014)	0.478	0.230	$Fe^{2+}_{0.02}+Fe^{3+}_{0.04}$	$Fe^{2+}_{0.02}+Fe^{3+}_{0.08}$	$Fe^{2+}_{0.07}$

had analysed three tremolite samples, had already advanced the idea that the presence of OH ions was essential for the structure of amphibole. More recently, the main mass loss of 2.02% between 850 and 1050°C due to tremolite (Val d'Ala, Italy) dehydroxylation on the TG curve was shown by Bloise *et al.* (2016c) (see Bloise *et al.*, 2017, this volume).

2.6. Fluoro-edenite

Fluoro-edenite $NaCa_2Mg_5[Si_7AlO_{22}]F$ is an end-member of the calcic amphibole group (Gianfagna and Oberti, 2001) that occurs with both prismatic and fibrous morphologies. It has been found in altered volcanic products of Monte Calvario, Biancavilla, on the flanks of Mount Etna, Sicily, Italy. It represents a typical example of non-occupational exposure of an environmental nature to mineral fibres because an epidemiological study of the area around Biancavilla pointed to a significant incidence of malignant pleural mesothelioma (Paoletti *et al.*, 2000) that was subsequently attributed to the inhalation of fluoro-edenite (Comba *et al.*, 2003). Fluoro-edenite has been included recently by IARC in the Group 1 Human-Carcinogens list (IARC, 2017; *in press*). Multi-analytical investigations carried out on both prismatic and fibrous samples (Gianfagna and Oberti, 2001; Gianfagna *et al.*, 2003, 2007) have shown differences with respect to their crystal chemistry. In particular, samples with fibrous morphology are characterized by larger Fe_{total}, Fe^{2+} and Si contents, and smaller Mg^{2+} and Ca^{2+} contents compared to those with prismatic morphology (Gianfagna *et al.*, 2007). Iron partition, performed by Mössbauer spectroscopy, indicated a preferential partition of Fe^{2+} at *M*(2) [with a minor occurrence at *M*(3)] and of Fe^{3+} at *M*(3). The *C* sites of fibrous fluoro-edenite show an increased site scattering (s.s) compared to that of prismatic fluoro-edenite. This behaviour is driven mainly by the significant increase of s.s. at *M*(2). Cell parameters reflect the chemical differences as the *a* parameter of the fibrous samples is consistently shorter than that of the prismatic ones. On the contrary, the *b* parameter of fibres is slightly larger (Andreozzi *et al.*, 2009). As mentioned above, fibres sampled at Biancavilla show a marked chemical variability as they spread across four amphibole compositional fields *i.e.* edenite (prevailing), winchite, tremolite and richterite (rare) (Gianfagna *et al.*, 2007; Mazziotti *et al.*, 2008; Andreozzi *et al.*, 2009).

2.7. Winchite/fibrous riebeckite

Winchite is the most abundant fibrous amphibole occurring at the Rainy Creek complex, Libby, Montana, USA. However, following the classification criteria of Leake *et al.* (1997), as many as five further amphiboles have been reported: richterite, tremolite and possibly magnesioriebeckite (magnesio-riebeckite: Hawthorne and Oberti, 2006), edenite and magnesio-arfedsonite (Meeker *et al.*, 2003). Libby is famous for hosting the biggest vermiculite mine in the world (peaking at ~80% of the total world production) that closed in 1990. However, the vermiculite produced was found to be contaminated by traces of amphibole-asbestos initially referred to incorrectly as tremolite or actinolite (USEPA, 2000) and subsequently identified as winchite (Wylie and Verkouteren, 2000; Gunter *et al.*, 2003). Structure refinement was performed on a single crystal selected from the bulk sample characterized by a formula

$$(K_{0.19}Na_{0.32})(Na_{0.85}Ca_{1.12}Mn_{0.03})(Mn_{0.01}Mg_{4.43}Fe^{3+}_{0.34}Fe^{2+}_{0.19}Ti_{0.01}Al_{0.02})(Al_{0.03}Si_{7.97}O_{22})(OH_{1.63}F_{0.37})$$

as determined by EPMA and Mössbauer spectroscopy. The refinement, carried out in *C2/m*, a = 9.879(2) Å, b = 18.024(3) Å, and c = 5.288(1) Å, β = 104.377(3)°, outlined clearly the difficulty in describing correctly and classifying those minerals. In fact, according to the *A*-site occupancy based on EPMA data (0.51 a.p.f.u.) and that based on X-ray diffraction data (0.48 a.p.f.u.) the sample would be classified as richterite or winchite, respectively! Mg was distributed among *M*(1), *M*(2) and *M*(3), all available Fe^{3+} was located at *M*(2), whereas Fe was segregated over *M*(2) and *M*(3). Calcium and Na occupied *M*(4) whereas the remaining Na and K resided in *A*. A further study (Bandli *et al.*, 2003) reported optical, compositional, morphological and X-ray data of 11 particles of winchite from Libby. The results indicated that some of them were polycrystalline but in spite of being characterized by similar chemical composition they possessed significantly different cell parameters ranging from $9.77 < a < 9.93$ Å, $17.96 < b < 18.27$ Å, $5.27 < c < 5.31$ Å, $104.38 < \beta < 105.27°$.

In a recent paper, fibrous riebeckite from Libby was characterized in detail (Pacella and Ballirano, 2016). The most relevant information was that the chemistry of the riebeckite was very similar to that of prismatic winchite, the main difference being a significantly higher F content (0.82 a.p.f.u. *vs*. 0.37 a.p.f.u.) of riebeckite compared to winchite. This difference has produced smaller cell parameters for richterite than for winchite. The iron partition was found to be similar: all Fe^{3+} in *M*(2) and Fe^{2+} distributed among *M*(1), *M*(2) and *M*(3) sites. The cation partition was obtained by combining the chemical data with the Rietveld refinement results using the conventional assignment to the *C* group sites following Hawthorne (1981). The partition produced a regular discrepancy of −5% of the s.s. from structure refinement with respect to those from site occupancy for *A*, *B* and *C* group sites. This result is in qualitative agreement with the findings of Vignaroli *et al.* (2014) who hypothesized that such behaviour was related to an imperfect absorption-effects correction. Those results support their hypothesis because an increase in the capillary diameter compared to that used in previous investigations (1 mm instead of 0.7 mm) produced similar effects but distributed in a different way among the various *A*, *B* and *C* group sites. The difference between refined s.s. and those calculated from chemical data for the *A* site reaches ~4% relative and may reflect the simplified model used for the *A* site (a single site located at 0, ½, 0), which does not take into account properly the possible presence of electron density at *A*(*m*) and/or *A*(2) sites. The $\langle M(1){-}O \rangle$ (2.073 Å), $\langle M(2){-}O \rangle$ (2.074 Å) and $\langle M(3){-}O \rangle$ (2.068 Å) distances observed were found to be in good agreement with those calculated using the equations of Hawthorne and Oberti (2007) (2.066 Å, 2.075 Å and 2.064 Å, respectively). The mean discrepancy of ~0.004 Å was found to be at the same level as the mean standard error of estimate of the three regression formulae

(0.0046 Å) used for modelling the $^{[6]}M$ sites (Hawthorne and Oberti, 2007). The same behaviour is shown by prismatic winchite, the structure of which was refined by single-crystal X-ray diffraction, which surprisingly has a mean discrepancy (0.007 Å) greater than that observed for fibrous richterite (XRPD data). The slight increase in the <*M*(4)–O> distance observed for the fibrous richterite with respect to prismatic winchite was attributed to the presence of a larger $^{M(4)}$Na content. For both richterite and winchite the calculated <*M*(4)–O> distance is smaller than that refined (0.011–0.013 Å) but the discrepancy is at the same level as the mean standard error of estimate of the regression formula (0.011 Å) for $^{[8]}M$ (Hawthorne and Oberti, 2007).

3. Chrysotile

Chrysotile belongs to the serpentine group, together with lizardite and antigorite. Although chrysotile is the least abundant of the three serpentine minerals, it accounts for most of the world's asbestos production to date and hence is of key importance when considering the health effects of industrial mineral fibres. It occurs most commonly in mildly prograde metamorphosed serpentinite deposits, as slip fibre along shear zones, and less frequently as mass fibre formed by the replacement of the host serpentine in asbestos deposits (Wicks and O'Hanley, 1988). It is a minor component in retrograde lizardite serpentinites (Cressey, 1979) and in prograde antigorite serpentinites (Mellini *et al.*, 1987).

At a basic level, all serpentine minerals are layer silicates composed of Si-centred tetrahedral (T) sheets in a pseudo-hexagonal network joined to Mg-centred octahedral (O) sheets in units with a 1:1 (TO) ratio (Fig. 3a). The ideal chemical formula of serpentine minerals is $Mg_3(OH)_4Si_2O_5$. In chrysotile, substitutions may occur in both T and O sheets but are limited. Al^{3+} may substitute for both Si^{4+} and Mg^{2+} in the T and O sheets, respectively. Wicks and Plant (1979) determined that most chrysotile samples contain <0.9 wt.% Al_2O_3. Fe^{2+} and Fe^{3+} may also substitute for Mg^{2+} in the O sheet while replacement for Si^{4+} in the T sheet is minor and infrequent. Wicks and Plant (1979) found that FeO may be present in natural chrysotile at <6 wt.%. Both Fe^{2+} and Fe^{3+} ions can replace Mg in the octahedral sheet (Stroink *et al.* 1980; Hardy and Aust, 1995) and Fe^{3+} ions may eventually replace Si ions although this position may preferentially host Al^{3+} (Blaauw *et al.* 1979; O'Hanley and Dyar, 1998). In a recent combined X-ray absorption spectroscopy (XAS) and Mössbauer study, Pollastri *et al.* (2015) found that Fe^{3+} is prevailing over Fe^{2+} and both are in octahedral coordination in standard UICC chrysotile. Equivalent amounts of Fe^{3+} and Fe^{2+} are found mainly in octahedral environments, with the possible minor presence of tetrahedral iron, in the chrysotile from the Balangero mine (Italy). Finally, iron in chrysotile from Val Malenco (Italy) shows comparable fractions of Fe^{2+} and Fe^{3+} almost exclusively in 6-fold coordination.

In chrysotile samples, trace metals such as Cr, Mn, Co, Ni and Cu represent an almost exclusively isomorphous substitute for magnesium (Wicks and O'Hanley, 1988; Bloise *et al.*, 2009a, 2010). Trace-metal substitution in chrysotile is usually more restricted than in the other serpentine minerals (*i.e.* lizardite and antigorite). Moreover, the

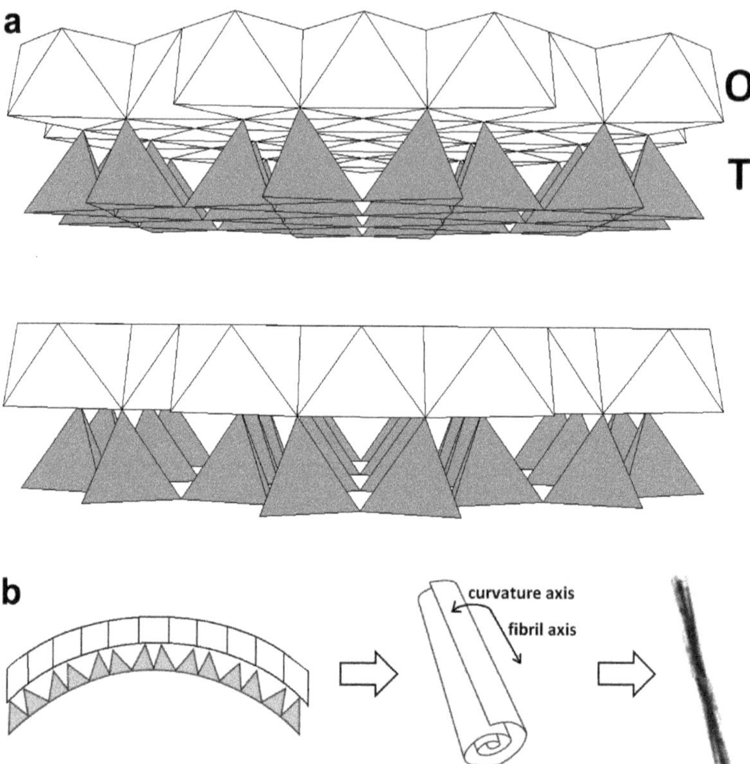

Figure 3. (a) The ideal layer unit of serpentine with Si-centred tetrahedral (T) sheets joined to Mg-centred octahedral (O) sheets in units with a 1:1 (TO) ratio of ~7 Å along the *c* axis. (b) The rolling of the TO layers in serpentine to form a papyrus like structure (cylindrical spiral lattice) which may make up single chrysotile fibrils.

concentration of trace elements is highly variable among the different samples, due to the different geochemical processes involved in their formation (Upreti *et al.*, 1984; Bloise *et al.*, 2016a). Bloise *et al.* (2016a), who compared the sum of metals (Co, Cr, Cu, Mn, Zn, Ni) content in three different chrysotile samples, have shown the following decreasing order of abundance: chrysotile from Balangero (Italy), chrysotile from Val Malenco (Italy) and, finally, UICC chrysotile from Canada (Table 7). However, some authors have shown that the concentrations of Cr, Zn, Cu and Ni in some samples were often >1000 ppm (Upreti *et al.*, 1984; Bowes and Farrow, 1997) and, therefore, they were considered as minor elements in the calculations of their chemical formulae. For example, analyses of chrysotile from Balangero and Val Malenco (Italy) by EPMA, showed the substitution of Mg by Cr and Ni, which were both >1000 ppm (Pollastri *et al.*, 2014). This finding is in agreement with the data from Harington and Roe (1965) who found Ni (0.5 wt.%) and Cr (0.1 wt.%) in relatively high concentrations in standard chrysotile. Other trace metals, such as Co and Mn, were always present at concentrations <1000 ppm (Morgan and Cralley, 1973; Bowes and Farrow, 1997; Bloise *et al.*, 2016a) and, therefore, considered as trace elements in the calculations of their chemical formulae. Because trace metals cause lung cancer, many authors (Harington and Roe, 1965; Cralley *et al.*, 1968; Dixon *et al.*, 1970) have hypothesized that chrysotile carcinogenesis effects may be due to their presence.

The lateral size of an ideal Mg-occupied O sheet (b = 9.43 Å) is larger than the lateral size of an ideal T Si-occupied sheet (b = 9.1 Å). As a result of the misfit between the T and O sheets (Bailey, 1988) and because of the polarity of the TO unit, a differential stress occurs between the two sides of the layer. In chrysotile, the stress is released by rolling the TO layer around the fibril axis (Fig. 3b) to end up with a cylindrical structure. Historically, it is curious to note that Linus Pauling (1930), at the dawn of structural crystallography, observed that: "The non-existence of a magnesium analogue of kaolinite is accounted for by the large values of the fundamental translations in the brucite layer (with a = 5.40 Å), which would cause the kaolinite-type layer to curve". Note that the cylindrical structure does not completely compensate for the O and T sheet misfit. Only one layer will be at the ideal radius of curvature (namely at 88 Å: Whittaker, 1957) for a perfect match of the sheets along the circumferential direction. Most of the O sheets are outside the ideal radius and will experience greater and greater compression along the circumferential direction, as the radius increases and misfit relief through curvature becomes less efficient. Across the tube wall, there is a radial gradient of stored elastic energy (Baronnet and Devouard, 1996) and the tubular structure as a whole obeys five-fold symmetry (Cressey and Whittaker, 1993). It would be desirable in the future to solve the misfit issue in a quantitative way. To this end, the use of quantum mechanical *ab initio* calculations (*e.g.* D'Arco *et al.*, 2009) offer some promise in terms of accurate quantitative solutions.

The earlier X-ray diffraction studies on chrysotile showed a remarkable distortion (curvature) of the unit cell with respect to the conventional crystal structures, so that a new theory specially formulated for cylindrical lattices had to be developed (*e.g.* Whittaker, 1954, 1955a,b,c,d, 1956a,b,c; Devouard and Baronnet, 1995).

In earlier selected area electron diffraction (SAED) studies, when viewing fibres perpendicular to the fibre axis, patterns of regular cylindrical fibres showed that the central part of a fibre has 4.5 Å fringes, representing the (020) planes lying parallel to the fibre axis. On the other hand, in fibres with helical cylindrical structures, the fringes were inclined a few degrees to the fibre axis and the $0k0$ and $hk0$ reflections were split into pairs that represent diffraction from either side of the twisted helical fibre (Yada, 1971). In addition, spiral growth fibres were revealed to be composed of a single layer, a double layer or multilayer spirals.

The development of a theory for the description of diffraction effects from cylindrical structures allowed definition of cylindrical circular lattices (concentric) and cylindrical spiral lattices and classification of the different chrysotile polymorphs/ polytypes. Both cylindrical circular and cylindrical spiral lattices can have a translational component along the fibre axis. In this case, they are classified as helical (H). If that component does not exist, they are referred to as regular (R). Single-crystal structure refinements of clino-, ortho- and parachrysotile fibres based on the theory of diffraction for cylindrical lattices were restricted by the circumferential disorder of the fibres to two-dimensional refinements projected onto (010) (Whittaker, 1956a,b,c). The earlier diffraction studies showed that the circular stacking arrangement has three effects: (1) it involves a shift of 0.4 Å (~a/13) away from the position of normal hydrogen bonding found in most 1:1 layer structures; (2) it introduces disorder along the circumferential direction (usually b); and (3) one of the basal oxygen atoms (O_1 in Whittaker, 1953) not only projects out of the layer in the radial direction but is displaced by 0.10 Å from its ideal position parallel to the cylinder axis. This shift with respect to the upper part of its own layer occurs in either the positive or negative direction, and has been designated as δ by Wicks and Whittaker (1975). In terms of diffraction, the fact that chrysotile fibrils are formed by a concentric radial sequence of n layers (with a geometrical relationship of $nb = 2\pi c$, where n is an integer, b = the circumferential repeat distance and c = the interlayer spacing) or by a spiral roll are not so different (Whittaker, 1954). Jagodzinski and Kunze (1954a,b,c) pointed out that either type of spiral or helical structures may be described in terms of radial and axial dislocations introduced into a normal 3-dimensional lattice.

Fibres are achiral and most commonly, the fibril axis is parallel to the crystallographic a axis and the chrysotile polymorph is referred to as clinochrysotile or orthochrysotile. More rarely, the fibril axis is parallel to the crystallographic b axis and the chrysotile polymorph is referred to as parachrysotile (Whittaker, 1956c). There are different polytypes associated with each polymorphic configuration:

(1) 2-layer clinochrysotile (Whittaker, 1956a) identified as chrysotile-$2M_{c1}$ (2 stands for two 1:1 layers in the unit cell, M stands for monoclinic, c indicates a cylindrical lattice, and 1 indicates that this is the first of the possible monoclinic 2-layer structures), with a monoclinic helical cylindrical lattice of the third kind and $\beta = \gamma = 90°$ (Whittaker, 1955);
(2) 1-layer clinochrysotile (Zvyagin, 1967) identified as chrysotile-$1M_{c1}$;
(3) 2-layer orthochrysotile (Or) identified as chrysotile-$2Or_{c1}$ (Whittaker, 1956b), with an orthorhombic symmetry, regular or helical cylindrical lattice of the first kind and $\gamma = \beta = 90°$ (Whittaker, 1955);
(4) disordered sequences identified as chrysotile-D_c.

Parachrysotile has a helical cylindrical lattice of the second kind, with orthorhombic symmetry and $\gamma = \beta = 90°$. As for clinochrysotile, there are basal oxygen anion corrugations but at intervals of $n\underline{b}/6$ along the fibre axis, probably with considerable disorder. Alternate layers may be displaced by b/12 shifts (Whittaker, 1956c).

The single cylinder or spiral arrangement is called a 'fibril' and is composed of concentrically or spirally curved layers, forming a tubular structure (Yada, 1971). By this mechanism, a layer silicate assumes a fibrous crystal habit (see Fig. 3b). Because the layers cannot energetically withstand too tight a curvature, the rolls possess hollow cores with a diameter of ~5–8 nm (Cressey *et al.*, 1994) that may be empty or filled with silica or an Fe-rich amorphous phase (Bloise *et al.*, 2009a, 2012). The cross sections of the cylinders can be circular, spiral, eliptical or deformed (Yada, 1971). In chrysotile synthesized in the laboratory (Bloise *et al.*, 2017, this volume), curious shapes have been observed especially for the nanophasic products: cone-in-cone shapes were observed by Yada and Iishi (1977), tubular and "tube in tube" types were synthesized by Falini *et al.* (2004) and Roveri *et al.* (2006), conical and cone-in-cone types (see for example Fig. 4, modified after Smolikov *et al.*, 2013) characterized the synthetic nano-chrysotiles obtained in distilled water by Smolikov *et al.* (2013). Such peculiar microstructures have also been observed recently in natural samples: conical serpentine nanofibres were recognized in optically isotropic veins of serpentinite (Andreani *et al.* 2007).

Another aspect to consider is that the presence of impurities in ideal chemical compositions in chrysotile fibres, even in trace amounts, affects their chemical/physical properties and size and have been invoked to explain the different variety of shapes observed: hollowed cylinder, non-hollowed cylinder, tube-in-tube, conically wrapped, spiralized (helical) and cone-in-cone fibres (Yada and Iishi, 1977; Amelinckx *et al.*, 1996; Roveri *et al.*, 2006; Bloise *et al.*, 2009a,b, 2010). Other trace elements such as Be, As, V and Pb were scarcely present in chrysotile samples compared to other asbestos samples (Bloise *et al.*, 2016a), whereas they are toxic elements of great concern (IARC 2012). Bloise *et al.* (2016a) showed that chrysotile from Val Malenco (Italy) contains a high concentration of *LREE* (Sc, La, Ce, Pr, Nd, Sm, Eu and Gd) while chrysotile from Balangero (Italy) has the largest amount of *HREE* (Y, Tb, Dy, Ho, Er, Tm, Yb and Lu) and UICC chrysotile from Canada contains the smallest amount of *REE*. *REE*-associated health effects have been associated mainly with redox reactivity (Pagano *et al.*, 2015a,b), involving ROS formation, analogous to the recognized redox

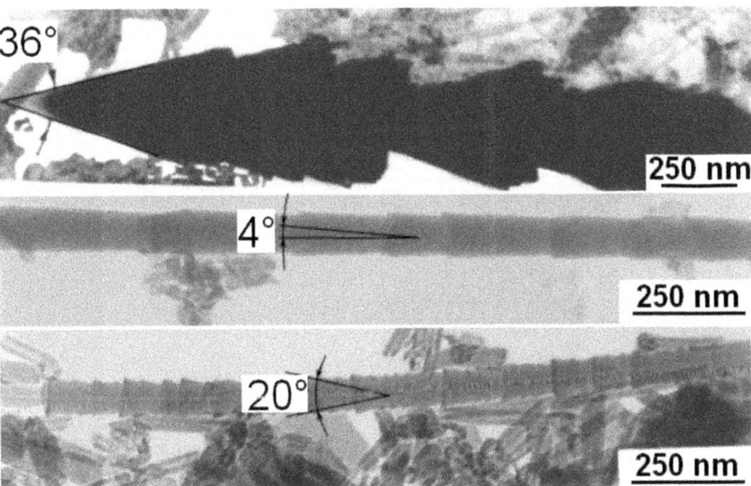

Figure 4. Outstanding examples of conical and cone-in-cone types of synthetic nano-chrysotiles (from Smolikov *et al.*, 2013, reproduced with the permission of Elsevier).

mechanisms observed for other transition elements such as iron and nickel. According to their relatively large ionic radius, *REE* are not incorporated as substitutes for Mg in chrysotile structures. In previous studies, *REE* in chrysotile were assigned to the hollow (core) centre present in its structure. Indeed, as previously mentioned, in the chrysotile structure, the TO layers are curved to form cylindrical fibres having generally empty central tubes of diameter ~5–8 nm that act as trapping locations for elements with large ionic radii (Bloise *et al.*, 2012; Lafay *et al.*, 2014), such as *REE*. In addition to the polymorphs/polytypes described above, some microstructures have been recognized only recently using TEM (Evans *et al.*, 2013). A serpentine form described as lath-like and sometimes referred to as Povlen chrysotile or more commonly polygonal serpentine, was first observed by Zussman *et al.* (1957). Later, this phase was investigated using both X-ray and electron diffraction/microscopy (Cressey, 1979; Cressey *et al.*, 1994, 2008, 2010). It occurs naturally and comprises a cylindrical core and polyhedral overgrowths of platy lizardite radiating around the core (Guggenheim, 2015). The core may also be composed of planar layers, antigorite and partially curved layers, and groups of fibres. Polygonal serpentine reported by Mitchell and Putnis (1988) contains curved and flat parts along individual layers, resulting in a microstructure with 15 or 30 sectors (see its beautiful microstructure in Fig. 5, modified after Yada and Wei, 1987). Polygonal serpentine may be viewed as a less metastable microstructure than that of chrysotile (Mitchell and Putnis, 1988). According to Wicks and O'Hanley (1988), it is not correct to consider polygonal serpentine as one of the serpentine polymorphs as it is a specific configuration of the known structures of chrysotile, lizardite and antigorite.

Polyhedral serpentine has also been observed as faceted, onion-like nanospherules (Baronnet *et al.* 2007) and recently associated with organic matter intimately

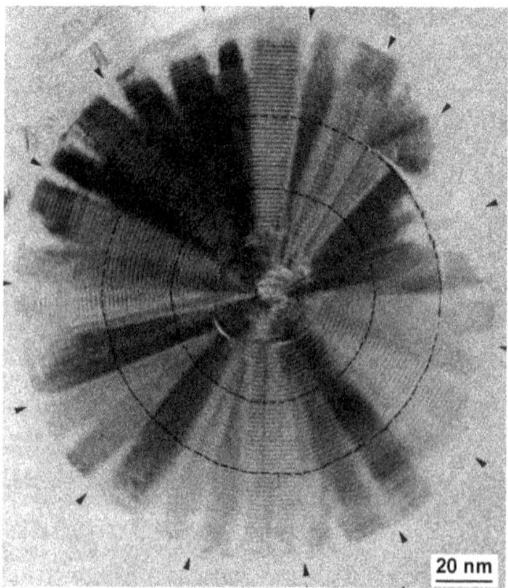

Figure 5. Cross section of a polygonal serpentine with a cylindric chrysotile core and polygonal overgrowth of 30 sections (reproduced from Yada and Wei, 1987, with the permission of Sociedad Española de Arcillas).

associated with serpentine hosted hydro-garnets in serpentinized peridotites from the Mid-Atlantic Ridge (Ménez *et al.*, 2012).

The fast development of the structure refinement from X-ray powder diffraction data made it possible in recent times to attempt a Rietveld structure refinement of chrysotile phases. The nano-chrysotile (Falini *et al.*, 2004) was refined using the *GSAS* software (Larson and Von Dreele, 1994), in the monoclinic space group *Cc*. The calculated cell parameters were $a = 5.340(1)$, $b = 9.241(1)$, $c = 14.689(2)$ Å, and $\beta = 93.66(3)°$. A more robust Rietveld structure refinement of the same nano-chrysotile described by Falini *et al.* (2004) was obtained using the *DIFFaX+* software (Leoni *et al.*, 2004) to reproduce empirically the curved lattice distortions. The best fit was obtained with a model composed of a statistical ensemble of six layers (an ideal and 1/4*b*, 1/3*b*, 1/2*b*, 2/3*b* and 3/4*b* shifted layers), which is a reasonable approximation of the cylindrical stacking with random disorder along the *b* axis. Additional disorder along *a*, reproducing a helical structure, did not improve the fit. The refined unit-cell parameters of the ideal orthogonal layer are a = 5.330 (1), b = 9.220 (1) and c = 7.3508 (9) Å (Leoni *et al.*, 2004). It would be desirable in the future to obtain structure models of greater accuracy by taking into account the lattice curvatures and distortions in the Rietveld-based algorithms. A possible way to do this would be to implement the new theoretical models for the description of diffraction effects from spiral nanotubes (see for example Figovsky *et al.*, 2014) in the Rietveld based codes.

Extensive experiments on chrysotile dehydration mechanisms and high-*T* crystallization have been carried out over the past couple of decades (*e.g.* Ball and Taylor 1963; Brindley and Hayami, 1965; Martin, 1977; MacKenzie and Meinhold, 1994; Bloise *et al.*, 2017, this volume). The water in the chrysotile was driven out of the structure over the temperature range 500–800°C and, when measured by thermogravimetric (TG) analysis, the values of water were between ~14% and 11% (*e.g.* Bush and Schumacher, 1982; Khorami *et al.*, 1984; Viti, 2010; Viti *et al.*, 2011; Punturo *et al.*, 2015; Bloise *et al.*, 2014; 2016b,c). Note that in serpentines the percentage of water loss increases progressively from antigorite to lizardite, polygonal serpentine and chrysotile (Viti, 2010). Moreover, in many chrysotile samples an anomalously high value of mass loss (>14%), due to impurities such as calcite or/and siderite was detected (Viti, 2010; Bloise *et al.*, 2016c). Khorami *et al.* (1984) reported that when chrysotile was heated in air, it underwent an endothermic weight loss of ~14% between ~600 and 720°C, representing the removal of structural water. Recently, Bloise *et al.* (2016c) showed three different dehydration processes in chrysotile samples, using TG analyses. The results of their research showed that water loss of chrysotile from Val Malenco (Italy) was 12.01%; chrysotile from Balangero (Italy) in the temperature range of 480–800°C showed a structural water loss of 11.8%; UICC chrysotile from Canada had a structural water loss of 12.22% (Bloise *et al.*, 2016c). These latter data are in agreement with those from Bush and Schumacher (1982) who reported a structural water loss of 12.58% in a similar sample of chrysotile coming from Canada. Moreover, these authors showed a TG curve in which the structural water loss of chrysotile from Rhodesia (Zimbabwe) was 11.74%. The water loss by synthetic chrysotile is not discussed here as it is covered by Bloise *et al.* (2017, this volume).

4. Palygorskite

The name palygorskite was used first by von Saftschenkov in 1862 to describe the occurrence of a fibrous mineral from the Palygorsk Range, Ural Mountains, Russia (Hey, 1975). Due to its fibrous crystal habit (see an example in Fig. 6), when found in the Palygorsk Division Mine near Popovka River, Perm Province (Russia), this unusual mineral was assumed to represent a variety of asbestos (Jones and Galán, 1988). As a matter of fact, palygorskite has been called mountain leather, mountain wood or wool although most of the earlier occurrences of fibrous palygorskite were later discovered to be essentially chrysotile. For some time, the name attapulgite was used in place of palygorskite. Attapulgite was a term introduced by de Lapparent (1935) for a fibrous clay found near Attapulgus, Georgia, USA, but the name was subsequently discredited by the International Mineralogical Association (IMA) because palygorskite, reported from Palygorsk has precedence.

Palygorskite forms either in lacustrine (Chahi *et al.*, 1997) or in perimarine (Singer, 1979; Velde, 1985) environments (Guggenheim and Krekeler, 2011). Crystallization may also occur during diagenesis (Couture, 1977), or hydrothermally (Imai and Otsuka, 1984) although alteration from precursors such as smectite is also known (Singer, 1979). Deep-ocean authigenic palygorskite near active ridge zones was described by Bowles *et al.* (1971). Jones and Galán (1988) summarized the occurrences of palygorskite (and sepiolite) in soils. At lower pH, palygorskite may form from amorphous silica and dioctahedral smectite, whereas at slightly higher pH, sepiolite, amorphous silica and palygorskite can precipitate.

Palygorskite together with sepiolite is included in the group of phyllosilicates because it contains a continuous 2-dimensional tetrahedral sheet (T) of composition

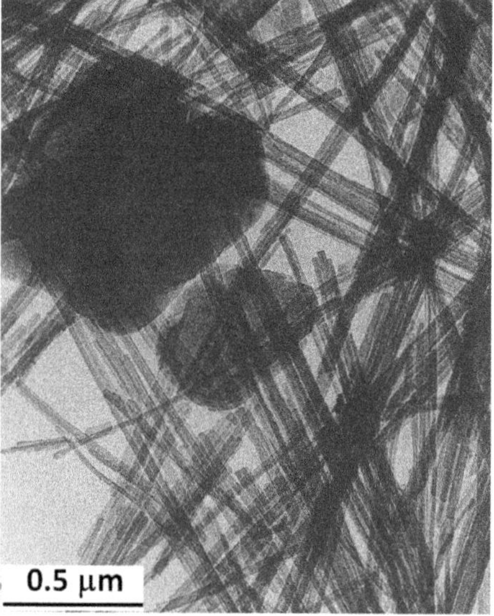

Figure 6. TEM image of fibrous palygorskite, together with illite, from Spain (modified after Galán, 1996; reproduced with the permission of the Mineralogical Society of Great Britain & Ireland).

T_2O_5 but it differs from the other layer silicates in lacking continuous octahedral (O) sheets (Jones and Galán, 1991). The palygorskite structure contains ribbons of 2:1 layer silicates, one ribbon being linked to the next by inversion of SiO_4 tetrahedra along a set of Si–O–Si bonds. Ribbons extend parallel to the *x* axis and have an average width along *y* of two pyroxene-like single chains. Hence, the linked ribbons form a continuous 2:1 layer along *x* but of limited extent along *y*. The T sheet is continuous across ribbons but with apices pointing in different directions in adjacent ribbons while the O sheet is discontinuous so that rectangular channels running parallel to the *x* axis are formed between opposing 2:1 ribbons (see Fig. 7). As the O sheet is discontinuous at each inversion of the T sheet, oxygen atoms of the O sheet are coordinated to cations on the ribbon side only and coordination is completed along the channel by hydrogen, (zeolitic) water molecules and eventually exchangeable cations (Jones and Galán, 1991). Possible exchange reactions with organic molecules show that exchange is dependent on the size of the organic cations because of the steric constraints of the channels. One synthetic pigment (Maya Blue, used extensively by the Mayan people because of its beautiful blue colour) involves the adsorption of the indigo molecule in palygorskite. Larger molecules may also be adsorbed by the structure, probably because of the existence of defects. As already noted, the O strips are terminated at the channel by four OH_2 per formula unit to form a part of the octahedral coordination polyhedron around Mg or Al (Guggenheim and Krekeler, 2011). Guggenheim and Eggleton (1988) proposed a 'modulated' phyllosilicates classification scheme, with palygorskite as a member because of the inverted tetrahedral arrangement and the formation of the channel where the octahedral sheet becomes discontinuous. This classification scheme was extended and adopted by The Clay Minerals Society (Martin

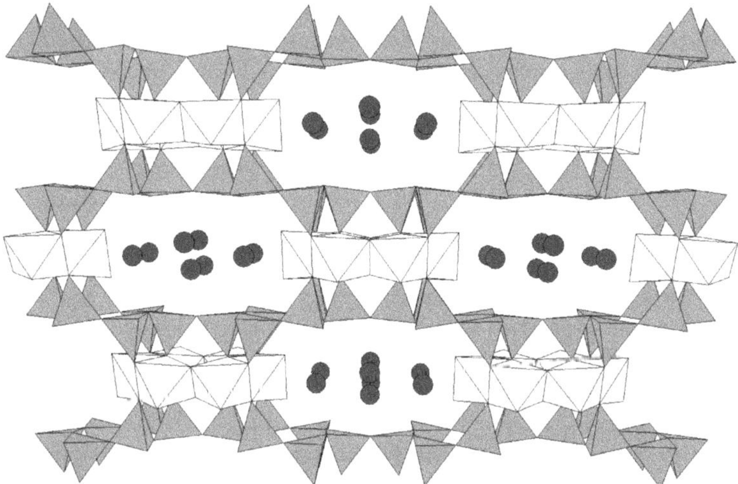

Figure. 7. The structure of palygorskite in the *a-b* plane. The grey spheres represent water molecules.

et al., 1991) and included in a summary of recommendations of nomenclature committees relevant to clay mineralogy presented by Guggenheim *et al.* (2006).

Regarding the chemical formula of trioctahedral palygorskite, we report herewith the approximate formula by Drits and Sokolova (1971): $(Mg_{5-y-z}R_y^{3+}\square_z)(Si_{8-x}R_x^{3+})O_{20}(OH)_2(OH_2)_4R^{2+}_{(x-y+2z)/2}\cdot(H_2O)_4$. T occupancies for Al ranges from 0.01 to 0.69 for eight positions (Weaver and Pollard, 1973) or from 0.12 to 0.66 (Newman and Brown, 1987). O cations (Mg, Al, Fe) are ordered. The exchangeable cations *R* can be Ca, K or Na atoms.

Single-crystal X-ray analyses of palygorskite are not available, but the overall structure has been obtained from powder data studies (generally Rietveld refinements). These studies (Christ *et al.*, 1969; Drits and Sokolova, 1971; Chisholm, 1992; Artioli and Galli, 1994; Chiari *et al.*, 2003; Giustetto and Chiari, 2004; Post and Heaney, 2008) confirmed the basic structure of Bradley (1940) for palygorskite (Guggenheim and Krekeler, 2011). The known structural modifications are monoclinic (*C*2/*m*) and orthorhombic (*Pbmn*); these modifications are commonly intergrown.

According to the structural studies on palygorskite available, there are orthorhombic and monoclinic polymorphs:

(1) *Orthorhombic*: *Pnmb*/*Pbmn* (Preisinger, 1963; Artioli and Galli, 1994; Giustetto and Chiari, 2004), *Pn* (Christ *et al.*, 1969), *Pncn* (Post *et al.*, 2007)

(2) *Monoclinic*: *A*2/*m* (Bradley, 1940; Drits and Sokolova, 1971), *P*2/*a* (Christ *et al.*, 1969), *C*2/*m* (Artioli and Galli, 1994; Artioli *et al.*, 1994; Chiari *et al.*, 2003; Giustetto and Chiari, 2004; Post and Heaney, 2008)

Bailey (1980) related the two varieties of stacking (monoclinic, orthorhombic) of the palygorskite structures to the atomistic approach he used to derive the standard polytypes for the 2:1 layer silicates (Guggenheim and Krekeler, 2011). Within a polysome, the direction of shift of the upper tetrahedral strip may be either $+c/3$ or $-c/3$, depending on whether set I or set II octahedral cations occupy possible sites. The direction of shift of the upper tetrahedral strip in the polysome relative to the lower tetrahedral strip is based on closest packing within the polysome. Where the same set of octahedral cations is occupied from polysome to polysome (regardless if it is set I or set II, Bailey, 1980), the direction of shift always remains the same and a monoclinic structure results (with an ideal $\beta = 105.2^o$). Where alternation of shift directions occurs because set I alternates with set II in adjacent strips, an orthorhombic structure results ($\beta = 90^o$).

Cation sites in palygorskite are usually defined from the centre of the O strip, where there is often a special position such as a mirror plane, to the outer edge. For example, *M*1 is the central O site on the special position, with *M*2 adjacent to *M*1, and *M*3 adjacent to *M*2 but further from the special position. T cation sites are defined in a similar fashion. From spectroscopic studies, Serna *et al.* (1977) and Heller-Kalai and Rozenson (1981) concluded that vacancies are ordered into *M*1 in palygorskite, and that Mg (and Fe) orders preferentially into *M*3. Chryssikos *et al.* (2009) found that regions of the palygorskite structure were dioctahedral (with $AlAlOH$, $AlFe^{3+}OH$, $Fe^{3+}Fe^{3+}OH$

interactions) and trioctahedral (MgMgOH) regions, with these interactions implying that Al and Fe^{3+} order into *M*2. Post and Heaney (2008) found vacancies ordered in *M*1, Al in *M*2, and Mg in *M*3, in accord with the spectroscopy studies and other Rietveld refinement studies (Artioli and Galli, 1994; Chiari *et al.*, 2003; Giustetto and Chiari, 2004).

There is controversial evidence of possible health effects due to inhalation of palygorskite fibres (Galan, 1996; Alvarez *et al.*, 2011). Oscarson *et al.* (1986) studied the effect of sepiolite and palygorskite on erythrocyte lysis and found the two minerals to be lysing agents. Experiments carried out in Florida and Georgia on men employed in mining and factories led the IARC (1987) to conclude that there was inadequate evidence for the cancinogenicity of palygorskite in humans. Pott *et al.* (1990) and Wagner *et al.* (1987) studied the effects of sepiolite and palygorskite inhalation and injection in rats. The results proved the production of mesotheliomas on exposure to a significant number of fibres >5–6 μm in length. Following such studies, IARC has classified short palygorskite fibres (<5 μm) in Group 3 (not classifiable as carcinogenic in humans), meanwhile long palygorskite fibres (>5 μm) are classified in Group 2B (possibly carcinogenic in humans) because of evidence from the experiments on animals.

5. Balangeroite

Balangeroite is a fibrous silicate, ideally $(Mg,Fe)_{42}O_6(Si_4O_{12})_4(OH)_{40}$ (Bonaccorsi *et al.*, 2012), that was found first in the serpentinites of the asbestos mine of Balangero (Turin, Italy), in the Dora-Maira Massif (Compagnoni *et al.*, 1983). Commonly it forms bundles of long brown fibres albeit a prismatic variety has been reported from Balangero and other localities of the same Massif (Groppo *et al.*, 2005). In spite of being an unregulated asbestiform mineral it has been shown to be potentially harmful because of its high durability, an *in vitro* cytotoxic effect on human epithelial cell linings and an oxidant activity similar to that of UICC crocidolite (Groppo *et al.*, 2005; Turci *et al.*, 2005). Balangeroite has been shown to possess two maximum degree of order (MDO) polytypes, balangeroite-2*M* (MDO_1: a = 19.179(6) Å, b = 9.601(6) Å, c = 19.218(6) Å, β = 90.52(1)°; space group $P2/n$) and balangeroite 1*A* (MDO_2: a = 9.602(6) Å, b = 13.891(6) Å, c = 14.012(6) Å, α = 86.99(1)°, β = 76.79(1)°, γ = 76.67(1)°; space group $P\bar{1}$). That it is isostructural with gageite, a fibrous manganese silicate, from Franklin, New Jersey (Phillips, 1910), was immediately apparent from XRPD (Compagnoni *et al.*, 1983) and subsequently confirmed from transmission electron microscopy (TEM) and electron diffraction (ED) (Ferraris *et al.*, 1987). The structure of both polymorphs is built from the packing of three different structural modules, namely 3 × 1 octahedral walls, 2 × 2 octahedral bundles, and four-repeat silicate chains (Fig. 8). The modules are aligned along the 9.6 Å axis (the b axis for MDO_1 and the a axis for MDO_2, respectively). A different relative positioning of the four-repeat silicate chains represents the sole distinguishing feature between the two MDOs. Structure refinements pointed to a larger Fe content of MDO_2 compared to that of MDO_1 (Bonaccorsi *et al.*, 2012).

Figure 8. The crystal structure of balangeroite-2*M* (MDO_1) highlighting the various modules. Red octahedra indicate a prevailing Fe content with respect to yellow ones.

6. Fibrous erionite

Erionite is a zeolite that occurs, often under a fibrous morphology, as a diagenetic alteration product of sediments (Sheppard and Gude, 1973), in cavities of altered basalts (Tschernich, 1992), or as a hydrothermal alteration product (Bargar and Keith, 1995). It pertains to the so-called ABC-6 family (Gottardi and Galli, 1985), the members of which are described conventionally as originating from stacking along the *c* axis, following an ABC scheme, of layers of six-membered rings made of $(Si,Al)O_4$ tetrahedra (Fig. 9). Erionite is characterized by a six-layer repetition. It has been pointed out (Meier and Groner, 1981; Smith and Bennett, 1981) that 10 different stacking sequences are possible for a period of six-layers, allowing for the occurrence of both single- and double-rings, but only four correspond to observed minerals: AABAAC, erionite (Staples and Gard, 1959), AABBCC, chabazite (Smith *et al.*, 1963), ABBACC, bellbergite (Rüdinger *et al.*, 1993) and ABABAC, liottite (Ballirano *et al.*, 1996b). Intergrowth of erionite with offretite (OFF) (Gard and Tait, 1972) may occur,

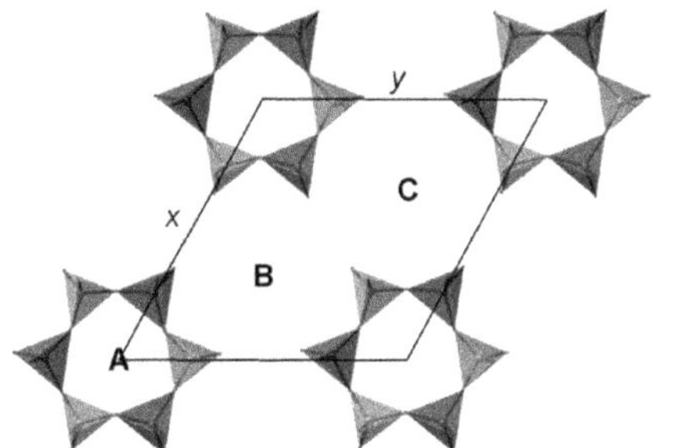

Figure 9. The plane of six-member rings is the common unit block for all ABC-6 minerals (*e.g.* sodalite and cancrinite groups). The three A, B and C positions in correspondence of which the stacking of the layers occurs is labelled.

due to the close similarity between the two zeolites. In fact, offretite has an AAB sequence, and stacking faults corresponding to partial substitution of *C* with *B* 6-rings have been observed (Kokotailo *et al.*, 1972; Schlenker *et al.*, 1977) (Fig. 10). Passaglia *et al.* (1998) pointed out that a simple discrimination between the two zeolites is difficult as they display very similar X-ray diffraction patterns owing to similar structures and unit-cell parameters. However, the compositional fields built into the chemical space describing the extra-framework (EF) cation content may be exploited to identify unambiguously the correct mineralogical species, in particular using a Mg-Ca(+Na)-K(+Sr+Ba) triangular diagram (Passaglia *et al.*, 1998).

Erionite is hexagonal, space group $P6_3/mmc$ (Kawahara and Curien, 1969), $a \approx$ 13.2 Å, $c \approx$ 15.1 Å, and has an average formula $K_2(Na,Ca_{0.5})_8[Al_{10}Si_{26}O_{72}]{\cdot}30H_2O$ (Coombs *et al.*, 1997). A large chemical variability is typical of this mineral, and for this reason three different species are identified, according to the most abundant extra-

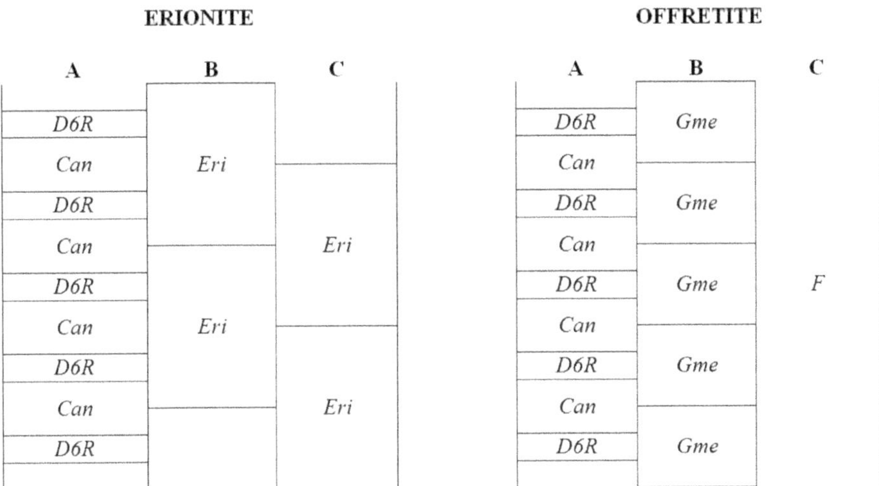

Figure 10. Sub-units-like representation (Ballirano *et al.*, 1996a) of the framework of erionite and offretite depicting the various cavities. *D6R*: double 6-rings; *Can*: cancrinite cage; *Eri*: erionite cage; *Gme*: gmelinite cage; *F*: free channel.

framework (EF) cation: erionite-Na, erionite-K and erionite-Ca (Coombs *et al.*, 1997; Gualtieri *et al.*, 1998; Passaglia *et al.*, 1998; Dogan and Dogan, 2008). The framework of erionite contains two cancrinite (ε) cages, two double-6 rings (D6R), and two erionite (23-hedron) cages per unit cell. Cancrinite and erionite cages host the EF cations. The former cage contains a K^+ ion (site K1) located at the centre of the cavity, whereas several EF cation sites are placed along the axis of the erionite cavity (Kawahara and Curien, 1969; Gard and Tait, 1973; Alberti *et al.*, 1997; Gualtieri *et al.*, 1998; Ballirano *et al.*, 2009; Cametti *et al.*, 2013). In addition, a further site, labelled K2 and located at the centre of the boat-shaped 8-member rings (8MR) forming the walls of the erionite cage, has been found in erionite-K and attributed to the presence of extra-K^+ ions (Schlenker *et al.*, 1977; Ballirano *et al.*, 2009; Cametti *et al.*, 2013) or Ca^{2+} and Na^+ ions (Gualtieri *et al.*, 1998). Besides, water molecules are arranged around the axis of the erionite cavity providing coordination to the EF cations.

The structure of fibrous samples of erionite has been investigated by laboratory XRPD on samples from the localities of Rome and Durkee, both in Oregon, USA (Ballirano *et al.*, 2009; Cametti *et al.*, 2013). The species analysed included erionite-K from Rome and erionite-Na from both Rome (fibrous) and Durkee (woolly). The studies have faced severe experimental difficulties, as it is very difficult to obtain a detailed chemistry of the fibres because of their significant instability under the electron beam (Carbone *et al.*, 2011; Pacella *et al.*, 2016) possibly magnified by partial surface protonation/damage occurring during the enrichment routine (Ballirano *et al.*, 2015). However, an optimization of both experimental set-up and analytical procedures may provide reliable results after filtering the results by the balance error formula E% (Passaglia, 1970), Mg-content test (Dogan and Dogan, 2008) and K test (Cametti *et al.*, 2013). A further point to be addressed is that the surface of erionite fibres may be coated by iron oxides/oxyhydroxides (Ballirano *et al.*, 2009; Croce *et al.*, 2015), Fe sulfates (Croce *et al.*, 2015) and Fe-bearing phyllosilicates (Ballirano *et al.*, 2015; Matassa *et al.*, 2015) contributing to the final chemical analysis. As proof of this contribution, Fe_2O_3 contents up to 3 wt.% have been reported for this nominally Fe-free zeolite (Eberly, 1964; Dogan *et al.*, 2006; Dogan and Dogan, 2008; Dogan, 2012). XRPD data have been collected using a laboratory focusing-beam θ/θ diffractometer operating in transmission mode and equipped with a PSD detector (Ballirano and Melis, 2009). It is worth noting that two problems need to be dealt with during the refinements, *i.e.* the occurrence of minor phases and in particular of nontronite, an Fe-bearing smectite and, in the case of woolly erionite-Na from Durkee, the strongly anisotropic peak-broadening arising mainly from the curved lattice of the bent fibrils. The first one has been solved by using a mixed Pawley/Rietveld approach and the second by applying the ellipsoid model of Katerinopoulou *et al.* (2012) describing the diffraction-vector dependent broadening of diffraction maxima. The Rietveld and combined mixed Pawley/Rietveld refinement strategy have been detailed in Ballirano and Melis (2009), Cametti *et al.* (2013) and Ballirano and Cametti (2015), whereas the modelling of small preferred orientation effects and the choice of the maximum $\sin\theta/\lambda$ for an optimum data collection were performed according to Ballirano (2003, 2014).

Results indicate very minor differences as far as the framework is concerned. In all cases Al^{3+} shows a preferential partition at T2, as deduced from the Jones (1968) determinative curves and by comparison with chemical data. On the contrary, EF cations are arranged dissimilarly as a function of the chemical composition of the fibres (Table 6). Erionite samples from Rome are characterized by the occurrence of extra K^+ ions located at K2. The relevant difference between the two samples is related to the fact that EF ions are located at Ca1 and Ca2 in erionite-K whereas they are distributed among three sites: Ca1, Ca2 and Ca3 in erionite-Na. Erionite-Na from Durkee, with a smaller K_2O content, has an empty K2 site. In all cases, water molecules are distributed among six sites the positions of which are almost unaffected by the moderate EF site variability, whereas differences in site populations are observed.

Ca1 is able to host all the available Mg^{2+} plus Na^+ ions. Mg^{2+} occurs in a fairly regular six-fold coordination with water molecules. It has been shown that this cation is the first one to be mobilized during heating-induced dehydration in both Na and K species (Ballirano and Cametti, 2012; Ballirano and Pacella, 2016). It is expected that the two ions are occupying, in effect, sites very close together and not resolvable with the available data (Ballirano and Cametti, 2012). This hypothesis is confirmed by the observation that, in the case of erionite-K from Rome, the Ca1 site occupied by Mg^{2+} only, is slightly displaced off-axis in order to provide a suitable six-fold coordination for water molecules. Moreover, it has been pointed out that the occurrence of Mg^{2+} at Ca1 and that of the K2 site are mutually excluded because the presence of the OW9 water molecule site, required for providing a six-fold coordination for Mg^{2+}, is incompatible with that of K2 due to a very short contact (Ballirano and Cametti, 2012). This restraint is confirmed by a check of the EMPA data by Passaglia *et al.* (1998) revealing that, invariably, a high K^+ content correlates with a low Mg^{2+} content.

Table 6. EF content of erionite fibres. Partition derived from combining site scattering (*s.s.*) at the relevant sites and chemical analyses.

	Erionite-K (Rome)[a]	Erionite-Na (Rome)[c]	Erionite-Na (Rome)[b]	Erionite-Na (Durkee)[d]
Ca1	$Mg_{0.80}$	$Mg_{0.53}Na_{1.60}$	$Mg_{0.42}Na_{0.04}$	$Mg_{0.24}$
Ca2	$Na_{2.00}$-$Ca_{0.86}Mn_{0.07}$	$Na_{2.74}$	$Na_{2.39}$	$Na_{3.48}$
Ca3	–	–	$Na_{1.32}$	$Na_{1.90}$
K1	$K_{2.00}$	$K_{2.00}$	$K_{2.00}$	$K_{1.99}$
K2	$K_{0.87}$	$K_{0.23}$	$K_{0.45}$	–

Data from: [a]Ballirano *et al.* (2009); [b]Ballirano and Cametti (2015); [c]Ballirano *et al.* (2015); [d]Cametti *et al.* (2013).

Ca2 and Ca3 offer different possible coordination for both Ca^{2+} and Na^{+} ions. However, a further restraint, involving the Mg^{2+} occurrence at Ca1, has been devised. In fact, the OW8 water molecule site, which is fundamental in providing a nine-fold coordination for Na^{+} at Ca2, has a very short OW8–OW9 contact indicating the mutual exclusion of Mg^{2+} at Ca1 and the Ca2 site (Ballirano and Cametti, 2015).

Because of the importance of the topics being discussed, significant effort has been made in recent years to determine the structure modifications of erionite fibres induced by leaching in simulated lung fluids (SLFs) (Ballirano and Cametti, 2015) and in Fe-containing solutions (Ballirano *et al.*, 2015; Pacella *et al.*, 2017). Various tests have been performed to mimic the behaviour of the fibres within the human body and to check their behaviour whenever in contact with a source of both Fe^{2+} and Fe^{3+} because it has been claimed that erionite toxicity is due mainly to the effect of ion-exchanged Fe, acquired after inhalation. This process occurs because protein injury can lead to metal-ion release (Carr and Frei, 1999) and Fe may participate in Fenton chemistry, eventually generating reactive oxygen species (ROS) inducing DNA damage (Eborn and Aust 1995; Hardy and Aust 1995).

Investigation of the chemical and structural modifications occurring in fibrous erionite leached with SLFs has been carried out on erionite-Na from Rome, Oregon, USA. To simulate different interstitial conditions occurring in the lung, artificial lysosomal fluid (ALF: pH ~4.5) and Gamble's solution (pH ~7.4) were used (Ballirano and Cametti, 2015). From a chemical point of view, compositional variations were apparent after just leaching with Gamble's solution, as a result of Ca^{2+} substitution for Na^{+}. Structure refinements indicated modifications of both the *a* and *c* cell parameters of erionite after leaching, consistently leading to a contraction of the unit-cell volume. The leaching with both SLFs results in depletion of the Ca1 site (occupied preferentially by Mg^{2+}, as discussed above) but, in the case of ALF it is coupled with an increase in s.s. at the Ca2 site, whereas in the case of the Gamble's solution with the fixing of Ca^{2+} ions, coming from the SLF, at Ca3. The relevant variations in the sites occupancy of individual water molecules did not cause notable modification of the overall water content, as indicated by the final total water s.s. ranging from ~259 e^{-} to ~272 e^{-}. Such discrepancy has been attributed to the partial migration of EF cations towards alternative sites close to water molecules. Moreover, a further effect of the leaching with ALF was a partial dissolution of erionite fibres because of the acidic pH of the solution.

As far as the iron-loading process by erionite fibres is concerned, two samples (Durkee and Rome, Oregon, USA) were suspended in $FeCl_2$ solutions at different concentrations, at pH ~5, in anaerobic conditions (Ballirano *et al.*, 2015). Comparison between released and acquired charges in the form of Fe indicates unequivocally that erionite binds Fe^{2+} by ion exchange with the EF cations, mainly Na^{+}. The process, common to both samples, testifies to a direct correlation existing between the degree of ion-exchange and the Fe^{2+} concentration of the solution used for fibre incubation. X-ray photo-electron spectroscopy (XPS) revealed no Fe enrichment of the fibre surface when comparing surface and bulk composition of the samples. Moreover, by combining chemical and structural data it was established that, for both samples, Fe^{2+} is fixed at the Ca3 site being six-fold coordinated to water molecules. The occurrence of Fe^{2+} within the erionite cavity causes a gradual migration of the other EF cations and in addition, induces a small rearrangement of the water molecules positions. A recent quantitative re-evaluation of XPS data provided strong support for the hypothesis that a fraction of charged Fe was, in effect, ferric and located exclusively at the fibre surface (Pacella *et al.*, 2017).

One sample of fibrous erionite-K from Rome (Oregon, USA) was studied after incubation in $FeCl_3$ solution (Ballirano *et al.*, 2016). Inductively coupled plasma-optical emission spectrometry results, supported by XPS investigation, showed unambiguously that Fe^{3+} is fixed mainly at the fibre surface, in agreement with the results of Eborn and Aust (1995). However, a fraction of Fe^{3+}, which is prevalent in the case of very diluted Fe^{3+} solutions and decreases rapidly as concentration increases, is segregated by an ion-exchange mechanism in the erionite cavity at the Ca3 site, similarly to Fe^{2+} (Ballirano *et al.*, 2015). The maximum amount of Fe^{3+} residing at Ca3 is 0.17(7) a.p.f.u., a significantly smaller iron content than that ion-exchanged in the form of Fe^{2+}. Therefore, Fe^{2+} and Fe^{3+} are both fixed at the same crystallographic site, albeit with a significantly different efficiency with respect to iron available in solution. In addition, it has been shown that ascorbic acid is able to reduce Fe^{3+} residing at the fibres' surface fairly easily, whereas this reaction does not occur (or possibly occurs only very marginally) in the case of the ion-exchanged metal. Gutteridge *et al.* (1996) reported the total iron content being as little as 0.21(4) μmol/L (12(2) μg/L) from bronchoalveolar lavage of normal humans, suggesting strongly that under physiological conditions iron tends to be mainly ion-exchanged.

It has been demonstrated that, in Fe-loaded zeolites with rather small Fe contents, Fe sites with very low nuclearity and residing at well defined crystallographic positions, act as active sites for the formation of ROS (Zecchina *et al.*, 2007). In this case, the identification of the segregation of both Fe^{2+} and Fe^{3+} at Ca3, coupled with the large surface area of erionite, represents a very important piece of information in the understanding of the molecular mechanism/s causing its carcinogenicity.

Erionite offers a large retention capacity for positively and negatively charged ions and has been used to remove, for example, toxic elements from industrial waste waters (*e.g.* Ahmed *et al.*, 1998), the cation exchange capacities of many zeolites being greater, in general, than those of most other minerals. In a recent work, Bloise *et al.* (2016a) showed that erionite from Jersey (Nevada, USA), contains a lot of As, Pb and Be. The presence of large amounts of As in fibrous erionite is justified by the high selectivity of this zeolite species for As (Fukuda *et al.*, 2004). In erionite, Be^{2+} may replace Al^{3+} or Si^{4+} in the framework (Grice *et al.*, 2013) and this leads to compensation of the associated increased negative charge of the framework by introduction of further extraframework cations into the cavities. However, the trace metals content in erionite from Jersey (Nevada, USA), was much lower than in the asbestos (Table 7) (Bloise *et al.*, 2016a). On the other hand, all the *REE* have higher concentrations in erionite than in the asbestos samples (*i.e.* crocidolite, tremolite, anthophyllite and chrysotile)

Table 7. Levels of trace metals in samples of UICC asbestos.

Sample	Laboratory	Cr (ppm)	Co (ppm)	Mn (ppm)	Ni (ppm)
UICC chrysotile, Canada	AERE[a]	490	50	480	820
	DHEW[b]	317	45	444	802
	DGAG[c]	900	82	700	975
	DBEES[d]	360	45	458	866
UICC amosite, South Africa	AERE[a]	35	7.0	>1000	<100
	DHEW[b]	31	11	>1000	33
	DGAG[c]	<15	25	>1000	33
	DBEES[d]	12	1.6	>1000	16
UICC crocidolite, South Africa	AERE[a]	16	2.0	880	<100
	DHEW[b]	20	10	833	8
	DGAG[c]	<15	20	>1000	18
	DBEES[d]	20	1.6	829	13
UICC anthophyllite, Finland	AERE[a]	870	50	>1000	>1000
	DHEW[b]	584	24	986	414
	DGAGc	>1000	70	>1000	>1000
	DBEES[d]	818	25	>1000	704

[a]Atomic Energy Research Establishment (AERE), Harwell, UK (Holmes *et al.*, 1971);
[b]Department of Health Education and Welfare (DHEW), Cincinnati, USA (Morgan and Cralley, 1973);
[c]Department of Geology and Applied Geology (DGAG), Glasgow, UK (Bowes and Farrow, 1997);
[d]Department of Biology, Ecology and Earth Sciences (DBEES), Rende, Italy (Bloise *et al.*, 2016a).

(Bloise *et al.*, 2016a). For example, erionite from Jersey (Nevada, USA), has a high concentration of La (65.6 ppm) and Ce (192.9 ppm). The *REE* are adsorbed in the different channel systems present in erionite.

7. Concluding remarks

Regulated and non-regulated fibrous minerals are intrinsically complex materials from both the structural and crystal chemical points of view. This makes it difficult to link those features with the chemical and biological reactivity of fibrous minerals. In fact, as highlighted in the present chapter, detailed structural and chemical characterization is made very difficult by the small dimensions, the instability under different analytical probes and the extremely variable chemical composition even for neighbouring fibres.

All these drawbacks can be overcome only by combining several techniques aimed at exploring the structure and chemical composition both at the bulk and at the surface levels. In fact, it should be clear that the surface may behave quite differently from the bulk and its diversity plays a potentially crucial role in modulating the chemical and biological reactivity of fibrous minerals. Nevertheless, only a relatively limited number of studies dealing with those topics has been published so far despite the importance of the expected results. This fact reflects possibly not only the inherent complexity of the task but also the amount of time and (human/financial) resources required to carry out those studies that are in striking contrast with the current general reduction in funding and the exponential increase in pressure to publish. Our hope is that this chapter will encourage young researchers to take an interest in these fascinating topics.

References

Ahmed, S., Chughtai, S. and Keane, M.A. (1998) The removal of cadmium and lead from aqueous solution by ion exchange with Na-Y zeolite. *Separation and Purification Technology*, **13**, 57–64.

Alberti, A., Martucci, A., Galli, E. and Vezzalini, G. (1997) A reexamination of the crystal structure of erionite. *Zeolites*, **19**, 349–352.

Alvarez, A., Santarén, J., Esteban-Cubillo, A. and Aparicio, P. (2011) Current industrial applications of palygorskite and sepiolite. Pp. 281–298 in: *Developments in Palygorskite-Sepiolite Research: A New Outlook on these Nanomaterials* (E. Galán and A. Singer, editors). Developments in Clay Science, **3**, Elsevier, Amsterdam.

Amelinckx, S., Devouard, B. and Baronnet, A. (1996) Geometrical aspects of the diffraction space of serpentine rolled microstructures: their study by means of electron diffraction and microscopy. *Acta Crystallographica Section A: Foundations of Crystallography*, **52**, 850–878.

Andreani, M., Baronnet, A., Boullier, A.M. and Gratier, J.P. (2007) A microstructural study of a "crack-seal" type serpentine vein using SEM and TEM techniques. *European Journal of Mineralogy*, **16**, 585–595.

Andreozzi, G.B., Ballirano, P., Gianfagna, A., Mazziotti-Tagliani, S. and Pacella, A. (2009) Structural and spectroscopic characterization of a suite of fibrous amphiboles with high environmental and health relevance from Biancavilla (Sicily, Italy). *American Mineralogist*, **94**, 1333–1340.

Artioli, G. and Galli, E. (1994) The crystal structures of orthorhombic and monoclinic palygorskite. *Materials Science Forum*, **166–169**, 647–652.

Artioli, G., Galli, E., Burattini, E., Cappuccio, G. and Simeoni, S. (1994) Palygorskite from Bolca, Italy: a characterization by high-resolution synchrotron radiation powder diffraction and computer modeling. *Neues Jahrbuch für Mineralogie Monatshefte*. **5**, 217–229.

Bailey, S.W. (1980) Structures of layer silicates. Pp. 1–124 in: *Crystal Structures of Clay Minerals and their X-ray Identification* (G.W. Brindley and G. Brown, editors). Monograph **5**, Mineralogical Society of Great Britain & Ireland, London.

Bailey, S.W. (editor) (1988) *Hydrous Phyllosilicates (Exclusive of Micas).* Reviews in Mineralogy, **19**, Mineralogical Society of America, Chantilly, Virginia, USA. 725 pp.

Ball, M.C. and Taylor, H.F.W. (1963) The dehydration of chrysotile in air and under hydrothermal conditions. *Mineralolgical Magazine*, **33**, 467–82.

Ballirano, P. (2003) Effects of the choice of different ionisation level for scattering curves and correction for small preferred orientation in Rietveld refinement: the $MgAl_2O_4$ test case. *Journal of Applied Crystallography*, **36**, 1056–1061.

Ballirano, P. (2014) Dependence of structural data from $\sin\theta/\lambda$ extension in Rietveld refinement of virtually texture-free laboratory X-ray powder-diffraction data. *Periodico di Mineralogia*, **83**, 41–53.

Ballirano, P. and Cametti, G. (2012) Dehydration dynamics and thermal stability of erionite-K: experimental evidence of the "internal ionic exchange" mechanism. *Microporous and Mesoporous Materials*, **163**, 160–168.

Ballirano, P. and Cametti, G. (2015) Crystal chemical and structural modifications of erionite fibers leached with simulated lung fluids. *American Mineralogist*, **100**, 1003–1012.

Ballirano, P. and Melis, E. (2009) Thermal behaviour and kinetics of dehydration of gypsum in air from *in situ* real-time laboratory parallel-beam X-ray powder diffraction. *Physics and Chemistry of Minerals*, **36**, 391–402.

Ballirano, P. and Pacella, A. (2016) Erionite-Na upon heating: dehydration dynamics and exchangeable cations mobility. *Scientific Reports*, **6**, 22786.

Ballirano, P., Maras, A. and Buseck, P.R (1996a) Crystal chemistry and IR spectroscopy of Cl- and SO_4-bearing cancrinite-like minerals. *American Mineralogist*, **81**, 1003–1012.

Ballirano, P., Merlino, S., Bonaccorsi, E. and Maras, A. (1996b) The crystal structure of liottite, a six-layer member of the cancrinite group. *The Canadian Mineralogist*, **34**, 1021–1030.

Ballirano, P., Andreozzi, G.B. and Belardi, G. (2008) Crystal chemical and structural characterization of fibrous tremolite from Susa Valley, Italy, with comments on potential harmful effects on human health. *American Mineralogist*, **93**, 1349–1355.

Ballirano, P., Andreozzi, G.B., Dogan, M. and Dogan, A.U. (2009) Crystal structure and iron topochemistry of erionite-K from Rome, Oregon, U.S.A. *American Mineralogist*, **94**, 1262–1270.

Ballirano, P., Pacella, A., Cremisini, C., Nardi, E., Fantauzzi, M., Atzei, D., Rossi, A. and Cametti, G., (2015) Fe (II) segregation at a specific crystallographic site of fibrous erionite: A first step toward the understanding of the mechanisms inducing its carcinogenicity. *Microporous and Mesoporous Materials*, **211**, 49–63.

Bandli, B.R., Gunter, M.E., Twamley, B., Foyt Jr., F.F. and Cornelius, S.B. (2003) Optical, compositional, morphological, and X-ray data of eleven particles of amphibole from Libby, Montana, U.S.A. *The Canadian Mineralogist*, **41**, 1241–1255.

Bargar, K.E. and Keith, T.E.C. (1995) Calcium zeolites in rhyolitic drill cores from Yellowstone National Park. Pp. 69–86 in: *Natural Zeolites '93* (D.W. Ming and F.S. Mumpton, editors). International Committee on Natural Zeolites, Brockport, New York.

Baronnet, A. and Devouard, B. (1996) Topology and crystal growth of natural chrysotile and polygonal serpentine. *Journal of Crystal Growth*, **166**, 952–960.

Baronnet, A., Andréani, M., Grauby, O., Devouard, B., Nitsche, S. and Chaudanson, D. (2007) Onion morphology and microstructure of polyhedral serpentine. *American Mineralogist*, **92**, 687–690.

Blaauw, C., Stroink, G., Leiper, W. and Zentilli, M. (1979) Mössbauer analysis of some Canadian chrysotiles. *The Canadian Mineralogist*, **17**, 713–717.

Bloise, A., Belluso, E., Barrese, E., Miriello, D. and Apollaro, C. (2009a) Synthesis of Fe-doped chrysotile and characterization of the resulting chrysotile fibers. *Crystal Research and Technology*, **44**, 590–596.

Bloise, A., Barrese, E. and Apollaro, C. (2009b) Hydrothermal alteration of Ti-doped forsterite to chrysotile and characterization of the resulting chrysotile fibers. *Neues Jahrbuch für Mineralogie-Abhandlungen: Journal of Mineralogy and Geochemistry*, **185**, 297–304.

Bloise, A., Belluso, E., Fornero, E., Rinaudo, C., Barrese, E. and Capella, S. (2010) Influence of synthesis conditions on growth of Ni-doped chrysotile. *Microporous and Mesoporous Materials*, **132**, 239–245.

Bloise, A., Belluso, E., Catalano, M., Barrese, E., Miriello, D. and Apollaro, C. (2012) Hydrothermal alteration of glass to chrysotile. *Journal of the American Ceramic Society*, **95**, 3050–3055.

Bloise, A., Critelli, T., Catalano, M., Apollaro, C., Miriello, D., Croce, A., Barrese, E., Liberi F., Piluso, E., Rinaudo, C. and Belluso E. (2014) Asbestos and other fibrous minerals contained in the serpentinites of the Gimigliano-Mount Reventino unit (Calabria, S-Italy). *Environmental Earth Sciences*, **71**, 3773–3786.

Bloise, A., Barca, D., Gualtieri, A.F., Pollastri, S. and Belluso E. (2016a) Trace elements in hazardous mineral fibers. *Environmental Pollution*, **216**, 314–323.

Bloise, A., Punturo, R., Calatano, M., Miriello, D. and Cirrincione, R. (2016b) Naturally occurring asbestos (NOA) in rock and soil and relation with human activities: the monitoring example of selected sites in Calabria (southern Italy). *Italian Journal of Geosciences*, **135**, 268–279.

Bloise, A., Catalano, M., Barrese, E., Gualtieri, A.F., Gandolfi, N.B., Capella, S. and Belluso, E. (2016c) TG/DSC study of the thermal behaviour of hazardous mineral fibres. *Journal of Thermal Analysis and Calorimetry*, **123**, 2225–2239.

Bloise, A., Kusiorowski, R., Lassinantti Gualtieri, M. and Gualtieri, A.F. (2017) Thermal behaviour of mineral fibres. Pp. 215–260 in: *Mineral Fibres: Crystal Chemistry, Chemical-physical Properties, Biological Interaction and Toxicity* (A.F. Gualtieri, editor). EMU Notes in Mineralogy, **19**. European Mineralogical Union and Mineralogical Society of Great Britain & Ireland, London.

Bonaccorsi, E., Ferraris, G. and Merlino, S. (2012) Crystal structure of 2M and 1A polytypes of balangeroite. *Zeitschrift für Kristallographie*, **227**, 460–467.

Bottazzi, P., Tiepolo, M., Vannucci, R., Zanetti, A., Brumm, R., Foley, S.F. and Oberti, R. (1999) Distinct site preferences for heavy and light REE in amphibole and the prediction of Amph/LDREE. *Contributions to Mineralogy and Petrology*, **137**, 36–45.

Bowes, D.R. and Farrow, C.M. (1997) Major and trace element compositions of the UICC standard asbestos samples. *American Journal of Industrial Medicine*, **32**, 592–594.

Bowles, A., Angino, E.A., Hosterman, J.W. and Galle, O.K. (1971) Precipitation of deep-sea palygorskite and sepiolite. *Earth and Planetary Science Letters*, **11**, 324–332.

Bradley, W.F. (1940) The structural scheme of attapulgite. *American Mineralogist*, **25**, 405–410.

Brindley, G.W. and Hayami, R. (1965) Mechanism of formation of forsterite and enstatite from serpentine. *Mineralogical Magazine*, **35**, 189–195.

Burragato, F., Ballirano, P., Fiori, S., Papacchini, L. and Sonno, M. (2001) Segnalazione di tremolite asbestiforme nel Lazio. *Il Cercapietre*, **1-2/2001**, 33–35.

Bush, D.G. and Schumacher, R.A. (1982) Analysis of chrysotile asbestos in bulk materials. *Analytical Chemistry*, **54**, 1792–1795.

Cametti, G., Pacella, A., Mura, F., Rossi, M. and Ballirano, P. (2013) New morphological, chemical, and structural data of woolly erionite-Na from Durkee, Oregon, USA. *American Mineralogist*, **98**, 2155–2163.

Carbone, M., Baris, I., Bertino, P., Brass, B., Corertpay, S., Dogan, A., Gaudino, G., Jube, S., Kanodia, S., Partridge, C., Pass, H., Rivera, Z., Steele, I., Tuncer, M., Way, S., Yang, H. and Miller, A. (2011) Erionite exposure in North Dakota and Turkish villages with mesothelioma. *Proceedings of the National Academy of Sciences USA*, **108**, 13618–13623.

Carr, A. and Frei, B. (1999) Does vitamin C act as a pro-oxidant under physiological conditions? *The FASEB Journal*, **13**, 1007–1023.

Chahi, A., Fritz, B., Duplay, J., Weber, F. and Lucas, J. (1997) Textural transition and genetic relationship between precursor stevensite and sepiolite in lacustrine sediments (Jbel Rhssoul, Morocco). *Clays and Clay Minerals*, **45**, 378–389.

Chiari, G., Giustetto, R. and Ricchiardi, G. (2003) Crystal structure refinements of palygorskite and Maya Blue from molecular modeling and powder synchrotron diffraction. *European Journal of Mineralogy*, **15**, 21–33.

Chisholm, J.E. (1973) Planar defects in fibrous amphiboles. *Journal of Materials Science*, **8**, 475–483.

Chisholm, J.E. (1992) Powder diffraction patterns and structural models for palygorskite. *The Canadian Mineralogist*, **30**, 61–73.

Christ, C.L., Hathaway, J.C., Hostetler, P.B. and Shepard, A.O. (1969) Palygorskite: new X-ray data. *American Mineralogist*, **54**, 198–205.

Chryssikos, G.D., Gionis, V., Kacandes, G.H., Stathopoulou, E.T., Suarez, M. and Garcia-Romero, E. (2009). Octahedral cation distribution in palygorskite. *American Mineralogist*, **94**, 200–203.

Comba, P., Gianfagna, A. and Paoletti, L. (2003) The pleural mesothelioma cases in Biancavilla are related to the new fibrous amphibole fluoro-edenite. *Archives of Environmental Health*, **58**, 229–232.

Compagnoni, R., Ferraris, G. and Fiora, L. (1983) Balangeroite, a new fibrous silicate related to gageite from Balangero, Italy. *American Mineralogist*, **68**, 214–219.

Coombs, D.C., Alberti, A., Armbruster, T., Artioli, G., Colella, C., Galli, E., Grice, J.D., Liebau, F., Mandarino, J.A., Minato, H., Nickel, E.H., Passaglia, E., Peacor, D.R., Quartieri, S., Rinaldi, R., Ross, M., Sheppard, R.A., Tillmanns, E. and Vezzalini, G. (1997) Recommended nomenclature for zeolite minerals: Report of the subcommittee on zeolites of the International Mineralogical Association, Commission on New Minerals and Mineral Names. *The Canadian Mineralogist*, **35**, 1571–1606.

Couture, R.A. (1977) Composition of palygorskite-rich and montmorillonite-rich zeolite-containing sediments from the Pacific Ocean. *Chemical Geology*, **19**, 113–130.

Cralley, L.J., Key, M.M., Groth, D.H., Lainhart, W.S. and Ligo, R.M. (1968) Fibrous and mineral content of cosmetic talcum products. *American Industrial Hygiene Association Journal*, **29**, 350–354.

Cressey, B.A. (1979) Electron microscope studies of serpentine textures. *The Canadian Mineralogist*, **17**, 741–756.

Cressey, B.A. and Whittaker, E.J.K. (1993) Five-fold symmetry in chrysotile asbestos revealed by transmission electron microscopy. *Mineralogical Magazine*, **57**, 729–732.

Cressey, B.A., Cressey, G. and Cernik, R.J. (1994) Structural variations in chrysotile asbestos fibers revealed by synchrotron X-ray diffraction and high resolution transmission electron microscopy. *The Canadian Mineralogist*, **32**, 257–270.

Cressey, B.A., Cressey, G. and Wicks, F.J. (2008) Polyhedral serpentine: a spherical analogue of polygonal serpentine? *Mineralogical Magazine*, **6**, 1229–1242.

Cressey, B.A., Cressey, G., Wicks, F.J. and Yada., K. (2010) A disc with five-fold symmetry: the proposed fundamental seed structure for the formation of chrysotile asbestos fibres, polygonal serpentine and polyhedral lizardite spheres. *Mineralogical Magazine*, **74**, 29–37.

Croce, A., Allegrina, M., Rinaudo, C., Gaudino, G., Yang, H. and Carbone, M. (2015) Numerous iron-rich particles lie on the surface of erionite fibers from Rome (Oregon, USA) and Karlik (Cappadocia, Turkey). *Microscopy and Microanalysis*, **21**, 1341–1347.

D'Arco, P., Noel, Y., Demichelis, R. and Dovesi, R. (2009) Single-layered chrysotile nanotubes: a quantum mechanical *ab initio* simulation. *Journal of Chemical Physics*, **131**, 204701.

de Lapparent, J. (1935) Sur un constituent essential des terres a foulon. *Comptes Rendues de l'Academie des Sciences de Paris*, **201**, 481–482.

Della Ventura, G., Robert, J.L., Raudsepp, M., Hawthorne, F.C. and Welch, M.D. (1997) Site occupancies in synthetic monoclinic amphiboles: Rietveld structure refinement and infrared spectroscopy of (nickel, magnesium, cobalt)-richterite. *American Mineralogist*, **82**, 291–301.

Devouard, B. and Baronnet, A. (1995) Axial diffraction of curved lattices: geometrical and numerical modeling. Application to chrysotile. *European Journal of Mineralogy*, **7**, 835–846.

Dixon, J.R., Lowe, D.B., Richards, D.E., Cralley, L.J. and Stokinger, H.E. (1970) The role of trace metals in chemical carcinogenesis: Asbestos cancers. *Cancer Research*, **30**, 1068–1074.

Dogan, M. (2012) Quantitative characterization of the mesothelioma-inducing erionite series minerals by transmission electron microscopy and energy dispersive spectroscopy. *Scanning*, **34**, 37–42.

Dogan, A.U. and Dogan, M. (2008) Re-evaluation and re-classification of erionite series minerals. *Environmental Geochemistry and Health*, **30**, 355–366.

Dogan, A.U., Baris, Y.I., Dogan, M., Emri, S., Steele, I., Elmishad, A.G. and Carbone, M. (2006) Genetic predisposition to fiber carcinogenesis causes a mesothelioma epidemic in Turkey. *Cancer Research*, **66**, 5063–5068.

Drits, V.A. and Sokolova, G.V. (1971) Structure of palygorskite. *Soviet Physics, Crystallography*, **16**, 183–185.

Eberly, P.E. (1964) Adsorption properties of naturally occurring erionite and its cationic-exchanged forms. *American Mineralogist*, **49**, 30–40.

Eborn, S.K. and Aust, A.E. (1995) Effect of iron acquisition on induction of DNA single-strand breaks by erionite, a carcinogenic mineral fiber. *Archives of Biochemistry and Biophysics*, **316**, 507–514.

Evans, B.W., Ghiorso, M.S., Yang, H. and Medenbach, O. (2001) Thermodynamics of the amphiboles: Anthophyllite-ferroanthophyllite and the ortho-clino phase loop. *American Mineralogist*, **86**, 640–651.

Evans, B.W., Hattori, K. and Baronnet, A. (2013) Serpentinite: what, why, where? *Elements*, **9**, 99–106.

Falini, G., Foresti, E., Gazzano, M., Gualtieri, A.F., Leoni, M., Lesci, I.G. and Roveri, N. (2004) Tubular-shaped stoichiometric chrysotile nanocrystals. *Chemistry - A European Journal*, **10**, 3043–3049.

Ferraris, G., Mellini, M. and Merlino, S. (1987) Electron-diffraction and electron-microscopy study of balangeroite and gageite: crystal structures, polytypism, and fiber texture. *American Mineralogist*, **72**, 382–391.

Figovsky, O., Pashin, D., Khalitov, Z. and Khadiev, A. (2014) The quantitative theory of diffraction by spiral nanotubes. *Chemistry, Chemical Technology*, **8**, 41–50.

Finger, L.W. (1969) The crystal structure and cation distribution of a grunerite. *Mineralogical Society of America Special Paper*, **2**, 95–100.

Finger, L.W. (1970) Refinement of the crystal structure of an anthophyllite. *Carnegie Institute of Washington*

Year Book, **68**, 283–288.
Fischer, K.F. (1966) A further refinement of the crystal structure of cummingtonite, $(Mg,Fe)_7(Si_4O_{11})_2(OH)_2$. *American Mineralogist*, **51**, 814–818.
Freeman, A.G. (1966) The dehydroxylation behaviour of amphiboles. *Mineralogical Magazine*, **35**, 953–957.
Fubini, B. and Otero Aréan, C. (1999) Chemical aspects of the toxicity of inhaled mineral dusts. *Chemical Society Review*, **28**, 373–381.
Fujishige, M., Kuribara, A., Karasawa, I. and Kojima, A. (2007) Low-temperature pyrolysis of crocidolite and amosite using calcium salts as a flux. *Journal of the Ceramic Society of Japan*, **115**, 434–439.
Fukuda, H., Ebara, M., Yamada, H., Arimoto, M., Okabe, S., Obu, M., Yoshikawa, M., Sugiura, N. and Saisho, H. (2004) Trace elements and cancer. *Japan Medical Association Journal*, **47**, 391–395.
Galán, E. (1996) Properties and applications of palygorskite-sepiolite clays. *Clay Minerals*, **31**, 443–453.
Gard, J.A. and Tait, J.M. (1972) The crystal structure of the zeolite offretite $K_{1.1}Ca_{1.1}Mg_{0.7}[Si_{12.8}Al_{5.2}O_{36}] \cdot 15.2H_2O$. *Acta Crystallographica*, **B28**, 825–834.
Gard, J.A. and Tait, J.M. (1973) Refinement of the crystal structure of erionite. Pp. 94–99 in: *Proceedings of the 3rd International Conference on Molecular Sieves*. University Press, Leuven, Belgium.
Ghose, S. (1961) The crystal structure of a cummingtonite. *Acta Crystallographica*, **14**, 622–627.
Ghose, S. and Ganguly, J. (1982) Mg-Fe order-disorder in ferromagnesian silicates. Pp. 3–99 in: *Advances in Physical Geochemistry* (S.K. Saxena, editor). Springer, New York.
Ghose S. and Hellner E. (1959) The crystal structure of grunerite and observations on the Mg-Fe distribution. *The Journal of Geology*, **67**, 691–701.
Gianfagna, A. and Oberti, R. (2001) Fluoro-edenite from Biancavilla (Catania, Sicily, Italy): crystal chemistry of a new amphibole end-member. *American Mineralogist*, **86**, 1489–1493.
Gianfagna, A., Ballirano, P., Bellatreccia, F., Bruni, B., Paoletti, L. and Oberti, R. (2003) Characterization of amphibole fibres linked to mesothelioma in the area of Biancavilla, Eastern Sicily, Italy. *Mineralogical Magazine*, **67**, 1221–1229.
Gianfagna, A., Andreozzi, G.B., Ballirano, P. and Mazziotti-Tagliani, S. (2007) Structural and chemical contrast between prismatic and fibrous fluoro-edenite from Biancavilla, Sicily, Italy. *The Canadian Mineralogist*, **45**, 249–262.
Giustetto, R. and Chiari, G. (2004) Crystal structure refinement of palygorskite from neutron powder diffraction. *European Journal of Mineralogy*, **16**, 521–532.
Gottardi, G. and Galli, E. (1985) *Natural Zeolites*. Springer-Verlag, Heidelberg, 409 pp.
Gottschalk, M., Andrut, M. and Melzer, S. (1999) The determination of the cummingtonite content of synthetic tremolite. *European Journal of Mineralogy*, **11**, 967–982.
Grice, J.D., Kristiansen, R., Friis, H., Rowe, R., Poirier, G.G., Selbekk, R.S., Cooper, M.A. and Larsen, A.O. (2013) Ferrochiavennite, a new beryllium silicate zeolite from syenite pegmatites in the Larvik plutonic complex, Oslo region, southern Norway. *The Canadian Mineralogist*, **51**, 285–296.
Groppo, C., Tomatis, M., Turci, F., Gazzano, E., Ghigo, D., Compagnoni, R. and Fubini, B. (2005) Potential toxicity of nonregulated asbestiform minerals: balangeroite from the Western Alps. Part 1: Identification and characterization. *Journal of Toxicology and Environmental Health, Part A*, **68**, 1–19.
Gualtieri, A., Artioli, G., Passaglia, E., Bigi, S., Viani, A. and Hanson, J.C. (1998) Crystal structure-crystal chemistry relationships in the zeolites erionite and offretite. *American Mineralogist*, **83**, 590–606.
Guggenheim, S. (2015) Phyllosilicates used as nanotube substrate in engineering materials: structures, chemistries, and textures. Pp 3–50 in *Natural Mineral Nanotubes Properties and Applications* (P. Pasbakhsh and G. Jock Churchman, editors). CRC Press, Boca Raton, Florida, USA.
Guggenheim, S. and Eggleton, R.A. (1988) Crystal chemistry, classification, and identification of modulated layer silicates. Pp. 675–725 in: *Hydrous Phyllosilicates (Exclusive of Micas)* (S.W. Bailey, editor), Reviews in Mineralogy and Geochemistry, **19**. Mineralogical Society of America, Chantilly, Virginia, USA.
Guggenheim, S. and Krekeler, M.P.S. (2011) The structures and microtextures of the palygorskite–sepiolite group minerals. Pp. 3–32 in: *Developments in Palygorskite-Sepiolite Research: A New Outlook on these Nanomaterials* (E. Galán and A. Singer, editors). Developments in Clay Science **3**, Elsevier, Amsterdam.
Guggenheim, S., Adams, J.M., Bain, D.C., Bergaya, F., Brigatti, M.F., Drits, V.A., Formoso, M.L.L., Galán, E., Kogure, T. and Stanjek, H. (2006) Summary of recommendations of nomenclature committees relevant to

clay mineralogy: report of the Association Internationale pour l'Etude des Argiles (AIPEA) Nomenclature-Committee for 2006. *Clays and Clay Minerals*, **54**, 761–772 and *Clay Minerals*, **41**, 863–877.

Gunter, M.E., Darby Dyar, M., Twamley, B., Foit, F.F. Jr. and Cornelius, S. (2003) Composition, $Fe^{3+}/\Sigma Fe$, and crystal structure of non-asbestiform and asbestiform amphiboles from Libby, Montana, U.S.A. *American Mineralogist*, **88**, 1970–1978.

Gutteridge, J.M.C., Mumby, S., Quinlan, G.J., Chung, K.F. and Evans, T.W. (1996) Pro-oxidant iron is present in human pulmonary epithelial lining fluid: implications for oxidative stress in the lung. *Biochemical and Biophysical Research Communications*, **220**, 1024–1027.

Hardy, J.A. and Aust, E.A. (1995) Iron in asbestos chemistry and carcinogenicity. *Chemical Reviews*, **95**, 97–118.

Harington, J.S. and Roe, F. (1965) Studies of carcinogenesis of asbestos fibers and their natural oils. *Annals of the New York Academy of Sciences*, **132**, 439–450.

Hawthorne, F.C. (1978) The crystal chemistry of the amphiboles. VIII. The crystal structure and site chemistry of fluor-riebeckite. *The Canadian Mineralogist*, **16**, 187–194.

Hawthorne, F.C. (1981) Crystal chemistry of the amphiboles. Pp. 1–102 in: *Amphiboles and Other Hydrous Pyriboles – Mineralogy* (D.R. Veblen, editor). Reviews in Mineralogy, **9A**, Mineralogical Society of America, Chantilly, Virginia, USA.

Hawthorne, F.C. and Oberti, R. (2006) On the classification of amphiboles. *The Canadian Mineralogist*, **44**, 1–21.

Hawthorne, F.C. and Oberti, R. (2007) Amphiboles: Crystal chemistry. Pp. 1–54 in: *Amphiboles: Crystal Chemistry, Occurrence, and Health Issues* (F.C. Hawthorne, R. Oberti, G. Della Ventura and A. Mottana, editors). Reviews in Mineralogy and Geochemistry, **67**, Mineralogical Society of America and Geochemical Society, Chantilly, Virginia, USA.

Hawthorne, F.C., Oberti, R., Della Ventura, G. and Mottana, A. (editors) (2007) *Amphiboles: Crystal Chemistry, Occurrence, and Health Issues*. Reviews in Mineralogy and Geochemistry **67**, Mineralogical Society of America and Geochemical Society, Chantilly, Virginia, USA, 546 pp.

Hawthorne, F.C., Oberti, R., Zanetti, A. and Nayak, V.K. (2008a) The crystal chemistry of alkali amphiboles from the Kajlidongri manganese mine, India. *The Canadian Mineralogist*, **46**, 455–466.

Hawthorne, F.C., Schindler, M., Abdu, Y., Sokolova, E., Evans, B.W. and Ishida, K. (2008b). The crystal chemistry of the gedrite-group amphiboles. II. Stereochemistry and chemical relations. *Mineralogical Magazine*, **72**, 731–745.

Heller-Kallai, L. and Rozenson, I. (1981) Mössbauer studies of palygorskite and some aspects of palygorskite mineralogy. *Clays and Clay Minerals*, **29**, 226–232.

Hey, M.H. (1975) *Chemical Index of Minerals*. 2nd edition. British Museum (Natural History) London, 728 pp.

Hodgson, A.A. (1979) Chemistry and physics of asbestos. Pp. 67–114 in: *Asbestos – Properties, Applications and Hazards* (L. Michaels and S.S. Chissick, editors). John Wiley & Sons, New York.

Hodgson, A.A., Freeman, A.G. and Taylor, H. (1965a) The thermal decomposition of crocidolite from Koegas, South Africa. *Mineralogical Magazine*, **35**, 5–30.

Hodgson, A.A., Freeman, A.G. and Taylor, H. (1965b) The thermal decomposition of amosite. *Mineralogical Magazine*, **35**, 445–463.

Holmes, A., Morgan, A. and Sandalls, F.J. (1971) Determination of iron, chromium, cobalt, nickel, and scandium in asbestos by neutron activation analysis. *The American Industrial Hygiene Association Journal*, **32**, 281–286.

Hutchison, J.L., Irusteta, M.C. and Whittaker, E.J.W. (1975) High-resolution electron microscopy and diffraction studies of fibrous amphiboles. *Acta Crystallographica*, **A31**, 794–801.

IARC (2011) IARC Monographs on the evaluation of the carcinogenic risk to humans. *Arsenic, Metals, Fibres and Dusts*, Vol. 100 C, 311–316.

IARC (2012) IARC Monographs on the evaluation of the carcinogenic risk to humans. *Arsenic, Metals, Fibres and Dusts*, Vol. 100 C, 401–435.

IARC (2014) Carcinogenicity of fluoro-edenite, silicon carbide fibres and wiskers, and carbon nanotubes. *Lancet Oncology*, **15**, 1427–1428.

IARC (2017) IARC Monographs on the evaluation of the carcinogenic risk to humans. Fluoro-edenite, silicon carbide fibres and wiskers, and single-walled and multi-walled carbon nanotubes. Vol. 111 (in press).

Imai, N. and Otsuka, R. (1984) Sepiolite and palygorskite in Japan. Pp. 211–232 in: *Palygorskite-Sepiolite. Occurrences, Genesis, and Uses* (A. Singer and E. Galán, editors), Developments in Sedimentology, **37**. Elsevier, Amsterdam.

Jagodzinski, H. and Kunze, G. (1954a) Die Rollchen-struktur des Chrysotils. I Allgemeine Beugungstheorie und Kleinwinkelstreuung. *Neues Jahrbuch für Mineralogie Monatshefte*, 95–108.

Jagodzinski, H. and Kunze, G. (1954b) Die Rollchen-struktur des Chrysotils. II Weitwinkelinterferenzen. *Neues Jahrbuch für Mineralogie Monatshefte*, 113–130.

Jagodzinski, H. and Kunze, G. (1954c) Die Rollchen-struktur des Chrysotils. III Versetzungswachtum der Rollchen. *Neues Jahrbuch für Mineralogie Monatshefte*, 137–150.

Jones, J.B. (1968) Al-O and Si-O tetrahedral distances in aluminosilicate framework structures. *Acta Crystallographica*, B**24**, 355–358.

Jones, B.F. and Galán, E. (1988) Sepiolite and palygorskite. Pp. 631–674 in: *Hydrous Phyllosilicates (Exclusive of Micas)* (S.W. Bailey, editor), Reviews in Mineralogy, **19**. Mineralogical Society of America, Chantilly, Virginia, USA.

Katerinopoulou, A., Balic-Zunic, T. and Lundegaard, L.F. (2012) Application of the ellipsoid modeling of the average shape of nanosized crystallites in powder diffraction. *Journal of Applied Crystallography*, **45**, 22–27.

Kawahara, A. and Curien, H. (1969) La structure crystalline de l'érionite. *Bulletin de la Société Française de Minéralogie et de Cristallographie*, **92**, 250–256.

Khorami, J., Choquette, D., Kimmerle, F.M. and Gallagher, P.K. (1984) Interpretation of EGA and DTG analyses of chrysotile asbestos. *Thermochimica Acta*, **76**, 87–96.

Kohyama, N., Shinohama, Y. and Suzuki, Y., (1996) Mineral phases and some re-examined characteristics of the international union against cancer standard asbestos samples. *American Journal of Industrial Medicine*, **30**, 515–543.

Kokotailo, G.T., Sawruk, S. and Lawton, S.L. (1972) Direct observation of stacking faults in the zeolite erionite. *American Mineralogist*, **57**, 439–444.

Lafay, R., Montes-Hernandez, G., Janots, E., Auzende, A.L., Chiriac, R., Lemarchand, D. and Toche, F. (2014) Influence of trace elements on the textural properties of synthetic chrysotile: complementary insights from macroscopic and nanoscopic measurements. *Microporous and Mesoporous Materials*, **183**, 81–90.

Larson, A.C. and Von Dreele, R.B. (1994) GSAS Generalized structure analysis system. *Laur 86-748*. Los Alamos National Laboratory, Los Alamos, New Mexico.

Leake, B.E., Woolley, A.R., Arps, A.R., Birch, W.D., Gilbert, M.C., Grice, J.D., Hawthorne, F.C., Kato, A., Kisch, H.J., Krivovichev, V.G., Linthout, K., Laird, J., Mandarino, J.A., Maresch, W.V., Nickel, E.H., Rock, N.M.S., Schumacher, J.C., Smith, D.C., Stephenson, N.C.N., Ungaretti, L., Whittaker, E.J.W. and Youzhi, G. (1997) Nomenclature of the amphiboles: report of the subcommittee on amphiboles of the International Mineralogical Association, Commission on New Mineral Names. *American Mineralogist*, **82**, 1019–1037.

Leoni, M., Gualtieri, A.F. and Roveri, N. (2004) Simultaneous refinement of structure and microstructure of layered materials. *Journal of Applied Crystallography*, **37**, 166–173.

Lindemann, W. (1964) Beitrag zur Struktur des Anthophyllits. *Fortschritte der Mineralogie*, **42**, 205 (abstract).

Lyon, G.L., Brooks, R.R. and Peterson, P.J. (1970) Some trace elements in plants from serpentine soils. *New Zealand Journal of Science*, **13**, 133–139.

Lyon, G.L., Brooks, R.R., Peterson, P.J. and Butler, G.W. (1968) Trace elements in a New Zealand serpentine flora. *Plant and soil*, **29**, 225–240.

MacKenzie, K.J.D. and Meinhold, R.H. (1994) Thermal reactions of chrysotile revisited: a ^{29}Si and ^{25}Mg MAS NMR study. *American Mineralogist*, **79**, 43–50.

Mancuso, T.F. (1970) Relation of duration of employment and prior respiratory illness to respiratory cancer among beryllium workers. *Environmental Research*, **3**, 251–275.

Martin, C.J. (1977) The thermal decomposition of chrysotile. *Mineralogical Magazine*, **41**, 453–462.

Martin, R.T., Bailey, S.W., Eberl, D.D., Fanning, D.S., Guggenheim, S., Kodama, H., Pevear, D.R., Środoń, J. and Wicks, F.J. (1991) Revised classification of clay materials: report of The Clay Minerals Society Nomenclature Committee for 1986–1988. *Clays and Clay Minerals*, **39**, 333–334.

Matassa, R., Famigliari, G., Relucenti, M., Battaglione, E., Downing, C., Pacella, A., Cametti, G. and Ballirano, P. (2015) A deep look into erionite fibres: an electron microscopy investigation of their self-assembly. *Scientific Reports*, **5**, 16757.

Mazziotti-Tagliani, S., Andreozzi, G.B., Bruni, B.M., Gianfagna, A., Pacella, A. and Paoletti, L. (2009) Quantitative chemistry and compositional variability of fibrous amphiboles from Biancavilla (Sicily, Italy). *Periodico di Mineralogia*, **78**, 65–75.

Meeker, G.P., Bern, A.M., Brownfield, I.K., Lowers, H.A., Sutley, S.J., Hoefen, T.M. and Vance, J.S. (2003) The composition and morphology of amphiboles from the Rainy Creek Complex, near Libby, Montana. *American Mineralogist*, **88**, 1955–1969.

Meier, W.M. and Groner, M. (1981) Zeolite structure type EAB: crystal structure and mechanism for the topotactic transformation of the Na, TMA form. *Journal of Solid State Chemistry*, **37**, 204–218.

Mellini, M, Trommsdorff, V. and Compagnoni, R. (1987) Antigorite polysomatism: behaviour during progressive metamorphism. *Contributions to Mineralogy and Petrology*, **97**, 147–155.

Ménez, B., Pasini, V. and Brunelli, D. (2012) Life in the hydrated suboceanic mantle. *Nature Geoscience*, **5**, 133–137.

Mitchell, R.H. and Putnis, A. (1988) Polygonal serpentine in segregation-textured kimberlite. *The Canadian Mineralogist*, **26**, 991–997.

Miyano, T. and Beukes, N.J. (1997) Mineralogy and petrology of the contact metamorphosed amphibole asbestos-bearing Penge iron formation, Eastern Transvaal, South Africa. *Journal of Petrology*, **5**, 651–676.

Morgan, A. and Cralley, L. (1973) Chemical characteristics of asbestos and associated trace elements. Pp. 113–132 in: *Biological Effects of Asbestos*. Scientific Publication, IARC, Lyon, France.

Nackerdien, Z., Kasprzak, K.S., Rao, G., Halliwell, B. and Dizdaroglu, M. (1991) Nickel (II)-and cobalt (II)-dependent damage by hydrogen peroxide to the DNA bases in isolated human chromatin. *Cancer Research*, **51**, 5837–5842.

Newman, A.C.D. and Brown, G. (1987) The chemical constitution of clays. Pp. 1–128 in: *Chemistry of Clays and Clay Minerals* (A.C.D. Newman, editor). Monograph **6**, Mineralogical Society, London.

Oberti, R., Hawthorne, F.C., Cannillo, E. and Cámara, F. (2007) Long-range order in amphiboles. Pp. 125–171 in: *Amphiboles: Crystal Chemistry, Occurrence, and Health Issues* (F.C. Hawthorne, R. Oberti, G. Della Ventura and A. Mottana, editors). Reviews in Mineralogy and Geochemistry, **67**, Mineralogical Society of America and Geochemical Society, Chantilly, Virginia, USA.

O'Hanley, D.S. and Dyar, M.D. (1998). The composition of chrysotile and its relationship with lizardite. *The Canadian Mineralogist*, **36**, 727–740.

Oscarson, D.W., van Scoyoc, G.E. and Ahlrichs, J.L. (1986) Lysis of erythrocytes by silicate minerals. *Clays and Clay Minerals*, **34**, 74–86.

Pacella, A. and Ballirano, P. (2016) Chemical and structural characterization of fibrous richterite with high environmental and health relevance from Libby, Montana (USA). *Periodico di Mineralogia*, **85**, 169–177.

Pacella, A., Andreozzi, G.B., Ballirano, P. and Gianfagna, A. (2008) Crystal chemical and structural characterization of fibrous tremolite from Ala di Stura (Lanzo Valley, Italy). *Periodico di Mineralogia*, **77**, 51–62.

Pacella, A., Andreozzi, G.B. and Fournier, J. (2010) Detailed crystal chemistry and iron topochemistry of asbestos occurring in its natural setting: a first step to understanding its chemical reactivity. *Chemical Geology*, **277**, 197–206.

Pacella, A., Fantauzzi, M., Turci, F., Cremisini, C., Montereali, M.R., Nardi, E., Atzei, D., Rossi, A. and Andreozzi, G.B. (2014) Dissolution reactions and surface iron speciation of UICC crocidolite in buffered solutions at pH 7.4: a combined ICP-OES, XPS and TEM investigation. *Geochimica et Cosmochimica Acta*, **127**, 221–232.

Pacella, A., Ballirano, P. and Cametti, G. (2016) Quantitative chemical analysis of erionite fibres using a micro-analytical SEM-EDX method. *European Journal of Mineralogy*, **28**, 257–264.

Pacella, A., Fantauzzi, M., Atzei, A., Cremisini, C., Nardi, E., Montereali, M.R., Rossi, A. and Ballirano, P. (2017) Iron within the erionite cavity and its potential role in inducing its toxicity: Evidence of Fe(III) segregation as extra-framework cation. *Microporous and Mesoporous Materials*, **237**, 168–179.

Pagano, G., Aliberti, F., Guida, M., Oral, R., Siciliano, A., Trifuoggi, M. and Tommasi, F. (2015a) Rare earth

elements in human and animal health: State of art and research priorities. *Environmental Research*, **142**, 215–220.
Pagano, G., Guida, M., Tommasi, F. and Oral, R. (2015b) Health effects and toxicity mechanisms of rare earth elements knowledge gaps and research prospects. *Ecotoxicology and Environmental Safety*, **115**, 40–48.
Paoletti, L., Battisti, D., Bruno, C., Di Paola, M., Gianfagna, A., Mastrantonio, M., Nesti, M. and Comba, P. (2000) Unusually high incidence of malignant pleural mesothelioma in a town of eastern Sicily: an epidemiological and environmental study. *Archives of Environmental Health*, **55**, 392–398.
Passaglia, E. (1970) The crystal chemistry of chabazites. *American Mineralogist*, **55**, 1278–1301.
Passaglia, E., Artioli, G. and Gualtieri, A. (1998) Crystal chemistry of the zeolites erionite and offretite. *American Mineralogist*, **83**, 577–589.
Pauling, L. (1930) The structure of chlorites. *Proceedings of the National Academy of Science USA*, **16**, 578–582.
Phillips, A.H. (1910) Gageite, a new mineral from Franklin, New Jersey. *American Journal of Science*, **30**, 1035–1045.
Pollastri, S., Gualtieri, A.F., Gualtieri, M.L., Hanuskova, M., Cavallo, A. and Gaudino, G. (2014) The zeta potential of mineral fibres. *Journal of Hazardous Materials*, **276**, 469–479.
Pollastri, S., D'Acapito, F., Trapananti, A., Colantoni, I., Andreozzi, G.B. and Gualtieri, A.F. (2015) The chemical environment of iron in mineral fibres. A combined X-ray absorption and Mössbauer spectroscopic study. *Journal of Hazardous Materials*, **298**, 282–293.
Pollastri, S., Perchiazzi, N., Gigli, L., Ferretti, P., Cavallo, A., Bursi Gandolfi, N., Pollok, K. and Gualtieri, A.F. (2017) The crystal structure of mineral fibres. 2. Amosite and fibrous anthophyllite. *Periodico di Mineralogia*, **86**, 55–65.
Posnjak, E. and Bowen, N.L. (1931) The role of water in tremolite. *American Journal of Science*, **222**, 203–214.
Post, J.E. and Heaney, P.J. (2008) Synchrotron powder X-ray diffraction study of the structure and dehydration behavior of palygorskite. *American Mineralogist*, **93**, 667–675.
Post, J.E., Bish, D.L. and Heaney, P.J. (2007) Synchrotron powder X-ray diffraction study of the structure and dehydration behavior of sepiolite. *American Mineralogist*, **92**, 91–97.
Pott, F., Bellmann, B., Mohle, H., Rodelsperger, K., Rippe, R.M., Roller, M. and Rosenbruch, M. (1990) Intraperitoneal injection studies for the evaluation of the carcinogenicity of fibrous phyllosilicates. Pp. 319–331 in: *Health Related Effects of Phyllosilicates* (J. Bignon, editor). NATO ASI Series G. Ecological Science, Vol. **G21**, Springer-Verlag, Heidelberg, Germany.
Preisinger, A. (1961) Sepiolite and related compounds: its stability and application. *Clays and Clay Minerals*, **10**, 365–371.
Punturo, R., Bloise, A., Critelli, T., Catalano, M., Fazio, E. and Apollaro, C. (2015) Environmental implications related to natural asbestos occurrences in the ophiolites of the Gimigliano-Mount Reventino Unit (Calabria, Southern Italy). *International Journal of Environmental Research*, **9**, 405–418.
Roveri, N., Faligi, G., Foresti, E., Fracasso, G., Lesci, I.G. and Sabatino, P. (2006) Geoinspired synthetic chrysotile nanotubes. *Journal of Materials Research*, **21**, 2711–2725.
Rüdinger, B., Tillmanns, E. and Hentschel, G. (1993) Bellbergite – a new mineral with the structure type EAB. *Mineralogy and Petrology*, **48**, 147–152.
Scambelluri, M., Piccardo, G.B., Philippot, P., Robbiano, A. and Negretti, L. (1997) High salinity fluid inclusions formed from recycled seawater in deeply subducted alpine serpentinite. *Earth and Planetary Science Letters*, **148**, 485–499.
Schaller, W.T. (1916) The chemical composition of tremolite. *Geological Survey Mineralogical Notes Series 3 Bullettin*, **610**, 133–136.
Schindler, M., Sokolova, E., Abdu, Y., Hawthorne, F.C., Evans, B.W. and Ishida, K. (2008) The crystal chemistry of the gedrite-group amphiboles. I. Crystal structure and site populations. *Mineralogical Magazine*, **72**, 703–730.
Schlenker, J.L., Pluth, J.J. and Smith, J.V. (1977) Dehydrated natural erionite with stacking faults of the offretite type. *Acta Crystallographica*, **B33**, 3265–3268.
Serna, C., Van Scoyoc, G.E. and Ahlrichs, J.L. (1977) Hydroxyl groups and water in palygorskite. *American Mineralogist*, **62**, 784–792.

Shannon, R.D. (1976) Revised effective ionic radii and systematic studies of interatomic distances in halides and chalcogenides. *Acta Crystallographica*, **A32**, 751–767.

Sheppard, R.A. and Gude, A.J. 3rd (1973) Zeolites and associated authigenic silicate minerals in tuffaceous rocks of the Big Sandy Formation, Mohave County, Arizona. *U.S. Geological Survey Professional Paper*, **830**, 35 pp.

Singer, A. (1979) Palygorskite in sediments: detrital, diagenetic, or neoformed. A critical review. *Geologische Rundschau*, **68**, 996–1008.

Smith, I.M., Hall, K.J., Lavkulich, L.M. and Schreier, H. (2007) Trace metal concentrations in an intensive agricultural watershed in British Columbia, Canada. *Journal of the American Water Resources Association*, **43**, 1455–1467.

Smith, J.V. and Bennett, J.M. (1981) Enumeration of 4-connected 3-dimensional nets and classification of framework silicates; the infinite set of ABC-6 nets; the Archimedean and sigma-related nets. *American Mineralogist*, **66**, 777–788.

Smith, J.V., Rinaldi, F. and Dent Glasser, L.S. (1963) Crystal structures with a chabazite framework. II. Hydrated Ca-chabazite at room temperature. *Acta Crystallographica*, **16**, 45–53.

Smolikov, A., Vezentsev, A., Beresnev, V., Kolesnikov, D. and Solokha, A. (2013) Morphology of synthetic chrysotile nanofibers (Mg-hydro silicate). *Journal of Materials Science and Engineering*, **A3**, 523–530.

Staples, L.W. and Gard, J.A. (1959) The fibrous zeolite erionite; its occurrence, unit cell and structure. *Mineralogical Magazine*, **247**, 261–281.

Stroink, G., Blaauw, C., White, C.G. and Leiper, W. (1980) Mössbauer characteristics of UICC standard reference asbestos samples. *The Canadian Mineralogist*, **18**, 285–290.

Thilo, E. and Rogge, G. (1939) Chemisches Untersuchungen von Silikaten, VIII. Thermische Umwandlung des anthophyllite; polymorphie des $MgSiO_3$. *Berichte der deutschen chemischen Gesellschaft*, **72**, 341–362.

Tiepolo, M., Oberti, R., Zanetti, A., Vannucci, R. and Foley, S.F. (2007) Trace-element partitioning between amphibole and silicate melt. Pp. 417–452 in *Amphiboles: Crystal Chemistry, Occurrence, and Health Issues* (F.C. Hawthorne, R. Oberti, G. Della Ventura and A. Mottana, editors). Reviews in Mineralogy and Geochemistry, **67**, Mineralogical Society of America and Geochemical Society, Chantilly, Virginia, USA.

Toby, B.H. (2001) EXPGUI, a graphical user interface for GSAS. *Journal of Applied Crystallography*, **34**, 210–213.

Tschernich, R.W. (1992) *Zeolites of the World*. Geoscience Press, Phoenix, Arizona, USA, 563 pp.

Turci, F., Tomatis, M., Gazzano, E., Riganti, C., Martra, G.. Bosia, A., Ghigo, D. and Fubini, B. (2005) Potential toxicity of nonregulated asbestiform minerals: balangeroite from the Western Alps. Part 2: Oxidant activity of the fibers. *Journal of Toxicology and Environmental Health, Part A*, **68**, 21–39.

United States Environmental Protection Agency (USEPA) (2000) *Sampling and analysis of consumer garden products that contain vermiculite*. Report 2001-S-7. United States Environmental Protection Agency, Washington, DC.

Upreti, R.K., Dogra, R., Shanker, R., Murti, C.K., Dwivedi, K.K. and Rao, G.N. (1984) Trace elemental analysis of asbestos with an X-ray fluorescence technique. *Science of the Total Environment*, **40**, 259–267.

Veblen, D.R. (1980) Anthophyllite asbestos: microstructures, intergrown sheet silicates, and mechanisms of fiber formation. *American Mineralogist*, **65**, 1075–1086.

Veblen, D.R. and Ribbe, P.H. (editors) (1981) *Amphiboles and Other Hydrous Pyriboles – Mineralogy*. Reviews in Mineralogy, **9A**. Mineralogical Society of America, Chantilly, Virginia, USA, 372 pp.

Velde, B. (1985) *Clay Minerals: A Physico-chemical Explanation of their Occurrence*. Developments in Sedimentology, vol. **40**. Elsevier, New York, 427 pp.

Verkouteren, J.R. and Wylie, A.G. (2000) The tremolite-actinolite-ferro-actinolite series: systematic relationship among cell parameters, composition, optical properties, and habit, and evidence of discontinuities. *American Mineralogist*, **85**, 1239–1254.

Vignaroli, G., Ballirano, P., Belardi, G. and Rossetti, F. (2014) Asbestos fibre identification vs. evaluation of asbestos hazard in ophiolitic rock mélanges, a case study from the Ligurian Alps (Italy). *Environmental Earth Sciences*, **72**, 3679–3698.

Vils, F., Pelletier, L., Kalt, A., Mntener, O. and Ludwig, T. (2008) The lithium, boron and beryllium content of serpentinized peridotites from ODP Leg 209 (Sites 1272A and 1274A): implications for lithium and boron

budgets of oceanic lithosphere. *Geochimica et Cosmochimica Acta*, **72**, 5475–5504.
Virta, R.L. (2005) *Mineral commodity profiles – Asbestos*. Circular 1255-KK U.S. Department of Interior and U.S. Geological Survey, 63 pp.
Viti, C. (2010) Serpentine minerals discrimination by thermal analysis. *American Mineralogist*, **95**, 631–638.
Viti, C., Giacobbe, C. and Gualtieri, A.F. (2011) Quantitative determination of chrysotile in massive serpentinites using DTA: implications for asbestos determinations. *American Mineralogist*, **96**, 1003–1011.
Wagner, J.C., Griffiths, D.M. and Munday, D.E. (1987) Experimental studies with palygorskite dust. *British Journal of Industrial Medicine*, **44**, 749–763.
Walitzi, E.M., Walter, F. and Ettinger, K. (1989) Verfeinerung der kristallstruktur von anthophyllit vom Ochsenkogel/Gleinalpe, Osterreich. *Zeitschrift für Kristallographie*, **188**, 237–244.
Warren, B.E. (1930a) The structure of tremolite $H_2Ca_2Mg_5(SiO_3)_8$. *Zeitschrift für Kristallographie*, **72**, 42–57.
Warren, B.E. (1930b) The crystal structure and chemical composition of the monoclinic amphiboles *Zeitschrift für Kristallographie*, **72**, 493–516.
Warren, B. and Modell, D. (1930) The structure of Anthophyllite $H_2Mg_7(SiO_3)_8$. *Zeitschrift für Kristallographie*, **75**, 161–179.
Weaver, C.E. and Pollard, L. (1973) *The Chemistry of Clay Minerals*. Elsevier, Amsterdam, 213 pp.
Whittaker, E.J.W. (1949) The structure of Bolivian crocidolite. *Acta Crystallographica*, **2**, 312–317.
Whittaker, E.J.W. (1953) The structure of chrysotile. *Acta Crystallographica*, **6**, 747–748.
Whittaker, E.J.W. (1954) The diffraction of X-rays by a cylindrical lattice I. *Acta Crystallographica*, **7**, 827–832.
Whittaker, E.J.W. (1955a) The diffraction of X-rays by a cylindrical lattice II. *Acta Crystallographica*, **8**, 261–265.
Whittaker, E.J.W. (1955b) The diffraction of X-rays by a cylindrical lattice III. *Acta Crystallographica*, **8**, 265–271.
Whittaker, E.J.W. (1955c) A classification of cylindrical lattices. *Acta Crystallographica*, **8**, 571–574.
Whittaker, E.J.W. (1955d) The diffraction of X-rays by a cylindrical lattice IV. *Acta Crystallographica*, **8**, 726–729.
Whittaker, E.J.W. (1956a) The structure of chrysotile II. Clinochrysotile. *Acta Crystallographica*, **9**, 855–862.
Whittaker, E.J.W. (1956b) The structure of chrysotile III. Orthochrysotile. *Acta Crystallographica*, **9**, 862–864.
Whittaker, E.J.W. (1956c) The structure of chrysotile IV. Parachrysotile. *Acta Crystallographica*, **9**, 865–867.
Whittaker, E.J.W. (1957) The structure of chrysotile V. Diffuse reflections and fibre texture. *Acta Crystallographica*, **10**, 149–156.
Whittaker, E.J.W., Cressey, B.A. and Hutchinson, B.L. (1981) Edge dislocations in fibrous grunerite. *Mineralogical Magazine*, **44**, 287–291.
Wicks, F.J. and O'Hanley, D.S. (1988) Serpentine minerals: structures and petrology. Pp. 91–168 in: *Hydrous Phyllosilicates (Exclusive of Micas)* (S.W. Bailey, editor). Reviews in Mineralogy, **19**, Mineralogical Society of America, Chantilly, Virginia, USA.
Wicks, F.J. and Plant, A.G. (1979) Electron microprobe and X-ray microbeam studies of serpentine minerals. *The Canadian Mineralogist*, **17**, 785–830.
Wicks, F.J. and Whittaker, E.J.W. (1975) A reppraisal of the structures of the serpentine minerals. *The Canadian Mineralogist*, **13**, 227–243.
Wylie, A.G. and Verkouteren, J.R. (2000) Amphibole asbestos from Libby, Montana: aspects of nomenclature. *American Mineralogist*, **85**, 1540–1542.
Yada, K. (1971) Study of microstructure of chrysotile asbestos by high-resolution electron microscopy. *Acta Crystallographica*, **A27**, 659–664.
Yada, K. and Iishi, K. (1977) Growth and microstructure of synthetic chrysotile. *American Mineralogist*, **62**, 958–965.
Yada, K. and Wei, L. (1987) Polygonal microstructures of Povlen chrysotile observed by high resolution electron microscopy. *Euroclay 87 (Sevilla, Spain)*, abstract.
Yang, H. and Evans, B.W. (1996) X-ray structure refinements of tremolite at 140 and 295 K: crystal chemistry and petrologic implications. *American Mineralogist*, **81**, 1117–1125.
Zecchina, A., Rivallan, M., Berlier, G., Lamberti, C. and Ricchiardi, G. (2007) Structure and nuclearity of active

sites in Fe-zeolites: Comparison with iron sites in enzymes and homogeneous catalysts. *Physical Chemistry Chemical Physics*, **9**, 3483–3499.

Zussman, J., Brindley, G.W. and Comer, J.J (1957) Electron diffraction studies of serpentine minerals. *American Mineralogist*, **42**, 133–153.

Zvyagin, B.B. (1967) *Electron-Diffraction Analysis of Clay Mineral Structures*. Plenum Press, New York.

Crystal habit of mineral fibres

ELENA BELLUSO[1], ALESSANDRO CAVALLO[2] and DON HALTERMAN[3]

[1]Dipartimento di Scienze della Terra - Centro Interdipartimentale per lo studio degli amianti e di altri particolati nocivi "Giovanni Scansetti", Università degli Studi di Torino, Via Valperga Caluso 35, I-10125, Turin, Italy, e-mail: elena.belluso@unito.it
[2]Dipartimento di Scienze dell'Ambiente e del Territorio e di Scienze della Terra, Università degli Studi di Milano Bicocca, Piazza della Scienza 1, 20126 Milan, Italy
[3]Occupational Safety and Health Administration, Salt Lake Technical Center, 8660 South Sandy Parkway, Sandy, UT 84070, USA

This chapter describes the crystal habit of mineral fibres. This topic is so complex and multifaceted that it deserves a detailed and systematic description in order to guide the reader through the jungle of the intricate, overlapping, sometimes contradictory definitions of 'mineral fibre' and the complex relationship between crystal structure and growth and overall crystal habit. The definition of 'mineral fibre' or 'fibre' itself is intrinsically chameleon-like because it assumes different meanings and relevance in different areas and scientific fields. As a matter of fact, if you ask ten different people to define what 'fibre' means to them, you may receive ten different answers. Even among specialists in the field of 'mineral fibres', there is no clear agreement on what we call a 'fibre'. It is important to be able to rely upon indisputable and unanimous definitions and share a common language because the crystal habit of 'mineral fibres', especially asbestos, is considered one of the major characteristics affecting their toxicity, pathogenicity and carcinogenicity and hence it has a paramount importance for the social and legal issues.

1. Introduction

If you ask ten different people to define what 'fibre' means to them, you may receive ten different answers. A botanist, a tailor and a mineralogist will certainly have widely variant ideas of what a 'fibre' is. However, even among mineralogists, there is no clear agreement on what we call a 'fibre'. To some mineral scientists, any mineral with a growth habit that is very thin and long and with a high aspect ratio (length-to-width ratio) can be called fibrous. This includes such species and varieties as millerite, filiform pyrite, artinite, natrolite and mesolite, as well as the regulated asbestos minerals such as chrysotile and crocidolite (asbestiform riebeckite). Crystallographers, geochemists and mineralogists are typically more concerned with the structure, chemistry, and phase diagrams of minerals, and less concerned with the qualitative descriptive terms we give to their habits and macroscopic appearance. However, to a health scientist or a regulatory official, the terms in use are critical to defining what is, and is not, asbestiform.

DOI: 10.1180/EMU-notes.18.3

We must also remember that there is no formal mineralogical definition of 'asbestos' and the word should be avoided in scientific literature, unless referring to one of the six regulated species. The term 'asbestos' derives from the ancient Greek for 'inextinguishable', referring to the fire-resistant properties of asbestos fibres. We can assume they referred to chrysotile, but we do not know all the other species that the ancients used for practical purposes and called 'asbestos'. There was no International Mineralogical Association in the time of ancient civilization! Therefore, in the modern era, it has been necessary to form a consensus definition of what we mean by the colloquial term 'asbestos' and especially by the more mineralogically useful term 'asbestiform'.

We will not discuss the regulatory definitions of asbestos, such as those used by the United States Environmental Protection Agency (EPA) or the Occupational Safety and Health Administration (OSHA). We will observe only those properties of minerals that make them asbestiform, apart from any of the many and sometimes conflicting definitions of 'asbestos' used by regulatory agencies around the world.

In Fig. 1a, you may recognize the fragments as cheese flavoured crackers. Let us imagine they are single crystals of minerals. If you look carefully, you can see the parallel vertical striations. Let us imagine they are crystallographic planes. Now let us apply mechanical stress to our crackers, and imagine we are applying crushing, grinding, or impact forces to some mineral crystals as we do so. In Fig. 1b we see that they have broken into elongate pieces along the striations. However, as we see detailed in Fig. 1c, there is also fracturing that is not along crystallographic planes. This behaviour is called 'parting'. Most minerals are inflexible and tend to be brittle, like these crackers. If these had truly been mineral crystals, the fractured particles would be called 'cleavage fragments'. However, cleavage fragments are not fibres. They did not grow in this habit. They are broken pieces of a crystal that can be blocky, irregular or have a high aspect ratio with parallel sides. Having a high aspect ratio with parallel sides does not make a cleavage fragment a fibre, and certainly does not make it asbestiform. Figure 2 illustrates a cleavage fragment of altered prismatic diopside from Globe, Arizona, USA. Note the elongate appearance and high aspect ratio, as well as the parting on the left end. We would not call this a fibre.

Figure 1. (a) Crackers representing single mineral crystals; (b) breakage representing cleavage fragments; (c) breakage representing parting.

Figure 2. Cleavage fragment of altered prismatic diopside (SEM image).

Now let us consider another food product – this time, we will enjoy some cherry flavoured liquorice candy. Upon careful observation of Fig. 3a, you will see that this is not a single, compact structure as it might appear from a distance. Rather, it is composed of many long, thin strands. These strands are held together loosely and pull apart easily along their length. There is one critical difference in this analogy; these liquorice strands are formed in a helical shape whereas in a mineral the strands or fibrils would be essentially straight and parallel. What is a mineral 'fibril'? A fibril is defined in ISO 22262-2:2014 as a "single fibre of asbestos which cannot be further separated longitudinally into smaller components without losing its fibrous properties or appearances" (see the next paragraph for a systematic description). As we see in Figs 3b and 3c, this bundle can be subdivided into smaller bundles, and ultimately into fibrils. Notice the frayed ends of some the bundles.

Next, let us observe a scanning electron microscope (SEM) photomicrograph of actual chrysotile fibres from the vicinity of Superior, Arizona, USA (Fig. 4). You can see how the fibres occur in bundles that can be subdivided into smaller bundles and then into fibrils. Some of the other characteristics of asbestiform fibres are not only being long and thin with a high aspect ratio, but also flexible, having great tensile strength,

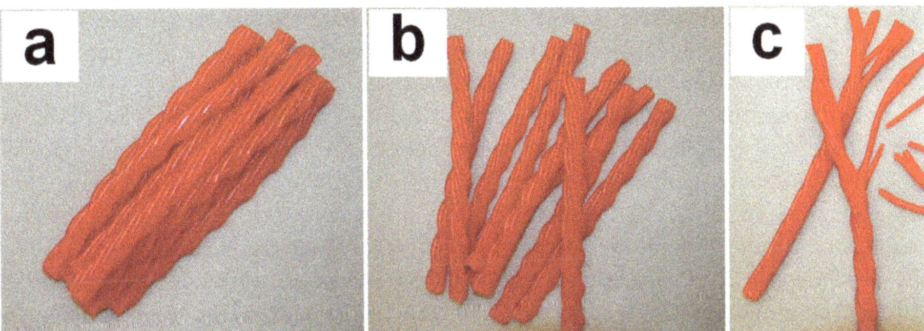

Figure 3. (a) Liquorice candy representing an asbestiform mineral; (b) liquorice candy strands subdivided into smaller bundles along their lengths; (c) the analogy of fibre bundles being subdivided into smaller bundles and then into fibrils.

Figure 4. SEM image of chrysotile fibre bundles and fibrils.

and having bundles with frayed or splayed ends. Asbestiform minerals grow in nature this way. Their asbestiform habit is not an artifact of mechanical action, but rather it is a growth habit. These analogies and illustrations should make clear the differences between a cleavage fragment and an asbestiform habit. Remember, cleavage fragments are portions of minerals that have broken primarily along crystallographic planes, though some of the breakage may be along non-crystallographic planes of weakness called 'parting planes'. They are not flexible, they do not exhibit frayed or splayed ends, and they cannot be easily subdivided into smaller fragments without substantial mechanical stress. Asbestiform minerals, on the other hand, are flexible and often curved, exhibit frayed or splayed ends, and can be easily subdivided until they are isolated as 'fibrils'. Asbestiform minerals under magnification typically do not appear compact, but rather have visible divisions and gaps in their structures.

Does all this mean that asbestiform mineral particles are harmful, and cleavage fragments are not? The answer is negative. Science does not yet have all the answers regarding the potential adverse health effects of the wide range of elongate mineral particles. We do know that the six commercial asbestos minerals – chrysotile, amosite (asbestiform cummingtonite-grunerite), crocidolite (asbestiform riebeckite), asbestiform tremolite, asbestiform actinolite, and asbestiform anthophyllite are defined as Group I carcinogens by the International Agency for Research on Cancer (IARC). However, respirable quartz (known to the health community as 'respirable silica') is also a Group I carcinogen. We also know that any respirable particles that persist in the lungs can have various adverse health effects.

For the reasons given above, the crystal habit of asbestos as well as fibres not classified as asbestos is still today considered one of the major characteristics affecting their toxicity, pathogenicity and carcinogenicity (*e.g.* Committee on Asbestos, 2006; Boulanger *et al.*, 2014). As a matter of fact, it has been demonstrated that different habits of the same mineral species provoke different responses during both *in vitro* (*e.g.* Mossman *et al.*, 2011) and *in vivo* testing (*e.g.* Davis *et al.*, 1991). Subordinate to the fibrous morphology, the length and diameter are considered as major influencing factors for the cells' engulfment mechanism and clearance of the fibres from the respiratory system. The chemical composition of both the bulk and the surface of the fibres can be considered as factors contributing to their toxicity (*e.g.* Aust *et al.*, 2011).

For example, Brown and Gunter (2003) stated that interaction between lung cells and the (100) surface of amphibole asbestos could be more harmful than the cell interaction with the (110) surface of the same mineral.

This chapter describes the habit of asbestos and minerals classified as asbestiform (interpreted as minerals not classified as asbestos, but 'asbestos-like') and therefore it is appropriate to introduce first the different points of view concerning the meaning of several terms used by the scientific community and regulators. In particular, the characteristics of the crystal habits of the six minerals commercially classified asbestos (and regulated) will be presented in addition to several other fibrous minerals not classified as asbestos. Regarding the latter, we will focus on the habit of two groups of mineral fibres: the first regarded as carcinogenic following high-dose exposures (as in the case of asbestiform winchite, richterite, fluoredenite and erionite); the second includes minerals that may have been confused with asbestos or are naturally mixed with chrysotile. These latter may have contaminated product from a mine, the so-called asbestos-containing material. Examples are asbestiform antigorite, polygonal serpentine, sepiolite, carlosturanite and balangeroite. In some cases, these minerals are also classified as toxic on the basis of *in vitro* studies.

2. Terminology issues

The terms 'asbestos' and 'asbestiform minerals', according to regulators all over the world, refer only to those minerals (in most cases silicates) that occur in poly-filamentous bundles and that are composed of very long fibres with a relatively small diameter, most of them showing a high degree of flexibility. These fibre bundles have splaying ends and the fibres are easily separated from one another. The individual fibres, called fibrils, have a tendency to grow along the fibre axis in parallel orientation to the bundle length (*e.g.* Ross *et al.*, 2008).

It is interesting to note that the origin of the word 'fibre' comes from Latin *fibra* perhaps related to Latin *filum* 'a thread, string' or from the root *findere* that is 'to split'. During the 14th century, it was adopted to describe 'a lob of liver' and also 'entrails'.

The meaning 'thread-like structure in animal bodies', applied to various tissues such as muscles, nerves and tendons, is from the 16th century. In the same period, it was used in reference to force or toughness (1630). Later, the word was also adopted to define fine and threadlike organic material such as cotton or jute. 'Fibreglass', a U.S. registered trademark name, dates back to 1937 and the general term 'fibreglass' is from 1941 (Online Etymology Dictionary: http://www.etymonline.com/index.php). Therefore, the application of this word to natural minerals is rather recent.

As far as the definition of asbestos is concerned, the different mineral names are attributed on the basis of structural and chemical characteristics but, except for chrysotile, which crystallizes with curved morphology only, the habit of the other asbestos amphiboles is variable.

2.1. Size matters

Let us start with an example: can a crystal, which is chemically and structurally identified as anthophyllite, being 9 μm long and 3 μm wide with parallel sides, be classified as anthophyllite asbestos? According to the dimensional definition of the World Health Organization (WHO, 1997) and many regulatory agencies, the answer is positive. In fact, for these entities the criteria are the following: length ⩾5 μm, diameter ⩽3 μm, length/diameter (aspect) ratio ⩾3:1. On the other hand, some scientists claim that only fibres with an aspect ratio ⩾20:1 should be classified as asbestos (*e.g.* Chatfield, 2008) wherefore the answer is negative in this case.

It is important to highlight that the WHO (1997) criteria concern health issues and the dimensional standards for asbestos are based on the respirability of the specific particles (fibres). That is, the particle/fibre should not only be inhalable but also be able to penetrate down to the alveolar space.

Nevertheless, fibres shorter than 5 μm, abundantly diffused in air and consequently respirable can represent a health problem as well (Lemen, 2004; Suzuki *et al.*, 2005; Boulanger *et al.*, 2014). These recent considerations thus indicate that the definition of asbestos, as far as the dimensions are concerned, is no longer adequate.

Another problem is that the definition of the term 'fibre' (and therefore the consequent classification as asbestos or non-asbestos) for several regulators is based on the investigation methods used. A specific particle can be considered a fibre when its aspect ratio is >3 according to light microscopy images or >5 when based on images collected with electron microscopy (Gunter *et al.*, 2007).

An additional problem is the fact that parallel aggregates of fibrous minerals (named fibre bundles), fitting the WHO criteria and/or the more restrictive ones (*e.g.* Chatfield, 2008), are actually an intermix of fibres with different sizes (see the National Institute for Occupational Safety and Health – NIOSH, 2011). In this case, particles should be classified as asbestos or non-asbestos from a dimensional point of view by considering the particle-size distribution within the investigated population (Lee and Van Orden, 2016).

In light of all these problems and interpretations, prior to any investigation, one should state what the selected dimensional criteria are, and all together agree upon unequivocal definitions.

2.2. Cleavage fragments

Minerals with crystals that grow in two or three dimensions and that cleave into fragments rather than separate longitudinally into fibrils are classified as minerals with a 'non-asbestiform' habit and are identified as 'cleavage fragments', from the morphological point of view. These minerals may have the same chemical formula of the asbestiform varieties (*e.g.* IARC, 2012). Their distinction from fibres of asbestos is very important in that they appear to be less harmful (*e.g.* Gunter and Brown, 2003; Gamble and Gibbs, 2008) and thus deregulated in 1992 by one out of two principal federal government agencies in the USA that deal with asbestos regulation (OSHA, 1992).

In practice, the cleavage fragments generated from the massive habits of the non-asbestiform analogues of the asbestos minerals are not true mineralogical fibres (*e.g.* NIOSH, 2011). Nevertheless, they can be classified as asbestos if they fit the WHO (1997) criteria.

One consequence of the possible failed recognition is that workers can be erroneously classified as exposed to certain concentrations of asbestos (*i.e.* asbestiform amphiboles) when in fact they are not exposed to asbestos at all but to non-asbestiform amphibole cleavage fragments (Gamble and Gibbs, 2008). Moreover, because in some cases it is not possible to know if crystals are fragments from a larger crystal or pristine individual crystals, it can be difficult to distinguish between prismatic crystals and cleavage fragments (*e.g.* Buck *et al.*, 2013) (Fig. 5).

The identification and classification of amphibole asbestos is still a matter of debate and the application of the most recent criteria may give rise to different classifications. As an example, according to the procedure proposed by Chatfield (2008), more restrictive than the definition by WHO (1997), the amphiboles classified as asbestos are only those thinner than 1.5 μm and with an aspect ratio exceeding 20:1. By applying this rule, it follows that many amphibole fibres previously defined as asbestos are instead classified as 'non-asbestos' cleavage fragments.

As a matter of fact, the distinction between 'true fibre' and cleavage fragment is not so easy, notably when the particles are very thin.

2.3. Elongate particles

Confusion still exists in the use of various terms dealing with 'fibres'. To reach a widespread shared consensus on the definitions among the various parts of the scientific community (mineralogists, chemists, medics, biologists, *etc.*), and to avoid the misclassification of minerals with non-asbestiform habit but with a length >5 μm and an aspect ratio ⩾3:1 as asbestos, NIOSH (2011) indicates the use of the term 'elongate mineral particles' (EMP). This term refers to any mineral particle with a minimum aspect ratio of 3:1, therefore comprising both asbestiform and non-asbestiform habits (see Gualtieri, 2017, this volume). The term EMP thus substitutes for 'mineral fibre' and beyond asbestiform particles. It also includes elongated particles with non-

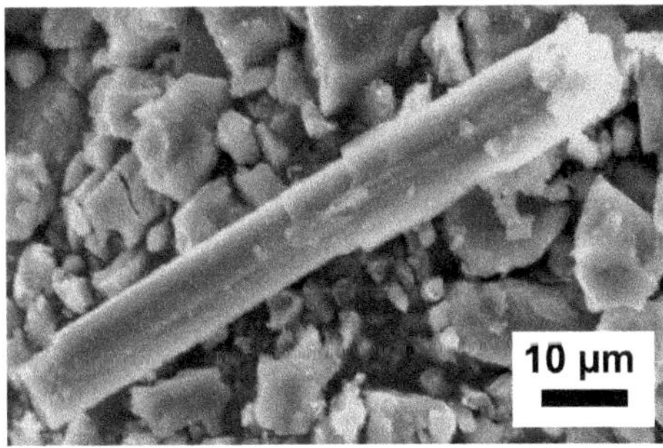

Figure 5. SEM image of a prismatic individual crystal of amphibole with irregular, blunt edges, from a soil. Because it is not known if it has broken away from a larger crystal fibre, its nature is unknown and it could be a cleavage fragment (modified from Buck *et al.*, 2013).

asbestiform habit such as acicular, needle-like and prismatic in addition to cleavage fragments. The term EMP also includes the family of the 'fibres classified as non-asbestos'. At present the nomenclature issue is still open to debate.

2.4. Adopted definitions

A long list of general mineralogical terms and their definitions is given in the NIOSH *Current Intelligence Bulletin* (2011). In this publication, the different definitions referring to amphiboles, given by the U.S. Bureau of Mines (1996) in the *Dictionary of Mining, Mineral, and Related Terms* of the American Geological Institute, in *Glossary of Geology* of the American Geological Institute (Neuendorf *et al.*, 2005), in NIOSH (1990) and by Leake *et al.* (1997), are compared.

The meanings of words that are used in this chapter to describe the characteristics of different fibrous minerals are shown below.

> *Habit* is the shape assumed by a single crystal or an aggregate of crystals. In the case of asbestos and asbestiform minerals, they are fibrils and bundles of fibres respectively.
> *Fibril* is the thinnest component (single crystal) of a fibre bundle. In contrast, a bundle is composed of many fibrils with their elongation axes disposed in parallel, and it can be longitudinally reduced to thinner aggregates or bundles. As we have seen above, a fibril can be defined as a single fibre of asbestos that cannot be further separated longitudinally into smaller components without losing its fibrous properties or appearance.
> *Fibre* is a word indicating a generically elongated particle with length $\geqslant 5$ µm, diameter $\leqslant 3$ µm, length/diameter ratio $\geqslant 3:1$ (WHO, 1997), uniform parallel sides and geometrical faces (*e.g.* rectangular faces elongated along the fibre axis).
> *Fibrous* is a general adjective and a broad term applied to particles that occur as fibres: acicular, columnar, prismatic and asbestiform are specific fibrous types. When crystals show a fine needle-like habit, the adjective used is acicular.
> *Asbestiform* is a word applied to mineral fibres which are not classified as asbestos but have the same dimensional parameters (length $\geqslant 5$ µm, diameter $\leqslant 3$ µm, length/diameter ratio $\geqslant 3:1$: WHO, 1997) and at least one of the asbestos properties (*e.g.* flexibility). Minerals classified as asbestos have an asbestiform habit and therefore, tremolite asbestos is synonymous, for example, with asbestiform tremolite.

The terms describing the particular habit of fibres can be used at different scales, depending on the possibility of observing and distinguishing fibrils and fibre bundles, the latter sometimes being visible with the naked eye whereas the former require an enhanced optical microscope or an electron microscope (SEM or transmission electron microscope, TEM, for micrometric to nanometric dimensions, respectively) to be visualized.

3. Habits of serpentine asbestos minerals

3.1. Chrysotile

The only serpentine mineral classified as asbestos is chrysotile (see Ballirano *et al.*, 2017, this volume) the habit of which is the most investigated among asbestos minerals and has been studied over the past 60 years (*e.g.* Yada, 1971; Veblen and Buseck, 1979).

Chrysotile is the only one of the six minerals classified as asbestos for which the mineralogical name fits with the regulated name, showing asbestiform habit only. Chrysotile fibres consist of aggregates or bundles of long, very thin, flexible fibrils that resemble scrolls or cylinders; the dimensions of each individual chrysotile fibre (fibril) depend on the mode and kinetics of crystal growth (Fig. 6).

Chrysotile forms irregular, curvilinear, curly aggregates of fibril bundles with splayed ends (see some examples in the SEM images of Fig. 7 for chrysotile from Valmalenco, Italy). The diameter of each single fibril is very small (<0.1 μm), and the fibre length ranges between 0.2 and 200 μm or even longer.

The chrysotile habit, as can be seen in TEM images taken along the fibre axis (usually [100], but [010] in the case of parachrysotile), appears to have a concentric or spiral-like core and a concentric or spiral-like wrapping, and a cylindrical, helical or conical overall microstructure (Fig. 8) with [001] and [010] directions radial and tangent respectively (for normal chrysotile). Fibrils with different habits may be present in the same bundle and a single fibril may be the result of the combination of two or more

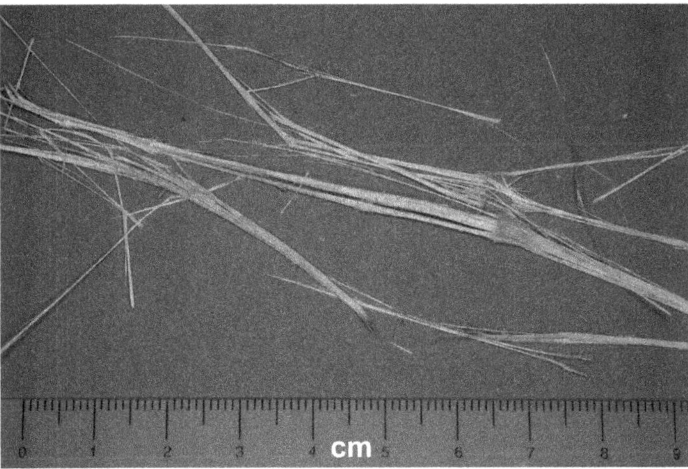

Figure 6. Macrophotograph of very long chrysotile bundles from Emarese mine site (Valle d'Aosta Region, Italy). Their flexibility is medium to high.

Figure. 7. SEM images of chrysotile fibres from Valmalenco (Italy).

different shapes (*e.g.* Amelinckx *et al.*, 1996). Generally, the conical kind does not appear in natural samples. Infrequently, fibrils are made of a single layer.

Natural chrysotile fibres generally have outer and inner diameters of 22–27 nm and 7–10 nm, respectively (*e.g.* Whittaker, 1957; Cressey and Whittaker, 1993; Baronnet and Belluso, 2002; Vigliaturo, 2015). In several cases, the outer diameter can be as small as 5 nm and the inner diameter can be as large as 50 nm (Wagner, 2015). In any

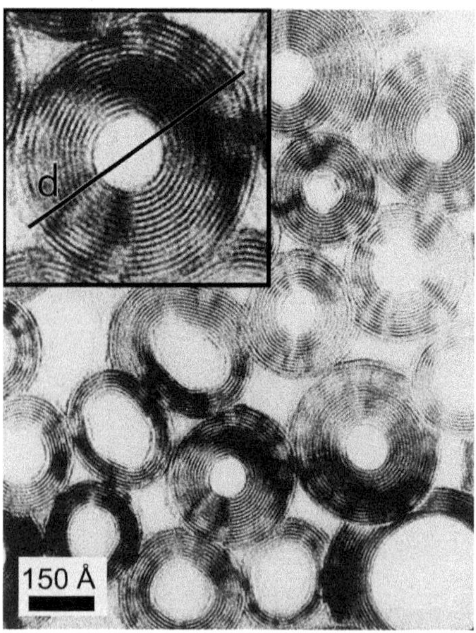

Figure 8. HRTEM image of cylindrical chrysotile fibres from Susa Valley (Italy) as seen along the fibre axes. The outer diameters vary between 200 and 500 Å.

case, the diameter invariably does not exceed 100 nm. The thinnest fibrils consist of a single wall.

The central cores of all kinds of fibres are usually empty along the entire length of the fibres (Fig. 9). The only mineral showing a similar habit (fibrous with a central core) when observed by TEM is halloysite (see for example Shu *et al.*, 2015).

Basically the diameter of chrysotile fibres tends to increase with increased fibre length. Long fibres tend to be curvilinear (*e.g.* Sporn, 2014).

Depending on the formation mechanisms, the length of chrysotile fibres in rocks is extremely variable, from a few micrometers to several millimetres (*e.g.* Bloise *et al.*, 2014). TEM investigations showing a curved character of fibrils and bundles, thus proving their high flexibility, have been reported by Kühn (1941) and later by others (*e.g.* Vigliaturo *et al.*, 2016).

Many synthetic fibres may also show two other kinds of habit: cylinder-in-cylinder and cone-in-cone (Fig. 10), sometimes also spiral-like and helical (*e.g.* Yada and Iishi, 1977; Roveri *et al.*, 2006; Bloise *et al*, 2014).

3.1.1. Chrysotile in air

As shown by TEM investigations, chrysotile may be dispersed in air either as single fibrils or in thin bundles, having mean diameters of 0.07 μm and 0.85 μm, respectively. The mean lengths are ~8.1 μm and 9.5 μm, respectively (Cattaneo *et al.*, 2012).

3.1.2. Chrysotile in the human body and in animals

Chrysotile fibres detected in lung tissues from humans and animals show the same asbestiform habit as that observed in rocks (Fig. 11, modified after Vigliaturo, 2015). Nonetheless, the true nature of chrysotile is identified unequivocally by TEM (both by direct imaging and electron diffraction: *e.g.* Vigliaturo *et al.*, 2016). Even though the fibrils are partially or totally amorphized, they preserve their tubular habit (Vigliaturo, 2015) and are distinguished from the similar (tubular) halloysite by chemical composition.

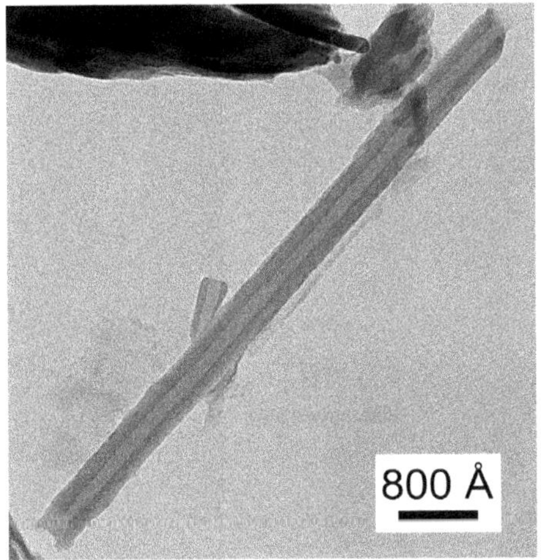

Figure 9. TEM image of a chrysotile fibre as viewed perpendicular to fibre axes with the empty core running along the fibre.

Figure 10. TEM image of cone-in-cone fibrils of synthetic chrysotile as viewed perpendicular to the fibre axes. As for the cylindrical fibre in Fig. 9, the vacant central core runs along the fibre axis. Where it appears partially filled, it is an amorphization effect due to the electron beam.

Because chrysotile fibres can be <5 μm long (*e.g.* IARC, 2011; Fornero *et al.*, 2009; Vigliaturo, 2015), they should not be classified as asbestos according to the dimensional definition by WHO (length ⩾5 μm, diameter ⩽3 μm, aspect ratio ⩾3:1).

Furthermore, a recent TEM study by Gilham *et al.* (2016) conducted on asbestos fibres extracted from the lungs of a large population of people (395 samples) with a

Figure 11. TEM image of a thin bundle chrysotile detected in human lungs (from after Vigliaturo, 2015, reproduced with the permission of the author).

history of occupational exposure to asbestos and experience of malignant mesothelioma or lung cancer, showed that most of the chrysotile fibres were 5–10 μm long with a small proportion >20 μm in length. The average width was 0.09 μm.

3.1.3. Chrysotile after in vitro and in vivo treatments

Experiments involving instillation of chrysotile in rats showed no alteration of the fibril habit, whereas changes in length and width appeared after a few days, as observed by TEM (*e.g.* Chung and Han, 2015).

A TEM study of chrysotile fibres, following interactions with bronchoalveolar and mesothelial cell cultures showed that the fibre habit suffered several changes which appeared after only 48 h of *in vitro* treatment and were subsequently enhanced after 96 h. In particular, the side and apex of the fibrils showed incipient dissolution whereas the inner channel was modified and became undistinguishable, possibly in correlation with the amorphization of the structure (Vigliaturo, 2015).

4. Habits of amphibole asbestos minerals

Amphibole asbestos occurs commonly together with a non-asbestiform mineral counterpart within the same area and the same deposits. Hence, the identification and quantification of asbestos is difficult.

The matter is even more problematic when the exposure suffered by people needs to be evaluated, as in the cases of the pathology diagnosis or trials. The same problem occurs when the risk of asbestos exposures during processing of rocks containing amphiboles has to be predicted so that necessary safety precautions can be taken. The NIOSH (2011) highlights the difficulty in ascertaining the source of exposure in the case of mixed exposures for some mining operations and downstream users of their mined commodities. As a matter of fact, identifying tremolite asbestos, actinolite asbestos and anthophyllite asbestos and determining their location in deposits of their non-asbestiform mineral counterparts can be very difficult.

Besides, the issue becomes even more complicated by the presence of other fibrous minerals frequently intergrown with asbestos, as illustrated below.

4.1. Tremolite and actinolite asbestos

Natural crystals of tremolite and actinolite exhibit a prismatic habit from stocky to acicular or asbestiform, the two last being elongated crystals. When the dimensions fit the WHO criteria (1997), these amphiboles are classified as asbestos.

Besides the issue regarding discrimination between crystals fragments and fibres for these minerals and amphiboles in general, it is also very hard to determine unequivocally their chemical composition and structural characteristics. Detailed studies have presented the criteria for distinguishing the different amphibole species. In general, the classification is based on crystal-chemical data (Hawthorne and Oberti, 2007; Hawthorne *et al.*, 2012). According to accurate chemical characterization, amphiboles previously classified as tremolite were reclassified as winchite after the IMA nomenclature definition (Leake *et al.*, 1997). In addition, as a consequence of boundaries now existing in continuous compositional fields, amphiboles with different names have chemical compositions that are more similar than minerals with the same name (*e.g.* Gunter *et al.*, 2007).

In terms of the evaluation of the fibre dimensions, note an important point regarding the characterization techniques themselves. Because of the poor flexibility of tremolite fibres, the preparation procedure for TEM specimens includes gentle grinding which may reduce their length. Hence, the actual measurement of the fibre size should be evaluated by SEM. To determine the width by TEM, it is possible to use an alternative sample-preparation procedure such as ion-thinning preparation of fibres sectioned perpendicular to the fibre axis.

Tremolite and actinolite asbestos typically show a long, thin and quite rigid columnar habit (Fig. 12), elongated along the [001] fibre axis and generally flattened along the (100) plane (*e.g.* Brown and Gunter, 2003). A qualitative tendency to exhibit a strong preferential orientation of the (100) planes has been observed for asbestiform minerals (excluding those with curved habit), whereas their massive counterparts show a less pronounced orientation (Bandli and Gunter, 2014).

Tremolite and actinolite, as observed by SEM, form long, straight, parallel sided and lath-shaped aggregates, fairly flexible to brittle, with single fibrils having a diameter in the range 0.5–0.15 μm (Fig. 13), as well as bundles of fibres with distinct longitudinal aggregation.

TEM images show quite straight and rigid fibres sometimes with partial defibrillation. Fibrils splitting from fibres show a certain degree of flexibility (*e.g.* Langer *et al.*, 1974).

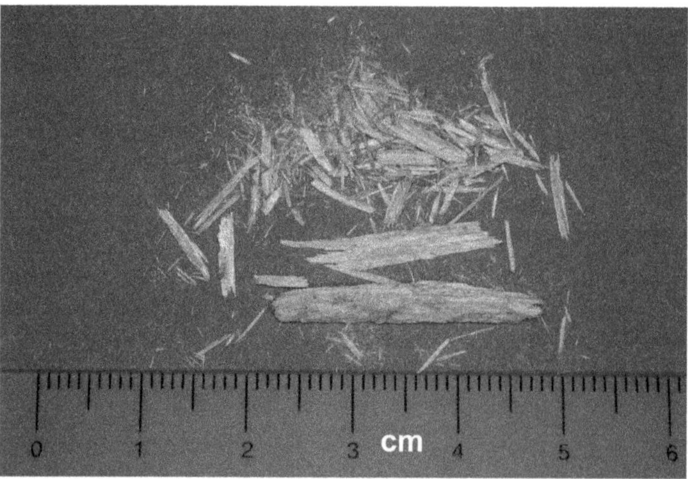

Figure 12. Macrophotograph of tremolite asbestos from Val d'Ala (Piemonte Region, Italy). Bundles of fibres show a long and quite rigid habit.

Figure 13. SEM images of tremolite fibres showing different widths and stiffness.

The fibre length is normally >8 μm and generally falls in the range of 60 to 100 μm although it can reach some tens of millimetres (*e.g.* Nolan *et al.*, 1991; Bloise *et al.*, 2014). The width is normally <0.5 μm, but it varies from <0.1 μm to 0.3 μm (*e.g.* Langer *et al.*, 1974; Davis *et al.*, 1991; Vignaroli *et al.*, 2014; Gaffney *et al.*, 2016).

According to TEM analyses, the fibres show parallel sides, regular terminations and straight shapes (*e.g.* Fig. 14).

TEM investigations along the [001] fibre axes show fibrils of different widths, with typical 120° angles between (110) faces when the defibrillation comes from primary crystals. The partial transformation of fibres to talc or sheet silicates in general, originates as layers from the fibre boundaries at the (110) cleavage planes, thus

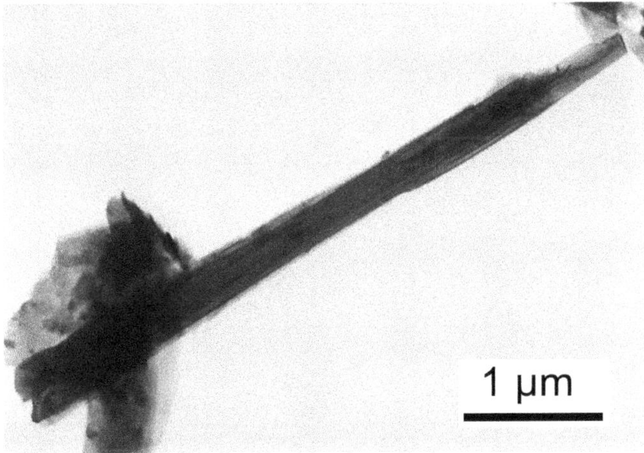

Figure 14. TEM image of a single fibre of tremolite asbestos as viewed perpendicular to the fibre axis. The lower end displays an initial split into two fibrils.

determining the formation of steps and indentation (*e.g.* Fig. 15). Consequently, the original parallelogram habit loses its regularity in the cross sections.

4.1.2. Tremolite and actinolite asbestos in soils

Investigations of tremolite and actinolite asbestos present in soil show that their habits are not substantially different from their counterparts in pristine rocks. In fact, the fibres eroded from the rocks, transported and finally deposited in the soil, display little or no change in shape. A change in size is observed as the various processes induce granulometric separation (Buck *et al.*, 2013).

4.1.3. Tremolite and actinolite asbestos in the human body

Germine and Puffer (2015) performed a TEM investigation of tremolite and actinolite asbestos fibres recovered from lungs of miners and millers working in Canadian asbestos mines. They reported regular fibres, fibres split at the end, fibres with jagged sides and cleavage fragments. All fibres were monocrystalline and showed a silica-rich amorphous surface coating due to partial dissolution of fibres in contact with the biological environment. These authors also found one fibre consisting, from the

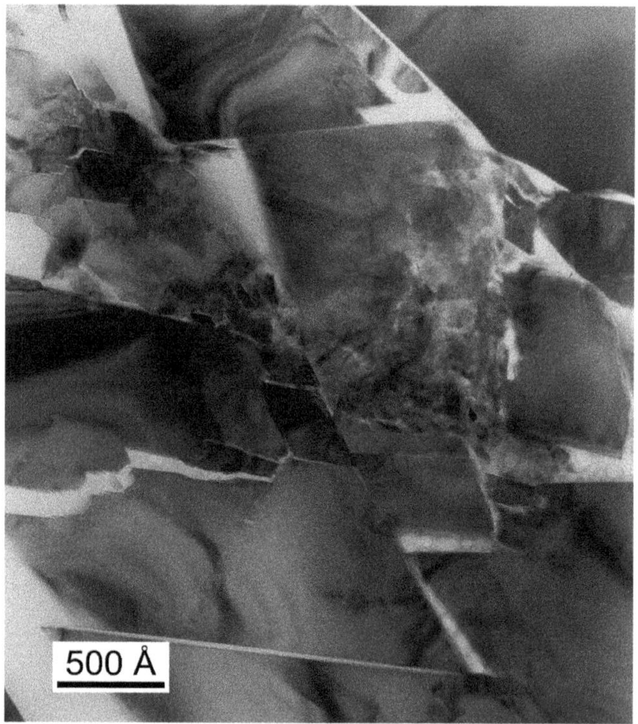

Figure 15. TEM image of tremolite asbestos fibres as viewed along the [001] fibre axes. The long, thin fibres are aggregated longitudinally in columnar prisms.

structural and compositional point of view, of tremolite-actinolite and anthophyllite intergrown lamellae. This situation confirms the complexity of fibre identification and the need for an accurate TEM investigation. Moreover, only 21% of the fibres were >5 μm long. Hence, according to the WHO criteria, the majority of the fibres should not be defined as asbestos. However, their length may not correspond to that before inhalation because the fibres were probably shortened due to transformation in the lungs (*i.e.* partial dissolution) or by comminution (*e.g.* milling) during mineral processing. Germaine and Puffer (2015) concluded that most of the fibres detected in the lungs are the product of cleavage or parting fragments. Nevertheless, given that the samples are from workers deceased because of mesothelioma, this observation demonstrates that cleavage fragments are carcinogenic, contrary to what has been assumed recently.

For fibres ⩾5 μm long, the study by Gilham *et al.* (2016) described above for chrysotile, showed that most tremolite and actinolite asbestos fibres are 5–10 μm long with a small proportion >20 μm in length; with an average width of 0.49 and 0.61 μm, respectively.

4.2. Anthophyllite asbestos

Unlike other amphibole asbestos, anthophyllite asbestos has not been investigated extensively. This may be due to its limited use. Anthophyllite asbestos fibres may occur together with their non-asbestos counterpart in the same rock sample. Generally, the fibres are not longer than several millimeters (Fig. 16). The morphology of asbestiform anthophyllite under SEM is similar to that of other amphibole asbestos varieties (needle-like, fibrous, acicular or columnar aggregates) (Fig. 17).

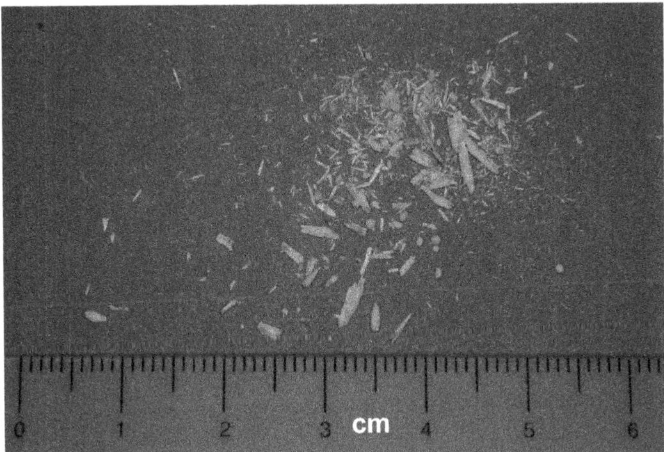

Figure 16. Macrophotograph of anthophyllite asbestos from Bresimo (Trentino Alto Adige Region, Italy).

Figure 17. SEM of anthophyllite asbestos.

The habit of anthophyllite asbestos has been described thoroughly using TEM analyses (*e.g.* Veblen, 1980; Cressey *et al.*, 1982). The bundles consist of fibres aligned along the [001] fibre axis and more or less rotated with respect to each other around the bundle axis. Fibres are quite rigid and straight, but fibrils splitting from fibres show a certain degree of flexibility (*e.g.* Langer *et al.*, 1974).

The main production of the fibres occurs through the separation along grain boundaries from a primary crystal. The pervasive transformation in sheet silicates (slabs of antigorite, talc and chlorite, partial to nearly complete chrysotile tubes) along the grain boundaries enhances the separation process.

When alteration is at an advanced state, very thin fibrils, <0.1 µm wide, are preserved as islands with an irregular contour, slightly flattened on sheet silicate (100) planes, as for anthophyllite asbestos in talc, where the intergrowth is well oriented along the [001] and [010] directions of talc and [100] and [010] directions of anthophyllite, respectively (Cressey *et al.*, 1982).

The mechanism of fibre formation is influenced by (100) stacking faults and (010) chain-width errors, which are widespread in amphiboles. Regarding these structural errors, various reports (*e.g.* Chisholm, 1973; Alario Franco *et al.*, 1977; Veblen *et al.*, 1977; Veblen and Buseck, 1980) suggested that amphibole asbestos specimens contain a greater concentration of multiple-chain lamellae than their massive counterparts.

The most diffused chain-width errors consist of regularly alternating double and triple chains, named chesterite (Veblen and Burnham, 1978), and therefore the fibre bundles may be considered as an aggregate of anthophyllite asbestos and asbestiform chesterite, the latter being very narrow (up to 0.05 µm, but mostly <0.02 µm).

Thinner fibres are generated by splitting along parallel planes. Their thickness along [100] is <2 μm, frequently ~0.4 μm (*e.g.* Gaffney *et al.*, 2016). The size in the (001) plane varies between <0.1 μm and 0.8 μm with most falling between <0.1 and 0.5 μm; occasionally the width exceeds 1 μm (*e.g.* Langer *et al.*, 1974).

The length of fibrils can exceed 50 μm, with an average of ~9 μm (Vigliaturo, 2015).

As for the other fibrous amphiboles, fibres outside the WHO classification accompany 'WHO' fibres in the same sample.

4.2.1. Anthophyllite asbestos in the human body

Regarding fibres >5 μm long, the study of Gilham *et al.* (2016) (see above), showed that the highest and the lowest percentage of anthophyllite fibres had lengths in the range 5 to 10 μm and >20 μm, respectively, with an average width of 0.58 μm.

4.3. Grunerite asbestos (amosite)

Amosite fibres can be 30 μm long (Fig. 18). Under the SEM, amosite appears as fibrous, columnar or acicular aggregates, and can be distinguished easily by energy dispersive spectrometry (EDS) from other amphibole asbestos varieties due to its large iron content. The fibre diameter is generally in the range 0.15 to 15 μm, whereas the fibre length is in the 0.4–40 μm range (Fig. 19). TEM observations perpendicular to the fibre axes, reveal quite straight and rigid fibres with initial to advanced defibrillation at the ends. Fibrils splitting at the fibre ends show a certain degree of flexibility (*e.g.* Langer *et al.*, 1974). As seen by TEM investigations in sections cut perpendicular to elongation fibres, grunerite asbestos fibres are [001] elongated. They constitute bundles where

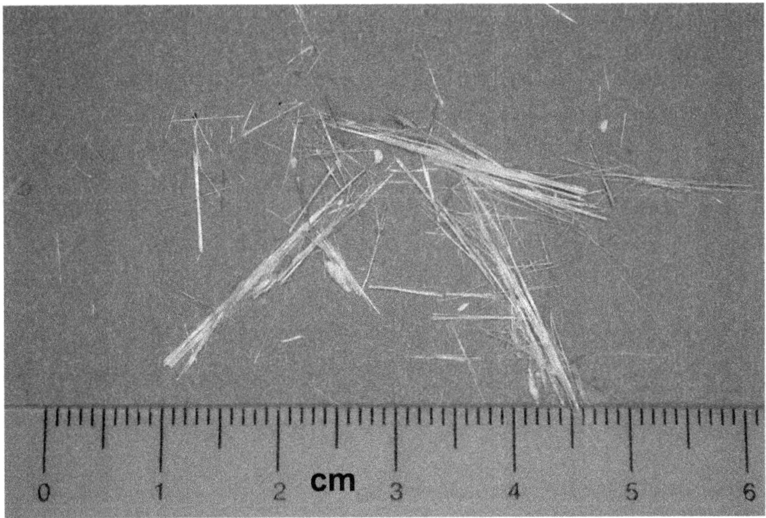

Figure 18. Macrophotograph of amosite from South Africa.

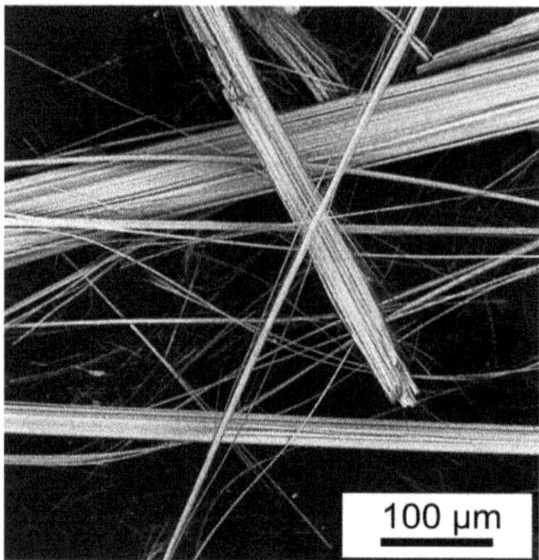

Figure 19. SEM image of very long, thin fibres of amosite. Several of the thinnest fibres show some flexibility.

fibre axes are rotated with respect to each other in certain areas of the specimen, whereas in other areas they show a preferred (100) orientation. In this last case, the sections of the fibres are more developed along the [010] direction and the fibres appear slightly flattened along the (100) plane (*e.g.* Cressey *et al.*, 1982).

The fibril boundaries are generally irregular in shape with faces parallel to the fibre axis; only occasionally do they have crystallographic indices coincident with the low-index faces (010) and (110) (Cressey *et al.*, 1980). The shape of their section is irregular and the dimensions are small, usually varying from <0.1 µm to 0.5 µm (*e.g.* Vigliaturo, 2015; Gaffney *et al.*, 2016). Less frequently, they are >1 µm (*e.g.* Langer *et al.*, 1974).

Amosite fibres, as anthophyllite asbestos, contain several multiple-chain lamellae (*e.g.* Whittaker *et al.*, 1981).

Similarly to anthophyllite asbestos but to a lesser extent, amosite fibres also show alteration to talc and a curved serpentine-like mineral, the first being well oriented, with the grunerite asbestos structure. The intergrowth between talc and amosite is well oriented with the [001] direction of talc parallel to the [100] direction of amosite (*e.g.* Cressey *et al.*, 1982).

Grunerite as well as other amphibole asbestos, exhibits heterogeneous fibre size-distribution.

4.3.1. Grunerite asbestos (amosite) in the human body

A recent TEM investigation of asbestos fibres (see the study of Gilham *et al.*, 2016, described above) showed that most amosite fibres were between 5 and 10 µm with a much smaller proportion >20 µm long; they have a mean width of 0.3 µm.

4.4. Crocidolite (asbestiform riebeckite)

Crocidolite fibres are elongated along [001] and form bundles. They may reach 15 cm in length (Fig. 20).

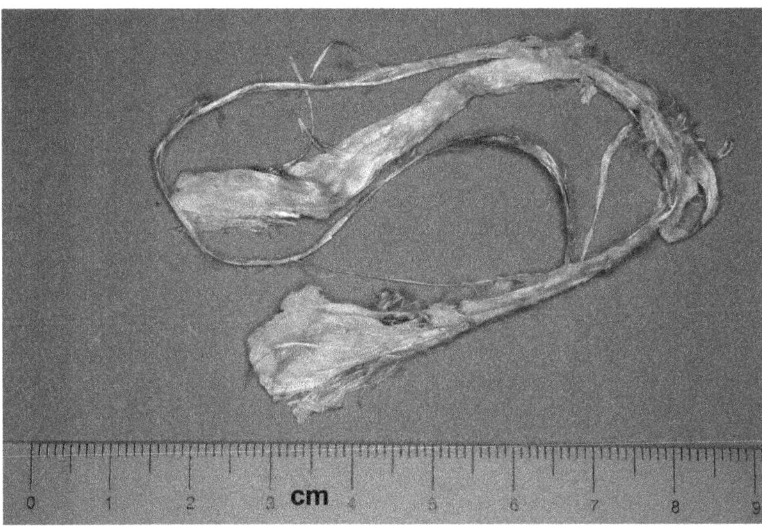

Figure 20. Macrophotograph of very long crocidolite fibres from South Africa.

Crocidolite, as seen by SEM, appears in fibrous, columnar, felt-like aggregates, with fibre diameters between 0.06 and 1.2 μm and between 0.2 and 17 μm long (Fig. 21).

The axes of the fibres can either be rotated by several degrees with respect to each other (Alario Franco *et al.*, 1977) or have similar orientations (*e.g.* Ahn and Buseck, 1991) as shown in TEM images obtained from a section cut perpendicular to the fibre axes. In the areas where fibres have similar orientations, thin fibrils are aggregated into larger clusters. In this case, the formation of fibrils from large crocidolite fibres is evident (Fig. 22, modified from Ahn and Buseck, 1991).

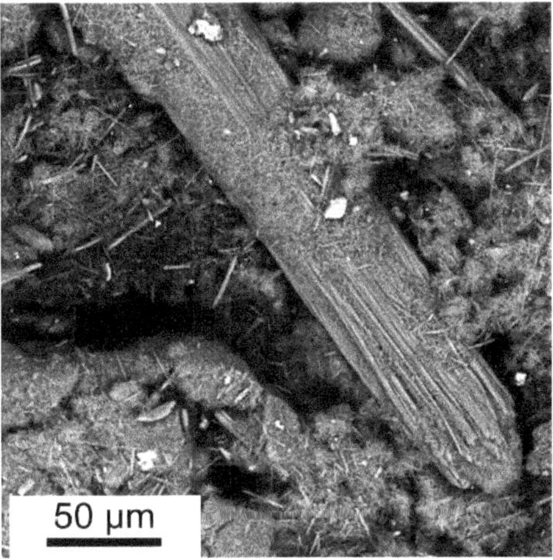

Figure 21. SEM image of crocidolite fibres. The shortest fibres make up wool-like aggregates.

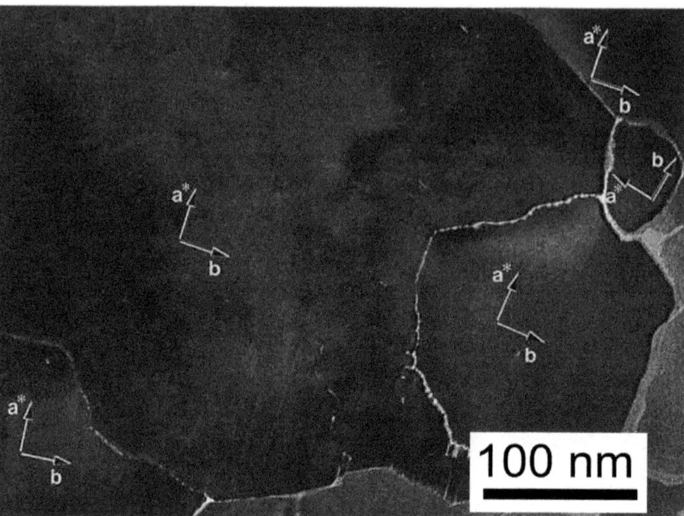

Figure 22. TEM image, as seen along the bundle axis, of a thin fibril of crocidolite surrounding a larger fibre, all having similar orientation. The white contrasted strips inside the grains are wide chain silicates (from Ahn and Buseck, 1991, reproduced with the permission of the Mineralogical Society of America).

Many fibre sections display an irregular shape. They exhibit a range of widths: from a minimum of 0.04 μm up to >1 μm in rare cases (*e.g.* (Vigliaturo, 2015). The most frequently occurring width is between 0.1 μm and 0.3 μm (*e.g.* Langer *et al.*, 1974; Ahn and Buseck, 1991; Gaffney *et al.*, 2016).

The (110) grain boundaries generally coincide with the cleavage planes of amphibole.

Crocidolite contains intergrown wide-chain silicates, most of which are three, four or five chains wide (referred to as pyriboles after Veblen and Buseck, 1980); they appear with a white contrast in the cross section TEM images (Fig. 22).

The presence of various planar defects and wide-chain pyriboles leads to the polygonization of the fibres. In this case, the formation of fibrils is driven within or at the edges of larger crystals through the disaggregation of polygonized fibres, a process enhanced by geological stresses (*e.g.* Ahn and Buseck, 1991).

Other investigations suggest that each fibre nucleates separately and grows laterally until it comes in contact with neighbouring fibres (*e.g.* Whittaker, 1979).

Crocidolite fibres alter to sheet silicate (probably mica), developing lamellae usually 100 Å thick. This sort of fibrous slab develops coherently with crocidolite fibres, and as a consequence, their [100] elongation axis is parallel to the [001] crocidolite fibre axis.

As seen from TEM images, fibres are quite rigid and straight perpendicular to fibre axes, (Fig. 23), with initial-to-advanced defibrillation at the ends; fibrils splitting at the end of fibres show a certain degree of flexibility (*e.g.* Langer *et al.*, 1974). Their length can exceed 50 μm (Vigliaturo, 2015).

As for the other fibrous amphiboles, within the same specimen crocidolite fibres can coexist with others that are not classified as WHO fibres.

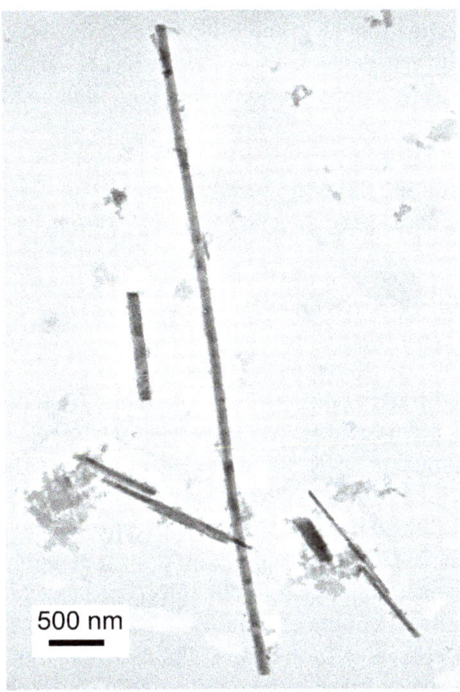

Figure 23. TEM image of thin and quite rigid fibrils of crocidolite as viewed perpendicular to fibre axes.

4.4.1. Crocidolite after in vitro treatments

TEM investigations of crocidolite following interactions with bronchoalveolar and mesothelial cell cultures, showed that the asbestiform habit was kept after 96 h although an amorphous layer was detected around some fibres (Vigliaturo, 2015; Fig. 24).

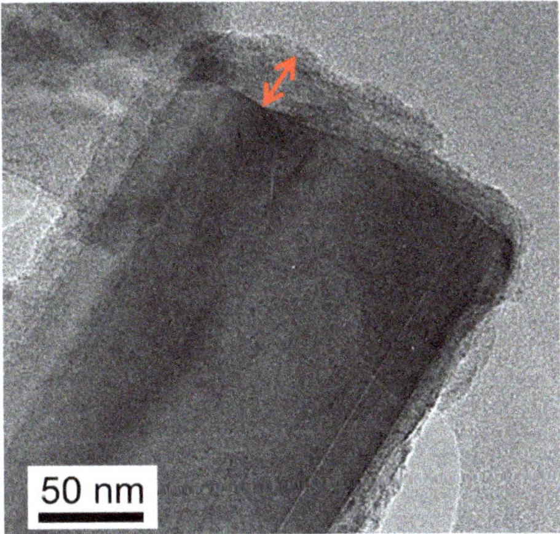

Figure 24. TEM image, as viewed perpendicular to the fibre axis, of a crocidolite fibre after 96 h of *in vivo* treatment. The red arrows indicate the thickness of the amorphous layer (modified from Vigliaturo, 2015, and reproduced with the permission of the author).

Crawford (1980) reported a TEM investigation of crocidolite fibres exposed to biological material (human blood serum in that case), showing evidence of surface dissolution and attack at (100) planar defects (previously observed in large numbers).

4.4.2. Crocidolite in the human body

Gilham *et al.* (2016) showed that most crocidolite fibres are between 5 and 10 μm long and the smallest group is >20 μm long; the fibres have a mean width of 0.17 μm.

5. Habits of asbestiform minerals

5.1. Asbestiform polygonal serpentine

The polygonal serpentine (PS) (see also Ballirano *et al.*, 2017, this volume), initially named 'Povlen' serpentine (Kristanovic and Pavlovic, 1964), is a type of serpentine consisting of flat sheets (like lizardite) arranged in 15 or 30 ordered sectors (Fig. 25), surrounding an empty hollow or a thin chrysotile core (*e.g.* Cressey and Zussman, 1976; Wagner, 2015). The PS crystals appear as fibres elongated along [100], stockier than chrysotile, forming bundles parallel to their [100] fibre axis. TEM investigations showed that PS could be the results of polygonization of chrysotile rolls (Baronnet and Belluso, 2002), as evidenced also by synthetic samples (Grauby *et al.*, 1998). The synthetic PS showed that the mineral counterpart may form from cylindrical chrysotile when a critical diameter of ~50–100 nm is exceeded (Grauby *et al.*, 1998).

The length of the PS fibre is variable: it can be longer than several micrometres but is generally shorter than that of chrysotile (Baronnet and Belluso, 2002); the diameter is usually >100 nm (Papp, 1990; Fig. 26) but may be 50–400 nm wide, with irregular, thin canals and often irregular fibre terminations (*e.g.* Andreani *et al.*, 2004; Baronnet and Belluso, 2002).

As seen in TEM images along the [100] fibre axis, the polygonal serpentine fibres may show elliptical sections (Fig. 27). Crushing of these fibres probably represents the response to deformation strain suffered by the rock. In fact, chrysotile fibres are thinner

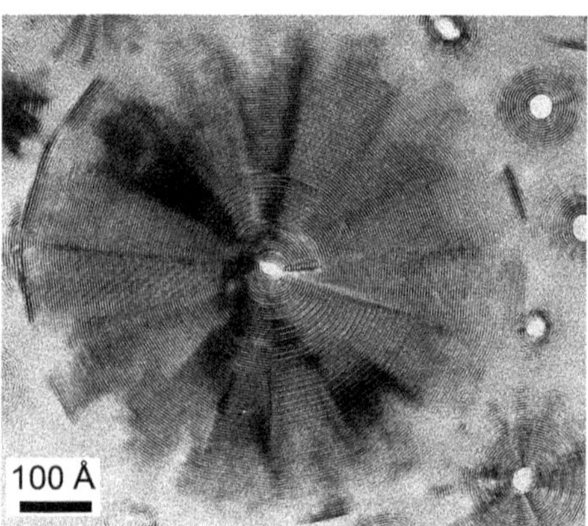

Figure 25. TEM image of 15-sector asbestiform polygonal serpentine as viewed along the fibre axis.

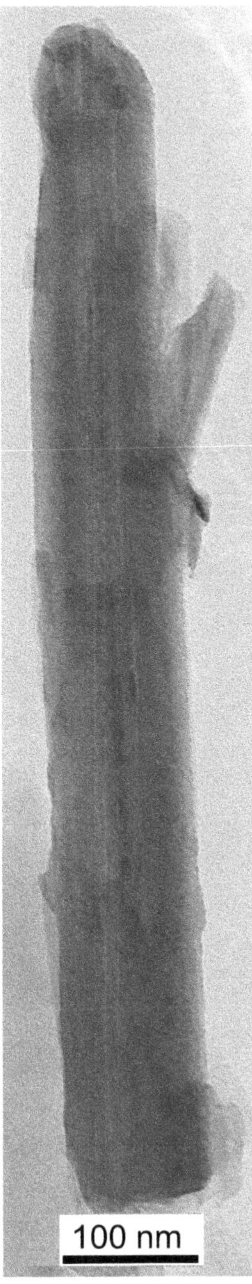

Figure 26. TEM image of a polygonal serpentine fibre (PS) observed perpendicular to the [100] zone axis. The inner tube is scarcely visible due to its very small dimension.

than PS and therefore able to arrange themselves inside the bundle, at least under weak strain. On the contrary, the larger PS fibres cannot absorb the strain in the same way and, because of the angular arrangement of their flat layers (and the empty central space), they can bend perpendicular to the fibre axis.

Chrysotile and asbestiform PS are usually mixed in the same bundles (Fig. 28). Surprisingly, PS is widespread in nature. In some rock types, especially serpentinites, it may constitute a remarkable percentage. For example, as natural occurrence of asbestos (NOA) it is abundant in serpentinite rocks of the Italian Western Alps (Belluso *et al.*, 1995) and southwestern Italy (Bloise *et al.*, 2014), and along the coast in northern California (USA) where it is associated with various minerals belonging to the serpentine mineral group (Wagner, 2015).

To date, no data on the detection of PS in human or animal tissue or in other biological samples have been reported in the literature. This may be due to the very limited number of TEM investigations performed on inorganic fibres in such materials.

5.2. Asbestiform antigorite

The structural characteristics of antigorite, that is an alternating-wave arrangement of octahedral and tetrahedral sheets (see Ballirano *et al.*, 2017, this volume), give rise to different habits, from flattened prismatic to asbestiform crystals (*e.g.* Keeling *et al.*, 2008). When the dimensions of the elongated laths fit the dimensional criteria applied to asbestos (WHO, 1997), fibres appear flexible, and separate longitudinally into thinner fibres. The antigorite fibres may be

Figure 27. HRTEM image of asbestiform carlosturanite (a) and polygonal serpentine with elliptic section (b), viewed along the fibre axis.

classified as 'asbestiform'. Despite papers dealing with this kind of antigorite, many authors still ignore its nature (*e.g.* Glenn *et al.*, 2008). In the past, the word 'picrolite' was used for antigorite generically defined as fibrous (Chapman and Zussman, 1959).

Figure 28. TEM image of chrysotile and asbestiform polygonal serpentine intergrowth, as viewed along the fibre axes. Almost all fibril sections are cylindrical and the outer diameters vary between 200 and 400 Å. The largest fibrils correspond to polygonal serpentine. The two sections that are not circular correspond to two asbestiform carlosturanite fibrils.

The asbestiform crystals constitute bundles partially or totally filling veins and cracks in serpentinite rocks (*e.g.* Belluso *et al.*, 1995; Viti and Mellini, 1996) and in calc-silicate marbles more or less altered to serpentine and talc (Fitz Gerald *et al.*, 2010).

Using only macroscopic (Fig. 29) and optical microscope observations, asbestiform antigorite may be confused with chrysotile, especially when the fibre bundles are thin. In these cases, only TEM-EDS analyses can resolve the issue (*e.g.* Belluso *et al.*, 1995; Fitz Gerald *et al.*, 2010). Owing to this problem, it is known that these fibres have been used commercially as asbestos (see for example Keeling *et al.*, 2006).

Misidentification is even more likely because of the intergrowth of asbestiform antigorite with various asbestiform and asbestos phases, as in the case of chrysotile, and asbestiform carlosturanite bundles in serpentine of the Italian Western Alps (Belluso *et al.*, 1995; Baronnet and Belluso, 2002) or chrysotile, tremolite asbestos and subordinate asbestiform antigorite aggregates in metamorphosed ultramafic rocks of the Ankara and Eskişehir areas in northwestern Turkey (Dogan and Emri, 2000).

The fibres, lath-shaped and elongated along the [010] direction, vary in length from several micrometres to some hundreds of millimetres, with smallest diameters of 0.08 μm, and usually <1 μm, a large aspect ratio (exceeding even 100:1), with length and thickness being the most and the least developed dimensions, respectively (*e.g.* Cardile *et al.*, 2007; Keeling *et al.*, 2008; Bloise *et al.*, 2014). Their flexibility varies from stiff to moderate, but it is very high in fibres having a diameter <1 μm (Keeling *et al.*, 2008). The fibres, aggregated in bundles, have their fibre axes parallel to each other, and are more or less randomly rotated about [010]. However, owing to their habit, when the (001) faces are developed extensively along the [100] direction (the fibrils are much wider than they are thick, with the short direction near to [001]), they tend to be oriented with parallel (001) faces (Fig. 30).

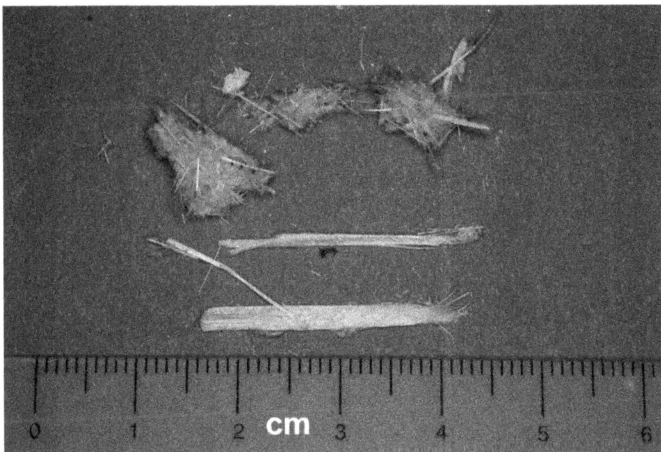

Figure 29. Macrophotograph of long bundles of asbestiform antigorite from Australia. The right end of the lowest bundle shows an opening longitudinal splitting in flexible fibres. The macroscopic characters are very similar to those of chrysotile.

Figure 30. TEM image of asbestiform antigorite as seen along the [010] fibre axis. Fibres are flattened along the (001) plane.

TEM imaging performed along the fibre axis invariably reveals the lath morphology and permits asbestiform antigorite to be distinguished from chrysotile tubes and from asbestiform polygonal serpentine prisms. The absence of a core in TEM images perpendicular to the fibre axis represents another tool to distinguish antigorite fibres from chrysotile, but not from polygonal serpentine because the latter generally has a small core (sometimes filled) scarcely visible when fibres are observed perpendicular to their elongation axis (Fig. 26).

5.2.1. Asbestiform antigorite from air

Asbestiform antigorite is dispersed in air in single fibrils. TEM investigations showed that its elongated prismatic habit has a mean diameter of 1.42 μm and length of 8.2 μm (Cattaneo *et al.*, 2012).

The discrimination between chrysotile and asbestiform antigorite with the aid of SEM-EDS investigations may be done on the basis of morphometric criteria. An investigation by Cattaneo *et al.* (2012) showed that the diameters of antigorite and chrysotile single fibres occur in two distinct dimensional regions. Thus, based on a combined SEM-EDS and

TEM-SAED observations on airborne samples, a diameter cut-off of 0.25 µm was proposed as a threshold marker for the differentiation between chrysotile and antigorite fibres.

5.3. Asbestiform erionite

Asbestiform erionite fibres occur mainly as two varieties: thin and flexible or quite rigid and straight. Both form bundles where the fibres are grouped together with their fibre axes parallel or sub-parallel to each other.

The first type of fibre shows medium to high flexibility and looks like wool (*e.g.* Van Gosen *et al.*, 2013). The length may exceed 200 µm; the minimum width of each fibril is ~0.01–0.015 µm (*e.g.* Mattioli *et al.*, 2016). Bundles, with a width of 10–20 µm, consist of parallel or sub-parallel fibrils (*e.g.* Cametti *et al.*, 2013).

The fibres of the second type are thicker than 2 µm and are up to ~30 µm long (Vigliaturo, 2015) as for the sample from Oregon (Fig. 31). They are generally grouped together in irregular to parallel bundles with a variable diameter and are up to 100 µm long, with a great tendency to separate into thin fibrils with widths in the range 0.03 to 0.5 µm (*e.g.* Van Gosen *et al.*, 2013; Vigliaturo, 2015; Mattioli *et al.*, 2016).

TEM observations perpendicular to fibres that are elongated parallel to [001] show isolated linear defects developed across the fibres (along the width) and may be interpreted as offretite slabs (Vigliaturo, 2015).

As for the amphibole asbestos, erionite samples display heterogeneous size distribution of the fibres.

In SEM images, erionite may occur in straight acicular particles, elongated hexagonal prisms, or in thin and fibril aggregates (Fig. 31).

5.3.1. Asbestiform erionite after in vitro treatments

TEM investigation of fibres treated *in vitro* show some changes. After 48 h of treatment, different degrees of morphological and dimensional changes appear with dissolution

Figure 31. SEM image of asbestiform erionite from Oregon (USA). Fibres thicker than 2 µm are grouped as squat aggregates.

along the boundaries and/or at the ends, with rounded edges. Observations after 96 h of treatment revealed that the dissolution process for some fibres appeared more advanced (Fig. 32, modified after Vigliaturo, 2015).

5.4. Asbestiform offretite

Asbestiform offretite fibres form quite rigid bundles, generally <30 μm long and <1 μm wide. Only some of the thinnest fibrils show flexibility, but in most cases they are rigid (Mattioli *et al.*, 2016).

5.5. Asbestiform winchite and richterite

In general, the asbestiform amphiboles richterite and winchite are intermixed and only careful chemical investigation allows us to distinguish them.

The habits of the two amphiboles are similar and appear in a range of morphologies from more or less prismatic crystals to very long, thin and flexible fibrils. The prismatic fibres may show splayed ends and often cleavage steps across the entire fibre. The other kinds of fibres are aggregated in bundles, with their elongation axes parallel or sub-parallel (Meeker *et al.*, 2003).

In all cases, they are elongated on [001], not less than 5 μm long and 0.1–1 μm wide.

5.5.1. Asbestiform winchite and richterite from air

TEM images of asbestiform winchite and richterite detected in air samples collected around the closed vermiculite mine of Libby, Montana, USA, show near perfect euhedral amphibole fibrils (displaying growth of {110} faces), being up to 10 μm long and <0.5 μm wide (Fig. 33, modified after Gunter *et al.*, 2008).

Harris *et al.* (2007) investigated the morphology of >3700 fibrous amphiboles collected from outdoor air in residential zones and former vermiculite screening plant locations in the Libby area. Regarding the amphibole richterite, TEM-EDS and FEG-SEM analyses showed four different habits:

- fibrous or asbestiform habit, thin particles (⩽0.5 μm wide) with parallel sides and smooth surfaces, no visible crystal faces (*i.e.* an

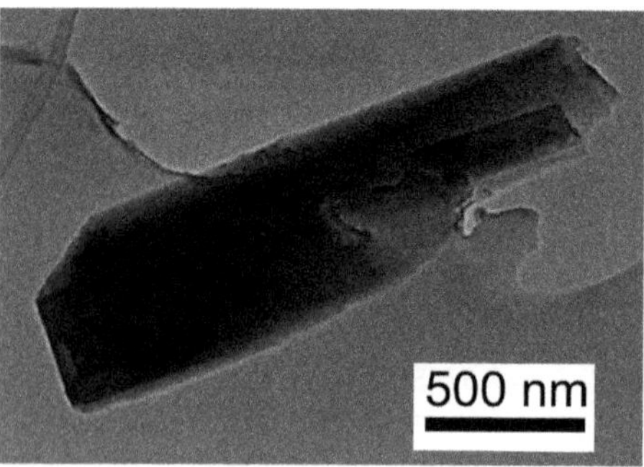

Figure 32. TEM image of fibrous erionite after 96 h of cell treatment, as viewed perpendicular to the fibre axis (modified from Vigliaturo, 2015 and reproduced with the permission of the author).

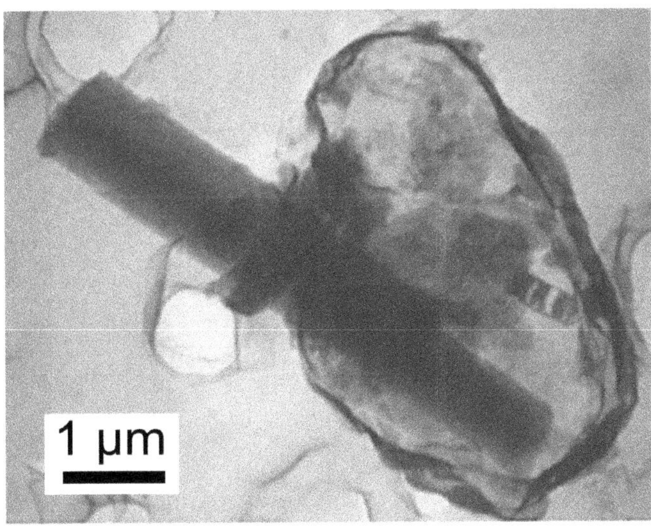

Figure 33. TEM image of asbestiform winchite or richterite fibre from the air, as viewed perpendicular to fibre axes (modified from Gunter *et al.*, 2008 and reproduced here with the permission of Schweizerbart'sche www.schweizerbart.de/journals/ejm).

apparently round cross-section), very large aspect ratios (20:1–100:1, or higher), often curved;

- acicular habit, thin particles ($\leqslant 0.5$ µm wide), generally with moderate (10:1 to 20:1) to high (>20:1) aspect ratios, visible crystal faces, sides generally (but not always) parallel, often needle-like or tapered ends (sometimes perpendicular ends were also observed), smooth, textured, mottled, or rough surface;
- prismatic habit, particles having well developed crystal faces and generally small (<10:1) to moderate (10:1 to 20:1) aspect ratios; sides typically parallel (but sometimes non-parallel sides were observed), a well defined corner or edge and a crystalline termination, smooth and rough surface texture;
- blade-like habit, particles with well developed crystal faces as found for those with prismatic habit, generally having small (<10:1) to moderate (10:1 to 20:1) aspect ratios, the widths typically 4 to 5 times larger than the thickness, sides typically parallel (although non-parallel sides are observed occasionally), typically wedge-shaped cross-section, surface texture more or less smooth.

5.6. Asbestiform fluoro-edenite

The asbestiform fluoro-edenite (see the macroscopic picture in Fig. 34) fibres are generally <1 µm wide and up to 100–150 µm long (*e.g.* Bruni *et al.*, 2014). The maximum length is the shortest among the fibres described in this chapter and reported in the literature.

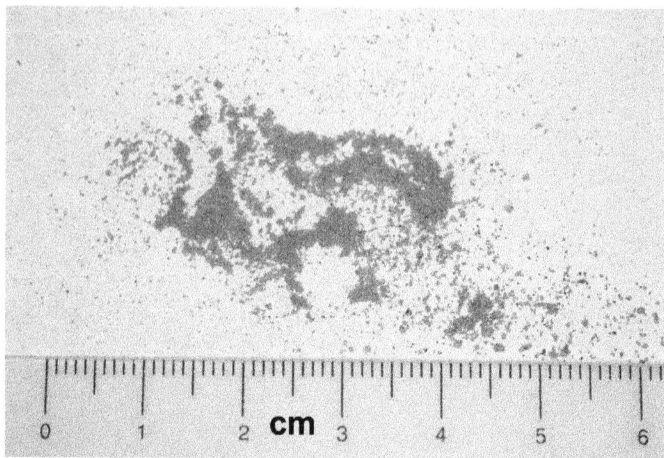

Figure 34. Macrophotograph of asbestiform fluoro-edenite from Biancavilla (Sicilia Region, Italy). Neither fibres nor bundles are discernible at the macro level.

The shorter fibres appear straight and stiff; those with a diameter of 0.5 μm and longer than 10 μm, appear flexible (*e.g.* Comba *et al.*, 2003). Some fibrous crystals show different degrees of longitudinal parting, with fibrils splayed and curved (Gianfagna *et al.*, 2003).

The habit of asbestiform fluoro-edenite is similar to that of asbestiform winchite, richterite and tremolite asbestos. Fluoro-edenite coexists with the other asbestiform amphiboles described above and the distinction of the different amphibole species is achieved only by detailed chemical analysis (*e.g.* Bruni *et al.*, 2014).

The asbestiform fluoro-edenite has recently been declared as carcinogenic to humans (Group 1) by IARC after *in vivo* and *in vitro* studies concerning fibres found in some volcanic rocks on the slopes of the Etna Volcano in Sicily, Italy (Grosse *et al.*, 2014). Only one other occurrence of this fibrous species has been found to date, namely in the lavas of the Kimpo volcano, Kumamoto Prefecture, Japan (Makino *et al.*, 1996).

The apparent rarity could also be due to the difficulty in distinguishing tremolite asbestos (as well as asbestiform winchite) without in-depth chemical analyses (Paoletti *et al.*, 2000).

SEM investigations show that fluoro-edenite fibres generally have a diameter <1 μm (200 to 600 nm), are up to 100–150 μm long, and are characterized by an acicular-fibrous morphology up to filamentous-asbestiform (Fig. 35).

5.6.1. *Asbestiform fluoro-edenite from animal lungs*

Fibres of asbestiform fluoro-edenite detected in the lungs and in the lymph-nodes of sheep living on the slopes of Mount Etna appear long (up to 180 μm), straight and stiff with regular and parallel sides (De Nardo *et al.*, 2004). The fibres are >~20 μm long, with diameters between 0.7 μm and 1 μm (Fig. 35) (Ledda *et al.*, 2016).

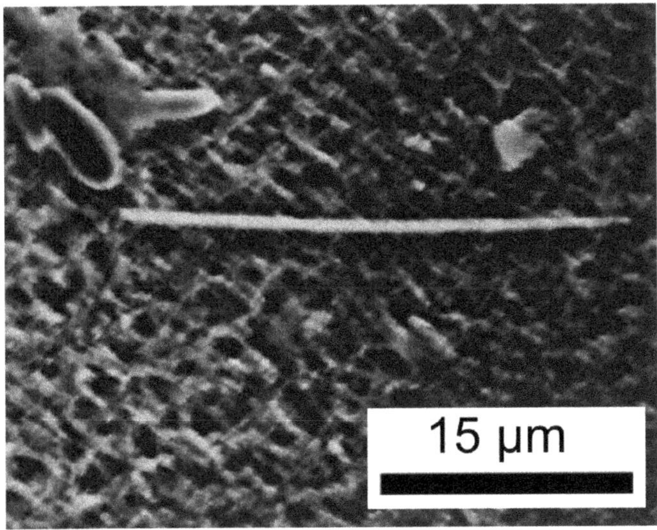

Figure 35. SEM image of an asbestiform fluoro-edenite fibre extracted from the lung of a sheep which lived in the Biancavilla area (Sicily, Italy).

5.6.2. Asbestiform fluoro-edenite from in vivo and after the in vitro test

Fibres of fluoroedenite were used to test their carcinogenicity by *in vivo* tests in rats (*e.g.* Soffritti *et al.*, 2004) and their toxicity by *in vitro* studies (*e.g.* Travaglione *et al.*, 2006). Unfortunately, the habit and size of the fibres detected in the neoplasms and after the cell treatments were not described and therefore it is not possible to evaluate possible changes that occurred in contact with the biologic media.

5.7. Other minor asbestiform minerals

Several minor mineral species may display asbestiform habit, as reported in the first detailed report on this topic (Skinner *et al.*, 1988). Some of these are not widespread, although they may be present in some localities in very large amounts (see for example asbestiform carlosturanite and balangeroite in the Italian Western Alps (Belluso *et al.*, 1995). Others may be found as minor phases in association with major fibrous species in the filling of veins or covering of surface cracks in rocks.

In other cases, such fibrous minerals may occur as the result of the transformation of primary fibrous species or be derived from longitudinal splitting of elongate and large primary crystals.

Several fibrous minerals with potentially harmful dimensions, although not diffused to date, are naturally intermixed with their short counterparts and mined for some industrial applications, as in the case of asbestiform sepiolite and palygoskite.

For all the above-listed cases, the fibres can be detected and distinguished from the associated fibres only by SEM observations (and EDS investigations) and in certain cases only by TEM and EDS.

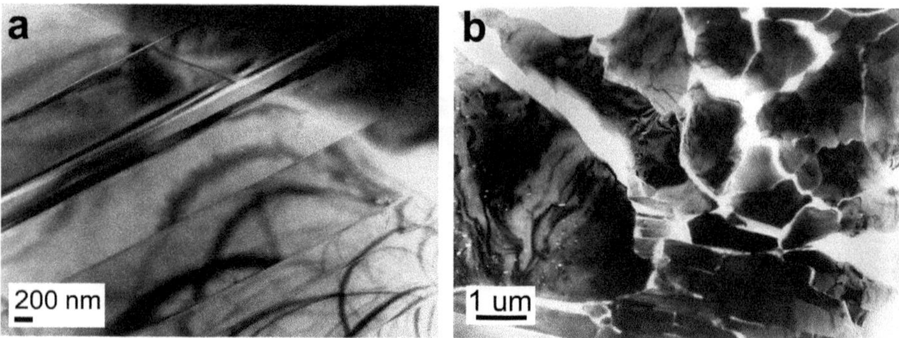

Figure 36. TEM image of asbestiform diopside as viewed perpendicular to the fibre axis (a) and along it (b).

A minor mineral species is asbestiform diopside, arranged in bundles with the straight-shaped fibres elongated along the [001] direction and usually aggregated in parallel to chrysotile (Belluso and Ferraris, 1991, Baronnet and Belluso, 2002). The fibres are usually ~10 μm long but may reach several millimetres; the width ranges from ~0.07 μm to 2.5 μm. As seen by TEM along the fibre axes, some fibres show pristine to complete segmentation into thinner fibres (Fig. 36a). Many fibres appear to have rectangular sections, flattened on the (100) plane, as displayed typically by asbestos and asbestiform amphiboles (Fig. 36b).

Another example is asbestiform gedrite found in serpentinite rocks in southern Italy (Bloise *et al.*, 2014). Due to defibrillation of the flattened crystals along the direction parallel to the crystal elongation, the split fibres are mostly <3 μm wide and >70 μm long (Fig. 37, modified after Bloise *et al.*, 2014). In the same rocks, fibres of magnesio-hornblende have been detected, but only during TEM investigations. They are slightly larger than ~5 μm, but very thin (<0.22 μm) and therefore should be classified as WHO fibres.

Asbestiform magnesio-hornblende aggregated to actinolite asbestos has also been detected in veins filling fractures in granitoid rocks of southern Nevada (Buck *et al.*, 2013).

Figure 37. SEM image of a large, flattened gedrite crystal splitting longitudinally into thinner fibrils (reproduced from Bloise *et al.*, 2014 with the permission of Springer).

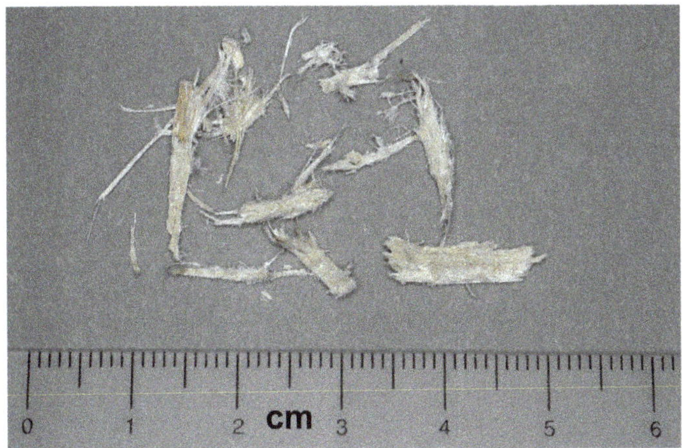

Figure 38. Macrophotograph of asbestiform sepiolite bundles (Piemonte Region, Italy). The terminations of bundles invariably split into thin fibrils.

5.7.1. Asbestiform sepiolite and palygorskite

Only in a few localities worldwide does asbestiform sepiolite occur with a fibre length of >5 μm (Fig. 38) and hence it is generally considered to be non-carcinogenic (WHO carcinogenic fibres: WHO, 1997). As illustrated in Fig. 39, fibres longer than 20 μm have been detected in Finland, China and a few other geographic areas such as northwestern Italy (*e.g.* Lopez-Galindo and Navas, 1989; Santarén and Alvarez, 1994; Giustctto *et al.*, 2014). In Argentina, the fibres are longer than 10 cm (Lescano *et al.*, 2015). Instead, according to Garcia-Romero *et al.* (2007) and Hong *et al.* (2007), sepiolite fibres are invariably shorter than 5 μm.

As regards the aforesaid threshold, TEM investigations show that the fibres are indeed flexible bundles of very thin and long fibrils, <0.1 μm wide and >150 μm long, with parallel [001] elongation axes and longitudinally separable. TEM images from sections cut perpendicular to the bundle elongation show clusters of (001) regular rhomboidal to parallelogram-like cross-sections similarly oriented (Fig. 40). Different clusters are

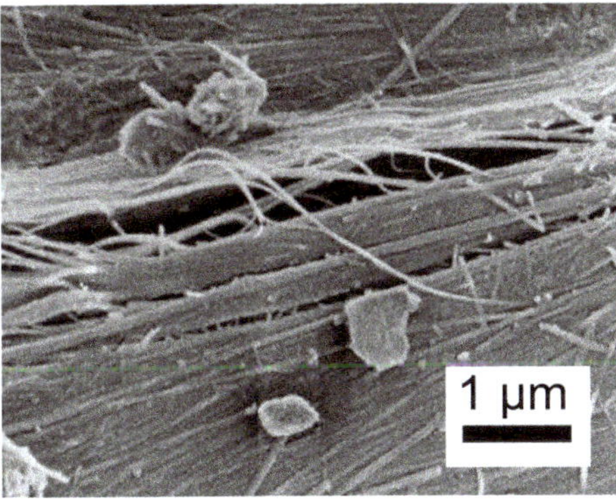

Figure 39. SEM image of asbestiform and flexible sepiolite fibres.

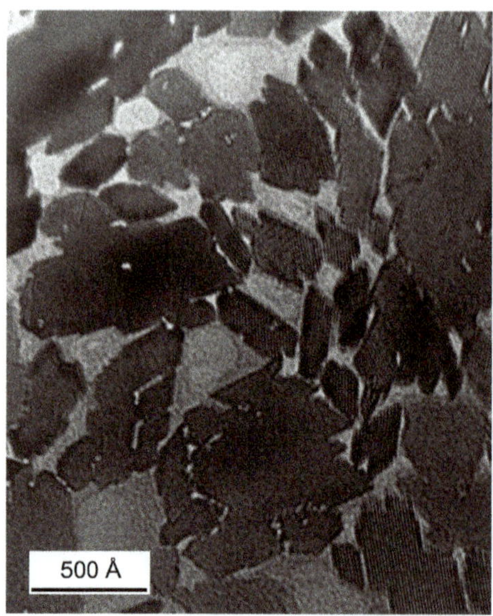

Figure 40. HRTEM images of asbestiform sepiolite fibrils as seen along the fibre axis from Aosta Valley Region (Italy). The sections of the fibre appear as more or less regular lozenges showing incipient partition in smaller rhombs.

variously oriented around the bundle axes, as also shown by electron diffraction rings (Giustetto *et al.*, 2014). The maximum dimension (base of parallelogram) is not larger than 0.075 μm and the mean minimum dimension is <0.02 μm. Parallelograms with larger

Figure 41. Hand specimen of asbestiform palygorskite (courtesy of M. Gunter).

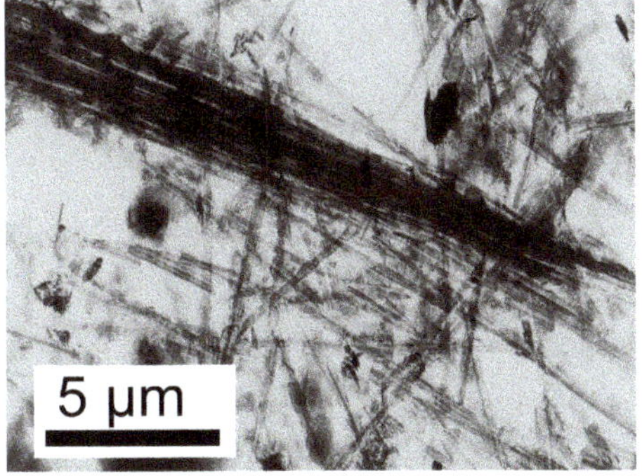

Figure 42. TEM image of bundled and isolated fibrils of palygorskite, as viewed perpendicular to the elongation axis (reproduced from Knidiri *et al.*, 2014, with the permission of the Mineralogical Society of Great Britain & Ireland).

dimensions show an initial defibrillation, *i.e.* an incipient separation into thinner fibres as observed for many amphibole fibres. The process occurs mainly across the (110) cleavage planes. The aperture mechanism is enhanced by the presence of aliphatic hydrocarbons as also observed for flexible fibrous sheet silicates (Sciré *et al.*, 2011).

The association of sepiolite with palygorskite (known also as attapulgite) fibres is not rare. Both sepiolite and palygorskite are lath-shaped on a microscopic scale, with the laths clumping together to form frayed bundles between 0.2 and 5 mm long, exceptionally up to several cm in length (Figs 38 and 42).

As for sepiolite fibres, palygorskite fibres longer than 5 μm are not widespread. In fact, the predominant length is <1.0 μm. Nevertheless, fibres of different sizes generally occur in the same site and may be intermixed (*e.g.* IARC, 1997) and appear as a wool felt (Fig. 41).

Long fibres (up to 20 μm) are flexible and occur either isolated or in bundles, as seen by TEM investigations (Knidiri *et al.*, 2014: Fig. 42).

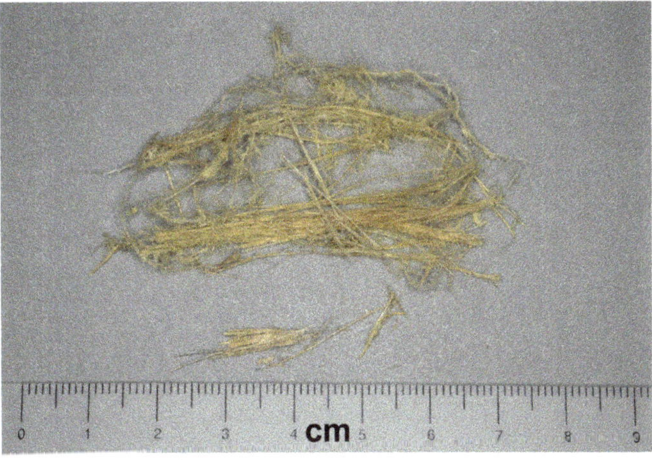

Figure 43. Macrophotograph of asbestiform carlosturanite from Valle Varaita (Piemonte Region, Italy).

Figure 44. TEM image of asbestiform carlosturanite and the intergrown chrysotile, asbestiform polygonal serpentine and asbestiform antigorite, as seen along their fibre axis.

5.7.2. *Asbestiform carlosturanite*

Carlosturanite is an asbestiform silicate (Fig. 43) found mainly in the Italian Western Alps (*e.g.* Compagnoni *et al.*, 1985; Belluso and Ferraris, 1991; Belluso *et al.*, 1995). To date, only one other locality has been reported, namely Taberg in Sweden (Mellini

Figure 45. Macrophotograph of asbestiform balangeroite from the former chrysotile mine at Balangero (Piemonte Region, Italy).

Figure 46. SEM image of asbestiform balangeroite fibres from the former chrysotile mine at Balangero (Piemonte Region, Italy).

and Zussman, 1986). It occurs with fibrous habit, mostly in association with antigorite and chrysotile. The fibres are flexible and long (>150 µm), with a width varying generally between 0.5 µm and several µm. They are aggregated with the fibre axes parallel to each other, thus forming bundles.

TEM investigations performed along the [010] fibre axis show that the fibres are rotated around the [010] direction.

Larger fibres commonly show longitudinal defibrillation into thinner fibres. The cross sections of fibres are not regular, rarely resembling a parallelogram. As a matter of fact, they show saw-shapes and indentations (Fig. 44) due to various degrees of transformation into other fibrous silicates such as chrysotile, asbestiform polygonal serpentine, antigorite, diopside and brucite, all having their fibre axes parallel to each other (Compagnoni *et al.*, 1985; Baronnet and Belluso, 2002).

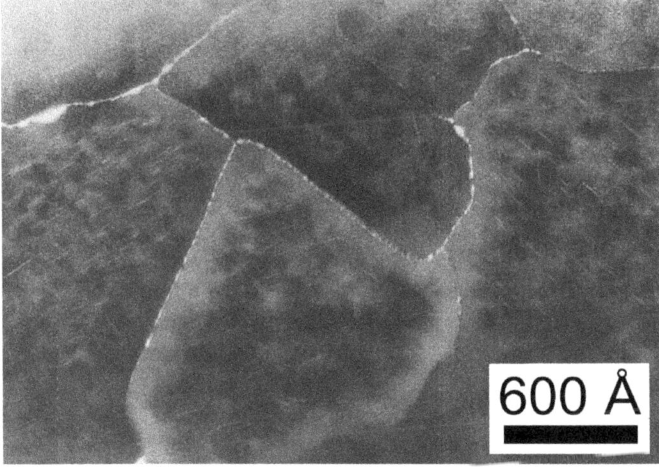

Figure 47. TEM image of asbestiform balangeroite fibres as seen along their fibre axes (modified from Ferraris *et al.*, 1987 and reproduced with the permission of the Mineralogical Society of America).

5.7.3. Asbestiform balangeroite

Balangeroite is a silicate mineral, detected to date only in a small area of the Italian Western Alps (Compagnoni *et al*, 1983; Belluso *et al.*, 1995). This mineral generally forms bundles of long brown fibres, but a prismatic variety from three localities in the same area has also been reported (Groppo, 2005).

Like carlosturanite, balangeroite fibres are macroscopically very similar to chrysotile. For this reason, fibrous balangeroite has been confused with chrysotile for a long time (Fig. 45).

The fibres are long, from a few tens to several thousands of micrometres (Groppo *et al.*, 2005), elongated along the [001] direction and aggregated in bundles. Only the longer ones are slightly flexible (Fig. 46).

TEM investigations along the fibre axes show that the fibres are randomly rotated around the [001] direction, the width of the sections being between 0.05 and 0.5 μm and rarely up to 1 μm (Fig. 47). The larger fibres separate into thinner ones apparently along faulted regions. The cross sections appear as irregular parallelograms, indented forms or even residual islands owing to the transformation in chrysotile and asbestiform polygonal serpentine, either along the boundaries between balangeroite fibres or at the interconnection between three fibrils (Ferraris *et al.*, 1987).

Acknowledgements

E.B. is grateful to A. Baronnet and S. Nietsche for their invaluable help and collaboration for many years with HRTEM at the CINaM (Campus de Luminy, France) and to E. Costa (DST, University of Torino, Italy) for the macrophotography.

References

Alario Franco, M., Hutchison, J.L., Jefferson, D.A. and Thomas, J.M. (1977) Structural imperfection and morphology of crocidolite (blue asbestos). *Nature*, **266**, 520–521.

Amelinckx, S., Devouard, B. and Baronnet, A. (1996) Geometrical aspects of the diffraction space of serpentine rolled microstructures: their study by means of electron diffraction and microscopy. *Acta Crystallographica*, **A52**, 850–878.

Ann, H. and Buseck, P.R. (1991) Microstructures and fibre-formation mechanisms of crocidolite asbestos. *American Mineralogist*, **76**, 1467–1478.

Andreani, M., Baronnet, A., Boullier, A.M. and Gratier, J.P. (2004) A microstructural study of a "crack-seal" type serpentine vein using SEM and TEM techniques, *European Journal of Mineralogy*, **16**, 585–595.

Atkinson, A.W., Gettins, R.B. and Rickards, A.L. (1970) Estimation of fibril lengths in chrysotile asbestos fibres. *Nature*, **226**, 937–938.

Aust, A., Cook, P. and Dodson, R. (2011) Morphological and chemical mechanisms of elongated particle toxicities. *Journal of Toxicology and Environmental Health, Part B*, **14**, 40–75.

Ballirano, P., Bloise, A., Gualtieri, A.F., Lezzerini, M., Pacella, A., Perchiazzi, N., Dogan, M. and Dogan, A.U. (2017) Crystal structure of mineral fibres. Pp. 17–64 in: *Mineral Fibres: Crystal Chemistry, Chemical-physical Properties, Biological Interaction and Toxicity* (A.F. Gualtieri, editor). EMU Notes in Mineralogy, **18**. European Mineralogical Union and Mineralogical Society of Great Britain and Ireland, London.

Bandli, B.R. and Gunter, M.E. (2014) Scanning electron microscopy and transmitted electron backscatter diffraction examination of asbestos standard reference materials, amphibole particles of differing morphology, and particle phase discrimination from talc ores. *Microscopy and Microanalysis*, **20**, 1805–1816.

Baronnet, A. and Belluso, E. (2002) Microstructures of the silicates: key information about mineral reactions and a link with the Earth and materials sciences. *Mineralogical Magazine*, **66**, 709–735.

Belluso, E. and Ferraris, G. (1991) New data on balangeroite and carlosturanite from alpine serpentinites. *European Journal of Mineralogy*, **3**, 559–566.

Belluso, E., Compagnoni, R. and Ferraris, G. (1995) Occurrence of asbestiform minerals in the serpentinites of the Piemonte Zone,Western Alps. Pp. 57–66 in: *Giornata di studio in ricordo del prof. Stefano Zucchetti* (Politecnico di Torino Editor), Turin, Italy.

Bloise, A., Critelli, T., Catalano, M., Apollaro, C,. Miriello, D., Croce, A., Barrese, E., Liberi, F., Piluso, E., Rinaudo, C. and Belluso, E. (2014) Asbestos and other fibrous minerals contained in the serpentinites of the Gimigliano-Mount Reventino Unit (Calabria, S-Italy). *Environmental Earth Sciences*, **71**, 3773–3786.

Boulanger, G., Andujar, P., Pairon, J.C., Billon-Galland, M.A., Dion, C., Dumortier, P., Brochard, P., Sobaszek, A., Bartsch, P., Paris, C. and Jaurand, M.C. (2014) Quantification of short and long asbestos fibers to assess asbestos exposure: a review of fiber size toxicity. *Environmental Health*, **13**, 59.

Brown, B.M. and Gunter, M.E. (2003) Morphological and optical characterization of amphiboles from Libby, Montana U.S.A. by spindle stage assisted polarized light microscopy. *Microscope*, **51**, 121–140.

Buck, B.J., Goossens, D., Metcalf, R.V., McLaurin, B. and Freudenberger M.R.F. (2013) Naturally occurring asbestos: Potential for human exposure, southern Nevada, USA. *Soil Science Society of America Journal*, **77**, 2192–2204.

Cardile, V, Lombardo, L., Belluso, E., Panico, A., Silvana, C. and Balazy, M. (2007) Toxicity and carcenogenicity mechanisms of fibrous antigorite. *International Journal of Environmental Research and Public Health*, **4**, 1–9.

Cametti, G., Pacella, A., Mura, F., Rossi, M. and Ballirano, P. (2013) New morphological chemical, and structural data of woolly erionite-Na from Durkee, Oregon, USA. *American Mineralogist*, **98**, 2155–2163.

Cattaneo A., Somigliana A., Gemmi, M., Bernabeo, F., Savoca, D., Cavallo, D.M. and Bertazzi, P.A. (2012) Airborne concentrations of chrysotile asbestos in serpentine quarries and stone processing facilities in Valmalenco, Italy. *The Annals of Occupational Hygiene*, **56**, 671–683.

Chapman, J.A. and Zussman, J. (1959) Further electron optical observation on crystals of antigorite. *Acta Crystallographica*, **12**, 550–552.

Chatfield, E.J. (2008) A procedure for quantitative description of fibrosity in amphibole minerals. In: *2008 Johnson Conference: critical issues in monitoring asbestos.* ASTM International, Burlington, Vermont, USA, July 14–18, 2008.

Chisholm, J.E. (1973) Planar defects in fibrous amphiboles. *Journal of Materials Science*, **8**, 475–483.

Chung, Y.H. and Han, J.H. (2015) Evaluation of biodurability of Korean chrysotile within the lung of rats. *Journal of Korean Society of Occupational and Environmental Hygiene*, **25**, 20–26.

Comba, P., Gianfagna, A. and Paoletti, L. (2003) Pleural mesothelioma cases in Biancavilla are related to a new fluoro-edenite fibrous amphibole. *Archives of Environmental Health: An International Journal*, **58**, 229–232.

Committee on Asbestos (2006) *Asbestos: Selected Cancers.* The National Academies Press, NW Washington, D.C., pp. 1–327.

Compagnoni, R., Ferraris, G. and Fiora, L. (1983) Balangeroite, a new fibrous silicate related to gageite from Balangero, Italy. *American Mineralogist*, **68**, 214–219.

Compagnoni, R., Ferraris, G., and Mellini, M. (1985) Carlosturanite, a new asbestiform rock-forming silicate from Yal Varaita, Italy. *American Mineralogist*, **70**, 767–772.

Crawford, D. (1980) Electron microscopy applied to studies of the biological significance of defects in crocidolite asbestos. *Journal of Microscopy*, **120**, 181–192.

Cressey, B.A. and Whittaker, E.J.W. (1993) Five-fold symmetry in chrysotile asbestos as revealed by transmission electron microscopy. *Mineralogical Magazine*, **57**, 729–732.

Cressey, B.A. and Zussman, J. (1976) Electron microscopic studies of serpentinites. *The Canadian Mineralogist*, **14**, 307–313.

Cressey, B.A., Whittaker, E.J.W. and Hutchison, J.L. (1982) Morphology and alteration of asbestiform grunerite and anthophyllite. *Mineralogical Magazine*, **46**, 77–87.

Davis, J.M.G., Addison, J., McIntosh, C., Miller, B.G. and Niven, K. (1991) Variations in the carcinogenicity of tremolite dust samples of differing morphology. *Annals of the New York Academy of Sciences*, **643**, 473–490.

DeNardo, P., Bruni, B., Paoletti, L., Pasetto, R. and Sirianni, B. (2004) Pulmonary fibre burden in sheep living in the Biancavilla area (Sicily): preliminary results. *Science of the Total Environment*, **325**, 51–58.

Doğan, M. and Emri, S. (2000) Environmental health problems related to mineral dusts in Ankara and Eskişehir, Turkey. *Bulletin of Earth Sciences Application and Research Centre of Hacettepe University*, **22**, 149–161.

Falini, G., Foresti, E., Gazzano, M., Gualtieri, A.F., Leoni, M., Lesci, I.G. and Roveri, N. (2004) Tubular-shaped stoichiometric chrysotile nanocrystals. *Chemistry – A European Journal*, **10**, 3043–3049.

Ferraris, G., Mellini, M. and Merlino, S. (1987) Electro-diffraction and electron microscopy study of balangeroite and gageite: crystal structures, polytypism and fiber textures. *American Mineralogist*, **72**, 382–391.

Fitz Gerald, J.D., Eggleton, R.A. and Keeling, J.L. (2010) Antigorite from Rowland Flat, South Australia: asbestiform character. *European Journal of Mineralogy*, **22**, 525–533.

Fornero, E., Belluso, E., Capella, S. and Bellis, D. (2009) Environmental exposure to asbestos and other inorganic fibres by animal lung investigation. *Science of the Total Environment*, **407**, 1010–1018.

Gaffney, S.H., Grespin, M., Garnick, L., Drechsel, D.A., Hazan, R., Paustenbach, D.J. and Simmons, B.D. (2016) Anthophyllite asbestos: state of the science review. *Journal of Applied Toxicology*, on line, DOI 10.1002/jat.3356

Gamble, J.F. and Gibbs, G.W. (2008) An evaluation of the risks of lung cancer and mesothelioma from exposure to amphibole cleavage fragments. *Regulatory Toxicology and Pharmacology*, **52**, S154–S186.

Garcia-Romero, E., Suarez, M., Santaren, J. and Alvarez, A. (2007) Crystallochemical characterization of the palygorskite and sepiolite from the Allou Kagne deposit, Senegal. *Clays and Clay Minerals*, **55**, 606–617.

Germine, M. and Puffer, J.H. (2015) Analytical transmission electron microscopy of amphibole fibers From the lungs of Quebec miners. *Archives of Environmental and Occupational Health*, **70**, 323–331.

Gianfagna, A., Andreozzi, G.B., Ballirano, P. and Mazziotti-Tagliani, S. (2007) Structural and chemical contrasts between prismatic and fibrous fluoro-edenite from Biancavilla, Sicily, Italy. *The Canadian Mineralogist*, **45**, 249–262.

Gianfagna, A., Ballirano, P., Bellatreccia, F., Bruni, B.M., Paoletti, L. and Oberti, R. (2003) Characterization of amphibole fibres linked to mesothelioma in the area of Biancavilla, eastern Sicily, Italy. *Mineralogical Magazine*, **67**, 1221–1229.

Gilham, C., Rake, C., Burdett, G., Nicholson, A.G., Davison, L., Franchini, A., Carpenter, J., Hodgson, J., Darnton, A. and Peto, J. (2016) Pleural mesothelioma and lung cancer risks in relation to occupational history and asbestos lung burden. *Occupational and Environmental Medicine*, **73**, 290–299.

Giustetto, R., Seenivasan, K. and Belluso, E. (2014) Asbestiform sepiolite coated by aliphatic hydrocarbons from Perletoa, Aosta Valley Region (Western Alps, Italy): characterization, genesis and possible hazards. *Mineralogical Magazine*, **78**, 919–940.

Glenn, R.E., Lee, R.J., Jastrem, L.M., Bunker, K.L., Van Orden, D.R. and Strohmeier, B.R. (2008) Asbestos: by any other name, is it still? *Occupational Safety and Health Reporter*, **38**, 428–439.

Grauby, O., Baronnet, A., Devouard, B., Schumaker, K. and Demirdjian, L. (1998) The chrysotile–polygonal serpentine–lizardite suite synthesized from a 3MgO–2SiO_2–excess H_2O gel. Abstract, Seventh International Symposium on Experimental Mineralogy, Petrology and Geochemistry, Orléans, France, April 1998, *Terra Nova*, **10**, 24.

Groppo, C. (2005) *Rischio amianto nelle Alpi Occidentali*. PhD Thesis, University of Torino, Italy.

Groppo, C., Tomatis, M., Turci, F., Gazzano, E., Ghigo, D., Compagnoni, R. and Fubini, B. (2005) Potential toxicity of nonregulated asbestiform minerals: balangeroite from the Western Alps. Part 1: identification and characterization. *Journal of Toxicology and Environmental Health, Part A*, **68**, 1–19.

Grosse, Y., Loomis, D., Guyton, K.Z., Lauby-Secretan, B., El Ghissassi, F., Bouvard, V., Benbrahim-Tallaa, L., Guha, N., Scoccianti, C., Mattock, H. and Straif, K. (2014) Carcinogenicity of fluoro-edenite, silicon carbide fibres and whiskers, and carbon nanotubes. *Lancet Oncology*, **15**, 1427–1428.

Gunter, M.E., Belluso, E. and Mottana, A. (2007) Amphiboles: environmental and health concerns. Pp. 453–516 in: *Amphiboles: Crystal Chemistry, Occurrence and Health Issues* (F.C. Hawthorne, R. Oberti, G. Della Ventura and A. Mottana, editors). Reviews in Mineralogy and Geochemistry, **67**, Mineralogical Society of America and Geochemical Society, Chantilly, Virginia, USA.

Gunter, M.E., Harris, K.E., Bunker, K.L., Wyss, R.K. and Lee, R.J. (2008) Amphiboles between the sheets:

observations of interesting morphologies by TEM and FESEM. *European Journal of Mineralogy*, **20**, 1035–1041.

Harris, K.E., Bunker, K.L., Strohmeier, B.R., Hoch, R. and Lee, R.J. (2007) Discovering the true morphology of amphibole minerals: Complementary TEM and FESEM characterization of particles in mixed mineral dust. Pp. 643–650 in: *Modern Research and Educational Topics in Microscopy* (A. Méndez-Vilas and J. Díaz, editors). FORMATEX 2007.

Hawthorne, F.C., Oberti, R., Harlow, G.E., Maresch, W.V., Martin, R.F., Schumacher, J.C. and Welch, M.D. (2012) Nomenclature of the amphibole supergroup. *American Mineralogist*, **97**, 2031–2048.

Hong H.L., Yu N., Xiao P., Zhu Y.H., Zhang K.X., and Xiang S.Y. (2007) Authigenic palygorskite in Miocene sediments in Linxia basin, Gansu, northwestern China. *Clay Minerals*, **42**, 45–58.

IARC (1997) Asbestos. Pp. 1–106 in: *Evaluation of Carcinogenic Risks of Chemicals to Man*, Vol. **14**, Lyon, France.

IARC (2012) Asbestos (chrysotile, amosite, crocidolite, tremolite, actinolite, and anthophyllite). Pp. 219–309 in: *Evaluation of Carcinogenic Risks to Humans. A Review of Human Carcinogens.* International Agency for Research on Cancer, Monographs, Vol. **100**, Part C: Arsenic, Metals, Fibres, and Dusts, Lyon, France.

Keeling, J.L., Raven, M.D. and McClure, S.G. (2006) *Identification of Fibrous Minerals from Rowland Flat Area, South Australia.* Report Book 2006/02, Department Primary Industries and Resources, South Australia.

Keeling, J.L., Raven, M.D., Self, P.G. and Eggleton, R.A. (2008) Asbestiform antigorite occurrence in South Australia. Pp. 329–336 in: *Proceedings of the 9th International Congress for Applied Mineralogy*, ICAM08, Brisbane, September 8–10, 2008.

Knidiri, A., Daoudi, L., El Ouahabi, M., Rhouta, B., Rocha, F. and Fagel, N. (2014) Palaeogeographic controls on palygorskite occurrence in Maastrichtian-Palaeogene sediments of the Western High Atlas and Meseta Basins (Morocco). *Clay Minerals*, **49**, 595–608.

Kristanovic, I. and Pavlovic, S. (1964) X-ray study of chrysotile. *American Mineralogist*, **49**, 1769–1771.

Kühn, J. (1941) Übermikroskopische untersuchungen an asbeststaub und asbestlungen. *Archiv für gewerbepathologie und gewerbehygiene*, **10**, 473–485.

Langer, A.M., Mackler, A.D. and Pooley, F.D. (1974) Electron microscopical investigation of asbestos fibers. *Environmental Health Perspectives*, **9**, 63–80.

Leake, B.E., Woolley, A.R., Arps, C.E.S., Birch, W.D., Gilbert, M.C., Grice, J.D., Hawthorne, F.C., Kato, A., Kisch, H.J., Krivovichev, V.G., Linthout, K., Laird, J., Mandarino, J.A., Maresch, W.V., Nickel, E.H., Rock, N.M.S., Schumacher, J.C., Smith, D.C., Stephenson, N.C.N., Ungaretti, L., Whittaker, E.J.W. and Youzhi, G. (1997) Nomenclature of the amphiboles: Report of the subcommittee on amphiboles of the International Mineralogical Association, Commission on New Minerals and Mineral Names. *American Mineralogist*, **82**, 1019–1037.

Ledda, C., Loreto, C., Pomara, C., Rapisarda, G., Ferrante, M., Fiore, M., Bracci, M., Santarelli, L., Fenga, C. and Rapisarda, V. (2016) Sheep lymph-nodes as a biological indicator of environmental exposure to fluoro-edenite. *Environmental Research*, **147**, 97–101.

Lee, R. and Van Orden, D. (2016) Asbestos in commercial cosmetic talcum powder as a cause of mesothelioma in women. *International Journal of Occupational and Environmental Health*, **21**, 337–341.

Lemen, R.A. (2004) Asbestos in brakes: Exposure and risk of disease. *American Journal of Industrial Medicine*, **45**, 229–237.

Lescano, L., Gandini, N.A., Marfil, S.A. and Maiza, P.J. (2015) Biological effects of argentine asbestos: mineralogical and morphological characterization. *Environmental Earth Sciences*, **73**, 3433–3444.

Lopez-Galindo, A. and Sanchez Navas, A. (1989) Morphological, crystallographical and geochemical criteria of the differentiation between sepiolites of sedimentary and hydrothermal origin. *Boletín de la Sociedad Española de Mineralogía*, **12**, 375–384 (in Spanish).

Makino, K, Yamaguchi, Y and Tomita, K. (1996) Fluor edenite from the Ishigamiyama lava dome of the Kimpo volcano, Kumamoto, southwest Japan. *Journal of Mineralogy, Petrology and Economic Geology*, **91**, 419–423 (in Japanese).

Mattioli, M., Giordani, M., Dogan, M., Cangiotti, M., Avella, G., Giorgi, R., Dogan, A.U. and Ottaviani, M.F. (2016) Morpho-chemical characterization and surface properties of carcinogenic zeolite fibers. *Journal of*

Hazardous Materials, **306**, 140–148.

Meeker, G.P., Bern, A.M., Brownfield, I.K., Lowers, H.A., Sutley, S.J., Hoefen, T.M. and Vance, J.S. (2003) The composition and morphology of amphibole from the Rainy Creek Complex, near Libby, Montana. *American Mineralogist*, **88**, 1955–1969.

Mellini, M. and Zussman, J. (1986) Carlosturanite (not picrolite) from Taberg, Sweden. *Mineralogical Magazine*, **50**, 675–679.

Middleton, A.P. and Whittaker, E.J.W (1976) The structure of Povlen-type chrysotile. *The Canadian Mineralogist*, **14**, 301–306.

Mossman, B.T., Lippmann, M., Hesterberg, T.W., Kelsey, K.T., Borchowsky, A. and Bonner, J.C. (2011) Pulmonary endpoints (lung carcinomas and asbestosis) following inhalation exposures to asbestos. *Journal of Toxicology and Environmental Health Part B*, **14**, 76–121.

Neuendorf, K.K.E., Mehl, J.P. and Jackson, J.A. (editors) (2005) *Glossary of Geology*. American Geological Institute, 5th edition. Alexandria, Virginia, USA.

NIOSH (1976) *Revised recommended asbestos standard*. Department of Health, Education, and Welfare, Center for Disease Control, National Institute for Occupational Safety and Health, Cincinnati, Ohio, U.S.A. DHEW (NIOSH) Publication No. 77-169.

NIOSH (1990) *Comments of the National Institute for Occupational Safety and Health on the Occupational Safety and Health Administration's Notice of Proposed Rulemaking on Occupational Exposure to Asbestos, Tremolite, Anthophyllite, and Actinolite*. OSHA Docket No. H–033d.

NIOSH (2011) *Asbestos Fibers and Other Elongate Mineral Particles: State of the Science and Roadmap for Research*. Revised edition. Department of Health and Human Services. DHHS (NIOSH) Publication No. 2011-159, *Current Intelligence Bulletin*, **62**, 1–159.

Nolan, R.P., Langer, A.M., Oechsle, G.E., Addison, J. and Colflesh, D.E. (1991) Association of tremolite habit with biological potential. Pp. 231–251 in: *Mechanisms in Fibre Carcinogenesis* (R.C. Brown, J.A. Hoskins and N.F. Johnson, editors). Plenum Press, New York.

OSHA (1992) *Occupational Exposure to Asbestos, Tremolite, Anthophyllite and Actinolite*. Occupational Safety and Health Administration Federal Registers, 24310–24311.

Paoletti, L., Batisti, D., Bruno, C., Di Paola, M., Gianfagna, A., Mastrantonio, M., Nesti, M. and Comba, P. (2000) Unusually high incidence of malignant pleural mesothelioma in a town of eastern Sicily: an epidemiological and environmental study. *Archives of Environmental Health: An International Journal*, **55**, 392–398.

Papp, G. (1990) A review of the multi-layer lizardite polytypes. *Annals of the Natural History National Museum, Hungary*, **82**, 9–17.

Roveri, N., Falini, G., Foresti, E., Fracasso, G., Lesci, I.G. and Sabatino, P. (2006) Geoinspired synthetic chrysotile nanotubes. *Journal of Materials Research*, **21**, 2711–2725.

Ross, M., Langer, A.M., Nord, G.L., Nolan, R.P., Lee, R.J., Orden, D. Van and Addison, J. (2008) The mineral nature of asbestos. *Regulatory Toxicology and Pharmacology*, **52**, S26–S30.

Santarén, J. and Alvarez, A. (1994) Assessment of the health effects of mineral dusts. The sepiolite case. *Industrial Minerals*, **April**, 1–12.

Shu, Z., Chena, Y., Zhou, J., Li, T., Yu, D. and Wang, Y. (2015) Nanoporous-walled silica and alumina nanotubes derived from halloysite: controllable preparation and their dye adsorption applications. *Applied Clay Science*, **112–113**, 17–24.

Skinner, H.C.W., Ross, M. and Frondel, C. (1988) *Asbestos and Other Fibrous Materials. Crystal Chemistry and Health Effects*. Oxford University Press, New York, USA.

Soffritti, M., Minardi, F., Bua, L., Degli Esposti, D. and Belpoggi, F. (2004) First experimental evidence of peritoneal and pleural mesotheliomas induced by fluoro-edenite fibres present in Etnean volcanic material from Biancavilla (Sicily, Italy). *European Journal of Oncology*, **9**, 169–175.

Sporn, T.A. (2014) The mineralogy of asbestos. Pp. 1–10 in: *Pathology of Asbestos Associated Diseases* (T.D. Oury, T.A. Sporn and V.L. Roggli, editors). Springer-Verlag, Berlin, Heidelberg.

Straif, K., Benbrahim-Tallaa, L., Baan, R., Grosse, Y., Secretan, B., El Ghissassi, F., Bouvard, V., Guha, N., Freeman, C., Galichet, L. and Cogliano, V. (2009) WHO International Agency for Research on Cancer Monograph Working Group. A review of human carcinogens. Part C: metals, arsenic, dusts, and fibres. *The*

Lancet Oncology, **10**, 453–454.

Strohmeier, B.R., Huntington, J.C., Bunker, K.L., Sanchez, M.S., Allison, K. and Lee, R.J. (2010) What is asbestos and why is it important? Challenges of defining and characterizing asbestos. *International Geology Review*, **52**, 801–872.

Suzuki, Y., Yuen, S.R. and Ashley, R. (2005) Short, thin asbestos fibers contribute to the development of human malignant mesothelioma: pathological evidence. *International Journal of Hygiene and Environmental Health*, **208**, 201–210.

Travaglione, S., Bruni, B.M., Falzano, L., Filippini, P., Fabbri, A., Paoletti, L. and Fiorentini, C. (2006) Multinucleation and pro-inflammatory cytokine release promoted by fibrous fluoro-edenite in lung epithelial A549 cells. *Toxicology in Vitro*, **20**, 841–850.

U.S. Bureau of Mines (1996) *Dictionary of Mining, Mineral, and Related Terms*. American Geological Institute, 2nd edition. Alexandria, Virginia, USA.

Van Gosen, B.S., Blitz, T.A., Plumlee, G.S., Meeker, G.P. and Pierson, M.P. (2013) Geologic occurrences of erionite in the United States: an emerging national public health concern for respiratory disease. *Environmental Geochemistry and Health*, **35**, 419–430.

Veblen, D.R. and Burnham, C.W. (1978) New biopyriboles from Chester, Vermont: I. Descriptive mineralogy. *American Mineralogist*, **63**, 1000–1009.

Veblen, D.R. and Buseck, P.R. (1979) Serpentine minerals: intergrowths and new combination structures. *Science*, **206**, 1398–1400.

Veblen, D. R. and Buseck, P.R. (1980) Microstructures and reaction mechanisms in biopyriboles. *American Mineralogist*, **65**, 599–623.

Veblen, D.R., Buseck, P.R. and Burnham, C.W. (1977) Asbestiform chain silicates: new minerals and structural groups. *Science*, **198**, 359–365.

Vigliaturo, R. (2015) *Microstructures of potentially harmful fibrous minerals*. PhD Thesis, University of Torino, Italy

Vigliaturo, R., Capella S., Rinaudo, C. and Belluso, E. (2016) "Rinse and trickle": a protocol for TEM preparation and investigation of inorganic fibers from biological material. *Inhalation Toxicology*, **8**, 537–363.

Vignaroli, G., Ballirano, P., Belardi, G. and Rossetti, F. (2014) Asbestos fibre identification vs. evaluation of asbestos hazard in ophiolitic rock melanges, a case study from the Ligurian Alps (Italy). *Environmental Earth Sciences*, **72**, 3679–3698.

Viti, C. and Mellini, M. (1996) Vein antigorites from Elba Island, Italy. *European Journal of Mineralogy*, **8**, 423–434.

Wagner, J. (2015) Analysis of serpentine polymorphs in investigations of natural occurrences of asbestos. *Environmental Science: Processes and Impacts*, **17**, 985–996.

Whittaker, E.J.W. (1957) The structure of chrysotile. V. Diffuse reflexions and fibre texture. *Acta Crystallographica*, **10**, 149–156.

Whittaker, E.J.W. (1979) Mineralogy, chemistry, and crystallography of amphibole asbestos. Pp. 1–34 in: *Short Course in Mineralogical Techniques of Asbestos Determination*. (R.L. Ledoux, editor). Mineralogical Association of Canada, Toronto, Canada.

Whittaker, E.J.W., Cressey, B.A. and Hutchison, J.L. (1981) Terminations of multiple-chain lamellae in grunerite asbestos. *Mineralogical Magazine*, **44**, 27–36.

WHO (1997) *Determination of airborne fibre number concentrations; a recommended method, by phase contrast optical microscopy (membrane filter method)*. World Health Organization, Geneva, Switzerland.

Yada, K. (1971) Study of microstructure of chrysotile asbestos by high resolution electron microscopy. *Acta Crystallographica*, **27**, 659–664.

Yada, K. and Iishi, K. (1977) Growth and microstructure of synthetic chrysotile. *American Mineralogist*, **62**, 958–965.

EMU Notes in Mineralogy, Vol. 18 (2017), Chapter 4, 111–134

Bulk spectroscopy of mineral fibres

GIOVANNI B. ANDREOZZI[1] and SIMONE POLLASTRI[2]

[1]*Department of Earth Sciences, Sapienza University of Rome, Piazzale Aldo Moro 5, I-00185 Rome, Italy, e-mail: gianni.andreozzi@uniroma1.it*
[2]*Department of Chemical and Geological Sciences, University of Modena and Reggio Emilia, Via Campi 103, I-41125, Modena, Italy*

Spectroscopic methods are utilized widely for characterizing minerals and other geomaterials in terms of electronic, vibrational and nuclear properties. The basics and applications of spectroscopic methods in mineralogy were reported comprehensively by Hawthorne (1988), and later discussed carefully and updated by Burns (1993) and Clark (1999), by Beran and Libowitzky (2004) and more recently by Henderson *et al.* (2014). These esteemed books and reviews focused generally on topics of immediate mineralogical interest, but nevertheless contain stimulating parallel excursions into the fields of geology and materials sciences. This chapter is built on the shoulders of those giants and is devoted specifically to exploring spectroscopic investigations of electronic and nuclear properties of mineral fibres, a topic not reviewed previously. A number of spectroscopies (though not all) will be mentioned without covering in detail their physical bases (which can be found easily in the books and reviews mentioned above), because this chapter is intended to serve as a review of their contribution to increasing comprehension of the bulk properties of mineral fibres.

1. Characterization of the electronic properties of mineral fibres

1.1. XAS

1.1.1. General introduction to the technique

Absorption spectroscopy refers to a family of experimental techniques used to gain information on the electronic, structural and magnetic properties of matter. Such techniques are based on the study of the variation of the linear absorption coefficient with the energy of the incident photons. When X-rays are used, the technique is known as X-ray absorption spectroscopy (XAS). The linear attenuation coefficient depends on both the scattering phenomena (elastic or inelastic) and on the photoelectric absorption, but in the range of energies used to perform XAS (1–40 keV), the component due to absorption dominates that caused by diffusion (Buffiere and Baruchel, 2015). Therefore, the attenuation coefficient can be approximated as the coefficient of photoelectric absorption and this approximation applies in general for energies of X-rays <200 keV (Buffiere and Baruchel, 2015).

When the energy of the beam is sufficient to extract or induce transition of the electrons from one energy level to another, photons are absorbed by the sample, decreasing dramatically the number of transmitted particles. This causes the

DOI: 10.1180/EMU-notes.18.4

appearance of a discontinuity in the vicinity of that particular value of energy in the absorption spectrum. The analysis of the energy distribution yields information on the innermost electron shells of the target element. The analysis of an XAS spectrum can convey different figures depending on the range of energies considered; for this reason, it is convenient to divide the spectrum into different regions (Fornasini, 2015):

> XANES (X-ray Absorption Near Edge Structure) and pre-edge region. This is the part of the spectrum from a few eV before the absorption edge to 50 eV above it; the combined study of XANES and pre-edge provides information on local geometric and electronic configuration. It is particularly useful for obtaining information on the valence state, local structure, coordination chemistry and ligand symmetry around the absorber atom (Benfatto and Meneghini, 2015);
>
> EXAFS (Extended X-ray Absorption Fine Structure). This defines the region of the spectrum between 100 and 1000 eV beyond the absorption edge. This region contains the sinusoidal oscillations of the absorption coefficient, discovered by Hugo Fricke and Gustav Hertz (Fricke, 1920; Hertz, 1920) which turned out to be related to the local structure of the sample (Sayers *et al.*, 1969, 1971). Therefore, EXAFS analysis allows us to determine the geometric arrangement in the immediate vicinity of the absorber atom (~10 Å).

Considering that XANES has a much larger signal than EXAFS, spectra can be collected even in samples containing smaller concentrations of the target element and in less than perfect sample conditions (Newville, 2004), thus making it a more suitable region for investigating difficult samples such as mineral fibres.

The chemical information in the XANES region (formal valence and coordination environment) makes it possible to use it as a fingerprint to identify phases (Newville, 2004). In fact, a common application of XANES is to use the shift of the edge position to determine the valence state and the heights and positions of pre-edge peaks to determine empirically oxidation states and coordination chemistry (Newville, 2004). The shape of the edge is also characteristic of the chemical environment and bonding geometry (Benfatto and Meneghini, 2015). For many systems (such as the mineral fibres) XANES analysis based on linear combinations of known spectra from the 'model compounds' is sufficient to determine the ratios of valence states and/or phases (Newville, 2004). Model compounds are represented by minerals having the target element in a known chemical environment, such as siderite, hercynite and hematite (having octahedral Fe^{2+}, tetrahedral Fe^{2+} and octahedral Fe^{3+}, respectively) for iron.

For a detailed description of the XAS techniques see for example: Sayers *et al.*, 1969, 1971; Lytle, 1999; Mottana, 2004; Newville, 2004; Vlaic and Olivi, 2004; Benfatto and Meneghini, 2015; Fornasini, 2015.

1.1.2. Application to mineral fibres

A survey of the literature indicates that XAS techniques have been used rarely for the study of minerals in fibrous form. Only a handful of recent studies have attempted to apply the XAS technique to mineral fibres: Giacobbe *et al.* (2010) used XANES pre-edge peak analysis to study standard NIST chrysotile asbestos fibres. Gunter *et al.* (2011) used XANES pre-edge peaks analysis to investigate differences in Fe-redox for 'asbestiform and non-asbestiform' amphiboles. Pascolo *et al.* (2013) characterized asbestos fibres in human lung tissues of naval shipyard workers with synchrotron X-ray fluorescence (XRF) mapping and micro-XANES. Pollastri *et al.* (2016) used the same approach to study the structural variations in representative mineral fibres (chrysotile UICC, crocidolite UICC and erionite) after being in contact with human cell cultures. The only systematic characterization of mineral fibres using XAS was reported by Pollastri *et al.* (2015), where eight mineral fibres of socio-economic and industrial importance were characterized combining XANES and EXAFS investigations with Mössbauer data.

All these works used XAS for the study of the chemical environment of iron exclusively, and no other elements. This is justified basically by the abundance of iron in mineral fibres and its key role in determining the toxicity of mineral fibres (see for example Hardy and Aust, 1995).

Data from the systematic characterization reported by Pollastri *et al.* (2015) on major mineral fibres will be presented in the following paragraphs. The XANES data are reported in Table 1. These data show the presence of both Fe^{2+} and Fe^{3+} oxidation states. This is seen clearly in Fig. 1 where the parameters of XANES pre-edge peaks of the mineral fibres investigated are plotted against those of standard reference compounds.

Chrysotile. The samples investigated by Pollastri *et al.* (2015) were three chrysotile samples of different origins: (1) UICC standard chrysotile asbestos "B" Canadian (NB #4173-111-1) from Quebec, Canada; (2) chrysotile asbestos from Balangero (Turin, Italy); and (3) chrysotile asbestos from Valmalenco (Sondrio, Italy).

Regarding chrysotile UICC, XANES data provide evidence of major fractions of Fe^{3+} with respect to Fe^{2+}, with a pre-edge peaks centroid position at 7114.42 eV (Table 1). The total pre-edge peaks area of 0.129 indicates that iron is hosted in both octahedral and tetrahedral environments. This is confirmed by the EXAFS data (Pollastri *et al.*, 2015) showing two different site positions for iron: one having a first shell of four oxygen atoms at 1.860 Å and another one with a first shell of six oxygen atoms at 2.028 Å. The first shell is compatible with an Fe^{3+} tetrahedral environment (theoretical distance 1.865 Å) while the second is compatible with an octahedral environment for mainly Fe^{3+} with a small presence of Fe^{2+} (theoretical distances 2.015 Å and 2.140 Å, respectively). The presence of tetrahedral iron is justified widely by the previous Mössbauer study on this sample (Stroink *et al.*, 1980) showing that ~60% of total iron belongs to magnetite (or maghemite). Hence, assigning all tetrahedral iron to magnetite/maghemite and taking into account its contribution for octahedral Fe^{2+} and Fe^{3+}, the UICC chrysotile sample displays ferric iron prevailing over ferrous iron in octahedrally coordinated environments (Fig. 1).

Table 1. XANES pre-edge peaks parameters of samples investigated by Pollastri *et al.* (2015). All reported data are averages obtained from a minimum of three analyses.

Sample	Component position (eV)	Component area	Total area	r^2	Centroid position (eV)
Anthophyllite	7112.28	0.019	0.054 (5)	0.9997	7113.13 (3)
	7113.07	0.016			
	7114.03	0.019			
Amosite	7112.37	0.019	0.044 (5)	0.9997	7113.17 (4)
	7113.45	0.017			
	7114.51	0.008			
Tremolite	7112.63	0.012	0.032 (8)	0.9996	7113.56 (7)
	7113.75	0.010			
	7114.83	0.008			
Crocidolite	7112.36	0.011	0.054 (7)	0.9993	7114.05 (5)
	7113.87	0.024			
	7115.16	0.019			
Chrysotile Balangero	7112.81	0.018	0.074 (7)	0.9991	7114.04 (9)
	7114.19	0.042			
	7115.27	0.014			
Chrysotile Valmalenco	7112.72	0.013	0.072 (5)	0.9990	7114.13 (5)
	7114.02	0.035			
	7115.10	0.024			
Chrysotile UICC	7112.96	0.016	0.129 (8)	0.9997	7114.42 (5)
	7114.22	0.077			
	7115.54	0.036			
Erionite	7112.85	0.006	0.076 (2)	0.9994	7114.65 (3)
	7114.17	0.036			
	7115.45	0.034			

In the Balangero chrysotile, XANES data suggest comparable fractions of Fe^{2+} and Fe^{3+}, with a pre-edge peaks centroid position at 7114.04 eV (Table 1). The total pre-edge peaks area of 0.074 points to iron in mainly octahedrally coordinated environments, with a possible minor presence of tetrahedrally coordinated iron. EXAFS refinements show two site positions: a first shell of four oxygen atoms at a distance of 1.770 Å (attributed to tetrahedral ferric iron from magnetite) and another with six oxygen atoms at 2.054 Å (Pollastri *et al.*, 2015), consistent with an octahedral environment halfway between Fe^{2+} and Fe^{3+} (theoretical distances of 2.140 Å and

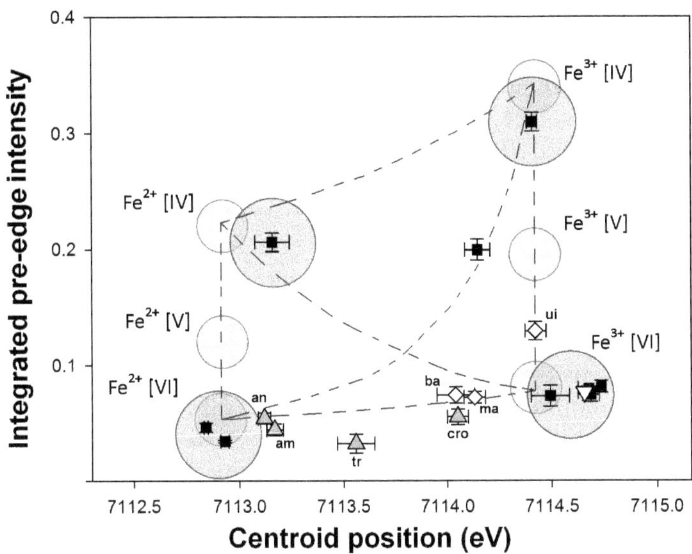

Figure 1. Pre-edge parameters of samples investigated by Pollastri et al. (2015) and reference compounds plotted in the modified variogram from Wilke *et al.* (2005). The small grey fields designate pre-edge parameters for the Fe coordination and oxidation state whereas dashed lines between fields indicate the variation of pre-edge parameters assuming binary mixtures of respective end-members. The larger grey fields designate the pre-edge parameters of Pollastri *et al.* (2015). Legend: black squares = reference compounds; grey triangles = amphibole; white diamonds = chrysotile; white triangle = erionite. Modified after Pollastri *et al.* (2015).

2.015 Å, respectively). Hence, iron in chrysotile from Balangero is half ferrous and half ferric, both hosted in octahedrally coordinated sites (Fig. 1).

Concerning the Valmalenco sample, the XANES pre-edge centroid position at 7114.13 eV and total pre-edge peaks area of 0.072 indicate comparable fractions of Fe^{2+} and Fe^{3+} almost exclusively in 6-fold coordination (Table 1 and Fig. 1). EXAFS results are in agreement with XANES data and are very similar to those of Balangero chrysotile, with a first shell of four oxygen atoms at a distance of 1.810 Å and another with six oxygen atoms at 2.033 Å (Pollastri *et al.*, 2015) amongst the octahedral theoretical Fe^{2+}–O distance of 2.140 Å and the theoretical Fe^{3+}–O distance of 2.015 Å. Hence, the Valmalenco chrysotile displays comparable amounts of Fe^{2+} and Fe^{3+} in octahedrally coordinated environments, very similar to what was observed for the Balangero chrysotile.

Even though they investigated a different sample (chrysotile NIST), Giacobbe *et al.* (2010) found the pre-edge peaks parameters to be in line with our results; in particular they are very similar to those of chrysotile UICC, with a centroid position of 7114.30 eV (rescaled value with respect to a threshold of iron at 7112.00 eV, to be compared with our data) and a total pre-edge peaks area of 0.143.

Fibrous amphiboles. Fibrous amphiboles investigated by Pollastri *et al.* (2015) were: (1) UICC standard amosite asbestos (South African, NB #4173-111-4) from Penge mine, Northern Province (South Africa); (2) UICC standard anthophyllite asbestos (Finnish NB #4173-111-5) from Paakkila (Finland); (3) UICC standard crocidolite asbestos (South African, NB #4173-111-3) from Koegas Mine, Northern Cape (South Africa); and (4) fibrous tremolite from Val d'Ala, Turin (Italy).

In amosite, a pre-edge peaks centroid position of 7113.17 eV and total area of 0.044 indicate predominance of Fe^{2+} hosted in an octahedral position only (Table 1 and Fig. 1). EXAFS data (Pollastri *et al.*, 2015) confirm this picture.

For anthophyllite, the pre-edge peaks centroid position at 7113.13 eV (Table 1) indicates the presence of Fe^{2+} and the total pre-edge peaks area of 0.054 points to iron hosted in octahedrally coordinated environment only (Table 1).

Regarding crocidolite, XANES data indicated comparable fractions of Fe^{2+} and Fe^{3+}, with a pre-edge centroid position at 7114.05 eV (Table 1), halfway between that of the Fe^{2+} and Fe^{3+} standards. The total area of the pre-edge peaks of 0.054 indicated that the iron has an octahedral coordination. EXAFS data confirm a first shell of six oxygen atoms, indicating comparable amounts of ferrous and ferric iron, both in octahedral coordination (Fig. 1, modified after Pollastri *et al.*, 2015).

Tremolite displays a pre-edge centroid position at 7113.56 eV (Table 1) close to that of Fe^{2+} (Fig. 1) whereas the total pre-edge area of 0.032 is the smallest of the samples investigated and indicates that iron is in octahedral positions only. The best fit of the EXAFS data (Pollastri *et al.*, 2015), collected in fluorescence mode, was obtained with a single shell of six oxygen atoms. This is consistent with the coexistence of Fe^{2+} and Fe^{3+}, with Fe^{2+} dominant.

Erionite. Pollastri *et al.* (2015) investigated a fibrous erionite-Na from Jersey (Nevada, USA). XANES data showed clear evidence for the presence of Fe^{3+} only, having a pre-edge peaks centroid position at 7114.65 eV (Table 1). The pre-edge peaks area of 0.076 reveals that iron is hosted in octahedrally coordinated sites only, in perfect agreement with the EXAFS results, having a first shell of six oxygen atoms at a mean distance of 2.011 Å (Pollastri *et al.*, 2015). It is important to note that (XAS being a bulk spectroscopy) these results refer to the whole iron content of the sample, and do not enable us to determine whether it is part of the zeolite structure or is in iron-bearing impurities. This particular issue has been addressed recently by Gualtieri *et al.* (2016).

1.2. EELS and ELNES

1.2.1. General introduction to the technique

Electron energy-loss spectroscopy (EELS) is an analytical technique that measures the change in kinetic energy of electrons after they have interacted with a specimen (Egerton, 2008). Thanks to modern transmission electron microscopes, EELS in favourable conditions is capable of giving information with a spatial resolution at the atomic level (Egerton, 2008). An EELS spectrum provides information equivalent to that obtainable from XAS using a dedicated synchrotron source (Brown, 1997), and the study of its near-edge structures (ELNES: Electron energy Loss Near Edge Structure) is similar to that of XANES. Moreover, the study of the 'EXtended Energy-Loss Fine Structure' (EXELFS) corresponds to EXAFS in X-ray spectroscopy.

The possibility of combining transmission imaging, electron diffraction and X-ray emission spectroscopy makes this technique an important tool for investigating the physics and chemistry of solid matter (Egerton, 2008). EELS is particularly sensitive for the low-atomic-number elements such as C, N and O as well as other elements with favourable ionization cross-sections, such as Fe (Leapman *et al.*, 1987). Considering the experimental difficulties in dealing with soft X-rays and the fact that, for light elements, there is a high probability that the extended fine structures (in both EXAFS and EXELFS) are obscured by edges from other elements, EELS appears to be superior to X-ray absorption (Taftø and Zhu, 1982).

1.2.2. Application to mineral fibres

One of the earliest studies to use EELS to investigate samples containing mineral fibres was that of Jeanguillaume *et al.* (1984), where EELS elemental mapping was used to characterize human tissue sections containing asbestos fibres (the type of fibre is not specified) and asbestos bodies. Leapman *et al.* (1987) used EELS elemental imaging to study alveolar macrophages containing respirable particles. Later, several works using EELS and Electron Spectroscopic Imaging (ESI) to study chemically modified chrysotile fibres were published: Xoffer *et al.* (1991) characterized chrysotile fibres treated with organosilane; Berghmans *et al.* (1992, 1994) and Berghmans and Adams (1992) used EELS and ESI to investigate organosilane-coated chrysotile and for the localization of titanium in $TiCl_3$-treated chrysotile fibres; Wirth (1997) reported EELS spectra at the oxygen *K*-edge of synthetic tremolite, although it is not reported if the sample was fibrous or not. Similarly, Xu (2000) used EELS to determine the oxidation state of iron in amphiboles, although he did not specify the asbestos species and whether it was fibrous or massive. In addition, an Fe $L_{2,3}$-ELNES of a 'generic' calcic amphibole was reported by Calvert *et al.* (2001).

The poor application of this technique for the study of mineral fibres is probably due to the fact that it is used mainly to investigate low-atomic-number elements ($Z < 15$), which are present commonly in small amounts in raw mineral fibres.

1.3. EPR and luminescence spectroscopy used for the study of mineral fibres

Electron paramagnetic resonance (EPR) spectroscopy is of considerable importance for the characterization of chemical reactivity of mineral fibres (*e.g.* radical-adduct detection and quantification: Pacella *et al.*, 2010; Turci *et al.*, 2011; Turci *et al.*, 2017, this volume), but not really for bulk fibre characterization. However, EPR has been used to reveal the presence of Mn^{2+} and Fe^{3+} in fibrous actinolite (Gopal *et al.*, 2004). Moreover, EPR and optical absorption spectroscopies (OAS) were performed on chrysotile by Anbalagan *et al.* (2008). Optical absorption indicated the presence of Fe^{2+} and Fe^{3+} ions, with the bands at 10989, 11111, 15873, 19608, 22727 and 26316 cm^{-1} assigned to Fe^{3+} and the other bands at 9260, 10526, 11628, 12048 and 21739 cm^{-1} assigned to Fe^{2+}. More recently, laser-induced breakdown spectroscopy (LIBS) has been performed on asbestos samples with analysis of optical emission in the visible region of the spectrum (Benton, 2013). The principal elements were identified and the six types of asbestos were determined.

Luminescence spectroscopy is used widely to reveal cell damage induced by fibres (*e.g.* cell fluorescence, Pacella *et al.*, 2012), but Fluorescence Microscopy has been applied recently to the detection of airborne asbestos fibres of nanometric size (Ishida *et al.*, 2012).

2. Characterization of nuclear properties of mineral fibres

2.1. Mössbauer spectroscopy

The chemical activity of mineral fibres has received considerable attention from the biomedical community and has been related frequently to the presence and bioavailability of iron (*e.g.* Kell, 2009, 2010 and references therein). For asbestos the most robust mechanism-based structure-activity relationship includes generation of iron-mediated reactive oxygen species (ROS), and both the oxidation state and the structural coordination of Fe were shown to be crucial factors for asbestos toxicity (Fubini and Otero-Areán, 1999; Shukla *et al.*, 2003; Gazzano *et al.*, 2007; Turci *et al.*, 2011; Pacella *et al.*, 2012). Mineral characterization in terms of Fe oxidation state and structural coordination is obtained principally *via* Mössbauer spectroscopy, a versatile technique used to study the nuclear structure of an atom in a solid by means of the absorption and re-emission of γ rays. Mössbauer spectroscopy requires the use of solids or crystals that have a probability of absorbing an emitted photon in a recoilless manner. The technique uses a combination of the Mössbauer effect and Doppler shift to probe the hyperfine transitions between the excited and ground states of the nucleus. Rudolf L. Mössbauer (1929–2011) provided experimental evidence for recoilless resonant absorption of γ rays in the nucleus in 1958, later to be called the Mössbauer Effect. In 1961, at the age of 32, Rudolf Mössbauer was awarded the Nobel Prize in Physics. By exploiting the Doppler effect (the change in frequency due to a moving source and/or a moving observer), Mössbauer spectroscopy allows the researcher to probe nuclear details in several ways, measuring electric monopole, electric quadrupole and magnetic interactions. Together they are known as hyperfine interactions and may generate absorption in the form of singlets, doublets or sextets, with a Mössbauer spectrum being likely to contain a combination of all of them. Many isotopes have displayed the possibility of providing a Mössbauer spectrum (*e.g.* ^{119}Sn, ^{121}Sb, ^{197}Au, *etc.*) The isotope studied most commonly using Mössbauer spectroscopy is ^{57}Fe (using ^{57}Co as a source) due to the relevance iron in a number of fields, from physics to medicine.

^{57}Fe is a stable isotope representing only 2.1% of natural iron (Coplen *et al.*, 2002). The various stable isotopes of iron are considered to have approximately fixed ratios and very close partitioning behaviour, so that results obtained by ^{57}Fe Mössbauer spectroscopy can be extended to all Fe nuclei of any investigated material. Noteworthy, with respect to the total content of Fe in a sample (Fe_{tot}), ^{57}Fe Mössbauer spectroscopy can determine only the relative abundance of the various types of Fe in terms of valence state (*e.g.* Fe^{2+} and Fe^{3+}), spin state (high/low) and likely coordination number (*e.g.* IV and VI). It cannot determine Fe_{tot} itself (*i.e.* Fe content of the sample relative to the other elements) because the presence of the other elements has no effect on the Mössbauer spectrum (except when they alter the local Fe environment). Consequently, ^{57}Fe Mössbauer spectroscopy provides no information on samples that do not contain Fe in their structures (or in their impurities) and it should be used in combination with other analytical techniques.

Basic principles of ^{57}Fe Mössbauer spectroscopy, together with several examples of its application in mineralogy, petrology and geochemistry, may be found in the excellent reviews of Amthauer *et al.* (2004) and McCammon (2004). In the present chapter, applications of ^{57}Fe Mössbauer spectroscopy to the study of Fe ions in the structure of fibrous minerals will be described so that only the principal aspects of the method will be recalled.

2.1.1. Methodology

^{57}Fe Mössbauer spectroscopy is an absorption spectroscopy, because some of the 14.4 keV γ-rays emitted by the source (^{57}Co which decays to ^{57}Fe in 270 days) are absorbed in a resonant manner by the sample (the absorber). The source is moved by an electromechanical drive system relative to the absorber with different velocities (measured in mm/s), and this modulates the energy of the emitted γ-rays *via* the Doppler effect. When resonant absorption occurs, a decrease in the count rate is registered by the detector (usually a proportional counter). The count rate registered as a function of the source velocity in a multi-channel analyser ultimately represents the raw Mössbauer spectrum.

Fibrous sample preparation for a Mössbauer absorber requires special care due to the possibility of fibre iso-orientation, which would affect the quadrupole doublet components as well as magnetic sextet components. To avoid (or reduce) preferred orientations and texture effects, fibres are usually cut into small pieces, ground gently in an agate mortar with acetone and mixed with powdered acrylic resin, Vaseline gel or other inert material (such as sucrose or graphite) to assist in spreading evenly across the diameter of the sample holder. Furthermore, orientation effects may be minimized by measurements performed with an angle of 54.7 degrees ("magic angle") between the surface of a pressed absorber disc and the incident γ-rays (*e.g.* Ericsson and Wäppling, 1976). The sample is then placed in a 'Plexiglas' sample-holder or held in place with a plastic film non-absorbent of γ-rays. The absorber has a conventional diameter of ~1 cm, equal to that of the window in the detector, and the amount of sample used to make it must fit the limits for the thin absorber thickness described by Long *et al.* (1983). If the chemical composition of the sample to be analysed is known, then the optimal amount of sample can be calculated on the basis of the mass absorption coefficients for the 14.4 keV γ radiation, and the absorber should contain ~5 mg Fe/cm^2 (a typical thickness used in Mössbauer experiments to minimize data collection time). If too small a sample is used, then Fe atoms in the absorber will be too few and the resonance absorption will be too weak. In this case a very long registration time will be needed to improve statistics and obtain a reliable absorption spectrum (*e.g.* several millions of counts in the background channels must be collected). Notably, the use of too much sample (*i.e.* a quantity leading to Fe content exceeding the limits for the thin absorber thickness) should also be avoided as it can affect area, intensity, width and shape details of the Mössbauer absorption peaks.

Collection of Mössbauer spectra may be done in air at room temperature or with the absorber cooled to liquid nitrogen (77 K) or less frequently liquid helium (4 K) temperatures, heated in an inert or reducing atmosphere, kept under high pressure or exposed to an external magnetic field. The velocity range of the source drive system is usually −4 to +4 mm/s for samples with paramagnetic behaviour (*e.g.* silicates, the spectra of which exhibit only quadrupole doublets), while the velocity range −10 to +10 mm/s is adopted for samples with magnetic behaviour (*e.g.* Fe oxides, the spectra of which exhibit magnetic sextets). In specific cases the larger velocity interval was adopted also for fibrous silicates to reveal and quantify magnetic oxide impurities (Ballirano *et al.*, 2008; Pollastri *et al.*, 2015). Raw data, after velocity calibration against a standard spectrum – a high-purity α-iron foil 25 μm thick is commonly adopted and will be used as a reference hereafter – are folded to reconcile back and forth collection paths. The mirror-symmetric absorption spectrum that is obtained is deconvoluted *via* computer fitting programs based on least-squares minimization methods, generally assuming Lorentzian line shape of singlet, doublet or sextet. The hyperfine interaction parameters isomer shift (IS or δ_0), quadrupole splitting (QS or ΔE_Q, only for doublets), and magnetic splitting (H or ΔE_M, only for sextets) are refined. Several fitting methods have been proposed in the literature, sometimes also modelling Gaussian distributions of hyperfine parameters when single Lorentzian lineshapes were no longer appropriate (*e.g.* Lagarec and Rancourt, 1997). The examples in this paragraph were obtained using *RECOIL* software (written by Lagarec and Rancourt and distributed by Intelligent Scientific Applications Inc, Canada).

Isomer shift, which measures the electric monopole interaction between the Fe nucleus and the electronic charge, decreases with increasing density of the *s* electrons at the nucleus. Going from Fe^{2+} to Fe^{3+}, the density of *s* electrons at the Fe nucleus increases inversely with the number of 3*d* electrons ($3d^6$ for Fe^{2+} and $3d^5$ for Fe^{3+}). Consequently, going from Fe^{2+} to Fe^{3+} the δ_0 value decreases (from roughly 1.1 mm/s to roughly 0.3 mm/s at room temperature), while it may take intermediate values (roughly 0.7 mm/s) for intervalence charge-transfer between the two oxidation states. In addition to the oxidation state, the spin state also influences the δ_0 value, and the high-spin state, commonly assumed for both Fe^{2+} and Fe^{3+} in silicate minerals, is associated with the highest δ_0 values. Moreover, coordination number and bond nature operate in opposition, because increasing coordination number of iron ions makes δ_0 increase, whereas increasing covalent bonding makes δ_0 decrease. The isomer shift, which applies to singlet, doublet and sextet components, is therefore very informative and takes values very close to (and sometime confused with) what is referred to as 'chemical shift', 'centre shift' or 'centroid shift' (CS). Actually, this latter is the real (and experimentally measurable) shift of the signal centre from zero velocity – and is known as CS or δ – and corresponds to the value of the isomer shift (δ_0) decreased by the second-order Doppler shift (δ_{SOD}), a negative term the modulus of which increases with temperature (~0.07 mm/s for every 100 K). In a real ^{57}Fe Mössbauer spectrum, CS is the measured centroid shift of a doublet component with respect to zero, and the distance between the two lines of the doublet (in mm/s) is the quadrupole splitting QS. The two lines of the doublet are highly symmetric, but texture effects arising from incorrect sample preparation of fibrous minerals may alter the symmetry in terms of absorption intensity. Increasing QS values reflect deviations from cubic symmetry of the electric field gradient around Fe nuclei. Non-cubic electric charge distribution around the Fe nuclei is determined by: (1) anisotropic electron distribution and spin orientation in the valence shell of Fe ions, *i.e.* the valence term; and (2) deviation from cubic symmetry of ion charge-distribution in the nearest-neighbours (the first coordination sphere made of anions) and next-nearest-neighbours (the second coordination sphere made of cations) in the crystalline lattice, *i.e.* the lattice term. The two contributions are opposite in sign, with the valence term largely exceeding the lattice term. This explains why high-spin Fe^{3+} (which is highly symmetric for both electron distribution and spin orientation) shows generally lower QS values than high-spin Fe^{2+} (which has a strong asymmetry of electron distribution and spin orientation due to one unpaired electron). Moreover, as a consequence of general relationships between geometric/electronic site distortion and QS values, a direct correlation is expected for high-spin Fe^{3+} and the opposite for high-spin Fe^{2+} (McCammon, 2004). When there is a magnetic exchange interaction between Fe atoms in the absorber, the contribution of six absorption lines with relative intensities 3:2:1:1:2:3 occurs in the spectrum. The sextet centroid is shifted from zero velocity by the CS, and the magnitude of the hyperfine magnetic field is measured by the magnetic splitting H. Various Fe oxides and hydroxides have magnetic behaviour and characteristic (and diagnostic) H values, whereas Fe-bearing silicates have paramagnetic behaviour and show only quadrupole doublets. A special case is that of zeolite that may have several types of Fe, from framework to extra-framework Fe ions, and from Fe_xO_y nanoclusters within the pores to larger nanoparticles outside the channels. In this case Fe oxide nanoparticles show superparamagnetic behaviour at room temperature, and the characteristic sextet is relaxed into a doublet or a singlet (Ballirano *et al.*, 2009; Moretti *et al.*, 2014).

2.1.2. Case studies

^{57}Fe Mössbauer spectroscopy is applied to the study of fibrous minerals less frequently than other techniques; nonetheless the information retrieved may be very useful and sometimes unique, spreading from the Fe^{3+}/Fe_{tot} ratios to the coordination environment of Fe ions in the structure to the discovery of Fe oxide impurities.

Chrysotile. Bulk characterization of several Canadian chrysotile samples by ^{57}Fe Mössbauer spectroscopy was done by Blaauw *et al.* (1979), while Stroink *et al.* (1980) studied the two UICC standard samples, namely Rhodesian and Canadian chrysotiles.

The latter sample is a mixture of chrysotile fibres from eight Canadian mines, with half of the material coming from the mining area around the world-famous town of Asbestos. More recently, NIST standard chrysotile was studied by Giacobbe *et al.* (2010) and two natural samples from Balangero and Valmalenco (Italy) were characterized by Pollastri *et al.* (2015). The Mössbauer spectra of all the investigated chrysotile samples showed that they are far from being a single phase, as their spectra are typical of a paramagnetic material (*e.g.* silicate fibres) with accessory magnetic phases as, *e.g.* Fe oxides (Fig. 2).

The occurrence of an accessory magnetic phase was indicated by the appearance of two sextets in the external part of the spectra (CS ~0.30 and 0.67 mm/s, H ~49 and 46 Tesla, respectively), assigned to magnetite and accounting for 30–60% of Fe_{tot} (even if magnetite represents only 1–2 wt.% of the total sample). After subtraction of the magnetic contribution to the absorption spectrum, the chrysotile spectrum was modelled by a combination of quadrupole doublets (up to three): the first two showed CS values of ~1.1 mm/s, QS values in the range of 2.30–2.90 mm/s, and were assigned to octahedrally coordinated Fe^{2+}; the third showed CS values of ~0.30 mm/s, QS values of ~0.70 mm/s, and was assigned to octahedrally coordinated Fe^{3+} (Table 2).

Fibrous amphiboles. Many amphibole species have been studied by ^{57}Fe Mössbauer spectroscopy (Fig. 2) in order to measure Fe^{3+}/Fe_{tot} ratios and get insight into Fe^{2+} and Fe^{3+} distribution within the complex amphibole structure (*e.g.* Bancroft *et al.*, 1967;

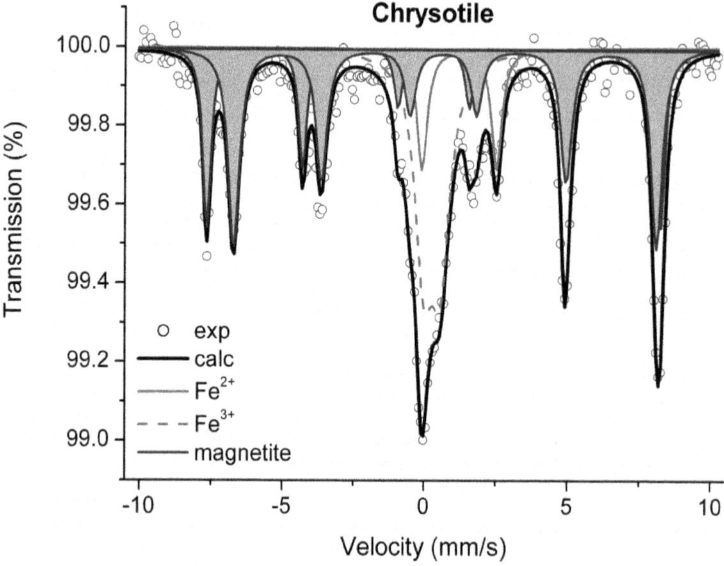

Figure 2. Room-temperature ^{57}Fe Mössbauer spectrum of NIST chrysotile. The grey-shadowed absorption area is due to two magnetic sextets assigned to magnetite (reproduced from Giacobbe *et al.*, 2010 with the permission of Schweizerbart, Germany, www.schweizerbart.de/journals/ejm).

Table 2. Room-temperature ^{57}Fe Mössbauer hyperfine parameters for the fibrous samples investigated compared to extant data from the literature (reproduced from Pollastri *et al.*, 2015, with the permission of Elsevier).

Sample	χ^2	Fe^{2+}				Fe^{3+}					Fe^{3+}_{raw} (% Fe_{tot})	Fe^{3+}_{corr} (% Fe_{tot})
		δ (δ_0) (mm/s)	ΔE_Q (Δ_0) (mm/s)	Γ (σ_Δ) (mm/s)	Area (%)	δ (δ_0) (mm/s)	ΔE_Q (Δ_0) (mm/s)	Γ (σ_Δ) (mm/s)	(H) Tesla	Area (%)		
Amosite	1.12	1.16	2.79	0.32	64	0.30	0.00	0.82		9	9	8
UICC standard		1.08	1.57	0.32	26							
Amosite		1.14	2.80	0.38	53	0.37	1.06	0.68		9	9	
UICC standard[b]		1.07	1.58	0.52	38							
Anthophyllite		1.11	2.64	0.48	38						0	
UICC standard[b]		1.10	1.82	0.32	62							
Crocidolite	3.16	1.07	2.92	0.03		0.37	1.00	1.86			57	52
UICC standard[a]		1.07	2.64	0.38		0.37	0.43	0.18				
Crocidolite		1.12	2.90	0.34	38	0.38	0.42	0.31		42	42	37
UICC standard[b]		1.13	2.42	0.34	20							
Tremolite	0.79	1.19	2.92	0.17	48	0.25	2.16	0.62		7	15	12
Ala di Stura[a]		1.19	1.89	0.20	37	0.25	1.02	0.32		8		
Chrysotile		1.12	2.65			0.34	0.75					75
UICC standard[b]						0.20	0.34					
Chrysotile	0.97	1.14	2.73	0.30	17	0.33	0.66	0.64		30	51	43
Balangero		1.16	2.32	0.44	23	*0.26*			*49*	*12*		
						0.68			*46*	*18*		
Chrysotile	0.88	1.13	2.88	0.24	10	0.35	0.69	0.62		37	56	47
Valmalenco		1.15	2.63	0.33	26	*0.26*			*49*	*10*		
						0.67			*46*	*17*		
Erionite	0.49					0.35	0.24	0.44		55	100	100
Jersey, Nevada						0.32	0.85	0.64		45		
Erionite	0.89					0.34	0.34	0.50		80	100	100
Rome, Oregon[c]						0.38	1.17	0.40		20		

Notes: Centre shift (δ) measured with respect to α-iron. Lorentzian site analysis: uncertainties were estimated at ~±0.02 mm/s for both δ, quadrupole splitting (ΔE_Q) and line width (Γ), ±0.5 Tesla for magnetic field (H), and no less than ±3% for absorption area (expressed as % of Fe_{tot}). In *italics*: parameters assigned to Fe^{3+} and $Fe^{2.5+}$ of accessory magnetite in chrysotile. Fe^{3+}_{raw}: area of absorption peaks assigned to Fe^{3+}; Fe^{3+}_{corr}: obtained from the raw value by applying the correction factor of Dyar *et al.* (1993) for amphibole and chrysotile, and that of De Grave and Van Alboom (1991) for magnetite.
[a] Original data from Fantauzzi *et al.* (2010): in this case the QSD parameters centre shift (δ_0), centre of Gaussian components (Δ_0) and Gaussian width (σ_Δ) are reported; [b] original data from Stroink *et al.* (1980); [c] original data from Ballirano *et al.* (2009). Symbols according to Rancourt and Ping (1991).

Hawthorne, 1983; Skogby and Annersten, 1985; Schmidbauer *et al.*, 2000; Hawthorne *et al.*, 2007). Goldman (1979), in particular, demonstrated the advantages of low-temperature measurements to improve the resolution of various quadrupole doublets in amphibole spectra. However, the Mössbauer technique has rarely been applied to the fibrous amphibole counterpart. The UICC asbestos standards amosite, fibrous anthophyllite and crocidolite were investigated by ^{57}Fe Mössbauer spectroscopy by Stroink *et al.* (1980) and Luys *et al.* (1983). More recently, Fantauzzi *et al.* (2010) re-examined UICC crocidolite in a combined study of bulk and surface Fe oxidation states of amphibole asbestos by ^{57}Fe Mössbauer and X-ray Photoelectron spectroscopies, and Pollastri *et al.* (2015) re-examined UICC amosite in a combined ^{57}Fe Mössbauer spectroscopic and X-ray absorption study. Fibrous tremolite, representative of asbestos in its natural setting, has been involved recently in several studies focused on Fe speciation because its toxic potential had been shown to be as high as that of crocidolite (Wagner *et al.*, 1982; Weill *et al.*, 1990). On the basis of those studies, samples from Val di Susa and Ala di Stura (Piedmont, Italy) were chemically and structurally characterized by ^{57}Fe Mössbauer, electron probe microanalysis and X-ray diffraction to determine Fe oxidation state and site partitioning (Ballirano *et al.*, 2008; Pacella *et al.*, 2008); samples from S. Mango (Calabria, Italy) and Montgomery County (Maryland, USA) were investigated by ^{57}Fe Mössbauer and X-ray Photoelectron spectroscopies to ascertain Fe oxidation state of bulk fibre and of surfaces (Fantauzzi *et al.*, 2010); and samples from Mt. Rufeno (Latium, Italy) and Castelluccio Superiore (Basilicata, Italy) underwent an interdisciplinary and multi-analytical study (by X-ray diffraction, electron microprobe, scanning electron microprobe, inductively coupled plasma-mass spectrometry (ICP-MS), ^{57}Fe Mössbauer, infrared and EPR spectroscopies) to unveil relationships among crystal chemistry, iron topochemistry and chemical reactivity (Pacella *et al.*, 2010). Moreover, Gunter *et al.* (2003) examined the crystal chemistry and the Fe oxidation state of asbestiform richterite-whinchite samples in comparison with coexisting non-asbestiform samples from Libby (Montana, USA). Asbestiform richterite from Libby was studied also by Fantauzzi *et al.* (2012) in an interdisciplinary and multi-analytical study of Fe speciation, surface chemistry and surface reactivity of fibrous amphiboles that are not regulated as asbestos. Similarly, Gianfagna *et al.* (2007) and Andreozzi *et al.* (2009) studied crystal chemistry and Fe oxidation states of a suite of fibrous samples of fluoro-edenite (another fibrous amphibole not regulated as asbestos) coexisting with crystals of the same species from Biancavilla (Sicily, Italy). In addition to natural samples, a number of crystallochemical studies have been performed on synthetic amphibole samples (produced hydrothermally using internally-heated pressure vessels) that grew with an asbestiform shape, *i.e.* high aspect ratio. Notably, only some of them are synthetic analogues of amphiboles having a natural asbestiform variety, *e.g.* riebeckite (= crocidolite) and richterite. The synthetic fibrous amphiboles were characterized fully *via* a multi-analytical approach including ^{57}Fe Mössbauer spectroscopy, used to determine Fe^{3+}/Fe_{tot} ratios and obtain information about Fe ions site distribution and short-range ordering (Iezzi *et al.*, 2003a,b, 2004, 2005; Della Ventura *et al.*, 2005a,b, 2016).

Mössbauer spectra of amphiboles (both asbestiform and not) are typical of paramagnetic materials and are composed of variable combinations of absorption doublets (Fig. 3).

In the early literature, amphibole spectra were fitted using Lorentzian line shapes, as usually done for other mineral spectra. In most cases a good fit was obtained, with satisfactory statistical parameters, reasonable hyperfine parameters CS, QS and, above all, full width at half maximum (FWHM or Γ) close to the 14.4 keV γ radiation natural linewidth (0.194 mm/s). In some cases, however, results were not satisfactory due to extremely large linewidths (up to 2 mm/s). Better results were obtained by fitting quadrupole splitting distributions (QSD), taking advantage of the approach proposed by Rancourt and Ping (1991) and Rancourt (1994) who showed convincingly that a pure Lorentzian approach is inadequate when analysing spectra with poorly resolved quadrupole doublets. QSD builds on fitting the unconstrained parameters isomer shift (δ_0), coupling parameter (δ_1), centre of a Gaussian component (Δ_0), Gaussian width (σ_D), and absorption area (A). It was noteworthy, in most cases, that the Fe^{3+}/Fe_{tot} ratios obtained for the same sample by applying QSD and Lorentzian fitting were not

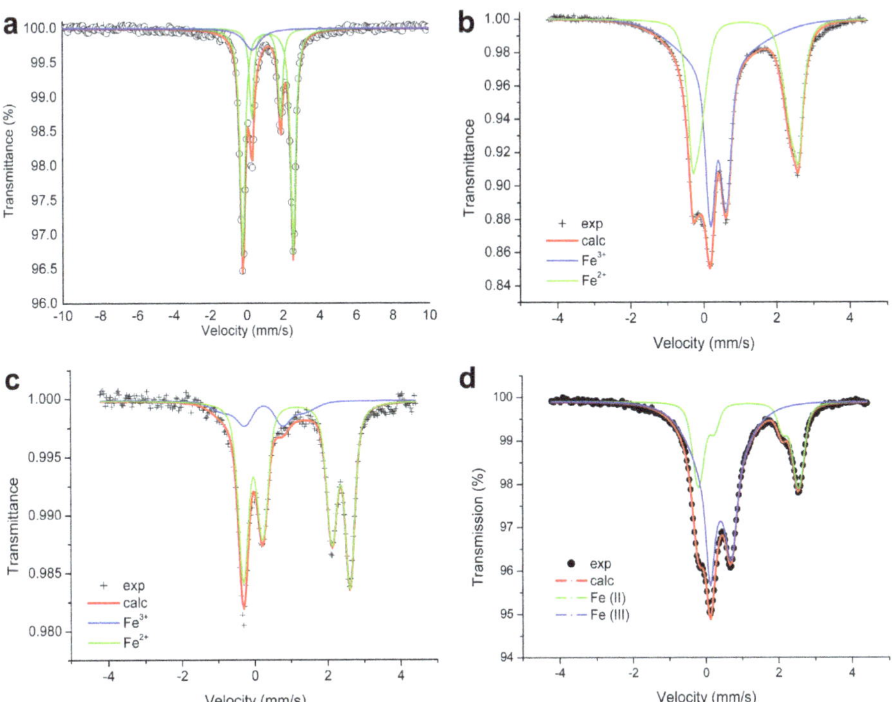

Figure 3. Room-temperature ^{57}Fe Mössbauer spectrum of: (a) UICC amosite (Lorentzian line fitting, from Pollastri *et al.*, 2015); (b) UICC crocidolite (QSD fitting, from Fantauzzi *et al.*, 2010); (c) Ala di Stura tremolite (QSD fitting, from Pacella *et al.*, 2008); and (d) Libby richterite (QSD fitting, from Fantauzzi *et al.*, 2012).

dramatically different (with the QSD values always greater than the Lorentzian ones), and this allows comparison between results of recent and early studies. For several fibrous amphiboles, for example, the difference in Fe^{3+}/Fe_{tot} values was found to be within 3% absolute or less (*i.e.* within experimental error), with the only exception being one sample from Libby (Montana, USA) for which the difference was 12% absolute (Gunter *et al.*, 2003; Gianfagna *et al.*, 2007; Andreozzi *et al.*, 2009; Pacella *et al.*, 2010). Furthermore, to account for the temperature influence on Fe^{3+}/Fe_{tot} quantification, the 'raw' absorption areas measured at room temperature (RT) conditions are usually corrected using an empirical value for the recoil-free fraction correction (1.22 was retrieved for amphiboles by Dyar *et al.*, 1993). This must be taken into account when data from different sources are compared, as fitting "raw" RT data leads to a systematic overestimation of Fe^{3+} proportions, up to twice the experimental error (Gunter *et al.*, 2003; Gianfagna *et al.*, 2007; Ballirano *et al.*, 2008; Pacella *et al.*, 2008, 2010; Andreozzi *et al.*, 2009; Fantauzzi *et al.*, 2010, 2012; Pollastri *et al.*, 2015).

In the amphibole structure, both Fe^{2+} and Fe^{3+} may be hosted: with reference to the structural formula $AB_2C_5T_8O_{22}W_2$, Fe^{2+} ions may occupy the 6- to 8-fold coordinated $M(4)$ site (as B cations) and the octahedrally coordinated $M(1)$, $M(2)$ and $M(3)$ sites (as C cations), whereas Fe^{3+} ions are mainly (or exclusively) ordered at $M(2)$ site (Hawthorne *et al.*, 2012).

When interpreting the Mössbauer spectrum of an amphibole (asbestiform or not), the distinction between Fe^{2+} and Fe^{3+} is quite easy thanks to extremely differentiated CS values, ~1.0–1.2 mm/s for Fe^{2+} and 0.2–0.4 mm/s for Fe^{3+} (Fig. 4). On the contrary, the assignment of the various absorptions to a specific site occupancy or local environment on the basis of QS values may be more problematic. In fibrous amphiboles the octahedrally coordinated Fe^{3+} shows QS values highly variable between zero and 2 mm/s, with a considerable cluster of data between 0.2 and 1.2 mm/s (Fig. 4). Because it is known that Fe^{3+} is hosted preferentially at $M(2)$ sites, and that for Fe^{3+} the QS is correlated positively with geometric/electronic distortion of the octahedral environment, increasing QS values have been interpreted as a reflection of increasing local distortion around ${}^{M(2)}Fe^{3+}$ (Hawthorne, 1983; Iezzi *et al.*, 2003a; Della Ventura *et al.*, 2016). For octahedrally coordinated Fe^{2+}, recorded QS values are generally higher than those of Fe^{3+} and range from 1.5 to 3.0 mm/s (Fig. 4, Table 2). Note that for Fe^{2+} an inverse correlation exists between QS values and distortion of the octahedral environment, due to opposite variations of lattice and valence terms as a function of site distortion (see above or McCammon, 2004, for more details). Accordingly, decreasing QS values should reflect an increase in the geometric/electronic distortion of the octahedral environment around Fe^{2+}. The commonly accepted site assignment – based largely on convergence between spectroscopic, structural and chemical data – is in agreement with theoretical predictions as it proposes that Fe^{2+} doublets with QS (or Δ_0) values that fall approximately in the range 3.0–2.7 mm/s are assigned to $M(1)$, those in the range 2.7–2.4 mm/s are assigned to $M(3)$, those in the range 2.4–1.8 mm/s are assigned to $M(2)$, and those in the range 1.8–1.5 mm/s are assigned to the peripheral and more distorted $M(4)$ sites (Burns and Greaves, 1971; Gunter *et al.*, 2003; Gianfagna *et al.*, 2007; Ballirano *et al.*, 2008; Pacella *et al.*, 2008; Andreozzi *et*

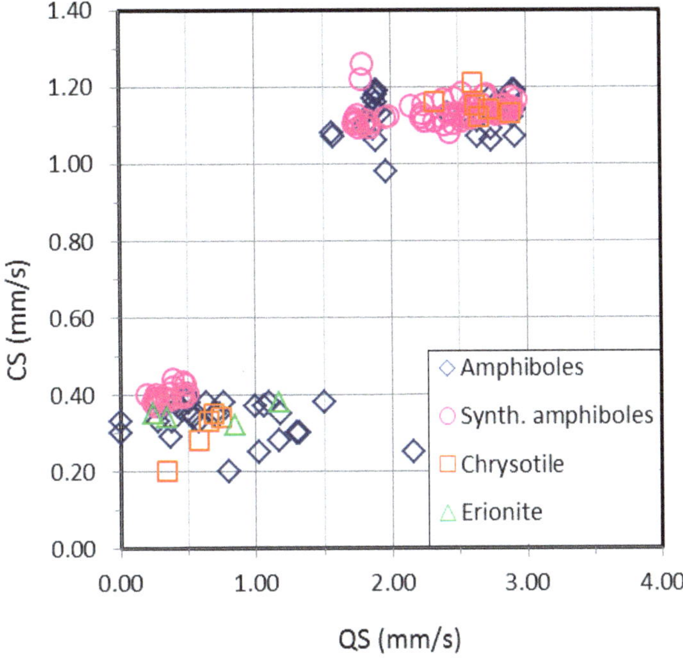

Figure 4. Quadrupole splitting (QS or Δ_0) *vs.* centre shift (CS or δ_0, with respect to α-iron) of fibrous minerals. Hyperfine parameters extracted from room-temperature ^{57}Fe Mössbauer spectra fitted with either Lorentzian lineshape or QSD. Data from Stroink *et al.* (1980), Gunter *et al.* (2003), Iezzi *et al.* (2003a,b, 2004, 2005), Della Ventura *et al.* (2005a,b, 2016), Gianfagna *et al.* (2007), Ballirano *et al.* (2008, 2009), Giacobbe *et al.* (2010); Pacella *et al.* (2008, 2010), Andreozzi *et al.* (2009), Fantauzzi *et al.* (2010, 2012) and Pollastri *et al.* (2015).

al., 2009; Pacella *et al.*, 2010). A certain ambiguity may exist for the attribution of Fe^{2+} doublets with QS in the range 1.7–1.9 mm/s, that were assigned to *M*(4) by Goldman (1979), Hawthorne *et al.* (1983), Skogby and Annersten (1985) and Schmidbauer *et al.* (2000). In these cases, the availability of low-temperature Mössbauer spectra may help to distinguish Fe^{2+} in *M*(2) from Fe^{2+} in *M*(4).

A further complication in fitting ^{57}Fe Mössbauer spectra of fibrous amphiboles arises when the sample is impure, as is the asbestiform tremolite collected in the Val di Susa (Piedmont, Italy), at the excavation front of the Musinè-Gravio tunnel on the planned Turin-Lyon railway (Ballirano *et al.*, 2008). A careful characterization of this fibrous amphibole was necessary because of its potentially high impact on health, related to the risk of lung pathologies of workers and local inhabitants after environmental exposure to airborne powders of this material. Similarly to what was observed previously for chrysotile, the Mössbauer spectrum of fibrous tremolite from Val di Susa was typical of a paramagnetic material with some accessory magnetic phases (Fig. 5).

The magnetic phases (very few wt.% and practically invisible to other techniques) were identified as hematite and magnetite and their contribution to the absorption was

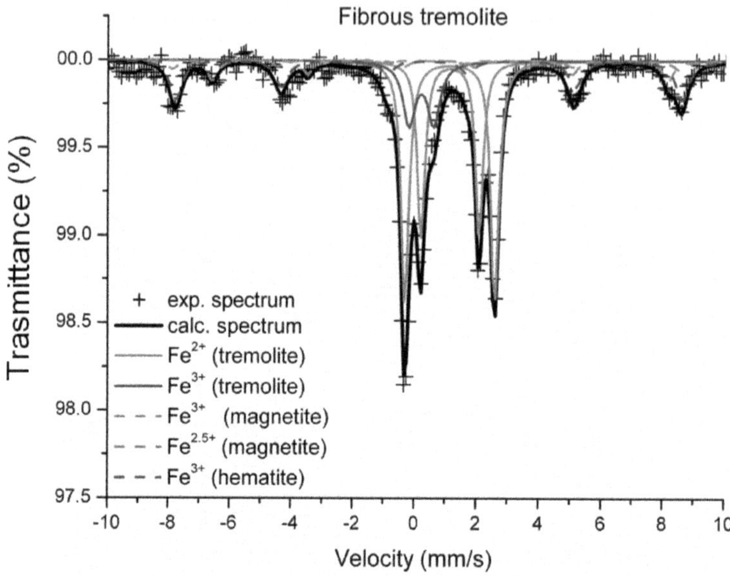

Figure 5. Room-temperature ^{57}Fe Mössbauer spectrum of impure fibrous tremolite from Val di Susa containing magnetic iron oxides (reproduced from Ballirano *et al.*, 2008 with the permission of the Mineralogical Society of America).

subtracted from the spectrum. The real absorption of fibrous tremolite was represented by three quadrupole doublets: the first with CS of ~0.2 mm/s and QS of ~0.8 mm/s was assigned to $^{M(2)}Fe^{3+}$ (17% of Fe_{tot}), the other two with CS values of ~1.1 mm/s and QS values of ~2.9 and 1.9 mm/s were assigned to $^{M(1)}Fe^{2+}$ and $^{M(2)}Fe^{2+}$ (47 and 36% of Fe_{tot}), respectively.

Erionite. The fibrous zeolite erionite was investigated by ^{57}Fe Mössbauer spectroscopy with the purpose of determining Fe oxidation state, topochemistry and aggregation state (Ballirano *et al.*, 2009; Pollastri *et al.*, 2015). Whenever iron is present in a zeolite sample, in addition to its oxidation state it is necessary to determine its position and partitioning, that is the quantities possibly located at framework sites (*i.e.* tetrahedrally coordinated, substituting for Si/Al), within extra-framework positions (*i.e.* in the cages) and on external surfaces. Closely correlated with Fe position, another crucial factor is its aggregation state (also called nuclearity) that can be represented by a single Fe ion, an Fe_xO_y dinuclear or oligonuclear cluster and an Fe_2O_3 hematite-like nanoparticle (Zecchina *et al.*, 2007; Moretti *et al.*, 2014). For erionite, both Fe position and nuclearity are very important because they strongly affect iron availability and bioactivity (Fach *et al.*, 2003; Ruda and Dutta, 2005; Gualtieri *et al.*, 2016).

In the study by Ballirano *et al.* (2009), a sample of erionite-K from Rome (Oregon, USA) was examined, and the oxidation and coordination state of Fe were clarified by exploiting room- and low-temperature ^{57}Fe Mössbauer spectroscopy. The spectrum collected at room temperature (Fig. 6a) showed a broad absorption signal, which was

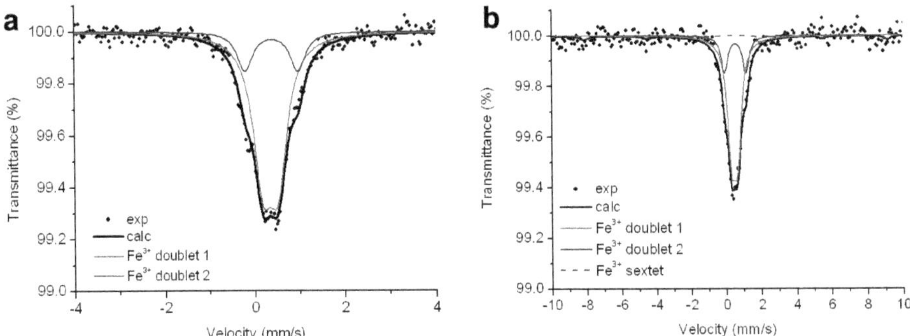

Figure 6. ^{57}Fe Mössbauer spectra of erionite-K from Rome, Oregon, USA collected at: (a) room temperature from −4 to 4 mm/s; and (b) 77 K, from −10 to 10 mm/s (reprinted from Ballirano *et al.*, 2009 with the permission of the Mineralogical Society of America).

modelled initially by a close quadrupole doublet with CS of 0.34 mm/s relative to α-iron and QS of 0.37 mm/s, indicating the presence of high-spin Fe^{3+} ions. However, the large value of linewidth (0.75 mm/s) suggested the presence of at least two overlapping signals. A second model with two doublets gave better resolution, with a first doublet characterized by CS = 0.34 mm/s, QS = 0.34 mm/s and linewidth of 0.50 mm/s, and a second doublet characterized by CS = 0.38 mm/s, QS = 1.17 mm/s and linewidth of 0.40 mm/s (Table 2). On the basis of CS values, both doublets were ascribed to Fe^{3+} coordinated octahedrally. However, because the limited size of erionite cages could hardly host octahedrally coordinated Fe^{3+}–O clusters and site scattering of extra-framework sites, retrieved by structural refinement, did not support the presence of iron in that position, the hypothesis of surface iron oxides was explored. The most common iron oxide and hydroxide with Fe^{3+} coordinated octahedrally are α-Fe_2O_3 (hematite) and α-FeO(OH) (goethite). Magnetically ordered materials such as hematite and goethite generally give a sextet absorption in the ^{57}Fe Mössbauer spectrum, which was not observed in the present case. However, it is known that in the case of very small particles (<10 nm) the magnetic sextet collapses into a doublet (the phenomenon is called superparamagnetic relaxation), and the influence of super-paramagnetism decreases with temperature, so another spectrum of erionite was collected at 77 K (Fig. 6b).

The low-temperature spectrum was similar to the room-temperature one, but it was modelled by two quadrupole doublets and, tentatively, one magnetic sextet. The CS of all doublets and sextet was ~0.5 mm/s confirming the presence of high-spin Fe^{3+} ions (the numerical difference with previous CS values being due to second-order Doppler shift). The magnetic field measured for the sextet was ~54 Tesla, a value typical of hematite, and its absorption area was estimated to be ~5% Fe_{tot}. On these bases, the largest proportion of the Fe (95%) was attributed to Fe^{3+}-bearing, superparamagnetic, oxide-like nanoparticles with dimensions between 1 and 10 nm, and the remaining 5% was attributed to hematite nanoparticles with sizes >10 nm, both located on the erionite surface.

In the study by Pollastri *et al.* (2015), a sample of fibrous erionite-Na from Jersey (Nevada, USA) was examined by a combined ^{57}Fe Mössbauer and XAS spectroscopic approach. In spite of the different origin of the sample, the ^{57}Fe Mössbauer parameters were closely comparable to those of the Oregon sample (Table 2). Also in this case the results of the multi-analytical study converged to attribute Fe signals to high-spin Fe^{3+} not contained in the erionite structure, but confined to external surfaces as Fe^{3+}–O clusters with increasing nuclearity bound by either weak physisorption or strong chemisorption, which was supported by high-resolution transmission electron microscopy.

2.2. NMR spectroscopy used for the study of mineral fibres

Proton, aluminium and silicon magic-angle-spinning nuclear magnetic resonance spectroscopy (^{1}H, ^{27}Al and ^{29}Si MAS NMR), together with infrared spectroscopy, were adopted to investigate erionite to obtain information about the location and concentration of Broensted acid sites, the distribution of aluminium in the framework and the concentration of non-framework species (Roessner *et al.*, 1989). ^{1}H-NMR spectroscopy was employed by Ozeki *et al.* (1991) to investigate water adsorbed on a synthetic chrysotile asbestos, and ^{29}Si-NMR was used by Habaue *et al.* (2006) to study the porous, disordered and fibrous silica derived from serpentine slag containing chrysotile asbestos by selective acid leaching.

References

Amthauer, G., Grodzicki, M., Lottermoser, W. and Redhammer, G. (2004) Mössbauer spectroscopy: Basic principles. Pp. 345–367 in: *Spectroscopic Methods in Mineralogy* (A. Beran and E. Libowitzky, editors). EMU Notes in Mineralogy, **6**. European Mineralogical Union, Eötvös University Press, Budapest, 661 pp.

Anbalagan, G., Sakthimurugesan, K., Balakrishnan, M. and Gunasekaran, S. (2008) Structural analysis, optical absorption and EPR spectroscopic studies on chrysotile. *Applied Clay Science*, **42**, 175–179.

Andreozzi, G.B., Ballirano, P., Gianfagna, A., Mazziotti-Tagliani, S. and Pacella, A. (2009) Structural and spectroscopic characterization of a suite of fibrous amphiboles with high environmental and health relevance from Biancavilla (Sicily, Italy). *American Mineralogist*, **94**, 1333–1340.

Ballirano, P., Andreozzi, G.B. and Belardi, G. (2008) Crystal chemical and structural characterization of fibrous tremolite from Susa Valley, Italy, with comments on potential harmful effects on human health. *American Mineralogist*, **93**, 1349–1355.

Ballirano, P., Andreozzi, G.B., Dogan, M. and Dogan, A.U. (2009) Crystal structure and iron topochemistry of erionite-K from Rome, Oregon, USA. *American Mineralogist*, **94**, 1262–1270.

Bancroft, G.M., Maddock, A.G. and Burns, R.G. (1967) Applications of the Mössbauer effect to silicate mineralogy. Iron silicates of known crystal structure. *Geochimica et Cosmochimica Acta*, **31**, 2219–2246.

Benfatto, M. and Meneghini, C. (2015) A close look into the low energy region of the XAS spectra: the XANES region. Pp. 213–240 in *Synchrotron Radiation* (S. Mobilio, F. Boscherini and C. Meneghini, editors). Springer-Verlag, Berlin, Heidelberg.

Benton, D.M. (2013) Identification of asbestos using laser-induced breakdown spectroscopy: A viable alternative to the conventional approach? *ISRN Spectroscopy*, Volume 2013, article ID 362694, 6 pp.

Beran, A. and Libowitzky, E. (editors) (2004) *Spectroscopic Methods in Mineralogy*. EMU Notes in Mineralogy, **6**. European Mineralogical Union, Eötvös University Press, Budapest, 661 pp.

Berghmans, P.A. and Adams, F.C. (1992) Electron energy loss spectroscopy (EELS) and electron spectroscopic imaging (ESI) for the localization of titanium in chrysotile asbestos. *Surface and Interface Analysis*, **19**, 439–444.

Berghmans, P., Jacob, W. and Adams, F. (1992) Ultrastructural localization of titanium in chemically modified chrysotile asbestos. *Journal of Microscopy*, **166**, 307–316.

Berghmans, P., Injuk, J., Van Grieken, R. and Adams, F. (1994) Microanalysis of atmospheric particles and fibres by electron energy loss spectroscopy, electron spectroscopic imaging and scanning proton microscopy. *Analytica Chimica Acta*, **297**, 27–42.

Blaauw, C., Stroink, G., Leiper, W. and Zenteilli, M. (1979) Mössbauer analysis of some Canadian chrysotiles. *The Canadian Mineralogist*, **17**, 713–717.

Brown, L.M. (1997) A synchrotron in a microscope. Pp. 17–22 in: *Institute of Physics Conference Series*. Adam Hilger, Ltd., Bristol, UK.

Buffiere, J.Y. and Baruchel, J. (2015) Hard X-ray synchrotron imaging techniques and applications. Pp. 389–408 in: *Synchrotron Radiation* (S. Mobilio, F. Boscherini and C. Meneghini, editors). Springer-Verlag, Berlin, Heidelberg.

Burns, R. (1993) *Mineralogical Applications of Crystal Field Theory*. Cambridge Topics in Mineral Physics and Chemistry (No. 5), Cambridge University Press, London, 551 pp.

Burns, R.G. and Greaves, C. (1971) Correlations of infrared and Mössbauer site population measurements of actinolites. *American Mineralogist*, **56**, 2010–2033.

Calvert, C.C., Brydson, R., Banks, D.A. and Lloyd, G.E. (2001) Quantification of Fe-oxidation state in mixed valence minerals: a geochemical application of EELS. Pp. 251–254 in: *Electron Microscopy and Analysis* (M. Aindow and C.J. Kiely, editors). Institute of Physics Publishing, Bristol and Philadelphia, USA.

Clark, R.N. (1999) Spectroscopy of rocks and minerals and principles of spectroscopy. Pp. 3–58 in: *Manual of Remote Sensing* (A.N. Rencz, editor). Remote Sensing for the Earth Sciences, **3**. 3rd Edition Wiley & Sons, New York.

Coplen, T.B., Böhlke, J.K., De Bièvre, P., Ding, T., Holden, N.E., Hopple, J.A., Krouse, H.R., Lamberty, A., Peiser, H.S., Révész, K., Rieder, S.E., Rosman, K. JR., Roth, E., Taylor, P.D.P., Vocke, R.D. Jr. and Xiao, Y.K. (2002) Isotope-abundance variations of selected elements. *Pure and Applied Chemistry*, **74**, 1987–2017.

De Grave, E. and Van Alboom, A. (1991) Evaluation of ferrous and ferric Mössbauer fractions. *Physics and Chemistry of Minerals*, **18**, 337–342.

Della Ventura, G., Iezzi, G., Redhammer, G.J., Hawthorne, F.C., Scaillet, B. and Novembre, D. (2005a) Synthesis and crystal-chemistry of alkali amphiboles in the system Na_2O-MgO-FeO-Fe_2O_3-SiO_2-H_2O as a function of f_{O_2}. *American Mineralogist*, **90**, 1375–1383.

Della Ventura, G., Redhammer, G.J., Iezzi, G., Hawthorne, F.C., Papin, A. and Robert, J.L. (2005b) A Mössbauer and FTIR study of synthetic amphiboles along the magnesioriebeckite – ferri-clinoholmquistite join. *Physics and Chemistry of Minerals*, **32**, 103–113.

Della Ventura, G., Redhammer, G., Robert, J.L., Sergent, J., Iezzi, G. and Cavallo, A. (2016) Crystal-chemistry of synthetic amphiboles along the join richterite–ferro-richterite: a combined spectroscopic (FTIR, Mössbauer), XRPD and microchemical study. *The Canadian Mineralogist*, **54**, 97–114.

Dyar, M.D., Mackwell, S.M., McGuire, A.V., Cross, L.R. and Robertson, J.D. (1993) Crystal chemistry of Fe^{3+} and H^+ in mantle kaersutite: Implications for mantle metasomatism. *American Mineralogist*, **78**, 968–979.

Egerton, R.F. (2008) Electron energy-loss spectroscopy in the TEM. *Reports on Progress in Physics*, **72**, 016502.

Ericsson, T. and Wäppling, R. (1976) Texture effects in 3/2-1/2 Mössbauer spectra. *Journal de Physique Colloques*, **37**, C6-719–C6-723.

Fach, E., Kristovich, R., Long, J.F., Waldman, W.J., Dutta, P.K. and Williams, M.V. (2003) The effect of iron on the biological activities of erionite and mordenite. *Environment International*, **29**, 451–458.

Fantauzzi, M., Pacella, A., Atzei, D., Gianfagna, A., Andreozzi, G.B. and Rossi, A. (2010) Combined use of X-ray photoelectron and Mössbauer spectroscopic techniques in the analytical characterization of iron oxidation state in amphibole asbestos. *Analytical and Bioanalytical Chemistry*, **396**, 2889–2898.

Fantauzzi, M., Pacella, A., Fournier, J., Gianfagna, A., Andreozzi, G.B. and Rossi, A. (2012) Surface chemistry and surface reactivity of fibrous amphiboles that are not regulated as asbestos. *Analytical and Bioanalytical Chemistry*, **404**, 821–833.

Fornasini, P. (2015) Introduction to X-ray absorption spectroscopy. Pp. 181–211 in: *Synchrotron Radiation* (S. Mobilio, F. Boscherini and C. Meneghini, editors). Springer-Verlag, Berlin, Heidelberg.

Fricke, H. (1920) The K-characteristic absorption frequencies for the chemical elements magnesium to

chromium. *Physical Review*, **16**, 202–215.
Fubini, B. and Otero-Areán, C. (1999) Chemical aspects of the toxicity of inhaled mineral dusts. *Chemical Society Review*, **28**, 373–381.
Gazzano, E., Turci, F., Foresti, E., Putzu, M.G., Aldieri, E., Silvagno, F., Lesci, I.G., Tomatis, M., Riganti, C., Romano, C., Fubini, B., Roveri, N. and Ghigo, D. (2007) Iron-loaded synthetic chrysotile: A new model solid for studying the role of iron in asbestos toxicity. *Chemical Research in Toxicology*, **20**, 380–387.
Giacobbe, C., Gualtieri, A.F., Quartieri, S., Rinaudo, C., Allegrina, M. and Andreozzi, G.B. (2010) Spectroscopic study of the product of thermal transformation of asbestos containing materials (ACM). *European Journal of Mineralogy*, **22**, 535–546.
Gianfagna, A., Andreozzi, G.B., Ballirano, P., Mazziotti-Tagliani, S. and Bruni, B.M. (2007) Structural and chemical contrasts between prismatic and fibrous fluoro-edenite from Biancavilla, Sicily, Italy. *The Canadian Mineralogist*, **45**, 249–262.
Goldman, D.S. (1979) A reevaluation of the Mössbauer spectroscopy of calcic amphiboles. *American Mineralogist*, **64**, 109–118.
Gopal, N.O, Narasimhulu, K.V. and Rao, J.L (2004) EPR, optical, infrared and Raman spectral studies of actinolite mineral. *Spectrochimica Acta Part A: Molecular and Biomolecular Spectroscopy*, **60**, 2441–2448.
Gualtieri, A.F., Bursi Gandolfi, N., Pollastri, S., Pollok, K. and Langenhorst, F. (2016) Where is iron in erionite? A multidisciplinary study on fibrous erionite-Na from Jersey (Nevada, USA). *Scientific Reports*, **6**, 37981.
Gunter, M.E., Dyar, M.D., Twamley, B., Foit, F.F. and Cornelius, C. (2003) Composition, $Fe^{3+}/\Sigma Fe$, and crystal structure of non-asbestiform and asbestiform amphiboles from Libby, Montana, U.S.A. *American Mineralogist*, **88**, 1970–1978.
Gunter, M.E., Dyar, M.D., Lanzirotti, A., Tucker, J.M. and Speicher, E.A. (2011) Differences in Fe-redox for asbestiform and nonasbestiform amphiboles from the former vermiculite mine, near Libby, Montana, USA. *American Mineralogist*, **96**, 1414–1417.
Habaue, S., Hirasa, T., Akagi, Y., Yamashita, K. and Kajiwara, M. (2006) Synthesis and property of silicone polymer from chrysotile asbestos by acid-leaching and silylation. *Journal of Inorganic and Organometallic Polymers and Materials*, **16**, 155–160.
Hardy, J.A. and Aust, E.A. (1995) Iron in asbestos chemistry and carcinogenicity. *Chemical Reviews*, **95**, 97–118.
Hawthorne, F.C. (1983) The crystal chemisty of the amphiboles. *The Canadian Mineralogist*, **21**, 173–480.
Hawthorne, F.C. (editor) (1988) *Spectroscopic Methods in Mineralogy and Geology*. Reviews in Mineralogy, **18**. Mineralogical Society of America, Chantilly, Virginia, USA, 512 pp.
Hawthorne, F.C., Oberti, R., Della Ventura, G. and Mottana, A. (editors) (2007) *Amphiboles: Crystal Chemistry, Occurrence, and Health Issues*. Reviews in Mineralogy and Geochemistry **67**, Mineralogical Society of America and Geochemical Society, Chantilly, Virginia, USA, 545 pp.
Hawthorne, F.C., Oberti, R., Harlow, G.E., Maresch, W.V., Martin, R.F., Schumacher, J.C. and Welch, M. (2012) Nomenclature of the amphibole supergroup. *American Mineralogist*, **97**, 2031–2048.
Henderson, G.S., Neuville, D.R. and Downs, R.T. (editors) (2014) *Spectroscopic Methods in Mineralogy and Materials Sciences*. Reviews in Mineralogy and Geochemistry, **78**. Mineralogical Society of America and Geochemical Society, Chantilly, Virginia, USA, 800 pp.
Hertz, G. (1920) Über die Absorptionsgrenzen in der L-Serie. *Zeitschrift für Physik A Hadrons and Nuclei*, **3**, 19–25.
Iezzi, G., Della Ventura, G., Pedrazzi, G., Robert, J.L. and Oberti, R. (2003a) Synthesis and characterisation of ferri-clinoferroholmquistite, $\square Li_2(Fe^{2+}_3Fe^{3+}_2)Si_8O_{22}(OH)_2$. *European Journal of Mineralogy*, **15**, 321–327.
Iezzi, G., Della Ventura, G., Cámara, F., Pedrazzi, G. and Robert, J.L. (2003b) The ^{B}Na- ^{B}Li exchange in A-site vacant amphiboles: synthesis and cation ordering along the ferri-clinoferroholmquistite – riebeckite join. *American Mineralogist*, **88**, 955–961.
Iezzi, G., Cámara, F., Della Ventura, G., Oberti, R., Pedrazzi, G. and Robert, J.L. (2004) Synthesis, crystal structure and crystal-chemistry of ferri-clinoholmquistite, $\square Li_2Mg_3Fe^{3+}_2Si_8O_{22}(OH)_2$. *Physics and Chemistry of Minerals*, **31**, 375–385.
Iezzi, G., Della Ventura, G., Hawthorne, F.C., Pedrazzi, G., Robert, J.L. and Novembre, D. (2005) The (Mg,Fe^{2+}) substitution in ferri-clinoholmquistite, $\square Li_2\ (Mg,Fe^{2+})_3\ Fe^{3+}_2\ Si_8\ O_{22}\ (OH)_2$. *European Journal*

of Mineralogy, **17**, 733–740.

Ishida T., Alexandrov M., Nishimura T., Minakawa K., Hirota R., Sekiguchi K., Kohyama N. and Kuroda, A. (2012) Evaluation of sensitivity of fluorescence-based asbestos detection by correlative microscopy. *Journal of Fluorescence*, **22**, 357–363.

Jeanguillaume, C., Berry, J.P., Colliex, C., Galle, P., Tence, M. and Trebbia, P. (1984) Recent results in EELS elemental mapping of thin biological sections. *Le Journal de Physique Colloques*, **45**, 577–580.

Kell, D.B. (2009) Iron behaving badly: inappropriate iron chelation as a major contributor to the aetiology of vascular and other progressive inflammatory and degenerative diseases. *BMC Medical Genomics*, **2**, 2.

Kell, D.B. (2010) Towards a unifying, systems biology understanding of large-scale cellular death and destruction caused by poorly liganded iron: Parkinson's, Huntington's, Alzheimer's, prions, bactericides, chemical toxicology and others as examples. *Archives of Toxicology*, **84**, 825–889.

Lagarec, K. and Rancourt, D.G. (1997) Extended Voigt-based analytical lineshape method for determining N-dimensional correlated hyperfine parameter distributions in Mössbauer spectroscopy. *Nuclear Instruments and Methods in Physic Research*, **B129**, 266–280.

Leapman, R.D., Fiori, C.E. and Gorlen, K.E. (1987) Elemental imaging by EELS and EDXS in the analytical electron microscope. *Biological Trace Element Research*, **13**, 89–102.

Long, G.J., Cranshaw, T.E. and Longworth, G. (1983) The ideal Mössbauer effect absorber thickness. *Mössbauer Effect Reference Data Journal*, **6**, 42–49.

Luys, M.J., De Roy, G.L., Adams, F. and Vansant, E.F. (1983) Cation site population in amphibole asbestos: A Mössbauer study. *Journal of the Chemical Society, Faraday Transactions*, **1**, 1451–1459.

Lytle, F.W. (1999) The EXAFS family tree: a personal history of the development of extended X-ray absorption fine structure. *Journal of Synchrotron Radiation*, **6**, 123–134.

McCammon, C. (2004) Mössbauer spectroscopy: Applications. Pp. 369–398 in: *Spectroscopic Methods in Mineralogy* (A. Beran and E. Libowitzky, editors). EMU Notes in Mineralogy, **6**. European Mineralogical Union, Eötvös University Press, Budapest.

Moretti, G., Fierro, G., Ferraris, G., Andreozzi, G.B. and Naticchioni, V. (2014) N_2O decomposition over [Fe]-MFI catalysts: influence of the Fe_xO_y nuclearity and the presence of framework aluminum on the catalytic activity. *Journal of Catalysis*, **318**, 1–13.

Mottana, A. (2004) X-ray absorption spectroscopy in mineralogy: Theory and experiment in the XANES region. Pp. 465–552 in: *Spectroscopic Methods in Mineralogy* (A. Beran and E. Libowitzky, editors). EMU Notes in Mineralogy, **6**. European Mineralogical Union, Eötvös University Press, Budapest.

Newville, M. (2004) *Fundamentals of XAFS. Consortium for Advanced Radiation Sources*. University of Chicago, Chicago, Illinois, USA.

Ozeki, S., Masuda, Y., Sano, H., Seki, H. and Ooi, K. (1991) ^1H NMR spectroscopy of water adsorbed on synthetic chrysotile asbestos: Microtubes with acidic and basic surfaces. *Journal of Physical Chemistry*, **95**, 6309–6316.

Pacella, A., Andreozzi, G.B., Ballirano, P. and Gianfagna, A. (2008) Crystal chemical and structural characterization of fibrous tremolite from Ala di Stura (Lanzo Valley, Italy). *Periodico di Mineralogia*, **77**, 51–62.

Pacella, A., Andreozzi, G.B. and Fournier, J. (2010) Detailed crystal chemistry and iron topochemistry of asbestos occurring in its natural setting: a first step to understanding its chemical reactivity. *Chemical Geology*, **277**, 197–206.

Pacella, A., Andreozzi, G.B., Fournier, J., Stievano, L., Giantomassi, F., Lucarini, G., Rippo, M.R. and Pugnaloni, A. (2012) Iron topochemistry and surface reactivity of amphibole asbestos: relations with in vitro toxicity. *Analytical and Bioanalytical Chemistry*, **402**, 871–881.

Pascolo, L., Gianoncelli, A., Schneider, G., Salomé, M., Schneider, M., Calligaro, C., Kiskinova, M., Melato, M. and Rizzardi, C. (2013) The interaction of asbestos and iron in lung tissue revealed by synchrotron-based scanning X-ray microscopy. *Scientific Reports*, **3**, 1123.

Pollastri, S., D'Acapito, F., Trapananti, A., Colantoni, I., Andreozzi, G.B. and Gualtieri, A.F. (2015) The chemical environment of iron in mineral fibres. A combined X-ray absorption and Mössbauer spectroscopic study. *Journal of Hazardous Materials*, **298**, 282–293.

Pollastri, S., Gualtieri, A.F., Vigliaturo, R., Ignatyev, K., Strafella, E., Pugnaloni, A. and Croce, A. (2016)

Stability of mineral fibres in contact with human cell cultures. An *in situ* μXANES, μXRD and XRF iron mapping study. *Chemosphere*, **164**, 547–557.

Rancourt, D.G. (1994) Mössbauer spectroscopy of minerals. Inadequacy of Lorentzian-line doublets in fitting spectra arising from quadrupole splitting distributions. *Physics and Chemistry of Minerals*, **21**, 244–249.

Rancourt, D.G. and Ping, J.Y. (1991) Voigt-based methods for arbitrary-shape static hyperfine parameter distributions in Mössbauer spectroscopy. *Nuclear Instruments and Methods in Physics Research*, **B58**, 85–97.

Roessner, F., Steinberg, K.-H., Freude, D., Hunger, M. and Pfeifer, H. (1989) NMR and IR studies of zeolites of the erionite type. *Studies in Surface Science and Catalysis*, **46**, 421–427.

Ruda, T.A. and Dutta, P.K. (2005) Fenton chemistry of Fe^{III}-exchanged zeolitic minerals treated with antioxidants. *Environmental Science & Technology*, **39**, 6147–6152.

Sayers, D.E., Lytle, F.W. and Stern, E.A. (1969) Point scattering theory of X-ray K-absorption fine structure. *BSRL Document D1-82- 0880*. Boeing, USA.

Sayers, D.E., Stern, E.A. and Lytle, F.W. (1971) New determination of amorphous germanium structure using X-ray absorption spectroscopy. *Bulletin of the American Physical Society*, **16**, 302.

Schmidbauer, E., Kunzmann, T.H., Fehr, T.H. and Hochleitner, R. (2000) Electrical resistivity and ^{57}Fe Mössbauer spectra of Fe-bearing calcic amphiboles. *Physics and Chemistry of Minerals*, **27**, 347–356.

Shukla, A., Gulumian, M., Hei, T.K., Kamp, D., Rahman, Q. and Mossman, B.T. (2003) Multiple roles of oxidants in the pathogenesis of asbestos-induced diseases. *Free Radical Biology and Medicine*, **34**, 1117–1129.

Skogby, H. and Annersten, H. (1985) Temperature dependent Mg-Fe cation distribution in actinolite-tremolite. *Neues Jahrbuch für Mineralogie, Monatshefte*, 193–203.

Stroink, G., Blaauw, C., White, C.G. and Leiper, W. (1980) Mössbauer characteristics of UICC standard reference asbestos samples. *The Canadian Mineralogist*, **18**, 285–290.

Taftø, J. and Zhu, J. (1982) Electron energy loss near edge structure (ELNES), a potential technique in the studies of local atomic arrangements. *Ultramicroscopy*, **9**, 349–354.

Turci, F., Tomatis, M., Lesci, I.G., Roveri, N. and Fubini, B. (2011) The iron-related molecular toxicity mechanism of synthetic asbestos nanofibres: a model study for high-aspect-ratio nanoparticles. *Chemistry*, **17**, 350–358.

Turci, F., Tomatis, M. and Pacella, A. (2017) Surface and bulk properties of mineral fibres relevant to toxicity. Pp. 171–214 in: *Mineral Fibres: Crystal Chemistry, Chemical-physical Properties, Biological Interaction and Toxicity* (A.F. Gualtieri, editor). EMU Notes in Mineralogy, **18**. European Mineralogical Union and Mineralogical Society of Great Britain & Ireland, London.

Vlaic, G. and Olivi, L. (2004) EXAFS spectroscopy: a brief introduction. *Croatica Chemica Acta*, **77**, 427–433.

Wagner, J.C., Chamberlain, M., Brown, R.C., Berry, G., Pooley, F.D., Davies, R. and Griffiths D.M. (1982) Biological effects of tremolite. *British Journal of Cancer*, **45**, 352.

Weill, H., Abraham, J.L., Balmes, JR., Case, B., Chrug, A., Hughes, J., Schenker, M. and Sebastien, P. (1990) Health effects of tremolite. *American Review of Respiratory Disease*, **142**, 453–1458.

Wilke, M., Partzsch, G.M., Bernhardt, R. and Lattard, D. (2005) Determination of the iron oxidation state in basaltic glasses using XANES at the K-edge. *Chemical Geology*, **220**, 143–161.

Wirth, R. (1997) Water in minerals detectable by electron energy-loss spectroscopy EELS. *Physics and Chemistry of Minerals*, **24**, 561–568.

Xhoffer, C., Berghmans, P., Muir, I., Jacob, W., Grieken, R. and Adams, F. (1991) A method for the characterization of surface-modified asbestos fibres by electron energy-loss spectroscopy and electron spectroscopic imaging. *Journal of Microscopy*, **162**, 179–184.

Xu, Hui-fang (2000) Using Electron Energy-Loss Spectroscopy (EELS) associated with Transmission Electron Microscopy to determine oxidation states of Ce and Fe in minerals. *Geological Journal of China Universities*, **6**, 137–144.

Zecchina, A., Rivallan, M., Berlier, G., Lamberti, C. and Ricchiardi, G. (2007) Structure and nuclearity of active sites in Fe-zeolites: Comparison with iron sites in enzymes and homogeneous catalysts. *Physical Chemistry Chemical Physics*, **9**, 3483–3499.

EMU Notes in Mineralogy, Vol. 18 (2017), Chapter 5, 135–169

The analysis of asbestos minerals using vibrational spectroscopies (FTIR, Raman): Crystal-chemistry, identification and environmental applications

GIANCARLO DELLA VENTURA

Dipartimento di Scienze, Università di Roma Tre, Largo S. Leonardo Murialdo 1, I-00146 Roma, e-mail: giancarlo.dellaventura@uniroma3.it

This chapter deals with the use of vibrational spectroscopies, *i.e.* FTIR and Raman, for the study of asbestos minerals. Both techniques have been applied extensively for studies on amphiboles and layer silicates. According to factor group analysis, FTIR and Raman spectra show the same number of bands in the OH-stretching region (4000–3000 cm^{-1}) whereas in the lower frequency range (<1200 cm^{-1}) the spectra are much more complex and significantly different. In this region, vibrations due to the tetrahedral double-chain, metal-oxygen modes and OH librations overlap, thus making the assignment of the observed bands difficult. FTIR spectroscopy has been used extensively in recent decades for crystal-chemical studies and as a tool for characterizing short-range and long-range arrangements in amphiboles and layer silicates. A large corpus of data, summarized and discussed in the text, shows that Raman spectroscopy in the low-frequency region is valuable as an identification tool for the regulated asbestos silicates; for this reason, Raman microscopy, also thanks to its intrinsic higher spatial resolution compared to FTIR, has been used extensively to study fibrous minerals in environmental and biomedical applications. Selected examples of these applications are described.

1. Introduction

The term 'asbestos' is used to define several naturally occurring silicates with fibrous crystal habit. As discussed in the first chapters of this book, six fibrous minerals are regulated as being dangerous to human health: five of them belong to the amphibole group while the sixth is the serpentine-group mineral chrysotile. While chrysotile shows relatively limited chemical variation, amphiboles typically display extremely wide ranges of composition, even in samples from the same outcrop or, commonly, within the same crystal (see review by Gunter *et al.*, 2007). Such chemical variations result in significant problems for their identification and classification, and imply difficulties for health and legal issues arising from exposure to the amphibole fibres (*e.g.* Gunter *et al.*, 2007; Mazziotti-Tagliani *et al.*, 2009).

A proper identification of asbestos fibres needs the combination of several analytical techniques; the most notable are optical microscopy, electron microscopy (both SEM and TEM-STEM), X-ray diffraction and vibrational spectroscopies (FTIR, Raman), but more sophisticated methods can be also applied (*e.g.* Benton, 2013). All these

DOI: 10.1180/EMU-notes.18.5

techniques have advantages and disadvantages, as discussed in various chapters of this book. However, none of these techniques alone is able to provide a complete answer to the needs of modern research on these materials.

This chapter deals with the use of vibrational spectroscopies, *i.e.* FTIR and Raman, for the study of asbestos minerals. Both techniques have been applied extensively to amphiboles and layer silicates, although it will be clear, reading the text, that, while FTIR has been used extensively for crystal-chemical research, in most biological/medical applications, where the identification of the fibre was the main goal, Raman spectroscopy has been preferred.

2. Vibrational spectroscopies: FTIR *vs.* Raman

FTIR and Raman spectroscopies are grouped conventionally into 'vibrational spectroscopies', although Raman should be considered more correctly as a 'scattering spectroscopy'. However, as will be shown later, both methods probe the vibrational states of a material and provide similar and complementary information. Among the different analytical methods used for asbestos studies, vibrational spectroscopies are probably the most powerful tools; they allow rapid identification (*e.g.* Bard *et al.*, 1997; Rinaudo *et al.*, 2004) as well as determination of crystal-chemical features, *e.g.* LRO (long-range order) and SRO (short-range order) (*e.g.* Hawthorne and Della Ventura, 2007; Leissner *et al.*, 2015) of the sample. Theoretical and technical details of these methods can be found in many books (*e.g.* Hawthorne, 1988; Smith, 1989; Ferraro *et al.*, 2003; Stuart, 2004; Beran and Libowitzky, 2004; Henderson *et al.*, 2014) and will not be covered here, except for a brief commentary on the concepts involved.

When light interacts with matter it can be absorbed, reflected or diffused. Infrared (IR) spectroscopy is based on the first process; the spectrum is obtained by passing IR radiation through a sample and determining what fraction of the incident light is absorbed at a particular energy. The energy at which any peak in the absorption spectrum appears, corresponds to the frequency of vibration of a bond in the sample. Raman spectroscopy is based on the scattering process; in this case, monochromatic light in the visible or, more rarely, IR or UV electromagnetic range, provided by a laser source, interacts with the sample and the scattered light is detected. Most of the incident light is scattered elastically (without any energy transfer), and this is referred to as Rayleigh scattering. The inelastically (with energy transfer) scattered light produces the Raman effect; only 1 in 10^6 photons is diffused inelastically, thus the Raman effect is extremely weak. Figure 1a shows schematically the different processes. Molecular vibrations have frequencies from $<10^{12}$ to $\sim10^{14}$ Hz; these frequencies correspond to the IR region of the electromagnetic spectrum. At room temperature, most of the molecules in a substance occupy the ground vibrational state; IR absorption occurs when a molecule is excited to a higher vibrational state by direct absorption of the incoming radiation (Fig. 1a, left). The resonant frequencies (ν) are related to the strength of the bond and the mass of the atoms involved, according to: $\nu = 1/2\pi*(k/\mu)^{1/2}$, where k is the force constant of the bond, and μ is the reduced mass of the atoms in the

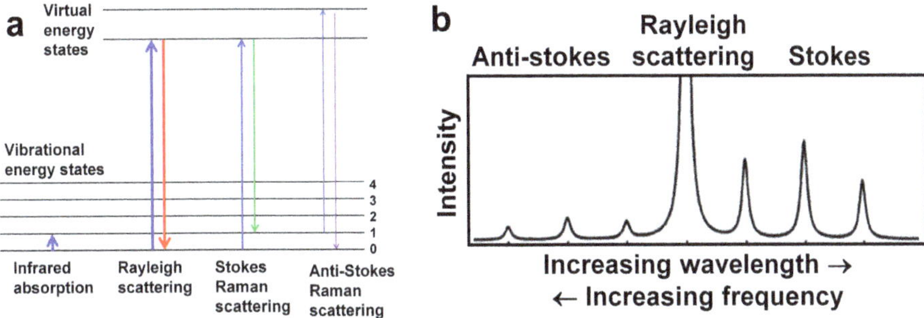

Figure 1. Schematic (a) energy-levels diagram showing the processes involved in the IR *vs.* Raman spectroscopy, and (b) Raman spectrum.

bond ($\mu = m_1 \cdot m_2/m_1+m_2$). This relationship is extremely useful for the interpretation of the band shifts observed in isotopic exchange experiments. Consider for example the O–H molecule and let ν_{OH} be the stretching wavenumber in a particular structural arrangement. If we replace H with D, the ν_{OD} wavenumber of the new O–D stretching can be calculated from the above equation to be $\nu_{OD} = \nu_{OH}/1.37$, where 1.37 is the ratio of the reduced masses of the two atoms couples: $(\mu_{OH}/\mu_{OD})^{1/2}$. This procedure has been used, for instance, by Della Ventura *et al.* (1991) to assign with confidence the tetrahedral Ti–O bands in the Raman spectra of $^{[4]}$Ti-substituted potassium-richterites; in this case, the shift expected considering the difference in reduced masses between the Si–O and Ti–O atom couples (0.92) was exactly that measured in the spectra. The higher-energy near-IR (NIR), at wavenumber ~14000–4000 cm^{-1} (wavelength 0.8–2.5 μm) can excite the overtone or harmonic vibrations. The mid-infrared (MIR), ~4000–400 cm^{-1} (2.5–25 μm) may be used to study the fundamental vibrations and associated rotational-vibrational structure. The far-infrared (FIR), ~400–10 cm^{-1} (25–1000 μm), has low energy and may sample rotational or optical modes.

In Raman spectroscopy, photons from the laser beam produce an oscillating polarization of the molecules, exciting them to a virtual energy state (Fig. 1a, right); if the oscillating polarization does not couple with the vibrational state of the molecule, the scattered photon will have the same energy as the incoming photon (Rayleigh scattering). If the polarization in the molecules couples to a vibrational state, then the incoming photon and the scattered photon will differ in energy by the amount required to excite the molecule. Such an inelastic scattering may occur in two ways: (1) the material absorbs energy and the emitted photon has a lower energy than the absorbed photon. This is called Stokes scattering; (2) the material loses energy and the emitted photon has a higher energy than the absorbed photon. This is called anti-Stokes scattering. In both effects, the energy difference between the absorbed and emitted photon corresponds to the energy difference between two vibrational states of the sample and is independent of the absolute energy of the beam. The Raman spectrum thus shows the intensity of the scattered (not absorbed!) light as a function of the difference in frequency with respect to the incident light; Stokes and anti-Stokes peaks occur symmetrically displaced on both sides of the Rayleigh peak (Fig. 1b). At room temperatures, the upper vibrational state will be less populated than the lower state, thus the anti-Stokes peaks are less intense than the Stokes peaks (Fig. 1b); for this reason the Stokes peaks are conventionally used in Raman work. Note that, unlike fluorescence, which occurs only for incident light able to raise the system to an excited state, the Raman effect may occur for any frequency of the incident light, because a virtual, non-quantized state, is involved.

The important point to note in Fig. 1 is that, although for different reasons (absorption *vs.* diffusion), both methods sample the same vibrational states of the material examined. Molecular bonds may be either IR or Raman active (*i.e.* may absorb/diffuse the incident light), and this depends on selection rules (*e.g.* Turrell, 1972) dictated by symmetry; accordingly, a molecule absorbs IR radiation if it possesses an electric dipole moment, and this must change during the vibration. This implies, for example, that homonuclear molecules are infrared-inactive because their dipole moment remains zero during the vibration. For a molecule to exhibit the Raman effect, a change in polarizability is required. A molecule can vibrate in different ways, called vibrational modes: symmetric and antisymmetric stretching, scissoring (bending), rocking, wagging and twisting. N atoms or molecules have $3N-6$ degrees of freedom (vibrational modes); this number increases to $3N-5$ for linear molecules. H_2O, for example, which is a non-linear molecule, will have $3 \times 3-6 = 3$ vibrational modes: symmetric stretching at 3657 cm^{-1}, antisymmetric stretching at 3755 cm^{-1} and symmetric bending at 1595 cm^{-1} (*e.g.* Nakamoto, 1986).

FTIR data can be collected from powders and from single crystals, and the analysis requires only a very small amount of material. For powders, as little as 0.5 mg is required; the mineral is typically embedded within a KBr (more rarely KCl, nujol or polyethylene) matrix, and the analysis takes only a few minutes. For single-crystal work, a FTIR microscope is attached to the optical bench and the data can be collected from areas as small as a few tenths of μm^2 when using a conventional IR source, or a few μm^2 when using a synchrotron source (Della Ventura *et al.*, 2014a). For asbestos materials, an interesting methods is diffuse reflectance IR spectroscopy (DRIFT); in this case, the powder is placed in a sample holder and the incoming light is scattered by the rough sample surface, collected by a mirror and sent to the detector (Fig. 2). This technique is interesting when studying cement-asbestos or similar materials, because the powder to be analysed could be simply scratched with an abrasive paper on a wall, a brick, a roof tile, *etc.* and placed on the DRIFT sample holder without any further preparation.

Raman data also can be collected from powders and single crystals; the greatest advantage of Raman spectroscopy compared to FTIR is that it allows data to be collected from a sample area as small as 1 μm^2 or less with modern instruments. Such spatial resolution cannot be achieved with FTIR except by using unconventional and not easily accessible techniques such as near-field infrared (*e.g.* Hermann *et al.*, 2013). Raman spectroscopy for hydrous minerals, such as asbestos, provides essentially the same kind of information as FTIR, *i.e.* both identification and crystal-chemical information (Leissner *et al.*, 2015), as will be shown below.

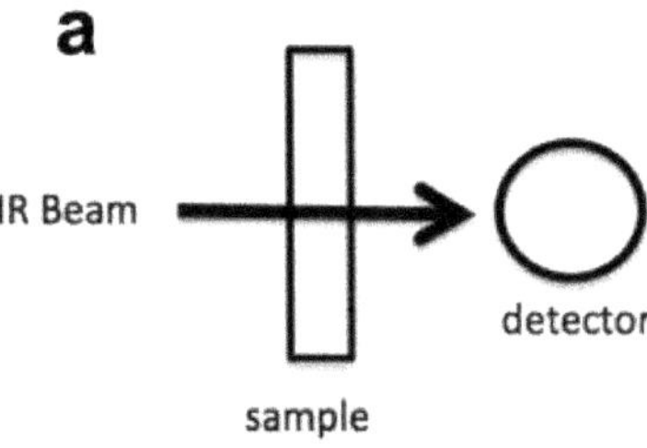

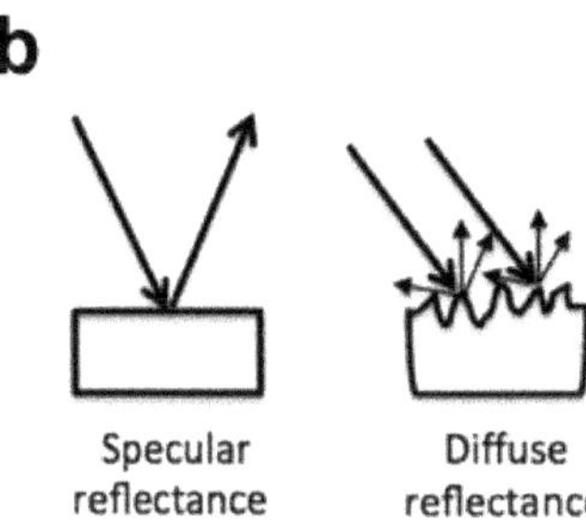

Figure 2. Schematic illustration of (a) transmission FTIR data acquisition, and (b) specular *vs.* diffuse reflectance.

The vibrational spectrum of hydrous silicates is divided conventionally into two ranges: the higher-wavenumber region (4000–3000 cm^{-1}) where bands due to water/OH stretching are observed, and the lower-wavenumber range (<1200 cm^{-1}) where bands due to longer bonds such as those of the tetrahedral skeleton, the cation-oxygen ($\geqslant$six-fold) polyhedra or the OH bending/libration modes are found. Additional spectral regions of interest are the NIR (near-infrared) at wavenumbers 10,000–4,000 cm^{-1}, where combination and overtone bands of H_2O/OH occur (*e.g.* Della Ventura *et al.*, 2009), and the FIR (far infrared) at wavenumbers <300 cm^{-1} down to the THz domain. The former spectral range, in particular, has been explored for hydrous minerals with a view to its possible use in remote sensing applications (*e.g.* Mustard 1992; Lewis *et al.*, 1996).

3. Vibrational spectroscopy applied to the study of asbestos silicates: Amphiboles

Amphiboles may crystallize under four different space groups; the *C2/m* is by far the most common, however (Hawthorne and Oberti, 2007); in this structure, the hydrogen atoms occupy the *4i* Wyckoff position (C_s site symmetry) and factor group analysis (Kroumova *et al.*, 2003) provides two internal vibrational modes for the OH^- ions: Γ_{int} $(OH^-) = A_g(R) + B_u(IR)$. The Raman active A_g species corresponds to the in-phase stretching motion of the O–H group, while the IR active B_u species represents the out-of-phase movement. Therefore, the same number of OH-stretching bands will be observed in the IR and Raman spectra of amphiboles (Wang *et al.*, 1988a,b; Leissner *et al.*, 2015). In Fig. 3a the FTIR and Raman patterns collected in the OH-stretching range of a synthetic richterite with composition $K(CaNa)Mg_5Si_8O_{22}OH_2$ are compared; the spectra are identical and consist of a main band with the same wavenumber, shape and

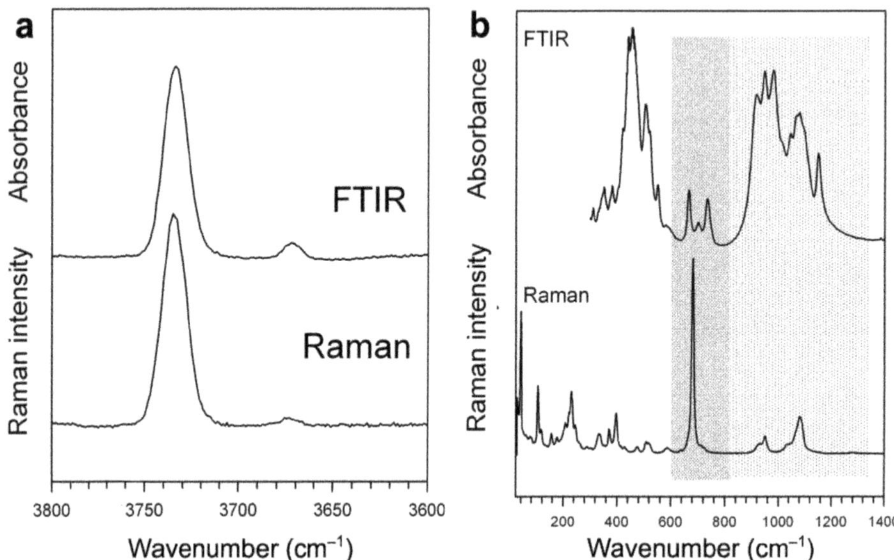

Figure 3. The Raman spectrum of potassium-richterite in (a) the OH-stretching region, and (b) the low-frequency region, compared with the FTIR spectra collected on the same sample. Data from Della Ventura (unpublished).

FWHM (full width at half maximum), plus a second, minor component at lower wavenumbers. Factor-group analysis of vibrations in the low-frequency (<1200 cm^{-1}) range is much more complex; according to Wang *et al.* (1988a), for a *C2/m* amphibole, up to 121 modes are calculated, *i.e.* $\Gamma_{vib} = 29\ A_g(R) + 30\ B_g(R) + 28\ A_u(IR) + 34\ B_u(IR)$. In this list, the purely *gerade*-species (59) are Raman active, while the purely *ungerade*-species (62) are IR active. The lower-frequency FTIR and Raman patterns for synthetic richterite are compared in Fig. 3b where it can be seen that, unlike the OH-stretching region (Fig. 3a) the spectra are now significantly different in terms of band multiplicity, width and intensity, and are both very complex. Therefore the observed bands are difficult to assign with certainty. As a rule of thumb, in the IR pattern the most intense bands are due to antisymmetric vibrations, while in the Raman pattern the most intense peaks (see the sharp peak at 681 cm^{-1} in Fig. 3b, for example) are due to the symmetric vibrations. However, vibrational modes are rarely purely symmetric or antisymmetric, and this increases the complexity of the spectra and their interpretation. Hopefully, for the goal of using spectroscopic data for mineral identification and crystal-chemical studies, we can ignore the theoretical analysis. We can simply take advantage of the large body of data collected on samples spanning the complete compositional space of amphiboles and layer silicates of environmental/biological interest. In the following, the relevant data are reviewed and it will be shown how we can use the spectroscopic techniques to augment the information obtainable with XRD or microanalytical/optical techniques.

3.1. The OH-stretching region: Basic concepts

In amphiboles, as well as in layer silicates, the OH group is bonded directly to three *M*(1,3) cations forming the base of a nearly equilateral triangle around the OH-group (Fig. 4a); the same local arrangement is also common to many other hydroxyl-bearing minerals, a notable case being provided by tourmaline (*e.g.* Robert *et al.*, 1996; Watenphul *et al.*, 2016). The O–H distance is extremely short (~0.96 Å, *e.g.* Hawthorne and Grundy, 1976), therefore the OH-stretching bands occur at very high wavenumbers (3740–3600 cm^{-1}), in a spectral region that is free from interferences from other absorptions, and this is a clear advantage when using O–H stretching spectra for analytical purposes.

The exact position (energy) of the principal OH-stretching band in the IR spectrum of an hydroxyl-bearing mineral is a function of the relative strength of the O–H bond (Huggins and Pimentel, 1956). As this strength depends on the degree of hydrogen bonding between the H atom and the surrounding O atoms, the relative strength of the O–H bond is sensitive to the local arrangement of atoms around the OH group, *i.e.* to the cation distribution in the immediate vicinity of the hydrogen. In amphiboles, these arrangements involve atoms at the NN (nearest-neighbour) *M*(1) and *M*(3) octahedra, but also at the NNN (next-nearest-neighbour) *M*(2), *M*(4), *T* and *A* sites, thus making IR and Raman spectroscopies very important tools for their crystal-chemical study. This concept is well represented in Fig. 5a where the IR spectra of two synthetic amphiboles are compared: tremolite [$Ca_2Mg_5Si_8O_{22}(OH)_2$] and richterite [$Na(NaCa)Mg_5Si_8O_{22}(OH)_2$]. These amphiboles have very close chemistries, the only difference being the presence of Na at the *A* site in richterite, balanced by an equal amount of ^{B}Na substituting for ^{B}Ca. Despite this, their spectra are extremely different: the spectrum of tremolite features a very sharp peak at 3674 cm^{-1} with a minor component at 3666 cm^{-1}, while the richterite spectrum shows a broader band at 3730 cm^{-1} with a minor component at 3674 cm^{-1}. The reasons for this difference will be discussed below to show how these spectra can be used to derive the amount of *A*-site occupancy in amphiboles, a feature that is crucial for the identification of the species. Another point

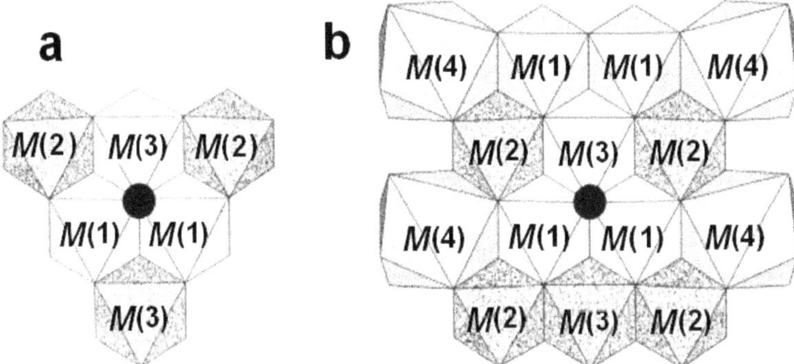

Figure 4. Nearest neighbour (a) and next-nearest neighbour arrangement of octahedra around the O(3) site of amphiboles. Reprinted with permission from Hawthorne and Della Ventura (2007).

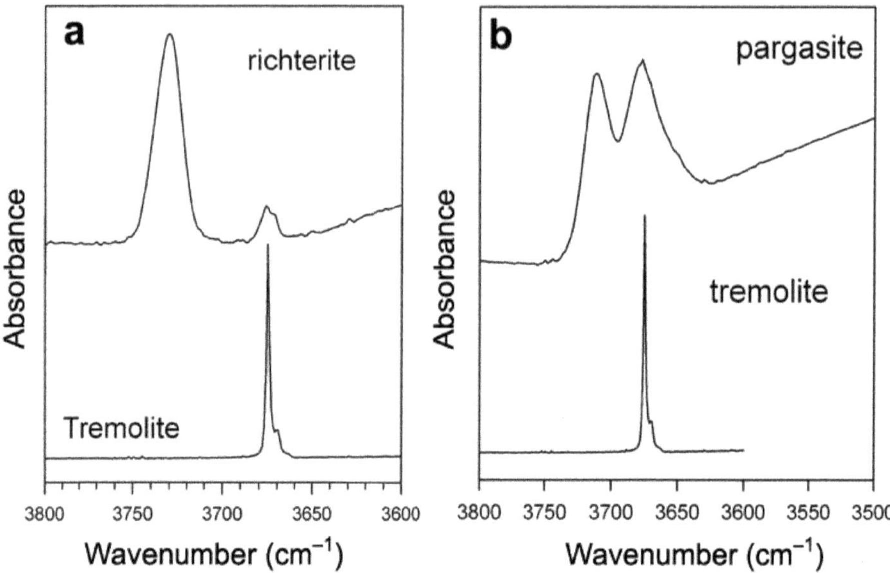

Figure 5. The FTIR OH-stretching spectrum of tremolite compared with that of (a) synthetic richterite, and (b) synthetic pargasite.

to note in Fig. 5a is the very different band widths: the FWHM is ~5 cm^{-1} for tremolite, while being ~25 cm^{-1} for richterite. This difference arises from the structural disorder around the OH group: end-member tremolite is a rare case of a perfectly ordered amphibole (Hawthorne 1997), whereas in richterite there is positional disorder at the *A* site (Hawthorne *et al.*, 1996a) coupled to a possibly disordered cation distribution at the *B* sites; this is sufficient to broaden significantly the OH-stretching band (Hawthorne *et al.*, 1997). These features are even more evident in Fig. 5b where the spectrum of synthetic tremolite is compared with the spectrum of synthetic pargasite [$NaCa_2(Mg_4Al)(Si_6Al_2)O_{22}(OH)_2$], in which additional disorder due to the distribution of Al at the *M* and *T* sites correlates with a band shift and with an evident and increasing band broadening.

This preliminary discussion suggests that a careful analysis of the OH-stretching spectrum can provide a full description of the sample studied; it is worth mentioning in this context the work of Petry *et al.* (2006) on the use of 244 and 257 nm argon-ion lasers to collect high-resolution OH-stretching Raman spectra. Their data show that UV Raman spectra measured on asbestos fibres are free from fluorescence and exhibit very well resolved and extremely sharp OH-stretching peaks allowing unambiguous identifications of the species.

Below, it will be described how we can disentangle the various effects of cationic/anionic substitutions at the different structural sites and use this information for the crystal-chemical characterization of the sample. In the following discussion the spectral features of tremolite (band position and width) will be used as a reference point to model band shifts as a function of cationic/anionic substitutions. Tremolite has a very simple chemical composition; the tetrahedral sites are all occupied by Si, the octahedral ribbon by Mg, the *M*(4) site by Ca and the *A* site is empty. In spectroscopic notation the local configuration in the immediate vicinity of the OH group can be expressed as: MgMgMg–OH–$^{A}\square$–SiSi (Della Ventura *et al.*, 1999). For layer silicates, the spectral features of talc are used as a reference point. The OH group in layer silicates has a local environment very similar to that in amphiboles, therefore the behaviour of the O–H stretching band with respect to varying chemistry can be modelled in the same way. Talc shares with tremolite the same NN environment around the hydroxyl ion and thus has an identical OH-stretching spectrum, consisting of a single, very sharp peak at 3677 cm^{-1} (Parry *et al.*, 2007).

3.2. OH as a chemical/structural probe in amphiboles

3.2.1. Substitutions at the NN M(1) and M(3) octahedra

Substitutions influencing directly the charge balance at the O(3) oxygen, and consequently the O–H bond strength/length, may involve only divalent cations (*e.g.* Mg, Fe^{2+}, Ni, Co, *etc.*) or atoms with different valence (*e.g.* Li, Mg, Al, Fe^{3+}, Ti). In layer silicates, one additional substitution that has significant effects on the O–H stretching frequency involves the presence of vacant *M* sites (for example in dioctahedral species, see Robert and Kodama, 1988). However, considering that this substitution occurs neither in amphiboles nor in chrysotile (a trioctahedral layer-silicate), it will be ignored in the following discussion.

Consider two cation species with the same valence (*e.g.* Mg and Fe^{2+}) at the *M*(1) and *M*(3) sites. Because O(3) bonds to these *M* sites, there are eight different possible short-range arrangements around an OH anion at O(3) (Hawthorne and Della Ventura, 2007). The spectrum that results from this situation (Fig. 6) shows only four absorption bands, as observed in early work on amphiboles (Strens, 1966; Burns and Strens, 1966) and later verified on synthetic analogues (Della Ventura *et al.*, 1996, 1997, 2005, 2016; Hawthorne *et al.*, 1996b; Iezzi *et al.*, 2005). This difference in number of bands arises from the pseudo-trigonal nature of the local arrangements at *M*(1,3) (see table 1 in Hawthorne and Della Ventura, 2007). The downward frequency shift of the different configurations is a function of the mean electronegativity of the cations at *M*(1,3); this feature was observed by Velde (1983) for a variety of layer silicates, and holds true also for amphiboles (Fig. 7). Similar patterns are also observed for layer silicates where (Mg, Fe^{2+}) distribute at the OH-coordinated octahedra; see, for example, Fig. 7 and the case of samples along the talc–minnesotaite series (Wang *et al.*, 2014). The important issue here is that, for simple binary cases, measurement of the relative intensities of the bands in the spectra allows derivation of the Mg/Fe^{2+} content at *M*(1,3) with a high level of accuracy (Hawthorne *et al.*, 1996b; Della Ventura *et al.*, 1996, 1997, 2016; Iezzi *et al.* 2005). Note that the same procedures are also valid for Raman spectroscopy in the OH-stretching region (Kloprogge *et al.*, 2001; Leissner *et al.*, 2015).

The presence of trivalent cations at the *M*(1,3) octahedra has a significant effect on the OH spectra of amphiboles (Hawthorne *et al.*, 2000; Della Ventura *et al.*, 2003,

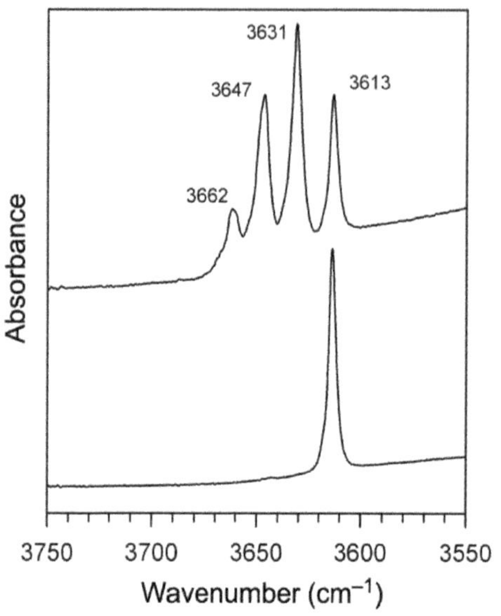

Figure 6. The spectrum of end-member ferri-clinoholmquistite compared with that of an intermediate (Mg,Fe) composition along the ferri-clinoholmquistite–ferri-clinoferro-holmquistite join.

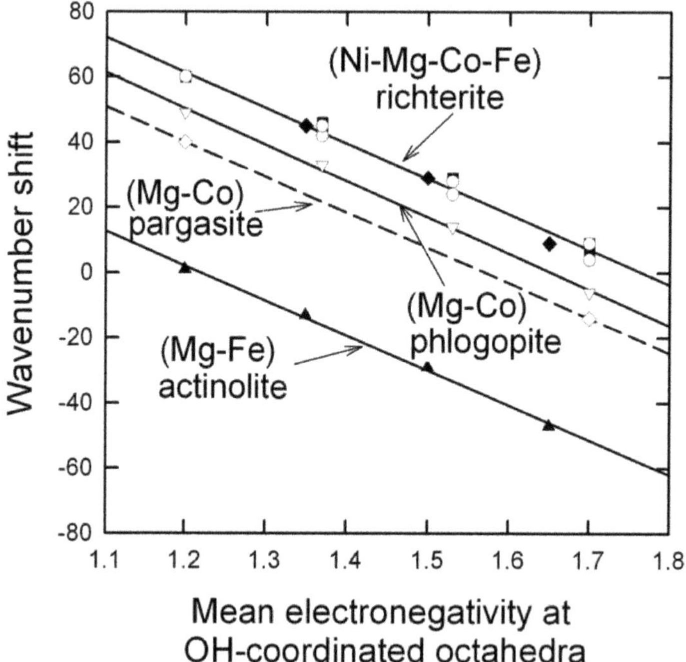

Figure. 7. The wavenumber shift (with respect to tremolite or talc) of the principal OH-stretching band of Mg-M^{2+} substituted amphiboles and layer silicates as a function of the mean electronegativity at the O–H coordinated octahedra. Data from Della Ventura *et al.* (1996, 1998), Burns and Strens (1966) and Robert (1981).

Hawthorne and Della Ventura, 2007) and micas (Robert and Kodama, 1988), and is correlated with the significant increase of incident bond-valence at O(3) when the formal charge at *M*(1,3) increases from +6 (like in the $^{M(1)}Mg^{M(1)}Mg^{M(3)}Mg$ local configurations) to +7 (*e.g.* $^{M(1)}Mg^{M(1)}Mg^{M(3)}Al$ local configurations). Semet (1973) first noted the presence of two broad bands at 3711 and 3678 cm^{-1} in synthetic pargasite and magnesiohastingsite, and suggested that these were due to Al-Mg and Fe^{3+}-Mg disorder over the *M*(1) and *M*(3) sites. Later, Raudsepp *et al.* (1987, 1991) presented similar spectra for pargasite and M^{3+}-substituted pargasite (M^{3+} = Cr, Sc, In). The frequency shift due to the replacement of Al for Mg at *M*(1,3), measured in pargasite, is ~30 cm^{-1} while for Fe^{3+} the frequency shift is ~50 cm^{-1} (Hawthorne and Della Ventura, 2007). Similar values are expected also for Mn^{3+}, although for manganese things are complicated by extreme polyhedral distortion due to the Jahn-Teller effect (Oberti *et al.*, 2016a). The presence of a monovalent cation at *M*(1,3), *e.g.* Li (Oberti *et al.*, 2007), is correlated to an increase (of ~30 cm^{-1}) in vibrational frequency. This effect was first recognized by Addison and White (1968) and is due to the decreasing electron valence (e.v.) on O(3) when the cation charge at *M*(1,3) is +5 (*i.e.* when the local configuration is MgMgLi). In such a case, the charge deficit is alleviated *via* a shortening of the O(3)–H bond distance, and by consequence, an increase in the O–H stretching wavenumber is observed.

3.2.2. Substitutions at the tetrahedral sites

In the amphibole structure, the O(7) atom acts as a hydrogen bond acceptor for the OH group at the adjacent O(3) site (Fig. 8). Where the O(7) atom occurs as Si–O(7)–Si linkage, the H...O(7) bond is weak; when O(7) is involved in Si–O(7)–Al linkages, the O–H bond is stronger (and consequently the O(3)–H bond is weaker), hence the principal stretching band shifts to lower wavenumbers. This being the case, the O–H stretching frequency may be used to derive the distribution of cations at the tetrahedral

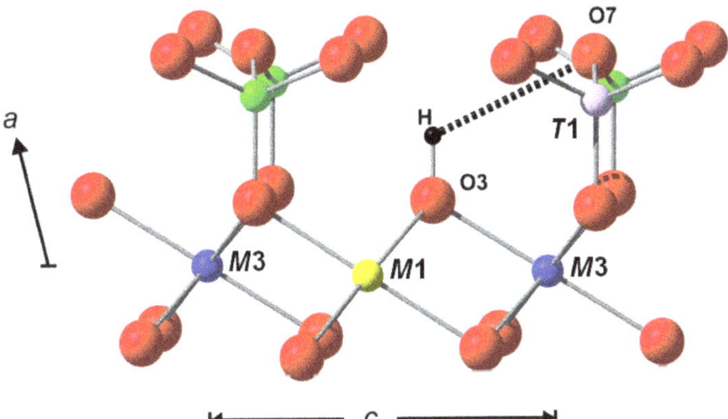

Figure 8. The local arrangement of the H atom in *C2/m* amphiboles viewed down [010]. Reprinted with permission from Della Ventura *et al.* (2014b).

double-chain of amphiboles (Della Ventura *et al.*, 2014b). Accordingly, in pargasite where all T–O(7)–T linkages are of the Si–O(7)–Al type, all OH groups are involved in significant O–H...O(7) hydrogen bonds. In richterite, conversely, all T–O(7)–T linkages are of the Si–O(7)–Si type, therefore the IR bands have lower wavenumbers in pargasite with respect to richterite (Della Ventura *et al.*, 1999, 2014b). In edenite, although the tetrahedral composition is (Si_7Al), thus intermediate between richterite and pargasite, all OH-bands are shifted to lower frequency, indicating complete Si–Al disorder at the tetrahedral double-chain (Della Ventura *et al.*, 2014b).

3.2.3. Substitutions at the NNN M(2) and M(4) octahedra

Cations occupying the *M*(2) and *M*(4) sites are not bonded directly to the hydroxyl oxygen (Fig. 4b); however, they have an indirect, although limited, effect on the OH-stretching frequency. The presence of ^{B}Mg in tremolite ('cummingtonite component') for example has a well known effect on the OH-spectrum with a negative shift of 8 cm^{-1} (Hawthorne *et al.*, 1997, Gottschalk *et al.*, 1999), while the presence of Li substituting for Na induces a negative shift of 4 cm^{-1} (Iezzi *et al.*, 2003a). Substitutions important for the study of asbestos amphiboles involve the presence of Fe^{3+} at *M*(2), common, for example, in Na-amphiboles of the riebeckite type ('crocidolites'). The presence of Fe^{3+} at *M*(2) is correlated to a significant negative shift (10 cm^{-1}) as determined in synthetic ferri-clinoferroholmquistite (Iezzi *et al.*, 2003b).

3.2.4. Substitutions at the O(3) anionic site

Solid-solutions at the anionic sites are very common in all hydroxyl-bearing minerals and typically involve the substitution of OH with F or Cl; a particular case involves the presence of O^{2-} at the anionic site, *i.e.* the removal of H compensated by the simultaneous substitution of a divalent cation at *M*(1,3) with a trivalent one. This latter case (oxi-substitution) may occur in minerals rich in multi-valent elements (*e.g.* Fe, Mn and Ti), and is well known in both amphiboles and micas (Scordari *et al.*, 2006; Della Ventura *et al.*, 2007; Oberti *et al.*, 2016b). Chemical variations at the anionic site have a significant effect on the vibrational frequency of the O–H groups. Robert *et al.* (1999, 2000) and Della Ventura *et al.* (2001) have shown that amphiboles with occupied *A* sites show 'two-mode behaviour' of the O–H stretching band, whereby the presence of F substituting for OH is correlated with the appearance of a second component in the spectrum; this component is due to the local OH–*A*–F configuration (Fig. 9a), and is shifted by 15–20 cm^{-1} towards lower wavenumbers (Fig. 9b). This behaviour is consistent with vibrational coupling between two facing O(3) anions across the Na or K that occupies the *A* cavity. Amphiboles with empty *A* sites (*e.g.* tremolite), conversely, show 'one-mode behaviour' of the O–H stretching band; in such a case the OH stretching band simply vanishes for decreasing OH. In the former case, provided there is OH–F disorder in the structure, the OH/F content of the sample can be quantified (Fig. 9c) on the basis of the band intensities (Robert *et al.*, 1999, 2000). The same concepts are valid for the oxi (OH–O^{2-}) substitutions (Della Ventura *et al.*, 2007, 2016).

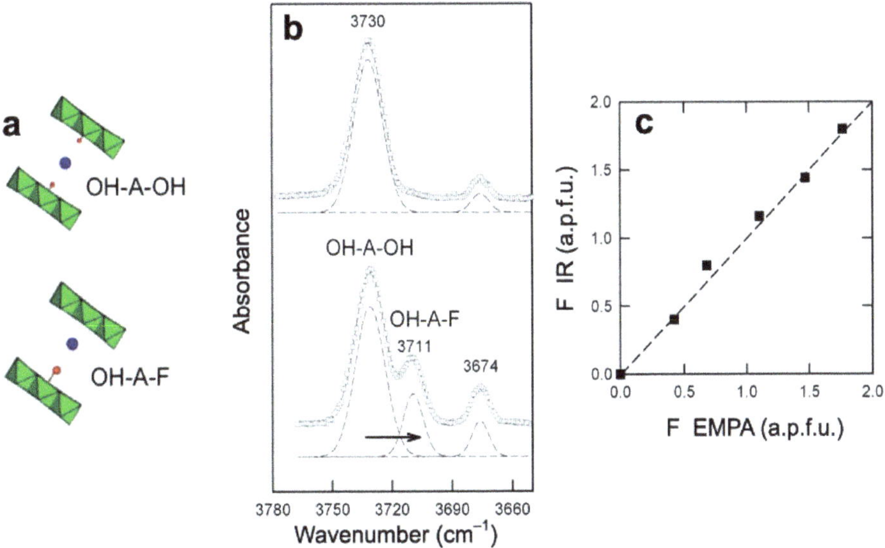

Figure 9. (a) Local environment in the amphibole structure involving facing O(3) sites. Red = H, blue = K, Na. (b) FTIR spectra of synthetic end-member OH richterite and an intermediate (OH-F) composition; the arrow indicates the wavenumber shift due to F. (c) The relationship between the F content (a.p.f.u.) in synthetic (OH,F) richterites analysed by EMP and that obtained by FTIR. Reprinted with permission from Robert *et al.* (1999).

3.2.5. Substitutions at the A site

The exact evaluation of the *A*-site occupancy of amphiboles is not easy because Na, which is the most common alkali cation, is also a common *B*-type cation, thus the sodium content analysed using microchemical techniques needs to be differentiated correctly between the available structural sites. This point is extremely important in asbestos studies because the *A*-site occupancy has significant implications for identification purposes (see for example Gunter *et al.*, 2007). Indeed, all regulated amphibole asbestos are *A*-site vacant species, therefore determining in a sample the *A*-site occupancy to be $A > 0.5$ a.p.f.u. implies that the sample is formally not an asbestos. The occupancy of the *A* site in amphiboles has a major effect on the OH-spectrum; this feature that was recognized initially by Rowbotham and Farmer (1974) is due to the strong H^+–A^+ repulsion that affects the force constant of the O–H bond. The shift is ~60 cm^{-1} for ^{A}K (Della Ventura 1992; Hawthorne *et al.*, 1997) while being slightly lower for ^{A}Na (~55 cm^{-1}: Robert *et al.*, 1989; Meltzer *et al.*, 2000) or ^{A}Li (~50 cm^{-1}: Iezzi *et al.*, 2004). Hawthorne *et al.* (1997) have shown that the molar fraction of the vacant ($^{\square}X$) *vs.* full *A* site can be derived from the relative intensity of the bands in the FTIR spectrum by using the equation $^{\square}X = R/[k+R(1-k)]$; R is the relative intensity ratio between the bands assigned to the *A* site filled and *A*-site vacant environments, and k is the correction factor for the difference in molar absorptivity (Skogby and Rossman, 1991). For the tremolite–richterite system k = 2.2 (Hawthorne *et al.*, 1997). A recent study (Della Ventura, unpublished) shows that measurements on the relative intensities of the OH-stretching bands in the Raman spectra provide directly the amount of *A*-site vacant *vs*. *A*-site filled environments, without any correction for the difference in molar absorptivity.

3.3. OH-spectra as a tool to identify and characterize the amphibole species

The notions presented above have been used during recent decades to characterize both synthetic and natural amphiboles; the most prominent applications in crystal-chemical studies have been summarized recently by Hawthorne and Della Ventura (2007) and Oberti *et al.* (2007). However, the method proved to be an extremely valuable tool also in studies concerning asbestos materials. Iezzi *et al.* (2007) studied a series of well characterized samples from the vermiculite mine of Libby (Montana, USA) and showed that FTIR powder spectroscopy, when combined with XRD and SEM methods, can provide detailed crystal-chemical features for extremely fibrous amphibole samples. The same kind of procedures and analytical protocols were later applied by Pacella *et al.* (2009) to characterize a series of asbestos amphiboles of interest for surface reactivity studies. Della Ventura *et al.* (2014c) examined bundles of very tiny, fibrous and wool-like crystal aggregates scattered within the pyroclastic deposits on the Alban Hills (Rome); based on powder FTIR spectra they were able to constrain the chemical composition of the fibres and identify them as tremolite. A recent study of synthetic Mg-Fe^{2+}-exchanged richterites (Della Ventura *et al.*, 2016) showed how, for iron-rich fibrous materials, the combined use of FTIR and Mössbauer spectroscopies is mandatory for characterizing the crystal-chemistry of the amphiboles. Interestingly, Della Ventura *et al.* (2016) observed a general increase in the crystal size and an evolution of the crystal morphologies, from extremely acicular at the Mg end-member composition to prismatic and stubby for the Fe^{2+}-richer compositions.

As an example of the working procedure that can be used to characterize asbestiform materials using powder FTIR, consider the spectrum given in Fig. 10a, collected from very acicular, fibrous amphibole crystals; the sample (MNHN31.73) is from the collection of the Muséum National d'Histoire Naturelle, Paris (courtesy of Dr P.J. Chiappero). Four peaks are resolved in the OH-stretching region, centred at 3667, 3652, 3637 and 3618 cm^{-1}, respectively. Note that: (1) the peaks are sharp (FWHM ~8 cm^{-1}), and (2) the splitting parameter between the four bands is constant and is Δ ~15 cm^{-1}. On the basis of the above discussion, we can make now a series of considerations:

(1) there are four bands in the spectrum, and thus two cations distributed in the vicinity of the OH group; the band splitting is consistent with both of these cations being divalent;
(2) the sample has a relatively ordered cation distribution (small FWHM), and totally empty *A* sites, because all bands have wavenumbers <3680 cm^{-1} (see Della Ventura *et al.*, 2003);
(3) the band frequencies are typical of amphiboles with no ^{T}Al (Della Ventura *et al.*, 2003).

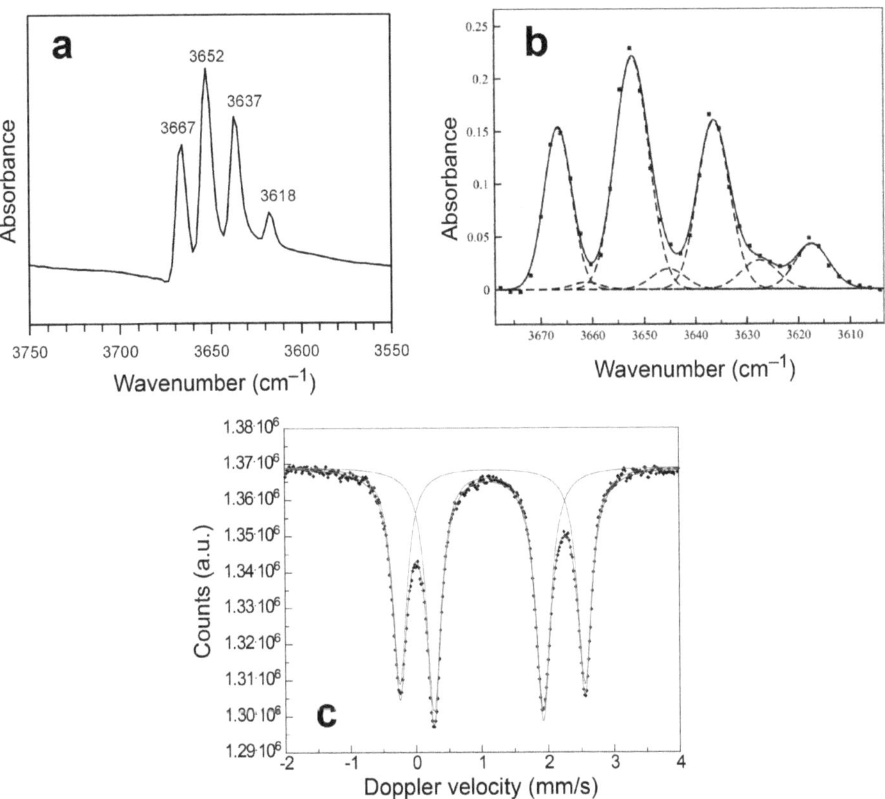

Figure 10. (a) FTIR OH-stretching powder spectrum of sample MNHN31.72; (b) the fitted spectrum; experimental data = filled squares, fitted component bands = dashed lines, sum of the fitted components = black line; and (c) the fitted Mössbauer spectrum of sample MNHN31.72.

We are thus left with three possibilities: actinolite, with composition $\square Ca_2(Mg,Fe^{2+})_5Si_8O_{22}(OH)_2$, cummingtonite, with composition $\square(Mg,Fe^{2+})_2(Mg,Fe^{2+})_5Si_8O_{22}(OH)_2$ or riebeckite, ideally $\square Na_2(Fe^{2+},Mg)_3Fe^{3+}_2Si_8O_{22}(OH)_2$. The higher-wavenumber band, assigned to the MgMgMg–OH–$\square$ configuration, is observed at 3667 cm^{-1}, thus the amphibole has a *B*-site composition = (Mg, Fe^{2+}), therefore we can exclude actinolite, the MgMgMg–OH–$\square$ configuration of which occurs at 3674 cm^{-1}. The choice between cummingtonite and riebeckite can be based on a quick EDS check for Na; sample MNHN31.73 does not contain Na, therefore we can definitively classify it as an intermediate (Mg, Fe^{2+}) cummingtonite. The quantitative determination of the Mg/Fe^{2+} ratio at *M*(1,3) can be obtained by using the equation of Burns and Strens (1966):

$$^{M(1,3)}Mg = 3I_A + 2I_B + I_C \text{ and } ^{(M1,3)}Fe^{2+} = I_B + 2I_C + 3I_D$$

where I_A to I_D are the measured intensities of the resolved quartet of bands (labelled from A to D, Fig. 10b). The calculation provides (a.p.f.u.): $^{M(1,3)}Mg$ = 1.86, $^{M(1,3)}Fe^2$

= 1.14. Obviously, for a complete characterization of the sample, additional data are needed, notably Mössbauer spectroscopy to check for any Fe^{3+}, and to quantify $^{M(4)}Fe^{2+}$. The Mössbauer spectrum of cummingtonite MNHN31.73 (courtesy of G. Redhammer, Salzburg) is given in Fig. 10c; it shows two doublets, one having the isomer shift = 1.09 mm/s and quadrupole splitting = 1.65, and a second having the isomer shift = 1.15 mm/s and quadrupole splitting = 2.85 mm/s. From the literature data (see Andreozzi and Pollastri, 2017, this volume), the former is assigned to $^{M(4)}Fe^{2+}$, while the latter is assigned to $^{M(1,2,3)}Fe^{2+}$. No peaks due to Fe^{3+} are present in the spectrum. From the relative peak areas, the ferric iron can be assigned 54% to *M*(4) and 46% to *M*(1,2,3). Note that a series of minor peaks is needed to model the FTIR spectrum of Fig. 10a; as will be shown below, these are due to the Mg/Fe^{2+} at *M*(4).

3.4. Quantitative measurements and worked examples

When the bands in the spectrum are well separated, as is the case for sample MNHN31.73 described above, the measure of the band intensity is relatively straightforward and can be done with the built-in tools in any spectrometer software. When the component bands in the spectrum overlap significantly, we need to treat the curve mathematically, and fit the proper number of bands to the spectrum. Technically, this task can be accomplished by many dedicated software packages, the most popular being *PeakFit*™ and *GRAMS*™; *Magic Plot* (http://magicplot.com) is a user-friendly app that can be downloaded from the web. The mathematical basis of the spectra treatments can be found in many papers, see for example Hawthorne and Waychunas (1988). The point to stress here is that, in general, there is no rule in this matter, a concept that is well represented by the statement "Curve fitting is an Art as well a Science". The main goal of any spectral resolution is not necessarily the minimization of the final agreement parameter, such as the R^2; it is rather the extraction of component bands hidden below broad and convolute signals, bearing in mind that the fitted components must be assigned finally to structurally realistic local configurations. The working philosophy, used by the writer during decades of spectroscopic work, has always involved an initial interactive optimization followed by least-squares refinement. As a rule of thumb, the minimum number of components needed for an accurate reconstruction of the experimental pattern has been introduced in the model; the number of peaks has, however, been based on what was expected by considering possible local configurations from the composition of the sample. This task has been eased by the study of several synthetic solid-solution series (see Hawthorne and Della Ventura, 2007 for a comprehensive list), where the presence of bands in the experimental curve could be constrained by precise knowledge of the chemical system, and supported by additional data from other techniques; a good example of this concept is provided by the complex case of amphiboles along the richterite–pargasite join described by Della Ventura *et al.* (1999). The starting point of any spectral resolution, however, is setting up the algorithms to model the background type and the peak parameters, *i.e.* the peak shape and FWHM (Hawthorne and Della Ventura, 2007).

The choice of the background type, *i.e.* a linear *vs.* a polynomial interpolation may affect the final intensity data; in general this choice depends on the pattern shape in the spectral region of interest (compare for example in Fig. 5 the spectra of tremolite and pargasite). The band shape is, however, the most critical point in curve-fitting procedures; a quick look in the literature shows that, in spectroscopic applications, Lorentzian and Gaussian are the most used functions. The large amount of work done on the OH-stretching region of hydrous minerals shows, however, that the Gaussian function is usually the best choice to model the experimental patterns. In principle, the OH-stretching bands of amphiboles are expected to be Lorentzian, however structural disorder produces Gaussian shapes and increasing band-widths (Strens, 1974); accordingly, the main band in the spectrum of end-member tremolite has an evident Lorentzian shape (see for example Gottschalk *et al.*, 1999), while even small amounts of ^{A}K in amphiboles induce significant band broadening (Hawthorne *et al.*, 1997) turning the shape into a Gaussian. In some cases, *e.g.* the spectra of amphiboles along the potassium-richterite–tremolite join, these parameters can be modelled easily (Hawthorne *et al.*, 1997); when there is significant peak overlap, some assumptions need to be made, and these assumptions can be based on known spectral parameters from similar systems. Consider the case of amphiboles along the richterite–ferro-richterite join (Della Ventura *et al.*, 2016). The spectrum of end-member richterite (Fig. 11a, top) shows an intense band at 3730 cm^{-1}, assigned to an OH group in the local configuration $^{M(1)}$Mg$^{M(1)}$Mg$^{M(3)}$Mg–OH–^{A}Na, and a minor component at 3674 cm^{-1} due to the presence of vacant *A* sites in the structure (band A*, Fig. 11a). This component indicates a slight departure of the synthesized amphibole from the nominal composition towards tremolite. As discussed above, for all intermediate (Mg, Fe)-amphiboles along the join, both bands split into four components, which are labelled A–D and A*–D*, respectively (Fig. 11). From the intensities of the A–D components we can derive the Mg/Fe^{2+} at *M*(1,3), while from the relative intensities of the A–D and A*–D* quadruplets we can calculate the number of empty *vs.* occupied *A* sites in the samples. In the present case, the spectra are very complex, thus the peak parameters of strongly overlapping peaks were refined where these peaks are most prominent, and then fixed during the refinement of the other samples. This is clearly the case of the band at 3670 cm^{-1} (band A* in Fig. 11a) that overlaps with the D band occurring at the same wavenumbers. At convergence, the peak positions were released and only the widths were constrained to be roughly constant in all spectra. A typical example of a fitted spectrum is displayed in Fig. 11b. Combination of the resulting data with Mössbauer spectroscopy allowed a detailed crystal-chemical study of these samples (Della Ventura *et al.*, 2016).

The Mg-Fe^{2+} exchange in ferri-clinoholmquistite has been studied by Iezzi *et al.* (2005) by combining XRD with FTIR and Mössbauer spectroscopies; Fig. 12 shows the fitted spectrum of an intermediate amphibole composition; a simple four-band fitting (Fig. 12b) gave significant residual intensity (Fig. 12c); a satisfactory reproduction of the experimental envelope could be achieved by introducing four additional peaks in the refinement (Fig. 12a), each associated with a main band. Peak-shape parameters for

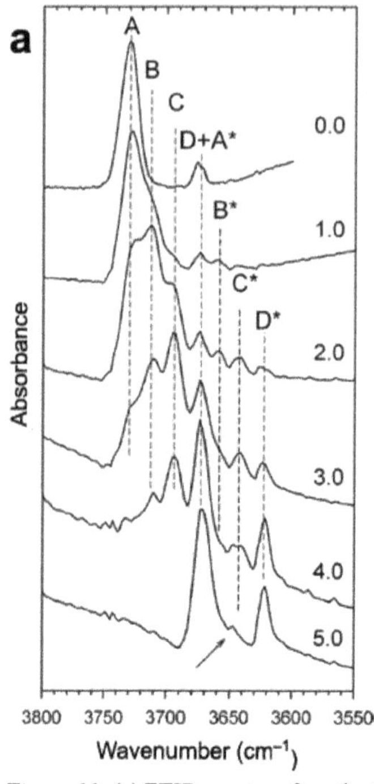

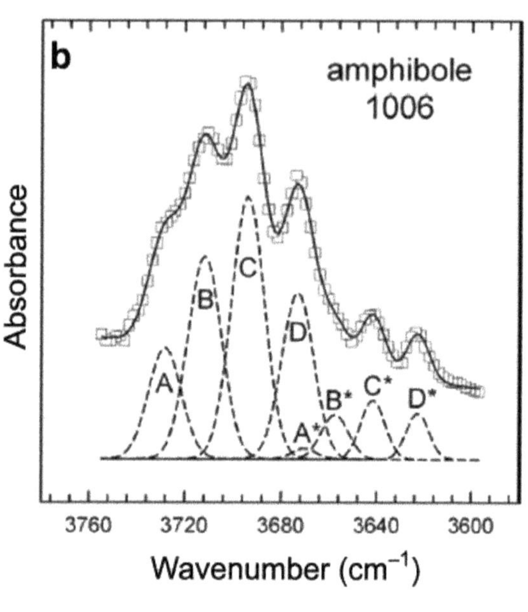

Figure 11. (a) FTIR spectra of synthetic amphiboles along the richterite–ferro-richterite join as a function of increasing (nominal) Fe^{2+} (a.p.f.u.) indicated on the right side. (b) Example of fitted spectrum: experimental data = empty boxes, fitted component bands = dashed lines, sum of the fitted components = black line. Reprinted with permission from Della Ventura *et al.* (2016).

these additional components were constrained initially to be similar to those of the associated main peak. On the basis of Mössbauer data, these weaker components could be assigned to minor Fe^{2+} at *M*(4), in agreement with the frequency shift observed in the FTIR spectra. The final Mg/Fe^{2+} ratio derived from the FTIR data was in excellent agreement with expected values (Iezzi *et al.*, 2005).

Finally, Fig. 13 displays the results obtained for a set of fibrous actinolites and riebeckites from different provenances (Della Ventura, unpublished data) using the above method, in comparison with the results from EMP analysis; the good agreement between the two sets of data suggests that, for relatively simple chemical compositions, and in particular for amphiboles with the anionic composition close to $(OH)_2$, FTIR spectroscopy can indeed be extremely reliable as an analytical tool.

3.5. The low-frequency region

The interpretation of the vibrational spectra of amphiboles in the region <1200 cm^{-1} and, in particular, in the <650 cm^{-1} range, is difficult due to the structural complexity of these minerals (Wang *et al.*, 1988b; Gillet *et al.*, 1989) and to the fact that, in this range, tetrahedral modes overlap with octahedral (M–O) modes and OH librations. The

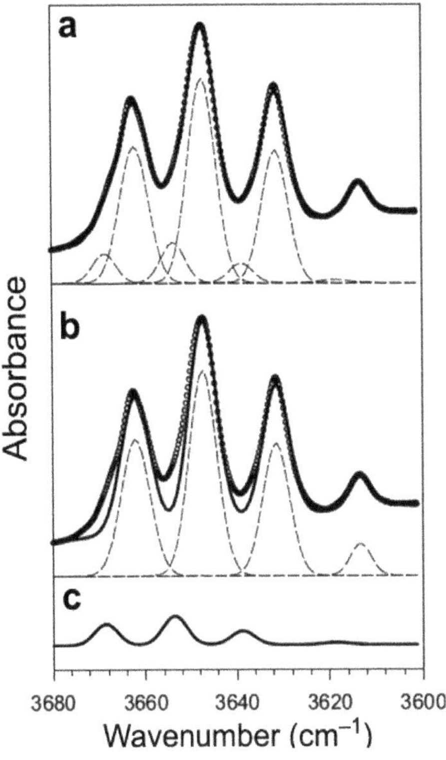

Figure 12. The FTIR spectrum of an intermediate Mg-Fe^{2+} ferri-clinoholmquistite resolved using (a) an eight-band or (b) a four-band model. The residual from the four-band fitting shown in (b) is given in (c), and corresponds to the additional component bands needed for a complete description of the experimental profile. Observed intensities are shown by the hollow squares, and the line following the observed intensities is the envelope of the sum of the fitted component bands. Reprinted with permission from Iezzi *et al.* (2005).

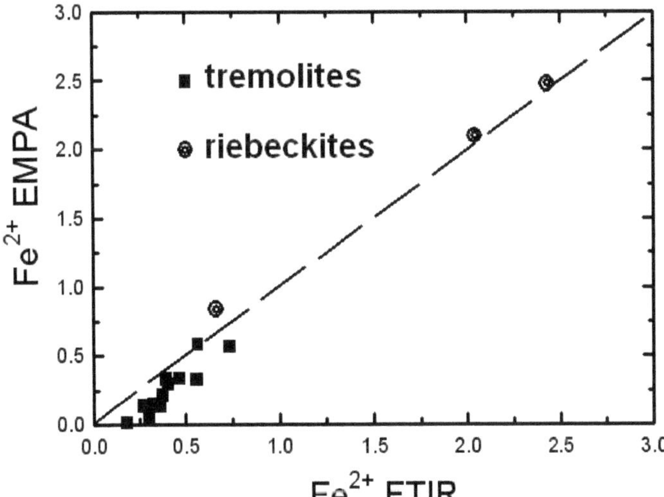

Figure 13. Relationship between the Fe^{2+} content at $M(1,3)$ (a.p.f.u.) analysed by EMP and that obtained by powder FTIR for a series of actinolitic and riebeckitic amphiboles. From Della Ventura (unpublished data).

internal vibrations of the T_4O_{11} double chain involve different types of T–O bonds, because of the occurrence of both bridging (O_b): O(5), O(6), O(7) and non-bridging (O_{nb}): O(1), O(2) oxygen atoms, as discussed by Apopei and Buzgar (2010) and displayed schematically in Fig. 14. Despite these difficulties, vibrational spectroscopy in this region is extremely useful (and used widely) for asbestos identification (*e.g.* Blaha and Rosasco, 1978; Bard *et al.*, 1997; Rinaudo *et al.*, 2004, 2005, 2006). For this reason, μ-Raman spectroscopy, in particular, is one of the most important analytical techniques in environmental and medical applications where high spatial resolution is needed.

Raman data for amphiboles in the framework phonon modes region have been provided in several studies (see Lazarev, 1972; Wang *et al.*, 1988a,b; Gillet *et al.*, 1989; Della Ventura *et al.*, 1991, 1993; Kloprogge *et al.*, 2001; Rinaudo *et al.*, 2004, 2006; Fornero *et al.*, 2008 among the others); a collection of spectra for a large set of compositions was provided by Apopei and Buzgar (2010). FTIR data for the <1200 cm^{-1} region are, on the other hand, relatively scarce; analysis of the IR spectra in this range has been published by Andrut *et al.* (2000) for synthetic tremolite, Sr-substituted tremolite and richterite–tremolite solid solutions, and by Ishida *et al.* (2008) for synthetic calcic amphiboles; data for natural samples have been given by Ishida (1989, 1990a,b, 1998).

According to this vast literature, Si–O–Si and O–Si–O antisymmetric stretching vibrations occur in the 950–1130 cm^{-1} range. As expected, these absorptions are weak in the Raman and intense in the IR spectra (Fig. 3b, lighter shading). The highest-

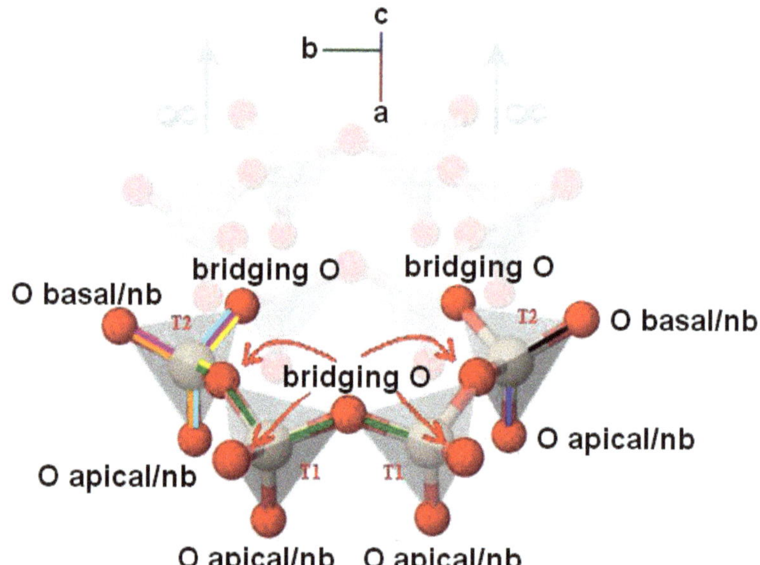

Figure 14. Schematic representation of the type of T–O bonds in the amphibole structure. Reprinted with permission from Apopei and Buzgar (2010).

wavenumber bands in this range can be assigned to T(1)–O(1), a very short bond in the amphibole structure (Della Ventura *et al.*, 1991). Si–O–Si symmetric stretching modes occur in the 800–600 cm^{-1} range; the most intense peak in the Raman spectra is observed in this region (Fig. 3b, darker shading) and it has been shown that the actual position (wavenumber) of this sharp peak provides a reliable identification of the amphibole species (Rinaudo *et al.*, 2004, 2006). At wavenumbers <650 cm^{-1}, following general considerations for silicates (*e.g.* Lazarev, 1972; Dowty, 1987) a mixture of T–O and M–O bending modes and OH librations have been proposed (see correlation tables in Kloprogge *et al.*, 2001 and Apopei and Buzgar, 2010). Ishida (1990a) reported linear shifts for most bands in the FTIR spectra for alkali amphiboles of the glaucophane–riebeckite series although no band assignments were proposed; in ^{T}Al-free amphiboles, OH-libration bands at 700 and 690–640 cm^{-1} could be identified *via* hydrothermal treatment with deuterium (Ishida, 1990b). Selected Raman spectroscopy data for the low-frequency region are provided in Table 1.

4. Vibrational spectroscopy applied to the study of asbestos silicates: Chrysotile and serpentine-group minerals

Chrysotile is a member of the serpentine group (see Ballirano *et al.*, 2017, this volume) comprising the rock-forming minerals lizardite, chrysotile and antigorite, plus additional rarer species, such as polygonal and polyhedral serpentine (Whittaker and Zussman 1956; Cressey *et al.* 2008). The members of this group share the common chemistry $Mg_3(OH)_4Si_2O_5$; this stoichiometry implies a significant structural mismatch between the tetrahedral and the octahedral sheets that can be alleviated *via* a series of chemical and crystallographic modifications, giving rise to the various polymorphs of the group (Wicks and O'Hanley, 1988). Compared to amphiboles, substitutions of other cations for magnesium and silicon are limited and typically include Al for Si at the *T* sites coupled with Al for Mg and/or Fe for Mg at the *M* sites (Viti and Mellini, 1997). Two kinds of OH groups can be distinguished in the *T*–O layer of serpentines (Fig. 15): (1) the outer OH groups groups, that are located at the top of the trioctahedral sheet, and are hydrogen-bonded with the basal plane of O atoms of the next layer; (2) the inner OH groups that are located inside the layer, between the trioctahedral and the tetrahedral sheets, pointing toward the centre of the ditrigonal cavity (Balan *et al.*, 2002). There has been significant interest in the H*T*-spectroscopy of serpentines in the last decade, particularly due to the widespread occurrence of these minerals in key geodynamic locations, such as subduction and suture zones, and to the interest for their thermal and mechanical properties (Trittschack and Grobéty, 2013; Petriglieri *et al.*, 2014).

4.1. The OH-stretching spectra of serpentines

The FTIR OH-stretching spectrum of chrysotile consists of two broad bands centred at 3692 and 3645 cm^{-1} (*e.g.* Farmer, 1974; Ristić *et al.*, 2011) and this pattern is used commonly as a finger-print for its identification in environmental or experimental

Table 1. Raman data for selected asbestos fibres in the literature; data for amphiboles are from Rinaudo *et al.* (2004), for chrysotile (*) are from Trittschack and Grobéty (2013). ° = fitted components. s = strong, m = medium, w = weak, v = very.

Amosite	Anthophyllite	Chrysotile*	Crocidolite	Tremolite
		3697, $368\bar{8}$, $368\bar{1}$, 3649		
1093 vw		1102	1082 s	
				1062 m
	1044 s			
1020 s			1030 w	1031 m
968 m			967 vs	950 w
	928 w			932 m
904 vw		709	889 m	
		705	771 w	751 w
		692	733 w	740 w
	699 w	629		
659 vs	674 vs	622	664 s	676 vs
		607	577 s	
528 m	539 m	466	537 m	531 w
507 w	503 vw	458	506 w	516 w
		432	470 w	
423 w	433 m		428 m	438 w
	410 m			418 m
400 w	387 m	388	374 s	396 s
368 w	364 w	374	360 sh	355 vw
348 m	342 vw	345		348 vw
		318	331 m	335 vw
307 vw	304 m	304	300 m	306 w
289 vw				290 w
	265 m		272 m	
252 w	254 w		246 m	245 m
				234 s
216 m	222 vw	231	211 m	225 s
182 vs	188 m	199	195 s	180 s
155 s			162 vs	162 s

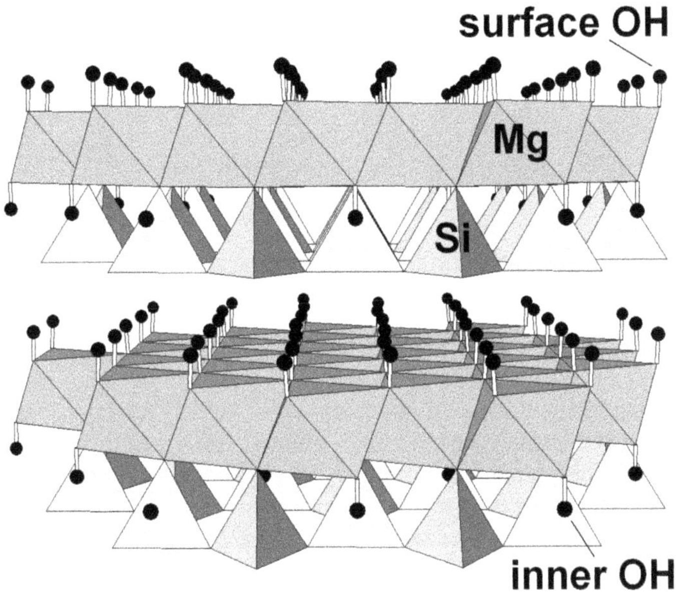

Figure 15. The crystal-structure of lizardite-1*T*, showing the inner O4–H4 hydroxyl groups, pointing towards the centre of the ditrigonal cavity, and the outer (inner-surface) O3–H3 hydroxyl groups. Reprinted with permission from Balan *et al.* (2002).

studies (*e.g.* Hiavay *et al.*, 1984; Liu *et al.*, 2007; Foresti *et al.*, 2009; Giacobbe *et al.*, 2010; Lafay *et al.*, 2012; Accardo *et al.*, 2014) or even in meteorites (Osawa *et al.*, 2005). The higher-frequency band is assigned to the inner OH groups, while the lower-frequency band is assigned to the outer (interlayer) OH groups (*e.g.* Titulaer *et al.*, 1993; Jolicoeur and Duchesne, 1981). However, on closer inspection, the chrysotile pattern is much more complicated; several components can in fact be fitted to the experimental envelope. Figure 16 compares the FTIR and Raman spectra in the OH-region of an almost pure chrysotile (from Trittschack and Grobety, 2013), where the different components are assigned on the basis of previous literature (Balan *et al.*, 2002; Auzende *et al.*, 2004; Petriglieri *et al.*, 2014). Examination of Fig. 16 shows that the spectra are similar but not identical; the Raman pattern consists of a relatively sharp component at 3697 cm^{-1} with a broad shoulder on the lower-frequency side, plus a weaker band at 3650 cm^{-1}. Two components can be fitted to the broad shoulder, at 3688 and 3681 cm^{-1}, respectively, while three components can be fitted to the higher-frequency broad band in the FTIR pattern, at 3702, 3695 and 3682 cm^{-1} (Fig. 16). These peaks are all assigned to the inner-surface OH (Trittschack and Grobety, 2013). Petriglieri *et al.* (2014) and Wang *et al.* (2014) showed that the different serpentine-group species can be well differentiated on the basis of the OH-stretching Raman spectra. The most notable difference was observed for the peak assigned to the outer O–H groups; this finding is consistent with a variable orientation of the outer O–H bonds, due to the waving of the layers in the different polymorphs (Wang *et al.*, 2014).

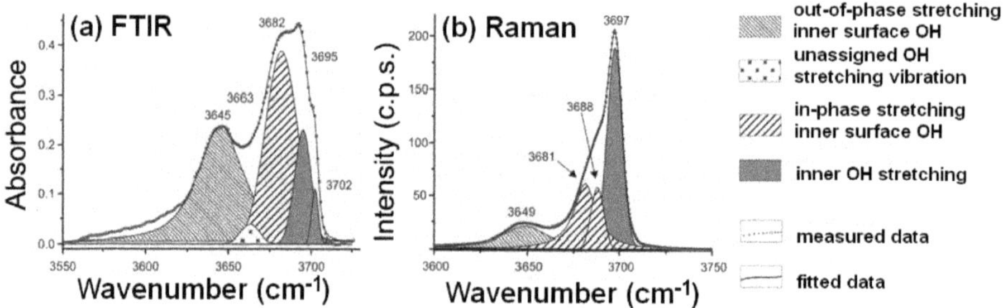

Figure 16. The FTIR spectrum of pure chrysotile compared with the Raman spectrum collected from the same sample. Reprinted with permission from Trittschack and Grobéty (2013).

According to the Raman work of Kloprogge *et al.* (1999) the intensities of the OH-bands are strongly dependent on the crystal orientation and the direction of polarization of the laser and the scattered beams. The chemically limited cationic substitutions apparently have little effect on the OH-spectra (Bishop *et al.*, 2002); in chrysotile, the band due to the inner OH-groups tends to shift toward lower wavenumbers, while that due to the outer OH-groups tends to shift towards higher wavenumbers for increasing ^{T}Al (Velde, 1980; Groppo *et al.*, 2006). The effect of the Mg-Fe exchange in chrysotile nanocrystals was examined experimentally by Foresti *et al.* (2009) who found that iron may substitute both octahedral Mg and tetrahedral Si, modifying the chrysotile structure to a variable extent as a function of the Fe content. Accordingly, the entry of Fe in the structure is correlated to minor shifts of both the OH-stretching (FTIR) and low-frequency (Raman) bands. The spectroscopic data also highlight that iron has a possible role influencing the health hazard of chrysotile in biological systems (Foresti *et al.*, 2009).

4.1.1. The low-frequency region

Extensive work has been done in the lower-frequency region of serpentine-group minerals; the Raman spectrum of chrysotile in this range is given in Fig. 17 together with the proposed band assignments (from Trittschack and Grobéty, 2013). Similarly to amphiboles (see above), the antisymmetric Si–O modes occur at higher frequency as very weak absorptions, while a very intense peak around 700 cm^{-1} is related to the symmetric Si–O vibrations. Rinaudo *et al.* (2003) and Petriglieri *et al.* (2014) have shown that chrysotile can be distinguished effectively from antigorite by considering that the frequencies of the $Si–O_b$-Si and $Si–O_{nb}$ stretchings of antigorite occur at lower wavenumbers with respect to the corresponding modes of chrysotile. Additional differences are found also in the 500–550 cm^{-1} region where *M*–OH librations occur. Serpentine-group minerals can be also distinguished by carlosturanite, a rare serpentine-like, asbestiform mineral, by the presence of a diagnostic peak at 774 cm^{-1} (Belluso *et al.*, 2007). A study of serpentinized ultramafic rocks of the Lanzo massif was presented by Groppo *et al.* (2006) where μ-Raman spectroscopy was applied as a tool to identify the different serpentine minerals.

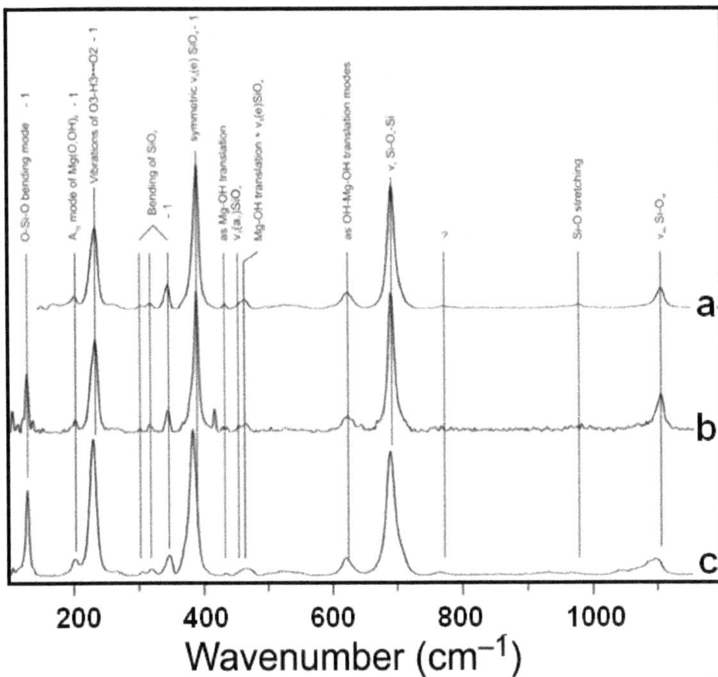

Figure 17. Raman spectra of reference chrysotile samples in the low-frequency (<1200 cm^{-1}) region. The band assignments are indicated. Legend: sample a: chrysotile R070088 785 nm RRUFF; sample b: chrysotile FR 488 nm (GFZ); sample c: lizardite Monte Fico 488 nm (GFZ). Reprinted with permission from Trittschack and Grobéty (2013).

IR spectra of serpentine-group minerals in the <1200 cm^{-1} region have also been shown to be sensitive to the structure type; Yariv and Heller-Kallai (1975) presented polarized-light spectra for the different species and proposed the assignment for the observed bands in the spectra collected from both KBr pellets and oriented crystals deposited on AgCl or polythene windows. The higher-wavenumber peaks were related to Si–O bonds perpendicular to the tetrahedral sheet, while absorption in the 600–650 cm^{-1} range were confirmed to be due to in-plane O–H librations, in agreement with the early work of Stubican and Roy (1961) done on synthetic chrysotile. The most striking feature in the IR spectra, useful for differentiating the various serpentine minerals, was, however, found to be a well resolved peak assigned to out-of-plane Mg–O vibrations, occurring at 570 cm^{-1} in antigorite and missing or occurring as a broad, convolute band, in both chrysotile and lizardite.

A method able to distinguish the serpentine minerals directly from petrographic thin sections, and possibly suitable also for histological sections, was presented by Petriglieri *et al.* (2014). The method is based on μ-Raman mapping, *i.e.* on the collection of spectra in a grid of points and subsequent integration of the absorbance, a procedure used extensively in FTIR imaging of geological materials (Della Ventura *et al.*, 2014a). Petriglieri *et al.* (2014) showed that with this method a fine-scale study of the structural relationships between different serpentines is possible, thus providing a clue for characterizing the serpentinization processes in complex geological sequences.

5. Environmental/medical applications

Vibrational spectroscopies, and μ-Raman spectroscopy in particular, have been used extensively as identification tools in medical/biological and environmental studies. The list below cannot be complete or exhaustive of the work done, but is intended to give some notable examples highlighting the use of these techniques.

Rinaudo *et al.* (2009) and Musa *et al.* (2012) presented methodological studies for the identification of particles and fibres in histological sections of patients affected by respiratory diseases or malignant mesothelioma (MM). They showed how the use of μ-Raman allowed the study of the inorganic materials while preserving the biological system. The same technique was extended successfully to the identification of erionite fibres in tissues from both mice injected with asbestos and slides of biopsies of MM-affected patients from Turkey (Croce *et al.*, 2012). Tomatis *et al.* (2010) examined the toxicological role of fibre morphologies *vs.* their surface chemistry. They used Raman spectroscopy, in conjunction with X-ray diffraction, to assess the crystallinity of the amosite fibres used for the experiments; IR spectroscopy was used to determine the adsorption of NO by the fibres as a probe for the coordination of surface iron. Gualtieri *et al.* (2013) studied the crystal-chemistry of mesothelioma-inducing crocidolite fibres injected into mice peritonea. Samples of IUCC crocidolite before the injection and those embedded in the mice tissues were examined by μ-Raman and *in situ* synchrotron X-ray diffraction. The fibres occurring in the histological sections after tumour development were found to be free from ferro-protein coating ('asbestos bodies', *e.g.* Stewart and Haddow, 1929); this allowed the examination of the early crystal-chemical changes due to the interaction of the mineral fibre with the biological medium, before the development of the asbestos bodies. Interestingly, combination of X-ray diffraction and spectroscopic data showed a partial degradation of the mineral structure, with intra-crystalline migration of the iron and sodium ions. Applications of μ-Raman spectroscopy combined with SEM-EDX and X-ray diffraction for the study of asbestos bodies were recently presented by Rinaudo *et al.* (2010) and Bursi Gandolfi *et al.* (2016). Rinaudo *et al.* (2010), in particular, observed a photo-crystallization of hematite from the iron-rich materials of the asbestos bodies induced by the 632.8 nm laser beam of the Raman spectrometer. This feature highlights one of the main drawbacks of Raman spectroscopy, *i.e.* possible modifications of the system due to the power of the laser beam during the analysis; this is indeed a problem of paramount importance, especially when working with organic or carbon-rich materials (Notingher *et al.*, 2002; Yoshida *et al.*, 2009; Hodkiewicz, 2010). For this reason, there is now a tendency, in biological studies, to develop spectroscopic methods based on FTIR spectroscopy (*e.g.* Seydou *et al.*, 2010, 2016) that employ beams with energies lower than those of the visible range, therefore harmless for organic systems. The important point here is that methods based on IR light would also allow *in vivo* studies (Marcelli *et al.*, 2012). Bursi Gandolfi *et al.* (2016) studied the development of iron coatings around crocidolite, chrysotile and erionite fibres injected in the intraperitoneal or intrapleural tissues of Sprague-Dowley rats. μ-Raman spectra were collected from the asbestos bodies to identify the nature of the coatings and showed that these consisted essentially of Fe-rich apatite. Interestingly, their study showed that only long fibres were coated and that no coatings developed on erionite, an iron-free zeolite (Bursi Gandolfi *et al.*, 2016). Seydou *et al.* (2013) used Raman spectroscopy to study the role of asbestos morphologies on their cellular toxicity. A novel fast-acquisition method for collecting optical and Raman images was developed to characterize *in vitro* the interaction between single cells and fibres. Reconstruction of successive images at different elevation steps (900 nm) allowed collection for the first time of a 3D μ-Raman image of a fibre embedded within the cell (Fig. 18).

The main result of this study was that cell–fibre interaction phenomena are heavily dependent on the fibre morphology; fibres <5 μm long were in fact completely internalized and vesiculated, while fibres 5 to 15 μm long were embedded in the extracellular matrix; longer fibres were found to be incompletely embedded and thus available for mechanical damage to lung tissues following lung movements during breathing (Seydou *et al.*, 2013).

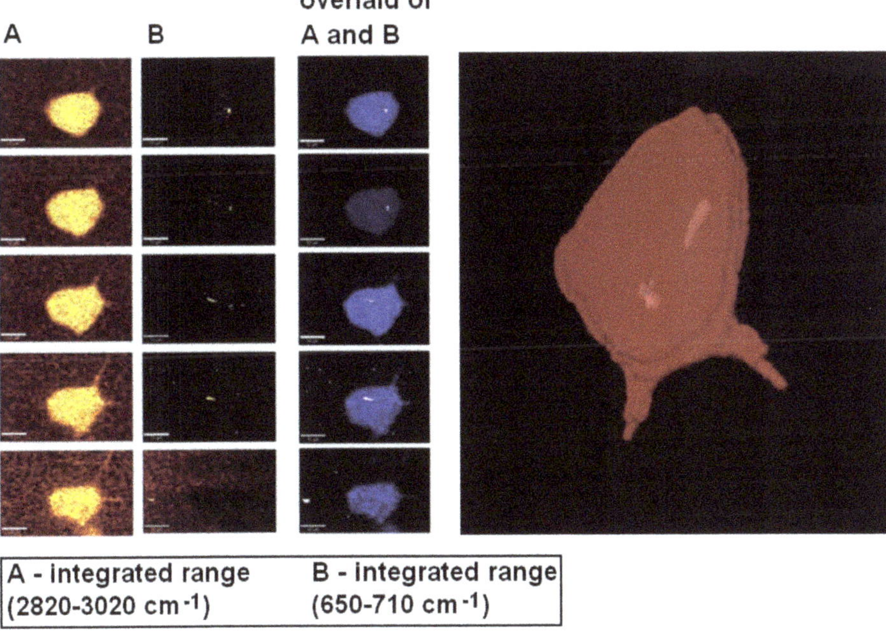

Figure 18. Optical and Raman images of an A549 cell interacted with a single crocidolite fibre. The Raman images (modified from Seydou *et al.*, 2013) have been taken at different elevation levels with a step of 0.9 μm and integrated in different spectra ranges for lipids (2820–3020 cm^{-1}), and the fibre (650–710 cm^{-1}), respectively. The resulting images (left and middle columns), can be combined in single images (right column), and eventually in a 3D rendering (right side, courtesy of Yao Seydou).

In environmental studies, vibrational spectroscopies have been used extensively to address the mineralogical composition and the transformation products, during heat or physical treatments, of cement-asbestos or related hazardous materials. Leviatan *et al.* (2009) studied the mine wastes at the Vermont Asbestos Group Mine (Vermont, USA), the second largest mine in USA, where chrysotile was mined from serpentinized ophiolites; the waste piles were characterized by combining μ-Raman with XRD, optical and micro-chemical (SEM-EDS, EMPA) techniques.

The study of cement-asbestos fragments of roof-slates from several localities in Italy was presented by Viani *et al.* (2013a). Complete mineralogical compositions of the samples have been obtained by combining different modern mineralogical techniques, including XRF, high-resolution SEM, quantitative Rietveld analysis of the powder diffraction data, electron-backscattered diffraction (EBSD) and μ-Raman spectroscopy. Although chrysotile was found to be the main asbestos constituents (up to >13 wt.%), minor but significant amphibole fibres were also identified. The annealing products of cement-asbestos were studied by Colangelo *et al.* (2011), Viani *et al.* (2013b) and Croce *et al.* (2014). Powdered and massive samples were studied before and after the thermal treatment using a combination of spectroscopic, X-ray diffraction and SEM techniques; in these studies Raman spectroscopy was revealed to be an essential tool to assess directly the transformation of needle-like crystals into non-hazardous minerals.

FTIR and Raman spectroscopies were applied by Kudu-Petersen *et al.* (2000) to study white fibrous material used in pot making by women in a village in Botswana; the identification of tremolite in the samples prompted recognition of occupational hygiene implications and the need to take measures improving the safety of the pot-making processes.

A final application worthy of mention is the work of Odziemkowski *et al.* (2001) who used confocal μ-Raman to identify asbestos fibres within the airborne particulate matter sampled by poly(dimethylsiloxane) filters.

References

Accardo, G., Cioffi, R., Colangelo, F., d'Angelo, R., De Stefano, L. and Paglietti, F. (2014) Diffuse reflectance infrared Fourier transform spectroscopy for the determination of asbestos species in bulk building materials. *Materials*, **7**, 457–470.

Addison, W.E. and White, A.D. (1968) Spectroscopic evidence for the siting of lithium ions in a riebeckite. *Mineralogical Magazine*, **36**, 743–745.

Andreozzi, G.B. and Pollastri, S. (2017) Bulk spectroscopy of mineral fibres. Pp. 111–134 in: *Mineral Fibres: Crystal Chemistry, Chemical-physical Properties, Biological Interaction and Toxicity* (A.F. Gualtieri, editor). EMU Notes in Mineralogy, **18**. European Mineralogical Union and Mineralogical Society of Great Britain & Ireland, London.

Andrut, M., Gottschalk, M., Melzer, S. and Najorka, J. (2000) Lattice vibrational modes in synthetic tremolite-Sr-tremolite and tremolite-richterite solid solutions. *Physics and Chemistry of Minerals*, **27**, 301–309.

Apopei, A. I. and Buzgar, N. (2010) The Raman study of amphiboles. *Analele Ştiinţifice ale Universităţii "Al. I. Cuza" Iaşi, Geologie*, **56**, 57–83.

Auzende, A.-L., Daniel, I., Reynard, B., Lemaire, C. and Guyot, F. (2004) High pressure behaviour of serpentine minerals: a Raman spectroscopic study. *Physics and Chemistry of Minerals*, **31**, 269–277.

Bard, D., Yarwood, J. and Tylee, B. (1997) Asbestos fiber identification by Raman microspectrometry. *Journal of Raman Spectroscopy*, **28**, 803–809.

Balan, E., Saitta, A.M., Mauri, F., Lemaire, C. and Guyot, F. (2002) First-principles calculation of the infrared spectrum of lizardite. *American Mineralogist*, **87**, 1286–1290.

Ballirano, P., Bloise, A., Gualtieri, A.F., Lezzerini, M., Pacella, A., Perchiazzi, N., Dogan, M. and Dogan, A.U. (2017) The crystal structure of mineral fibres. Pp. 17–64 in: *Mineral Fibres: Crystal Chemistry, Chemical-physical Properties, Biological Interaction and Toxicity* (A.F. Gualtieri, editor). EMU Notes in Mineralogy, **18**. European Mineralogical Union and Mineralogical Society of Great Britain & Ireland, London.

Belluso, E., Fornero, E., Cairo, S., Albertazzi, G. and Rinaudo, C. (2007) The application of micro-Raman spectroscopy to distinguish carlosturanite from serpentine-group minerals. *The Canadian Mineralogist*, **45**, 1495–1500.

Benton, D.M. (2013) Identification of asbestos using Laser-Induced Breakdown Spectroscopy: a viable alternative to the conventional approach? *International Scholarly Research Notices*, **2013**, article ID 362694, 1–6.

Beran, A. and Libowitzky, E. (editors) (2004) *Spectroscopic Methods in Mineralogy*. EMU Notes in Mineralogy, **6**. European Mineralogical Union and The Mineralogical Society of Great Britain & Ireland, Eötvös University Press, Budapest, 661 pp.

Bishop, J., Murad, E. and Dyar, M.D. (2002) The influence of octahedral and tetrahedral cation substitution on the structure of smectites and serpentines as observed through infrared spectroscopy. *Clay Minerals*, **37**, 617–628.

Blaha, J.J. and Rosasco, G.J. (1978) Raman microprobe spectra of individual microcrystals and fibers of talc, tremolite, and related silicate minerals. *Analytical Chemistry*, **50**, 892–896.

Burns, R.G. and Strens, R.G.J. (1966) Infrared study of the hydroxyl bonds in clinoamphiboles. *Science*, **153**, 890–892.

Bursi Gandolfi N., Gualtieri, A.F., Pollastri, S., Tibaldi, E. and Belpoggi, F. (2016) Assessment of asbestos body formation by high resolution FEG-SEM after exposure of Sprague-Dawley rats to chrysotile, crocidolite, or erionite. *Journal of Hazardous Materials*, **306**, 95–104.

Colangelo, F., Cioffi, R., Lavorgna, M., Verdolotti, R. and De Stefano, L. (2011) Treatment and recycling of asbestos-cement containing waste. *Journal of Hazardous Materials* **195**, 391–397.

Cressey, G., Cressey, B.A. and Wicks, F.J. (2008) Polyhedral serpentine: a spherical analogue of polygonal serpentine? *Mineralogical Magazine*, **72**, 1229–1242.

Croce, A., Musa, M., Allegrina, M., Rinaudo, C., Izzettin Baris, Y., Umran Dogan, A., Powers, A., Rivera, Z., Bertino, P., Yang, H., Gaudino, G. and Carbone, M. (2012) Micro-Raman spectroscopy identifies crocidolite and erionite fibers in tissue sections. *Journal of Raman Spectroscopy*, **44**, 1440–1445.

Croce, A., Allegrina, M., Trivero, P., Rinaudo, C., Viani, A., Pollastri, S. and Gualtieri, A.F. (2014) The concept of 'end of waste' and recycling of hazardous materials: in depth characterization of the product of thermal transformation of cement-asbestos. *Mineralogical Magazine*, **78**, 1177–1191.

Della Ventura, G. (1992) Recent developments in the synthesis and characterization of amphiboles. Synthesis and crystal-chemistry of richterites. *Trends in Mineralogy*, **1**, 153–192.

Della Ventura, G., Robert, J.-L. and Bény, J.-M. (1991) Tetrahedrally coordinated Ti^{4+} in synthetic Ti rich potassic richterite: Evidence from XRD, FTIR and Raman studies. *American Mineralogist*, **76**, 1134–1140.

Della Ventura, G., Robert, J.-L., Bény, J.-M., Raudsepp, M. and Hawthorne, F.C. (1993) The OH-F substitution in Ti-rich potassium-richterites: Rietveld structure refinement and FTIR and microRaman spectroscopic studies of synthetic amphiboles in the system K_2O–Na_2O–CaO–MgO–SiO_2-TiO_2-H_2O–HF. *American Mineralogist*, **78**, 980–987.

Della Ventura, G., Robert, J.-L. and Hawthorne, F.C. (1996) Infrared spectroscopy of synthetic (Ni,Mg,Co)-potassium-richterite. *Geochimica and Cosmochimica Acta special issue*, **5**, 55–63.

Della Ventura, G., Robert, J.-L., Raudsepp, M., Hawthorne, F.C. and Welch, M. (1997) Site occupancies in synthetic monoclinic amphiboles: Rietveld structure-refinement and infrared spectroscopy of (nickel, magnesium, cobalt)-richterite. *American Mineralogist*, **82**, 291–301.

Della Ventura, G., Robert, J.-L., Hawthorne, F.C., Raudsepp, M. and Welch, M.D. (1998) Contrasting $^{[6]}Al$ ordering in synthetic Mg- and Co-pargasite. *The Canadian Mineralogist*, **36**, 1237–1244.

Della Ventura, G., Hawthorne, F.C., Robert, J.-L., Delbove, F., Welch, M.D. and Raudsepp, M. (1999) Short-range order of cations in synthetic amphiboles along the richterite–pargasite join. *European Journal of Mineralogy*, **11**, 79–94.

Della Ventura, G., Robert, J.-L., Sergent, J., Hawthorne, F.C. and Delbove, F. (2001) Constraints on F vs. OH incorporation in synthetic [6]Al-bearing monoclinic amphiboles. *European Journal of Mineralogy*, **13**, 841–847.

Della Ventura, G., Hawthorne, F.C., Robert, J.-L. and Iezzi, G. (2003) Synthesis and infrared spectroscopy of amphiboles along the tremolite – pargasite join. *European Journal of Mineralogy*, **15**, 341–347.

Della Ventura, G., Iezzi, G., Redhammer, G.J., Hawthorne, F.C., Scaillet, B. and Novembre, D. (2005) Synthesis and crystal-chemistry of alkali amphiboles in the system Na_2O–MgO–FeO–Fe_2O_3-SiO_2-H_2O as a function of f_{O_2}. *American Mineralogist*, **90**, 1375–1383.

Della Ventura, G., Oberti, R., Hawthorne, F.C. and Bellatreccia, F. (2007) Single-crystal FTIR study Ti-rich pargasites from Lherz: the spectroscopic detection of $^{O3}O^{2-}$ in amphiboles. *American Mineralogist*, **92**, 1645–1651.

Della Ventura, G., Gatta, D., Redhammer, G., Bellatreccia, F., Loose, A. and Parodi, G.C. (2009) Single-crystal polarized FTIR spectroscopy and neutron diffraction refinement of cancrinite. *Physics and Chemistry of Minerals*, **36**, 193–206.

Della Ventura, G., Marcelli, A. and Bellatreccia, F. (2014a) SR-FTIR microscopy and FTIR imaging in the Earth Sciences. Pp. 447–479 in: *Spectroscopic Methods in Mineralogy and Materials Science* (G.S. Henderson, D.R. Neuville and R.T. Downs, editors). Reviews in Mineralogy and Geochemistry, **78**, Mineralogical Society of America and Geochemical Society, Chantilly, Virginia, USA.

Della Ventura, G., Bellatreccia, F., Cámara, F. and Oberti, R. (2014b) Crystal-chemistry and short-range order of fluoro-edenite and fluoro-pargasite: a combined X-ray diffraction and FTIR spectroscopic approach. *Mineralogical Magazine*, **78**, 293–310.

Della Ventura, G., Mottana, A., Caprilli, E., Bellatreccia, F. and De Benedetti, A. (2014c) Asbestiform tremolite within the late pyroclastic deposits of the Alban Hills volcano (Latium, Italy): FTIR spectroscopy and crystal chemistry. *Rendiconti Accademia dei Lincei*, **25**, 229–236.

Della Ventura, G., Redhammer, G., Robert, J.L., Sergent, J., Iezzi, G. and Cavallo, A. (2016) Crystal-chemistry of synthetic amphiboles along the join richterite–ferro-richterite: a combined spectroscopic (FTIR, Mössbauer), XRPD and microchemical study. *The Canadian Mineralogist*, **54**, 97–114.

Dowty, E. (1987) Vibrational interactions of tetrahedra in silicate glasses and crystals: I. calculations on ideal silicate aluminate-germanate structural units. *Physics and Chemistry of Minerals*, **14**, 80–93.

Farmer, V.C. (1974) The layer silicates. Pp. 331–363 in: *The Infrared Spectra of Minerals* (V.C. Farmer, editor). Monograph **4**, The Mineralogical Society, London.

Ferraro, J.R., Nakamoto, K. and Brown, C.W. (2003) *Introductory Raman Spectroscopy* (second edition). Elsevier, 434 pp.

Foresti, E., Fornero, E., Lescia, I.G., Rinaudo, C., Zuccheri, T. and Roveri, R. (2009) Asbestos health hazard: A spectroscopic study of synthetic geoinspired Fe-doped chrysotile. *Journal of Hazardous Materials*, **167**, 1070–1079.

Fornero, E., Allegrina, M., Rinaudo, C., Mazziotti-Tagliani, S. and Gianfagna, A. (2008) Micro-Raman spectroscopy applied on oriented crystals of fluoro-edenite amphibole. *Periodico di Mineralogia*, **77**, 2–14.

Giacobbe, C., Gualtieri, A.F., Quartieri, S., Rinaudo, C., Allegrina, M. and Andreozzi, G.B. (2010) Spectroscopic study of the product of thermal transformation of chrysotile-asbestos containing materials (ACM). *European Journal of Mineralogy*, **22**, 535–546.

Gillet, P., Reynard, B. and Tequi, C. (1989) Thermodynamic properties of glaucophane. New data from calorimetric and spectroscopic measurements. *Physics and Chemistry of Minerals*, **16**, 659–667.

Gottschalk, M., Andrut, M. and Melzer, S. (1999) The determination of cummingtonite content of synthetic tremolite. *European Journal of Mineralogy*, **11**, 967–982.

Groppo, C., Rinaudo, C., Cairo, S., Gastaldi, D. and Compagnoni, R. (2006) Micro-Raman spectroscopy for a quick and reliable identification of serpentine minerals from ultramafics. *European Journal of Mineralogy*, **18**, 319–329.

Gualtieri, A.F., Giacobbe, C., Rinaudo, C., Croce, A., Allegrina, M., Gaudino, G., Yang, H. and Carbone, M.

(2013) Preliminary results of the spectroscopic and structural characterization of mesothelioma inducing crocidolite fibers injected in mice. *Periodico di Mineralogia*, **82**, 299–312.

Gunter, M.E., Belluso, E. and Mottana, A. (2007) Amphiboles: environmental and health concerns. Pp. 453–516 in: *Amphiboles: Crystal Chemistry, Occurrence and Health Issues* (F.C. Hawthorne, R. Oberti, G. Della Ventura and A. Mottana, editors). Reviews in Mineralogy and Geochemistry, **67**, Mineralogical Society of America and Geochemical Society, Chantilly, Virginia, USA.

Hawthorne, F.C. (editor) (1988) *Spectroscopic Methods in Mineralogy and Geology*. Reviews in Mineralogy, **18**, Mineralogical Society of America. Chantilly, Virginia, USA, 698 pp.

Hawthorne, F.C. (1997) Short-range order in amphiboles: a bond-valence approach. *The Canadian Mineralogist*, **35**, 201–216.

Hawthorne, F.C. and Della Ventura, G. (2007) Short-range order in amphiboles. Pp. 173–222 in: *Amphiboles: Crystal Chemistry, Occurrence and Health Issues* (F.C. Hawthorne, R. Oberti, G. Della Ventura and A. Mottana, editors). Reviews in Mineralogy and Geochemistry, **67**, Mineralogical Society of America and Geochemical Society, Chantilly, Virginia, USA.

Hawthorne, F.C. and Grundy, H.D. (1976) The crystal chemistry of the amphiboles. IV. X-ray and neutron refinements of the crystal structure of tremolite. *The Canadian Mineralogist*, **14**, 334–345.

Hawthorne, F.C. and Oberti, R. (2007) Amphiboles: crystal chemistry. Pp. 1–54 in: *Amphiboles: Crystal Chemistry, Occurrence and Health Issues* (F.C. Hawthorne, R. Oberti, G. Della Ventura and A. Mottana, editors). Reviews in Mineralogy and Geochemistry, **67**, Mineralogical Society of America and Geochemical Society, Chantilly, Virginia, USA.

Hawthorne, F.C. and Waychunas, G.A. (1988) Spectrum-fitting methods. Pp. 63–98 in: *Spectroscopic Methods in Mineralogy and Geology* (F.C. Hawthorne, editor). Reviews in Mineralogy, **18**, Mineralogical Society of America. Chantilly, Virginia, USA.

Hawthorne, F.C., Oberti, R. and Sardone, N. (1996a) Sodium at the A site in clinoamphiboles: the effect of composition on patterns of order. *The Canadian Mineralogist* **34**, 577–593.

Hawthorne, F.C., Della Ventura, G. and Robert, J.-L. (1996b) Short-range order and long-range order in amphiboles: A model for the interpretation of infrared spectra in the principal OH-stretching region. *Geochimica and Cosmochimica Acta special issue*, **5**, 49–54.

Hawthorne, F.C., Della Ventura, G., Robert, J.-L., Welch, M.D., Raudsepp, M. and Jenkins, D.M. (1997) A Rietveld and infrared study of synthetic amphiboles along the potassium-richterite–tremolite join. *American Mineralogist*, **82**, 708–716.

Hawthorne, F.C., Welch, M.D., Della Ventura, G., Shuangxi Liu, Robert, J.-L. and Jenkins, D.M. (2000) Short-range order in synthetic alluminous tremolites: an infrared and triple-quantum MAS NMR study. *American Mineralogist*, **85**, 1716–1724.

Hermann, P., Hoehl, A., Patoka, P., Huth, F., Rühl, E. and Ulm, G. (2013) Near-field imaging and nano-Fourier-transform infrared spectroscopy using broadband synchrotron radiation. *Optics Express*, **21**, 2913–2919.

Henderson, G.S., Neuville, D.R. and Downs, R.T. editors (2014) *Spectroscopic Methods*. Reviews in Mineralogy and Geochemistry **78**, Mineralogical Society of America and Geochemical Society, Chantilly, Virginia, USA, 800 pp.

Hiavay, J., Antal, L., György-Poszonyi, I. and Inczédy, J. (1984) Determination of chrysotile content of asbestos cement dusts by IR-spectroscopy. *Fresenius' Zeitschrift für analytische Chemie*, **319**, 547–551.

Hodkiewicz, J. (2010) The importance of tight laser power control when working with carbon nanomaterials. *ThermoFisher Application Note*, 51948.

Huggins, C.M. and Pimentel, G.C. (1956) Systematics of the infrared spectral properties of hydrogen-bonded systems: frequency-shift, half width and intensity. *Journal of Physical Chemistry*, **60**, 1615-1619

Iezzi, G., Della Ventura, G., Cámara, F., Pedrazzi, G. and Robert, J.-L. (2003a) The ^{B}Na- ^{B}Li exchange in A-site vacant amphiboles: synthesis and cation ordering along the ferri-clinoferroholmquistite–riebeckite join. *American Mineralogist*, **88**, 955–961.

Iezzi, G., Della Ventura, G., Pedrazzi, G., Robert, J.-L. and Oberti, R. (2003b) Synthesis and characterisation of ferri-clinoferroholmquistite, $\square Li_2(Fe^{2+}_3Fe^{3+}_2)Si_8O_{22}(OH)_2$. *European Journal of Mineralogy*, **15**, 321–327.

Iezzi, G., Cámara, F., Della Ventura, G., Oberti, R., Pedrazzi, G. and Robert, J.-L. (2004) Synthesis, crystal structure and crystal-chemistry of ferri-clinoholmquistite, $\square Li_2Mg_3Fe^{3+}_2Si_8O_{22}(OH)_2$. *Physics and*

Chemistry of Minerals, **31**, 375–385.

Iezzi, G., Della Ventura, G., Hawthorne, F.C., Pedrazzi, G., Robert, J.-L. and Novembre, D. (2005) The (Mg,Fe^{2+}) substitution in ferri-clinoholmquistite, □Li_2 (Mg,Fe^{2+})$_3$ Fe^{3+}_2 Si_8 O_{22} $(OH)_2$. *European Journal of Mineralogy*, **17**, 733–740.

Iezzi, G., Della Ventura, G., Bellatreccia, F., Lo Mastro, S., Gunther, M. and Bandly, B.R. (2007) Site occupancy of natural richterite-winchite amphiboles from Libby, Montana, USA: a comparison between FTIR OH-stretching spectroscopy and micro-chemical (EPMA), X-ray diffraction (SREF) and Mössbauer data. *Mineralogical Magazine*, **71**, 93–104.

Ishida, K. (1989) Infrared study of manganoan alkali-calcic amphiboles. *Mineralogical Journal*, **14**, 255–265.

Ishida, K. (1990a) Infrared spectra of alkali amphiboles of the glaucophane-riebeckite series and their relation to chemical composition. *Mineralogical Journal*, **15**, 147–161.

Ishida, K. (1990b) Identification of infrared OH-librational bands of talc-willemseite solid-solutions and $Al^{(VI)}$-free amphiboles through deuteration. *Mineralogical Journal*, **15**, 93–104.

Ishida, K. (1998) Cation disordering in heat-treated anthophyllites through oxidation and dehydrogenation. *Physics and Chemistry of Minerals*, **25**, 160–167.

Ishida, K., Jenkins, D.M. and Hawthorne, F.C. (2008) Mid-IR bands of synthetic calcic amphiboles of tremolite-pargasite series and of natural calcic amphiboles. *American Mineralogist*, **93**, 1112–1118.

Jolicoeur, C. and Duchesne, D. (1981) Infrared and thermogravimetric studies of the thermal degradation of chrysotile asbestos fibers: evidence for matrix effects. *Canadian Journal of Chemistry*, **59**, 1521–1526.

Kloprogge, J.T., Frost, R.L. and Rintoul, L. (1999) Single crystal Raman microscopic study of the asbestos mineral chrysotile. *Physical Chemistry Chemical Physics*, **1**, 2559–2564.

Kloprogge, J.T., Case, M.H. and Frost, R.L. (2001) Raman microscopic study of the Li amphibole holmquistite, from the Martin Marietta Quarry, Bessemer City, NC, USA. *Mineralogical Magazine*, **65**, 775–785.

Kroumova, E., Arojo, M.I., Perez Mato, J.M., Kirov, A., Capillas, C., Ivanchev, S. and Wondratschek, H. (2003) Bilbao Crystallographic Server: useful data-bases and tools for phase transition studies. *Phase Transitions*, **76**, 155–170.

Khudu-Petersen, K., Bard, D., Garriton, N., Yarwood, J. and Tylee, B. (2000) Microscopic identification of asbestos fibres associated with african clay crafts manufacture. *Annals of Occupational Hygiene*, **44**, 137–141.

Lafay, R., Montes-Hernandez, G., Janots, E., Chiriac, R., Findling, N. and Toche, F. (2012) Mineral replacement rate of olivine by chrysotile and brucite under high alkaline conditions. *Journal of Crystal Growth*, **347**, 62–72.

Lazarev, A.N. (1972) *Vibrational Spectra and Structure of Silicates*. Consultants Bureau (Plenum Publishing Company Ltd.), New York, London, 302 pp.

Leissner, L., Schlüter, J., Horn, I. and Mihailova, B. (2015) Exploring the potential of Raman spectroscopy for crystallochemical analyses of complex hydrous silicates: I. Amphiboles. *American Mineralogist*, **100**, 2682–2694.

Levitan, D.M., Hammarstrom, J.M., Gunter, M.E., Seal, R.R., Chou, I.-M. and Piatak, N.M. (2009) Mineralogy of mine waste at the Vermont Asbestos Group mine, Belvidere Mountain, Vermont. *American Mineralogist*, **94**,1063–1066.

Lewis, I.R., Chaffin, N.C., Gunter, M.E. and Griffiths, P.R. (1996) Vibrational spectroscopic studies of asbestos and comparison of suitability for remote analysis. *Spectrochimica Acta part A*, **52**, 315–328.

Liu, K., Feng, Q., Yang, Y., Zhang, G., Ou, L. and Lu, Y. (2007) Preparation and characterization of amorphous silica nanowires rom natural chrysotile. *Journal of Non-crystalline Solids*, **353**, 1534–1539.

Marcelli, A., Cricenti, A., Kwiatek, W.M. and Petibois, C. (2012) Biological applications of synchrotron radiation infrared spectromicroscopy. *Biotechnology Advances*, **30**, 1390–1404.

Mazziotti-Tagliani, S., Andreozzi, G.B., Bruni, B.M., Gianfagna, A., Pacella, A. and Paoletti, L. (2009) Quantitative chemistry and compositional variability of fluorine fibrous amphiboles from Biancavilla (Sicily, Italy). *Periodico di Mineralogia*, **78**, 65–74.

Melzer, S., Gottschalk, M., Andrut, M. and Heinrich, W. (2000): Crystal chemistry of K-richterite-richterite-tremolite solid solutions: a SEM, EMP, XRD, HRTEM and IR study. *European Journal of Mineralogy*, **12**, 273–291.

Musa, M., Croce, A., Allegrina, M., Rinaudo, C., Belluso, E., Bellis, D., Toffalorio, F. and Veronesi, G. (2012) The use of Raman spectroscopy to identify inorganic phases in iatrogenic pathological lesions of patients with malignant pleural mesothelioma. *Vibrational Spectroscopy*, **61**, 66–71.

Mustard, J.F. (1992) Chemical analysis of actinolite from reflectance spectra. *American Mineralogist*, **77**, 345–358.

Nakamoto, K. (1986) *Infrared and Raman Spectra of Inorganic and Coordination Compounds*. John Wiley and Sons, New York.

Notingher, I., Verrier, S., Romanska, H., Bishop, A.E., Polak, J.M. and Hench, L.L. (2002) *In situ* characterisation of living cells by Raman spectroscopy. *Spectroscopy*, **16**, 43–51.

Oberti, R., Hawthorne, F.C., Cannillo, E. and Cámara, F. (2007) Long-range order in amphiboles. Pp. 125–172 in: *Amphiboles: Crystal Chemistry, Occurrence and Health Issues* (F.C. Hawthorne, R. Oberti, G. Della Ventura and A. Mottana, editors). Reviews in Mineralogy and Geochemistry, **67**, Mineralogical Society of America and Geochemical Society, Chantilly, Virginia, USA.

Oberti, R., Della Ventura, G., Boiocchi, M., Zanetti, M. and Hawthorne, F.C. (2016a) The crystal chemistry of oxo-mangani-leakeite and mangano-mangani-ungarettiite from the Hoskins mine and their apparent but impossible solid-solution–An XRD and FTIR study. *Mineralogical Magazine*, **80**, 1013–1021.

Oberti, R., Boiocchi, M., Zema, M. and Della Ventura, G. (2016b) Synthetic potassic-ferro-richterite: 1. Composition, crystal structure refinement and HT behavior by in operando single-crystal X-ray diffraction. *The Canadian Mineralogist*, **54**, 311–335.

Osawa, T., Kagi, H., Nakamura, T. and Noguchi, T. (2005) Infrared spectroscopic taxonomy for carbonaceous chondrites from speciation of hydrous components. *Meteoritics and Planetary Science*, **40**, 71–86.

Odziemkowski, M., Koziel, J.A., Irish, D.E. and Pawliszyn, J. (2001) Sampling and Raman confocal microspectroscopic analysis of airborne particulate matter using poly(dimethylsiloxane) solid-phase microextraction fibers. *Analytical Chemistry*, **73**, 3131–3139.

Pacella, A., Andreozzi, G.B. and Fournier, J. (2009) Detailed crystal chemistry and iron topochemistry of asbestos occurring in its natural setting: A first step to understanding its chemical reactivity. *Chemical Geology*, **277**, 197–206.

Parry, S.A., Pawley, A.R., Jones, R.L. and Clark, S.M. (2007) An infrared spectroscopic study of the OH stretching frequencies of talc and 10-Å phase to 10 GPa. *American Mineralogist*, **92**, 525–531.

Petry, R., Mastalerz, R., Zahn, S., Mayerhöfer, T.G., Völksch, G., Viereck-Götte, L., Kreher-Hartmann, B., Holz, L., Lankers, M. and Popp, J. (2006) Asbestos mineral analysis by UV Raman and energy-dispersive X-ray spectroscopy. *ChemPhysChem*, **7**, 414–420.

Petriglieri, J.R., Salvioli-Mariani, E., Mantovani, L., Tribaudino, M., Lottici, P.P., Laporte-Magonib, C. and Bersani, D. (2014) Micro-Raman mapping of the polymorphs of serpentine. *Journal of Raman Spectroscopy*, **46**, 953–958.

Raudsepp, M., Turnock, A.C., Hawthorne, F.C., Sherriff, B.L. and Hartman J.S. (1987) Characterization of synthetic pargasitic amphiboles ($NaCa_2Mg_4M^{3+}Si_6Al_2O_{22}(OH,F)_2$; M^{3+} = Al, Cr, Ga, Sc, In) by infrared spectroscopy, Rietveld structure refinement and ^{27}Al, ^{29}Si, and ^{19}F MAS NMR spectroscopy. *American Mineralogist*, **72**, 580–593.

Raudsepp, M., Turnock, A.C. and Hawthorne, F.C. (1991) Amphiboles synthesis at low-pressure: what grows and what doesn't. *European Journal of Mineralogy*, **3**, 983–1004.

Rinaudo, C., Gastaldi, D. and Belluso, E. (2003) Characterization of chrysotile, antigorite and lizardite by FT-Raman spectroscopy. *The Canadian Mineralogist*, **41**, 883–890.

Rinaudo, C., Belluso, E. and Gastaldi, D. (2004) Assessment of the use of Raman spectroscopy for the determination of amphibole asbestos. *Mineralogical Magazine*, **68**, 455–465.

Rinaudo, C., Gastaldi, D., Belluso, E. and Capella, S. (2005) Application of Raman spectroscopy on asbestos fibre identification. *Neues Jahrbuch für Mineralogie Abhandlungen*, **182**, 31–36.

Rinaudo, C., Cairo, S., Gastaldi, D., Gianfagna, A., Mazziotti Tagliani, S., Tosi, G. and Conti, C. (2006) Characterization of fluoro-edenite by μ-Raman and μ-FTIR spectroscopy. *Mineralogical Magazine*, **70**, 291–298.

Rinaudo, C., Allegrina, M., Fornero, E., Musa, M., Croce, A. and Belli, D. (2009) Micro-Raman spectroscopy and VP-SEM/EDS applied to the identification of mineral particles and fibres in histological sections.

Journal of Raman Spectroscopy, **41**, 27–32.

Rinaudo, C., Croce, A., Musa, M., Fornero, E., Allegrina, M., Trivero, P., Bellis, D., Sferch, D., Toffalorio, F., Veronesi, G. and Pelosi, G. (2010) Study of inorganic particles, fibers, and asbestos bodies by variable pressure scanning electron microscopy with annexed energy dispersive spectroscopy and micro-Raman spectroscopy in thin sections of lung and pleural plaque. *Applied Spectroscopy*, **64**, 571–577.

Ristić, M., Czakó-Nagy, I., Musić, S. and Vértes, A. (2011) Spectroscopic characterization of chrysotile asbestos from different regions. *Journal of Molecular Structure*, **993**,120–126.

Robert, J.-L. (1981) *Etude cristallochimiques sur les micas et les amphiboles: application à la petrographie et à la géochimie*. Thèse d'Etat, Univerity of Paris XI, Paris (France).

Robert, J.-L. and Kodama, H. (1988) Generalization of the correlation between hydroxyl-stretching wavenumbers and composition of micas in the system K_2O–MgO–Al_2O_3-SiO_2-H_2O: a single model for trioctahedral and dioctahedral micas. *American Journal of Science*, **288-A**, 196–212.

Robert, J.-L., Della Ventura, G. and Thauvin, J.-L. (1989) The infrared OH stretching region of synthetic richterites in the system Na_2O–K_2O–CaO–MgO–SiO_2-H_2O–HF. *European Journal of Mineralogy*, **1**, 203–211.

Robert, J.-L., Gourdant, J.-P. and Fuchs, Y. (1996) Characterization of tourmalines by FTIR absorption spectrometry. *Physics and Chemistry of Minerals*, **23**, 309.

Robert, J.-L., Della Ventura, G. and Hawthorne, F.C. (1999) Near-infrared study of short-range disorder of OH and F in monoclinic amphiboles. *American Mineralogist*, **84**, 86–91.

Robert, J.-L., Della Ventura, G., Welch, M. and Hawthorne, F.C. (2000) OH-F substitution in synthetic pargasite at 1.5 kbar, 850°C. *American Mineralogist*, **85**, 926–931.

Rowbotham, G. and Farmer, V.C. (1974) The effect of the "A" site occupancy upon the hydroxyl stretching frequency in clinoamphiboles. *Contributions to Mineralogy and Petrology*, **38**, 147–149.

Scordari, F., Ventruti, G., Sabato, A., Bellatreccia, F., Della Ventura, G. and Pedrazzi, G. (2006) Ti-rich phlogopite from Mt. Vulture (Potenza, Italy) investigated by a multianalytical approach: substitutional mechanisms and orientation of the OH dipoles *European Journal of Mineralogy*, **18**, 379–391.

Semet, M.P. (1973) A crystal-chemical study of synthetic magnesiohastingsite. *American Mineralogist*, **58**, 480–494.

Seydou, Y., Della Ventura, G. and Petibois, C. (2010) Analytical characterization of cell-asbestos fibers interactions in lung pathogenesis. *Analytical and Bioanalytical Chemistry*, **397**, 2079–2089.

Seydou, Y., Chen, H.H., Harte, E., Della Ventura, G. and Petibois, C. (2013) The role of asbestos morphology on their cellular toxicity: an in vitro 3D Raman/Raley imaging study. *Analytical and Bioanalytical Chemistry*, **405**, 8701–8707.

Seydou, Y., Cestelli Guidi, M., Delugin, M., Della Ventura, G., Marcelli, A. and Petibois, C. (2016) Methodology for FTIR imaging of individual cells. *Acta Physica Polonica A*, **129**, 255–259.

Skogby, H. and Rossman, G.R. (1991) The intensity of amphibole OH bands in the infrared absorption spectrum. *Physics and Chemistry of Minerals*, **18**, 64–68.

Smith, B.C. (1989) *Fundamentals of Fourier Transform Infrared Spectroscopy*. CRC Press, New York, 202 pp.

Strens, R.G.J. (1966) Infrared study of cation ordering and clustering in some (Fe,Mg) amphibole solid solutions. *Chemical Communications*, **15**, 519–520.

Strens, R.G.J. (1974) The common chain, ribbon, and ring silicates. Pp. 305–329 in: *The Infrared Spectra of Minerals* (V.C. Farmer, editor). Monograph **4**, Mineralogical Society, London.

Stewart, M.J. and Haddow, A.C. (1929) Demonstration of the peculiar bodies of pulmonary asbestosis (asbestosis bodies) in material obtained by lung puncture and in the sputum. *The Journal of Pathology and Bacteriology*, **32**, 172.

Stuart, B. (2004) *Infrared Spectroscopy: Fundamentals and Applications*. Wiley, New Jersery, USA, 203 pp.

Stubican, V. and Roy, R. (1961) Isomorphous substitution and infra-red spectra of layer lattice silicates. *American Mineralogist*, **46**, 32–51.

Titulaer, M.K., van Miltenburg, J.C., Jansen, J.B.H. and Geus, J.W. (1993) Characterization of tabular chrysotile by thermoporometry, nitrogen sorption, DRIFT and TEM. *Clays and Clay Minerals*, **41**, 496–513.

Tomatis, M., Turci, F., Ceschino, R., Riganti, C., Gazzano, E., Martra, G., Ghigo, D. and Fubini, B. (2010) High aspect ratio materials: role of surface chemistry vs. length in the historical "long and short amosite asbestos fibers". *Inhalation Toxicology*, **22**, 984–998.

Trittschack, R. and Grobéty, B. (2013) The dehydroxylation of chrysotile: A combined in situ micro-Raman and micro-FTIR study. *American Mineralogist*, **98**, 1133–1145.

Turrell, G. (1972) *Infrared and Raman Spectra of Crystals*. Academic Press, London, 384 pp.

Velde, B. (1980) Ordering in synthetic aluminous serpentines: infrared spectra and cell-dimensions. *Physics and Chemistry of Minerals*, **6**, 209–220.

Velde, B. (1983) Infrared OH-stretch bands in potassic micas, talcs and saponites; influence of electronic configuration and site of charge compensation. *American Mineralogist*, **68**, 1169–1173.

Viani, A., Gualtieri, A.F., Secco, M., Peruzzo, L., Artioli, G. and Cruciani, G. (2013a) Crystal chemistry of cement-asbestos. *American Mineralogist*, **98**, 1095–1105.

Viani, A., Gualtieri, A.F., Pollastri, S., Rinaudo, C., Croce, A. and Urso, G. (2013b) Crystal chemistry of the high temperature product of transformation of cement-asbestos. *Journal of Hazardous Materials*, **248–249**, 69–80.

Viti, C. and Mellini, M. (1997) Contrasting chemical compositions in associated lizardite and crysotile in veins from Elba, Italy. *European Journal of Mineralogy*, **9**, 585–596.

Wang, A., Dhamelincourt, P. and Turrell, G. (1988a) Infrared and low-temperature mico-Raman spectra of the OH stretching vibrations in cummingtonite. *Applied Spectroscopy*, **42**, 1451–1457.

Wang, A., Dhamelincourt, P. and Turrell, G. (1988b) Raman microspectroscopic study of the cation distribution in amphiboles. *Applied Spectroscopy*, **42**, 1441–1450.

Wang, A., Freeman, J.J. and Jolliff, B.L. (2014) Understanding the Raman spectral features of phyllosilicates. *Journal of Raman Spectroscopy*, **46**, 829–845.

Watenphul, A., Burgdorf, M., Schlüter, J., Horn, I., Malcherek, T. and Mihailova, B. (2016) Exploring the potential of Raman spectroscopy for crystallochemical analyses of complex hydrous silicates: II. Tourmalines. *American Mineralogist*, **101**, 970–985.

Wicks, F.J. and O'Hanley, D.S. (1988) Serpentines: structure and petrology. Pp. 91–168 in: *Hydrous Phyllosilicates Exclusive of Micas* (S.W. Bailey, editor). Reviews in Mineralogy, **19**, Mineralogical Society of America, Chantilly, Virginia, USA.

Whittaker, E.J.W. and Zussman, J. (1956) The characterization of serpentine minerals by X-ray diffraction. *Mineralogical Magazine*, **31**, 107–126.

Yariv, S. and Heller-Kallai, L. (1975) The relationship between the I.R. spectra of serpentines and their structures. *Clays and Clay Minerals*, **23**, 142–152.

Yoshida, M., Tanabe, T., Ohno, N., Yoshimi, M. and Takamura, S. (2009) High temperature irradiation damage of carbon materials studies by laser Raman spectroscopy. *Journal of Nuclear Materials*, **388**, 841–843.

EMU Notes in Mineralogy, Vol. 18 (2017), Chapter 6, 171–214

Surface and bulk properties of mineral fibres relevant to toxicity

Francesco TURCI[1,*], Maura TOMATIS[1] and Alessandro PACELLA[2]

[1]*"G. Scansetti" Center for Studies on Asbestos and Other Toxic Particulates and Dipartimento di Chimica, Università di Torino, Via P. Giuria 7, I-10125 Torino, Italy, e-mail: francesco.turci@unito.it*
[2]*Dipartimento di Scienze della Terra, Sapienza Università di Roma, Piazzale Aldo Moro 5, I-00185 Roma, Italy*
**Corresponding author*

Many physicochemical properties of fibrous minerals concur – often simultaneously – in determining the fate of a particle within the human body. A complex chain of physicochemical transformations and biological reactions occur when a mineral fibre comes in contact with biological material and occasionally this results in inhaled material being held in an organism for a very long time. During that time, which might last for decades in the case of highly bio-persistent minerals such as amphibole asbestos, the fibre evolves and reacts with the human body, initially with body fluids and immune system cells, and dynamically interacts with its biological surroundings. To understand the molecular mechanisms of interaction, both bulk and surface properties of toxic – and potentially toxic – mineral fibres have to be considered. Far from being an exhaustive compendium of the enormous numbers of works dealing with the toxicological properties of minerals, this chapter is devoted to the discrimination of the bulk and surface – often interrelated – properties that impart toxic potential to fibrous minerals. To allow the establishment of a common background for a multidisciplinary audience, some general considerations of the factors influencing the health effect of mineral fibres aimed at clarifying some key toxicological concepts, including dose, exposure, clearance and molecular mechanisms, are provided in the initial paragraphs. Because asbestos accounts undoubtedly for the vast bulk of toxicological research on minerals, an essential introduction of the key toxicological properties of non-asbestos mineral fibres is also provided. The chapter is divided into two main sections dealing with bulk and surface properties of mineral fibres variously involved in toxicological response. Specifically, fibre morphology, biopersistence and fibre surface properties, including the generation mechanisms of particle-induced reactive oxygen species (ROS), the investigation of surface active sites and the surface modification induced by the biological environment, are reported and discussed with the support of more than 200 bibliographic references.

1. Introduction

The investigation of the toxicological mechanisms of particles and fibres is an active area of environmental and biomedical research, with broad implications for geochemistry, medicine, occupational health and safety assessment. It is far beyond the scope of this

DOI: 10.1180/EMU-notes.18.6

chapter to describe, even partially, the vast scientific literature on this topic. Nonetheless, we try here to focus on the key toxicological concepts that will provide the necessary background to appreciate how mineral complexity may induce adverse effect on some well defined biochemical mechanisms. The crystal structural and the chemical composition or compositional range of a material define, in nature, a mineral species, which can be classified further into sub-species according to specific mineral characteristics such as crystal habit. It has been known for a long time that exposure to minerals may increase the risk of various diseases. It was only around the second half of the 20th century that systematic investigation into the health effects of minerals on human health began (Guthrie, 1997). During those years, Wagner *et al.* (1960) reported on the potential hazard to human health from exposure to asbestos fibres and several cases of pleural mesothelioma in miners working in South African mines contaminated with asbestos were described. Shortly afterwards, non-occupational cases were also reported (Newhouse and Thompson, 1965). Since 1977, the International Agency for Research on Cancer (IARC) of the United Nations recognized asbestos as a human carcinogen, able to induce a significant incidence of pleural and peritoneal mesothelioma, and lung, gastrointestinal and larynx cancer in groups exposed occupationally to the minerals. However, asbestos is the generic commercial designation for six naturally occurring mineral fibres belonging to the serpentine and amphibole series and showing quite different bulk and surface properties. The role of specific fibre parameters are still controversial, mainly the mineralogical species and size, in modulating the risk associated with exposure to asbestos. The serpentine group contains a single asbestiform variety, namely chrysotile; five asbestiform varieties of amphiboles are known: anthophyllite asbestos, grunerite asbestos (amosite), riebeckite asbestos (crocidolite), tremolite asbestos and actinolite asbestos. These fibrous minerals share several properties that qualify them as asbestiform fibres: they are found in bundles of fibres that can be separated easily from the host matrix or cleaved into thinner fibres; the fibres exhibit high tensile strengths, they show large length/diameter (aspect) ratios; they are sufficiently flexible to be spun; and macroscopically, they resemble organic fibres such as cellulose. As asbestos fibres are all silicates, they exhibit several other common properties, such as incombustibility, thermal stability, resistance to biodegradation, chemical inertia towards most chemicals, and low electrical conductivity. Interestingly, they all exhibit iron ions both as structural or substitutional cations. The term asbestos has been attributed traditionally only to those varieties that are commercially exploited, but other fibrous minerals (we will discuss ambiguous terms denoting elongated mineral particles in the following paragraphs), including the amphiboles winchite, richterite (Meeker *et al.*, 2003) and fluoro-edenite (Gianfagna *et al.*, 2003), or the asbestiform balangeroite (Compagnoni *et al.*, 1983) and carlosturanite (Compagnoni *et al.*, 1985) from the Western Alps are treated conveniently along with asbestos. These elongated minerals share many properties with asbestos, often some toxic effects as well, but they are not considered specifically in any current asbestos regulations. US National Institute of Occupational Health and Safety (NIOSH, 2011) has recently proposed to extend the definition of asbestos to all elongated mineral particles (EMP). The microscopic and macroscopic properties of asbestos fibres stem from their intrinsic, and sometimes unique, crystalline features. As with all silicate minerals, the basic building blocks of asbestos fibres are the silicate tetrahedral, which may occur as double chains $(Si_4O_{11})^{6-}$, as in the amphiboles, or in sheets $(Si_4O_{10})^{4-}$, as in chrysotile. In the case of chrysotile, an octahedral brucite layer having the formula $(Mg_6O_4(OH)_8)^{4-}$ is intercalated between each silicate tetrahedral sheet (Virta, 2002). The relationship between workplace exposure to airborne asbestos fibres and respiratory diseases is one of the most widely studied subjects in modern epidemiology (Becklake, 1976; Hessel *et al.*, 2005). The research effort has resulted in significant consensus in some areas, although controversies remain in others (Stayner *et al.*, 1996). It is recognized broadly that the inhalation of long (considered usually longer than 5 μm), thin and durable fibres can induce or promote cancer in rats (Stanton *et al.*, 1981). It is also widely accepted that asbestos fibres can be associated with three types of diseases: asbestosis, lung cancer and mesothelioma. A further consensus has developed within the scientific community regarding the relative carcinogenicity of the different types of asbestos fibres. There is strong evidence that the genotoxic and carcinogenic potentials of asbestos fibres are not identical; in particular, mesothelial cancer is associated mostly with amphibole fibres (Churg, 1988), although it is accepted generally that chrysotile can induce lung cancer and, in some cases, also mesothelioma (Smith and Wright, 1996; Nicholson, 2001; Lemen, 2004). Since the beginning of toxicological studies on mineral fibres, scientists recognized that some well defined mineralogical properties were involved directly in the biological responses to minerals, and detailed knowledge of such properties could be the key to understanding and ultimately predicting the positive or negative influence of minerals on human health. When considering the impact of minerals on human health, it is worth remembering that past exposure to asbestos is currently the leading cause of occupational cancer, followed by crystalline silica (quartz dust). The United Nations World Health Organization estimated that >100 million people per year are still exposed occupationally to asbestos and >100,000 die each year from asbestos-related lung cancer, mesothelioma and asbestosis resulting from occupational exposures (WHO, 2006). Nonetheless, mineral fibres differing from asbestos occur both as natural and man-made fibres. For a more comprehensive description of the mechanisms involved in the toxicity of mineral fibres, a brief summary of our knowledge of natural and man-made fibres is provided here. In all cases, inhalation is considered to be the most relevant route for exposure to mineral particles. However, ingestion and dermal contact have been considered in some cases, particularly when mineral particulate matter is suspended in water used for drinking and watering (WHO, 2003 Asbestos in drinking-water).

1.1. Key toxicological facts about non-asbestos mineral fibres

1.1.1. Naturally occurring fibres

Palygorskite, also known as attapulgite, is a hydrated magnesium aluminium silicate with magnesium replaced partially by aluminium or iron and is common in clay deposits and in calcareous soils. Palygorskite exhibits an elongated morphology and a structure similar to that of amphibole minerals. Fibre lengths are very variable, depending on geographic origin. Studies of long-term inhalation by rodents suggest that carcinogenicity is dependent on the proportion of fibres longer than 5 μm in the samples. Accordingly, several studies on rats showed no increase in the incidence of tumours after inhalation, intrapleural or intraperitoneal injection of fibres $\leqslant 5$ μm in length (IARC, 1997). On the basis of these studies IARC included long palygorskite (>5 μm long) in Group 2B (possibly carcinogenic to humans) and short palygorskite in Group 3 (not classifiable as to their carcinogenicity to humans). In 2005, the WHO workshop on mechanisms of fibre carcinogenesis confirmed this evaluation (WHO, 2005).

Sepiolite is a fibrous hydrated magnesium silicate. Although the characteristics of sepiolite vary with the source, fibres in commercial samples are usually short (<5 μm long). No data on carcinogenicity in humans are available, and no significant reduction of lung function was found in workers with high exposure to sepiolite dust. Evidence for carcinogenic effects on experimental animals is limited. An increase in the incidence of pleural and peritoneal mesothelioma was observed in rats and mice only after intrapleural or intraperitoneal administration of long fibres (up to 100 μm) (IARC, 1997). On the basis of the inadequate evidence in humans and on the limited evidence in rodents for the carcinogenicity of fibres longer than 5 μm, IARC included sepiolite in Group 3.

Wollastonite is a needle-like calcium silicate mineral. Occupational exposure to wollastonite causes reduction of lung functionality and pneumoconiosis (Maxim and McConnell, 2005). *In vivo* studies demonstrated that wollastonite has low biopersistence and induces only a transient inflammatory response in rats after inhalation. No development of tumour or interstitial fibrosis was found in laboratory animals after inhalation, or intratracheal, intrapleural and intraperitoneal administration of wollastonite (Maxim *et al.*, 2014). In 1997, on the basis of inadequate evidence for carcinogenicity in humans and animals, IARC included wollastonite in Group 3. In 2005, an IARC working group on mechanisms of fibre carcinogenesis confirmed that the hazard was likely to be low (WHO, 2005).

Erionite is a hydrated aluminosilicate that occurs naturally in fibrous form. Several epidemiological studies on people living in Cappadocia (Turkey), where erionite is largely diffuse, demonstrated that exposure to erionite is associated with a high incidence of pleural and peritoneal mesothelioma (IARC, 2012a). Its mesotheliogenic potency seems to be much higher than that of asbestos fibres. On the basis of this evidence, in 1987, IARC classified erionite in Group 1 'carcinogenic to humans'. The evaluation was confirmed by IARC in the monograph 100C (IARC, 2012a).

Fluoro-edenite is a mineral of volcanic origin belonging to the amphibole group that occurs as asbestiform fibres around the Etna volcano (Sicily, Italy). Exposure to fibrous fluoro-edenite was associated with an increased incidence of pleural mesothelioma in the small town of Biancavilla (Catania, Italy) where this mineral was present in the material used in the local building industry. Experimental evidence of carcinogenicity was obtained also in rats after intraperitoneal and intrapleural injection. Fibrous fluoro-edenite has been classified by IARC as carcinogenic to humans (Group 1) (Gross *et al.*, 2014; IARC, in press).

Balangeroite and other asbestiform minerals are associated commonly with chrysotile asbestos over large natural areas. They may modulate toxicity and in some cases act as hallmarks of asbestos exposure (Finkelstein and Dufresne, 1999). Balangeroite is a fibrous mineral found mainly in the now disused chrysotile mine in Balangero (Italy) and in its surroundings (Compagnoni *et al.*, 1983). Balangeroite is not an amphibole, nor does it belong to the serpentine family, but has a gageite-like structure (Ferraris *et al.*, 1987). The abundance of balangeroite varies from one site to the other, so that any claimed percentage of it is just a coarse estimate. As for carlosturanite (Compagnoni *et al.*, 1985) and other accessory fibrous minerals, no one has ever been exposed to balangeroite alone, nor is it feasible to perform sufficient *in vivo* experimental tests with chrysotile-free balangeroite (Turci *et al.*, 2009). Nevertheless, the bulk and surface properties of these minerals, including the high aspect ratio and the presence of reactive surfaces, often allows the formulation of hypotheses on their contribution to the overall toxicity of associated chrysotile (Gazzano *et al.*, 2005; Groppo *et al.*, 2005; Turci *et al.*, 2005).

1.1.2. Man-made fibres

The terms man-made mineral fibres (MMMF) and man-made vitreous fibres (MMVF) refer mainly to fibrous inorganic substances made from rock, clay, slag or glass. MMVF include glass fibres (comprising glasswool and glass filament), rockwool, slagwool and ceramic fibres. MMVF have an amorphous structure, and their diameters and lengths vary greatly depending upon the method of production and use (TIMA, 1991).

Continuous glass filament and insulation glass wool: epidemiological studies on workers exposed to these glass fibres did not provide evidence of excess in respiratory cancers (trachea, bronchus and lung), respiratory symptoms or decrease in lung function (IARC, 1988, 2002; Baan *et al.*, 2004). *In vivo* studies did not provide adequate evidence for carcinogenicity in experimental animals. Intraperitoneal administration in rats of continuous glass filament of diameter >3μm did not result in tumour development. No significant increase in lung tumours and no mesotheliomas were observed in rats and hamsters after inhalation of insulation glass wool (IARC, 1988, 2002). Continuous glass filament and insulation glass wool have been classified in Group 3 by IARC (IARC, 2002).

Special-purpose glass fibres are small-diameter fibres designed mainly for filtration. Chronic inhalation studies on rats showed pleural fibrosis and an increase in lung tumours and mesotheliomas in rats and, although few in number, in hamsters. Intraperitoneal studies also reported an increase in peritoneal tumours, whereas more contradictory data were obtained in intratracheal studies (IARC, 2002). On the basis of these data, the IARC Working Group in 2001 concluded that there is 'sufficient evidence' in experimental animals for the carcinogenicity of special-purpose glass fibres and included them in Group 2B 'possibly carcinogenic to humans'.

Slag and rockwool: large cohort studies and case-control studies from the USA and Europe indicated an overall elevated risk for lung cancer in workers exposed to slag and rockwool, but without correlation with duration of exposure or with time since first exposure (IARC, 1988, 2002). No increase in lung tumour and mesothelioma incidence was found in studies conducted after inhalation or intratracheal instillation. An increased incidence of mesothelioma was observed in rodents only after intraperitoneal injection of large doses of slag and rockwool (IARC, 2002). Considering the inadequate evidence for carcinogenicity in humans and the limited evidence in experimental animals, IARC included glasswool and rockwool in Group 3.

Refractory ceramic fibres (RCF): epidemiological data are limited and do not allow an adequate evaluation of the cancer risk for workers exposed to RCF (reviewed in Mast *et al.*, 2000; IARC, 2002; Utell and Maxim, 2010; McKay *et al.*, 2011; Greim *et al.*, 2014). The studies available provided no evidence of increase in the incidence of lung cancer, mesothelioma or pulmonary fibrosis and found no significant loss of lung function, even following long exposure time. Conversely, evidence of a causal relationship between exposure to RCF and occurrence of pleural plaques was found in some studies (IARC, 2002). *In vivo* studies (Mast *et al.*, 2000; IARC, 2002; Maxim *et al.*, 2003; Brown *et al.*, 2005) showed interstitial and pleural fibrosis in rats and hamsters, and lung cancer and mesothelioma in rats after inhalation of large doses of RCF. Intratracheal instillation did not induce development of lung tumours in rats and hamsters, whereas intrapleural and intraperitoneal administration produced variable incidences of mesotheliomas. Considering the sufficient evidence for carcinogenicity in experimental animals, the Working Group in 1987 included the refractory ceramic fibres in Group 2B (IARC, 1988), and confirmed this classification in 2002 (IARC, 2002).

Among synthetic crystalline fibres increasing attention is being paid to silicon carbide (SiC) fibres and whiskers and to carbon fibre. These were the object of two IARC workshops (2005, 2014).

Silicon carbide (SiC) fibres are poly-crystalline fibres of variable length and diameter. A significant amount of lung cancer mortality was found in workers operating the Acheson process for SiC production (Grosse *et al.*, 2014; IARC, in press). Because workers were exposed simultaneously to fibrous and non-fibrous SiC, quartz and cristobalite, the Acheson process was classified in Group 1 and SiC fibres in Group 2B.

SiC whiskers are single crystals a few microns in diameter and hundreds of microns in length. No epidemiological data are available for SiC whiskers, while experimental data provided evidence for carcinogenicity in rats after intrapleural implantation (Stanton *et al.*, 1981) or injection (Johnson *et al.*, 1996). Due to their size being similar to asbestos amphiboles, the IARC Working Group classified SiC whiskers in Group 2A 'probably carcinogenic to humans' (Grosse *et al.*, 2014).

Carbon fibres are made of at least 92 wt.% carbon and are prepared from heating of polymeric precursors. The term carbon fibres is commonly applied interchangeably for both fibres obtained by pyrolysis of organic polymers at 1100–1200°C, consisting essentially of amorphous carbon networks, and fibres produced by heating their precursor at 2000–2700°C and having a graphitic-like structure (Frank *et al.*, 2014).

Carbon fibres have diameter usually ranging from 5 to 15 μm, outside the respirable range. No pulmonary function abnormalities were observed in workers in a carbon fibre production plant (Jones *et al.*, 1982). Long-term inhalation studies on rats (Owen *et al.*, 1986; Martin *et al.*, 1989; Waritz *et al.*, 1998) and on guinea pigs (Holt and Horne, 1978; Holt, 1982) found no signs of lung inflammation or fibrosis, whereas a short-term inhalation assay (Warheit *et al.*, 1994) showed a dose-dependent transient inflammatory response in the lungs of exposed rats. On the basis of these data and considering that "workplace exposure in production and processing is mostly to non-respirable fibres", the IARC workshop considered that the hazard from inhalation exposure to these fibres to be low (WHO, 2005).

2. General considerations of the factors influencing the health effect of mineral fibres

2.1. Dose of toxic mineral

When describing the toxicological effect of substances, including mineral fibres, one should consider that the amount of toxicant taken up by the organism over a given period of time is defined as 'the dose'. The dose should not be confused with 'the exposure', which is any condition which provides an opportunity for an external environmental agent to enter the body (Bernstein *et al.*, 2005). During the design of toxicological investigations it is important to consider that the same exposure to a toxicant can result in largely different doses, *e.g.* in different respiratory systems (rodents *vs.* humans) (Hofmann *et al.*, 1989) the dose depositing in the lower respiratory tracts will be quite different because fibre respirability differs significantly between species.

The higher the dose of a toxic mineral, the greater the effects. For carcinogenic endpoints, the dose–response relationship is monotone, increasing and without a threshold below which the adverse effect will not occur (Moolgavkar *et al.*, 2014). For amphibole fibres, for example, the peritoneal mesothelioma risk is proportional to the square of cumulative exposure, whereas lung cancer risk lies between a linear and square relationship. Conversely, for non-carcinogenic endpoints a threshold below which the adverse effect will not occur can be observed (Moolgavkar *et al.*, 2014), *e.g.* the development of asbestosis requires workers to be exposed to asbestos concentration usually >25 fibre/mL/y (Mossman and Churg, 1998). Toxic effects engendered by the same dose of different mineral dusts depend on a number of physicochemical characteristics of each dust, such as size, *e.g.* long amosite fibres induce higher levels of pulmonary fibrosis and mesothelioma in rats than the same dose of short fibres (Davis *et al.*, 1986), shape, *e.g.* amphibole cleavage fragments show lower toxicity than their asbestiform analogues (Gamble and Gibbs, 2008) and durability, *e.g.* a certain dose of a rapidly dissolved fibre acts like a smaller dose of a durable fibre (Eastes and Hadley, 1996). The toxic effects also depend on the exposure route (Drummond *et al.*, 2016) and individual susceptibility (IARC, 2012b). The dose is usually expressed as particle mass per unit volume, even if other metrics are adopted under specific circumstances.

Particle volume (Morrow, 1988), particle surface area (Oberdörster, 1996) and particle number have often been used in fact. Fibre number per cm^3 is generally used to describe fibre exposure in occupational settings.

2.2. Exposure routes

The main ways by which people come into contact with mineral fibres are inhalation and ingestion. The latter pathway mainly refers to fibres in potable water supplies (*e.g.* from erosion of rocks or of cement-asbestos pipes) and from swallowing fibres cleared from the respiratory tract by the mucociliary system. However, no significant adverse effects have been observed following fibre ingestion (IARC, 2012b). Other exposure routes such as direct pulmonary deposition (*e.g.* intratracheal instillation or intrapleural injection) are relevant only for experimental studies. Mineral fibres can have different effects depending upon the exposure route. Inhalation of glass microfibres, for example, induced some fibrosis in rats, but no mesothelioma, whereas intraperitoneal administration increased the incidence of mesothelioma (Cullen *et al.*, 2000). Inhalation, but not ingestion, of asbestos fibres may induce pulmonary fibrosis, lung cancer and mesothelioma, and may result in an increased risk of development of stomach and colorectal cancer (IARC, 2012b). The difference in the effect produced is related to both physiological processes and chemical conditions to which the dust is subjected, and such differences should be taken into account during experimental studies for assessing potential dust toxicity. Inhalation is the most important exposure route for mineral fibres. Inhalation, as well as deposition, distribution, translocation and clearance, depends on particle size and shape. Particle inhalation and deposition are affected by mode of breathing and airway conditions too. Oral breathing increases the deposition of large particles in the tracheobronchial area, and in the alveoli with respect to nasal breathing; obstructive diseases usually reduce deposition in the alveolar region. Moreover, the amount deposited depends on duration of exposure, the subject's ventilation rate and anatomy of the respiratory tract. Differences in the structure of the lower respiratory tract within species explain the different depositional efficiency of inhaled particles in rodents and humans (Miller *et al.*, 1993). For example, the maximum diameter of particles able to reach the alveoli is ~5 μm in rodents and 10 μm in humans (Oberdörster, 2000).

2.3. Clearance mechanisms

Particles deposited in the airways may be cleared by several mechanisms (Lippmann *et al.*, 1980) depending upon the site of deposition and physical properties of the particles. Large particles deposited in the nose are rapidly removed by nose wiping, blowing, sneezing, *etc.*, while particles deposited in the pharynx are usually swallowed. In the upper airways covered by ciliated epithelium, clearance takes place by means of the mucociliary system. Particles are trapped in the mucus, which is then swept up to the mouth for swallowing to the gut. Mucociliary clearance takes place within the first 24 h of exposure (Sturm and Hoffman, 2009). Mucociliary transport rate is variable, both along the respiratory tract of a given individual and between individuals, and may be reduced by cigarette smoke, environmental pollutants and inflammatory conditions (Lippmann *et al.*, 1980), which decrease the pH of the airway surface liquid, a mucus gel-aqueous sol complex which covers the conducting airways, and increases mucus viscosity. Retardation of mucus flow has been observed also where smaller airways open into larger airways. Such retardation increases the contact time between particles and epithelial cells, increasing the opportunity of particle internalization. Removal of small particles that reach the alveolar region occurs through macrophages that move around on the lung surface, engulf the particles and then move upwards onto the mucociliary escalator for upward clearance. Alveolar macrophage-mediated clearance is slower than mucociliary clearance and particle retention half-time is several dozens of days. Major factors that determine macrophage clearance efficiency are particle size and amount of deposited particles (lung burden). Macrophages have diameters of ~10–15 μm in rodents and 14–21 μm in humans (Krombach *et al.*, 1997) and can engulf successfully particles smaller than their size. In particular, phagocytosis of fibres is strictly dependent on their length. Fibres shorter than 5 μm are completely phagocytized, whereas fibres longer than 30 μm cannot be engulfed (Allison, 1974; Schinwald *et al.*, 2012) and two or more macrophages were observed to attach to the same fibre. Long fibres not completely enclosed by macrophages lead to frustrated phagocytosis with leakage of the macrophage cells and release of fibres onto the epithelial surface. Autolysis of macrophages may also occur following internalization of particles with cytotoxic activity. Particles/fibres released from the macrophages and fibres not phagocytized because of their length may penetrate the epithelial cell barrier or be re-phagocytized by other macrophages. Some interstitialized fibres may translocate to capillary endothelial cells (Brody *et al.*, 1981), migrate along lymphatic drainage pathways (this pathway is limited by size with a maximal fibre length of ~9 μm: Oberdörster *et al.*, 1988), or to pleural space (Donaldson *et al.*, 2011).

Increase in particle lung burden, as a consequence of a high dose or long-term exposure, may also result in the reduction of the macrophage-mediated clearance and increase of the particle retention half-time (Oberdörster, 2002). Inhalation studies on rats have shown that clearance impairment starts when the volume of phagocytized particles exceeds 6% of the normal macrophage volume, and clearance ceases when this volume reaches 60% (Morrow, 1988).

2.4. Molecular pathogenesis of human cancers related to mineral dust exposure

Fibre inhalation may cause several health effects, ranging from a transient inflammation to fibrosis and cancer development. Some cell–fibre interactions taking place *in vivo* are recognized to be relevant to the overall pathogenicity. Once in the alveolar regions, fibres are engulfed by alveolar macrophages (AMs), which clear the particles out of the lungs. This clearance process may either succeed or fail, depending mainly on fibre length and intrinsic surface reactivity. In the latter case, macrophages will become activated and release transcription factors, reactive oxygen species (ROS) and reactive nitrogen species (RNS), chemotactic factors, hydrolases, cytokines, growth factors, with eventual cell death, release of the fibres in the alveolar space and onset of ingestion-reingestion cycles. Bronchiolar and alveolar epithelial cells will then be affected by products of both activated AM and generated by the fibre itself. Moreover, as mentioned above, fibres can be internalized by epithelial cells or translocate to interstitial space or migrate along lymphatic drainage pathways (Fig. 1).

2.5. Risk enhancing factors

Several factors are known to increase the risk associated with exposure to mineral fibres, most of them have been investigated extensively for asbestos.

2.5.1. Tobacco smoke

Tobacco smoke is one of the major causes of lung cancer *per se* and is classified in Group 1 (carcinogenic to humans) by IARC (2012c). Simultaneous exposition to tobacco smoke and asbestos fibres greatly increases the risk of developing lung cancer with respect to exposure to asbestos alone (IARC, 2012b). *In vitro* and *in vivo* studies showed that tobacco smoke increases the uptake of asbestos in lung epithelial cells. The

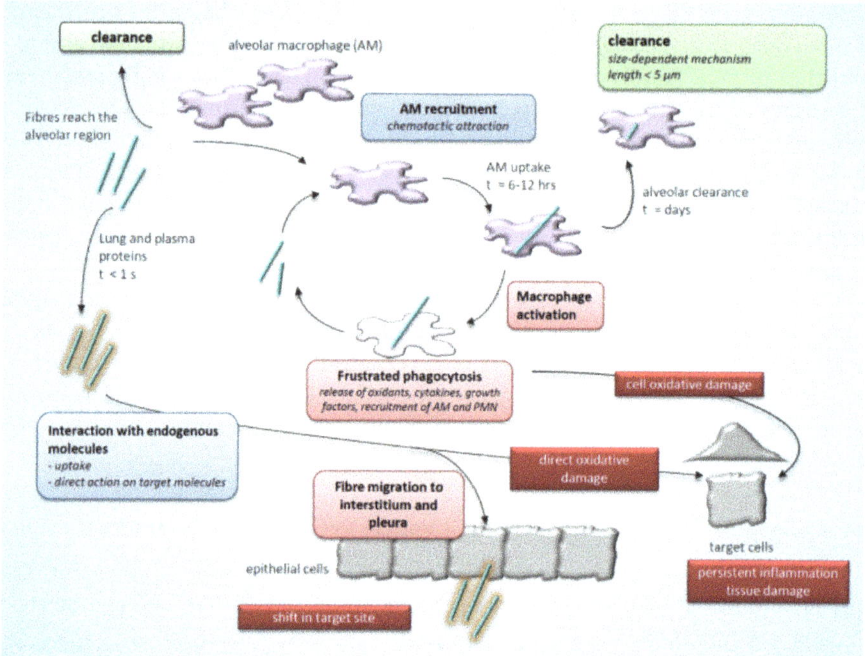

Figure 1. Schematic diagram of the multiple interactions of a mineral fibre in alveolar space. In the alveolar regions, fibres are engulfed by alveolar macrophages (AMs). If clearance fails, a frustrated phagocytosis cycle begin with the release of oxidant species, pro-inflammatory molecules. Bronchiolar and alveolar epithelial cells will then be affected by toxic products of both activated AM and generated by the fibre itself. Fibres can also be internalized by epithelial cells or translocate to different target sites.

mechanism responsible for enhanced uptake is still partially unclear; however, in smokers mucociliary clearance is significantly reduced with subsequent extension of the contact time with epithelial cells (Hammond *et al.* 1979) and increase of the amount of retained dusts (Cohen *et al.*, 1979). Moreover, reactive oxygen species derived from tobacco smoke may damage cellular membranes, thus increasing membrane permeability and enhancing fibre uptake (Churg *et al.*, 1989; Hobson *et al.*, 1990).

2.5.2. Genetic susceptibility

Genetic predisposition to malignant mesothelioma has been hypothesized in several studies reviewed by Neri *et al.* (2008). A plausible association with GSTM1 polymorphism has been observed for mesothelioma and other asbestos-related diseases, possibly linked to the role of GSTM1 in conjugation with reactive oxygen species, an important step in cellular response to oxidative damage.

The neurofibromatosis type *2 Nf2* gene product, merlin, is a tumour suppressor frequently inactivated in malignant mesothelioma and mutations in this gene have been found in 40% of mesothelioma (Thurneysen *et al.*, 2009). Mice exposed to asbestos fibres develop mesothelioma at a higher frequency after inactivation of one *Nf2* allele as compared to the wild-type (Fleury-Feith *et al.*, 2003). Down-regulation of *Nf2* in mouse embryonic fibroblasts results in enhanced cell spreading and invasion, whereas re-expression of *Nf2* inhibits markedly cell motility, spreading and invasiveness, properties connected with the malignant phenotype of malignant mesothelioma cells (Poulikakos *et al.*, 2006).

The genetic predisposition was supposed to be involved also in malignant mesothelioma observed in people exposed to erionite (Dogan *et al.*, 2006), a zeolite present in stones used to build some villages in Cappadocia (Turkey).

2.5.3. Simian virus 40 (SV40)

SV40 is a polyomavirus that is found in both monkeys and humans. SV40 is a DNA tumour virus with high oncogenic activity in hamsters in which it induces development of mesotheliomas, osteosarcomas, lymphomas or brain tumours depending on the route of inoculation. However, biological effects vary according to the species of origin (Carbone *et al.*, 2003). Hamster has been used as a preferential animal model to evaluate SV40 effects and there is a lack of studies on other animals and of epidemiologic studies on humans (Gazdar *et al.*, 2002; Carbone *et al.*, 2003). Humans have been exposed to SV40 through contaminated poliovaccines in the mid 1950s and an association between SV40 and mesothelioma was proposed on the basis of the large carcinogenic potential observed in experimental studies (reviewed by Rizzo *et al.*, 2001; Gazdar *et al.*, 2002; Poulin and DeCaprio, 2006) and the presence of SV40 in several human mesotheliomas (Carbon *et al.*, 2003). The role of SV40 in developing human cancer is still debated (Rizzo *et al.*, 2001; Poulin and DeCaprio, 2006); however, the Vaccine Safety Committee of the Institute of Medicine (IOM), National Academy of Sciences, USA, state that "the evidence is of moderate strength that SV40 exposure could lead to cancer in humans under natural conditions" (Stratton *et al.*, 2002).

3. Bulk structure of mineral fibres relevant to toxicology

The bulk structure of mineral fibres is usually investigated *via* a multidisciplinary approach. Conventional solid-state spectroscopic and diffraction techniques are employed commonly, even if more exotic approaches have been used to unveil subtle structural/compositional aspects that may impart special toxicological properties to the fibres. Among the several properties that may be involved in the toxicological outcome of mineral fibres the most investigated one is the effect of particle size and shape. Fibre morphology is considered universally to be a key parameter in pathogenesis and there are several mechanisms in which size and shape clearly influence toxicity and carcinogenicity. Fibre morphology determines deposition, translocation, and clearance in the lung, ultimately driving the fibre fate following inhalation. Bulk properties also control the solubility of a mineral, which in turn determines the biodurability, hence the residence time in the organism. The longer the residence time, the greater is the probability that the adverse interaction between mineral fibre and biological surroundings takes place. The following paragraphs will discuss in detail these aspects and their toxicological relevance.

3.1. Fibre morphology: size parameters and their relationship with adverse health effects

3.1.1. Definitions

As described by Belluso *et al.* (2017, this volume), the term 'mineral fibres' is used currently by the scientific community to indicate particles that occur in both asbestiform and other elongated-fibrous habits, *e.g.* needle-like (acicular) single crystals. The term "mineral fibres" is even used, under some circumstances, to define prismatic crystals and cleavage fragments that meet specified dimensional criteria, *e.g.* aspect ratio $\geqslant 3$, from the World Health Organization's definition of fibre (WHO, 1986). It is well known that cleavage fragments are generated by fracturing minerals, including the non-asbestiform polymorphs of asbestos minerals and the scientific community has clarified that cleavage fragments from non-asbestiform analogues are not as potent in terms of toxicity as asbestos fibres (Mossman, 2008). However, the use of non-standard terminology relative to dimensional aspects of mineral fibres has led to some ambiguity in exposure-response relationships, and could also lead to imprecise recommendations to protect human health. The terms 'asbestos', 'asbestiform', 'fibre' and 'fibrous' are often interpreted differently depending on the scientific discipline or the regulatory context. There is no current standard terminology that is complete and unambiguous. As described below, the definition of 'asbestos' is discipline-dependent and the elongation parameter, the length/width ratio or 'aspect ratio', is usually considered as the main discriminating factor between asbestos and non-asbestos minerals. A detailed tabulation of asbestos-related terminology, including acicular, asbestiform, cleavage, cleavage fragment, fibre, fibril, fibrous and parting, can be found in the work by Lowers and Meeker (2002). The term 'fibrous' in mineralogy is applied to elongated particles with a large aspect ratio ($\geqslant 10$) (Skinner *et al.*, 1988). NIOSH defines 'fibre' as a particle with aspect ratio >3 and length >5 μm as determined by the phase-contrast optical microscope at a magnification of 450 × to 500 × (NIOSH, 1984). WHO applies the term "fibre" to any elongated structure with $L > 5$ μm, $D < 3$ μm and aspect ratio >3 (WHO, 1999). Furthermore, the term "asbestos" refers to six fibrous silicates macroscopically readily separable into long, thin, flexible fibres when crushed or processed, but such straightforward differentiation is often blurred at the microscopic scale and it is often hard or even impossible to discriminate among specimens, *e.g.* fibrous antigorite *vs.* chrysotile asbestos. 'Asbestiform' refers to a subset of fibrous minerals having a macroscopic habit similar to that of asbestos. Usually, asbestiform fibres are small in diameter (generally <0.5 μm), have a large aspect ratio (>20) and exhibit two or more of the following properties: (1) parallel fibres occurring in bundles; (2) fibre bundles displaying splayed ends; (3) matted masses of individual fibres; and (4) fibres showing curvature. A prismatic crystal is a structure with one elongated dimension and two other dimensions nearly equal. A needle-shaped structure with aspect ratio ranging from 3 to 20 is called 'acicular'. This term is usually applied in mineralogy to designate straight, elongated, individual structures terminated by crystal faces. The aspect ratio of acicular crystals is similar to that of fibres, but the thickness is usually greater and may extend to several millimetres (Skinner *et al.*, 1988). Acicular crystals do not have the strength and flexibility typical of asbestiform fibres. A particle formed by breakage of a mineral is called 'cleavage fragment', but usually the term 'cleavage fragment' refers to elongated particles generated by breakage of non-asbestiform amphibole minerals. Some cleavage fragments are acicular or needle-shaped because of the tendency of amphibole to cleave along two dimensions. However, their aspect ratio is usually less than that of asbestos and rarely exceeds 10. Furthermore, cleavage fragments are commonly >0.5 μm wide and do not show the strength and flexibility of asbestiform particles (Harper *et al.*, 2008 and references therein).

3.1.2. Regulatory effect of fibre definition

One of the main difficulties in fibre analysis is the translation of macroscopic properties of minerals that can be observed in natural outcrops or with a light microscopy, to microscopically identifiable characteristics, such the micro and nanostructures visible with scanning and transmission electron microscopy. Furthermore, minerals sharing the same chemistry in an asbestiform crystal habit are not given a separate name (either chemical or common) from species in a massive habit. The above-mentioned lack of consistency in mineral definition generates uncertainty in the identification of asbestos fibres in both toxicological experiments and occupational measurements. In the latter setting, difficulties are greater when occurrences of fibres from natural sources are investigated. Further differentiation is induced by the matrix in which asbestos fibres have to be analysed. A more precautionary aspect ratio is generally considered a valid limit for airborne asbestos ($L/D \geqslant 3$, $L > 5$ μm, $D < 3$ μm), while, in drinking water, the US EPA (Method 100.2) proposes to include in the calculation only fibres with $L > 10$ μm. In soil and excavated rock the dimensional differentiation among asbestos and non-asbestos polymorphs may be even more subtle and often morphological criteria from mineralogy prevail, defining as asbestos those fibres with aspect ratio of 1:10, 1:20 or even 1:100.

3.1.3. Preparation of size-differentiated fibres for toxicological studies

The vast majority of toxicological investigations into the role of size parameters in *in vivo* and *in vitro* models are based on a mechanical preparation of a set of fibres sharing similar chemistry and diversification by size. Mechanical comminution is the reduction of solid materials from one average particle size to a smaller average particle size, by crushing, grinding, cutting, vibrating, or other. Size distribution and shape of the final product depends on grinding conditions, type of mill and the original size and nature of the mineral being ground. The most common forces applied during comminution are compression (particle disintegration by two rigid forces), impact (concussion by a single rigid force), shear (compression between two hard surfaces moving tangentially) and attrition (particles scraping against another or against a rigid surface). The finest particles with a narrow size distribution profile are usually obtained with shear forces, but shear on the crystalline lattice structure commonly causes a modification of the physicochemical properties of the ground particles. Application of compressive stress results in wide particles with a broad size distribution whereas impact gives intermediate results. During comminution, part of the mechanical energy is used to cause fractures and part is converted to heat. Size reduction may thus be accompanied not only by a change of solid surface area, but also of solid surface state and sometimes of bulk structure. Local heating and mechanical action may cause strain and defect generation and promote polymorphic transformation, formation of amorphous phases, surface activation, *etc*. For example, grinding causes the progressive amorphization of quartz and of some phyllosilicates, such as kaolin or talc (Aglietti *et al.*, 1986a,b; Liao and Senna, 1993), and transformation of magnetite in maghemite and then into hematite (Plescia *et al.*, 1995). The structural changes induced in the milled product depend upon milling conditions (time and energy of milling), grinding atmosphere (dry or wet environment), and type of solid being milled. Langer *et al.* (1978) showed that prolonged milling of chrysotile causes a decrease in the fibre crystallinity, alteration of Si–O and Mg–O interlayer bonding, and changes in the brucite layer. This modification was probably responsible for the smaller hemolityc potency of milled sample compared with pristine ones. Energy intensive milling induces structural changes of serpentine minerals enhancing Mg and Fe leaching (Kim and Chung, 2002) and may transform chrysotile, crocidolite and amosite into non-fibrous amorphous solids. This procedure was proposed for recycling asbestos-containing waste (Plescia *et al.*, 2003). High-energy milling also modifies drastically carbon nanotubes, causing nanotubes to break, formation of carbon onion-like particles and, when prolonged, formation of amorphous carbon nanoparticles (Li *et al.*, 1999). Wet grinding preserves asbestos fibre integrity and generates individual fibrils with a higher aspect ratio, whereas dry grinding produces mainly fibre bundles with a smaller aspect ratio (Salamatipour *et al.*, 2016). Chrysotile ground in dry conditions induced higher ROS production in alveolar macrophages than chrysotile ground in wet conditions, probably because of the larger amount of iron on the surface of dry-ground fibres. However, no details of the materials used in the ball-milling procedure were provided; thus contamination with iron from materials used in the mill construction may not be excluded. Finally, Assuncao and Corn (1975) demonstrated that ball milling reduces the lengths of glass fibres without affecting their diameters and produces also non-fibrous particles, whereas it reduces both diameters and lengths of chrysotile asbestos. Mechanical milling of a sample of long amosite (LFA) fibres was used to produce a batch of short fibres (SFA), a technique employed widely, in the past three decades, to investigate the effect of length on asbestos toxicity. Ball milling resulted in a significant shortening of fibres that retained a diameter close to the diameter of the pristine sample, with a subsequent marked decrease in the aspect ratio (Davis *et al.*, 1986), and formation of several small fragments characterized by rough profiles and irregular shape. Ball milling also resulted in a change in surface properties, in particular in the amount and oxidation state of surface iron. Surfaces of short fibres contain mainly poorly coordinated, thus quickly removable, iron ions in the oxidized state (Tomatis *et al.*, 2010). This explains the ability of SFA to release larger amounts of oxidized iron per unit surface compared to LFA (Graham *et al.*, 1999) and the higher potency for generating carbon-centred radicals in aqueous solution (Tomatis *et al.*, 2010). Indeed, reactive surface sites involved in the generation of carbon-centred radicals are poorly coordinated ferrous ions (Tomatis *et al.*, 2002; Martra *et al.*, 2003; Turci *et al.*, 2011). Accordingly, unlike LFA, SFA were reactive in this mechanism only when ascorbic acid, which reduces the highly uncoordinated Fe^{3+} to Fe^{2+}, was introduced into the reaction medium. Alternative methods, *e.g.* ultrasonic treatment, for preparation of short fibres that limit surface modifications were thus proposed (Spurny *et al.*, 1979; Hanton *et al.*, 1988; Turci *et al.*, 2012).

3.1.4. Relationship between size parameters and lung deposition

The penetration of a particle into the lung depends primarily upon a size parameter defined as "aerodynamic diameter", D_{ae} the diameter of a unit density sphere (1.0 g cm^{-3}) with identical aerodynamic characteristics as the particle of interest (WHO 1999; Sturm and Hofmann, 2009). D_{ae} of a fibre is dominated by the fibre width and only slightly by the length. For instance, D_{ae} of a fibre 0.25 μm wide and 10 μm long is 0.67 μm, and if the diameter is doubled (width = 0.50 μm) D_{ae} becomes 1.47 μm, whereas if length is doubled (L = 20 μm) a 10% increase of D_{ae} (= 0.74 μm) is observed (Gross, 1981). Interestingly, with an aspect ratio of ~20 a fibre exhibits D_{ae} about only three times greater than its physical diameter and thus can flow through airways while acting like a much smaller particle. Due to its extensive use in inhalation toxicology, the aerodynamic diameter concept has found a rather wide acceptance, and alternative – possibly more accurate – concepts with similar reliability have not been introduced so far. Large particles (inhalable fraction, D_{ae} < 100 μm) are deposited in the nasopharyngeal tract, smaller particles (up to 10 μm) pass the larynx (thoracic fraction), and particles with D_{ae} < 4 μm (respirable fraction) may reach the alveoli (European Committee for Standardization, 1993). Once inhaled, the particles will deposit in the respiratory tract principally by inertial impaction (extrathoracic and upper bronchial regions), gravitational sedimentation (lower bronchial and alveolar region) and Brownian diffusion (alveolar region). The largest particles deposit principally by impaction, which occurs when the inertia of a particle leads to a deviation from the airstream. Sedimentation occurs for particles with $D_{ae} < 2.5$ μm when air velocity is slow, whereas the smallest particles ($D_{ae} < 0.5$ μm) deposit as a result of random collisions with airway walls by Brownian diffusion (Tsuda *et al.*, 2013). The behaviour of particles of elongated shape is more complex than that of isometric particles and is depicted in several theoretical models (Sturm and Hoffmann, 2009; Hoffmann, 2011; Roshchenko *et al.*, 2011). The depositional efficiency of fibres in the nasal region is usually less than that of spherical particles of comparable size. This increases the probability of fibrous particles penetrating the lung. Fibre deposition depends heavily on the orientation of fibres in the flow field. Fibre aggregates behave as non-fibrous particles, whereas single fibres tend to flow aerodynamically. A fibre whose flow orientation differs from axial alignment may deposit by interception (Lippmann, 1990). This mechanism is more important than the others in controlling fibre deposition in the lungs. Interception increases with fibre length and is more relevant for curly fibres than for straight ones (Roshchenko *et al.*, 2011). For small fibres, which may be assumed to be oriented randomly, deposition takes place in the small airways mainly by Brownian motion.

3.1.5. Relationship between size parameters and fibre carcinogenicity

The first evidence of the high carcinogenicity of long and thin fibres was provided by Stanton *et al.* (1981). Examining the effects engendered in rats by a large number of fibrous materials after intraperitoneal administration, Stanton and colleagues found that the probability of developing pleural sarcoma correlated best with the number of fibres with $D < 0.25$ μm and $L > 8$ μm ('Stanton's hypothesis'). The role of fibre length in induction of lung cancer was subsequently demonstrated by *in vivo* studies on asbestos (Davis *et al.*, 1985, 1986; Wagner *et al.*, 1985), erionite (Wagner *et al.*, 1985) and glass fibres (Muhle *et al.*, 1987) that revealed a large carcinogenic potential for fibres longer than 10 μm. Conversely, fibres shorter than 5 μm are regarded as having small potentcy for inducing development of lung cancer or mesothelioma. Long-term inhalation assays showed that short fibres of amosite (median length 2.7 μm) (Davis *et al.*, 1986), erionite ($L < 5$ μm) (Wagner., 1988) and chrysotile ($L < 5$ μm) (Platek *et al.*, 1985; Davis and Jones, 1988, Stettler *et al.*, 2008) did not induce development of lung cancer or mesotheliomas in laboratory animals. A large set of intratracheal and intraperitoneal studies (IARC, 1988, 1997, 2002) confirmed the low carcinogenic and fibrotic potential of short fibres. Long and thin fibres are suspected to be more potent than short and thick fibres in the induction of lung cancer in humans also (IARC, 2012b). However, there is a lack of information on fibre size-distributions in most epidemiological data. Two studies on workers at chrysotile textile mills in South and North Carolina (USA) showed that, even if exposure to chrysotile of all sizes increases the risk of cancer in humans, the best correlation is with fibres longer than 10 μm and thinner than 0.25 μm (Stayner *et al.*, 2008) or 1 μm (Loomis *et al.*, 2010). A successive reanalysis of the pooled data from both studies provided strong correlation with fibres having length ranging from 5 to 10 μm and D < 0.25 μm (Loomis *et al.*, 2012). In a meta-analysis of data from 15 asbestos cohort studies, Berman and Crump (2008) found weak evidence that fibres longer than 10 μm are more carcinogenic than fibres with length ranging from 5 to 10 μm, regardless of the width. On the basis of the available data, the World Health Organization rates breathable asbestos fibres (diameter <3 μm) as a health hazard under worldwide regulations only if they are longer than 5 μm. Nevertheless, a new hypothesis on the role of short fibres in the development of mesothelioma has recently surfaced in the medical literature (Dodson *et al.*, 2003; Suzuki *et al.*, 2005; Lemen, 2006; Roggli, 2015). In a critical review of asbestos fibre length and pathogenicity, Dodson *et al.* (2003) claimed that shorter fibres also may contribute to the pathological response since the most abundant fibres in the lung and mesothelial tissues of mesothelioma patients were reported to be shorter than 5 μm and thinner than 0.3 μm (Dodson *et al.*, 2003; Adib *et al.*, 2013). It is still unclear if these short fibres are the result of the splitting up of long fibres into sub-micrometric fibrils that may take place in the lung or if they were inhaled as short fibres. Despite these differences, however, all studies suggested that thin mineral fibres (diameter <1 μm) may be involved in the occurrence of lung cancer or mesothelioma, provided that they reside in the organism long enough (see Section 3.2. for more details on biopersistence).

3.1.6. Relationship between size parameters and fibre translocation

Because the pleura and peritoneum are two main target organs for asbestos-induced malignancy, the translocation from the alveolar region is a highly investigated topic. A model of alveolar retention of deposited fibres (Timbrell, 1983) showed that fibres thicker than 0.15 μm are retained easily in the lung and that the retention increases with diameter up to 0.8 μm. This leads to fibre accumulation and allows the long-fibre dose to build up, contributing to lung cancer development. Conversely, fibres thinner than 0.15 μm are readily translocated to the pleura and peritoneum and are involved in the induction of mesothelioma. The pathway for fibres to reach the pleural space is still unclear. Fibres can penetrate directly across injured alveolar epithelial cells or pass through the intercellular space between the cells (paracellular route) in the same manner as interstitial fluid (Miserocchi *et al.*, 2008). Once entered in the pleural space, fibres can pass through the stomata, pore-like structures with diameter between 3 and 10 μm in the parietal pleura linking the peritoneal cavity to the underlying lymphatic capillaries, and into the lymphatic system. This pathway depends on both fibre diameter and fibre length. Long asbestos fibres accumulate preferentially at the peritoneal face of the diaphragm around the stomata and could initiate inflammation, proliferation and granuloma formation (Donaldson *et al.*, 2010). However, the critical dimensions that govern fibre translocation as well mesothelioma induction are still debated (Broaddus *et al.*, 2011; Mossman *et al.*, 2011). Lentz *et al.* (2003), in a study on workers manufacturing refractory ceramic fibres, identified the critical size for translocation in the pleura as $D < 0.4$ µm and $L < 10$ µm, whereas Lippmann (1990, 2014) concluded that mesothelioma and development of pleural plaques is related to fibres thinner than 0.1 µm and longer than 5 µm. Clearly, ultrafine fibres (<0.25 µm) have the greatest chance of reaching the pleural space. Among these, short fibres (<5 µm) may cross the stomata and reach the lymphatic system, whereas long fibres are expected to block the stomata physically and accumulate forming "black spots" (Donaldson *et al.*, 2010). Black spots containing several amphibole and chrysotile fibres longer than 5 µm, were observed in some patients with a history of asbestos exposure (Boutin *et al.*, 1996). Accumulation of small fibres may also occur at these sites and several studies (Dodson *et al.*, 2003) have also demonstrated the large presence of short asbestos fibres in the pleura of asbestos-exposed workers. The accumulation leads to cell damage, recruitment of pleural macrophages which undergo frustrated phagocytosis in attempting to internalize the longer fibres, and subsequent inflammation, fibrosis and genotoxicity (Donaldson *et al.*, 2010). The studies of pleural translocation, focused mainly on asbestos, have been extended to other fibrous materials and a length-dependent inflammation, ascribed to the limited clearance through stomata of long fibres, has been reported recently for carbon nanotubes after intrapleural administration in mice (Murphy *et al.*, 2011).

3.1.7. Relationship between size parameters and cellular toxicity

Many cellular tests, reviewed by Boulanger *et al.* (2014) were performed with fibres of different length. The majority of these studies refer to asbestos and two sets of amosite fibres were prepared specifically, by milling from the same batch, in order to differ only in the dimensional parameters. Such samples were used widely to verify the Stanton hypothesis *in vitro*. The milled sample contained only 37% of particles with aspect ratio >3 with average length 2.7 µm (Davis *et al.*, 1986). 75% of these fibres had $L < 2$ µm and ~2% were longer than 5 µm (Riganti *et al.*, 2003). Short amosite fibres did not significantly inhibit the pentose phosphate pathway activity, did not modify the expression of glucose 6-phosphate dehydrogenase, did not induce lipid peroxidation (Riganti *et al.*, 2003), did not stimulate NO release (Tomatis *et al.*, 2010) and caused less epithelial detachment injury (Donaldson *et al.*, 1993) in alveolar epithelial cells compared to the long fibres. Moreover, short amosite fibres induced less release of superoxide anion (Hill *et al.*, 1995) and cytokine tumour necrosis factor (TNF) (Donaldson *et al.*, 1992; Dogra and Donaldson, 1995) in alveolar macrophages, and less chromosomal aberrations in Chinese hamster ovary cells (Donaldson and Golyasnya, 1995) with respect to the long ones. Goodglick and Kane (1990) evaluated the cytotoxicity of two crocidolite samples containing 72.4% and 98.5% of fibres with $L \leqslant 5$ µm. Cytotoxic effects were less for the sample with many short fibres; however, the authors considered that the observed effect was more dependent on the amount of iron on the fibre surface than on the fibre length, because the pre-treatment of fibres with a chelating agent reduced the cytotoxicity. Two samples of short chrysotile fibres (75% $L < 2.4$ µm and $L < 1.5$ µm, aspect ratio >10 for 87% and 49.6% of fibres, respectively) failed to cause LDH release, NO generation and lipid peroxidation in alveolar epithelial cells. However, both samples induced ROS production in a dose-dependent way, even if to a lesser extent than chrysotile samples containing fibres longer than 5 μm (Turci *et al.*, 2012). A study of different fibrous samples (asbestos and glass fibres) showed a positive correlation between fibre length greater than 3 μm in lung fibroblast or 4 μm in alveolar epithelial cells and cytotoxic effects (Brown *et al.*, 1986).

The transformation potency of several fibres as a function of their length was examined by Hesterberg and Barrett (1984) on Syrian hamster embryo cells. The ability of chrysotile to cause morphological transformation was decreased significantly by milling. The reduction of average length of thin glass fibres from 9.5 to 1.7 μm led to a 10-fold decrease in transforming activity, and the transforming activity was stopped completely following further reduction to 0.95 μm. Moreover, long glass fibres (17 μm) were twice as potent in stimulation of TNF-α production in murine peritoneal macrophages than short fibres (7 μm) (Ye *et al.*, 1999).

3.1.8. Health effect of cleavage fragments

The number of studies on asbestos cleavage fragments is limited, but they suggest that non-asbestiform amphiboles might pose lower risks than actual asbestos fibres (Ilgren, 2004; Addison and McConnell, 2008; Mossman, 2008; NIOSH, 2011; Williams *et al.*, 2013). Epidemiological studies (Gamble and Gibbs, 2008) on workers exposed to non-asbestiform amphiboles demonstrated no increased risk for mesothelioma development. *In vivo* studies showed that non-asbestiform amphiboles are non-carcinogenic in animals (Ilgren *et al.*, 2004; Addison and McConnell, 2008; Williams *et al.*, 2013). Intrapleural administration of non-asbestiform tremolite in hamsters and rats (and intraperitoneal injection of non-asbestiform actinolite in hamsters) failed to induce mesothelioma. Cellular studies have demonstrated also that low aspect-ratio cleavage fragments are less toxic than their asbestiform analogues (Mossman, 2008 and references therein). Riebeckite (non-fibrous crocidolite) and antigorite cleavage fragments, both with elongated shape but large diameters, failed to cause squamous metaplasia and to increase DNA synthesis in tracheal explant cultures, and to induce ornithine decarboxylase, an enzyme associated with cell proliferation, in hamster tracheal epithelial cells. Riebeckite and antigorite were less cytotoxic than crocidolite and chrysotile, and riebeckite was less potent than crocidolite in stimulating cell proliferation. Riebeckite induced a smaller release of superoxide than crocidolite in alveolar macrophages and did not induce apoptosis or DNA synthesis in mesothelial cells. The biological reactivity of cleavage fragments, however, tends to approach that of asbestos when the size of the acicular crystals approaches that of asbestiform analogues. A fibrous antigorite (14–20 μm long and 0.08–0.4 μm wide) was cytotoxic in macrophages, mesothelial (Cardile *et al.*, 2007) and alveolar epithelial cells (Pugnaloni *et al.*, 2010), stimulated ROS and NO generation in macrophages and mesothelial cells (Cardile *et al.*, 2007), and increased synthesis of vascular endothelial growth factor (VEGF) and of two proteins (Cdc42 and β-catenin) involved in tumour development (Pugnaloni *et al.*, 2010), even if to a lesser extent than crocidolite.

However, cleavage fragments with a large aspect ratio of >10 are not very common (Harper *et al.*, 2008). Even if some splinter-like cleavage fragments can be mistaken for asbestos, thin fragments are rarely found and an inter-laboratory study aimed at determining the best procedure for discriminating amphibole asbestos from cleavage fragments by phase-contrast microscopy suggested merely the use of a width-based criterion (Harper *et al.*, 2012). On the other hand, additional features could contribute to the difference in the biological reactivity between cleavage fragments and asbestos (Ilgren, 2004). Under pressure, asbestiform minerals tend to split longitudinally. This behaviour is also observed *in vivo*, after deposition in the lung. The longitudinal splitting is responsible for the increase in the total number of fibres retained in the lung. Conversely, cleavage fragments tend to break transversely resulting in shorter particles that can be cleared easily through phagocytosis. Moreover, cleavage fragments are less stable in acid medium than equivalent asbestos because of the abundance of surface defects, steps and cracks (Ilgren, 2004) that are preferred sites for chemical attack. Indeed, in contrast to asbestos, which attains its shape by crystal growth, cleavage fragments originate from fracturing processes that lead to defective surfaces. These surface differences could result in a higher solubility in lung fluids. This point, however, should be further investigated (Plumlee *et al.*, 2006).

3.2. Biopersistence and biodurability of mineral fibres

For a detailed discussion of fibrous mineral durability, both in terms of residence time in the organism and *in vitro*/cell-free evaluation of biopersistence, the reader should refer to Chapter 10 (Rozalén *et al.*, 2017, this volume). Here, the main aspects of these mineral properties are reported and discussed only in terms of their potential impact on human health. The importance of biopersistence in determining the health effect of inhaled fibres is widely recognized and has been affirmed by several government and advisory agencies (IARC, 2002; Bernstein *et al.*, 2005). The European Community, for example, adopted in 1997 a directive (97/69/EEC) on hazardous substance classification, in which it was stated that respirable man-made vitreous fibres (MMVF) may be excluded from classification as a carcinogen if low biopersistence is demonstrated through *in vivo* measurements. As far as asbestos is concerned, the most discussed question is about the relatively short persistence in the organism of chrysotile and the consequences for the carcinogenic potential of these fibres. The question of whether chrysotile is less potent than amphiboles in inducing lung cancer has been defined as the 'amphibole hypothesis' and discussed over the years in several lung biopersistence studies that compare chrysotile carcinogenicity with that of crocidolite and amosite (Cullen, 1996; Stayner *et al.*, 1996; McDonald and McDonald, 1997; Bernstein and Hoskins, 2006). However, recently, the IARC working group (IARC, 2012b) settled the dispute by pointing out that the lower persistence of chrysotile with respect to amphibole does not imply that chrysotile is a less potent aetiological agent for lung cancer than crocidolite and amosite. In fact, it is important to recognize that the mechanistic basis for carcinogenicity of asbestos, and xenobiotic minerals in general, derives from the complex interactions between the mineral surface and the biological hosting environment (body fluids, membranes, cells, tissues). The structural and physicochemical differentiation between mineral species and the surface transformations that may occur *in vivo* accounts for an endless number of permutations and a number of multiple direct and indirect mechanisms have been proposed to drive the multiple stages in the development of adverse health effects. The term biopersistence refers to the extent to which a fibre is able to resist chemical, physical and other physiological clearance mechanisms in the body (Bernstein *et al.*, 2005), whereas (bio)durability is the ability to resist (bio)chemical alteration. Solubility is the amount of a substance that dissolves in a unit volume of a liquid to form a saturated solution under specified conditions of temperature and pressure. The (bio)solubility of a mineral fibre is the extent to which the fibre dissolves in body fluids. Dissolution is the process in which one substance is dissolved in another and, in the case of mineral fibres, indicates the release of ions from the surface (Utembe *et al.*, 2015). Chemical dissolution clearly plays an important role in the biopersistence of several fibres, mainly of the synthetic ones, and the dissolution rate of fibres is often extended to apply to fibre biodurability/biosolubility. The dissolution may take place through a congruent (same rate of dissolution for all components) or an incongruent (preferential leaching of certain components) process (Hesterberg and Hart, 2001). In the first case, fibre length and diameter decrease at a constant rate and fibre bulk composition does not vary over time, though subtle alterations may occur at the fibre surface. Basalt and high-Al silicate fibres have been found to undergo congruent dissolution, but inorganic fibres generally dissolve incongruently. The different leaching of some components during incongruent dissolution may cause the formation of non-crystalline layers on the fibre surface and promote precipitation of newly formed mineral species, often in the nanometric range (Favero-Longo *et al.*, 2013; Pacella *et al.*, 2015). Moreover, leaching may weaken the structure causing fibre breakage. The transverse fragmentation of long fibres into short fragments, in turn, enhances macrophage clearance. Once inhaled, the fibres may encounter different chemical environments. Indeed, inhaled fibres come in contact with both extracellular and intracellular fluids. One of the most important extracellular fluids is the lung lining fluid (LLF), a thin layer of fluid which covers the surface of the epithelium from nasal mucosa to alveoli. The LLF composition varies along the respiratory tract. LLF contains phospholipids (mainly dipalmitoylphosphatidylcholine), glycomucoproteins, antioxidants (uric and ascorbic acid, and reduced glutathione) and an electrolyte phase (major components Na, K, Cl). Its pH is close to neutrality. Ascorbate and glutathione were observed to be greatest in broncho-alveolar space, conversely significantly higher concentrations of uric acid were found in the nasal region (Cross *et al.*, 1994). Interstitial fluid (pH = 7.4) contains proteins, organic acids, aminoacids, and sodium, chloride and bicarbonate ions. Intracellular fluids (pH 7.1) contain proteins, organic acids, antioxidants such as glutathione and are rich in potassium, magnesium, bicarbonate and phosphate ions. Finally, macrophage lysosomes contain acid proteases, acid phosphatases, and other enzymes and are slightly acidic pH (5.5–4.5) (Plumlee *et al.*, 2006). The interaction with these fluids promotes physicochemical transformation of the fibres. The physicochemical change depends upon the characteristics of the fluid, mainly composition and pH, and upon structure (amorphous *vs.* crystalline) and chemical composition of the fibres. Many comparative investigations on the bio-solubility of mineral fibres rely on data collected with simulated body fluids that mimic the different chemical environments that particles may encounter once inhaled. Amorphous solids are more soluble than crystalline ones. Asbestos is poorly soluble compared to synthetic vitreous fibres. Among the asbestos materials, chrysotile exhibits a faster dissolution rate than amphiboles, due to its crystal structure (see Ballirano *et al.*, 2017; Rozalén *et al.*, 2017, this volume). For silicate minerals the structure stability increases with the covalent character, which, in turn, is related directly to the silicon-to-oxygen molar ratio. Indeed, when the ratio increases only a few cations are required to balance the negative charge imparted to the structure by oxygen (Bernstein, 2007). The solubility of amorphous fibres, *e.g.* MMVF, depends on the nature and the proportion of network formers (compounds, such as SiO_2 and B_2O_3, that in their pure form can be melted and quenched into the glassy state) and network modifiers (compounds such as Na_2O, K_2O, Li_2O, CaO, MgO that modify the glass structure reducing the melt temperature of raw materials). Alkali and alkaline earth metal oxides enhance the dissolution rate of vitreous fibres, whereas Al_2O_3 has the opposite effect. Fibres with high aluminium or iron contents tend to dissolve more easily in acid than in neutral environment, whereas the reverse is true for fibres with high silicon and low calcium and magnesium contents (Hesterberg and Hart, 2001; Maxim *et al.*, 2006). Accordingly, with their composition (IARC, 2002), glass wools are more soluble in extracellular fluids than in phagolysosomal fluid, unlike rock wools, slag wools and refractory ceramic fibres (Hesterberg and Hart, 2001). For these fibres phagocytosis will aid dissolution and clearance. The dissolution process often induces fibre morphological changes. Crater-like surface pits, rounded or blunt ends, dissolved cores were observed on the surface of several MMVF (Hesterberg and Hart, 2001; Campopiano *et al.*, 2014). Leaching-induced weakness of the structure promotes fibre breakage (Hesterberg and Hart, 2001) that is an important mechanism in the biopersistence of MMVF. Several studies (reviewed by Maxim *et al.*, 2006) reported that fibres with a high dissolution rate in *in vitro* systems are also cleared rapidly during *in vivo* experiments. Moreover, Hesterberg and Hart (2000) found that fibres that dissolved quickly caused only transient lung inflammation in rodents. To enhance fibre dissolution, the industry has extended the traditional compositional ranges of MMVF by increasing the content of alkali oxides in glass wools and by developing high-alkaline earth silicate wools as an alternative to Al-rich refractory ceramic fibres (IARC, 2002). Some mathematical models have been developed to predict the dissolution rate of MMVF (Maxim *et al.*, 2006). The dissolution rate is related directly to fibre surface area and dissolution rate constant (k_{dis}), which in turn depends on fibre chemical composition and on the pH of the dissolution medium. For a wide set of borosilicate glasses, calculated k_{dis} are in agreement with k_{dis} values measured in experimental conditions. Some k_{dis} value ranges for MMVF are reported in Table 1. If the biodurability of MMVF depends upon chemical composition, the biodurability of

Table 1. Dissolution rate constant (k_{dis}) measured experimentally for different fibre types (adapted from Maxim *et al.*, 2006).

Fibre type	k_{dis} @ pH 7.4
Glass wool	70–300
Alkaline earth silicate wools (AES)	100–250
Rockwool	17–29
Slagwool	50–400
Ceramic refractory fibres (RCF)	7–10
Amphibole asbestos	0.1–1.3

carbon nanotubes depends on surface functionalization. Carbon nanotubes are quite stable in biological media, but surface functionalization with carboxyl groups promotes nanotube degradation in phagolysosomal simulant fluids, leading to length reduction and accumulation of ultrafine solid carbonaceous debris (Liu *et al.*, 2010a). The lower stability of carboxylated carbon nanotubes is probably due to damage to the tubular graphenic backbone following acid functionalization. The presence of surface defects plays a role also in the enzymatic degradation of CNT observed in an oxidative environment (H_2O_2-containing medium) in the presence of peroxidase (Allen *et al.*, 2008, 2009; Zhao *et al.*, 2011). Hypochlorite and reactive radical intermediates produced by myeloperoxidase, a complex heme protein abundant in neutrophils and monocytes, catalyse the biodegradation of CNT both in acellular and cellular (neutrophils and macrophages) tests. After biodegradation, nanotubes were not effective at generating an inflammatory response in the mice lung (Kagan *et al.*, 2010).

4. Surface properties of mineral fibre relevant to toxicology

The mineral surface is a dynamic entity that interacts with biological media, cells and tissues. The toxicity of a mineral fibre depends heavily on the combination of mechanical/physical properties (described in the previous paragraphs) and some key surface chemical properties, discussed here. The main surface factors involved in toxicity are described conveniently in term of reactivity, *i.e.* the potency of a mineral surface to alter chemically biomolecules, and the dissolution/re-precipitation reactions that either modulate dynamically the surface chemistry or segregate ions in neoformed minerals – often at the nanoscale – as new chemical entities. It is well known from studies of asbestos that interaction between fibrous material and a biological environment is strongly dependent on both the morphology and the crystal chemistry of mineral fibres. In particular, both the presence and the structural coordination of Fe have been shown to be crucial factors in fibre toxicity (Bonneau *et al.*, 1986; Turci *et al.*, 2011). Furthermore, only Fe exposed at the fibre surface, especially Fe(II), is considered to play a primary role in the production of reactive oxygen species (ROS) (Fantauzzi *et al.*, 2010, 2012; Pacella *et al.*, 2010, 2012; Turci *et al.*, 2005; Hardy and Aust, 1995).

4.1. Generation mechanisms of particle-induced reactive oxygen species (ROS)

Several mineral particles are able to stimulate ROS and RNS production after inhalation. Moderate concentrations of reactive species are produced during normal cellular metabolism and some of these play an important role as regulatory mediators in signalling processes. Endogenous antioxidants are generally capable of counterbalancing the effects of reactive species, but when the amount of ROS and RNS produced exceeds the cellular ability to remove them, cellular damage ensues. Oxidative stress plays an important role in inflammation and other diseases, including cancer (Azad *et al.*, 2008). Reactive species may be produced by the particles themselves or by the cells after contact with the particles. Specifically, on the fibre surface, production of ROS (superoxide anion, $O_2^{-\bullet}$; hydrogen peroxide, H_2O_2; hydroxyl radical, $HO^\bullet$; singlet oxygen, $^1O_2$) may be catalysed by inhaled particles through several different mechanisms (Schoonen *et al.*, 2006), which involve: (1) metal transition ions (Jomova and Valko, 2011); (2) surface defects (Fubini and Hubbard, 2003; Jiang *et al.*, 2008a); or (3) polycyclic aromatic hydrocarbons (Xue and Warshawsky, 2005). Specifically, (1) redox-active metals (iron, copper, *etc.*) may either be released into body fluids as a consequence of mineral dissolution or take part in redox cycles at the fibre surface. In both cases, mineral surface interaction with endogenous molecules which may act as metal chelators (Hardy and Aust, 1995) may enhance or suppress metal reactivity. In the former case, mineral particles act as a source of metals and are not involved directly in redox reactions. In the latter case, surface-bound metals may catalyse free radical release if appropriate coordination and oxidation states occur. For example, rhombic-distorted isolated iron ions in low oxidation states (Turci *et al.*, 2011) are found to be responsible for generation of carbon-centred radicals from chrysotile asbestos. (2) Surface defects may be present on the mineral surface both because of the mineral's intrinsic structure (*e.g.* anatase and rutile) and mechanical treatment such as crushing (*e.g.* quartz). These defects may react with water or oxygen to produce reactive oxygen species. For example, when quartz is fractured, both homolytic ("dangling bonds" $Si^\bullet$, $SiO^\bullet$) and heterolytic (Si^+, SiO^-) cleavage of the silicon-oxygen bond takes place. The reaction of surface $Si^\bullet$ and $SiO^\bullet$ with molecular oxygen yields other "surface ROS" such as $SiO_2^\bullet$ and $SiO_3^\bullet$. Surface radicals are probably responsible for the higher inflammogenic and fibrogenic activity of freshly fractured silica dust compared with aged dust (Fubini and Hubbard, 2003). (3) Particle-associated polycyclic aromatic hydrocarbons (PAHs) are transformed by cytochrome P-450 into epoxide and diol-epoxides, which can react with DNA or be converted into redox-active quinones that may produce reactive oxygen species (Valavanidis *et al.*, 2009). PAHs are present in tobacco smoke together with a number of other carcinogens and their ability to produce reactive oxygen species, which damage epithelial and endothelial cells and impair their function, helps to explain the increase in lung cancer incidence in people exposed to asbestos (IARC, 2012b).

4.2. Surface active sites at the inorganic bio-interface of mineral fibres

Transition metal ions, both structural and substituent, are important in accounting for fibre toxicity (Aust *et al.*, 2000). All natural asbestos fibres contain iron, either structural forming a consistent part of the crystal framework (crocidolite, grunerite/amosite), or as a substituting element for Mg^{2+} ions, which share with Fe^{2+} size and charge (such as in chrysotile and tremolite). Iron in asbestos fibres can be present in both ferrous (Fe^{2+}) and ferric (Fe^{3+}) form within the asbestos crystal structure. The role of iron in the induction of oxidative stress and toxicity upon exposure to asbestos fibres was confirmed by a number of studies (Hardy and Aust, 1995; Kamp and Weitzman, 1999; Shukla *et al.*, 2003). Fibre-derived free radicals (and also cell-derived free radicals) contribute to the overall asbestos-induced oxidative stress. Asbestos minerals have been shown to induce ROS generation in cell cultures as a consequence of both the occurrence of highly reactive surface iron ions and of frustrated phagocytosis (Manning *et al.*, 2002). Iron ions involved in free radical generation are those present at the fibre surface in a poor coordination state (Fubini and Mollo, 1995; Martra *et al.*, 2003; Turci *et al.*, 20011) that happen also to be the most easily removable during incubation with iron-chelators (bio-available iron) (Hardy and Aust, 1995; Aust *et al.*, 2000). The topmost layers in every solid, including minerals, are structurally and chemically different from the bulk, with ions exhibiting a partial coordinative unsaturation (Hochella and Banfield, 1995). Generally, structural ligands are lost due to the mineral micro-topography and the location of the ions at the surface (*e.g.* extended faces, edge and corner positions) further differentiate the partially unsaturated ions (Fig. 2). It is accepted that transition metal ions exposed at the mineral surface may show a redox reactivity, due to their high valence electron number. Those electrons are also involved in the definition of the metal coordination state and consequently transition ions at the surface will differentiate in redox potential according to their coordination state. In air or in an aqueous medium (*e.g.* in the lung), the structural coordinative vacancies of these ions are totally or partially replaced by molecular water or hydroxyl groups. However, those non-covalent bonds allow for a ligand-displacement mechanism in which water/OH may be replaced by endogenous target molecules.

Such surface ions are involved in redox or charge-transfer reactions which may occur at the fibre–living matter interface, and are responsible for chemical reactivity towards a variety of endogenous molecules, such as H_2O_2, NO, proteins, lipids and nucleic acids (Hardy and Aust, 1995; Fubini and Otero-Arean, 1999; Kamp and Weitzman, 1999; Shukla *et al.*, 2003). The evaluation of the structural state of iron, and transition metals in general, in asbestos and other mineral fibres may provide unique molecular information about the chemical environment and hence reactivity of surface active sites. The nature and abundance of surface active sites have been evaluated by means of the adsorption of site-specific molecular probes (*e.g.* nitric oxide) which exhibit a significant affinity for iron ions (Martra *et al.*, 1999, 2003; Turci *et al.*, 2005, 2007).

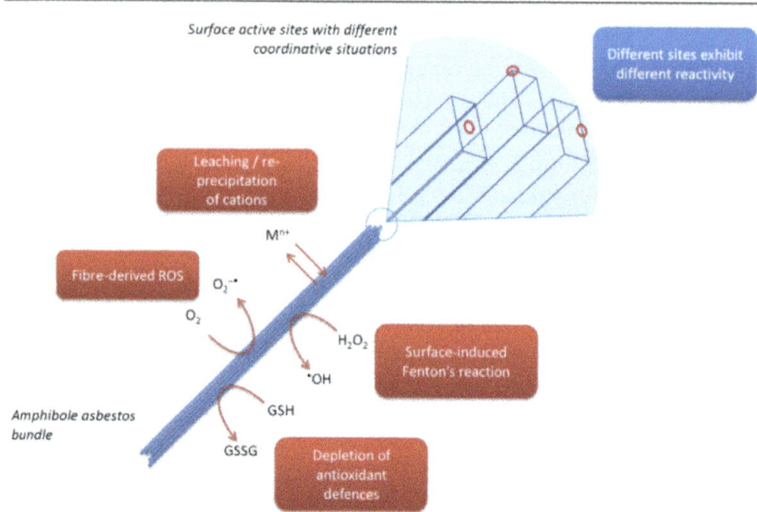

Figure 2. Schematic diagram of the surface reactions and the different coordinative situations (extended faces, edge or corner positions) a surface ion may experience at asbestos bio-interfaces.

4.3. Surface modification by biological environment: Leaching, re-precipitation, and ion exchange

The interaction of inhaled fibres with body fluids may alter both the biological environment and the surface properties of the fibres themselves. Chemical reactivity tests performed on both pristine and leached asbestos samples showed that fibre surface modifications due to dissolution processes affect significantly both hydroxyl and carbon-centred radical generation in well established reactivity tests. On these bases, Pacella *et al.* (2016) showed that the production of $HO^\bullet$ almost doubled for leached crocidolite after 48 h of leaching time in H_2O_2 solution buffered at pH 7.4. At the opposite end, leaching rapidly reduced the radical reactivity of tremolite after 24 h of leaching time. The continuous promotion of highly coordinated Fe(II) – but also Fe(III) – sites from the bulk to poorly coordinated Fe centres at the fibre surface, provided by the relatively rapid dissolution of crocidolite (Pacella *et al.*, 2014, 2015), actually fosters the reactivity in terms of $HO^\bullet$ production. In addition, during amphibole asbestos dissolution both Fe oxidation state and nuclearity increase with time and this leads to the formation of insoluble Fe nanoparticles (Pacella *et al.*, 2014, 2015) that are known to be only weakly active in triggering surface chemical processes (Fierro *et al.*, 2011; Freyria *et al.*, 2012; Moretti *et al.*, 2014). The formation of oxide/phosphate coating on the crocidolite surface probably inhibits the interaction with ascorbic acid, which is essential to promote Fe(III) reduction or prevent the rapid oxidation of Fe(II) from which carboxyl radical release is dependent. Also several molecules able to extract metal ions (chelators) may produce surface modifications. The role of iron chelators in asbestos toxicity has been investigated widely (Hardy and Aust, 1995; Shukla *et al.*,

2003; Liu *et al.*, 2010b). Mobilization of iron from different mineral fibres depends on the crystalline structure, surface area, iron content of the fibres, iron oxidation state and pH of the reaction medium, whereas the stability constants of the complex do not correlate well with the rate of iron mobilization which can be influenced by the geometry and size of the chelator (Hardy and Aust, 1995). Removal of iron has been used extensively in the past to demonstrate the role of iron in the asbestos toxicity. Pre-incubation of asbestos with several iron chelators reduced hydroxyl radical release (Fubini *et al.*, 1995; Martra *et al.*, 2003) and DNA single-strand breaks (Chao and Aust, 1994) in cell-free systems, cytotoxicity (Goodglick and Kane, 1990) and induction of TNF-alpha in rat alveolar macrophages (Simeonova and Luster, 1995). The addition of deferoxamine or phytic acid in the culture medium attenuated asbestos-induced injury in alveolar macrophages, lung epithelial and mesothelial cells (Hardy and Aust, 1995; Baldys and Aust, 2005, Dostert *et al.*, 2008; Liu *et al.*, 2010b). Several molecules present in lung fluids can mobilize iron from the surface of inhaled fibres. Mobilization by endogenous chelators may induce 'iron overload' in cells which have phagocytized the fibres and may be responsible for asbestos-dependent cytotoxicity (Chao *et al.*, 1994). Iron overload was recently supposed to be involved also in mesothelioma development and the use of iron chelators in mesothelioma prevention is under investigation (Toyokuni, 2014). The effect of iron mobilization depends upon both fibre and chelator. The addition of deferoxamine to the reaction medium decreased the reactivity of crocidolite, amosite and chrysotile in hydroxyl radical release, whereas other chelators such as citrate, adenosine 5′-triphosphate and guanosine 5′-triphosphate, did not inhibit this reaction, which was enhanced by nitrilotriacetic acid (Jiang *et al.*, 2008b). Intraperitoneal administration of crocidolite, amosite and chrysotile to rats caused mesothelioma development and repeated administration of nitrilotriacetic acid, which promotes the Fenton reaction, significantly reduced the period required for carcinogenesis (Jiang *et al.*, 2012). In addition, iron bound to low-molecular-weight chelators such as citrate, adenosine diphosphate (ADP), or ethylenediaminetetraacetic acid (EDTA) is redox active (Hardy and Aust, 1995). The ability of an iron chelate to catalyse hydroxyl radical release is related to its redox potential. Iron-catalysed hydroxyl radical formation also requires at least one iron coordination site occupied by water or a readily dissociable group allowing access to superoxide and/or hydrogen peroxide. Thus chelators which allow water, or other small molecules such as O_2, to occupy an available coordination site give rise to iron complexes able to generate highly reactive species (Graf *et al.*, 1984). Modification of mineral surface properties and chemical reactivity by chelators has been observed also in the natural environment. Lichens and fungi secrete a large variety of primary and secondary metabolites with acidic and chelating functions that act as potent weathering agents on natural and man-made substrates. Several fungi induced iron release from chrysotile (Daghino *et al.*, 2008; 2009) and crocidolite (Martino *et al.*, 2003; 2004), inhibiting the reactivity in free radical release (Martino *et al.*, 2003; Daghino *et al.*, 2005) and in inducing oxidative damage to DNA (Daghino *et al.*, 2006) in cell free tests. Ion mobilization by soil microorganisms may also involve magnesium. Indeed selective Mg depletion, due to two lichen metabolites (oxalic and pulvinic acid), was observed in chrysotile where the fibres were in intimate contact with the lichen hyphae (Favero-Longo *et al.*, 2005, 2007). Once metal ions are mobilized from the mineral framework, a mechanism of ion exchange may occur. This is of particular importance for erionite. The role of iron in inducing erionite toxicity is still controversial because erionite is a nominally Fe-free zeolite $K_2(Na,Ca_{0.5})_8[Al_{10}Si_{26}O_{72}]\cdot30H_2O$ (Coombs *et al.*, 1997). However, some authors (Eberly, 1964; Dogan *et al.*, 2006, 2008; Dogan, 2012) reported for this zeolite an Fe_2O_3 content close to 3 wt.%. In contrast, studies performed by Ballirano and co-workers (2009, 2015) on erionite samples from Oregon (USA) ruled out the presence of Fe within the structure of the zeolite, suggesting the adhesion to the surface of iron-rich nanoparticles, the occurrence of which has been confirmed by Croce *et al.* (2013). Recently, Matassa *et al.* (2015) showed unambiguously, by high-resolution transmission electron microscopy (HRTEM) and electron diffraction (ED), that iron is located at the fibre surface as a thin coating of an iron-bearing phyllosilicates (nontronite), which is carried into the human body by erionite fibres then becoming potential sources of iron. So far, erionite toxicity is ascribed partly to ion-exchanged and/or surface-deposited Fe participating in Fenton chemistry (Eborn and Aust, 1995; Fach *et al.*, 2002, 2003). In particular, Fe(III) is supposed to be fixed mainly at the fibre surface, with Fe(II) being ion-exchanged (Eborn and Aust, 1995). Although erionite contains little or no iron, it could acquire Fe *via* ion-exchange from intra- and extracellular media. Very recently, Ballirano *et al.* (2015) provided a detailed picture of the crystal chemical modifications occurring in erionite fibres when kept in contact with a source of both Fe(II) and Fe(III) chloride solutions at different concentrations, by combining chemical (scanning electron microscopy energy dispersive X-ray spectroscopy), structural (X-ray powder diffraction) and surface (X-ray photoelectron spectroscopy) data. The results highlighted that Fe(II) is segregated within the erionite cage, at the Ca3 site with six-fold coordination to water molecules (Ballirano *et al.*, 2015), whereas Fe(III) is fixed mainly at the fibre surface (Pacella *et al.*, 2016), in perfect agreement with previous results of Eborn and Aust (1995). Nevertheless, in very diluted Fe (III) solutions (<50 μM $FeCl_3$) a significant fraction of Fe(III) is segregated by an ion-exchange mechanism in the erionite cavity at the Ca3 site, albeit with a significantly lower efficiency with respect to Fe(II). In addition, it has been shown that ascorbic acid is able to reduce fairly easily Fe(III) residing at the fibre surface whereas this reaction does not occur (or possibly occurs only very marginally) in the case of the ion-exchanged metal (Pacella *et al.*, 2016). Considering the very low concentration of the total iron in lung fluids (~0.21 μM), Pacella *et al.* (2016) suggested strongly that under physiological conditions iron may be mainly ion-exchanged.

4.4. Surface modification *in vivo*: The ferruginous bodies

Alteration of iron homeostasis has been proposed to be involved in acute and chronic diseases related to mineral inhalation (Ghio *et al.*, 2008; Ghio, 2009). Such alteration may be due not only to mobilization of iron from inhaled particles but also to sequestration of host iron by the particles themselves. Indeed, mineral fibres can acquire iron during residence in the lung leading to the formation of a ferruginous body (Ghio *et al.*, 2004; Roggli, 2014). Surface minerals often expose oxygen-containing functional groups, mainly hydroxyl and silanol groups, which have an acidic character and are dissociated at physiologic pH. The dissociation imparts a negative charge to the mineral surface and allows adsorption of metal ions. Fe^{3+} has a large affinity for silanol groups at the silicate surfaces (Ghio *et al.*, 2004) and forms a silicate-iron coordination complex easily. The complexed iron eventually has the capacity to mediate electron exchange *via* the Fenton reaction after its reduction from Fe^{3+} to the Fe^{2+}. In the lower respiratory tract, both particles and fibres may accumulate iron from available sources in a cell and tissue, *e.g.* haemoglobin and plasma transferrin (Roggli, 2014). The formation of ferruginous bodies has been considered a protective reaction of the host to diminish injury mediated by inhaled fibres, making redox active iron at the fibre surface less available. However, some studies have suggested that the coated material may still be cytotoxic, increasing free radical production (Governa *et al.*, 1999; Ghio *et al.*, 2004: Borelli *et al.*, 2007) and inducing the formation of single DNA strand breaks in the presence of low molecular chelator or ascorbate (Lund *et al.*, 1994). Ferruginous bodies consist of the original fibres with a complex coating (Roggli, 2014) of iron, proteins such as ferritin, hemosiderin and albumin, mucopolysaccharides and organic acids such as oxalate with an heterogeneous distribution along the fibre (see Capella *et al.*, 2017, this volume, for more details). This heterogeneous distribution imparts to the ferruginous body a beaded, segmented or lancet shape and is probably due to adsorption of iron ions and of quaternary ammonium groups of proteins on the segments of the surface with greater silanol concentrations, thus with greater negative charge, as mentioned above (Ghio *et al.*, 2004). Not all fibres deposited in the distal lung will be coated, the tendency to form ferruginous bodies being higher for durable silicate fibres and for long fibres (Ghio *et al.*, 2004). Asbestos materials are the most effective in the formation of these structures, which are commonly defined as 'asbestos bodies' (see Capella *et al.*, 2017, this volume). Asbestos bodies were rarely formed on fibres <10 μm long and are often formed on straight fibres. They were found to occur on all the varieties of asbestos but less frequently on chrysotile. *In vivo* experiments demonstrated that inhalation of fibrous aluminium silicate, silicon carbide, rockwool (Roggli, 2014) and glass fibres (Ghio *et al.*, 2006; Roggli, 2014) resulted in structures identical to asbestos bodies in humans. Ferruginous bodies with cores of sheet silicate (talc, mica, or kaolinite) (Roggli, 2014), carbon and metal oxide (titanium, aluminium, chromium) (Roggli, 2014; Ghio *et al.*, 2006) have also been detected in human lung.

4.5. Surface charge

Among the surface properties affecting the interaction of particles with a biological medium, surface charge is receiving increasing attention, mainly in the field of synthesis of nanoparticles for biomedical application. Surface charge plays an important role in particle aggregation, is involved in particle interactions with phospholipid or protein domains on cell surfaces, therefore regulating particle uptake, and it seems to play a role also in particle cytotoxicity. Indeed, for several nanoparticles positively charged surfaces correlate with large cellular uptake and greater cytotoxicity in non-phagocytic cells, whereas negatively charged surfaces are associated with better internalization and higher cytotoxicity in phagocytic cells (Frohlich, 2012). Conversely, few studies on the role of surface charge in fibre toxicity are available and refer mostly to carbon nanotubes. CNT have a highly hydrophobic surface that limits their dispersion in aqueous solutions. Surface functionalization is often used to overcome these problems. Introduction of surface groups also improves biocompatibility and alters cellular interaction pathways, often resulting in reduced cytotoxic effects. Li (2013) reported that surface charge plays a role in induction of pulmonary fibrogenic effects by MWCNTs. Multiwall carbon nanotubes (MWCNT) functionalized with anionic groups (COOH and PEG) were less fibrogenic in mice than pristine MWCNTs, whereas positively charged MWCNTs induced more pulmonary fibrosis. Moreover, anionic MWCNTs were less internalized and induced less production of pro-fibrogenic cytokines and growth factors than cationic MWCNTs in human monocytes (Li *et al.*, 2013). Interaction of asbestos fibres with cellular membrane was studied by Light and Wei (1977). They demonstrated the role of surface charge, represented by the ξ potential, in haemolytic activity of chrysotile and amphibole asbestos. In distilled water at physiological pH, chrysotile and amphibole asbestos exhibit ξ potential values similar in magnitude but opposite in polarity, namely chrysotile surfaces being positively charged and amphibole surfaces negatively charged. Chrysotile was more haemolytic than crocidolite. After leaching, the surface charge of chrysotile decreased as did its haemolytic activity, whereas the haemolytic potential of crocidolite increased proportionately as it became more negative in charge. Electrostatic forces may contribute to adsorption of proteins, lipids and other biologically important macromolecules on inhaled dusts. Molecule adsorption may lead to alteration of fibre bioactivity. Selective protein adsorption depended on both asbestos surface charge and net protein charge, represented by its isoelectric point (pI). When guinea pig IgG (pI = 8.5) and bovine serum albumin (BSA; pI = 4.8) were present together, IgG was adsorbed selectively on negatively charged crocidolite. Similarly, IgG was adsorbed preferentially on the positively charged chrysotile surface when cytochrome C (pI = 10) was present together (Scheule and Holian, 1990). Pre-incubation of chrysotile with IgG resulted in an enhanced production of superoxide anion in alveolar macrophages (Scheule and Holian, 1990). Asbestos, quartz and titanium dioxide adsorb serum proteins to a different extent, depending on surface area and surface chemistry (Desai and Richards, 1978). Pre-incubation of asbestos with serum led to diminution of haemolysis (Sykes *et al.*, 1980), of release of inflammatory mediators from macrophage-like cells (Brown *et al.*, 1985), or of necrosis of tracheal-organ cultures (Mossman *et al.*, 1980). Chang *et al.* (1990) reported that several proteins, as well DNA and RNA, adsorb more on chrysotile and amosite than on glass-fibre surface as a result of coulombic interaction between the surface charge of the fibre and the hydrophilic groups on the macromolecule. The degree of macromolecule adsorption correlated positively with fibre cytotoxicity on human fibroblasts. Another study (Valerio *et al.*, 1987) showed that chrysotile has greater adsorption capability toward albumin and cytochrome C than crocidolite and found a good correlation between protein charge density and affinity for fibres. In this case, the sign of charge on fibres and proteins seemed to play a minor role in adsorption. Adsorption of vitronectin, one of the major cell adhesion proteins in plasma, on crocidolite surfaces increased fibre uptake by mesothelial cells, and enhanced intracellular oxidation, DNA strand breakage, and apoptosis with respect to uncoated fibres (Liu *et al.*, 2000). Vitronectin binds preferentially to negatively charged materials and was adsorbed on chrysotile at a smaller amount per surface area than on crocidolite. Despite the low adsorption, however, vitronectin-coated chrysotile fibres were more easily phagocytized than the uncoated version (Wu *et al.*, 2000). Finally, surface charge and other surface properties, such as hydrophilicity and the presence of functional groups able to form hydrogen bonds, also affect the adsorption of molecules from the external environment. The binding capacity of various mineral fibres for environmental carcinogens such as aromatic compounds was studied in the early 1980s (Kandaswami and O'Brien, 1980; Harvey *et al.*, 1984) mainly in an attempt to explain synergistic effects of tobacco smoke and asbestos in lung cancer induction. Chrysotile was found to bind significantly greater amounts of carcinogens than amphibole asbestos and glasswool, and a strong correlation between binding ability and fibre surface charge was observed by Harvey and colleagues (Harvey *et al.*, 1984).

5. Final remarks

It is widely accepted that a large aspect ratio (length/width), the biodurability and the surface chemistry of fibres are the three main factors which determine the pathological response to asbestos, with thin fibres being able to reach the alveolar region, the deepest portion of the lung. Both the presence and the structural coordination of Fe have been shown to be crucial factors in chrysotile-asbestos toxicity. In particular, poorly coordinated Fe centres (single ions, dimers or oligomers) exposed on particle surfaces have the greatest probability of being involved in several redox reactions and hence to rule the surface chemical reactivity of inorganic particulates. Furthermore, only the Fe on the fibre surface, and in particular Fe(II), was considered to play a primary role in ROS production from amphibole asbestos. It is now clear that the properties of minerals interacting with biological molecules change with time in a dynamic way and much effort will be required to explain the fundamental mechanisms of such interactions. General structure-activity relationships should be found and general mechanism of action proposed. The increasing predictive power of theoretical models may foster the development of these studies, to add to the large body of experimental knowledge acquired so far.

References

Addison, J. and McConnell, E.E. (2008) A review of carcinogenicity studies of asbestos and non-asbestos tremolite and other amphiboles. *Regulatory Toxicology and Pharmacology*, **52**, S187–S199.

Adib, G., Labreche, F., De Guire, L., Dion, C. and Dufresne, A. (2013) Short, fine and WHO asbestos fibers in the lungs of Quebec workers with an asbestos-related disease. *American Journal of Industrial Medicine*, **56**, 1001–1014.

Aglietti, E.F., Lopez, J.M.P. and Pereira, E. (1986a) Mechanochemical effects in kaolinite grinding. 1. Textural and physicochemical aspects. *International Journal of Mineral Processing*, **16**, 125–133.

Aglietti, E.F., Lopez, J.M.P. and Pereira, E. (1986b) Mechanochemical effects in kaolinite grinding. 2. Structural aspects. *International Journal of Mineral Processing*, **16**, 135–146.

Allen, B.L., Kichambare, P.D., Gou, P., Vlasova, I.I., Kapralov, A.A., Konduru, N., Kagan, V.E. and Star, A. (2008) Biodegradation of single-walled carbon nanotubes through enzymatic catalysis. *Nano Letters*, **8**, 3899–3903.

Allen, B.L., Kotchey, G.P., Chen, Y.N., Yanamala, N.V.K., Klein-Seetharaman, J., Kagan, V.E. and Star, A. (2009) Mechanistic investigations of horseradish peroxidase-catalyzed degradation of single-walled carbon nanotubes. *Journal of the American Chemical Society*, **131**, 17194-17205.

Allison, A.C. (1974) Pathogenic effects of inhaled particles and antigens. *Annals of the New York Academy of Sciences*, **221**, 299–308.

Assuncao, J. and Corn, M. (1975) The effects of milling on diameters and lengths of fibrous glass and chrysotile asbestos fibers. *American Industrial Hygiene Association Journal*, **36**, 811–819.

Aust, A.E., Lund, L.G., Chao, C.C., Park, S.H. and Fang, R.H. (2000) Role of iron in the cellular effects of asbestos. *Inhalation Toxicology*, **12**, 75–80.

Azad, N., Rojanasakul, Y. and Vallyathan, V. (2008) Inflammation and lung cancer: Roles of reactive oxygen/nitrogen species. *Journal of Toxicology and Environmental Health – Part B – Critical Reviews*, **11**, 1–15.

Baan, R.A. and Grosse, Y. (2004) Man-made mineral (vitreous) fibres: evaluations of cancer hazards by the IARC Monographs Programme. *Mutation Research*, **553**, 43–58.

Baldys, A. and Aust, A.E. (2005) Role of iron in inactivation of epidermal growth factor receptor after asbestos treatment of human lung and pleural target cells. *American Journal of Respiratory Cell and Molecular Biology*, **32**, 436–442.

Ballirano, P. and Cametti, G. (2015) Crystal chemical and structural modifications of erionite fibers leached with simulated lung fluids. *American Mineralogist*, **100**, 1003–1012.

Ballirano, P., Andreozzi, G.B., Dogan, M. and Dogan, A.U. (2009) Crystal structure and iron topochemistry of erionite-K from Rome, Oregon, USA. *American Mineralogist*, **94**, 1262–1270.

Ballirano, P., Pacella, A., Cremisini, C., Nardi, E., Fantauzzi, M., Atzei, D., Rossi, A. and Cametti, G. (2015) Fe (II) segregation at a specific crystallographic site of fibrous erionite: A first step toward the understanding of the mechanisms inducing its carcinogenicity. *Microporous and Mesoporous Materials*, **211**, 49–63.

Ballirano, P., Bloise, A., Gualtieri, A.F., Lezzerini, M., Pacella, A., Perchiazzi, N., Dogan, M. and Dogan, A.U. (2017) Crystal structure of mineral fibres. Pp. 17–64 in: *Mineral Fibres: Crystal Chemistry, Chemical-physical Properties, Biological Interaction and Toxicity* (A.F. Gualtieri, editor). EMU Notes in Mineralogy, **18**. European Mineralogical Union and Mineralogical Society of Great Britain & Ireland, London.

Becklake, M.R. (1976) Asbestos-related diseases of the lung and other organs: their epidemiology and implications for clinical practice. *American Review of Respiratory Disease*, **114**, 187–227.

Belluso, E., Cavallo, A. and Halterman, D. (2017) Crystal habit of mineral fibres. Pp. 65–109 in: *Mineral Fibres: Crystal Chemistry, Chemical-physical Properties, Biological Interaction and Toxicity* (A.F. Gualtieri, editor). EMU Notes in Mineralogy, **18**. European Mineralogical Union and Mineralogical Society of Great Britain & Ireland, London.

Berman, D.W. and Crump, K.S. (2008) A meta-analysis of asbestos-related cancer risk that addresses fiber size and mineral type. *Critical Reviews in Toxicology*, **38**, 49–73.

Bernstein, D.M. (2007) Synthetic vitreous fibers: A review toxicology, epidemiology and regulations. *Critical Reviews in Toxicology*, **37**, 839–886.

Bernstein, D.M. and Hoskins, J.A. (2006) The health effects of chrysotile: current perspective based upon recent data. *Regulatory Toxicology and Pharmacology*, **45**, 252–264.

Bernstein, D., Castranova, V., Donaldson, K., Fubini, B., Hadley, J., Hesterberg, T., Kane, A., Lai, D., McConnell, E.E., Muhle, H., Oberdörster, G., Olin, S. and Warheit, D.B. (2005) Testing of fibrous particles: Short-term assays and strategies – Report of an ILSI Risk Science Institute working group. *Inhalation Toxicology*, **17**, 497–537.

Bonneau, L., Malard, C. and Pezerat, H. (1986) Studies on surface properties of asbestos. II. Role of dimensional characteristics and surface properties of mineral fibers in the induction of pleural tumors. *Environmental Research*, **41**, 268–275.
Borelli, V., Brochetta, C., Melato, M., Rizzardi, C., Polentarutti, M., Busatto, C., Vita, F., Abbate, R., Gotter, R. and Zabucchi, G. (2007) A procedure for the isolation of asbestos bodies from lung tissue by exploiting their magnetic properties: A new approach to asbestos body study. *Journal of Toxicology and Environmental Health – Part A – Current Issues*, **70**, 1232–1240.
Boulanger, G., Andujar, P., Pairon, J.-C., Billon-Galland, M.-A., Dion, C., Dumortier, P., Brochard, P., Sobaszek, A., Bartsch, P. and Paris, C. (2014) Quantification of short and long asbestos fibers to assess asbestos exposure: a review of fiber size toxicity. *Environmental Health*, **13**, 1.
Boutin, C., Dumortier, P., Rey, F., Viallat, J.R. and De, V.P. (1996) Black spots concentrate oncogenic asbestos fibers in the parietal pleura. Thoracoscopic and mineralogic study. *American Journal of Respiratory and Critical Care Medicine*, **153**, 444–449.
Broaddus, V.C., Everitt, J.I., Black, B. and Kane, A.B. (2011) Non-neoplastic and neoplastic pleural endpoints following fiber exposure. *Journal of Toxicology and Environmental Health – Part B – Critical Reviews*, **14**, 153–178.
Brody, A.R., Hill, L.H., Adkins Jr, B. and O'Connor, R.W. (1981) Chrysotile asbestos inhalation in rats: Deposition pattern and reaction of alveolar epithelium and pulmonary macrophages 1, 2. *American Review of Respiratory Disease*, **123**, 670–679.
Brown, R., Poole, A. and Johnson, N. (1985) The release of inflammatory mediators from cells treated with leached asbestos. Pp. 313–319 in: *In Vitro Effects of Mineral Dusts* (E.G. Beck and J. Bignon, editors). Springer, Berlin.
Brown, G.M., Cowie, H., Davis, J.M. and Donaldson, K. (1986) In vitro assays for detecting carcinogenic mineral fibres: a comparison of two assays and the role of fibre size. *Carcinogenesis*, **7**, 1971–1974.
Brown, R.C., Bellmann, B., Muhle, H., Davis, J.M.G. and Maxim, L.D. (2005) Survey of the biological effects of refractory ceramic fibres: Overload and its possible consequences. *Annals of Occupational Hygiene*, **49**, 295–307.
Campopiano, A., Cannizzaro, A., Angelosanto, F., Astolfi, M.L., Ramires, D., Olori, A., Canepari, S. and Iavicoli, S. (2014) Dissolution of glass wool, rock wool and alkaline earth silicate wool: Morphological and chemical changes in fibers. *Regulatory Toxicology and Pharmacology*, **70**, 393–406.
Capella, S., Belluso, E., Bursi, N., Tibaldi, E. and Belpoggi, F. (2017) *In vivo* biological activity of mineral fibres. Pp. 307–345 in: *Mineral Fibres: Crystal Chemistry, Chemical-physical Properties, Biological Interaction and Toxicity* (A.F. Gualtieri, editor). EMU Notes in Mineralogy, **18**. European Mineralogical Union and Mineralogical Society of Great Britain & Ireland, London.
Carbone, M., Pass, H.I., Miele, L. and Bocchetta, M. (2003) New developments about the association of SV40 with human mesothelioma. *Oncogene*, **22**, 5173–5180.
Cardile, V., Lombardo, L., Belluso, E., Panico, A., Capella, S. and Balazy, M. (2007) Toxicity and carcinogenicity mechanisms of fibrous antigorite. *International Journal of Environmental Research and Public Health*, **4**, 1–9.
Chang, M.J., Joseph, L.B., Stephens, R.E. and Hart, R.W. (1990) Modulation of biological processes by mineral fiber adsorption of macromolecules in vitro. *Journal of Environmental Pathology, Toxicology and Oncology*, **10**, 89–93.
Chao, C.C. and Aust, A.E. (1994) Effect of long-term removal of iron from asbestos by desferrioxamine B on subsequent mobilization by other chelators and induction of DNA single-strand breaks. *Archives of Biochemistry and Biophysics*, **308**, 64–69.
Chao, C.C., Lund, L.G., Zinn, K.R. and Aust, A.E. (1994) Iron mobilization from crocidolite asbestos by human lung carcinoma cells. *Archives of Biochemistry and Biophysics*, **314**, 384–391.
Churg, A. (1988) Chrysotile, tremolite, and malignant mesothelioma in man. *Chest*, **93**, 621–628.
Churg, A., Warnock, M.L. and Green, N. (1979) Analysis of the cores of ferruginous (asbestos) bodies from the general population. II. True asbestos bodies and pseudoasbestos bodies. *Laboratory Investigation*, **40**, 31–38.
Churg, A., Hobson, J., Berean, K. and Wright, J. (1989) Scavengers of active oxygen species prevent cigarette

smoke-induced asbestos fiber penetration in rat tracheal explants. *American Journal of Pathology*, **135**, 599–603.

Cohen, D., Arai, S.F. and Brain, J.D. (1979) Smoking impairs long-term dust clearance from the lung. *Science*, **204**, 514–517.

Compagnoni, R., Ferraris, G. and Fiora, L. (1983) Balangeroite, a new fibrous silicate related to gageite from Balangero. *American Mineralogist*, **68**, 214–219.

Compagnoni, R., Ferraris, G. and Mellini, M. (1985) Carlosturanite, a new asbestiform rock-forming silicate from Val Varaita, Italy. *American Mineralogist*, **70**, 767–772.

Coombs, D.S., Alberti, A., Armbruster, T., Artioli, G., Colella, C., Galli, E., Grice, J.D., Liebau, F., Mandarino, J.A., Minato, H., Nickel, E.H., Passaglia, E., Peacor, D.R., Quartieri, S., Rinaldi, R., Ross, M., Sheppard, R.A., Tillmanns, E. and Vezzalini, G. (1997) Recommended nomenclature for zeolite minerals: Report of the subcommittee on zeolites of the International Mineralogical Association, Commission on New Minerals and Mineral Names. *The Canadian Mineralogist*, **35**, 1571–1606.

Croce, A., Musa, M., Allegrina, M., Rinaudo, C., Baris, Y.I., Dogan, A.U., Powers, A., Rivera, Z., Bertino, P., Yang, H.N., Gaudino, G. and Carbone, M. (2013) Micro-Raman spectroscopy identifies crocidolite and erionite fibers in tissue sections. *Journal of Raman Spectroscopy*, **44**, 1440–1445.

Cross, C.E., van der Vliet, A., O'Neill, C.A., Louie, S. and Halliwell, B. (1994) Oxidants, antioxidants, and respiratory tract lining fluids. *Environmental Health Perspectives*, **102**, 185.

Cullen, M.R. (1996) The amphibole hypothesis of asbestos-related cancer – gone but not forgotten. *American Journal of Public Health*, **86**, 158–159.

Cullen, R., Searl, A., Buchanan, D., Davis, J., Miller, B. and Jones, A. (2000) Pathogenicity of a special-purpose glass microfiber (E glass) relative to another glass microfiber and amosite asbestos. *Inhalation Toxicology*, **12**, 959–977.

Daghino, S., Martino, E., Fenoglio, I., Tomatis, M., Perotto, S. and Fubini, B. (2005) Inorganic materials and living organisms: surface modifications and fungal responses to various asbestos forms. *Chemistry – A European Journal*, **11**, 5611–5618.

Daghino, S., Turci, F., Tomatis, M., Favier, A., Perotto, S., Douki, T. and Fubini, B. (2006) Soil fungi reduce the iron content and the DNA damaging effects of asbestos fibers. *Environmental Science & Technology*, **40**, 5793–5798.

Daghino, S., Martino, E., Vurro, E., Tomatis, M., Girlanda, M., Fubini, B. and Perotto, S. (2008) Bioweathering of chrysotile by fungi isolated in ophiolitic sites. *FEMS Microbiology Letters*, **285**, 242–249.

Daghino, S., Turci, F., Tomatis, M., Girlanda, M., Fubini, B. and Perotto, S. (2009) Weathering of chrysotile asbestos by the serpentine rock-inhabiting fungus Verticillium leptobactrum. *FEMS Microbiology Ecology*, **69**, 132–141.

Davis, J.M. and Jones, A.D. (1988) Comparisons of the pathogenicity of long and short fibres of chrysotile asbestos in rats. *British Journal of Experimental Pathology*, **69**, 717–737.

Davis, J.M., Addison, J., Bolton, R.E., Donaldson, K., Jones, A.D. and Miller, B.G. (1985) Inhalation studies on the effects of tremolite and brucite dust in rats. *Carcinogenesis*, **6**, 667–674.

Davis, J.M., Addison, J., Boltob, R.E., Donaldson, K., Jones, A.D. and Smith, T. (1986) The pathogenicity of long versus short fibre samples of amosite asbestos administered to rats by inahlation and intraperitoneal injection. *British Journal of Experimental Pathology*, **67**, 415–430.

Desai, R. and Richards, R.J. (1978) The adsorption of biological macromolecules by mineral dusts. *Environmental Research*, **16**, 449–464.

Dodson, R.F., Atkinson, M.A. and Levin, J.L. (2003) Asbestos fiber length as related to potential pathogenicity: a critical review. *American Journal of Industrial Medicine*, **44**, 291–297.

Dogan, M. (2012) Quantitative characterization of the mesothelioma-inducing erionite series minerals by transmission electron microscopy and energy dispersive spectroscopy. *Scanning*, **34**, 37–42.

Dogan, A.U., Baris, Y.I., Dogan, M., Emri, S., Steele, I., Elmishad, A.G. and Carbone, M. (2006) Genetic predisposition to fiber carcinogenesis causes a mesothelioma epidemic in Turkey. *Cancer Research*, **66**, 5063–5068.

Dogan, A.U., Dogan, M. and Hoskins, J.A. (2008) Erionite series minerals: mineralogical and carcinogenic properties. *Environmental Geochemistry and Health*, **30**, 367–381.

Dogra, S. and Donaldson, K. (1995) Effect of long and short fibre amosite asbestos on in vitro TNF production by rat alveolar macrophages: the modifying effect of lipopolysaccharide. *Industrial Health*, **33**, 131–141.

Donaldson, K. and Golyasnya, N. (1995) Cytogenetic and pathogenic effects of long and short amosite asbestos. *Journal of Pathology*, **177**, 303–307.

Donaldson, K., Li, X.Y., Dogra, S., Miller, B.G. and Brown, G.M. (1992) Asbestos-stimulated tumor-necrosis-factor release from alveolar macrophages depends on fiber length and opsonization. *Journal of Pathology*, **168**, 243–248.

Donaldson, K., Miller, B.G., Sara, E., Slight, J. and Brown, R.C. (1993) Asbestos fibre length-dependent detachment injury to alveolar epithelial cells in vitro: role of a fibronectin-binding receptor. *International Journal of Experimental Pathology*, **74**, 243–250.

Donaldson, K., Murphy, F., Duffin, R. and Poland, C.A. (2010) Asbestos, carbon nanotubes and the pleural mesothelium: a review of the hypothesis regarding the role of long fibre retention in the parietal pleura, inflammation and mesothelioma. *Particle and Fibre Toxicology*, **7**, 5–22.

Donaldson, K., Murphy, F., Schinwald, A., Duffin, R. and Poland, C.A. (2011) Identifying the pulmonary hazard of high aspect ratio nanoparticles to enable their safety-by-design. *Nanomedicine: Nanotechnology, Biology, and Medicine*, **6**, 143–156.

Dostert, C., Petrilli, V., Van, B.R., Steele, C., Mossman, B.T. and Tschopp, J. (2008) Innate immune activation through Nalp3 inflammasome sensing of asbestos and silica. *Science*, **320**, 674–677.

Drummond, G., Bevan, R. and Harrison, P. (2016) A comparison of the results from intra-pleural and intra-peritoneal studies with those from inhalation and intratracheal tests for the assessment of pulmonary responses to inhalable dusts and fibres. *Regulatory Toxicology and Pharmacology*, **81**, 89–105.

Eastes, W. and Hadley, J.G. (1996) A mathematical model of fiber carcinogenicity and fibrosis in inhalation and intraperitoneal experiments in rats. *Inhalation Toxicology*, **8**, 323–343.

Eberly, P.E. (1964) Adsorption properties of naturally occurring erionite and its cationic-exchanged forms. *American Mineralogist*, **49**, 30–40.

Eborn, S.K. and Aust, A.E. (1995) Effect of iron acquisition on induction of DNA single-strand breaks by erionite, a carcinogenic mineral fiber. *Archives of Biochemistry and Biophysics*, **316**, 507–514.

European Committe for Standarization (CEN) (1993) Workplace atmospheres – size fraction definitions for measurement of airborne particles. Report BS EN 481. British Standards Institute, London.

Fach, E., Waldman, W.J., Williams, M., Long, J., Meister, R.K. and Dutta, P.K. (2002) Analysis of the biological and chemical reactivity of zeolite-based aluminosilicate fibers and particulates. *Environmental Health Perspectives*, **110**, 1087–1096.

Fach, E., Kristovich, R., Long, J.F., Waldman, W.J., Dutta, P.K. and Williams, M.V. (2003) The effect of iron on the biological activities of erionite and mordenite. *Environment International*, **29**, 451–458.

Fantauzzi, M., Pacella, A., Atzei, D., Gianfagna, A., Andreozzi, G.B. and Rossi, A. (2010) Combined use of X-ray photoelectron and Mössbauer spectroscopic techniques in the analytical characterization of iron oxidation state in amphibole asbestos. *Analytical and Bioanalytical Chemistry*, **396**, 2889–2898.

Fantauzzi, M., Pacella, A., Fournier, J., Gianfagna, A., Andreozzi, G.B. and Rossi, A. (2012) Surface chemistry and surface reactivity of fibrous amphiboles that are not regulated as asbestos. *Analytical and Bioanalytical Chemistry*, **404**, 821–833.

Favero-Longo, S.E., Turci, F., Tomatis, M., Castelli, D., Bonfante, P., Hochella, M.F., Piervittori, R. and Fubini, B. (2005) Chrysotile asbestos is progressively converted into a non-fibrous amorphous material by the chelating action of lichen metabolites. *Journal of Environmental Monitoring*, **7**, 764–766.

Favero-Longo, S.E., Girlanda, M., Honegger, R., Fubini, B. and Piervittori, R. (2007) Interactions of sterile-cultured lichen-forming ascomycetes with asbestos fibres. *Mycological Research*, **111**, 473–481.

Favero-Longo, S.E., Turci, F., Fubini, B., Castelli, D. and Piervittori, R. (2013) Lichen deterioration of asbestos and asbestiform minerals of serpentinite rocks in Western Alps. *International Biodeterioration and Biodegradation*, **84**, 342–350.

Ferraris, G., Mellini, M. and Merlino, S. (1987) Electron-diffraction and electron-microscopy study of balangeroite and gageite: Crystal structures, polytypism, and fiber texture. *American Mineralogist*, **72**, 382–391.

Fierro, G., Moretti, G., Ferraris, G. and Andreozzi, G.B. (2011) A Mössbauer and structural investigation of Fe-

ZSM-5 catalysts: Influence of Fe oxide nanoparticles size on the catalytic behaviour for the NO-SCR by C3H8. *Applied Catalysis B: Environmental*, **102**, 215–223.

Finkelstein, M.M. and Dufresne, A. (1999) Inferences on the kinetics of asbestos deposition and clearance among chrysotile miners and millers. *American Journal of Industrial Medicine*, **35**, 401–412.

Fleury-Feith, J., Lecomte, C., Renier, A., Matrat, M., Kheuang, L., Abramowski, V., Levy, F., Janin, A., Giovannini, M. and Jaurand, M.C. (2003) Hemizygosity of Nf2 is associated with increased susceptibility to asbestos-induced peritoneal tumours. *Oncogene*, **22**, 3799–3805.

Frank, E., Steudle, L.M., Ingildeev, D., Sporl, J.M. and Buchmeiser, M.R. (2014) Carbon Fibers: Precursor Systems, Processing, Structure, and Properties. *Angewandte Chemie-International Edition*, **53**, 5262–5298.

Freyria, F.S., Bonelli, B., Tomatis, M., Ghiazza, M., Gazzano, E., Ghigo, D., Garrone, E. and Fubini, B. (2012) Hematite nanoparticles larger than 90 nm show no sign of toxicity in terms of lactate dehydrogenase release, nitric oxide generation, apoptosis, and comet assay in murine alveolar macrophages and human lung epithelial cells. *Chemical Research in Toxicology*, **25**, 850–861.

Frohlich, E. (2012) The role of surface charge in cellular uptake and cytotoxicity of medical nanoparticles. *International Journal of Nanomedicine*, **7**, 5577–5591.

Fubini, B. and Otero Arean, C. (1999) Chemical aspects of the toxicity of inhaled mineral dusts. *Chemical Society Reviews*, **28**, 373–381.

Fubini, B. and Hubbard, A. (2003) Reactive oxygen species (ROS) and reactive nitrogen species (RNS) generation by silica in inflammation and fibrosis. *Free Radical Biology and Medicine*, **34**, 1507–1516.

Fubini, B. and Mollo, L. (1995) Role of iron in the reactivity of mineral fibers. *Toxicological Letters*, **82**, 951–960.

Fubini, B., Mollo, L. and Giamello, E. (1995) Free radical generation at the solid/liquid interface in iron containing minerals. *Free Radical Research*, **23**, 593–614.

Gamble, J.F. and Gibbs, G.W. (2008) An evaluation of the risks of lung cancer and mesothelioma from exposure to amphibole cleavage fragments. *Regulatory Toxicology and Pharmacology*, **52**, S154–S186.

Gazdar, A.F., Butel, J.S. and Carbone, M. (2002) SV40 and human tumours: myth, association or causality? *Nature Reviews Cancer*, **2**, 957–964.

Gazzano, E., Riganti, C., Tomatis, M., Turci, F., Bosia, A., Fubini, B. and Ghigo, D. (2005) Potential toxicity of nonregulated asbestiform minerals: balangeroite from the western Alps. Part 3: Depletion of antioxidant defenses. *Journal of Toxicology and Environmental Health: Part A*, **68**, 41–49.

Ghio, A.J. (2009) Disruption of iron homeostasis and lung disease. *Biochimica et Biophysica Acta*, **1790**, 731–739.

Ghio, A.J., Churg, A. and Roggli, V.L. (2004) Ferruginous bodies: implications in the mechanism of fiber and particle toxicity. *Toxicologic Pathology*, **32**, 643–649.

Ghio, A.J., Funkhouser, W., Pugh, C.B., Winters, S., Stonehuerner, J.G., Mahar, A.M. and Roggli, V.L. (2006) Pulmonary fibrosis and ferruginous bodies associated with exposure to synthetic fibers. *Toxicologic Pathology*, **34**, 723–729.

Ghio, A.J., Stonehuerner, J., Richards, J. and Devlin, R.B. (2008) Iron homeostasis in the lung following asbestos exposure. *Antioxidants and Redox Signaling*, **10**, 371–377.

Gianfagna, A., Ballirano, P., Bellatreccia, F., Bruni, B., Paoletti, L. and Oberti, R. (2003) Characterization of amphibole fibres linked to mesothelioma in the area of Biancavilla, Eastern Sicily, Italy. *Mineralogical Magazine*, **67**, 1221–1229.

Goodglick, L.A. and Kane, A.B. (1990) Cytotoxicity of long and short crocidolite asbestos fibers in vitro and in vivo. *Cancer Research*, **50**, 5153–5163.

Governa, M., Amati, M., Fontana, S., Visona, I., Botta, G.C., Mollo, F., Bellis, D. and Bo, P. (1999) Role of iron in asbestos-body-induced oxidant radical generation. *Journal of Toxicology and Environmental Health: Part A*, **58**, 279–287.

Graf, E., Mahoney, J.R., Bryant, R.G. and Eaton, J.W. (1984) Iron-catalyzed hydroxyl radical formation. Stringent requirement for free iron coordination site. *Journal of Biological Chemistry*, **259**, 3620–3624.

Graham, A., Higinbotham, J., Allan, D., Donaldson, K. and Beswick, P.H. (1999) Chemical differences between long and short amosite asbestos: differences in oxidation state and coordination sites of iron, detected by infrared spectroscopy. *Occupational and Environmental Medicine*, **56**, 606–611.

Greim, H., Utell, M.J., Maxim, L.D. and Niebo, R. (2014) Perspectives on refractory ceramic fiber (RCF) carcinogenicity: comparisons with other fibers. *Inhalation Toxicology*, **26**, 789–810.

Groppo, C., Tomatis, M., Turci, F., Gazzano, E., Ghigo, D., Compagnoni, R. and Fubini, B. (2005) Potential toxicity of nonregulated asbestiform minerals: balangeroite from the western Alps. Part 1: Identification and characterization. *Journal of Toxicology and Environmental Health: Part A*, **68**, 1–19.

Gross, P. (1981) Consideration of the aerodynamic equivalent diameter of respirable mineral fibers. *The American Industrial Hygiene Association Journal*, **42**, 449–452.

Grosse, Y., Loomis, D., Guyton, K.Z., Lauby-Secretan, B., El Ghissassi, F., Bouvard, V., Benbrahim-Tallaa, L., Guha, N., Scoccianti, C., Mattock, H. and Straif, K. (2014) Carcinogenicity of fluoro-edenite, silicon carbide fibres and whiskers, and carbon nanotubes. *The Lancet Oncology*, **15**, 1427–1428.

Guthrie, G.D., Jr. (1997) Mineral properties and their contributions to particle toxicity. *Environmental Health Perspectives*, **105**, 1003–1011.

Hammond, E.C., Selikoff, I.J. and Seidman, H. (1979) Asbestos exposure, cigarette smoking and death rates. *Annals of the New York Academy of Sciences*, **330**, 473–490.

Hanton, D.Y., Furtak, H. and Grimm, H.G. (1998) Preparation and handling conditions of MMVF for in-vivo experiments. *Aerosol Science and Technology*, **29**, 449–456.

Hardy, J.A. and Aust, A.E. (1995) Iron in asbestos chemistry and carcinogenicity. *Chemical Reviews*, **95**, 97–118.

Harper, M., Lee, E.G., Doorn, S.S. and Hammond, O. (2008) Differentiating non-asbestiform amphibole and amphibole asbestos by size characteristics. *Journal of Occupational and Environmental Hygiene*, **5**, 761–770.

Harper, M., Lee, E.G., Slaven, J.E. and Bartley, D.L. (2012) An inter-laboratory study to determine the effectiveness of procedures for discriminating amphibole asbestos fibers from amphibole cleavage fragments in fiber counting by phase-contrast microscopy. *Annals of Occupational Hygiene*, **56**, 645–659.

Hart, G., Hesterberg, T. and Bauer, J. (1999) Physical and chemical degradation of synthetic vitreous fibres in the lung and in vitro. *Toxicologist*, **48**, 133.

Harvey, G., Page, M. and Dumas, L. (1984) Binding of environmental carcinogens to asbestos and mineral fibres. *British Journal of Industrial Medicine*, **41**, 396–400.

Hessel, P.A., Gamble, J.F. and McDonald, J.C. (2005) Asbestos, asbestosis, and lung cancer: a critical assessment of the epidemiological evidence. *Thorax*, **60**, 433–436.

Hesterberg, T.W. and Barrett, J.C. (1984) Dependence of asbestos- and mineral dust-induced transformation of mammalian cells in culture on fiber dimension. *Cancer Research*, **44**, 2170–2180.

Hesterberg, T.W. and Hart, G.A. (2000) Lung biopersistence and in vitro dissolution rate predict the pathogenic potential of synthetic vitreous fibers. *Inhalation Toxicology*, **12**, 91–97.

Hesterberg, T.W. and Hart, G.A. (2001) Synthetic vitreous fibers: a review of toxicology research and its impact on hazard classification. *Critical Reviews in Toxicology*, **31**, 1–53.

Hesterberg, T.W., Miller, W.C., Musselman, R.P., Kamstrup, O., Hamilton, R.D. and Thevenaz, P. (1996) Biopersistence of man-made vitreous fibers and crocidolite asbestos in the rat lung following inhalation. *Fundamental and Applied Toxicology*, **29**, 267–279.

Hill, I.M., Beswick, P.H. and Donaldson, K. (1995) Differential release of superoxide anions by macrophages treated with long and short fibre amosite asbestos is a consequence of differential affinity for opsonin. *Occupational and Environmental Medicine*, **52**, 92–96.

Hobson, J., Wright, J.L. and Churg, A. (1990) Active oxygen species mediate asbestos fiber uptake by tracheal epithelial cells. *The FASEB Journal*, **4**, 3135–3139.

Hochella, M. and Banfield, J. (1995) Chemical weathering of silicates in nature; a microscopic perspective with theoretical considerations. *Reviews in Mineralogy and Geochemistry*, **31**, 353–406.

Hofmann, W. (2011) Modelling inhaled particle deposition in the human lung – A review. *Journal of Aerosol Science*, **42**, 693–724.

Hofmann, W., Koblinger, L. and Martonen, T.B. (1989) Structural differences between human and rat lungs – implications for monte-carlo modeling of aerosol deposition. *Health Physics*, **57**, 41–47.

Holt, P. (1982) Submicron carbon dust in the guinea pig lung. *Environmental Research*, **28**, 434–442.

Holt, P.F. and Horne, M. (1978) Dust from carbon fibre. *Environmental Research*, **17**, 276–283.

IARC (International Agency for Research on Cancer) (1977) *Asbestos* Vol. **14**, Monographs on the evaluation of the carcinogenic risk of chemicals to man, World Health Organization (WHO), Lyon (France), 106 pp.

IARC (International Agency for Research on Cancer) (1988) *Man-made Mineral Fibres and Radon* Vol. **43**, Monographs on the evaluation of carcinogenic risks to humans, World Health Organization (WHO), Lyon (France), 300 pp.

IARC (International Agency for Research on Cancer) (1997) *Silica, some Silicates, Coal Dust and Para-aramid Fibrils.* Vol. **68**, Monographs on the evaluation of carcinogenic risks to humans, World Health Organization (WHO), Lyon (France), 521 pp.

IARC (International Agency for Research on Cancer) (2002) *Man-made Mineral Fibres* Vol. **81**, Monographs on the Evaluation of Carcinogenic Risks to Humans, World Health Organization (WHO), Lyon (France), 381 pp.

IARC (International Agency for Research on Cancer) (2012a) Erionite. Pp. 311–316 in: *Arsenic, Metals, Fibres, and Dusts.* Monographs on the Evaluation of Carcinogenic Risks to Humans, Vol. **100C**. World Health Organization (WHO), Lyon (F), France.

IARC (International Agency for Research on Cancer) (2012b) Asbestos (chrysotile, amosite, crocidolite, tremolite, actinolite and anthophyllite). Pp. 219–309 in: *Arsenic, Metals, Fibres, and Dusts.* Monographs on the evaluation of carcinogenic risks to humans, Vol. **100C**. World Health Organization (WHO), Lyon (F), France.

IARC (International Agency for Research on Cancer) (2012c) Tobacco smoking. Pp. 43–211 in: *Personal Habits and Indoor Combustions.* Monographs on the Evaluation of Carcinogenic Risks to Humans, Vol. **100E**. World Health Organizaziont (WHO), Lyon (F), France.

IARC (International Agency for Research on Cancer) (in press) *Fluoro-edenite, Silicon Carbide Fibres and Whiskers, and Single-walled and Multi-walled Carbon Nanotubes* Vol. **111**, Monographs on the evaluation of carcinogenic risks to humans, World Health Organization (WHO), Lyon (F), France.

Ilgren, E.B. (2004) The biology of cleavage fragments: A brief synthesis and analysis of current knowledge. *Indoor and Built Environment*, **13**, 343–356.

Jiang, J., Oberdörster, G., Elder, A., Gelein, R., Mercer, P. and Biswas, P. (2008a) Does nanoparticle activity depend upon size and crystal phase? *Nanotoxicology*, **2**, 33–42.

Jiang, L., Nagai, H., Ohara, H., Hara, S., Tachibana, M., Hirano, S., Shinohara, Y., Kohyama, N., Akatsuka, S. and Toyokuni, S. (2008b) Characteristics and modifying factors of asbestos-induced oxidative DNA damage. *Cancer Science*, **99**, 2142–2151.

Jiang, L., Akatsuka, S., Nagai, H., Chew, S.H., Ohara, H., Okazaki, Y., Yamashita, Y., Yoshikawa, Y., Yasui, H., Ikuta, K., Sasaki, K., Kohgo, Y., Hirano, S., Shinohara, Y., Kohyama, N., Takahashi, T. and Toyokuni, S. (2012) Iron overload signature in chrysotile-induced malignant mesothelioma. *Journal of Pathology*, **228**, 366–377.

Johnson, N.F. and Hahn, F.F. (1996) Induction of mesothelioma after intrapleural inoculation of F344 rats with silicon carbide whiskers or continuous ceramic filaments. *Occupational and Environmental Medicine*, **53**, 813–816.

Jomova, K. and Valko, M. (2011) Advances in metal-induced oxidative stress and human disease. *Toxicology*, **283**, 65–87.

Jones, H., Jones, T. and Lyle, W. (1982) Carbon fibre: results of a survey of process workers and their environment in a factory producing continuous filament. *Annals of Occupational Hygiene*, **26**, 861–867.

Kagan, V.E., Konduru, N.V., Feng, W.H., Allen, B.L., Conroy, J., Volkov, Y., Vlasova, I.I., Belikova, N.A., Yanamala, N., Kapralov, A., Tyurina, Y.Y., Shi, J.W., Kisin, E.R., Murray, A.R., Franks, J., Stolz, D., Gou, P.P., Klein-Seetharaman, J., Fadeel, B., Star, A. and Shvedova, A.A. (2010) Carbon nanotubes degraded by neutrophil myeloperoxidase induce less pulmonary inflammation. *Nature Nanotechnology*, **5**, 354–359.

Kamp, D.W. and Weitzman, S.A. (1999) The molecular basis of asbestos induced lung injury. *Thorax*, **54**, 638–652.

Kandaswami, C. and O'Brien, P.J. (1980) Effects of asbestos on membrane transport and metabolism of benzo(a)pyrene. *Biochemical and Biophysical Research Communications*, **97**, 794–801.

Kim, D.J. and Chung, H.S. (2002) Effect of grinding on the structure and chemical extraction of metals from serpentine. *Particulate Science and Technology*, **20**, 159–168.

Krombach, F., Munzing, S., Allmeling, A.M., Gerlach, J.T., Behr, J. and Dorger, M. (1997) Cell size of alveolar macrophages: an interspecies comparison. *Environmental Health Perspectives*, **105**, 1261–1263.

Langer, A.M., Wolff, M.S., Rohl, A.N. and Selikoff, I.J. (1978) Variation of properties of chrysotile asbestos subjected to milling. *Journal of Toxicology and Environmental Health: Part A*, **4**, 173–188.

Lemen, R.A. (2004) Chrysotile asbestos as a cause of mesothelioma: application of the Hill causation model. *International Journal of Occupational and Environmental Health*, **10**, 233–239.

Lemen, R.A. (2006) Epidemiology of asbestos-related diseases and the knowledge that led to what is known today. Pp. 201–308 in: *Asbestos, Risk Assessment, Epidemiology, and Health Effects* (R.F. Dodson and S.P. Hammar, editors). CRC Press, Boca Raton, Florida, USA.

Lentz, T.J., Rice, C.H., Succop, P.A., Lockey, J.E., Dement, J.M. and Lemasters, G.K. (2003) Pulmonary deposition modeling with airborne fiber exposure data: a study of workers manufacturing refractory ceramic fibers. *Applied Occupational and Environmental Hygiene*, **18**, 278–288.

Li, Y.B., Wei, B.Q., Liang, J., Yu, Q. and Wu, D.H. (1999) Transformation of carbon nanotubes to nanoparticles by ball milling process. *Carbon*, **37**, 493–497.

Li, R.B., Wang, X., Ji, Z.X., Sun, B.B., Zhang, H.Y., Chang, C.H., Lin, S.J., Meng, H., Liao, Y.P., Wang, M.Y., Li, Z.X., Hwang, A.A., Song, T.B., Xu, R., Yang, Y., Zink, J.I., Nel, A.E. and Xia, T. (2013) Surface charge and cellular processing of covalently functionalized multiwall carbon nanotubes determine pulmonary toxicity. *ACS Nano*, **7**, 2352–2368.

Liao, J.F. and Senna, M. (1993) Mechanochemical dehydration and amorphization of hydroxides of Ca, Mg and Al on grinding with and without SiO_2. *Solid State Ionics*, **66**, 313–319.

Light, W.G. and Wei, E.T. (1977) Surface charge and hemolytic activity of asbestos. *Environmental Research*, **13**, 135–145.

Lippmann, M. (1990) Effects of fiber characteristics on lung deposition, retention, and disease. *Environmental Health Perspectives*, **88**, 311–317.

Lippmann, M. (2014) Toxicological and epidemiological studies on effects of airborne fibers: coherence and public [corrected] health implications. *Critical Reviews in Toxicology*, **44**, 643–695.

Lippmann, M., Yeates, D.B. and Albert, R.E. (1980) Deposition, retention and clearance of inhaled particles. *British Journal of Industrial Medicine*, **37**, 337–362.

Liu, W., Ernst, J.D. and Broaddus, V.C. (2000) Phagocytosis of crocidolite asbestos induces oxidative stress, DNA damage, and apoptosis in mesothelial cells. *American Journal of Respiratory Cell and Molecular Biology*, **23**, 371–378.

Liu, X.Y., Hurt, R.H. and Kane, A.B. (2010a) Biodurability of single-walled carbon nanotubes depends on surface functionalization. *Carbon*, **48**, 1961–1969.

Liu, G., Beri, R., Mueller, A. and Kamp, D.W. (2010b) Molecular mechanisms of asbestos-induced lung epithelial cell apoptosis. *Chemico-Biological Interactions*, **188**, 309–318.

Loomis, D., Dement, J., Richardson, D. and Wolf, S. (2010) Asbestos fibre dimensions and lung cancer mortality among workers exposed to chrysotile. *Occupational and Environmental Medicine*, **67**, 580–584.

Loomis, D., Dement, J.M., Elliott, L., Richardson, D., Kuempel, E.D. and Stayner, L. (2012) Increased lung cancer mortality among chrysotile asbestos textile workers is more strongly associated with exposure to long thin fibres. *Occupational and Environmental Medicine*, **69**, 564–568.

Lowers, H. and Meeker, G. (2002) *Tabulation of asbestos-related terminology* US Department of the Interior, US Geological Survey.

Lund, L.G., Williams, M.G., Dodson, R.F. and Aust, A.E. (1994) Iron associated with asbestos bodies is responsible for the formation of single strand breaks in phi X174 RFI DNA. *Occupational and Environmental Medicine*, **51**, 200–204.

Manning, C.B., Vallyathan, V. and Mossman, B.T. (2002) Diseases caused by asbestos: mechanisms of injury and disease development. *International Immunopharmacology*, **2**, 191–200.

Martin, T.R., Meyer, S.W. and Luchtel, D.R. (1989) An evaluation of the toxicity of carbon fiber composites for lung cells in vitro and in vivo. *Environmental Research*, **49**, 246–261.

Martino, E., Prandi, L., Fenoglio, I., Bonfante, P., Perotto, S. and Fubini, B. (2003) Soil fungal hyphae bind and attack asbestos fibers. *Angewandte Chemie International*, **42**, 219–222.

Martino, E., Cerminara, S., Prandi, L., Fubini, B. and Perotto, S. (2004) Physical and biochemical interactions of

soil fungi with asbestos fibers. *Environmental Toxicology and Chemistry*, **23**, 938–944.
Martra, G., Chiardola, E., Coluccia, S., Marchese, L., Tomatis, M. and Fubini, B. (1999) Reactive sites at the surface of crocidolite asbestos. *Langmuir*, **15**, 5742–5752.
Martra, G., Tomatis, M., Fenoglio, I., Coluccia, S. and Fubini, B. (2003) Ascorbic acid modifies the surface of asbestos: possible implications in the molecular mechanisms of toxicity. *Chemical Research in Toxicology*, **16**, 328–335.
Mast, R.W., Maxim, L.D., Utell, M.J. and Walker, A.M. (2000) Refractory ceramic fiber: Toxicology, epidemiology, and risk analyses – A Review. *Inhalation Toxicology*, **12**, 359–399.
Matassa, R., Familiari, G., Relucenti, M., Battaglione, E., Downing, C., Pacella, A., Cametti, G. and Ballirano, P. (2015) A deep look into erionite fibres: An electron microscopy investigation of their self-assembly. *Scientific Reports*, **5**.
Maxim, L.D. and McConnell, E.E. (2005) A review of the toxicology and epidemiology of wollastonite. *Inhalation Toxicology*, **17**, 451–466.
Maxim, L.D., Yu, C.P., Oberdörster, G. and Utell, M.J. (2003) Quantitative risk analyses for RCF: survey and synthesis. *Regulatory Toxicology and Pharmacology*, **38**, 400–416.
Maxim, L.D., Hadley, J.G., Potter, R.M. and Niebo, R. (2006) The role of fiber durability/biopersistence of silica-based synthetic vitreous fibers and their influence on toxicology. *Regulatory Toxicology and Pharmacology*, **46**, 42–62.
Maxim, L.D., Niebo, R., Utell, M.J., McConnell, E., LaRosa, S. and Segrave, A.M. (2014) Wollastonite toxicity: an update. *Inhalation Toxicology*, **26**, 95–112.
McDonald, J.C. and McDonald, A.D. (1997) Chrysotile, tremolite and carcinogenicity. *Annals of Occupational Hygiene*, **41**, 699–705.
McKay, R.T., LeMasters, G.K., Hilbert, T.J., Levin, L.S., Rice, C.H., Borton, E.K. and Lockey, J.E. (2011) A long term study of pulmonary function among US refractory ceramic fibre workers. *Occupational and Environmental Medicine*, **68**, 89–95.
Meeker, G., Bern, A., Brownfield, I., Lowers, H., Sutley, S., Hoefen, T. and Vance, J. (2003) The composition and morphology of amphiboles from the Rainy Creek Complex, near Libby, Montana. *American Mineralogist*, **88**, 1955–1969.
Miller, F.J., Mercer, R.R. and Crapo, J.D. (1993) Lower respiratory-tract structure of laboratory-animals and humans – dosimetry implications. *Aerosol Science and Technology*, **18**, 257–271.
Miserocchi, G.A., Sancini, G.A., Mantegazza, F. and Chiappino, G. (2008) Translocation pathways for inhaled asbestos fibers. *Environmental Health*, **7**, 4.
Moolgavkar, S.H., Anderson, E.L., Chang, E.T., Lau, E.C., Turnham, P. and Hoel, D.G. (2014) A review and critique of US EPA's risk assessments for asbestos. *Critical Reviews in Toxicology*, **44**, 499–522.
Moretti, G., Fierro, G., Ferraris, G., Andreozzi, G.B. and Naticchioni, V. (2014) N_2O decomposition over [Fe]-MFI catalysts: Influence of the FexOy nuclearity and the presence of framework aluminum on the catalytic activity. *Journal of Catalysis*, **318**, 1–13.
Morrow, P.E. (1988) Possible mechanisms to explain dust overloading of the lungs. *Fundamental and Applied Toxicology*, **10**, 369–384.
Mossman, B.T. (2008) Assessment of the pathogenic potential of asbestiform vs. nonasbestiform particulates (cleavage fragments) in in vitro (cell or organ culture) models and bioassays. *Regulatory Toxicology and Pharmacology*, **52**, S200–S203.
Mossman, B. and Churg, A. (1998) *Asbestos-related disorders*, 3rd edition, *Occupational Lung Disorders*. Butterworth-Heinemann, Oxford, UK, 504 pp.
Mossman, B.T., Adler, K.B. and Craighead, J.E. (1980) Cytotoxic and proliferative changes in tracheal organ and cell cultures after exposure to mineral dusts. Pp. 241–250 in: *The In Vitro Effects of Mineral Dusts* (R.C. Brown, I.P. Gormley, M. Chamberlain and R. Davies, editors). Academic Press, London.
Mossman, B.T., Lippmann, M., Hesterberg, T.W., Kelsey, K.T., Barchowsky, A. and Bonner, J.C. (2011) Pulmonary endpoints (lung carcinomas and asbestosis) following inhalation exposure to asbestos. *Journal of Toxicology and Environmental Health – Part B – Critical Reviews*, **14**, 76–121.
Muhle, H., Pott, F., Bellmann, B., Takenaka, S. and Ziem, U. (1987) Inhalation and injection experiments in rats to test the carcinogenicity of MMMF. *Annals of Occupational Hygiene*, **31**, 755–764.

Murphy, F.A., Poland, C.A., Duffin, R., Al-Jamal, K.T., Ali-Boucetta, H., Nunes, A., Byrne, F., Prina-Mello, A., Volkov, Y., Li, S.P., Mather, S.J., Bianco, A., Prato, M., MacNee, W., Wallace, W.A., Kostarelos, K. and Donaldson, K. (2011) Length-dependent retention of carbon nanotubes in the pleural space of mice initiates sustained inflammation and progressive fibrosis on the parietal pleura. *American Journal of Pathology*, **178**, 2587–2600.

NIOSH (National Institute for Occupational Safety and Health) (1984) *Manual of Analytical Methods. Method #7400* Government Printing Office, Washington, D.C.

NIOSH (National Institute for Occupational Safety and Health) (2011) Current Intelligence Bulletin 62: asbestos fibers and other elongate mineral particles: state of the science and roadmap for research. Report 2011-159. Department of health and human services, Cincinnati (OH).

National Research Council (NRC) (1984) *Asbestiform fibers: Nonoccupational health risks* National Academies Press, Washington (DC), 334 pp.

Neri, M., Ugolini, D., Dianzani, I., Gemignani, F., Landi, S., Cesario, A., Magnani, C., Mutti, L., Puntoni, R. and Bonassi, S. (2008) Genetic susceptibility to malignant pleural mesothelioma and other asbestos-associated diseases. *Mutation Research*, **659**, 126–136.

Newhouse, M.L. and Thompson, H. (1965) Mesothelioma of pleura and peritoneum following exposure to asbestos in the London area. *British Journal of Industrial Medicine*, **22**, 261–269.

Nicholson, W.J. (2001) The carcinogenicity of chrysotile asbestos – a review. *Industrial Health*, **39**, 57–64.

Oberdörster, G. (1996) Significance of particle parameters in the evaluation of exposure-dose-response relationships of inhaled particles. *Particulate Science and Technology*, **14**, 135–151.

Oberdörster, G. (2000) Determinants of the pathogenicity of man-made vitreous fibers (MMVF). *International Archives of Occupational and Environmental Health*, **73 Suppl**, S60–S68.

Oberdörster, G. (2002) Toxicokinetics and effects of fibrous and nonfibrous particles. *Inhalation Toxicology*, **14**, 29–56.

Oberdörster, G., Morrow, P.E. and Spurny, K. (1988) Size dependent lymphatic short term clearance of amosite fibers in the lung. *Annals of Occupational Hygiene*, **32**, 149–156.

Owen, P.E., Glaister, J.R., Ballantyne, B. and Clary, J.J. (1986) Subchronic Inhalation Toxicology of Carbon-Fibers. *Journal of Occupational and Environmental Medicine*, **28**, 373–376.

Pacella, A., Andreozzi, G.B. and Fournier, J. (2010) Detailed crystal chemistry and iron topochemistry of asbestos occurring in its natural setting: A first step to understanding its chemical reactivity. *Chemical Geology*, **277**, 197–206.

Pacella, A., Andreozzi, G.B., Fournier, J., Stievano, L., Giantomassi, F., Lucarini, G., Rippo, M.R. and Pugnaloni, A. (2012) Iron topochemistry and surface reactivity of amphibole asbestos: relations with in vitro toxicity. *Analytical and Bioanalytical Chemistry*, **402**, 871–881.

Pacella, A., Fantauzzi, M., Turci, F., Cremisini, C., Montereali, M.R., Nardi, E., Atzei, D., Rossi, A. and Andreozzi, G.B. (2014) Dissolution reaction and surface iron speciation of UICC crocidolite in buffered solution at pH 7.4: A combined ICP-OES, XPS and TEM investigation. *Geochimica et Cosmochimica Acta*, **127**, 221–232.

Pacella, A., Fantauzzi, M., Turci, F., Cremisini, C., Montereali, M.R., Nardi, E., Atzei, D., Rossi, A. and Andreozzi, G.B. (2015) Surface alteration mechanism and topochemistry of iron in tremolite asbestos: A step toward understanding the potential hazard of amphibole asbestos. *Chemical Geology*, **405**, 28–38.

Pacella, A., Fantauzzi, M., Atzei, D., Cremisini, C., Nardi, E., Montereali, M.R., Rossi, A. and Ballirano, P. (2016) Iron within the erionite cavity and its potential role in inducing its toxicity: Evidence of Fe (III) segregation as extra-framework cation. *Microporous and Mesoporous Materials*, **237**, 168–179.

Platek, S.F., Groth, D.H., Ulrich, C.E., Stettler, L.E., Finnell, M.S. and Stoll, M. (1985) Chronic inhalation of short asbestos fibers. *Fundamental and Applied Toxicology*, **5**, 327–340.

Plescia, P., Belardi, G. and Marruzzo, G. (1995) X-ray diffraction measurement of stress and strain induced in magnetite and chromite by ultrafine grinding. *Minerals and Metallurgical Processing*, **12**, 178–183.

Plescia, P., Gizzi, D., Benedetti, S., Camilucci, L., Fanizza, C., De Simone, P. and Paglietti, F. (2003) Mechanochemical treatment to recycling asbestos-containing waste. *Waste Management*, **23**, 209–218.

Plumlee, G.S., Morman, S.A. and Ziegler, T.L. (2006) The toxicological geochemistry of earth materials: an overview of processes and the interdisciplinary methods used to understand them. Pp. 5–57 in: *Medical*

Mineralogy and Geochemistry (N. Sahai and M.A.A. Schoonen, editors) Reviews in Mineralogy and Geochemistry, **64**. Mineralogical Society of America and Geochemical Society, Chantilly, Virginia, USA.

Poulikakos, P.I., Xiao, G.H., Gallagher, R., Jablonski, S., Jhanwar, S.C. and Testa, J.R. (2006) Re-expression of the tumor suppressor NF2/merlin inhibits invasiveness in mesothelioma cells and negatively regulates FAK. *Oncogene*, **25**, 5960–5968.

Poulin, D.L. and DeCaprio, J.A. (2006) Is there a role for SV40 in human cancer? *Journal of Clinical Oncology*, **24**, 4356–4365.

Pugnaloni, A., Giantomassi, F., Lucarini, G., Capella, S., Belmonte, M.M., Orciani, M. and Belluso, E. (2010) Effects of asbestiform antigorite on human alveolar epithelial A549 cells: a morphological and immunohistochemical study. *Acta Histochemica*, **112**, 133–146.

Riganti, C., Aldieri, E., Bergandi, L., Tomatis, M., Fenoglio, I., Costamagna, C., Fubini, B., Bosia, A. and Ghigo, D. (2003) Long and short fiber amosite asbestos alters at a different extent the redox metabolism in human lung epithelial cells. *Toxicology and Applied Pharmacology*, **193**, 106–115.

Rizzo, P., Bocchetta, M., Powers, A., Foddis, R., Stekala, E., Pass, H.I. and Carbone, M. (2001) SV40 and the pathogenesis of mesothelioma. *Seminars in Cancer Biology*, **11**, 63–71.

Roggli, V.L. (2014) Asbestos bodies and nonasbestos ferruginous bodies. Pp. 34–70 in: *Pathology of Asbestos-Associated Diseases* (V.L. Roggli, T.D. Oury and T.A. Sporn, editors). 3rd edition. Springer, New York.

Roggli, V.L. (2015) The so-called short-fiber controversy literature review and critical analysis. *Archives of Pathology and Laboratory Medicine*, **139**, 1052–1057.

Roshchenko, A., Finlay, W.H. and Minev, P.D. (2011) The aerodynamic behavior of fibers in a linear shear flow. *Aerosol Science and Technology*, **45**, 1260–1271.

Rozalén, M., Huertas, F., Pacella, A. and Ballirano, P. (2017) Dissolution and biodurability of mineral fibres. Pp. 347–366 in: *Mineral Fibres: Crystal Chemistry, Chemical-physical Properties, Biological Interaction and Toxicity* (A.F. Gualtieri, editor). EMU Notes in Mineralogy, **18**. European Mineralogical Union and Mineralogical Society of Great Britain & Ireland, London.

Salamatipour, A., Mohanty, S.K., Pietrofesa, R.A., Vann, D.R., Christofidou-Solomidou, M. and Willenbring, J.K. (2016) Asbestos Fiber Preparation Methods Affect Fiber Toxicity. *Environmental Science and Technology Letters*, **3**, 270–274.

Scheule, R.K. and Holian, A. (1990) Modification of asbestos bioactivity for the alveolar macrophage by selective protein adsorption. *American Journal of Respiratory Cell and Molecular Biology*, **2**, 441–448.

Schinwald, A., Chernova, T. and Donaldson, K. (2012) Use of silver nanowires to determine thresholds for fibre length-dependent pulmonary inflammation and inhibition of macrophage migration in vitro. *Particle and Fibre Toxicology*, **9**, 47.

Schoonen, M.A.A., Cohn, C.A., Roemer, E., Laffers, R., Simon, S.R. and O'Riordan, T. (2006) Mineral-induced formation of reactive oxygen species. *Medical Mineraology and Geochemistry*, **64**, 179–221.

Shukla, A., Gulumian, M., Hei, T.K., Kamp, D., Rahman, Q. and Mossman, B.T. (2003) Multiple roles of oxidants in the pathogenesis of asbestos-induced diseases. *Free Radical Biology and Medicine*, **34**, 1117–1129.

Simeonova, P.P. and Luster, M.I. (1995) Iron and reactive oxygen species in the asbestos-induced tumor necrosis factor-alpha response from alveolar macrophages. *American Journal of Respiratory Cell and Molecular Biology*, **12**, 676–683.

Skinner, H., Ross, M and Frondel, C. (1988) *Asbestos and other Fibrous Materials – Mineralogy, Crystal-chemistry and Health Effects* Vol. **14**, Community Health Studies, Oxford University Press, Oxford, 206 pp.

Smith, A.H. and Wright, C.C. (1996) Chrysotile asbestos is the main cause of pleural mesothelioma. *American Journal of Industrial Medicine*, **30**, 252–266.

Spurny, K., Opiela, H. and Weiss, G. (1979) On the milling and ultrasonic treatment of fibres for biological and analytical applications. *IARC Scientific Publications*, 931–933.

Stanton, M.F., Layard, M., Tegeris, A., Miller, E., May, M., Morgan, E. and Smith, A. (1981) Relation of particle dimension to carcinogenicity in amphibole asbestoses and other fibrous minerals. *Journal of the National Cancer Institute*, **67**, 965–975.

Stayner, L.T., Dankovic, D.A. and Lemen, R.A. (1996) Occupational exposure to chrysotile asbestos and cancer risk: a review of the amphibole hypothesis. *American Journal of Public Health*, **86**, 179–186.

Stayner, L., Kuempel, E., Gilbert, S., Hein, M. and Dement, J. (2008) An epidemiological study of the role of chrysotile asbestos fibre dimensions in determining respiratory disease risk in exposed workers. *Occupational and Environmental Medicine*, **65**, 613–619.

Stettler, L.E., Sharpnack, D.D. and Krieg, E.F. (2008) Chronic inhalation of short asbestos: lung fiber burdens and histopathology for monkeys maintained for 11.5 years after exposure. *Inhalation Toxicology*, **20**, 63–73.

Stratton, K., Almario, D.A. and McCormick, M. (2002) *Immunization safety review. SV40 contamination of poliovaccine and cancer.* Report. Institute of Medicine (IOM). National Academies Press, Washington, D.C.

Sturm, R. and Hofmann, W. (2009) A theoretical approach to the deposition and clearance of fibers with variable size in the human respiratory tract. *Journal of Hazardous Materials*, **170**, 210–218.

Suzuki, Y., Yuen, S.R. and Ashley, R. (2005) Short, thin asbestos fibers contribute to the development of human malignant mesothelioma: pathological evidence. *International Journal of Hygiene and Environmental Health*, **208**, 201–210.

Sykes, S., Morgan, A. and Holmes, A. (1980) The haemolytic activity of chrysotile asbestos. Pp. 113–119 in: *The In Vitro Effects of Mineral Dusts* (R.C. Brown, I.P. Gormley, M. Chamberlain and R. Davies, editors). Academic Press, London.

Thurneysen, C., Opitz, I., Kurtz, S., Weder, W., Stahel, R.A. and Felley-Bosco, E. (2009) Functional inactivation of NF2/merlin in human mesothelioma. *Lung Cancer*, **64**, 140–147.

TIMA (Thermal Insulation Manufacturers Association) (1991) Man-made vitreous fibers: Nomenclature, chemical and physical properties. Nomenclature Committee of TIMA, Stamford, Connecticut, USA.

Timbrell, V. (1983) Fibres and carcinogenesis. *Journal of the Occupational Health Society of Australia*, **3**, 3–12.

Tomatis, M., Prandi, L., Bodoardo, S. and Fubini, B. (2002) Loss of surface reactivity upon heating amphibole asbestos. *Langmuir*, **18**, 4345–4350.

Tomatis, M., Turci, F., Ceschino, R., Riganti, C., Gazzano, E., Martra, G., Ghigo, D. and Fubini, B. (2010) High aspect ratio materials: role of surface chemistry vs. length in the historical "long and short amosite asbestos fibers". *Inhalation Toxicology*, **22**, 984–998.

Toyokuni, S. (2014) Iron overload as a major targetable pathogenesis of asbestos-induced mesothelial carcinogenesis. *Redox Report*, **19**, 1–7.

Tsuda, A., Henry, F.S. and Butler, J.P. (2013) Particle transport and deposition: Basic physics of particle kinetics. *Comprehensive Physiology*, **3**, 1437–1471.

Turci, F., Tomatis, M., Gazzano, E., Riganti, C., Martra, G., Bosia, A., Ghigo, D. and Fubini, B. (2005) Potential toxicity of nonregulated asbestiform minerals: balangeroite from the western Alps. Part 2: Oxidant activity of the fibers. *Journal of Toxicology and Environmental Health: Part A*, **68**, 21–39.

Turci, F., Favero-Longo, S.E., Tomatis, M., Martra, G., Castelli, D., Piervittori, R. and Fubini, B. (2007) A biomimetic approach to the chemical inactivation of chrysotile fibres by lichen metabolites. *Chemistry -A-European Journal*, **13**, 4081–4093.

Turci, F., Tomatis, M., Compagnoni, R. and Fubini, B. (2009) Role of associated mineral fibres in chrysotile asbestos health effects: the case of balangeroite. *Annals of Occupational Hygiene*, **53**, 491–497.

Turci, F., Tomatis, M., Lesci, I.G., Roveri, N. and Fubini, B. (2011) The iron-related molecular toxicity mechanism of synthetic asbestos nanofibres: a model study for high-aspect-ratio nanoparticles. *Chemistry – A European Journal*, **17**, 350–358.

Turci, F., Colonna, M., Tomatis, M., Mantegna, S., Cravotto, G., Gulino, G., Aldieri, E., Ghigo, D. and Fubini, B. (2012) Surface reactivity and cell responses to chrysotile asbestos nanofibers. *Chemical Research in Toxicology*, **25**, 884–894.

Utell, M.J. and Maxim, L.D. (2010) Refractory ceramic fiber (RCF) toxicity and epidemiology: A review. *Inhalation Toxicology*, **22**, 500–521.

Utembe, W., Potgieter, K., Stefaniak, A.B. and Gulumian, M. (2015) Dissolution and biodurability: Important parameters needed for risk assessment of nanomaterials. *Particle and Fibre Toxicology*, **12**, 11.

Valavanidis, A., Vlachogianni, T. and Fiotakis, K. (2009) Tobacco smoke: involvement of reactive oxygen species and stable free radicals in mechanisms of oxidative damage, carcinogenesis and synergistic effects with other respirable particles. *International Journal of Environmental Research and Public Health*, **6**, 445–462.

Valerio, F., Balducci, D. and Lazzarotto, A. (1987) Adsorption of proteins by chrysotile and crocidolite – Role of molecular-weight and charge-density. *Environmental Research*, **44**, 312–320.

Virta, R.L. (2002) *Asbestos: geology, mineralogy, mining, and uses*. Report 2002-149. US Geological Survey.

Wagner, J. (1988) Significance of the fibre size of erionite. *Proceedings of the VIIth International Pneumoconioses Conference*, August 23, Pittsburgh, USA.

Wagner, J.C., Sleggs, C.A. and Marchand, P. (1960) Diffuse pleural mesothelioma and asbestos exposure in the North Western Cape Province. *British Journal of Industrial Medicine*, **17**, 260–271.

Wagner, J.C., Skidmore, J.W., Hill, R.J. and Griffiths, D.M. (1985) Erionite exposure and mesotheliomas in rats. *British Journal of Cancer*, **51**, 727–730.

Warheit, D.B., Hansen, J.F., Carakostas, M.C. and Hartsky, M.A. (1994) Acute inhalation toxicity studies in rats with a respirable-sized experimental carbon fibre: Pulmonary biochemical and cellular effects. *Annals of Occupational Hygiene*, **38**, 769–776.

Waritz, R.S., Ballantyne, B. and Clary, J.J. (1998) Subchronic inhalation toxicity of 3.5-μm diameter carbon fibers in rats. *Journal of Applied Toxicology*, **18**, 215–223.

Williams, C., Dell, L., Adams, R., Rose, T. and Van Orden, D. (2013) State-of-the-science assessment of non-asbestos amphibole exposure: Is there a cancer risk? *Environmental Geochemistry and Health*, **35**, 357–377.

World Health Organization (WHO) (1986) Asbestos and other natural mineral fibres. *Environmental Health Criteria*, **53**, 120–124.

World Health Organization (WHO) (1999) *Hazard prevention and control in the work environment: airborne dust*. Report WHO/SDE/OEH/99.14, Geneva.

World Health Organization (WHO) (2003) *Asbestos in drinking-water. Background document for development of WHO guidelines for drinking-water quality*. Report WHO/SDE/WSH/03.04/02, Geneva.

World Health Organization (WHO) (2005) *Workshop on Mechanisms of fibre carcinogenesis and assessment of chrysotile asbestos substitutes*. Report. Lyon, France.

World Health Organization (WHO) (2006) *Elimination of asbestos-related diseases*, Geneva.

Wu, J., Liu, W., Koenig, K., Idell, S. and Broaddus, V.C. (2000) Vitronectin adsorption to chrysotile asbestos increases fiber phagocytosis and toxicity for mesothelial cells. *American Journal of Physiology – Lung Cellular and Molecular Physiology*, **279**, L916–L923.

Xue, W.L. and Warshawsky, D. (2005) Metabolic activation of polycyclic and heterocyclic aromatic hydrocarbons and DNA damage: A review. *Toxicology and Applied Pharmacology*, **206**, 73–93.

Ye, J., Shi, X., Jones, W., Rojanasakul, Y., Cheng, N., Schwegler-Berry, D., Baron, P., Deye, G.J., Li, C. and Castranova, V. (1999) Critical role of glass fiber length in TNF-alpha production and transcription factor activation in macrophages. *American Journal of Physiology*, **276**, L426–434.

Zhao, Y., Allen, B.L. and Star, A. (2011) Enzymatic degradation of multiwalled carbon nanotubes. *Journal of Physical Chemistry A*, **115**, 9536–9544.

EMU Notes in Mineralogy, Vol. 18 (2017), Chapter 7, 215–260

Thermal behaviour of mineral fibres

ANDREA BLOISE[1,*], ROBERT KUSIOROWSKI[2],
MAGDALENA LASSINANTTI GUALTIERI[3],
ALESSANDRO F. GUALTIERI[4]

[1]Dipartimento di Biologia, Ecologia e Scienze della Terra, Università della Calabria, Via P. Bucci, IT-87036, Arcavacata di Rende (Cs), Italy, e-mail: andrea.bloise@unical.it
[2]Institute of Ceramics and Building Materials, Refractory Materials Division in Gliwice, ul. Toszecka 99, 44-100 Gliwice, Poland
[3]Dipartimento di Ingegneria "Enzo Ferrari", Università di Modena e Reggio Emilia, Via Vivarelli 10, I-41125, Modena, Italy
[4]Dipartimento di Scienze Chimiche e Geologiche, Università di Modena e Reggio Emilia, Via Campi 103, I-41125, Modena, Italy
**Corresponding author*

This chapter deals with the synthesis and thermal stability of mineral fibres. The different structural assemblages within mineral fibres and their resistance to high temperature changes from species to species. In general, the formation of such minerals takes place in hydrothermal environments. The thermal decomposition process consists of three main stages: the loss of water adsorbed on the surface of the fibre and the zeolitic water below 200–250°C; the removal of the structure water (the hydroxyl groups) in the range 500–1100°C and recrystallization into new stable crystalline phases. The thermal stability of chrysotile, amphiboles fibres and erionite will be described in detail and will be followed by specific sections describing how the concept of thermal decomposition is used for the remediation of wastes containing asbestos to produce secondary raw materials to be recycled in various industrial application.

1. Introduction

The thermal stability of asbestos fibres is generally relatively high and varies slightly between different species. Asbestos minerals belong to the group of hydrous silicates. When heated to sufficiently high temperatures, they decompose with accompanying release of water through the process of dehydroxylation (removal of hydroxyl groups from the structure).

The entire thermal decomposition process, much like in the case of other hydrous siliceous minerals, consists of three main stages (Fig. 1). The first is related to the loss of water adsorbed on the surface of the fibre and the zeolitic water. The next stage involves removal of the constitutional water (the hydroxyl groups) from the asbestos mineral's structure. The process of dehydroxylation is followed by the last stage,

DOI: 10.1180/EMU-notes.18.7

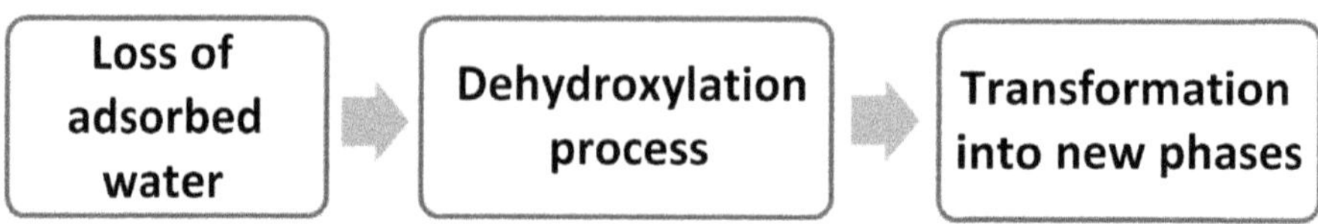

Figure 1. Thermal behaviour of mineral fibres.

responsible for transforming the material into new crystalline phases (Jeyaratnam and West, 1994; Bloise *et al.*, 2014).

In the process of heating asbestos, the adsorbed water starts to evaporate at ~100°C as it is released from the surface of the fibres. In this case, the water is trapped within the micropores of the structure as interlayer (zeolitic) water, the removal of which may not be complete until a temperature of ~250°C has been reached; this is related to the hindered migration of water molecules adsorbed inside the asbestos fibres. Constitutional water starts to exude from asbestos minerals at ~500°C. Complete dehydration takes place, depending on the type of asbestos, when the material is heated to even higher temperatures. The removal of structural water (hydroxyl groups) causes an irreversible decrease in mechanical strength; the fibres lose their flexibility and can be transformed easily into powdered material. During the last stage, new phases are created depending on the type of asbestos.

The loss in mass related to removal of water from the asbestos structure and the thermal decomposition of minerals which accompany asbestos varies. For chrysotile it reaches an average of 11 to 15%, while amphibole asbestos loses only a few percent (~1–4%).

2. Thermal behaviour of chrysotile

2.1. Hydrothermal synthesis of chrysotile

A significant amount of work has been done to synthesize chrysotile in the past 70 years, ever since the first experimental study by Bowen and Tuttle (1949) on the $MgO–SiO_2–H_2O$ system showed that chrysotile could be prepared under hydrothermal conditions. After an initial approach, and from the 1950s to the year 2000, several other authors synthesized chrysotile either from a petrological point of view or in order to define its stability field. An overview of these experimental works may be found in the papers by Evans (2004), Roveri *et al.* (2006) and Deer *et al.* (2009). In Table 1, details are given of the experimental conditions (T, P) and the starting materials used in the syntheses discussed below. In the past 15 years, several efforts have been made primarily to synthesize chrysotile nanotubes to be used for applications in nanotechnology, and also to obtain pure and chemistry-controlled chrysotile suitable for investigating its cytotoxicity. Chrysotile syntheses have been possible mainly under hydrothermal conditions at relatively high temperature and pressure. Bowen and Tuttle (1949) carried out experiments in the $MgO–SiO_2–H_2O$ system in the temperature range 300–1000°C and observed that pure Mg serpentine had a maximum temperature of

Table 1. Experimental conditions for the chrysotile reviewed in this chapter.

Reference	Temperature (°C)	Pressure (MPa)	Starting materials
Bowen and Tuttle (1949)	300–1000	13–275	SiO_2; MgO
Noll and Kircher (1952)	300	8	SiO_2; NiO
Roy and Roy (1954)	150–900	3–151	SiO_2; MgO; Ni; Mn; Zn; Al; Ge
Gillery (1959)	400–620	103–386	SiO_2; MgO; Al_2O_3
Noll *et al.* (1960)	300–350	9–16	SiO_2; MgO; NiO; CoO
Yang (1961)	300–350	8–12	SiO_2; MgO
Nesterchuk *et al.* (1966)	350–600	90–160	SiO_2; MgO
Johannes (1968)	350–430	50–650	SiO_2; MgO
Korytkova and Makarova (1971)	200–500	91	Olivine
Yada and Iishi (1974)	300–400	70	Olivine
Yamai and Saito (1974)	360–380	–	SiO_2; $Mg(OH)_2$; AlO(OH); $Fe(OH)_3$
Chernosky (1975)	363–647	100–1000	SiO_2; MgO; Al_2O_3
Jasmund *et al.* (1975)	250–350	6–25	Na_4SiO_4; Ni; Co; Fe
Moody (1976b)	302–361	50–200	Olivine
Yada and Iishi (1977)	300–400	70	Olivine
Hemley (1977)	200–660	100	SiO_2; MgO
Ozeki (1991)	300	7	SiO_2; $Mg(OH)_2$
Ueno *et al.* (1991)	300–400	20–51	SiO_2; $Mg(OH)_2$
Barrese *et al.* (1997)	250–390	100	$CaMgSi_2O_6$
Devouard *et al.* (1997)	300	70	$Si_2Mg_3O_7$
Normand *et al.* (2002)	300	30	Olivine
Falini *et al.*, (2002)	300	8.3	SiO_2; $MgCl_2$
Korytkova *et al.* (2002)	200–450	30–100	SiO_2; MgO
Korytkova *et al.* (2004)	100–450	30–100	SiO_2; MgO
Falini *et al.* (2004)	300	8	SiO_2; $MgCl_2$
Korytkova *et al.* (2005)	200–450	70	SiO_2; MgO; NiO
Foresti *et al.* (2005)	300	8	SiO_2; $FeCl_3$; $MgCl_2$
Xu *et al.* (2006)	400	–	SiO_2; $MgCO_3$; $Mg(OH)_2 \cdot 6H_2O$
Jancar and Suvorov (2006)	150–300	–	SiO_2; $Mg(OH)_2$
Korytkova *et al.* (2007)	200–450	30–100	MgO; SiO_2; $CoSiO_3 \cdot 9H_2O$
Ma *et al.* (2007)	160	–	SiO_2; MgO
Bloise *et al.* (2009a)	270–400	50–200	SiO_2; MgO; Fe; Fe_2O_3; Mg_2SiO_4
Bloise *et al.* (2009b)	300–320	50–200	Ti–Mg_2SiO_4
McDonald *et al.* (2009)	250	3.4	H_2SiO_3; $NiCl_2 \cdot 6H_2O$
Korytkova and Pivovarova (2010)	250–450	30–100	SiO_2; MgO; NiO; CoO; FeO; Fe_2O_3
Bloise *et al.* (2010)	300–350	15–200	SiO_2; MgO; NiO; Ni–Mg_2SiO_4
Ogorodova *et al.* (2010)	350–400	30–100	SiO_2; MgO; FeO; Fe_2O_3
Korytkova *et al.* (2011)	250–400	70	SiO_2; MgO
Krasilin *et al.* (2011)	300–400	50	SiO_2; $Mg(OH)_2$
Bloise *et al.* (2012)	300–400	100–200	SiO_2; MgO; TiO_2; CaO
Lafay *et al.* (2012)	150–200	0.5–1.6	Olivine
Lafay *et al.* (2013)	90–300	8.2–0.1	H_2SiO_3; $MgCl_2 \cdot 6H_2O$
Wu *et al.* (2013)	350	70	SiO_2; MgO
Smolikov *et al.* (2013)	390	98	SiO_2; MgO
Lafay *et al.* (2014)	300	8.2	H_2SiO_3; $MgCl_2 \cdot 6H_2O$; Li; Sb; Cs; B; As
Cheng *et al.* (2015)	240	–	SiO_2; MgO

existence below 500°C. Some years later, with the aim of investigating and determining the univariant reaction curve of serpentine (chrysotile) + brucite $\rightleftharpoons$ forsterite + water, many other authors have shown that chrysotile could be synthesized in the system $MgO–SiO_2–H_2O$ (*e.g.* Noll, 1950; Yoder, 1952; Gillery, 1959).

Chrysotile was subsequently prepared at higher temperatures by Nesterchuk *et al.* (1966) and Johannes (1968). Nesterchuk *et al.* (1966) reported the synthesis of chrysotile at temperatures between 350 and 600°C, and pressure ranging from ~90 MPa to ~160 MPa. In order to investigate and determine the univariant reaction curve of serpentine + brucite $\rightleftharpoons$ forsterite + water, Johannes (1968) carried out experiments in the system $MgO–SiO_2–H_2O$ in the pressure range 50–650 MPa and at temperatures between 350 and 430°C. Subsequently, Ueno *et al.* (1991) continued the work by Johannes (1968), investigating the univariant reaction curve of chrysotile + brucite $\rightleftharpoons$ forsterite + water, at relatively low H_2O pressure (20–50 MPa) and temperature ranging from 200 to 500°C. Submicron observations of serpentinitic rocks showed that, as a result of solid-state reactions, chrysotile fibres are intergrown with other minerals (*i.e.* olivine, diopside and brucite). On the basis of this petrological knowledge, since the 1970s most syntheses of chrysotile have been achieved using different starting minerals, such as brucite, quartz (Yamai and Saito, 1974) and mainly forsterite Mg_2SiO_4 (Yada and Iishi, 1977) in order to investigate the nucleation process and stability field of serpentine minerals (chrysotile, antigorite, lizardite). Korytkova and Makarova (1971) provided a first systematic study of chrysotile growth from olivine, with temperatures ranging from 200 to 500°C and pressure of 91 MPa. It was concluded that in a basic pH, olivine transforms first into antigorite, which then recrystallizes to chrysotile in further reactions. Yamai and Saito (1974) investigated the effect of different combinations of starting materials, such as sintered magnesia, quartz sand or silica gel, on the hydrothermal synthesis of chrysotile at 360°C. They showed that the starting materials affect the size and morphology (*i.e.* cylinder and cylinder or cone-in-cone) of the chrysotile thus obtained. Similar experiments were conducted by Yada and Iishi (1974, 1977) who subjected powdered olivine to hydrothermal alteration (at 300–400°C) to investigate the initial stages of chrysotile formation. By using high resolution transmission electron microscopy (HRTEM), they concluded that the initial nuclei of proto-chrysotile were formed as membranes of several thick unit layers and a few hundred Ångstroms wide, protruding from the surface of forsterite. Most proto-chrysotiles curled up with increased hydrothermal treatment time. Moreover, they were curved in various ways so as to form parts of a cone or cylinder, depending on the environmental conditions, especially on pH and temperature. Chernosky (1975) also reported the synthesis of chrysotile by hydrothermal reactions in the system $MgO–Al_2O_3–SiO_2–H_2O$ using stoichiometric mixtures of either gel or oxides as starting materials for the experiments. Chemical and morphological investigations of synthetic chrysotile showed that the tubes became shorter and thicker when the amount of alumina incorporated increased. Other studies have been based on experiments in the $MgO–SiO_2–H_2O$ system, with the aim of investigating the reactions involving the mineral genesis through appropriate mineral transformations in the attempt to define the chrysotile stability range inside the serpentine-group minerals (Moody, 1976a; Hemley *et al.*, 1977; Kalinin and Zubkov, 1981). More details of the serpentinization process are given by Moody (1976b) and Deer *et al.* (1982). In a later study, chrysotile was obtained by transformation of synthetic diopside $CaMg(Si_2O_6)$ by hydrothermal reactions in the temperature range 250–390°C, in the presence of MgO and $MgCl_2 \cdot 6H_2O$ used as mineralizing agents (Barrese *et al.*, 1997). However, the first synthesis of chrysotile, in which mineralizing agents were used successfully to catalyse its growth, was reported by Yang (1961). The experiments were carried out at temperatures between 300 and 350°C, from oxide mixtures in the presence of excess NH_4F aqueous solution and with pH in the range 10.3–10.7. Devouard *et al.* (1997) synthesized chrysotile at 300°C and 700 bar from a starting gel of $Mg_3Si_2O_7$ in an alkaline pH environment, with a synthesis duration set to 11 days. Normand *et al.* (2002) showed that chrysotile appeared to require greater fluid supersaturation for growth from olivine as compared to lizardite. In fact, the above-mentioned syntheses of chrysotile were carried out with large fluid/solid ratios, which appear to favour chrysotile growth. Indeed, in contrast to nature, chrysotile may be formed by hydrothermal alteration of olivine in the laboratory, also with modest supersaturation, provided that fluids are in excess (Berman, 1988).

The results of the above-mentioned work are summarized partially in the serpentinitic equilibria diagram by Chernosky *et al.* (1988). Evans (2004) highlighted the limits of this diagram, suggesting that the distribution of chrysotile and lizardite in serpentinites cannot be understood in terms of the temperature and pressure of their crystallization, but according to their modes of formation. More recently, thermogravimetric analyses were used to investigate the product of replacement reactions of olivine (serpentinization) for different starting grain sizes at 200°C. Based on this innovative approach, Lafay *et al.* (2012) were able to estimate time-dependent serpentinization advancement and they suggested that the process was heavily dependent on the grain size of the olivine used as starting material.

According to Roveri *et al.* (2006), only a few studies on chrysotile synthesis were published since the 1980s because the natural counterpart was widely available (*e.g.* Ozeki *et al.*, 1991). Since 2000, the synthesis of chrysotile has received renewed interest due to the possibility of using it as an alternative to carbon nanotubes. In fact, in recent years, many works have focused on the preparation of chrysotile with nanotube features. Moreover, chrysotile appears to be an excellent template for preparing innovative inorganic nanowires because, when filled with molten metals, semimetals and semiconductor materials, it attains magnetic and semiconducting properties. These syntheses are justified by the fact that natural chrysotile fibres contain typically varying amounts of chrysotile polytypes, and are often intergrown with polymorphs, such as lizardite and antigorite or other minerals that could degrade its performance.

Falini *et al.* (2002, 2004) provided a first solution to obtaining single-phase chrysotile nanotubes, by hydrothermal treatment of SiO_2 and $MgCl_2$ with the Si/Mg molar ratio equal to 0.66, temperature of 300°C, pressure of 82 atm and pH = 13. They pointed out that by varying the Si/Mg molar ratio, under the same condition of temperature, pressure, pH and time, the formation of secondary phases (*i.e.* brucite or talc) were observed together with chrysotile.

Extensive experiments on the preparation of $Mg_3Si_2O_5(OH)_4$ nanotubes were also conducted during the past decade by Korytkova and co-workers with the aim of synthesizing pure homogeneous chrysotile and improving knowledge of its physical properties (thermal stability, morphology and kinetic properties). This knowledge is of great importance because physical properties (morphology, size, specific surface area) influence significantly the reactivity of chrysotile.

Korytkova *et al.* (2002) described the synthesis of chrysotile nanotubes at temperatures between 200 and 450°C and pressure 10–100 MPa, from different starting materials (mixtures of magnesium oxide or magnesium hydroxide, silicon oxide, and silicates and hydrosilicates of magnesium). It was shown that the typology and the intensity of the formation of $Mg_3Si_2O_5(OH)_4$ nanotubes depend both on the type of initial compounds and their ratio, as well as on the temperature and duration of the process. Some years later, Korytkova *et al.* (2004) continued to investigate the synthesis conditions that led to the formation of nanotubular fibres of $Mg_3Si_2O_5(OH)_4$, by varying parameters such as temperature (100–450°C), as well as starting materials (synthetic enstatite, synthetic talc). They confirmed that the formation of nanotubular fibres of $Mg_3Si_2O_5(OH)_4$ was determined primarily by temperature and that it was preceded by a stage of formation of compounds with a layered structure irrespective of the initial type of starting materials. This finding is in agreement with the work of Jancar and Suvarov (2006), who showed that the initial stage of the chrysotile nanotube reaction was the formation of lamellar nanocrystals, which, after reaching a specific size, curled into nanotubes. They stressed the fact that chrysotile synthesis was favoured at a high pH (>13) and that the length and diameter of the nanotubes could be controlled by the reaction temperature and time. A similar study was conducted by Xu *et al.* (2006), who prepared, under solvothermal conditions, tubular $Mg_3Si_2O_5(OH)_4$ in a short period of time (4 h). They noted the importance of using $MgCO_3{\cdot}4Mg(OH)_2{\cdot}6H_2O$ as the magnesium source, and nanoporous silica as the silicon source to make the nanotube synthesis process faster. The possibility of synthesizing chrysotile at lower temperatures has been shown by Ma *et al.* (2007) who obtained nanotubes with optimum morphology by hydrothermal synthesis at 160°C and pH = 13.

Korytkova *et al.* (2011) and Lafay *et al.* (2013) showed that the formation of chrysotile nanotubes under hydrothermal conditions proceeds through step mechanisms. They suggested that chrysotile is nucleated spontaneously from suspension and growth through the simultaneous dissolution of proto-chrysotile and brucite. In agreement, Jancar and Suvarov (2006) also showed that the formation of the tubes occurred through the initial formation of thin silicate platelets (magnesium hydroxide, magnesium hydro-silicate) at low temperatures (100–150°C). These hydrated phases become laminar hydrosilicate modifications at higher temperatures which rapidly transformed into tubes of different morphologies (cones, scrolls, *etc.*) as a large mismatch in terms of size exists between the octahedral and tetrahedral structural sheets. Conversely, the crystallinity of chrysotile decreases at temperatures lower than 300°C (Jancar and Suvarov 2006; Kortykova *et al.*, 2011). These authors also defined the factors (long reaction time, high temperature, alkaline pH) which lead to an increase in tube size. Krasilin *et al.* (2011) carried out an interesting study on the effect of the structure of precursors on the formation of $Mg_3Si_2O_5(OH)_4$ nanotubes under hydrothermal conditions. The authors suggested that the nature of the spatial match between reactants has a significant effect on the rate of nanotube formation. When starting materials form interfacial chemical bonds and a structure similar to layered $Mg_3Si_2O_5(OH)_4$, the synthesis of nanotubes takes place at lower temperatures and the nanotubes are more uniform in length because of the higher nucleation rate. Similar conclusions were reached by Bloise *et al.* (2012) who studied the experimental conditions involved in the transformation of glass into chrysotile fibres. These authors concluded that the effect of starting materials and/or mineralizing agents ($MgCl_2 \cdot 6H_2O$) prevailed over growth parameters as regards increased chrysotile fibre production.

Smolikov *et al.* (2013) carried out a study on the structural and morphological features of chrysotile nanotubes, synthesized at a temperature of 390°C and pressure of 98 Pa. Their conclusions supported the idea that chrysotile synthesized in distilled water without additives crystallized into a great variety of fibril morphological forms: cylindrical, 'cone in cone' and 'cylinder in cylinder' types. Similar results had already been observed by Yada and Iishi (1977). Recently, chrysotile nanotubes have been prepared *via* hydrothermal methods, in view of their possible use as decontaminants of Pb(II) in the field of water treatment (Cheng *et al.*, 2015).

Experimental synthesis of doped chrysotile during the past decade had three main goals:

(1) to clarify whether it is possible to obtain an isomorphic series between chrysotile *s.s.* $Mg_3Si_2O_5(OH)_4$ and, for example, greenalite $Fe_3Si_2O_5(OH)_4$ or pecoraite $Ni_3Si_2O_5(OH)_4$, and to assess unambiguously the influence of Fe and Ni on the physical-chemical features of chrysotile (cell dimensions, FTIR and Raman spectra, XRPD line, *etc.*) by excluding the influence of other ions (Al, Cr, *etc.*) which are present commonly in natural samples.

(2) to exploit doped chrysotile in innovative technological applications, due to its physical and chemical properties. For these reasons, it is very important to study the conditions of formation, morphology and size modification in chrysotile fibres with respect to the various ion dopants. Natural chrysotile fibres are not suitable for use in nanoscience and technology, because of the presence of foreign ions (*e.g.* Al, Fe, Ni, *etc.*) in their structure and due to the intergrowth of different minerals (talc, antigorite, *etc.*).

(3) to support biological research using *in vitro* experiments, in order to obtain unambiguous results on the interaction between cells and asbestos, which have not yet been obtained through studies conducted

using natural fibres. The possibility of synthesizing fibres focusing on a limited set of variables (in terms of size, chemical compositions, *etc.*) could provide the key to the definition of the characteristic(s) of fibres that induce cytotoxicity. Until a few years ago, in an attempt to understand the pathological mechanisms of chrysotile, *in vitro* experiments were all performed using natural asbestos although natural samples do not have an ideal end-member composition. Moreover, the composition and size of the individual fibres may vary significantly.

The first experimental studies on the incorporation of ions into the chrysotile structure were conducted as early as the 1950s by many researchers. The influence of the substitution of Mg^{2+} with other cations, such as Ni^{2+} (Noll and Kircher, 1952; Roy and Roy, 1954), Co^{2+} (Noll *et al.*, 1958, 1960) (Fig. 2), Fe^{2+} (Jasmund and Sylla, 1970) Mn^{2+}, Zn^{2+}, Al^{3+} (Roy and Roy, 1954), and the substitution of Si^{4+} by Ge^{4+} and Al^{3+} (Roy and Roy, 1954) on the morphology of serpentine crystals were investigated.

Jasmund *et al.* (1975) reported the possibility of substituting Mg with Ni, Co and Fe in chrysotile at various temperatures (250–300°C), pH values (8–10) and pressures (60–250 bar). The formation of Mg-Ni, Mg-Co and Mg-Fe chrysotile solid solutions was firmly established by the continuous shift of certain infrared bands within the above-mentioned series. Korytkova *et al.* (2005) grew hydrothermally Ni-doped chrysotile nanotubes $(Mg,Ni)_3Si_2O_5(OH)_4$ using NiO and MgO as starting materials, at a relatively high temperature (400°C) and observed that: (1) $Ni_3Si_2O_5(OH)_4$ nanotubes could be obtained as a pure phase only by heating a mixture of $NiSiO_3$ and $Ni(OH)_2$; and (2) lower temperatures, shorter reaction times and/or the use of other starting mixtures produced only small amounts of short tubes, mixed with mostly small thin platelets. In the same year, McDonald *et al.* (2009) synthesized $Ni_3Si_2O_5(OH)_4$ nanotubes in a

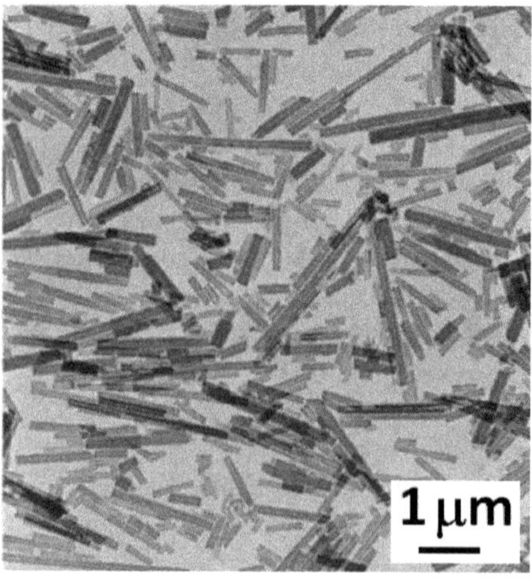

Figure 2. Electron micrograph of synthetic magnesium chrysotile. 37500 × (reproduced from Noll *et al.*, 1960 with the permission of Springer).

single phase using nickel chloride, silicic acid and sodium hydroxide as a starting mixture at relatively low temperature and pressure (250°C, 10 MPa). Bloise *et al.* (2010) studied the influence of various synthesis variables (amount of Ni-doping, starting materials) on the morphology, size and some chemical/physical features of grown Ni-doped chrysotile nanotubes (Fig. 3). They concluded that the best conditions for optimal fibre production was 350°C, and that the increase of Ni content in nanotubes produced a decrease in their length. Foresti *et al.* (2005) synthesized Fe-doped chrysotile nanotubes in the range 0.29–1.37 wt.% of Fe by hydrothermal treatment of the starting materials at 300°C. They concluded that the presence of Fe in the chrysotile lattice, if only to a limited extent, (1) reduced the degree of chrysotile crystallinity; and (2) led to changes from a fibrous to a flat morphology. Bloise *et al.* (2009a) suggested that the best conditions for the greatest lengthening of Fe-doped fibres (average length = 1 mm) with constant diameter, occurred at 380°C, 1 kbar with reaction times of 240 h, using synthetic forsterite as a starting material. Hydrothermally synthesized Fe-doped chrysotile $(Mg,Fe^{2+},Fe^{3+})_3Si_2O_5(OH)_4$ nanotubes were studied from a thermochemical point of view by Ogorodova *et al.* (2010). They showed that the enthalpy of formation of nanotubular hydrosilicates increases with increasing iron content. It follows that the formation of iron-containing nanotubes is less favourable energetically than the formation of $Mg_3Si_2O_5(OH)_4$ nanotubes. Bloise *et al.* (2009b) demonstrated that Ti can be incorporated into the chrysotile structure originating from the alteration of Ti-doped

Figure 3. TEM image of a cylinder-in-cylinder Ni-doped chrysotile nanotube (reproduced from Bloise *et al.*, 2010 with the permission of Elsevier.

synthetic forsterite (Fig. 4), and that the total hydrothermal alteration of Ti-doped forsterite to chrysotile requires long reaction times. Nanotubes of Co-doped chrysotile were synthesized under hydrothermal conditions at temperatures ranging between 200 and 450°C (Korytkova *et al.*, 2007). The starting mixtures were selected according to the stoichiometry of the $Co_3Si_2O_5(OH)_4$ and $(Co,Mg)_3Si_2O_5(OH)_4$ hydro-silicates with Mg:Co ratios of 1:2 and 2:1, respectively. Substitution of cobalt for magnesium led to a sharp decrease in the size of the nanotubes along both the axial and radial directions. In agreement with Korytkova *et al.* (2004), the stage of $(Co,Mg)_3Si_2O_5(OH)_4$ nanotube formation was consistently preceded by the formation of a lamellar structure. The incorporation of Fe (Foresti *et al.*, 2005) and Ni (Korytkova *et al.*, 2005) in the chrysotile structure has been shown to result in changes in morphology (*i.e.* cylinder-in-cylinder, cone-in-cone). Furthermore, as demonstrated recently (Lafay *et al.*, 2013), the morphology of synthetic chrysotile can be related also to the presence of trace elements in its structure (*i.e.* Li, As, Sb, B). However, other authors have emphasized the roles played by starting materials and experimental conditions on the various doped chrysotile morphologies obtained (Yamai and Saito, 1974; Bloise *et al.*, 2009a,b, 2010; Korytkova and Pivovarova, 2010; Korytkova *et al.*, 2011).

In summary, although chrysotile has been prepared from a variety of starting materials (oxides, gel, glass, natural and synthetic minerals), the most common synthesis procedures are either hydrothermal reaction of mixtures of cristobalite and periclase or autoclaving mixtures of magnesium hydroxide $Mg(OH)_2$ and silica gel with an appropriate $MgO:SiO_2$ ratio. Synthesis experiments provide large fluid/solid ratios, as well as large free-energy starting materials for the nucleation and growth of chrysotile. However, this approach leads the starting mixtures out of the expected stoichiometric proportions to form pure chrysotile. In fact, the syntheses of pure phase or doped chrysotile generally produce <100% chrysotile. Moreover, other phases, such as brucite, quartz or residual starting materials are often present in addition to

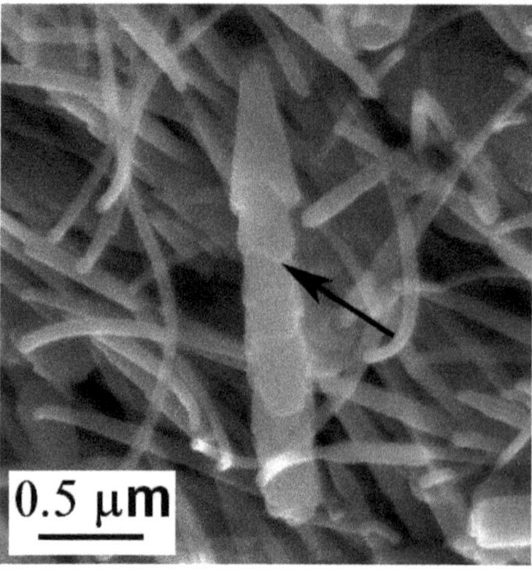

Figure 4. Secondary electron SEM image of cone-in-cone Ti-doped chrysotile fibre indicated by arrow (modified after Bloise *et al.*, 2009b; reproduced with the permission of Schweizerbart'sche, www.schweizerbart.de/journals/njma).

chrysotile. In most studies chrysotile was synthesized at a temperature of ~300°C, and many of these investigations revealed an optimal temperature between 300 and 400°C, and pressure of the order of 0.5 to 100 MPa. Long reaction time and high temperature are usually needed to increase the yield and size of the chrysotile nanotubes. The greatest crystallinity is obtained at a high pH (>13) by means of aqueous NaOH solution or using mineralizing agents. The presence of impurities (*i.e.* Fe, Ni, Ti, Co) in an ideal chemical composition in chrysotile fibres, even in trace amounts (*i.e.* Li, As, Sb, B), affects their chemical/physical properties, size and morphology.

2.2 Dehydration and breakdown of chrysotile

The thermal decomposition of chrysotile ($Mg_3(OH)_4Si_2O_5$) has been studied extensively over the last 60 years, since asbestos gained significant industrial interest. The main thermal reactions of the serpentine group (which includes chrysotile asbestos) are dehydroxylation and exothermic phase transformation. This process is similar to that for kaolinites. However, there is no clearly visible intermediate phase as in the case of kaolinite dehydroxylate (amorphous metakaolinite). In the case of chrysotile, the amorphous phase obtained after dehydroxylation exists over a very short temperature range and undergoes a rapid transformation into new phases.

Typical thermal analysis curves (DTA/TG/DTG) of selected natural chrysotile asbestos are given in Fig. 5. Within the temperature range 50–400°C minor endothermic mass losses are observed due to the loss of hygroscopic water adsorbed on the surface and from the inside of chrysotile fibres. The main endothermic effect related to mass loss between 500 and 800°C with the maximum rate at 710°C is caused by release of chemically bound water (dehydroxylation process of octahedrally coordinated hydroxyl groups from the crystal structure). In the context of chrysotile asbestos thermal inertization, this stage can be regarded as the most important. This process contributes to a complete breakdown of the mineral structure and creation of an anhydrous phase. In the next stage, the crystallization of this amorphous structure and the creation of forsterite (characteristic exothermic peak on the DTA curve at 850°C, without mass change) occur. Forsterite belongs to the family of nesosilicates and does not exhibit carcinogenic properties.

Sometimes, additional minor effects may be observed in the thermal analysis curves collected which are assigned to impurities accompanying chrysotile asbestos deposits. They may come from brucite, organic impurities and other compounds present in the sample (Zaremba *et al.*, 2010; Kusiorowski *et al.*, 2012; Bloise *et al.*, 2016). After being subjected to thermal treatment, the original habit is still evident, but the fibres are very fragile. This phenomenon is explained by the 'pseudomorphosis' effect. The overall original exterior crystal habit is preserved while the core is replaced by newly formed crystals of forsterite and/or enstatite grains along the fibre axis (Gualtieri *et al.*, 2012). The original cleavage parallel to the fibre axis is lost and fracture of the transformed pseudomorphic fibres occurs at the particle boundaries and not along the fibre axis (Bloise *et al.*, 2016). When subjected to a gentle mechanical action, this material is easily disintegrated into powder.

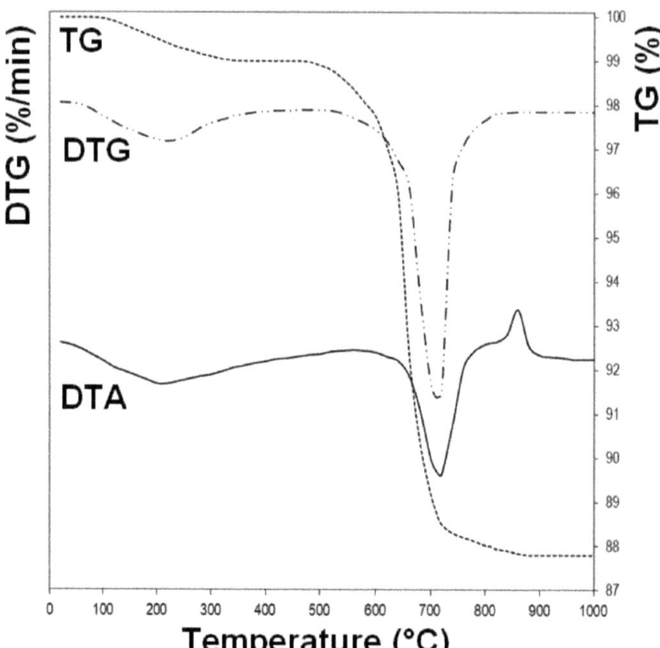

Figure 5. Thermoanalytical curves of chrysotile. Conditions: sample mass – 500 mg, heating rate – 10 K min^{-1}; inert material – α-Al_2O_3; air atmosphere.

A summary of thermo-analytical data (during non-isothermal treatment) based on different references from previous years is given in Table 2. Temperature differences for the dehydroxylation process and forsterite transformation reported by various researchers are associated with different measurement conditions, *i.e.* sample mass, heating rate, atmosphere. In turn, the results of experiments using isothermal conditions are shown in Table 3.

In the case of chrysotile asbestos, results reported in several studies show that with the loss of chemically bound water, chrysotile transforms into an anhydrous phase at ~500–800°C (with a maximum rate above 700°C during non-isothermal treatment). Rapid crystallization of the anhydrous phase occurs after this stage above 800°C. The first main product of this transformation is forsterite (Mg_2SiO_4) and amorphous silica. Enstatite ($MgSiO_3$) is generally formed at temperatures >1000°C (Brindley and Hayami, 1965). The above transformations could be simplified and represented by the following overall reaction:

$$Mg_3(OH)_4Si_2O_5 \text{ (chrysotile)} \rightarrow Mg_2SiO_4 \text{ (forsterite)} + MgSiO_3 \text{ (enstatite)} + 2H_2O$$

Despite a large number of studies on the dehydration of serpentine minerals (which include chrysotile), little is yet known about the mechanism of decomposition of this mineral subgroup (Zulumyan *et al.*, 2014). Consequently, the detailed course of the

Table 2. Characteristic temperature range and/or peak temperatures of chrysotile during thermal analysis experiment presented in various previous studies.

Reference	Loss of adsorbed water (°C)	Loss of the hydroxyl from the structure (°C) /Endothermic effect/	Transformation into forsterite (°C) /Exothermic effect/
Bloise *et al.* (2014)	<613 (most water loss <200)	636–646	819–826
Zaremba *et al.* (2010)	60–500 (most water loss <200)	600–800 (maximum rate at 725)	close to 840
Kusiorowski *et al.* (2012)	<500 (most water loss <200)	600–800 (maximum rate at 700–730)	820–830
Bloise *et al.* (2016)	<110	633–660	821–823
Virta (2012)	~90	640–810	>810
Belardi and Piga (2013)*	–	begins at ~600	840
Malkov *et al.* (2009)**	<300	maximum between 500–600 (complete at 800)	811
Dellisanti *et al.* (2001–2002)***	<250	450–700 (maximum at ~700)	~820
Gualtieri and Tartaglia (2000)	–	>500 (max ~600)	~870
Leonelli *et al.* (2006)	–	648–680	803
Cattaneo *et al.* (2003)	–	600–800	~800
Teixeira *et al.* (2013)	<200	500–750	800
Sprynsky *et al.* (2011)	<100	570–720 (maximum at 680)	~830
Vast *et al.* (2004)	–	~640	800–850

* commercial material containing ~60 wt.% chrysotile; ** synthetic chrysotile obtained by the hydrothermal method; *** commercial material containing ~80 wt.% chrysotile.

reaction sequence is yet to be revealed. Various detailed models for chrysotile dehydroxylation and breakdown mechanisms were proposed in the past by Ball and Taylor (1963), Brindley and Hayami (1965), Datta *et al.* (1986), MacKenzie and Meinhold (1994), Trittschack and Grobéty (2013) and Trittschack *et al.* (2014). According to the latter, the endothermic effect on the DTA curves connected with the dehydroxylation process took place in several stages. The authors explained this phenomenon by the presence of two types of particles with different reactivity in the chrysotile, which resulted in the creation of two reactive and X-ray amorphous

Table 3. Characteristic transformation temperature of a chrysotile sample during isothermal treatment based in various previous studies.

Reference	Conditions of isothermal experiment	Chrysotile disappearance /based on XRD measurement/	Forsterite presence
Jeyaratnam and West (1994)	0.2 g sample heated for 24 h at selected temperatures from 100–850°C	rapidly >500°C	>500°C (disordered) >800°C (well crystallized)
Zaremba *et al.* (2010)	100 g sample heated for 3 h at selected temperatures from 500–725°C	>650°C	>725°C (from 600°C weak peaks on XRD pattern)
Malkov *et al.* (2009)*	heated from 100–800°C	500–700°C	above 800°C
Dellisanti *et al.* (2001–2002)**	heated for 2 h at selected temperatures from 400–1100°C	700°C	from 700°C (also enstatite)
Cattaneo *et al.* (2003)	heated for 1 h at selected temperatures from 500–900°C	from 700°C	>600°C
Hashimoto and Yamaguchi (2006)	heated for 3 h at selected temperatures from 550–1000°C	from 600°C	from 700°C
Hashimoto *et al.* (2008)***	heated for 3 h at selected temperatures from 500–600°C	600°C	–

* synthetic chrysotile obtained by hydrothermal method; **commercial material containing ~80 wt.% chrysotile; *** measurement in a vacuum.

dehydroxylates with different Mg/Si ratios (dehydroxylate I with Mg/Si = 2 and dehydroxylate II with Mg/Si = 1). With further temperature increase, dehydroxylate I crystallizes into forsterite while dehydroxylate II is the precursor of enstatite (Fig. 6) (MacKenzie and Meinhold, 1994).

Studies conducted in recent years (Gualtieri *et al.*, 2012) indicate the correctness of the dehydroxylation and high-temperature reaction sequence of chrysotile postulated by Mackenzie and Meinhold (1994). The experimental data confirmed the formation of a Si-rich amorphous phase. Another product of this high-temperature reaction was forsterite (57 wt.% at 750°C), which was created from Mg-rich regions. At higher temperatures, Mg_2SiO_4 (forsterite) still crystallized in the Mg-rich regions, whereas $MgSiO_3$ (enstatite) began to form in the Si-rich regions.

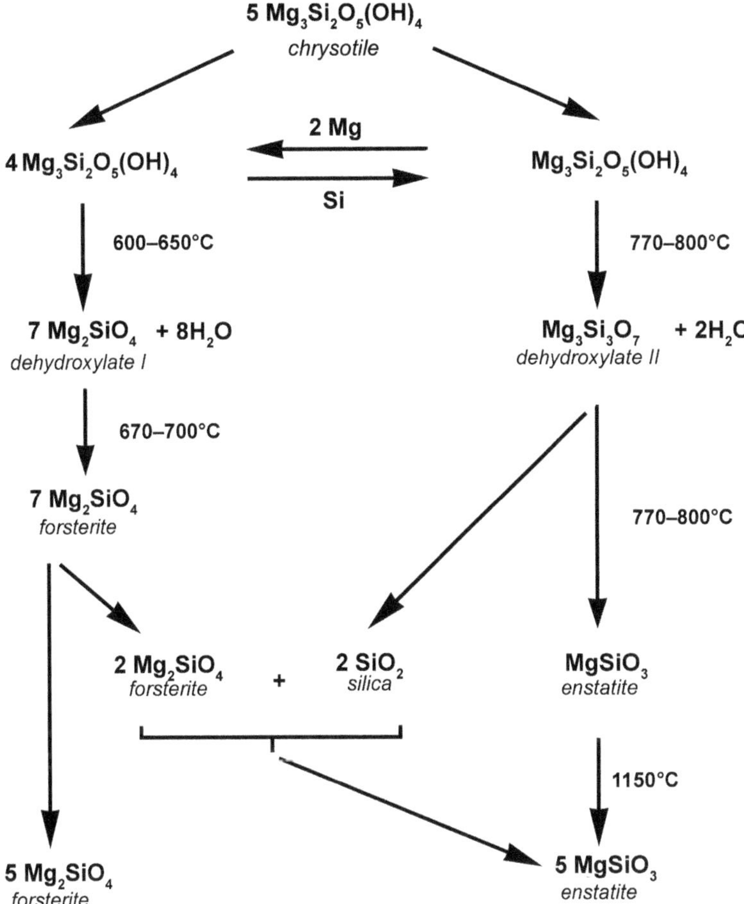

Figure 6. Schematic representation of proposed chrysotile reaction sequences according to MacKenzie and Meinhold (1994).

It can be concluded that the crystalline structure is stable up to approximately 550°C (depending on the heating period), after which the process of chrysotile dehydroxylation begins. This process is completed near 750°C. The resulting dehydroxylated magnesium silicate product undergoes a structural transformation (recrystallization) to form forsterite as a main product at a temperature of ~800°C. At higher temperatures enstatite is also formed.

Regarding the hazardous nature of asbestos, many researchers believe that the only point at which chrysotile changes into a substance that is not detrimental to human health is the temperature at which the phase transformation, *i.e.* the crystallization of forsterite takes place. According to Langer (2003), the complete transformation of an asbestos mineral during thermal treatment is not required for it to lose its toxic effect.

The process of dehydroxylation of asbestos leads to significant changes in the crystalline structure and the crystallization of products which are harmless (Langer, 2003).

2.3. Chrysotile remediation through thermal decomposition

Until the ban on amphibole asbestos, chrysotile accounted for >90% of the asbestos mineral fibres used in commerce worldwide. Since the ban on amphibole mineral fibres, chrysotile has been the only asbestos fibre still extracted, employed and marketed worldwide. Hence, remediation of asbestos-containing materials (ACMs) or wastes in practice means dealing with chrysotile-containing ACMs or wastes. Asbestos destruction *via* chemical-physical transformation (mostly temperature-induced recrystallization, sintering or melting) and recycling of the transformation product as secondary raw material is actually the only viable alternative to landfill disposal.

The preference for recycling over landfill disposal is indicated in the European Directive 2008/98/EC (November 19th, 2008). Although thermal transformation is the most common transformation process, other processes have also been developed in the last two decades (see for example Sugama *et al.*, 1998; Plescia *et al.*, 2003; Inoue *et al.*, 2007; Colangelo *et al.*, 2011) for the treatment of both pure asbestos minerals and ACMs.

At high temperature, chrysotile fibres in ACMs convert to stable crystalline silicates through solid-state reactions (Martin, 1977; MacKenzie and Meinhold, 1994; Gualtieri and Tartaglia, 2000; Cattaneo *et al.* 2003; Gualtieri *et al.*, 2008a; Kusiorowski *et al.*, 2012, 2013). As discussed above, dehydroxylation of chrysotile occurs in the range 650–750°C and is followed by solid-state recrystallization into forsterite (Mg_2SiO_4) first and enstatite ($MgSiO_3$) later. During the reaction, chrysotile fibres preserve the same overall crystal habit although a complete modification of the structure at the molecular scale has occurred ('pseudomorphosis', Giacobbe *et al.*, 2010 and Section 3.2.4, this chapter). Figure 7 shows a high-magnification SEM image of an original chrysotile fibre bundle fully recrystallized into forsterite and enstatite.

The number of research projects at laboratory and pilot scales and patents based on thermal transformation of ACM found in the literature is huge. The following paragraph presents the most relevant solutions.

Roberts and Stuart (1989) patented a method for the inertization of ACMs by vitrification/melting and subsequent recycling. The benefits of vitrification/melting derive from the complete destruction of the fibrous structure and the formation of an amorphous solid suitable for recycling as secondary glass material (see for example Osada *et al.*, 2013). Inaba *et al.* (1999) and Inaba and Iwao (2000) demonstrated that plasma torch vitrification permits conversion of asbestos into a rock-like structure with a Mohs hardness of 6. In the INERTAM-Europlasma process, plasma torch vitrification of ACMs at 1600°C takes place in a cylindrical furnace (Poiroux and Rollin, 1996; Borderes, 2000). The product of vitrification has been recycled as aggregate for road foundations. Fujishige *et al.* (2007) established an experimental protocol involving several annealing temperatures in the range 600–1300°C and chemical additives such as $CaCO_3$ and $CaCl_2$ to accomplish melting of chrysotile-rich wastes. Min *et al.* (2008)

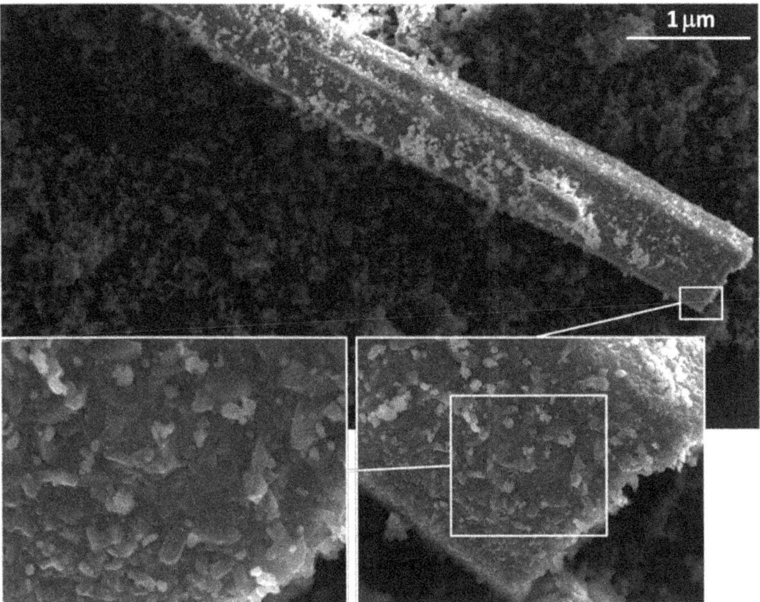

Figure. 7. An example of 'pseudomorphosis' of the high-temperature transformation product of chrysotile. The high-magnification SEM image shows a former chrysotile bundle fully transformed into forsterite and enstatite after thermal treatment.

proposed a melting furnace operating at 1450–1550°C that uses a mixture of hydrogen and oxygen (Brown's gas) as fuel to vitrify ACMs. The GeoMelt vitrification was designed by Finucane *et al.* (2008) while a method based on Joule-heating vitrification was developed by Dellisanti *et al.* (2009). A possible recycling solution for the product of plasma vitrification was reported by Bernardo *et al.* (2011). Dense sintered crystallized materials from vitrified asbestos-containing waste are obtained by fast heating processes and constituted the basis of glass ceramics and a new type of stoneware.

Melting-based processes have the disadvantage of being too energy consuming and expensive. Lower recrystallization (ceramization) temperatures (750–1300°C) are sufficient to denaturate ACMs, reducing the energy consumption and making the process economically more competitive. The CORDIAM project developed by Abruzzese *et al.* (1998) for the production of cordierite refractories is one of the pioneering ceramization processes of ACMs. Cordierite ceramics were also produced by thermal treatment at 1350°C of mixed chrysotile-rich tailing wastes (Zhu *et al.*, 2013). The Asbestex process (Johannes, 2003) includes a rotary kiln that converts ACMs thermally. Downey and Timmons (2005) proposed a thermochemical treatment (the so-called A.R.I. process) for destroying asbestos and for treating other hazardous and radioactive wastes.

Examples of microwave thermal treatment of ACMs (Leonelli *et al.*, 2006; Boccacini *et al.*, 2007; Yoshikawa *et al.*, 2015) are also well described in the existing literature. A combination of microwave, infra-red heating and chemical treatment is the core of the system installed on a mobile asbestos decontamination unit devised to disintegrate ACMs on-site (Gerdes *et al.*, 2007). Recently, a continuous microwave-driven treatment plant having a capacity of 2 t/day of ACMs has been developed in Japan, in the area affected by the earthquake disaster. The plant consists of a rotary kiln to burn the combustible waste (wood) and a microwave rotary kiln to treat asbestos-containing inorganic materials (Horikoshi *et al.*, 2014).

An industrial process, named KRY·AS, for the direct thermal conversion of cement-asbestos has been developed (Gualtieri *et al.*, 2008a). Sealed packages of cement-asbestos materials are treated thermally in a tunnel kiln in the temperature range 1200–1300°C. It was observed that the transformation sequence of the asbestos and cement-phases assemblage is different from that of the pure asbestos minerals, due to extra-crystalline reactions. The resulting product is composed mainly of SiO_2 and CaO and has a chemical nature comparable to that of a Mg-rich clinker (Gualtieri *et al.*, 2008a).

A hybrid method for making pulverized cement-asbestos inert has been described recently by Radvanec *et al.* (2013); this method includes thermal modification at 650°C followed by artificial carbonization. Following heating, the material is exposed to a mixture of CO_2 and water, during which the thermally modified chrysotile fibres recrystallize completely into hydromagnesite and magnesite.

Huang *et al.* (2013) studied *in situ* carbothermal reduction (CR) of chrysotile-rich tailing at high temperatures using coke (carbon) as reducing agent. The CR process resulted in the formation of stable composites consisting of forsterite, stishovite and silicon. In a similar vein, Teixeira *et al.* (2013) managed to develop a new heterogeneous catalyst for biodiesel synthesis by treatment of chrysotile with KOH followed by heating. In the process, the chrysotile fibres impregnated with KOH decompose at 700°C to produce sintered particles containing a catalytically active K-doped MgO dispersed phase. This K^+/MgO phase showed surface basic sites with very good activity for the transesterification of soyabean oil with methanol under various conditions.

It is possible to 'denaturate' chrysotile even at lower temperature in hydrothermal systems. Nam *et al.* (2014) investigated the feasibility of using a thermochemical technique for the inertization of chrysotile-containing roofing. Acid-digestive destruction (5 N sulfuric acid) was accomplished following treatment for 10–24 h at 100°C. A method based on hydrothermal conversion combined with acidic attack of friable asbestos has been developed by Yanagisawa *et al.* (2009). Chrysotile was decomposed by reacting with acidic gas, generated by the decomposition of $CHClF_2$ with superheated steam. The same authors proposed thermal inertization of asbestos minerals by treatment in a water vapour atmosphere at 800°C for 2 h (Kozawa *et al.*, 2010). On the same line, hydrothermal treatment of ACMs in temperature and pressure ranges of 300–700°C and 1.75–5.80 MPa, respectively, was described by Anastasiadou *et al.* (2010).

ACMs converted into End of Waste Materials (EoWM) offer attractive recycling solutions. There are plenty of examples in the literature describing recycling opportunities for EoWMs derived from inertization of ACMs. The product of transformation of cement-asbestos has been recycled successfully for the production of clay bricks (Kusiorowski *et al.*, 2014), glasses, glass-ceramics, ceramic frits, ceramic pigments and plastic materials (Gualtieri *et al.*, 2010). Recycling of thermally treated chrysotile-rich wastes in cement (Ca-sulfoaluminate clinker: Viani and Gualtieri, 2013; magnesium phosphate cement: Viani and Gualtieri, 2014; clinker bricks: Kusiorowski *et al.*, 2015a) and concrete (Gualtieri and Boccaletti, 2011) is a very attractive solution due to the high environmental impact of the cement industry (Bertolini *et al.*, 2004). Gualtieri and Boccaletti (2011), showed that the high-temperature transformation of cement-asbestos promoted the crystallization ('retro-clinkerization' process) of cement phases such as C_2S and ferrite (ideally $Ca_4Al_2Fe_2O_{10}$), and Al-, Ca-, Mg-rich silicates such as åkermanite (ideally $Ca_2MgSi_2O_7$) and merwinite (ideally $Ca_3MgSi_2O_8$). According to the UNI EN 206-1:2006 standards, a concrete product containing <20 wt.% of EoWM based on thermally transformed cement-asbestos may be used for: (1) indoor environments with very low moisture content; (2) reinforced concrete injected in chemically inert soil or unaggressive water; (3) unreinforced concrete under periodical dry/wet cycles but not subject to abrasion, frost nor chemical attack; and (4) liquid storage tanks or foundations. Recycling of this type of EoWM as cementitious material in concrete entails two major advantages: (a) reduction of the request for natural raw materials such as calcium carbonate, thus limiting the environmental impact caused by the extraction of the minerals from quarries and mines; and (b) substitution of ordinary Portland cement by EoWM-containing products enables a decrease in the emission of CO_2 (formed unavoidably during the clinkerization step of cement production) to the environment.

In the same way, studies of the thermochemical transformation of ACM wastes and subsequent recycling in cement products were conducted by Yvon and Sharrock (2011). The authors reported good mechanical properties of mortars containing up to 10 wt.% EoWM. The preparation of magnesium sulfate whiskers was possible by sulfuric acid hydrothermal leaching of the product of calcination of chrysotile asbestos tailings (Cunjin *et al.*, 2012). The same product was also used for the preparation of diopside-based glass ceramics (Ding *et al.*, 2012).

2.4. *In situ* electron microscopy studies of the thermal decomposition of chrysotile-rich wastes

Scanning electron microscopy (SEM) and energy dispersive X-ray spectroscopy (EDS) are basic tools for materials characterization in research communities as well as in industry. Of particular importance for the main subject of this book, regulatory analytical protocols for identification and quantitative assessment of asbestos fibre contaminants foresees the use of electron microscopy techniques (see for example the Italian Ministerial Decree no. 19 of 1994). High-resolution SEM imaging is also a valuable technique to verify complete thermal transformation of cement-asbestos in view of subsequent safe recycling. In fact, thermally transformed asbestos fibres can be distinguished from non-reacted material due to the appearance of sub-micron grains of newly formed crystalline phases (Gualtieri *et al.*, 2008a) as already discussed above. Unfortunately, the use of X-ray powder diffraction alone for quality assessment of thermally inertized cement-asbestos is not possible as the dilution of the asbestos phases in the cement matrix leads to peaks of low intensity which overlap with peaks from high-temperature phases.

The advent of SEM instruments which utilize a gas for image formation while also providing charge stabilization for dielectric materials, has opened up new possibilities for real-time imaging. These microscopes, commonly called 'environmental', may be accessorized with hot stages, allowing *in situ* high-temperature experiments up to 1500°C to be performed in a controlled environment (Podor *et al.* 2012). Direct imaging of thermally induced events such as crystal growth or decomposition is complementary to conventional *in situ* XRPD which instead provides average structural information of coherently diffracting domains. Considering pure or near-pure asbestos minerals, the latter technique offers several advantages over *in situ* ESEM not only due to higher statistics but also because resolution limitations of the ESEM permit visualization of the newly formed crystals only after they have reached a certain size whereas XRPD experiments allows the disappearance of asbestos diffraction peaks to be monitored and thus the initial dehydroxylation reactions. However, in light of the limitations of XRPD for the detection of fibrous minerals in cement-asbestos, Gualtieri *et al.* (2008b) applied *in situ* ESEM to observe the temperature-induced microstructural development of selected chrysotile fibres in this type of composite (Gualtieri *et al.*, 2008b). Additional advantages were offered by the possibility of controlling the gaseous atmosphere during the experiment and thus gain further insight into the transformation kinetics of chrysotile. Images were recorded in regular intervals during heating (20°C/min) up to a maximum temperature of 1300°C. The experiment was performed in both He (1.9–3.5 Torr) and water vapour (2.5–3.4 Torr) atmosphere.

Figures 8a and 8b show micrographs of chrysotile fibres collected in a He atmosphere during the heating ramp at temperatures of ~1000°C and 1150°C, respectively. Crystallization of newly formed enstatite and forsterite apparently started at ~1000°C, as indicated by the change in surface morphology from smooth to granular (Fig. 8a). Crystallization continued to a temperature of ~1150°C (Fig. 8b) after which no further morphology changes were observed. These results are in concert with those reported by Gregori and co-workers who performed similar *in situ* SEM experiments (Gregori *et al.*, 2002). Figure 9 shows micrographs of chrysotile fibres collected in a water vapour atmosphere during heating up to 1000°C (a) and following 3 h of isotherm at 1000°C following cooling from 1300°C (b). In the presence of water vapour, recrystallization was not observed during the non-isothermal run (Fig. 9a) but only following prolonged heating at 1000°C (Fig. 9b). The negative influence of water vapour on the reaction kinetics may be explained by a greather surface coverage of chemisorbed water, inhibiting the dehydroxylation reactions and thus recrystallization, as found by others for the structurally similar serpentine mineral (Brindley *et al.*, 1967).

Another interesting *in situ* ESEM experiment (water vapour atmosphere) performed in our laboratory was aimed at investigating the possibility of using a low-melting glass

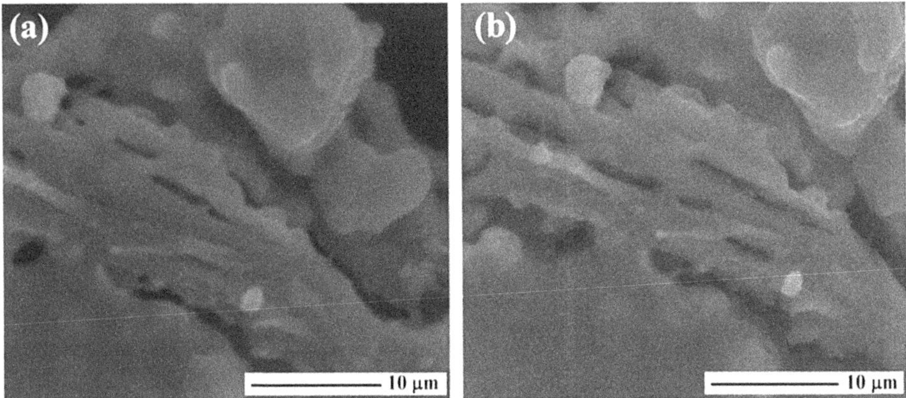

Figure 8. SE micrographs of chrysotile fibres, collected in He atmosphere at 1000°C (a) and 1150°C (b).

to decrease the inertization temperature of cement-asbestos (Gualtieri *et al.*, 2008b). In this experiment, a mixture of low-melting glass and cement-asbestos was heated to 850°C. Figure 10 shows micrographs of the mixture of glass and cement-asbestos collected at 155°C (a) and 750°C (b). As can be observed from this figure, complete melting was reached at 750°C. The chrysotile fibres as well as other crystalline phases were completely dissolved, thus forming a glassy phase as confirmed by *ex situ* X-ray diffraction. Hence, the addition of a low-melting glass to asbestos-containing materials during the abatement procedures would potentially reduce the firing temperature needed to obtain complete decomposition of the fibres.

3. Thermal behaviour of fibrous amphiboles

Dehydration and breakdown experiments on fibrous amphiboles were carried out extensively during the 1960s with the main aim of highlighting their thermal behaviour

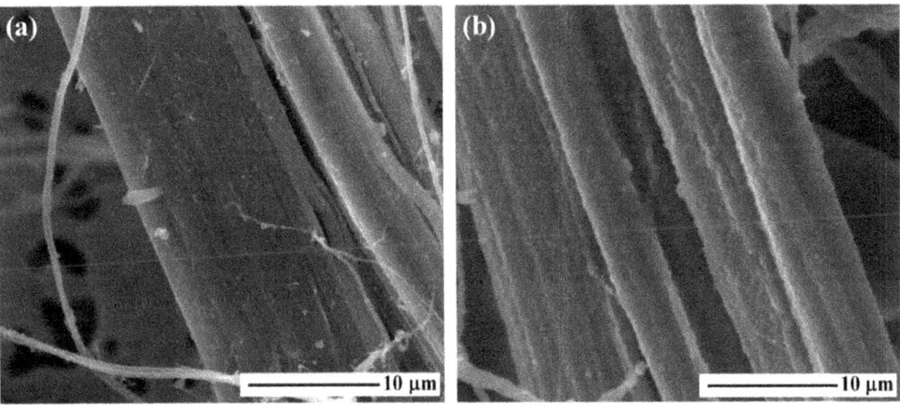

Figure 9. SE micrographs of chrysotile fibres collected in a water vapour atmosphere during heating to 1000°C (a) and following prolonged heating (3 h) at 1000°C (b).

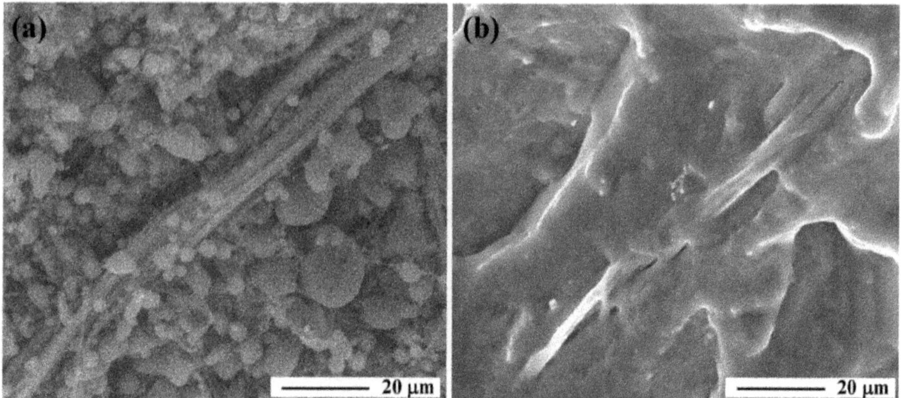

Figure 10. SE micrographs of a mixture of cement asbestos and glass, collected in a water-vapour atmosphere at 155°C (a) and 750°C (b).

and crystal chemistry. In the following years, a few studies addressed these topics. However, in the past decade, this issue has received renewed attention in view of the possibility of fibrous amphiboles being made inert and/or reused through the crystal chemical transformation induced by thermal treatment. The use of thermogravimetry provides much information for the investigation of the asbestos amphiboles. Important aspects include the detection of the structural water content or the temperature at which structural breakdowns of the fibrous amphiboles occur. The studies on thermal decomposition of fibrous amphiboles minerals have mainly been addressed to asbestos minerals of commercial value or health concern, such as crocidolite (*e.g.* Hodgson *et al.*, 1965a; Rouxhet *et al.*, 1972; Jeyaratnam and West 1994; Gualtieri *et al.*, 2004) and amosite (*e.g.* Hodgson *et al.*, 1965b; Jeyaratnam and West 1994). In recent years, the application of DTA, FTIR, SEM and XRPD, has allowed both the understanding of phase evolution during heating, and the gathering of information on the morphological changes of some fibrous amphiboles (Kusiorowski *et al.*, 2012, 2015b; Bloise *et al.*, 2016). Previous literature has shown that the investigation and interpretation of the thermograms of fibrous amphiboles containing ferrous iron are much more complex than those of other amphiboles, because the former are strongly oxidative and their oxidation behaviour is often complex. Indeed, the dehydroxylation of these minerals generally takes places in several steps. Besides, it is known that both the atmosphere of the furnace (N_2, air, vacuum) and sample-preparation conditions affect the shape of the thermogram curves.

Note that the presence of other phases (*e.g.* carbonates) leads to variations in the temperature at which breakdown of fibrous amphiboles occurs. The thermal transformations sequence of asbestos-containing materials (ACM), consisting of a variety of different crystalline and amorphous phases, is different from the transformations sequence of pure fibrous amphiboles.

The thermal analysis of synthetic amphiboles has not been discussed in this chapter as it has often shown different decomposition temperatures compared to the natural mineral, due to both low crystallinity and smaller grain size. What follows is not a review of all the fibrous amphiboles phase-equilibria stability obtained using calorimetric data, but a focus on thermal studies of fibres of social, health, economic and industrial relevance, carried out through calorimetric thermal analyses.

3.1. Hydrothermal synthesis of fibrous amphiboles

During the past 60 years, many studies have focused on the synthesis and characterization of amphiboles in order to study their crystal chemistry for petrological purposes or to define the kinetic processes allowing their growth as a function of the different growth parameters. Ernst (1968), Gilbert *et al.* (1982), Oberti *et al.* (2007), Evans (2007) and Maresch and Czank (2007) have provided an exhaustive overview of the early literature on synthetic amphiboles. In their conclusions, Maresch and Czank (2007) stress that for synthetic calcic and sodic-calcic amphiboles more promising results had been achieved than for other cases, such as for the members of the synthetic $(Mg,Fe^{2+})_7Si_8O_{22}(OH)$ series. However, a lot of work still has to be done to develop viable reaction paths that lead to acceptable products which can serve as analogues for natural ones. In this scenario, the main outcome of this subchapter is a review of the attempts that have been made to synthesize only fibrous-asbestos amphiboles:

-amosite: $Fe_7Si_8O_{22}(OH)_2$,
-crocidolite: $Na_2Fe_2^{3+}Fe_3^{2+}Si_8O_{22}(OH)_2$,
-anthophyllite: $Mg_7Si_8O_{22}(OH)_2$,
-tremolite: $Ca_2Mg_5Si_8O_{22}(OH)_2$,
-actinolite: $Ca_2Mg_3Fe_2^{2+}Si_8O_{22}(OH)_2$

Fibrous amphiboles were synthesized under hydrothermal conditions, varying temperature and pressure in the stability field of natural amphiboles (Deer *et al.*, 1966). To synthesize the various amphibole types, the preparation of starting material was required. The following types of starting material are commonly used in experimental investigations: glasses, gels, mixtures of oxides and natural minerals. Pre-treatment of the starting materials was sometimes necessary to improve the reaction time and to eliminate hygroscopic content. After pre-heating, the nutrients are mixed together according to the stoichiometry of the amphiboles. Most of the synthesis studies have been carried out with some variation in the three basic types of apparatus: (1) cold-seal pressure vessels (externally; internally); (2) piston-cylinder apparatus; and (3) internally heated gas-pressure vessels. The experimental conditions (temperature, pressure, reaction time and starting mixtures) chosen for the synthesis of end-member asbestos amphiboles aimed at:

(1) maintaining the ideal stoichiometry of the synthetic amphibole;
(2) avoiding the formation of additional phases.

In the past 25 years, the synthesis of tremolite and its reaction path have been studied accurately (Table 4), probably because it is an important mineral for deducing the conditions of medium-pressure metamorphism. An exhaustive overview of these experimental works may be found in the chapters by Evans (2007), Maresch and Czank

Table 4. Experimental conditions for the fibrous tremolite reviewed in this chapter. (Solid solutions radiating from synthetic end-member tremolite are not considered).

References	Pressure (kbar)	Temperature (ºC)	Time (h)
Skippen and McKinstry (1985)	2	780	336–504
Jenkins (1987)	2	796	285
Ahn *et al.* (1991)	2–5.88	701–1000	188–1152
Pawley *et al.* (1993)	2–6	850	160
Maresch *et al.* (1994)	2–15	600–875	50–1296
Zimmerman *et al.* (1996)	2	800	340–960
Hawthorne *et al.* (1997)	1	750	72–360
Gottschalk *et al.* (1998)	2	750	504
Gottschalk *et al.* (1999)	5–7	660–700	120–1224
Sharma and Jenkins (1999)	4–5	800	314
Della Ventura *et al.* (2003)	3	900	–
Bozhilov *et al.* (2004)	6.1	850	0.13–97.3
Bozhilov *et al.* (2007)	5	600	91–582
Bloise *et al.* (2008)	1.7	780	120

(2007), and some additional works reported below. The knowledge acquired to date about the synthesis and characterization of tremolite can be summarized as follows:

- in all preview experiments aiming to synthesize tremolite, the products contain ~10% of extraneous phases, such as quartz, diopside, talc, clinopyroxene and enstatite (*e.g.* Jenkins, 1987; Ahn *et al.*, 1991; Bozhilov *et al.*, 2004, Bloise *et al.*, 2008).
- all the previously synthesized tremolites shift from the ideal stoichiometry $Ca_2Mg_5Si_8O_{22}(OH)_2$ and show different amounts of cummingtonite component ($Mg_7Si_8O_{22}(OH)_2$) (~7%), depending on the thermodynamic growth conditions (Jenkins, 1987; Ahn *et al.*, 1991; Pawley *et al.*, 1993; Maresch *et al.*, 1994; Hawthorne *et al.*, 1997; Gottschalk *et al.*, 1999; Bozhilov *et al.*, 2004, 2007; Bloise *et al.*, 2008).
- high-resolution transmission electron microscope (HRTEM) images of synthetic tremolite reveal the presence of chain multiplicity faults (CMFs), the deviation from the ideal formula $Ca_2Mg_5Si_8O_{22}(OH)_2$ towards cummingtonite could be due to their structural defects.
- a more accurate evaluation of the cummingtonite component in the run products would require the gathering of HRTEM images to obtain the density of the CMFs (*i.e.* amounts and types of biopyriboles); these data should also be used to correct the results of the quantitative analyses performed by microprobe (Bozhilov *et al.*, 2004, 2007).

On the basis of the numerous experimental investigations carried out, it appears that stoichiometric tremolite cannot be synthesized. In fact, all synthetic tremolite contains a variable excess of Mg (3–7 wt.%) with respect to stoichiometry (Jenkins, 1987; Pawley *et al.*, 1993; Gottschalk *et al.*, 1999; Bozhilov *et al.*, 2004, 2007; Bloise *et al.*, 2008) probably due to cummingtonite substitution or chain-multiplicity defects (Maresch *et al.*, 1994). To summarize, the problems with synthetic stoichiometric tremolite come from two factors: (1) the growth during synthesis of precursor diopside or talc, which then require time to be dissolved in order to complete the growth of tremolite. Tremolite synthesis was studied in detail in time-series experiments by Bozhilov *et al.* (2004), who suggested that obtaining ~100% yields of tremolite with near end-member composition by purely hydrothermal methods would take several months, with intermediate grinding as well. However, in addition to unreasonably long synthesis durations (months rather than days), the product still tends to be cummingtonite–tremolite solid solutions with variable Ca/Mg ratios, most of them falling inside the cummingtonite–tremolite solvus; and (2) at high temperatures, for example >650°C, end-member tremolite is less stable than tremolite–cummingtonite solid solution; its synthesis above this temperature is unlikely if not impossible. However, it has been demonstrated (Zimmermann *et al.*, 1996; Gottschalk *et al.*, 1999) that the introduction of saline fluids ($CaBr_2$, $MgBr_2$, KOH, NaCl, *etc.*) in the reaction system can catalyse the growth of CMF-free tremolite of end-member composition. Besides, it is known that higher synthesis temperatures favour lower CMF concentrations (Ahn *et al.*, 1991).

We agree with Maresch and Czank (2007) when they say that "...many variable experimental factors combine to make generalizations difficult". It is worth remembering that, in an attempt to understand the pathological mechanisms of asbestos, all *in vitro* experiments have been performed using natural asbestos. These investigations have pointed out that asbestos toxicity may be related to two features of the fibres: dimension and chemistry. However, because these materials are natural, they may vary in size and chemistry thus not allowing a well defined correlation between their physicochemical properties and results from *in vitro* tests. With regard to the size of the tremolite fibres synthesized (*e.g.* Ahn *et al.*, 1991; Zimmermann *et al.*, 1996; Hawthorne *et al.*, 1997; Gottschalk *et al.*, 1999; Bozhilov *et al.*, 2004), they cover a large dimensional range (0.5–300 μm in length and 0.1–15 μm in width). In an attempt to understand the role of iron in the aetiology of the pathological effects of the tremolite fibres, Bloise *et al.* (2008) synthesized Fe-free tremolite under hydrothermal conditions, without using fluid reactants ($CaCl_2$, $CaBr_2$, $MgBr_2$, KOH, NaCl, *etc.*) to avoid their presence in the synthesized products. Furthermore, in agreement with all the previously reported data (Jenkins, 1987; Bozhilov *et al.*, 2004, 2007) also in Bloise *et al.* (2008), tremolite deviated from stoichiometry due to the presence of the cummingtonite component.

Whereas a wide literature exists on synthetic tremolite, only a few examples of synthetic fibrous crocidolite, amosite, anthophyllite and actinolite are known. Riebeckite was synthesized in the temperature range 408–700°C, and pressure up to

2 kbar by Ernst (1962) who studied its temperature and pressure stability and taking oxygen fugacity into account also. Some years later, Iezzi *et al.* (2003) synthesized hydrothermally fibrous riebeckite 0.2–1 μm wide and 1–5 μm long at 500°C and 4 kbar for a duration of 14 days. Lattard and Evans (1992) reported seeded synthesis experiments of grunerite at 1.2–2.0 GPa and 490–620°C with yields exceeding 90%. Amphibole compositions are estimated to contain between 87 and 98% grunerite end-member. The synthetic crystals were 5–10 μm long and 0.5–1 μm wide; HRTEM characterization showed almost no CMFs, but indicated the presence of numerous CAFs (polysynthetic micro-twinning). Veblen (1981) noted that, in the synthetic (Fe^{2+},Mg)-amphibole series, the density of CMFs decreases if Fe increases, while the number of CMFs decreases. Anthophyllite was obtained by Greenwood (1963) at 830°C, 0.1 GPa with 20 h of reaction of starting oxide mixture. However, synthetic anthophyllite produced in this way was poorly crystalline, and the yield of 100% anthophyllite was never attained. For reaction boundaries among anthophyllite and talc, Greenwood (1963) suggested that the reaction of talc to anthophyllite proceeded by breaking Si–O bonds in the tetrahedral sheet to produce strips of double chains. Skogby and Ferrow (1989) synthesized actinolite solid solutions at 500 to 700°C and 0.2 GPa. It was also shown that longer run times (2 to 6 months) with multiple interruptions of the run and intermittent grinding increased yield and decreased CMFs. Also Jenkins and Bozhilov (2003) used long reaction time (6 months) with intermitting grinding of the starting materials to synthesize ferro-actinolite. However, at 0.2–1.0 GPa and 420–600°C, the yields attained were <70 wt.%. The main problems in the synthesis of Fe-bearing amphiboles are related to the control of the oxidation state of Fe during the synthesis process.

3.2. Dehydration and breakdown of fibrous amphiboles

3.2.1. Crocidolite

Thermal studies conducted on crocidolite ($Na_2Fe_2^{3+}Fe_3^{2+}Si_8O_{22}(OH)_2$) have received more attention from researchers than similar studies on any other fibrous amphibole. Studies of the thermal behaviour of fibrous crocidolite are listed in Table 5. Vermaas (1952) provided the first systematic study of the thermal decomposition of crocidolite, using differential thermal analysis (DTA) and X-ray powder diffraction. He concluded that, in an oxidizing atmosphere, the exothermic effect at 400–500°C could be related to Fe^{2+} oxidation and that the decomposition to acmite ($NaFeSi_2O_6$), cristobalite and hematite occurred above 890°C. Similar results were obtained by Cilliers *et al.* (1961) who, analysing the crocidolite from Westerberg (South Africa), also observed an exothermic peak at 400°C and a dehydroxylation peak at 900°C (endothermic). The authors explained the thermal effect at 400°C as an oxidation process in which the conversion of Fe^{2+} to Fe^{3+} occurs. Addison *et al.* (1962) studied, in detail, the thermal oxidation of fibrous crocidolite in air and vacuum. They showed that when heating crocidolite in air at 450°C, partial oxidation occurred, leading to the conversion of Fe^{2+} to Fe^{3+}; the mechanism was considered to depend on the migration of protons and electrons through the crystal. They concluded that oxidation of ferrous iron within the lattice by

Table 5. Summary of literature studies of the crocidolite decomposition.

Reference	*T* (°C)	Analytical techniques used to characterize crocidolite
Vermaas (1952)	20–1200	DTA (air atmosphere), XRPD, PLOM
Cilliers *et al.* (1961)	20–1000	DTA (air atmosphere), XRD, PLOM
Addison *et al.* (1962)	350–480	Furnace (air atmosphere and vacuum)
Patterson (1964)	800–1000	Furnace (air atmosphere and vacuum), XRD
Hodgson (1965)	20–1000	DTA/TG (air atmosphere- N_2 atmosphere), EDS/SEM
Hodgson *et al.* (1965a)	20–1000	DTA/TG (air atmosphere- N_2 atmosphere), XRD, IR
Freeman (1966)	100–1200	TG/DTG (N_2 atmosphere), XRD
Rouxhet *et al.* (1972)	250–900	Furnace (vacuum and oxygen atmosphere), IR
Jeyaratnam and West (1994)	100–900	Furnace, XRPD
Kohyama *et al.* (1996)	20–1000	DTA/TG (air atmosphere-N_2 atmosphere), XRD, TEM/ EDS
Gualtieri *et al.* (2004)	25–1100	DTA/TG (air atmosphere), XRD
Fujishige *et al.* (2007)	20–1000	DTA/TG (air atmosphere), XRD, SEM
Kusiorowski *et al.* (2012)	20–1000	DTA/TG (air atmosphere), XRPD, SEM/EDS, FTIR
Kusiorowski *et al.* (2015b)	20–1000	DTA/TG (air atmosphere), XRPD, SEM/EDS, EGA, FTIR
Bloise *et al.* (2016)	25–1100	DSC/TG (air atmosphere), XRPD, SEM/EDS

* PLOM: polarized light optical microscopy

oxygen is unlikely, as oxygen molecules are far too big to spread through the lattice. The mechanism they proposed is that the oxidation of crocidolite is essentially a dehydrogenation, so that the mechanism of oxidation was considered to depend on migration of protons and electrons through the crystal, thus suggesting the following equation: $Fe^{2+} + (OH)^{-} = Fe^{3+} + O^{2-} + H^{+}$.

This reaction does not perturb the oxygen packing in the crystals and involves simply a migration of protons and electrons to the surface where hydrogen reacts with oxygen to form water (Hodgson, 1965). Patterson (1965) studied the thermal decomposition of crocidolite from Wittenoom Gorge (Australia) by XRPD in air and in vacuum. In air, the initial decomposition of the amphibole to acmite, spinel and cristobalite occurred at 850°C although this result had already been obtained by Vermaas (1952). After the initial decomposition, a further increase in temperature (>900°C) caused the transformation of the spinel to hematite in air and, at ~1000°C, acmite melted incongruently to hematite and glass. These findings were consistent with the data from Hodgson *et al.* (1965a), who concluded that the structure collapse of crocidolite from Koegas (South Africa) occurred between 800°C and 880°C, followed by the formation of magnetite, cristobalite and acmite. Moreover, through chemical analyses and DTA evidence, Hodgson *et al.* (1965a) concluded that the heating of crocidolite in air below 400°C and dehydroxylation took place within the range of 300–450°C, to form oxy-crocidolite which oxidized at >600°C. In the same conditions during dynamic heating at 400–600°C, hydrogen ions and electrons were lost, to form an oxy-crocidolite (crocidolite partially dehydrogenated).

At 600–950°C, the oxy-amphibole decomposed endothermically. They suggested that the decomposition of crocidolite took place in three steps: loss of physically combined water; loss of chemically combined water; and final breakdown into simpler phases (Hodgson, 1965). However, the temperature at which the three steps took place was not clear. Indeed, after 1965, an accurate evaluation of these range-steps has been the subject of much experimental investigation into the thermal behaviour of crocidolite. Using the maximum rate of mass loss due to the water (hydroxyls group) release during the heating of crocidolite at 1000°C, Freeman (1966) stated that the dehydroxylation temperature of crocidolite from Westerberg (South Africa) in an inert atmosphere was 520°C. That author concluded that crocidolite may be unique amongst the amphiboles in that the dehydroxylation process and the structural breakdown process were consecutive rather than concurrent (Freeman, 1966). Rouxhet *et al.* (1972) used infrared spectra (IR) to characterize the loss of constitutional hydroxyl of crocidolite at temperatures up to 900°C under both a vacuum and in air atmosphere. They confirmed that the oxidation of crocidolite took place progressively over a broad temperature range and was essentially a dehydrogenation that followed a dehydroxylation leading to the consequent formation of oxy-crocidolite. They also reported that the loss of tensile strength of the crocidolite with increasing temperature was due to the thermal decomposition and not to the oxidation and related dehydrogenation and/or dehydroxylation reactions. Jeyaratnam and West (1994) studied heat-degraded crocidolite from South Africa, focusing on the shifts of certain X-ray diffraction lines, and they concluded that oxy-amphiboles were formed at relatively low temperatures (400–500°C).

Subsequently, Gualtieri *et al.* (2004) studied the kinetics of crocidolite (NIST 1866a) decomposition using real-time conventional X-ray powder diffraction and DTA. These authors showed thermal analysis curves of crocidolite asbestos, and affirmed that the major mass loss occurs below 700°C. However, the structure did not show the collapses occurring at higher temperatures. Similar conclusions came from Fujishige *et al.* (2007), who did not observe any mass loss of crocidolite asbestos in oxygen atmosphere above 700°C, and the total mass loss recorded for this temperature was ~2%. Extensive experiments on the investigation of the thermal behaviour of crocidolite were also carried out during the last decade by Kusiorowski *et al.* (2012, 2015b), who aimed to determine a more accurate range of crocidolite asbestos (South Africa) dehydroxylation temperature, and thereby setting a minimum calcination temperature for its further thermal use. By running the data obtained through different analytical techniques (*i.e.* XRPD, FTIR, SEM-EDS, DTA-TG), they showed that the dehydroxylation process of crocidolite asbestos took place within the range of 400–700°C.

Recently, Bloise *et al.* (2016), using very sensitive DSC/DTG (differential scanning calorimetry and derivative thermogravimetry), have shown the full thermal behaviour of crocidolite from Koegas (South Africa) (Fig. 11) highlighting the endothermic peaks which were related to the dehydrogenation process. They concluded that crocidolite dehydroxylation took place in the range 350–700°C, and it was decomposed at ~850°C with the formation of magnetite, cristobalite and acmite.

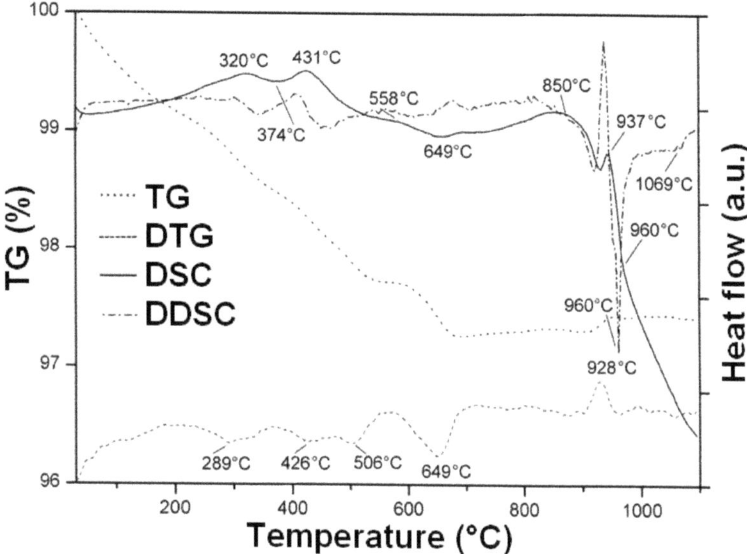

Figure 11. Thermal analysis of standard crocidolite UICC. Solid line: DSC (differential scanning calorimetry). Dashed line: DTG (derivative thermogravimetry). Dotted line: TG (thermogravimetry). Dashed and dotted line: DDSC (derivative differential scanning calorimetry) (modified after Bloise *et al.*, 2016; reproduced with the permission of Springer).

After heating at 1100°C, SEM characterization showed that the original morphology was strongly altered. The single fibres, originally arranged in bundles, now appeared as thick sticks, confirming that partial melting occurred during heating.

Summing up the results of the above-mentioned studies, we can conclude that the structural changes of crocidolite with increasing temperature, follow these steps: dehydrogenation and/or dehydroxylation (200–700°C), structure collapse (800–900°C) and crystallization of newly formed crystalline phases (850–950°C), early melting (1000–1070°C).

3.2.2. Tremolite

One of the first thermograms concerning the thermal behaviour of tremolite was published by Wittels (1951, 1952). That author studied tremolite in the temperature range 600 to 1100°C by differential thermal analysis, X-ray and optical methods. Combining the results from these different techniques showed that at ~1125°C, the theoretical water content (hydroxyl groups) in tremolite was driven off, and the following reaction took place:

$$Ca_2Mg_5Si_8O_{22}(OH)_2 + \Delta H = 2CaSiO_3{\cdot}5MgSiO_3 + SiO_2 + H_2O \quad (1)$$

When heated, tremolite decomposed into two pyroxenes, silica and water as suggested by equation 1. Considering the close relation between amphibole and pyroxene structures, the results revealed a mechanism of structural collapse whereby the double-chain units split into single-chain units.

Subsequently, using the weight-loss curve of tremolite from St Gotthard (Switzerland), Freeman (1966) stated that its dehydroxylation temperature in an inert atmosphere was at 790°C.

However, using the thermal analysis (DTA) for the identification of asbestos minerals present as impurities in talc, Luckewicz (1975) showed a thermogram in which an endothermic peak at 1075°C corresponding to the breakdown of tremolite was present. Kusiorowski *et al.* (2012) studied tremolite from Italy by DTA up to 1000°C, and showed that its thermal decomposition started at 950°C. In later studies, heating the tremolite asbestos up to 1100°C, Bloise *et al.* (2016) showed – through DSC thermal analysis – that the breakdown of tremolite asbestos from Val d'Ala (Italy) occurred at 1046°C. The temperature range of tremolite breakdown for the different above-mentioned samples was 1046–1125°C. Using characterizations of the reacted samples, kinetic data, and literature activation energy data for cation and oxygen diffusion, Johnson and Fegley (2003) suggested the following thermal decomposition mechanism for tremolite:

(1) the atomic bonds undergo thermal expansion and the *A* sites expand (Sueno *et al.*, 1973);
(2) two adjacent hydroxyls on the Mg-*M*(1) sites react to form water *via* $2OH^- \rightarrow H_2O + O^{2-}$;
(3) water escapes;
(4) the remaining cations undergo bond breaking/shifting/reforming with bridging oxygen; and
(5) silica tetrahedra shift to maintain electronic neutrality (single chains appear first, followed by the more complex silica phase).

Thanks to IR spectra and XRPD data, these authors demonstrated that tremolite decomposition occurred concurrently with dehydroxylation (Johnson and Fegley, 2000). However, this thermal behaviour had already been noted previously by Vermaas (1952) and Patterson and O'Conner (1966). As mentioned above, the final products of decomposition of tremolite above 1100°C were diopside ($CaMgSi_2O_6$), enstatite ($MgSiO_3$) and cristobalite. Bloise *et al.* (2016) showed a detailed set of SEM images in which the morphology of tremolite can be seen before and after heating at 1100°C. After thermal treatment, the crystal habit of tremolite was preserved, but almost all the individuals appear more brittle and fractured sub-perpendicular to the fibre axis (Fig. 12a). Furthermore, at a molecular scale, the presence of newly formed polyhedral crystals on the surface of the pristine tremolite asbestos was shown (Fig. 12b). Chemical EDS investigations confirmed that the pseudomorphic process involved a complete recrystallization of the original tremolite asbestos into diopside (Bloise *et al.*, 2016).

3.2.3. Amosite

One of the first syudies on amosite thermal stability was reported by Vermaas (1952), who used DTA and X-ray powder diffraction to investigate this subject. In the thermogram obtained, a broad exothermic effect is reported at 580–830°C, which the author attributed to oxidation and decomposition of amosite. Moreover, this author

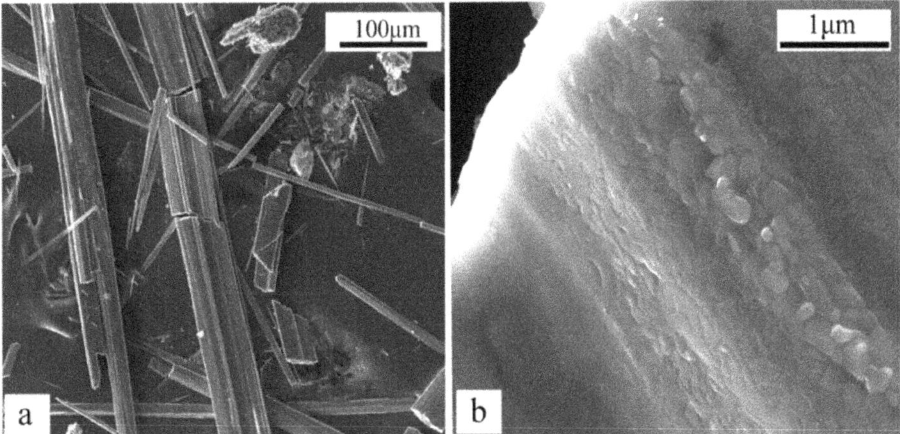

Figure 12. (a) Secondary electron SEM images of tremolite asbestos after heat treatment at 1100°C. (b) Image acquired using high-resolution SEM, showing the complete pseudomorphic recrystallization of fibres (modified after Bloise *et al.*, 2016; reproduced with the permission of Springer).

showed that magnetite and cristobalite were the newly formed phase after heating amosite above 1100°C. A significant effort was made by Hodgson *et al.* (1965b) to understand the thermal behaviour of amosite from Penge (South Africa), which was studied in argon or in air atmospheres mainly by differential thermal analysis (DTA), thermogravimetric analysis (TGA) and X-ray rotation photographs. Hodgson *et al.* (1965b) reported that the decomposition of amosite followed a very similar pattern to that of crocidolite, although in this case the formation of oxy-amosite (anhydrous amosite) involved a dehydroxylation process, as well as a dehydrogenation process. During thermal treatment (dynamic heating conditions in air atmosphere), amosite lost some of the hydroxyl water as hydrogen, with concurrent oxidation of the iron. Hydrogen migration through the amphibole structure to the surface, began rapidly at ~350°C, but slowed below ~800°C. The reaction products were partially amorphous to X-rays. Oxygenation processes occurred simultaneously, and it did not seem possible to isolate pure oxy-amosite. In the temperature range 350–800°C on the DTA curve, Hodgson *et al.* (1965b) recorded a broad exothermic effect to which the authors attributed a sequence of overlapping dehydrogenation, oxygen absorption and dehydroxylation reactions. The main products were an oxy-amosite at 350–800°C, and a pyroxene and amorphous material at 800–1000°C. The main reaction at 350–800°C can be represented, at least approximately, by the following equation:

$$Fe^{2+}_{5.5}Mg_{1.5}Si_8O_{24}H_2 + \tfrac{1}{2}\,O_2 = Fe^{3+}_2Fe^{2+}_{3.5}Mg_{1.5}\,Si_8O_{24} + H_2O$$

The conversion of amosite to pyroxene occurs without detectable formation of an amphibole anhydride. In this respect, amosite resembles tremolite and differs from crocidolite.

Rouxhet *et al.* (1972) studied the same amosite sample used by Hodgson *et al.* (1965b) and confirmed through infrared spectra and chemical analyses that, for amosite, dehydrogenation and oxygenation took place together. Fujishige *et al.* (2007) showed an endothermic effect at 865°C related to the structural breakdown of amosite. By characterizing the amosite (South Africa) after thermal treatment at 1000°C with IR spectra, Kusiorowski *et al.* (2012) concluded that the decomposition of amosite below 1000°C had already been initiated. A DSC/TG thermogram of amosite from Penge was reported by Bloise *et al.* (2016) who characterized amosite with XRPD and SEM/EDS before and after heating at 1100°C. They concluded that amosite decomposed into enstatite and hematite at 878°C although its decomposition started at 600°C. This finding was in agreement with the results of Freeman (1966) who had analysed the same sample of amosite from Penge (S. Africa).

After heating at 1100°C, the newly formed silicate (enstatite) preserved the original fibrous morphology (pseudomorphosis). In some cases, single fibres seemed to be fused together forming prismatic crystals (Fig. 13a), and when observed at greater magnification (Fig. 13b), they appeared partially covered by pseudo-spherical particles growing along the axial direction of the fibres (Bloise *et al.*, 2016).

3.2.4. Anthophyllite

Few thermal analyses have been carried out on natural anthophyllite asbestos, analogous to those carried out on other fibrous amphiboles. Wittels (1951, 1952) performed several differential thermal analyses of different members of the amphibole group. It was asserted that a close parallel could be drawn between the disintegration of anthophyllite and the collapse of tremolite. The thermogram obtained by that author showed a sharp endothermic peak at ~1020°C due to anthophyllite breakdown. Moreover, accurate thermal analyses, X-ray and optical investigation by Rabbit (1948), revealed that the transformation occurring in anthophyllite had already started at

Figure 13. (a) secondary electron SEM images of amosite after heating at 1100°C. (a) Magnification indicated by empty black square in micrograph. (b) High resolution SEM image showing the fibrous recrystallized forms (modified after Bloise *et al.*, 2016; reproduced with the permission of Springer).

830°C. Wittels (1952) showed that the products of structural disintegration were pyroxene, cristobalite and water, as suggested by the following equation.

$$Mg_7Si_8O_{22}(OH)_2 + \Delta H = 7MgSiO_3 + SiO_2 + H_2O$$

Using the maximum rate of mass loss due to the water (hydroxyl groups) release during heating of anthophyllite to 1200°C, Freeman (1966) stated that the dehydroxylation temperature of anthophyllite was 800°C. However, it is worth remembering that the morphology of the anthophyllite (Hillswick, Shetland) used by Freeman (1966) was not fibrous but tabular. Bush and Schumacher (1982), who analysed a sample of anthophyllite from Finland, showed a TG curve in which the water loss (hydroxyl groups) was observed between 700 and 900°C. Bloise *et al.* (2016) presented the result of DSC-TG analyses of anthophyllite asbestos from Paakkila (Finland) and reported that the structural breakdown of anthophyllite took place at ~900°C. Using also the maximum rate of mass loss due to the water (hydroxyl groups) release, they stated that the dehydroxylation temperature of anthophyllite was at 868°C. Enstatite and cristobalite were detected through an accurate examination of the samples heated at 1100°C using analytical electron microscopy and X-ray powder diffraction. SEM images collected at greater magnification showed that the new phase formed after heating (enstatite) preserved the original fibre morphology (pseudomorphosis), but the surface became rough (Bloise *et al.*, 2016).

Thermogravimetric analyses of fibrous amphiboles reveal that there are some differences in temperature (ranging from 800 to 1100°C) in the breakdown of fibrous amphiboles. Moreover, the dehydroxylation temperature has a range of ~350–850°C for the above mentioned fibrous amphiboles. The temperature at which dehydroxylation occurs in amphiboles increases proportionally with increasing $M(1)$ and $M(3)$ sites and with consequent decrease in Fe^{2+}, as the strength of the M–OH or MO–H bonds depends strongly on the type of cation that occupies these sites (Freeman, 1966; Kusiorowski *et al.*, 2015b; Bloise *et al.*, 2016). The decomposition temperatures of anthophyllite asbestos as well as tremolite asbestos were higher than those reported for amosite and crocidolite (Freeman, 1966; Bloise *et al.*, 2016). This behaviour depends on the correlation between Mg content, higher in the anthophyllite and tremolite as compared to amosite and crocidolite, and the decomposition peak temperature. It is worth noting that the thermal behaviour of amphiboles that contain iron 2+ and/or 3+ was affected by experimental conditions, such as inert or air atmosphere of the furnace and by static or dynamic heating.

The anhydrous material resulting from dehydroxylation of fibrous amphiboles, when heated at a temperature >800°C, recrystallized to other phases (*e.g.* pyroxene, cristobalite, plagioclase, hematite, *etc.*) depending on the composition of the amphibole (Hodgson *et al.*, 1965a,b; Kusiorowski *et al.*, 2015; Bloise *et al.*, 2016).

Despite the thermal treatment, all the fibrous amphiboles mentioned above preserved the same external fibrous habit, but the structure was completely changed at a molecular scale: this phenomenon, referred to as "pseudomorphosis" (Patterson, 1965; Giacobbe *et al.*, 2010, Bloise *et al.*, 2016), leads to the complete transformation of asbestos minerals into non-hazardous silicates.

3.2.5. Amphibole asbestos remediation through thermal decomposition

Like chrysotile, amphibole asbestos minerals convert into stable crystalline silicates through solid-state reactions at high temperature. Starting from the data reported in the previous paragraphs, this section is focused on defining the proper decomposition temperature for each amphibole species, leading to complete thermal transformation of the fibres and thus a safe EoWM. In addition, the existing literature data regarding their remediation is reviewed.

A preliminary remark that needs to be made is that the thermochemical transformation of fibrous amphibole species in hydrothermal systems combined with an acidic attack (*e.g.* Nam *et al.*, 2014) works for chrysotile but has little effect on amphiboles, as it is long known that their dissolution rate in contact with acidic solution is very sluggish (Speil and Leineweber, 1969).

The temperature of the thermal treatment to be set for successful decomposition of fibrous amphiboles depends upon the nature of the minerals.

According to the pioneering work of Freeman (1966), crocidolite asbestos undergoes dehydroxylation in the temperature range 350–800°C. The complex reaction involves iron oxidation (Gualtieri *et al.*, 2004) and leads to the formation of pyroxene ($NaFe^{3+}Si_2O_6$), enstatite ($MgSiO_3$), hematite (Fe_2O_3) and cristobalite (SiO_2). Recently, Kusiorowski *et al.* (2015b), who studied crocidolite samples from South Africa, Australia, Bolivia and Russia, found that the dehydroxylation temperature of crocidolite asbestos is dependent on the origin of the sample and takes place in the range 400–700°C. After annealing, the minimum temperature of thermal treatment decreases and the transformation process for all samples tested occurred in the temperature range 400–450°C. The material obtained, free of residual fibres, is brittle and easily reduced to a powder. If remediation is *via* thermal treatment of pure crocidolite, one of the crystalline phases that forms during the high-temperature reaction sequence is cristobalite due to segregation of amorphous silica. Milling of this high-temperature product may reduce the cristobalite particles to a size <10 μm, thus producing a respirable form of silica which is classified as a human carcinogen (group 1) by the International Agency for Research on Cancer (IARC) Monographs (1997). This side-effect makes this process unviable. However, if crocidolite is one of the minor phases forming the ACM (which is invariably the case for building materials), silica segregated during the transformation reaction of crocidolite readily reacts with calcium, magnesium and aluminium from the cementitious matrix to form stable silicates. For this reason, the thermal transformation of ACM like cement-asbestos is a viable remediation process.

As far as remediation techniques are concerned, the process described by Fujishige *et al.* (2007) involving heat treatment in the range 600–1300°C and chemical additives such as $CaCO_3$ and $CaCl_2$ was also applied to crocidolite-rich wastes. Dellisanti *et al.* (2009) applied successfully Joule-heating vitrification to crocidolite-rich ACM. Friable crocidolite was denaturized successfully by the method developed by Yanagisawa *et al.* (2009), which includes hydrothermal conversion at 500°C combined with acidic attack.

Tremolite is decomposed at very high temperature (1050°C) into clinopyroxene diopside ($CaMgSi_2O_6$), orthopyroxene enstatite and cristobalite (Gualtieri and Tartaglia, 2000). Freeman (1966) reported a kinetic curve of the dehydroxylation reaction of tremolite in the temperature range 500–1050°C. Accordingly, Kusiorowski *et al.* (2012), reported a DTA curve for tremolite where the endothermic dehydroxylation reaction started at 950°C.

The same mineral assemblage observed for tremolite has been reported for actinolite in metamorphic systems with the exception of quartz in place of cristobalite (Lledo and Jenkins, 2008). The presence of iron in the structure results in the formation of hematite at high temperature and oxidizing atmosphere. The FTIR study of Ishida *et al.* (2002) shows that two fibrous actinolite samples from Japan and Brazil still display the absorption band of structural hydroxyls following thermal treatment at 770 and 930°C, respectively. Hence, until conclusive experimental evidence is presented, a decomposition temperature >930°C has to be assumed.

According to Freeman (1966), amosite undergoes dehydroxylation in the temperature range 500–900°C. Amosite in air decomposes into spinel, hematite, pyroxene and an amorphous phase in the temperature interval 800–1100°C. Amorphous silica recrystallizes into cristobalite in the range 1100–1350°C (Hodgson *et al.*, 1965). On the other hand, according to Jeyaratnam and West (1994) amosite decomposition leads to the formation of spinel, hematite, magnetite and cristobalite. The issue raised for crocidolite regarding the presence of cristobalite in the EoWM obtained from pure samples also applies to amosite. The high temperature of dehydroxylation for amosite (>1000°C) is confirmed by Kusiorowski *et al.* (2012) who, surprisingly, observed only an amorphous phase as the final product of decomposition.

As far as remediation is concerned, the process described by Fujishige *et al.* (2007) was also applied with success to amosite-rich wastes. Yvon and Sharrock (2011) described a remediation process for spray-coated amosite which was thermochemically treated and subsequently recycled in the production of masonry products. The authors observed the formation of crystalline minerals belonging to the melilite group $(Ca,Na)_2(Mg,Fe,Al)(Al,Si)SiO_7$ and pyroxene group: (aegirine-augite) $(Ca,Na)(Mg,Fe,)Si_2O_6$. Apparently, no cristobalite was found in the thermally treated material. Like chrysotile and crocidolite, remediation through hydrothermal conversion at 600°C combined with acidic attack of friable asbestos developed by Yanagisawa *et al.* (2009) has also been applied successfully to friable amosite.

Anthophyllite is virtually the most stable fibrous amphibole species. Freeman (1966) reports a kinetic curve of its dehydroxylation reaction in the temperature range 350–1100°C. Temperature-induced recrystallization of anthophyllite results in forsterite and enstatite, eventually accompanied by cristobalite (Day and Halback, 1979). The issue raised for crocidolite and amosite regarding the presence of cristobalite in the EoWM obtained from pure samples also applies to this fibrous amphibole.

Table 6 reports a summary of the temperatures required for safe remediation of fibrous amphiboles and a comment concerning the possible presence of cristobalite in the reaction product from the pure phase.

Table 6. Heating temperature required for safe remediation of fibrous amphibole species and the presence of cristobalite in the reaction product from the pure phase.

Phase	$T_{remediation}$ (°C)	Presence of cristobalite in the EoWM
Actinolite	>930°C?	No
Amosite	900–1100?	Yes
Anthophyllite	1100	Yes
Crocidolite	800	Yes
Tremolite	1050	No

4. Thermal behaviour of erionite

Erionite is a zeolite known to have high thermal stability, a property generally attributed to its large silica content (Schlenker *et al.*, 1977; Cruciani, 2006). However, the effect of the framework Si/Al ratio on the thermal stability is best evaluated in a series of isostructural zeolites (Cruciani, 2006). Erionite has been classified as a zeolite which belongs to category 1 (Alberti and Vezzalini, 1984; Bish and Carey, 2001) which includes zeolites characterized by ''...Dehydration with a rearrangement of the extraframework cations and of residual water molecules without considerable changes in the geometry of the framework and in the cell volume...''

A handful of X-ray diffraction studies has been conducted to investigate the structural dehydration dynamics of erionite and its temperature stability. However, some open questions still remain. The unit-cell parameters as a function of temperature have been reported up to 300°C for an erionite sample from Eastgate, Nevada (Bish, 1989). The thermal behaviour of a fibrous sample from Jersey, Nevada (USA) is described by Papke (1972) who showed that the main endothermic event occurred at ~140°C. However, neither the thermogravimetry (TG) curve nor the temperature of the structural collapse was reported.

The most relevant temperature-dependent dynamic study was published by Ballirano and Cametti (2012) who investigated dehydration of fibrous erionite-K using *in situ* X-ray powder diffraction. The authors found that its structural breakdown occurred in the range 840–920°C. The dependence of cell parameters and volume on temperature is fairly complex, with the smallest cell volume observed at 820°C, just before the beginning of the structural breakdown (Fig. 14). The most relevant framework modifications with increasing temperature are: (1) the squeezing of the 8-member ring (8MR) leading to a reduction of the crystallographic free area (CFA) from ~14 to ~12.4 Å^2; (2) the expansion of the 6-member ring (6MR) shared between the cancrinite cage and the erionite cavity permitting cation diffusion between them; and (3) the expansion of the base of the erionite cavity to host the migrating Ca2 cation. Reduction of the site scattering of the K1 site during heating has been explained by invoking an

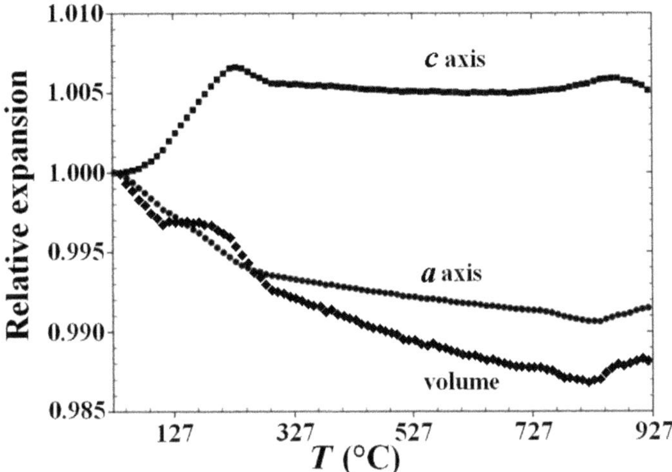

Figure 14. Relative expansion of unit-cell parameters and volume for erionite-K (modified after Ballirano and Cametti, 2012).

'internal ion exchange' mechanism the detailed description of which is difficult because of the simultaneous occurrence of three cationic species (Na, K and Ca). Depletion of the various water molecule sites is attained in the range 140–300°C although complete dehydration is only reached close to the temperature of structural collapse (700°C in Gottardi and Galli, 1985; 820–840°C in Ballirano and Cametti, 2012).

To the knowledge of the authors, there are no studies dealing with thermally induced remediation of erionite-rich ACMs or wastes. This is due to the fact that erionite has never been used as a commercial fibre as have chrysotile and amphiboles but simply as a rock-forming mineral of natural building stones.

Recently, Bloise *et al.* (2016) reported a thermal analysis study of fibrous erionite-Na from Jersey, Nevada. The TG curve of erionite exhibits a mass loss of 17.00 wt.% within the 25–450°C temperature range. The DTG curve shows two separate steps of dehydration: the former at 99°C corresponding to a water loss of 16.11 wt.% and the latter at 355°C associated with a water loss of 0.89 wt.%. Complete dehydration is attained without loss of crystallinity, which started at a temperature above 700°C and is followed by recrystallization of K-feldspar and plagioclase as revealed by the DSC exothermic peak at 911°C. The product of erionite recrystallization after heating to 1000°C is a mixture of K-feldspar (or), plagioclase (albite-anorthite ab-an), quartz (qtz) and possibly cristobalite (cr). No matter what the composition of pristine erionite is, the same mineral assemblage should be found after crystallization of the collapsed structure.

- Erionite-Ca (ideally $Ca_3K_2NaAl_9Si_{27}O_{72}{\cdot}29{-}32H_2O$) after dehydration gives: $Ca_3K_2NaAl_9Si_{27}O_{72} \rightarrow 3CaAl_2Si_2O_8$ (an) + $2KAlSi_3O_8$ (or) + $NaAlSi_3O_8$ (ab) + $12SiO_2$ (cr/qtz);

- Erionite-K (ideally $K_2Ca_2KNa_2Al_9Si_{27}O_{72}·29-32H_2O$) after dehydration gives: $K_2Ca_2KNa_2Al_9Si_{27}O_{72} \rightarrow 2CaAl_2Si_2O_8$ (an) + $3KAlSi_3O_8$ (or) + $2NaAlSi_3O_8$ (ab) + $8SiO_2$ (cr/qtz);
- Erionite-Na (ideally $Na_3Ca_2K_2Al_9Si_{27}O_{72}·29-32H_2O$) after dehydration gives: $Na_3Ca_2K_2Al_9Si_{27}O_{72} \rightarrow 2CaAl_2Si_2O_8$ (an) + $2KAlSi_3O_8$ (or) + $3NaAlSi_3O_8$ (ab) + $8SiO_2$ (cr/qtz).

The presence of Mg in the structure leads eventually to the formation of forsterite at the expense of the silica phases. Once again, in view of a safe reuse of the product of the thermal destruction of fibrous erionite, one should consider the issue raised for fibrous amphiboles regarding the possible presence of carcinogenic cristobalite in the EoWM which may rule out its recycling.

Acknowledgements

The authors are grateful to Professor V.V. Gusarov (Head of Department at the Russian Academy of Sciences, Moscow) for bibliographic data that he kindly provided to complete the chapter.

References

Abruzzese, C., Marabini, A.M., Paglietti, F. and Plescia. P. (1998) 'Cordiam' process: A new treatment for asbestos wastes. Pp. 563–577 in: *EPD Congress 1998*, San Antonio, Texas, USA.

Addison, C.C., Addison, W.E., Neal, G.H. and Sharp, J.H. (1962) Amphiboles. Part I. The oxidation of crocidolite. *Journal of the Chemical Society (Resumed)*, **278**, 1468–1471.

Ahn, J.H., Cho, M., Jenkins, D.M. and Buseck, P.R. (1991) Structural defects in synthetic tremolitic amphiboles. *American Mineralogist*, **76**, 1811–1823.

Alberti, A. and Vezzalini, G. (1984) Topological changes in dehydrated zeolites: breaking of T–O–T oxygen bridges. Pp. 834–831 in: *Proceedings of the Sixth International Zeolite Conference.* Reno, Nevada, USA.

Anastasiadou, K., Axiotis, D. and Gidarakos, E. (2010) Hydrothermal conversion of chrysotile asbestos using near supercritical conditions. *Journal of Hazardous Materials*, **179**, 926–932.

Ball, M.C. and Taylor, H.F.W. (1963) The dehydration of chrysotile in air and under hydrothermal conditions. *Mineralogical Magazine*, **33**, 467–482.

Ballirano, P. and Cametti, G. (2012) Dehydration dynamics and thermal stability of erionite-K: Experimental evidence of the "internal ionic exchange" mechanism. *Microporous and Mesoporous Materials*, **163**, 160–168.

Barrese, E., Belluso, E. and Abbona, F. (1997) On the transformation of synthetic diopside into chrysotile. *European Journal of Mineralogy*, **9**, 83–88.

Belardi, G. and Piga, L. (2013) Influence of calcium carbonate on the decomposition of asbestos contained in end-of-life products. *Thermochimica Acta*, **573**, 220–228.

Berman, R.G. (1988) Internally-consistent thermodynamic data for minerals in the system Na_2O-K_2O-CaO-MgO-FeO-Fe_2O_3-Al_2O_3-SiO_2-TiO_2-H_2O-CO_2. *Journal of Petrology*, **29**, 445–522.

Bernardo, E., Esposito, L., Rambaldi, E. and Tucci, A. (2011) Sintered glass ceramic articles from plasma vitrified asbestos containing waste. *Advances in Applied Ceramics*, **110**, 346–352.

Bertolini, L., Carsana, M., Cassago, D., Curzio, A.Q. and Collepardi, M. (2004) MSWI ashes as mineral additions in concrete. *Cement and Concrete Research*, **34**,1899–1906.

Bish, D.L. (1989) Determination of dehydration behaviour of zeolites using Rietveld refinement and high-temperature X-ray diffraction data. In: *Geological Society of America Annual Meeting, St. Louis, Missouri, USA, Abstract*, **2**, A73.

Bish, D.L. and Carey, J.W. (2001) Thermal behavior of natural zeolites. Pp. 403–452 in: *Natural Zeolites:*

Occurrence, Properties, Applications (D.L. Bish and D.W. Ming, editors). Reviews in Mineralogy and Geochemistry, **45**. Mineralogical Society of America and Geochemical Society, Chantilly, Virginia, USA.

Bloise, A., Fornero, E., Belluso, E., Barrese, E. and Rinaudo, C. (2008) Synthesis and characterization of tremolite asbestos fibres. *European Journal of Mineralogy*, **20**, 1027–1033.

Bloise, A., Belluso, E., Barrese, E., Miriello, D. and Apollaro, C. (2009a) Synthesis of Fe-doped chrysotile and characterization of the resulting chrysotile fibres. *Crystal Research and Technology*, **44**, 590–596.

Bloise, A., Barrese, E. and Apollaro, C. (2009b) Hydrothermal alteration of Ti-doped forsterite to chrysotile and characterization of the resulting chrysotile fibres. *Neues Jahrbuch für Mineralogie-Abhandlungen: Journal of Mineralogy and Geochemistry*, **185**, 297–304.

Bloise, A., Belluso, E., Fornero, E., Rinaudo, C., Barrese, E. and Capella, S. (2010) Influence of synthesis conditions on growth of Ni-doped chrysotile. *Microporous and Mesoporous Materials*, **132**, 239–245.

Bloise, A., Belluso, E., Catalano, M., Barrese, E., Miriello, D. and Apollaro, C. (2012) Hydrothermal alteration of glass to chrysotile. *Journal of the American Ceramic Society*, **95**, 3050–3055.

Bloise, A., Critelli, T., Catalano, M., Apollaro, C., Miriello, D., Croce, A., Barrese, E., Liberi, F. Piluso, E., Rinaudo, C. and Belluso, E. (2014) Asbestos and other fibrous minerals contained in the serpentinites of the Gimigliano-Mount Reventio Unit (Calabria, Italy). *Environmental Earth Sciences*, **71**, 3773–3786.

Bloise, A., Catalano, M., Barrese, E., Gualtieri, A.F., Gandolfi, N.B., Capella, S. and Belluso, E. (2016) TG/DSC study of the thermal behaviour of hazardous mineral fibres. *Journal of Thermal Analysis and Calorimetry*, **123**, 2225–2239.

Boccacini, D.N., Leonelli, C., Rivasi, M.R., Romagnoli, M., Veronesi, P., Pellacani, G.C. and Boccacini, A.R. (2007) Recycling of microwave inertised asbestos containing waste in refractory materials. *Journal of European Ceramic Society*, **27**, 1855–1858.

Borderes, A. (2000) Vitrification of the incineration residues. *Revue Verre*, **6**, 1–2.

Bowen, N.L. and Tuttle, O.F. (1949) The system $MgO–SiO_2–H_2O$. *Geological Society of America Bulletin*, **60**, 439–460.

Bozhilov, K.N., Jenkins, D.M. and Veblen, D.R. (2004) Pyribole evolution during tremolite synthesis from oxides. *American Mineralogist*, **89**, 74–84.

Bozhilov, K.N., Brownstein, D. and Jenkins, D.M. (2007) Biopyribole evolution during tremolite synthesis from dolomite and quartz in CO_2-H_2O fluid. *American Mineralogist*, **92**, 898–908.

Brindley, G.W. and Hayami, R. (1965) Mechanism of formation of forsterite and enstatite from serpentine, *Mineralogical Magazine*, **35**, 189-195

Brindley, G.W., Sharp, J.H., Patterson, J.H. and Narahari Achar, B.N. (1967) Kinetics and mechanism of dehydroxylation processes. II. Temperature and vapour pressure dependence of dehydroxylation of serpentine. *American Mineralogist*, **52**, 1697–1705.

Bush, D.G. and Schumacher, R.A. (1982) Analysis of chrysotile asbestos in bulk materials. *Analytical Chemistry*, **54**, 1792–1795.

Cattaneo, A., Gualtieri, A.F. and Artioli, G. (2003) Kinetic study of the dehydroxylation of chrysotile asbestos with temperature by in situ XRPD. *Physics and Chemistry of Minerals*, **30**, 177–183.

Cheng, L., Ren, X., Wei, X., Sun, X. and Yu, S. (2015) Facile synthesis and characterization of chrysotile nanotubes and their application for lead (II) removal from aqueous solution. *Separation Science and Technology*, **50**, 700–709.

Chernosky, J.V. (1975) Aggregate refractive indices and unit cell parameters of synthetic serpentine in the system MgO-Al_2O_3-SiO_2-H_2O. *American Mineralogist*, **60**, 200–208.

Chernosky, J.V., Berman, R.G. and Bryndzia, L.T. (1988) Stability, phase relations, and thermodynamic properties of chlorite and serpentine group minerals. Pp. 295–346 in: *Hydrous Phyllosilicates (Exclusive of Micas)* (S.W. Bailey, editor). Reviews in Mineralogy, **19**. Mineralogical Society of America, Chantilly, Virginia, USA.

Cilliers, J.J.R., Freeman, A.G., Hodgson, A. and Taylor, H. (1961) Crocidolite from the Koegas-Westerberg area, South Africa. *Economic Geology*, **56**, 1421–1437.

Colangelo, F., Cioffi, R., Lavorgna, M., Verdelotti, L. and De Stefano, L. (2011) Treatment and recycling of asbestos-cement containing waste. *Journal of Hazardous Materials*, **195**, 391–397.

Cruciani, G. (2006) Zeolites upon heating: Factors governing their thermal stability and structural changes.

Journal of Physics and Chemistry of Solids, **67**, 1973–1994.
Cunjin, T., Tongjiang, P., Haiyang, X., Jun, Z., Sheng, W. and Jingyong, T. (2012) Preparation of basic magnesium sulfate from chrysotile asbestos tailing. *Non Metallic Mines*, 01 (in Chinese).
Day, H. and Halbach, H. (1979) The stability field of anthophyllite: the effect of experimental uncertainty on pennissible phase diagram topologies. *American Mineralogist*, **64**, 809–823.
Datta, A.K., Samantaray, B.K. and Bhattacherjee S. (1986) Thermal transformation in a chrysotile asbestos. *Bulletin of Materials Science*, **8**, 497–503.
Deer, W.A., Howie, R.A. and Zussman, J. (1966) *An Introduction to the Rock-forming Minerals*. Longman, Green and Co. Ltd., London, 528 pp.
Deer, W.A., Howie, R.A. and Zussman, J. (1982) *Rock-forming Minerals. Vol. 1A, Orthosilicates*. Geological Society of London, London, 936 pp.
Deer, W.A., Howie, R.A. and Zussman, J. (2009) *Rock-forming Minerals. Vol. 3B, Layered Silicates Excluding Micas and Clay Minerals*. Geological Society of London, London, 320 pp.
Della Ventura, G., Hawthorne, F.C., Robert, J.L. and Iezzi, G. (2003) Synthesis and infrared spectroscopy of amphiboles along the tremolite-pargasite join. *European Journal of Mineralogy*, **15**, 341–347.
Dellisanti, F., Minguzzi, V. and Morandi, N. (2001–2002) Experimental results from thermal treatment of asbestos containing materials. *GeoActa*, **1**, 61–70.
Dellisanti, F., Rossi, P.L. and Valdrè, G. (2009) Remediation of asbestos containing materials by Joule heating vitrification performed in a pre-pilot apparatus. *International Journal of Mineral Processes*, **91**, 61–67.
Devouard, B., Baronnet, A., Van Tendeloo, G. and Amelinckx, S. (1997) First evidence of synthetic polygonal serpentines. *European Journal of Mineralogy*, **9**, 539–546.
Ding, W.J., Peng, T.J. and Chen, J.M. (2012) Diopside-based glass-ceramics from chrysotile asbestos tailings. *Advanced Materials Research*, **427**, 26–31.
Downey, A. and Timmons, D.M. (2005) Study into the applicability of thermo-chemical conversion technology to legacy asbestos wastes in the UK. *WM'05 Abstract Conference Book*, Tucson, Arizona, USA.
Ernst, W.G. (1962) Synthesis, stability relations, and occurrence of riebeckite and riebeckite-arfvedsonite solid solutions. *The Journal of Geology*, **70**, 689–736.
Ernst, W.G. (1968) *Amphiboles*. Springer, Heidelberg, Germany, 125 pp.
Evans, B.W. (2004) The serpentinite multisystem revisited: chrysotile is metastable. *International Geology Review*, **46**, 479–506.
Evans, B.W. (2007) The synthesis and stability of some end-member amphiboles. Pp. 261–286 in: *Amphiboles: Crystal Chemistry, Occurrence, and Health Issues* (F.C. Hawthorne, R. Oberti, G. Della Ventura and A. Mottana, editors). Reviews in Mineralogy and Geochemistry, **67**. Mineralogical Society of America and Geochemical Society, Chantilly, Virginia, USA.
Falini, G., Foresti, E., Lesci, G. and Roveri, N. (2002) Structural and morphological characterization of synthetic chrysotile single crystals. *Chemical Communications*, **14**, 1512–1513.
Falini, G., Foresti, E., Gazzano, M., Gualtieri, A. F., Leoni, M., Lesci, I. G. and Roveri, N. (2004) Tubular-shaped stoichiometric chrysotile nanocrystals. *Chemistry – A European Journal*, **10**, 3043–3049.
Finucane, K.G., Thompson, L.E., Abuku, T. and Nakauchi, H. (2008) Treatment of asbestos wastes using the GeoMelt® vitrification process. Pp. 1–10 in: *WM2008 Conference, Phoenix Arizona (USA), Abstract 8261*.
Foresti, E., Hochella, M.F., Kornishi, H., Lesci, I.G., Madden, A.S., Roveri, N. and Xu, H. (2005) Morphological and chemical/physical characterization of Fe-doped synthetic chrysotile nanotubes. *Advanced Functional Materials*, **15**, 1009–1016.
Freeman, A.G. (1966) The dehydroxylation behavior of amphiboles. *American Mineralogist*, **35**, 953–957.
Fujishige, M., Kuribara, A., Karasawa, I. and Kojima, A. (2007) Low-temperature pyrolysis of crocidolite and amosite using calcium salts as a flux. *Journal of the Ceramic Society of Japan*, **115**, 434–439.
Gerdes, T., Rosin, A. and Willert-Porada, M. (2007) Development of a mobile asbestos decontamination unit. *Nanotech 2007 Technical Proceedings*, **4**, 667–670.
Giacobbe, C., Gualtieri, A.F., Quartieri, S., Rinaudo, C., Allegrina, M. and Andreozzi, G.B. (2010) Spectroscopic study of the product of thermal transformation of chrysotile-asbestos containing materials (ACM). *European Journal of Mineralogy*, **22**, 535–546.

Gilbert, M.C. and Popp, R.K. (1982) Experimental studies of amphibole stability; End-member relations. Pp. 231–268 in: *Amphiboles: Petrology and Experimental Phase Relations* (D.R. Veblen and P.H. Ribbe, editors). Reviews in Mineralogy, **9B**. Mineralogical Society of America, Washington, USA.

Gillery, F.H. (1959) The X-ray study of synthetic Mg-Al serpentines and chlorites. *American Mineralogist*, **44**, 143–152.

Gottardi, G. and Galli, E. (1985) *Natural Zeolites*. Springer-Verlag, Heidelberg, Germany, 409 pp.

Gottschalk, M., Najorka, J. and Andrut, M. (1998) Structural and compositional characterization of synthetic (Ca, Sr)-tremolite and (Ca, Sr)-diopside solid solutions. *Physics and Chemistry of Minerals*, **25**, 415–428.

Gottschalk, M., Andrut, M. and Melzer, S. (1999) The determination of the cummingtonite content of synthetic tremolite. *European Journal of Mineralogy*, **11**, 967–982.

Gregori, G., Kleebe, H.-J., Siegelin, F. and Ziegler, G. (2002) In situ SEM imaging at temperatures as high as 1450ºC. *Journal of Electron Microscopy (Tokyo)*, **51**, 347–352.

Greenwood, H.J. (1963) The synthesis and stability of anthophyllite. *Journal of Petrology*, **4**, 317–351.

Gualtieri, A.F. and Boccaletti, M. (2011) Recycling of the product of thermal inertization of cement-asbestos for the production of concrete. *Construction and Building Materials*, **25**, 3561–3569.

Gualtieri, A.F. and Tartaglia, A. (2000) Thermal decomposition of asbestos and recycling in traditional ceramics. *Journal of the European Ceramic Society*, **20**, 1409–1418.

Gualtieri, A.F., Levy, D., Belluso, E. and Dapiaggi, M. (2004) Kinetics of the decomposition of crocidolite asbestos: a preliminary real-time X-ray powder diffraction study. *Materials Science Forum*, **443/444**, 291–294.

Gualtieri, A.F., Cavenati, C., Zanatto, I., Meloni, M., Elmi, G. and Lassinantti Gualtieri, M. (2008a) The transformation sequence of cement–asbestos slates up to 1200ºC and safe recycling of the reaction product in stoneware tile mixtures. *Journal of Hazardous Materials*, **152 (2)**, 563–570.

Gualtieri, A.F., Lassinantti Gualtieri, M. and Tonelli, M. (2008b) In situ ESEM study of the thermal decomposition of chrysotile asbestos in view of safe recycling of the transformation product. *Journal of Hazardous Materials*, **156**, 260–266.

Gualtieri, A.F., Giacobbe, C., Sardisco, L., Saraceno, M., Lassinantti Gualtieri, M., Lusvardi, G., Cavenati, C. and Zanatto, I. (2010) Recycling of the product of thermal inertization of cement–asbestos for various industrial applications. *Waste Management*, **31**, 91–100.

Gualtieri, A.F., Giacobbe, C. and Viti, C. (2012) The dehydroxylation of serpentine group minerals. *American Mineralogist*, **97**, 666–680.

Hashimoto, S. and Yamaguchi, A. (2006) Detoxification technique of asbestos using low-temperature heating and grinding. *Ceramics Japan*, **41**, 856–858.

Hashimoto, S., Takeda, H., Okuda, A., Kambayashi, A., Honda, S., Iwamoto, Y. and Fukuda, K. (2008) Detoxification of industrial asbestos waste by low-temperature heating in a vacuum. *Journal of the Ceramic Society of Japan*, **116**, 242–246.

Hawthorne, F.C., Ventura, G.D., Robert, J.-L., Welch, M.D., Raudsepp, M. and Jenkins, D.M. (1997) A Rietveld and infrared study of synthetic amphiboles along the potassium-richterite-tremolite join. *American Mineralogist*, **82**, 708–716.

Hemley, J.J., Montoya, J.W., Christ, C.L. and Hostetler, P.B. (1977) Mineral equilibria in the $MgO-SiO_2-H_2O$ system; I. Talc-chrysotile-forsterite-brucite stability relations. *American Journal of Science*, **277**, 322–351.

Hodgson, A.A. (1965) The thermal decomposition of miscellaneous crocidolites. *Mineralogical Magazine*, **35**, 291–305.

Hodgson, A.A., Freeman, A.G. and Taylor, H. (1965a) The thermal decomposition of crocidolite from Koegas, South Africa. *Mineralogical Magazine*, **35**, 5–30.

Hodgson, A.A., Freeman, A.G. and Taylor, H.F.W. (1965b) The thermal decomposition of amosite. *Mineralogical Magazine*, **35**, 445–463.

Horikoshi, S., Sumi, T., Ito, S., Dillert, R., Kashimura, K., Yoshikawa, N., Sato, M. and Shinohara, N. (2014) Microwave-driven asbestos treatment and its scale-up for use after natural disasters. *Environmental Science and Technology*, **48**, 6882–6890.

Huang, Li, W., Pan, Z.H., Liu, Y.G. and Fang, M.H. (2013) High-temperature transformation of asbestos tailings by carbothermal reduction. *Clays and Clay Minerals*, **61**, 75–82.

IARC (1997) Silica, some silicates, coal dust and *para*-aramid fibrils. *IARC Monographs on the Evaluation of the Carcinogenic Risk to Humans*, Vol. **68**, 1–475.

Iezzi, G., Della Ventura, G., Cámara, F., Pedrazzi, G. and Robert, J.-L. (2003) BNa–BLi solid-solution in A-site-vacant amphiboles: synthesis and cation ordering along the ferri-clinoferroholmquistite–riebeckite join. *American Mineralogist*, **88**, 955–961.

Inaba, T. and Iwao, T. (2000) Treatment of waste by DC arc discharge plasmas. *IEEE Transaction Dielectrics Electrical Insulation*, **7**, 684–692.

Inaba, T., Nagano, M. and Endo, M. (1999) Investigation of plasma treatment for hazardous wastes such as fly ash and asbestos. *Electrical Engineering in Japan*, **126**, 831–838.

Inoue, R., Kano, J., Shimme, K. and Saito, F. (2007) Safe decomposition of asbestos by mechano-chemical reaction. *Materials Science Forum*, **561–565**, 2257–2260.

Ishida, K., Hawthorne, F.C. and Ando, Y. (2002) Fine structure of infrared OH-stretching bands in natural and heat-treated amphiboles of the tremolite-ferro-actinolite series. *American Mineralogist*, **87**, 891–898.

Jancar, B. and Suvorov, D. (2006) The influence of hydrothermal-reaction parameters on the formation of chrysotile nanotubes. *Nanotechnology*, **17**, 25–29.

Jasmund, K. and Sylla, H.M. (1970) Synthese von Eisenchrysotil. *Naturwissenschaften*, **57**, 494–495.

Jasmund, K., Sylla, H.M. and Freund, F. (1975) Solid solution in synthetic serpentine phases. Pp. 267–274 in: *Proceedings of the International Clay Conference*. Applied Publishing, Wilmette, Illinois, USA.

Jenkins, D.M. (1987) Synthesis and characterization of tremolite in the system H_2O-CaO-MgO-SiO_2. *American Mineralogist*, **72**, 707–715.

Jenkins, D.M. and Bozhilov, K.N. (2003) Stability and thermodynamic properties of ferro-actinolite: a re-investigation. *American Journal of Science*, **303**, 723–752.

Jeyaratnam, M. and West, N.G. (1994) A study of heat-degraded chrysotile, amosite and crocidolite by X-ray diffraction. *The Annals of Occupational Hygiene*, **38**, 137–148.

Johannes, D. (2003) Method and device for safe and pollution-free disposal of asbestos-containing material. European Patent EP0484866.

Johannes, W. (1968) Experimental investigation of the reaction forsterite + $H_2O \rightleftharpoons$ serpentine + brucite. *Contributions to Mineralogy and Petrology*, **19**, 309–315.

Johnson, N.M. and Fegley, B. (2000) Water on Venus: New insights from tremolite decomposition. *Icarus*, **146**, 301–306.

Johnson, N.M. and Fegley, B. (2003) Tremolite decomposition on Venus II. Products, kinetics, and mechanism. *Icarus*, **164**, 317–333.

Kalinin, D.V. and Zubkov, M.Y. (1981) Kinetic investigation of the MgO-SiO_2-H_2O system, reaction: Serpentine = forsterite + talc + water. *Soviet Geology and Geophysics*, **22**, 61–68.

Kohyama, N., Shinohama, Y. and Suzuki, Y., (1996) Mineral phases and some re-examined characteristics of the international union against cancer standard asbestos samples. *American Journal of Industrial Medicine*, **30**, 515–543.

Korytkova, E.N. and Makarova, J.A. (1971) Experimental study of the serpentinization of olivine. *Doklady Akademii Nauk SSSR, Earth Science Sector*, **196**, 144–145.

Korytkova, E.N. and Pivovarova, L.N. (2010) Hydrothermal synthesis of nanotubes based on $(Mg,Fe,Co,Ni)_3Si_2O_5(OH)_4$ hydrosilicates. *Glass Physics and Chemistry*, **36**, 53–60.

Korytkova, E.N., Maslov, A.V. and Gusarov, V.V. (2002) Hydrothermal Synthesis of Inorganic Nanotubes. Pp. 54–59 in: *Collected Articles "Surface Chemistry and Synthesis of Low-Dimensional Systems"*. State Technological Institute (Technical University) St. Petersburg, (in Russian).

Korytkova, E.N., Maslov, A.V., Pivovarova, L.N., Drozdova, I.A. and Gusarov, V.V. (2004) Formation of $Mg_3Si_2O_5(OH)_4$ nanotubes under hydrothermal conditions. *Glass Physics and Chemistry*, **30**, 51–55.

Korytkova, E.N., Maslov, A.V., Pivovarova, L.N., Polegotchenkova, Y.V., Povinich, V.F. and Gusarov, V.V. (2005) Synthesis of nanotubular $Mg_3Si_2O_5(OH)_4$-$Ni_3Si_2O_5(OH)_4$ silicates at elevated temperatures and pressures. *Inorganic materials*, **41**, 743–749.

Korytkova, E.N., Pivovarova, L.N., Drosdova, I.A. and Gusarov, V.V. (2007) Hydrothermal synthesis of nanotubular Co-Mg hydrosilicates with the chrysotile structure. *Russian Journal of General Chemistry*, **77**, 1669–1676.

Korytkova, E.N., Brovkin, A.S., Maslennikova, T.P., Pivovarova, L.N. and Drozdova, I.A. (2011) Influence of the physicochemical parameters of synthesis on the growth of nanotubes of the $Mg_3Si_2O_5(OH)_4$ composition under hydrothermal conditions. *Glass Physics and Chemistry*, **37**, 161–171.

Kozawa, T., Onda, A., Yanagisawa, K., Chiba, O., Ishiwate, H. and Takanami, T. (2010) Thermal decomposition of chrysotile-containing wastes in a water vapor atmosphere. *Journal of the Ceramic Society of Japan*, **118**, 1199–1201.

Krasilin, A.A., Almjasheva, O.V. and Gusarov, V.V. (2011) Effect of the structure of precursors on the formation of nanotubular magnesium hydrosilicate. *Inorganic Materials*, **47**, 1111–1115.

Kusiorowski, R., Zaremba, T., Piotrowski, J. and Adamek. J, (2012) Thermal decomposition of different types of asbestos. *Journal of Thermal Analysis and Calorimetry*, **109**, 693–704.

Kusiorowski, R., Zaremba, T., Piotrowski, J. and Gerle, A. (2013) Thermal decomposition of asbestos-containing materials. *Journal of Thermal Analysis and Calorimetry*, **113**, 179–188.

Kusiorowski, R., Zaremba, T. and Piotrowski, J. (2014) The potential use of cement–asbestos waste in the ceramic masses destined for sintered wall clay brick manufacture. *Ceramics International*, **40**, 11995–12002.

Kusiorowski, R., Zaremba, T., Piotrowski, J. and Podwórny, J. (2015a) Utilisation of cement-asbestos wastes by thermal treatment and the potential possibility use of obtained product for the clinker bricks manufacture. *Journal of Materials Science*, **50**, 6757–6767.

Kusiorowski, R., Zaremba, T., Gerle, A., Piotrowski, J., Simka, W. and Adamek, J. (2015b) Study on the thermal decomposition of crocidolite asbestos. *Journal of Thermal Analysis and Calorimetry*, **120**, 1585–1595.

Lafay, R., Montes-Hernandez, G., Janots, E., Chiriac, R., Findling, N. and Toche, F. (2012) Mineral replacement rate of olivine by chrysotile and brucite under high alkaline conditions. *Journal of Crystal Growth*, **347**, 62–72.

Lafay, R., Montes-Hernandez, G., Janots, E., Chiriac, R., Findling, N. and Toche, F. (2013) Nucleation and growth of chrysotile nanotubes in $H_2SiO_3/MgCl_2/NaOH$ medium at 90 to 300°C. *Chemistry–A European Journal*, **19**, 5417–5424.

Lafay, R., Montes-Hernandez, G., Janots, E., Auzende, A.L., Chiriac, R., Lemarchand, D. and Toche, F. (2014) Influence of trace elements on the textural properties of synthetic chrysotile: complementary insights from macroscopic and nanoscopic measurements. *Microporous and Mesoporous Materials*, **183**, 81–90.

Langer, A.M. (2003) Reduction of the biological potential of chrysotile asbestos arising from conditions of service on brake pads. *Regulatory Toxicology and Pharmacology*, **38**, 71–77.

Lattard, D. and Evans, B.W. (1992) New experiments on the stability of grunerite. *European Journal of Mineralogy*, **4**, 219–238.

Leonelli, C., Veronesi, P., Boccaccini, D.N., Rivasi, M.R., Barbieri, L., Andreola, F., Lancellotti, I., Rabitti, D. and Pellacani, G.C. (2006) Microwave thermal inertisation of asbestos containing waste and its recycling in traditional ceramics. *Journal of Hazardous Materials*, **B135**, 149–155.

Lledo, H.L. and Jenkins, D.M. (2008) Experimental investigation of the upper thermal stability of Mg-rich actinolite; implications for Kiruna-type iron deposits. *Journal of Petrology*, **49**, 225–238.

Luckewicz, W. (1975) Differential thermal analysis of chrysotile asbestos in pure talc and talc containing other minerals. *Journal of the Society of Cosmetic Chemists*, **26**, 431–437.

Ma, G.H., Peng, T.J. and Duan, T. (2007) A study of chrysotile asbestos nanotube synthesized by hydrothermal reaction. *Journal of Mineralogy and Petrology*, **27**, 40–45 (in Chinese).

MacKenzie, K.J.D. and Meinhold, R.H. (1994) Thermal reactions of chrysotile revisited: a ^{29}Si and ^{25}Mg MAS NMR study. *American Mineralogist*, **79**, 43–50.

Malkov, A.A., Korytkova, E.N., Maslennikova, T.P., Shtykhova, A.M., and Gusarov, V.V. (2009) Effect of heat treatment on structural-chemical transformations in magnesium hydrosilicate $[Mg_3Si_2O_5(OH)_4]$ nanotubes. *Russian Journal of Applied Chemistry*, **82**, 2079–2086.

Maresch, W.V. and Czank, M. (2007) The significance of the reaction path in synthesizing single-phase amphibole of defined composition. Pp. 287–322 in: *Amphiboles: Crystal Chemistry, Occurrence and Health Issues* (F.C. Hawthorne, R. Oberti, G. Della Ventura and A. Mottana, editors). Reviews in Mineralogy and Geochemistry, **67**, Mineralogical Society of America and Geochemical Society, Chantilly, Virginia, USA.

Maresch, W.V., Czank, M. and Schreyer, W. (1994) Growth mechanisms, structural defects and composition of synthetic tremolite: what are the effects on macroscopic properties?. *Contributions to Mineralogy and Petrology*, **118**, 297–313.

Martin, C.J. (1977) The thermal decomposition of chrysotile. *Mineralogical Magazine*, **35**, 189–195.

McDonald, A., Scott, B. and Villemure, G. (2009) Hydrothermal preparation of nanotubular particles of a 1:1 nickel phyllosilicate. *Microporous and Mesoporous Materials*, **120**, 263–266.

Min, S.Y., Maken, S., Park, J.W., Gaur, A. and Hyun, J.S. (2008) Melting treatment of waste asbestos using mixture of hydrogen and oxygen produced from water electrolysis. *Korean Journal of Chemical Engineering*, **25**, 323–328.

Moody, J.B. (1976a) Serpentinization: a review. *Lithos*, **9**, 125–138.

Moody, J.B. (1976b) An experimental study on the serpentinization of iron-bearing olivines. *The Canadian Mineralogist*, **14**, 462–478.

Nam, S.N., Jeong, S. and Lim, H. (2014) Thermochemical destruction of asbestos-containing roofing slate and the feasibility of using recycled waste sulfuric acid. *Journal of Hazardous Materials*, **265**, 151–157.

Nesterchuk, N.I., Makarova, T.A. and Fedoseev, A.D. (1966) Hydrothermal synthesis of chrysotile. *Zapiski Vsesoyuznogo Mineralogicheskogo Obshchestva*, **95**, 75-79, (in Russian).

Noll, W. (1950) Synthesen im System $MgO/SiO_2/H_2O$. *Zeitschrift für anorganische und allgemeine Chemie*, **261**, 1–25.

Noll, W. and Kircher, H. (1952) Synthese des Garnierites. *Naturwissenschaften*, **39**, 233–234.

Noll, W., Kircher, H. and Sybertz, W. (1958) Ein weiteres Solenosilikat: Kobaltchrysotil. *Naturwissenschaften*, **45**, 489–489.

Noll, W., Kircher, H. and Sybertz, W. (1960) Über synthetischen Kobalt chrysotil und seine Beziehungen zu anderen Soleno silikaten. *Contributions to Mineralogy and Petrology*, **7**, 232–241.

Normand, C., Williams-Jones, A.E., Martin, R.F. and Vali, H. (2002) Hydrothermal alteration of olivine in a flow-through autoclave: Nucleation and growth of serpentine phases. *American Mineralogist*, **87**, 1699–1709.

Oberti, R., Della Ventura, G. and Cámara, F. (2007) New amphibole compositions: natural and synthetic. Pp. 89–124 in: *Amphiboles: Crystal Chemistry, Occurrence, and Health Issues* (F.C. Hawthorne, R. Oberti, G. Della Ventura and A. Mottana, editors). Reviews in Mineralogy and Geochemistry, **67**. Mineralogical Society of America and Geochemical Society, Chantilly, Virginia, USA.

Ogorodova, L.P., Kiseleva, I.A., Korytkova, E.N., Maslennikova, T.P. and Gusarov, V.V. (2010) The synthesis and thermochemical study of $(Mg,Fe)_3Si_2O_5(OH)_4$ nanotubes. *Russian Journal of Physical Chemistry A*, **84**, 44–47.

Osada, M., Takamiya, K., Manako, K., Noguchi, M. and Sakai, S. (2013) Demonstration study of high temperature melting for asbestos-containing waste (ACW). *Journal of Material Cycles and Waste Management*, **15**, 25–36.

Ozeki, S., Masuda, Y., Sano, H., Seki, H. and Ooi, K. (1991) Proton NMR spectroscopy of water adsorbed on synthetic chrysotile asbestos: microtubes with acidic and basic surfaces. *The Journal of Physical Chemistry*, **95**, 6309–6316.

Papke, K.G. (1972) Erionite and other associated zeolites in Nevada. *Nevada Bureau of Mines and Geology*, **79**, 1–31.

Patterson, J.H. and O'Connor, D.J. (1966) Chemical studies of amphibole asbestos. I. Structural changes of heat-treated crocidolite, amosite, and tremolite from infrared absorption studies. *Australian Journal of Chemistry*, **19**, 1155–1164.

Patterson, J.H. (1965) The thermal disintegration of crocidolite in air and in vacuum. *Mineralogical Magazine*, **35**, 31–37.

Pawley, A.R., Graham, C.M. and Navrotsky, A. (1993) Tremolite-richterite amphiboles: Synthesis, compositional and structural characterization, and thermochemistry. *American Mineralogist*, **78**, 23–35.

Plescia, P., Gizzi, D., Benedetti, S., Camilucci, L., Fanizza, C., De Simone, P. and Paglietti, F. (2003) Mechanochemical treatment to recycling asbestos-containing waste. *Waste Management*, **23**, 209–218.

Podor, R., Ravaux, J. and Brau, H.-P. (2012) In situ experiments in the Scanning Electron Microscope Chamber. Pp. 31–54 in: *Scanning Electron Microscopy* (V. Kazmiruk, editor). InTech, Rijeka, Croatia.

Poiroux, R. and Rollin, M. (1996) High temperature treatment of waste: from laboratories to the industrial stage. *Pure and Applied Chemistry*, **68**, 1035–1040.

Rabbitt, J.C. (1948) A new study of the anthophyllite series. *American Mineralogist*, **33**, 263–323.

Radvanec, M., Tucek, L., Derco, J., Cechovská, K. and Németh, Z. (2013) Change of carcinogenic chrysotile fibers in the asbestos cement (eternit) to harmless waste by artificial carbonatization: Petrological and technological results. *Journal of Hazardous Materials*, **252–253**, 390–400.

Roberts, D. and Stuart, J.H. (1989) *Vitrification of asbestos waste*, US Patent 4,820,328, Apr. 11, 1989.

Rouxhet, P.G., Gillard, J.L. and Fripiat, J.J. (1972) Thermal decomposition of amosite, crocidolite, and biotite. *Mineralogical Magazine*, **38**, 583–592.

Roveri, N., Falini, G., Foresti, E., Fracasso, G., Lesci, I. G. and Sabatino, P. (2006) Geoinspired synthetic chrysotile nanotubes. *Journal of Materials Research*, **21**, 2711–2725.

Roy, D.M. and Roy, R. (1954) An experimental study of the formation and properties of synthetic serpentines and related layer silicate minerals. *American Mineralogist*, **39**, 957–975.

Sharma, A. and Jenkins, D.M. (1999) Hydrothermal synthesis of amphiboles along the tremolite-pargasite join and in the ternary system tremolite-pargasite-cummingtonite. *American Mineralogist*, **84**, 1304–1318.

Schlenker, J.L., Pluth, J.J. and Smith, J.V. (1977) Dehydrated natural erionite with stacking faults of the offretite type. *Acta Crystallographica*, **B33**, 3265–3268.

Skippen, G. and McKinstry, B.W. (1985) Synthetic and natural tremolite in equilibrium with forsterite, enstatite, diopside and fluid. *Contributions to Mineralogy and Petrology*, **89**, 256–262.

Skogby, H. and Ferrow, E. (1989) Iron distribution and structural order in synthetic calcific amphiboles studied by Mössbauer spectroscopy and HRTEM. *American Mineralogist*, **74**, 360–366.

Smolikov, A., Vezentsev, A., Beresnev, V., Kolesnikov, D. and Solokha, A. (2013) Morphology of synthetic chrysotile nanofibers (Mg-hydro silicate). *Journal of Materials Science and Engineering A*, **3**, 523–530.

Speil, S. and Leineweber, J.P. (1969) Asbestos minerals in modern technology. *Environmental Research*, **2**, 166–208.

Sprynsky, M., Niedojadło, J. and Buszewski, B. (2011) Structural features of natural and acids modified chrysotile nanotubes. *Journal of Physics and Chemistry of Solids*, **72**, 1015–1026.

Sugama, T., Sabatini, R. and Petrakis, L. (1998) Decomposition of chrysotile asbestos by fluorosulfonic acid. *Industrial Engineering Chemical Research*, **37**, 79–88.

Sueno, S., Cameron, M., Papike, J.J. and Prewitt, C.T. (1973) The high temperature crystal chemistry of tremolite. *American Mineralogist*, **58**, 649–664.

Teixeira, A.P., Santos, E.M., Vieira, A.F.P. and Lago, R.M. (2013) Use of chrysotile to produce highly dispersed K-doped MgO catalyst for biodiesel synthesis. *Journal of Hazardous Materials*, **252–253**, 390–400.

Trittschack, R. and Grobéty, B. (2013) The dehydroxylation of chrysotile. A combined in situ micro-Raman and micro-FTIR study. *The American Mineralogist*, **98**, 1133–1145.

Trittschack, R., Grobéty, B. and Brodard, P. (2014) Kinetics of the chrysotile and brucite dehydroxylation reaction: a combined non-isothermal/isothermal thermogravimetric analysis and high-temperature X-ray powder diffraction study. *Swiss Journal of Geosciences*, **106**, 469–489.

Ueno, T., Furuta, Y., Koyama, T. and Imada, T. (1991) Phase relation among serpentine, brucite and forsterite from 200 to 500atm water pressure. *Mineralogical Journal*, **15**, 276–281.

Vast, P., Andries, V., Martines, M.A.U., Auffredic, J.P., Poulain, M. and Messaddeo, Y. (2004) Treatment and destruction of inorganic fibers wastes like asbestos by sodium polyphosphate. *Phosphorus Research Bulletin*, **15**, 68–82.

Veblen, D.R (editor) (1981) *Amphiboles and Other Hydrous Pyriboles – Mineralogy*. Reviews in Mineralogy, **9A**. Mineralogical Society of America, Chantilly, Virginia, USA.

Vermaas, F. (1952) The amphibole asbestos of South Africa. *Transactions of the Geological Society of South Africa*, **55**, 199–232.

Viani, A. and Gualtieri, A.F. (2013) Recycling the product of thermal transformation of cement-asbestos for the preparation of calcium sulfoaluminate clinker. *Journal of Hazardous Materials*, **260**, 813–818.

Viani, A. and Gualtieri, A.F. (2014) Preparation of magnesium phosphate cement by recycling the product of thermal transformation of asbestos containing wastes. *Cement and Concrete Research*, **58**, 56–66.

Virta, R.L. (2012) Asbestos. *Minerals Yearbook*. USGS, Virginia, USA.

Wittels, M.C. (1951) A thermo-chemical analysis of some amphiboles. *American Mineralogist*, **36**, 851–858.

Wittels, M.C. (1952) The structural disintegration of some amphiboles. *American Mineralogist*, **37**, 28–36.

Xu, J., Li, X., Zhou, W., Ding, L., Jin, Z. and Li, Y. (2006) The preparation of $Mg_3Si_2O_5(OH)_4$ nanotubes under solvothermal conditions. *Journal of Porous Materials*, **13**, 275–279.

Yada, K. and Iishi, K. (1974) Serpentine minerals hydrothermally synthesized and their microstructures. *Journal of Crystal Growth*, **24**, 627–630.

Yada, K. and Iishi, K. (1977) Growth and microstructure of synthetic chrysotile. *American Mineralogist*, **62**, 958–965.

Yamai, I. and Saito, H. (1974) The effects of starting components on the hydrothermal synthesis of chrysotile fibers. *Journal of Crystal Growth*, **24**, 617–620.

Yanagisawa, K., Kozawa, T., Onda, A., Kanazawa, M., Shinohara, J., Takanami, T. and Shiraishi, M. (2009) A novel decomposition technique of friable asbestos by $CHClF_2$-decomposed acidic gas. *Journal of Hazardous Materials*, **163**, 593–599.

Yang, J.C. (1961) The growth of synthetic chrysotile fiber. *American Mineralogist*, **46**, 748–752.

Yoder, H.S. (1952) The MgO-Al_2O_3-SiO_2-H_2O system and the related metamorphic facies. *American Journal of Science*, **250**, 569–627.

Yoshikawa, N., Kashimura, K., Hashiguchi, M., Sato, M., Horikoshi, S., Mitani, T. and Shinohara, N. (2015) Detoxification mechanism of asbestos materials by microwave treatment. *Journal of Hazardous Materials*, **284**, 201–206.

Yvon, Y. and Sharrock, P. (2011) Characterization of thermochemical inactivation of asbestos containing wastes and recycling the mineral residues in cement products. *Waste and Biomass Valorization*, **2**, 169–181.

Zaremba, T., Krząkała, A., Piotrowski, J. and Garczorz, D. (2010) Study on the thermal decomposition of chrysotile asbestos. *Journal of Thermal Analysis and Calorimetry*, **101**, 479–485.

Zimmermann, R., Heinrich, W. and Franz, G. (1996) Tremolite synthesis from $CaCl_2$-bearing aqueous solutions. *European Journal of Mineralogy*, **8**, 767–776.

Zhu, P., Wang, L.Y., Hong, D., Zhou, M. and Zhou, J. (2013) Investigative studies for inert transformation of toxic chrysotile tailing. *Journal of Material Cycles and Wastes Management*, **15**, 90–97.

Zulumyan, N., Mirgorodski, A., Isahakyan, A. and Beglaryan, H. (2014) The mechanism of decomposition of serpentines from peridotites on heating. *Journal of Thermal Analysis and Calorimetry*, **115**, 1003–1012.

EMU Notes in Mineralogy, Vol. 18 (2017), Chapter 8, 261–306

In vitro biological activity and mechanisms of lung and pleural cancers induced by mineral fibres

BROOKE T. MOSSMAN[1,*] and ARMANDA PUGNALONI[2]

[1]*Department of Pathology, University of Vermont College of Medicine, 89 Beaumont Avenue, Burlington, Vermont 05405, USA, e-mail: brooke.mossman@uvm.edu*
[2] *Department of Clinical and Molecular Sciences, Università Politecnica delle Marche, Via Tronto 10/A, Ancona 60020, Italy*
**Corresponding author*

The mechanisms of disease of the six naturally occurring asbestos fibres have been studied for decades although critical questions still remain. With the advent of cell and molecular biology, *in vitro* models have been developed that enable use of sophisticated analytical tools and important collaborations between biologists, geologists and experts in the chemical/physical characterization of not only asbestos, but other minerals in the environment that may have disease potential. This chapter stresses the fact that a number of cellular organelles participate in mineral fibre toxicity and describes the primary properties of pathogenic mineral fibres. In addition, it summarizes new directions in testing of suspect carcinogenic fibres in Italy.

1. Introduction

The cell is the structural unit of the body, and accordingly, studies evaluating the mechanisms of disease by mineral fibres have focused on cells maintained *in vitro*. Early studies focused on cells such as red blood cells (RBCs) (Woodworth *et al.*, 1982) and alveolar macrophages (AMs) (Quinlan *et al.*, 1998) that were isolated from humans or rodents with minimally invasive techniques. However, they could be maintained, generally in suspension, for only short periods of time. In the normal lung and pleura, there are more than 40 different cell types, each with their own specialized functions. *In situ*, epithelial cells of the lung that develop into lung cancers and mesothelial cells that line the body cavities occur in contiguous monolayers. Thus, a challenge to cell and molecular biologists has been to determine how to culture these cells in their differentiated states for prolonged periods of time. This alone has taken decades of research, especially for cultures of human cells. Historically, *in vitro* studies were first performed with transformed rodent and human cell epithelial cell lines because they have an infinite life span (A541, A549, HeLa, *etc.*). Studies then progressed to the use of rodent cell lines that were more normal in their growth features, but not necessarily of lung origin (CHO, SHE, 3T3). With the development of complex medium and supplements that encouraged normal differentiation of lung epithelial cells, the cell

DOI: 10.1180/EMU-notes.18.8

types giving rise to bronchial or pulmonary carcinomas, and mesothelial cells, the progenitor cells of mesotheliomas (MMs), a number of rodent lines flourished (RLE, HTE, C10, RPM) and were exchanged between laboratories. More recently, lines of human lung epithelial cells and mesothelial cells have been immortalized using viral vectors or telomerases to increase their life span (MeT5A, LP9-tert, BEAS-2B). Collaborations between thoracic surgeons, surgical pathologists, and cell biologists have also yielded primary isolates of normal human lung epithelial cells and mesothelial cells that can be maintained for a few passages. In some cases, certain organelles such as mitochondria (Bergamini *et al.*, 2004, 2007) and liposomes (artificial membrane preparations) (Woodworth *et al.*, 1982) have been isolated and studied *ex vivo*.

Other more complex *in vitro* models have been developed based upon the observations *in vivo* that lung and pleural reactions to mineral fibres involve a number of different cell interactions. These models include co-cultures of two or more cell types, spheroids or organ cultures derived from tracheobronchial tissues. These three-dimensional cultures maintained from rodent or human tissues can be cultured for months and implanted back into syngenetic or immunocompromised animals to determine their carcinogenicity (Mossman and Craighead, 1975, 1979). This model has been helpful in showing that mineral fibres adsorbing polycyclic aromatic hydrocarbons, combustion products and components of cigarette smoke, cause carcinomas while mineral fibres alone appear to act as cancer promoters or co-carcinogens (Topping and Nettesheim, 1980; Mossman and Craighead, 1982; Mossman *et al.*, 1984). Organ cultures have also been used to determine how different cell types in concert react to mineral fibres and other particles, how certain fibres and particles are cleared or removed by mucociliary action (Mossman *et al.*, 1978), and how fibres are taken up by various cell types, including epithelial cells and phagocytes (Mossman *et al.*, 1977). Lastly, because these 3-D models can be maintained for several months in culture, they can be evaluated over time for precancerous changes (cell proliferation, squamous metaplasia, dysplasia) that can be reversed using preventive or therapeutic approaches (Mossman *et al.*, 1976, 1980) (Fig. 1).

In this chapter, we provide first an introduction of *in vitro* models then discuss briefly the properties of mineral fibres with a focus on asbestos types. Next, we describe key organelles that may play a role in mineral fibre pathogenicity, *i.e.* disease development, excluding genes and gene products linked to causation or suppression of lung cancers and MMs that are covered in other chapters in this volume. We then describe novel and recent studies evaluating the toxicity, potential carcinogenicity, and cell injury of suspect minerals in Italy with relationship to the prevalence and possibly future episodes of lung cancers and MMs. Lastly, we discuss the toxic and carcinogenic properties of mineral fibres that have been linked to *in vitro* data in recent and historical studies with asbestos and other fibres.

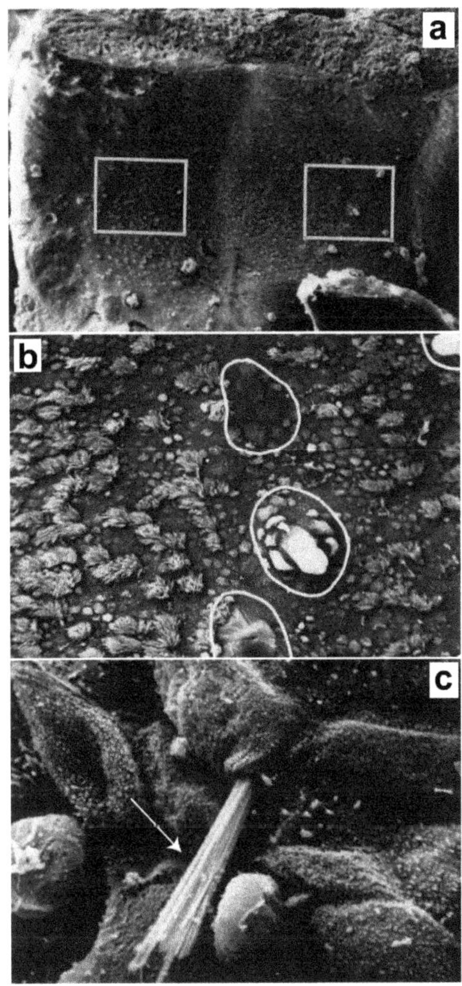

Figure 1. 3D organ culture (explants) models from a hamster trachea indicating by scanning electron microscopy (SEM) the epithelial surface: (a) approximately half of the explants surface area using low magnification (× 120); (b) an area within the white boxes indicated in 1a in which one can see the normal mucociliary epithelial cells with a predominance of cilia on the left and right, and areas of squamous metaplasia (circled) where cilia and cell surface features of mucin-producing cells are replaced by keratinizing cells (× 600); (c) a group of squamous epithelial cells that have lost contact inhibition and are proliferating around long crocidolite asbestos fibres that have been incompletely phagocytized (arrows) (× 5400). Reprinted from Woodworth *et al.* (1983b) with permission of *Environmental Perspectives*, **51**, 27–33.

2. Asbestos types and forms

'Asbestos' is a commercial term for a group of naturally occurring fibrous hydrated silicate minerals that have been used in a number of products for thousands of years. There are two sub-groups of asbestos minerals: amphibole and serpentine (see Gualtieri, 2017 and Ballirano *et al.*, 2017, this volume). Amphibole and serpentine minerals can exist in two different forms, non-asbestiform/non-asbestos or asbestiform that is used synonymously with 'asbestos' (Guthrie and Mossman, 1993; Finley *et al.*, 2012). In the past, the term 'asbestiform' has often been used inappropriately by biologists (IARC, 2012) to refer to fibres that look like asbestos morphologically, *i.e.* fibre glasses, *etc.*, but geologists have argued that asbestiform or asbestos fibres are distinctive in crystalline structure, growth habit and morphology from non-asbestos or non-asbestiform particles, sometimes called 'cleavage fragments' when they are crushed or milled from the parent rock. This has led to much confusion in interpretation of toxicologic and mechanistic data on asbestos and other fibres (see Belluso *et al.*, 2017, this volume for more details). Although six different asbestos minerals exist, the various potencies of three of these asbestos types in MMs are very different in that the amphiboles, (crocidolite > amosite > chrysotile asbestos) appear to be more potent than chrysotile asbestos (Wagner *et al.*, 1960; Mossman *et al.*, 1990; Hodgson and Darnton, 2000; Berman and Crump, 2008; Pierce *et al.*, 2016). While these estimations have been possible due to the commercial importance and documented health effects of workers using these types of asbestos, information on the disease potential of pure tremolite, actinolite and anthophyllite asbestos are more limited as they were not mined or used extensively in pure forms in industry (Finley *et al.*, 2012; Gaffney *et al.*, 2017). However, these minerals have been found in association with other asbestos types and mineral ores. While there appears to be a hierarchy of some asbestos fibres on their potency in development of MMs, it is more difficult to draw firm conclusions on their relative potencies in lung cancers, as smoking is a dominant contributing factor to this disease, and most asbestos workers historically were smokers.

Naturally occurring asbestos (NOA) is a term referring to asbestiform fibres present in rocks in the environment to distinguish them from asbestos contained in commercial asbestos-containing materials (ACM) such as asbestos cement (Agency for Toxic Substances and Disease Registry http://www.atsdr.cdc.gov/noa/) (Case *et al.*, 2011). To render appropriate precautions, it is necessary to know if other fibrous minerals (sometimes called elongated mineral particles or EMPs that can be asbestiform or non-asbestiform in their crystal formation) can be harmful alone or if found in association with asbestos fibres. Conceivably, their toxicity may enhance or attenuate that of asbestos fibres if they are dispersed in air. Limited data are currently available regarding human exposures to NOA, therefore, risk assessment also needs to take into account the effects of weathering on exposure to these fibres (Enrico Favero-Longo *et al.*, 2009). *In vitro* screening tests are valuable tools to identify and understand the molecular events that underpin pathogenicity of cancers and to compare the effects exerted by natural and synthetic analogue fibres to gain insights into the mechanisms by which these fibres induce cytotoxicity, cell injury and cell death at higher doses (Fubini *et al.*, 2010).

3. Key cell organelles involved in fibre interactions

The mammalian cell is orchestrated with many enzyme systems and organelles that control the balance between dose-related responses to various fibre types. It is clear in dose-response studies that fibres can elicit reversible repair and proliferation from fibre-induced damage at low concentrations and cell injury/death at higher concentrations of minerals. All of these reactions are a function of fibre type, dose, dimension and durability at sites of development of lung cancers and MMs. It has been concluded that asbestos fibres do not interact directly with DNA (Williams, 1979;

Eastman *et al.*, 1983; Mossman *et al.*, 1983) and are non-mutagenic in a number of cell types (Reiss *et al.*, 1982; Daniel, 1983; Denizeau *et al.*, 1985) (reviewed in Health Effects Institute-Asbestos Research, 1991), because of their inability to penetrate the nuclear membrane which acts as a barrier for protection of chromatin and DNA (Cole *et al.*, 1991; Ault *et al.*, 1995; Jensen and Watson, 1999). Moreover, whole genome sequencing of MMs has not revealed a uniformly common mutation or genetic or epigenetic change in oncogenes which drive the carcinogenic process or cancer suppressor genes that inhibit the development of cancers. However, mutations and epigenetic changes in certain cancer suppressor genes such as BAP1, NF2, P53 (Testa *et al.*, 2011) have been observed in some MMs. Because these genetic and epigenetic changes are covered in other chapters, and have been reviewed recently (IARC, 2012; Reid, 2015), they will not be discussed in detail here. Instead we focus on how asbestos fibres get into cells, trigger cell signalling pathways affect contemporary structures such as inflammasomes, and alter mitochondria.

3.1. Plasma membrane interactions and uptake of mineral fibres

After inhalation, the first site of contact of mineral fibres is the external plasma membrane of epithelial cells lining the airways and alveoli or phagocytes such as alveolar macrophages (AMs). Fibres may interact directly with the cell membrane or indirectly cause effects including lipid peroxidation and modification of membrane proteins and receptors *via* production of reactive oxygen or nitrogen species (ROS/RNS). Alternatively, these events and presentation of fibres or ROS/RNS to other cell types may occur through the accumulation and activation of AMs and other cells of the immune system including neutrophils. These dose-related membrane interactions include the stimulation of receptors on the cell membrane, initiation of kinase and other cascades of signalling events leading to cell uptake of fibres, proliferation, and transformation or cell injury and death at high concentrations. In addition to the epidermal growth factor receptor (EGFR) which is linked to activation of mitogen-activated protein kinases (MAPKs), and AKT/mTOR signalling, critical pathways in carcinogenesis (Altomare *et al.*, 2005; Heintz *et al.*, 2010), a number of other receptors including those for tumour necrosis alpha (TNF-α) are stimulated (Douvdevani *et al.*, 1996; Goldberg *et al.*, 1997). Nuclear Factor κB (NF-κB) (Janssen *et al.*, 1997) as well as chemokine receptors, cytokine receptors, Wnt signalling (Fox *et al.*, 2013) and tyrosine kinase receptors (Menges *et al.*, 2010) have been activated in both lung epithelial and mesothelial cells by asbestos fibres. Certain integrin receptors have also been linked to uptake of crocidolite fibres by cells and fibre incorporation into membrane-bound phagosomes which may merge with lysosomes (Boylan *et al.*, 1995). This latter defence mechanism may be particularly important with regard to short chrysotile (<5 μm long) fibres as the acidity of the phagolysosome is critical to leaching of Mg^{2+} from chrysotile asbestos fibres, their dissolution in the lung and pleura, and their lack of biodurability as compared to certain amphibole types of asbestos (see below).

3.2. Inflammasomes

3.2.1. Structure and function of inflammasomes

The inflammasome is a cytoplasmic complex sensing cell stress or danger signals and initiating inflammation. Priming and activation of these intracellular protein platforms trigger the maturation of pro-inflammatory chemokines and cytokines, most notably, interleukin-1β (IL-1β) and IL-18 (reviewed by Sayan and Mossman, 2016). As chronic inflammation is associated with lung cancers and MMs, mechanisms of action of inflammasomes have been elucidated recently in cells of the respiratory tract where they modulate the responses to inhaled pathogenic fibres including asbestos types and erionite.

The cytoplasmic, nucleotide binding and oligomerization domain (NOD)-like receptor (NLRP3 or NALP3) is one of a number of multi-protein pattern recognition receptors (PRRs) that are stimulated in the cytoplasm after exposure to inhaled pathogenic fibres (reviewed by Sayan and Mossman, 2016) and are thought to be critical to host resistance mechanisms and repair after cell damage. However, if inhaled materials elicit priming or activation of the NLRP3 inflammasome at high concentrations or continual exposures, the balance shifts from effective repair to chronic inflammation occurring with release of oxidants, tissue injury and release of chemokines, cytokines and angiogenic factors encouraging cancer initiation and development (Coussens and Werb, 2002). Three different models of inflammasome activation have been recognized classically and can be related to cellular effects of pathogenic fibres. The ion flux model shows that alterations in cytosolic levels of specific cations, such as K^+, Ca^{2+} and H^+, play a critical role in NLRP3 inflammasome activation (Franchi *et al.*, 2007; Petrilli *et al.*, 2007; Lee *et al.*, 2012; Murakami *et al.*, 2012; Rossol *et al.*, 2012). In a second model, release of reactive oxygen species (ROS) has been linked to NLRP3 sensing (Schroder and Tschopp, 2010). Mitochondrial ROS have been implicated in both NLRP3 priming (Bauernfeind *et al.*, 2011) and activation (Zhou *et al.*, 2011). A third model of NLRP3 activation is induced by lysosomal disruption which is partly attributable to release of cathepsin B (Mariathasan *et al.*, 2006; Martinon *et al.*, 2006; Halle *et al.*, 2008; Jin *et al.*, 2011). This model is particularly relevant to inhaled pathogenic fibres that are phagocytized by macrophages, lung epithelial cells and mesothelial cells. Upon activation of NLRP3, activation of caspase-1 leads to the maturation and secretion of the interleukins, IL-1β, IL-18 and IL-33. High mobility group box 1 protein (HMGB1), as well as some heat shock proteins that are produced after exposure of mesothelial cells or epithelial cells to crocidolite asbestos (Timblin *et al.*, 1998; Jube *et al.*, 2012) can also act as 'alarmins' or danger signals. Another important function of inflammasomes is the induction of caspase-1 dependent pyroptosis, a type of cell death characterized by both apoptotic and necrotic features (Bergsbaken *et al.*, 2009). Pyroptosis is characterized by nuclear DNA fragmentation, plasma membrane rupture, and release of additional inflammatory mediators that may function at low doses of mineral fibres to repair injury, but may contribute to chronic inflammation at high exposures (Bergsbaken *et al.*, 2009; Stegmann *et al.*, 2015).

3.2.2. Role of inflammation in lung cancers and MM

Acute inflammation is often reversible and efficacious in eliminating pathogenic fibres *via* accumulation of macrophages that can phagocytize or clear particles from sites of cancer development. In this regard, it is a fundamental tool in cell or tissue repair. However, chronic inflammation in response to high and persistent exposures can cause significant cell damage and disease, and IL-1β has been linked with the causation of cancers (Hu *et al.*, 2010; Grivennikov *et al.*, 2010; Grivennikov and Karin, 2010; Okamoto *et al.*, 2010; Hanahan and Weinberg, 2011) as well as fibrosis (Hanahan and Weinberg, 2011; Brusselle *et al.*, 2014; Rastrick and Birrell, 2014). One role of IL-1β in cancer is the recruitment of myeloid-derived suppressor cells (MDSCs) that impede natural killer (NK) cell development and function (Elkabets *et al.*, 2010). Alternatively, inflammasome-related cell death mechanisms and release of chemokines, cytokines or alarmins such as HMGB1 and tumour necrosis-alpha (TNF-α), have been linked to regression, resistance to toxicity by particle and fibres and cell growth (Luo *et al.*, 2004; Lin and Karin, 2007; Willingham *et al.*, 2009). Thus, inflammation can lead to compensatory cell proliferation and maturation of target cells of cancers as well as to immune suppression (Ben-Baruch, 2006).

3.2.3. Studies on regulation of the inflammasome by mineral fibres

Crocidolite asbestos-induced release of IL-1β is dependent upon activation of the NLRP3 inflammasome in human monocyte-derived macrophages and THP-1 cells *in vitro* and after inhalation of asbestos (Dostert *et al.*, 2008). Because monosodium urate crystals (MSU) and crystalline silica exhibited similar effects to crocidolite and were also phagocytized by these cell types *in vitro*, cell stimulation of phagocytosis and consequent release of oxidants as in activation models 2 and 3 above may be important mechanisms of inflammasome activation by particulates. The reinforcement of ROS as key players in inflammasome activation has been shown by knockdown of the NADPH subunit $p22^{Phox}$, and pre-treatment of cells with the antioxidant, N-acetylcysteine (NAC), or deferroxamine (an iron chelator). Treatment of cells with cytochalasin β blocks inflammasome activation by crocidolite asbestos and reduces IL-1β in the supernatants of these cell cultures, confirming the importance of phagocytosis and probable lysosomal instability that has been demonstrated after phagocytosis of crocidolite asbestos (Mossman *et al.*, 1977). The fact that high concentrations of long needle-like asbestos and carbon nanotubes (CNT) selectively activated the NLRP3 inflammasome despite phagocytosis of other materials including shorter, non-fibrous and tangled nanomaterials and glass beads that remained free in the cytoplasm (Palomaki *et al.*, 2011; Li *et al.*, 2012; Meunier *et al.*, 2012; Donaldson *et al.*, 2013; Cui *et al.*, 2014; Kanno *et al.*, 2015; Sun *et al.*, 2015), also suggests that frustrated phagocytosis and destabilization of lysosomes are key features of inflammasome activation in macrophages and monocytes. Cytotoxicity, production of oxidants, cathepsin B activation, $P2X_7$ receptor activation (a member of the family of purinoreceptors for extracellular ATP) and Src/Syk kinases were observed after exposures to crocidolite asbestos fibres and long needle-like CNT, but not by other materials. These observations suggest the importance of ion flux and ATP release as other models of inflammasome activation.

Although inflammasome activation by mineral fibres, particulates and microbial agents has been studied widely in cells of the immune system, it is only recently that it has been shown that epithelial and mesothelial cells have their own inflammasome systems. For example, NLRP3 inflammasomes in human mesothelial cells *in vitro* are primed and activated by crocidolite asbestos and erionite fibres, but not by glass beads (Hillegass *et al.*, 2013). It was noted that the 75×10^6 $\mu m/cm^2$ dish surface area dose of asbestos was toxic to cells, and erionite was non-toxic, yet dose-related increases in inflammasome priming and activation occurred with this potent mesotheliomagenic fibre in the absence of cytotoxicity. These studies also demonstrated that pre-treatment of crocidolite asbestos-induced mesothelial cells *in vitro* with anakinra, an IL-1 receptor antagonist, not only reduced IL-1β mRNA expression and protein secretion but also inhibited elevations of IL-6, IL-8, vascular endothelial growth factor (VEGF), and HMGB1 proteins, as well as NLRP3 mRNA expression (*i.e.* priming). These results suggest that inflammasome-related release and signalling by IL-1β and other cytokines are critical to inflammation and cancer development by long, thin mineral fibres.

3.3. Mitochondria

At non-lethal amounts of fibres, mitochondria play a role in asbestos toxicity in view of their important role in the production of ROS and RNS (Bergamini *et al.*, 2007). Moreover, toxic amounts of amphibole asbestos may lead to generation of oxidant-related dysfunction and apoptosis that may have therapeutic implications in treatment of MMs (Shukla *et al.*, 2003b; Cunniff *et al.*, 2013; Lim *et al.*, 2015). In addition, the mitochondrial and cytoplasmic antioxidant enzymes, MnSOD and CuZnSOD, which are increased in lung epithelial cells after inhalation of pathogenic minerals, and the efficacy of these naturally occurring antioxidants may determine the balance between reversible and permanent oxidative injury (Janssen *et al.*, 1990, 1992). At high concentrations of iron-rich amphibole fibres, cytochrome c from the inner mitochondrial space into the cytoplasm is a major step in the apoptotic cascade (Kamp *et al.*, 2002). This is accompanied by a decreased flux of electrons in the respiratory chain and enhanced production of ROS presumably by Complex I (Lenaz, 2001). Thus, ROS production may result not only from endogenous mechanisms or metal ions, but also from alterations in cellular metabolism. For example, Riganti *et al.* (2002) suggested that increased levels of ROS after exposure of lung epithelial cells to crocidolite asbestos resulted from direct inhibition of glucose-6-phosphate dehydrogenase activity and metabolic flux through the pentose phosphate pathway. This pathway represents the main antioxidant machinery of the cell by virtue of maintaining glutathione in its reduced form by continuous production of NADPH.

Asbestos fibres may generate ROS and RNS directly *via* a number of endogenous and exogenous mechanisms, including Haber-Weiss reactions by iron-containing fibres. Asbestos fibres can release different amounts of metal ions into buffer solutions that interfere with both mitochondrial function and integrity. In addition to their 'ideal' chemical formula, deposits of chrysotile and crocidolite can be contaminated with other minerals, elements and trace metals. The most common extraneous elements in crocidolite fibres are Ca, Co, Ni, Cr and Mn (Mossman and Churg, 1998). Al and Fe are the major impurities that vary in different chrysotile mines (Wicks and O'Hanley, 1988; Kamp and Weitzman, 1999).

Bergamini and coworkers (Bergamini *et al.*, 2007) investigated the effects of aqueous extracts of asbestos, from both natural and synthetic (SYN) fibres, on mitochondrial activities of isolated rat liver (RLM) mitochondria characterized by their bioenergetic and ultrastructural aspects (Pallotti and Lenaz, 2001). Chrysotile fibres were synthesized as a unique, morphological and structural phase under controlled hydrothermal conditions. They show a cylinder-intra-cylinder morphology and can be used as a reference samples with well defined chemical composition to investigate the factors responsible for chrysotile cytotoxicity and carcinogenicity (Falini *et al.*, 2002). To avoid mechanical mitochondrial damage due to direct contact with the asbestos fibres, a buffer in which the fibres had been incubated for at least 10 days was used to introduce these factors, *i.e.* an "asbestos buffer" (Bergamini *et al.*, 2007).

The comparative effects of aqueous extracts of natural and synthetic crocidolite asbestos fibres on mitochondrial activity showed that crocidolite fibres release substances that interfere directly with the mitochondrial cytochrome oxidase complex in solution. Crocidolite treatment caused a significant decrease in complex IV activity resulting in an increased reduced state of the respiratory chain. Moreover, the calcium ions released from these fibres induced mitochondrial swelling due to an opening of the permeability transition pore of the inner membrane leading to possible cytotoxic effects due to the release of apoptotic factors normally localized in the mitochondrial intermembrane space. In addition, crocidolite extracts enhanced the mitochondrial production of ROS. No significant biochemical effects were found with either natural or SYN chrysotile (Fig. 2a,b). However, all asbestos fibres tested induced morphological alterations, recognized using transmission electron microscopy (TEM) and morphometric analysis. In crocidolite-exposed samples, several destroyed and empty mitochondria ghosts as well as mitochondrial debris were detected by TEM, confirming the greater toxicity of these fibres. Exposure to chrysotile (both natural and SYN) induced more uniform morpho-structural impairments. In fact, mitochondria and ghosts which were destroyed completely were not found after this treatment (Fig. 2c–f). In this cell-free system, the toxic effect was related directly to the type and amount of ions released from the different asbestos fibres: crocidolite released more iron and calcium ions than either natural or SYN chrysotile, suggesting the role of these cations in mitochondrial effects of these fibres.

Panduri and co-workers (Panduri *et al.*, 2006) demonstrated that the mitochondria-regulated (intrinsic) death pathway mediated alveolar epithelial cell DNA damage and apoptosis and that amosite asbestos activated p53 promoter activity, mRNA levels, protein expression, Bax and p53 mitochondrial translocation. These results, which are detailed by Bernstein and Pavlisko (2017), Carbone and Yang (2017) and Jablonski *et al.* (2017), all this volume, suggest important interactive effects between p53 and mitochondria that may be involved in the pathogenesis of asbestos-induced pulmonary toxicity.

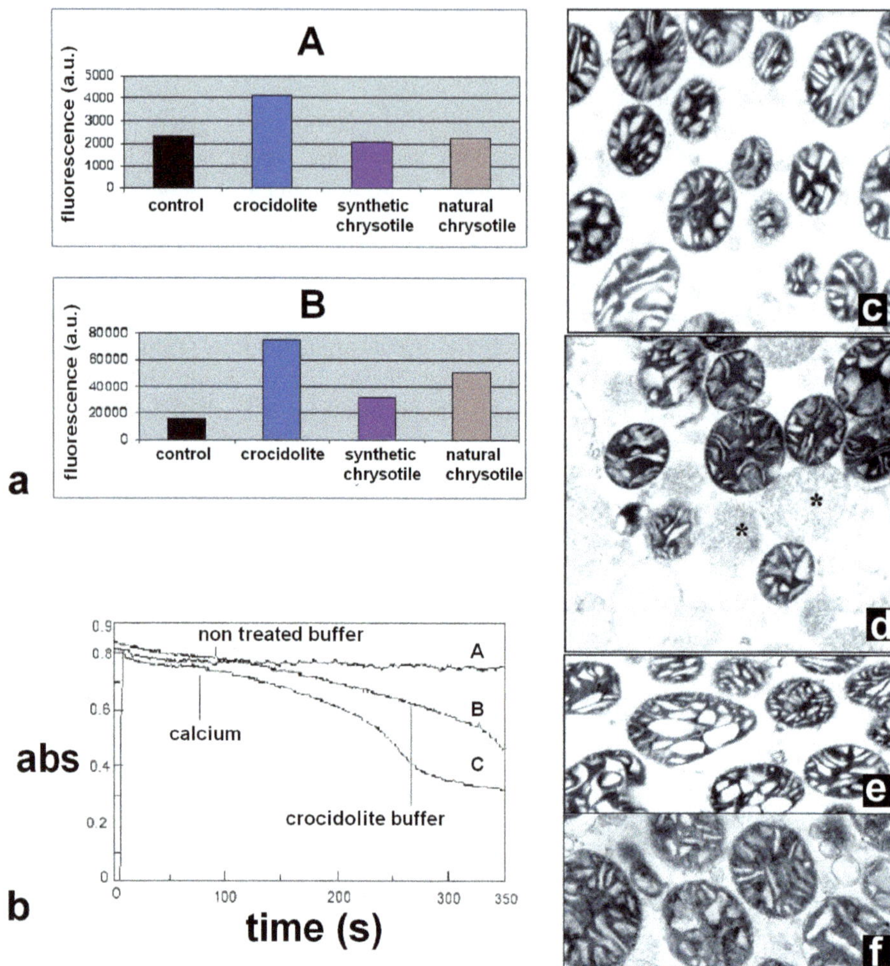

Figure 2. ROS production, opening of the permeability transition pore and morphological alterations occurring in rat liver mitochondria exposed to natural and synthetic (SYN) mineral fibres: (a) ROS production measured as increased DCF fluorescence in RLM treated with different kind of asbestos. Panel A = time 0 and Panel B = 90 min after exposure; (b) effects on permeability transition pore of untreated buffer alone (line A) or after incubation for 10 days with crocidolite asbestos on permeability transition pores in coupled RLM (line B), compared with 10 μM Ca^{2+} (line C). Transmission electron micrographs (TEM) showing ultrastructural aspects of RLM: (c) control; (d) crocidolite-treated; (e) NAT chrysotile; and (f) synthetic chrysotile-treated RLM. (× 18,500) (from Bergamini *et al.*, 2007; reprinted with the permission of *Cellular and Molecular Biology*).

4. Novel approaches for *in vitro* testing of Italian mineral fibres

Although most European communities have banned the industrial use of asbestos fibres, it persists at background levels in outdoor and indoor environments because of the presence of asbestos-containing minerals or from naturally occurring asbestos deposits. Health hazards resulting from para-occupational and environmental exposure to asbestos types including crocidolite in South Africa and Australia and tremolite asbestos in Corsica have been reported historically. However, most recently, the disease potential of non-asbestos classified fibrous minerals such as fluoro-edenite (FE) in Sicily has been reported (Comba *et al.*, 2003) (see Gualtieri, 2017; Ballirano *et al.*, 2017 and Carbone and Yang, 2017, all this volume).

4.1. Methodologies

For these reasons, FE, fibrous antigorite from the Western Italian Alps and asbestiform tremolite from Val d'Ala (NW Italian Alps) have been evaluated by *in vitro* tests using lung alveolar epithelial A549 (human pulmonary type II-like cancer cell line), mesothelial MeT-5A (SV40-immortalized human mesothelial cell line) and the mouse monocyte-macrophage J774 cell line (Fenoglio *et al.*, 2000; Pugnaloni *et al.*, 2007, 2010; Pacella *et al.*, 2012). The standard intermediate concentration of 50 μg/mL (equal to 14 $\mu g/cm^2$) was chosen for cell exposures based upon studies by others (Cardile *et al.*, 2007; Musumeci *et al.*, 2011).

The techniques used to evaluate specific bio-functional perturbations induced by Italian mineral fibres included:

- *MTT spectrophotometric assay* to evaluate cell viability as verified through mitochondrial succinate dehydrogenase enzyme activities as a consequence of mitochondrial functional state (Rapisarda *et al.*, 2015).
- *Time-lapse video microscopy* to observe cell motility over time (Mayer *et al.*, 2016).
- *Fluorescent phalloidin labelling* to detect cytoskeletal filamentous actin (F-actin) that plays important roles in several vital processes, including cell migration, endocytosis, vesicle trafficking, and cytokinesis (Ridley, 2001: Goley and Welch, 2006). Functional and morphological data obtained by time-lapse video microscopy were used to correlate cell functions and structural perturbations.
- *Immunolocalization of vascular endothelial growth factor (VEGF),* β-catenin and Cdc42. These are critical proteins in the signalling pathways in cancer development. For example, VEGF is a homodimeric glycoprotein with potent mitogenic effects on endothelial cells (Ferrara, 2000). It has an important role in development and sustaining angiogenesis, and prompts neoplastic cell growth and metastasis (Ishii *et al.*, 2004; Oppenheimer, 2006; Reinmuth *et al.*, 2008). β-catenin, a multifunctional protein crucial in cell fate (proliferation, differentiation, growth, and survival), links and

regulates cadherin-actin interactions (Fukumaru *et al.*, 2007). β-catenin activation is also considered a significant step in neoplastic transformation as it is frequently over-expressed in lung cancers (Sekine *et al.*, 2003; van Dekken *et al.*, 2007; Tennis *et al.*, 2007) and is associated with a high cell proliferative index, a key process in cancers (Hommura *et al.*, 2002). Lastly, Cdc42, a Ras-related GTP binding protein (Etienne-Manneville, 2006), plays a key role in controlling actin dynamics (Watanabe *et al.*, 2015), assembly and disassembly of F-actin in the cytoskeleton in response to extracellular signals. Moreover, it regulates the formation of membrane projections engaged in forward movements and migration polarity. Cdc42 may also be involved in lung cancer progression (Yao *et al.*, 2002), and is critical for the cell mobility and invasiveness of non-small cell lung cancers (NSCLC) (Chang *et al.*, 2015). It also modifies transcriptional programs in oncogenic transformation (Stengel and Zheng, 2011). Increased synthesis of VEGF and Cdc42 accompanies the development of malignant transformation (Ishii *et al.*, 2004; Tucci *et al.*, 2007).

4.2. Fluoro-edenite (FE)

The abnormal incidence of pleural MMs recorded among the inhabitants of Biancavilla, a town on the slopes of the Etna volcano (Catania in Sicily), in an epidemiological survey conducted from 1988 to 1997 (Gianfagna and Oberti, 2001; Grice and Ferraris, 2001; Travaglione *et al.*, 2003) has been linked to environmental exposure to FE. The mineral was extracted from the quarries in Monte Calvario where it was used widely as a building material for road paving and buildings (Bruno *et al.*, 2014). The association between FE and pleural MMs mandated more study of the amphibole, especially the salient aspects of its toxicity and role in neoplastic cell transformation.

Natural fibres collected directly from a stone quarry on M. Calvario, near Biancavilla were characterized using scanning electron microscopy and energy dispersive spectrometry (SEM-EDS) which showed the presence of edenite and FE-like fibres, the respective percent composition of which varied among bundles and even in contiguous areas. SEM analyses showed fibre bundles with fibres closely associated and parallel to the fibre axis and two types of chemical composition (Fig. 3a), both recognizable as calcic amphiboles. The original fibre widths (as fibres contained in outcropping rocks) ranged from <1 μm (mostly 0.5 μm) for the two fibre species to 2–3 μm and greater (Fig. 3b,c). Their lengths ranged from ~6 to 80 μm for edenite and to 100 μm for FE. The relative amount of the two fibre species varied between bundles. Both are respirable fibres and are found in lungs (human and sheep) and urine from people living in Biancavilla (Battaglia *et al.*, 5–7 December, 2005).

A549 and MeT-5A cells exposed for 24 h and 48 h to 50 μg/mL of FE fibres underwent functional modifications involving a number of cellular parameters (Pugnaloni *et al.*, 2007). In particular, cell viability was reduced, most markedly in mesothelial cells, and cells showed an irregular-fragmented staining of cytoskeletal

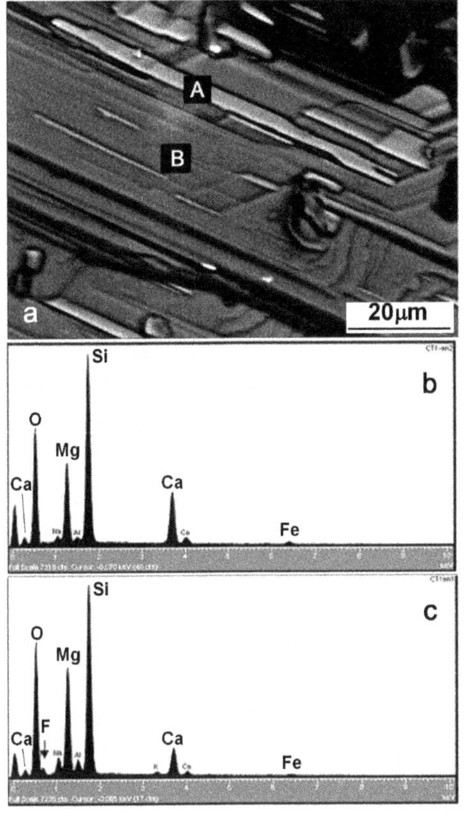

Figure 3. (a) Backscattered electron SEM image of fibre intergrowth of the two amphibole species, edenite and fluoro-edenite (FE); Dark fibrous lamellae (A) with the chemical composition of FE are larger and more numerous than light fibres (B) corresponding chemically to the edenite phase. EDS spectra of (b) edenite and (c) FE fibres. Different peak Mg and Ca height ratios compared to Si allow one to distinguish the two minerals, with F being found only in (c) (from Pugnaloni *et al.*, 2007; reprinted with the permission of *Cellular and Molecular Biology*).

actin, a possible consequence of a toxic depolymerizing activity (Boardman *et al.*, 2004) that is linked to impairment of cell motility. Exposure to FE fibres caused an almost total arrest of cell movement at 48 h in both cell lines. β-catenin and VEGF expression indicated upregulation of the Wnt-β-catenin signal transduction pathway and VEGF synthesis (Chen *et al.*, 2001; Beachy *et al.*, 2004). At 24 h, VEGF expression was increased significantly in A549 cells, whereas MeT-5A cells exhibited greater β-catenin expression (Fig. 4). At 48 h, increased VEGF and β-catenin expression was still evident in both cell lines. These studies suggest that FE fibres exert protein changes consistent with cell transformation, but impairment of cell motility and cell death over time.

FE also induced time-dependent cyclo-oxygenase (COX)-2 over-expression and prostaglandin (PGE2) increase in monocyte-macrophage J744 cells. COX-2 is an inducible enzyme with a central role in the conversion of arachidonic acid to a number of metabolites including prostaglandins (PGs) which have many roles in various diseases, including COX-2 participates in inflammation, cell proliferation and differentiation (Hla and Neilson, 1992). Its activity which is low in normal conditions, increases in fibroblasts, vascular endothelial cells, macrophages, neurons and other cells in response to various stimuli including cytokines, carcinogens, serum, hormones, and mitogens. COX-2 expression is upregulated in a variety of malignancies, including lung cancers (Xiao *et al.*, 2015) and may promote multiple events in the development of MMs (Cardillo *et al.*, 2005). Although recent studies show that COX-2 inhibitors are ineffectual in lowering the incidence of MMs in human cohorts and a mouse model of

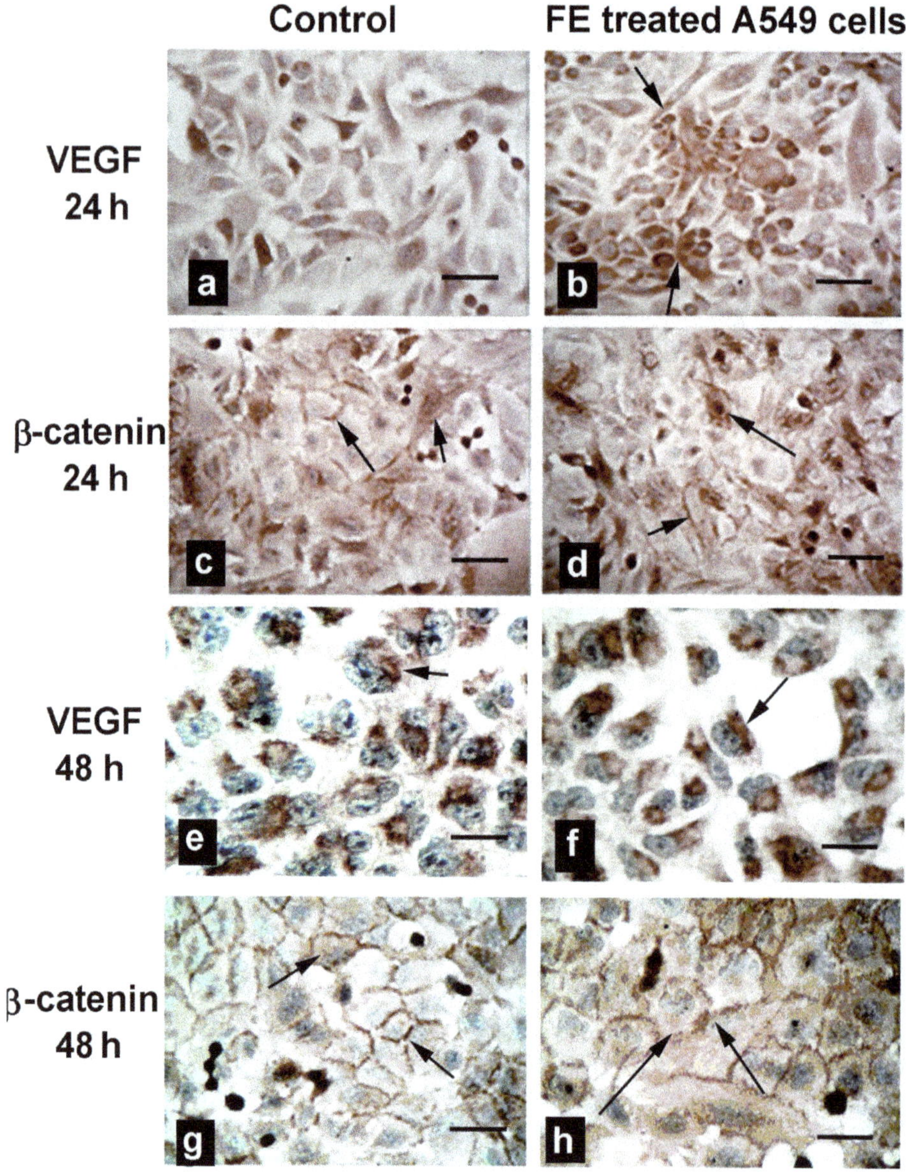

MM (Robinson *et al.*, 2014), they enhance the cytotoxic effects of chemotherapeutic agents in human MM cell lines (O'Kane *et al.*, 2010).

In summary, data using FE fibres show that they are capable of inducing *in vitro* functional modifications which may be involved in the development of cancer or cytotoxicity and eventually, cell death.

4.3. Antigorite

In the Italian Western Alps, as in many geographical locations worldwide, asbestos occurs in serpentine rocks. In addition to chrysotile asbestos, there are other species of mineral fibres found in veins of serpentine rocks that are more abundant than asbestos fibres (Belluso *et al.*, 28–30 June 2004). These include non-asbestiform antigorite that can be present with a fibrous crystal habit (Belluso *et al.*, 28–30 June 2004; Cardile *et al.*, 2007). This mineral also has structural and chemical analogies similar to those of chrysotile and lizardite. Thus, these three main minerals constitute the serpentine mineralogical group (Wicks and O'Hanley, 1988).

In contrast to chrysotile asbestos, exposure to fibrous, non-asbestiform antigorite has received little attention as a potential risk factor for the development and maintenance of chronic inflammation and neoplastic transformation (Wozniak *et al.*, 1988). For studies described below, serpentine rock samples bearing fibrous antigorite were collected from the Western Italian Alps (Belluso and Ferraris, 1991). The antigorite mineral exhibited a fibrous morphology when seen either in the rock fragment or by SEM (Fig. 5a,b) and a chemical-make up consistent with serpentine minerals (Fig. 5c). The single crystals showed a fibrous morphology also using TEM at the sub-micrometric scale (Fig. 5d) and high crystallinity by EDS-TEM (Fig. 5e). Size distributions from TEM investigations showed fibres from 14 to 20 μm long and from 0.08 to 0.4 μm wide. The chemical composition obtained using EDS-TEM (Fig. 5c) also revealed single fibres with the average formula $(Mg_{11.05}Fe^{2+}_{0.53}Al_{0.21})_{\Sigma 11.79}Si_{8.05}O_{20}(OH)_{16}$.

Fibrous antigorite exposure to A549 cells showed lower levels of viable cells detected by the MTT assay at 24 h and numerical cell density that became more marked

---

Figure 4 (facing page). Immunoperoxidase staining of VEGF and β-catenin expression in control and FE-treated A549 cell cultures at 24 h (a–d) and 48 h (e–h). (a) Control cultures at 24 h show moderate VEGF expression both in perinuclear zones and in cytoplasmic extensions (bar = 35 μm); (b) FE-treated cultures at 24 h show stronger VEGF expression with broad cytoplasmic diffusion (arrows) compared with control A549 cells (bar = 28 μm); (c) control cultures at 24 h exhibit focal β-catenin expression on the plasma membrane at intercellular contacts (arrows) and irregular cytoplasmic expression (arrows) (bar = 28 μm); (d) FE-treated A549 cell cultures at 24 h indicate increased β-catenin expression (arrows) compared to control A549 cells (bar = 28 μm); (e) control cells at 48 h indicate moderate expression of VEGF mainly at the perinuclear level (arrows), (bar = 12 μm); (f) FE-treated cells at 48 h show VEGF expression (arrows) is similar to control cultures (bar = 14 μm); (g) control cells at 48 h indicate that β-catenin expression (arrows) is evident on the cell membrane and faint in cytoplasm (bar = 14 μm); (h) FE-exposed A549 cells at 48 h show β-catenin positivity (arrows) similar to that noted in control cultures even though it is less faint in the cytoplasm (bar = 12 μm) (from Pugnaloni *et al.*, 2007; reprinted with the permission of *Cellular and Molecular Biology*).

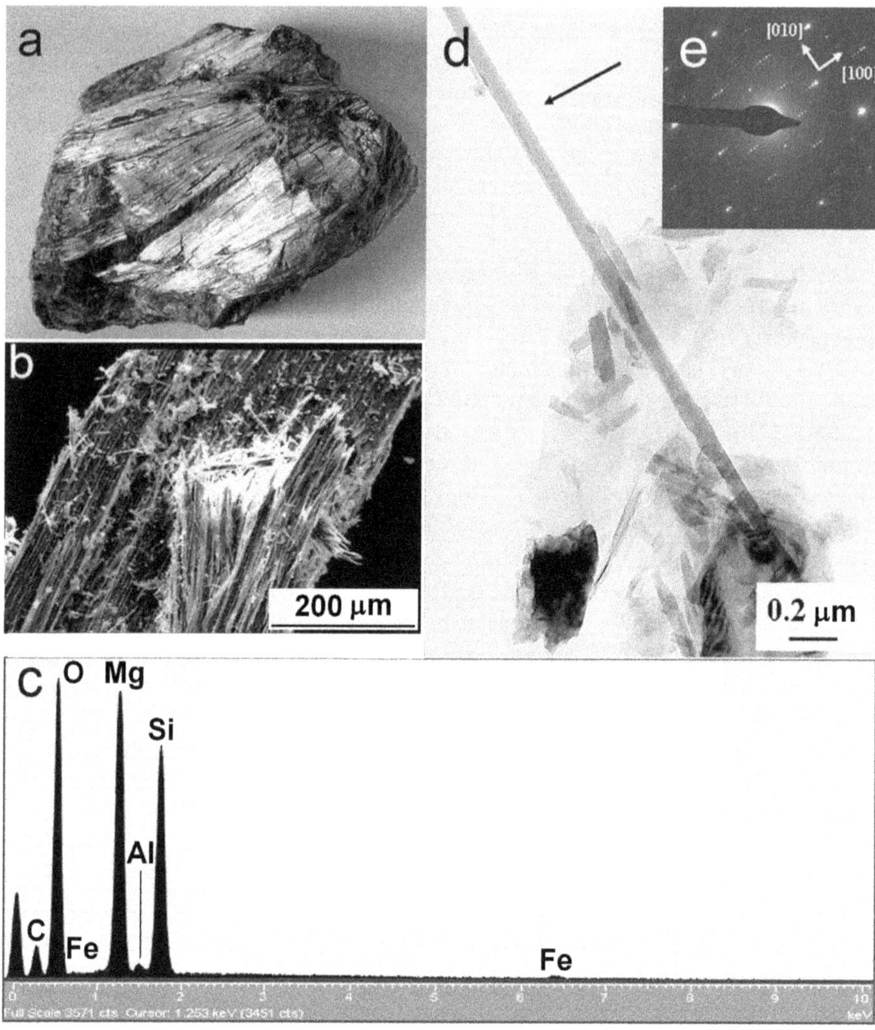

Figure 5. Fibres of antigorite: (a) on a serpentinite rock fragment; (b) secondary electron SEM image; (c) energy dispersive spectrum EDS-SEM; (d) TEM image of an antigorite single fibre observed in the plane of [010] fibre axis; and (e) electron diffraction of (d) along the [001] zone axis (from Pugnaloni *et al.*, 2010; reprinted with permission from *Acta Histochemica*).

at 48 h (Pugnaloni *et al.*, 2010). These findings correlated with morphometric data showing a greater incidence of necrotic and apoptotic cells over time (Fig. 6). Cell degeneration and necrotic death were frequently observed and characterized by cell debris at the external plasma membrane and mitochondrial alterations. Finger-like, irregular microvilli, decreased cell motility, swollen nuclei, karyolysis, cytosolic vacuolization, and membrane disruption with loss of cell contacts were detected in advanced necrosis. Fluorescent F-actin in exposed cells appeared irregular in the cytoplasm, assuming a more granular distribution than in the respective control cultures with loss of density at the level of the plasma membrane suggesting altered actin polymerization dynamics and impairment of cell motility.

VEGF, β-catenin and Cdc42 proteins showed increased labelling in A549 cells that was more evident in less confluent areas in fibrous antigorite-treated cultures at 24 h, although numbers of positive cells were not significantly different from untreated controls. These results at 24 h resembled cytoplasmic β-catenin immunelabelling without nuclear localization as observed in silica-induced rat lung dysplasias and carcinomas (Alvarez *et al.*, 2004; Blanco *et al.*, 2004).

Loss of β-catenin expression in membrane and cell junctions further supported the hypothesis that changes in cell-cell adhesion molecules are an early event in carcinogenic fibre interactions. Immunolabelling of Cdc-42, was also slightly increased at 24 h in the presence of fibrous antigorite (Fig. 7). However, at 48 h, β-catenin, Cdc42 and VEGF were almost undetectable in fibrous antigorite-exposed cultures, observations correlating with observations that the cells were unhealthy (Alvarez *et al.*, 2004; Blanco *et al.*, 2004).

In conclusion the significant morphological and functional changes supporting the cytotoxic effects exerted *in vitro* by fibrous antigorite should be evaluated in animal models to determine if these progress to cell transformation.

4.4. Tremolite

Pure asbestos tremolite and actinolite have not been used commercially, but they occur in metamorphic rocks used as construction materials for roads, railway beds, ornamental rocks, *etc.* In Italy, NOA deposits of tremolite, actinolite and chrysotile are widespread and abundant (as in the Piedmont region in NW Italy) whereas NOA of anthophyllite is less abundant and less common. Finally NOA of grunerite (the non-asbestos form of amosite) and crocidolite asbestos are absent from these areas (Belluso *et al.*, 1994).

Crocidolite and amosite were found to be far more potent than chrysotile asbestos in causing MMs because of their widespread occupational use, although epidemiological data on whether individuals with MMs are exposed occupationally to tremolite asbestiform or non-asbestiform fibres may be confusing without input and characterization by mineralogists (McDonald, 2010; Rudd, 2010; Finley *et al.*, 2012). In the Upper Susa Valley (Piedmont Region, NW, Italy) where NOA of asbestiform and non-asbestiform amphibole outcrops of rocks of tremolite and actinolite are still exploited, pleural and peritoneal malignancies were found to occur more frequently (Mirabelli and Cadum, 2002). The NOA tremolite tested *in vitro* (Pugnaloni *et al.*, 2013) come from an area close to the Susa Valley where specific epidemiological studies have not been reported.

Using *in vitro* assays, it was possible to compare cellular responses to natural tremolite (NAT) fibres (which have a complex chemical and crystalline composition) with synthetic fibres (characterized by more controlled chemical composition). NAT

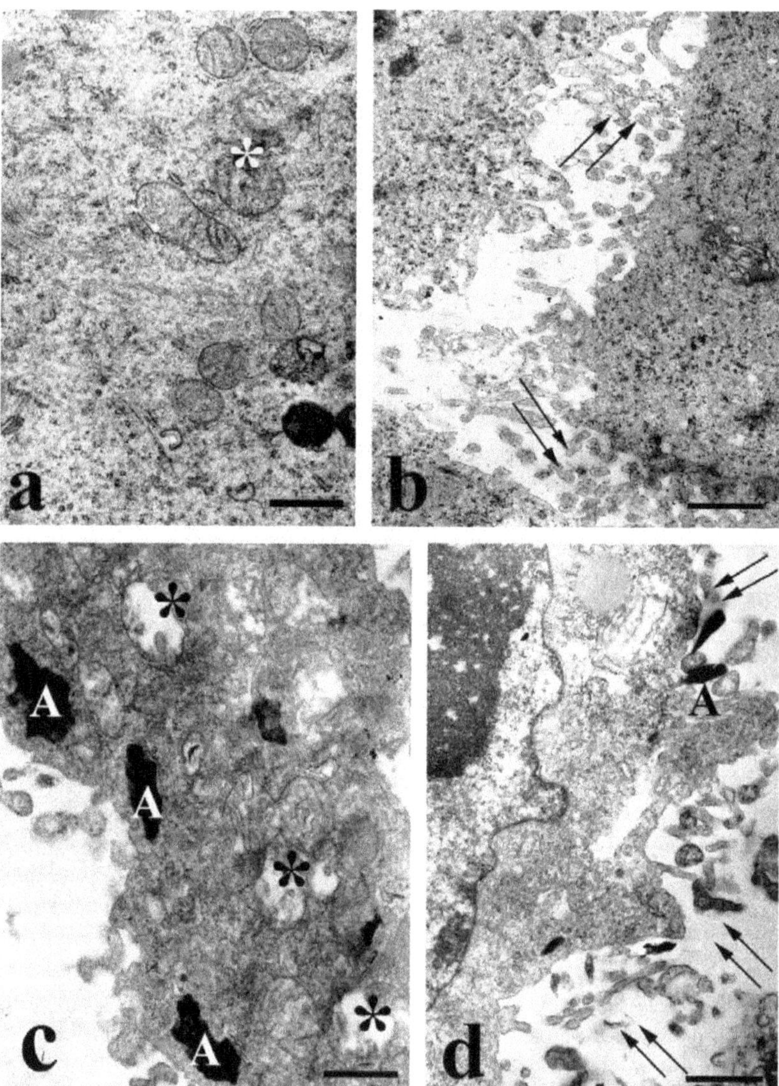

Figure 6 (above and facing page). TEM showing control (a, b) and treated (c, d) A549 cells at 24 h: (a) cytoplasmic mitochondria (*) showing membranous cristae (bar: 0.50 μm); (b) finger-like microvilli protrusions from cell membrane (arrows) (bar: 0.70 μm); (c) keratinocyte showing antigorite particles (A) after phagocytosis and swollen mitochondria with loss of matrical electron density (*) (bar: 0.70 μm); (d) finger-like microvilli (arrows) showing irregular shapes compared to controls (bar: 0.77 μm). (e) Control (e) and treated cells (f) at 48 h. Note that untreated control cells show normal cytoplasm rich in mitochondria (M) and cytoskeletal filaments (F) (bar: 0.85 μm) whereas the antigorite treated cell shows necrotic degeneration with a swollen nucleus, cytosolic vacuolization and membrane disruption. Antigorite = A. (bar: 1 μm) (from Pugnaloni *et al.*, 2010; reprinted with permission from *Acta Histochemica*).

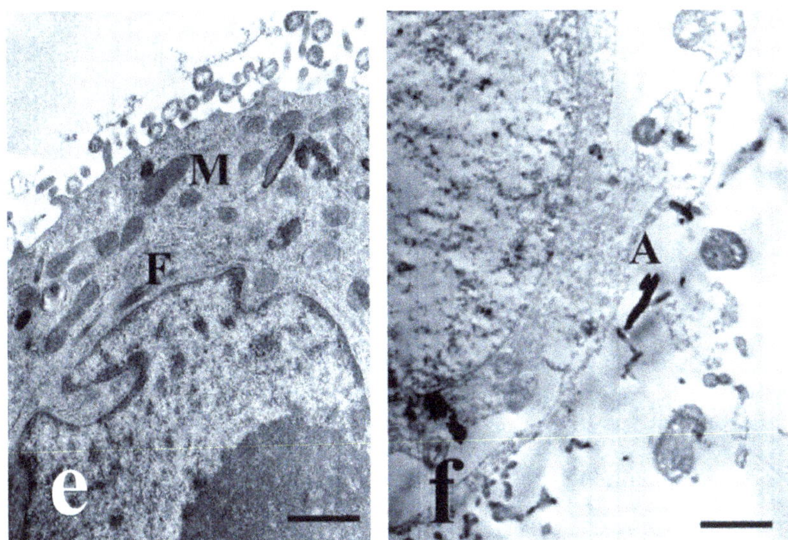

fibres were obtained in some serpentine rocks near Brachiello in Val d'Ala (NW Italian Alps) (Belluso *et al.*, 1994). Synthetic tremolite fibres (SAT) were prepared using standard cold-seal hydrothermal techniques (Bloise *et al.*, 2008). SAT was iron-free. SEM examination of NAT showed aggregates of long, rigid fibres the diameters of which ranged from 1 to 1.5 μm, and thicker fibre bundles measuring 150–200 μm in diameter (Fig. 8a,b). TEM and EDS of single fibres showed high crystallinity, the absence of structural defects, and a fibre axis running in the [001] direction, the latter feature being consistent with the amphibole structure of the mineral (Fig. 8c–e). NAT fibres had diameters ranging from 1 to 1.5 μm with a few thicker crystals ranging in diameter from 15 to 20 μm, therefore most were respirable. The formula of NAT fibres was: $Ca_{1.66}(Mg_{4.62}Fe^{2}_{0.27}Al_{0.18}Mn_{0.02})_{\Sigma 5.09}Si_{8.00}O_{22}(OH)_2$, consistent with the ideal theoretical formula of tremolite.

In contrast, SAT displayed traces of quartz and TEM observations showed thin, rigid fibres with an average length of 10 μm. Their diameters ranged from 0.04 to 0.68 μm with fibres measuring <0.25 μm in diameter accounting for 94% of all fibres. From a dimensional point of view, all the fibres could be classified as respirable fibres and did not appear as cleavage fragments (Addison and McConnell, 2008). Their chemical formula was $Ca_{1.71}Mg_{5.24}Si_{8.01}O_{22}(OH)_2$.

From the mineralogical studies, it was estimated that the total superficial areas of SAT fibres was ~10 times higher than NAT total superficial areas, in a ratio almost equal to 1 NAT/100 SAT fibres per cm^2.

Cytotoxic effects appeared at 24 h and 48 h in A549 cells, and decreased viability (MTT assay) was more evident with SAT. Significant decreases in healthy and mitotic cells were accompanied by increments in necrotic cells with broader variation after SAT exposures. Loss of cell motility was also evident in both NAT and SAT-treated

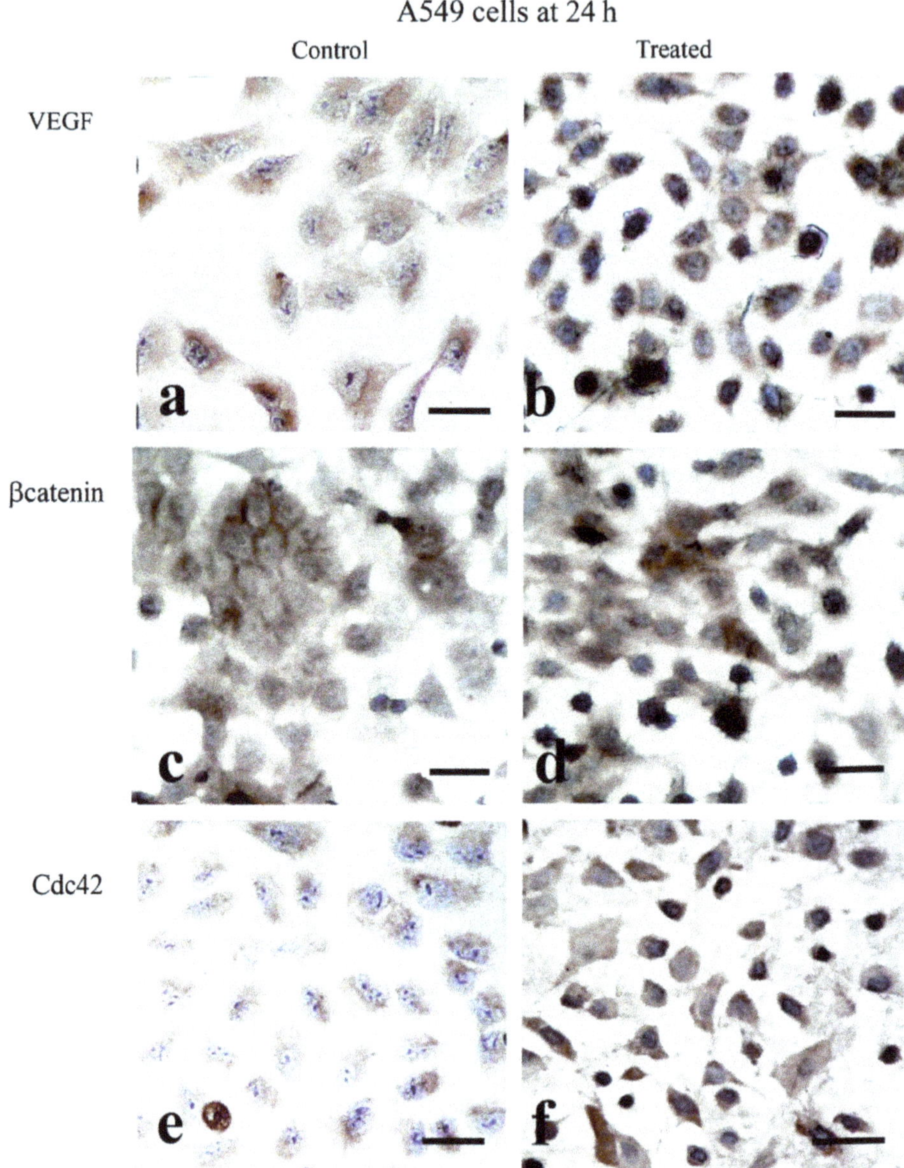

Figure. 7. Immunolocalization of VEGF, β-catenin and Cdc42 in control (a, c, e) and antigorite-treated A549 cells at 24 h (b, d, f). (bar in a = 10 μm; bar in b = 14 μm; bar in c = 16 μm; bar in d = 16 μm; bar in e = 12 μm; bar in f = 16 μm) (from Pugnaloni *et al.*, 2010; reprinted with permission from *Acta Histochemica*).

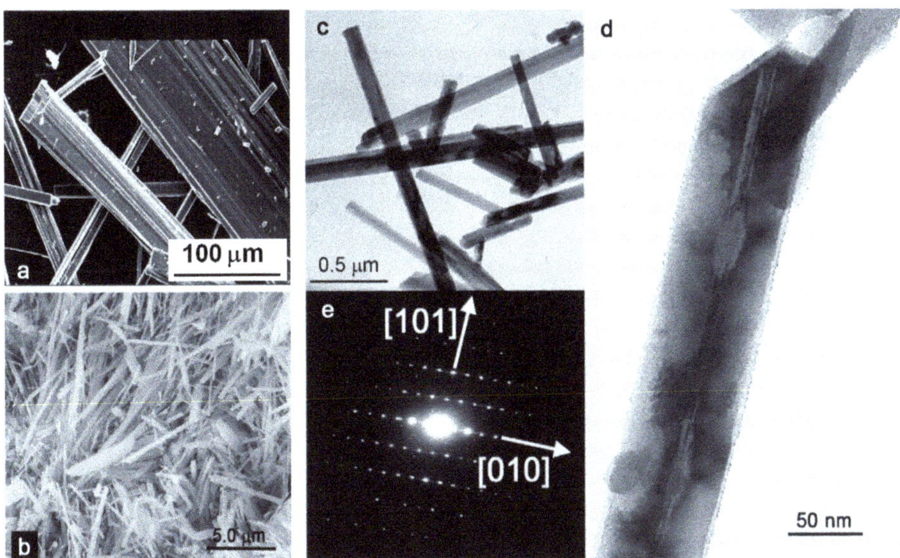

Figure 8. (a) Secondary electron SEM image of natural asbestos tremolite (NAT) and (b) secondary electron SEM image of synthetic asbestos tremolite (SAT); (c) TEM image of several SAT fibres; (d) high magnification; and (e) SAED pattern of a single SAT fibre (from Pugnaloni *et al.*, 2013; reprinted with permission from *Acta Histochemica*).

cells, with greater reductions at 48 h. After both NAT and SAT treatments, F-actin was no longer detected as cytoskeletal filaments with phalloidin staining but was irregularly structured with more focal distribution than in control cultures, suggesting impaired cell function. Abnormal F-actin organization was more strongly affected by SAT fibres. Cells that survived NAT treatment at 24 h were isolated and still engaged in mitotic activity despite the irregular organization of their intracellular F-actin (Fig. 9).

At 24 h, VEGF expression increased in both NAT- and SAT-exposed cells in terms of both numbers of positive cells and of staining intensity- the latter was stronger in NAT cells. Cdc42 staining was weak and the proportion of positive cells and staining intensity decreased after exposure to both fibre preparations (Fig. 10). β-catenin cytoplasmic and membrane expression decreased significantly after both exposures. Cell membrane expression was no longer detected after SAT treatment. At 48 h, VEGF and Cdc42 expression showed the same trends as at 24 h with both exposures. Loss of cell junctions in NAT and SAT cultures was associated with decreased expression of Cdc42 and membrane β-catenin that may destabilize these cell connections.

SAT is a well defined compound obtained from laboratory synthesis with smaller and more numerous fibres per unit weight, thus providing greater contact and uptake by cells. Moreover, because SAT fibres are iron-free, they can be used as standard mineral fibres in future investigations for achieving a better understanding of the role of iron in asbestos-induced toxicity and disease. As postulated by some authors (Dogan *et al.*, 2008) when comparing erionite or asbestos types (Mossman *et al.*, 1990), surface area

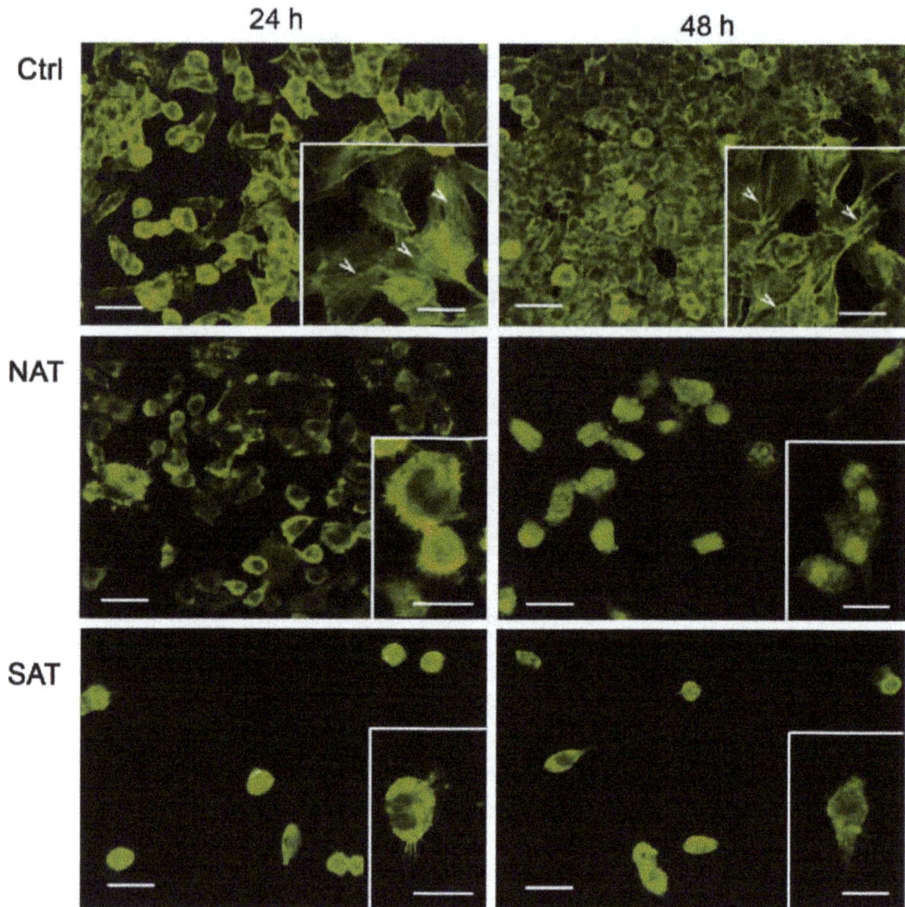

Figure 9. Fluorescence imaging of phalloidin staining for detection of F-actin at 24 h and 48 h. Control A549 cultures at 24 h and 48 h showing filamentous cytoskeletal actin (arrowheads). In NAT and SAT exposed cells, actin staining was irregular in the cytoplasm with a more granular, focal distribution. Insets show increased magnifications of cells from the same cultures. Scale bars in left corners = 20 μm; insert scale bars = 10 μm (from Pugnaloni *et al.*, 2013; reprinted with permission from *Acta Histochemica*).

or surface area-to-volume ratio may prove more relevant as dosimetry parameters as lower NAT cytotoxicity may be a function of fewer numbers of fibres per unit weight dose, providing a reduced surface area for cell uptake. These are questions that should be pursued in future studies.

4.5. Properties of mineral fibres linked to toxicity and cancers

It is clear that individual mineral fibres differ in their crystal patterns and habit, chemical composition, surface features, morphology, sizes and biodurability in the

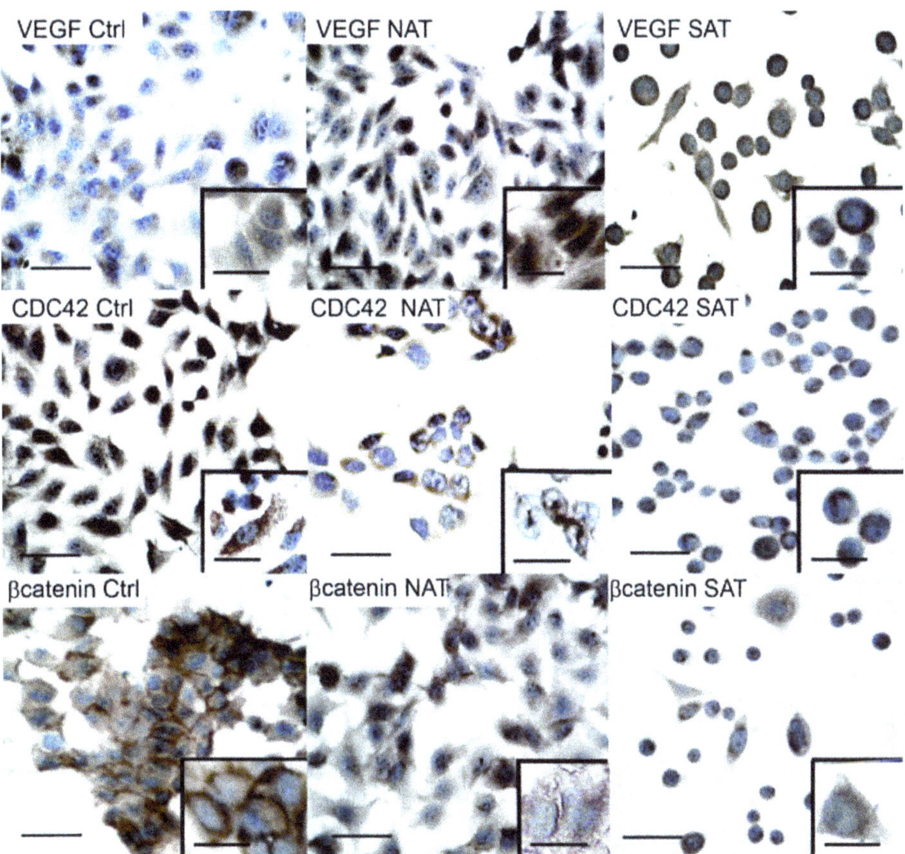

Figure 10. Immunohistochemical expression of VEGF, Cdc42 and β-catenin in control and tremolite-treated cultures at 24 h. VEGF and Cdc42 showed cytoplasmic staining. VEGF expression was increased in NAT and SAT cells both in terms of numbers of positive cells and staining intensity. Cdc42 staining was weak, and the proportion of positive cells and staining intensity decreased after both NAT and SAT treatments. β-catenin was detected in the cytoplasm and on the cell membrane of control cells. Cytoplasmic expression and staining intensity decreased significantly after both NAT and SAT treatments. After SAT treatment, cell membrane expression was no longer detected (scale bars on left = 25 μm). Inserts in boxes show magnification of the labelled cells (scale bars = 10 μm) (from Pugnaloni *et al.*, 2013; reprinted with permission from *Acta Histochemica*).

lung and pleura. Moreover, they may be modified in different natural outcrops or mining sites by association with other minerals, metals/elements and organic contaminants. All of these factors come into play when assessing dose-response relationships, cell damage or cytotoxicity, and carcinogenic potential (see Gualtieri *et al.*, 2017, this volume). We emphasize here both historical and new approaches to narrow in on the features and properties of materials that may be linked to human health effects.

4.5.1. Fibre size/dimensions

In vitro mechanistic studies and experimental animal studies involving inhalation, intra-tracheal or intra-pleural/peritoneal injection of fibres (reviewed in Health Effects Institute-Asbestos Research, 1991; Donaldson and Tran, 2002; Donaldson *et al.*, 2010; Lippmann, 2014; Roggli, 2015) have demonstrated that fibres >5 μm in length are associated with the development of asbestosis, lung cancers and both pleural and peritoneal MMs. Fibre analyses of human lung tissue samples provide further support for the increased pathogenicity of long amphibole fibres that are more durable in the lung than is chrysotile asbestos (reviewed in Roggli, 2015), and least durable, synthetic fibre glasses that have not been linked to MMs historically in epidemiological studies. Most noteworthy in the causation of MMs are studies showing that long (>5 μm) needle-like fibres of different types in the pleura (Murphy *et al.*, 2011; Schinwald *et al.*, 2012) and peritoneum (Moalli *et al.*, 1987) are not able to be cleared effectively through lymphatic channels because of their inability to go through stomata (pores connected to lymphatic channels) between parietal mesothelial cells, thus are trapped at sites of MM development. For many years, it has been shown in a number of studies that long, thin amphibole fibres are more cytotoxic, inflammatory and generate more ROS than shorter fibres due, in part, to their inability to be completely phagocytized by AMs (Woodworth *et al.*, 1983a; Dostert *et al.*, 2008; Murphy *et al.*, 2012). Short fibres are also cleared more effectively by these phagocytes and other cell types in the lung and pleura. These findings are in line with data described above from *in vitro* assays of Italian SAT and NAT (Pacella *et al.*, 2012).

Although primarily short chrysotile fibres can be found in the parietal pleura in MMs and lungs, leading to opinions that short fibres should not be ignored in lung-related pathogenicity (Suzuki and Yuen, 2002; Dodson *et al.*, 2003; Suzuki *et al.*, 2005; Lemen, 2006), there is currently no convincing scientific evidence that short fibres of 5 μm or less in length display carcinogenic effects *in vitro* or *in vivo*, even noting that the percentage of short fibres in pleural samples is greater than in lung tissue (reviewed by Roggli, 2015). It is most likely that these short chrysotile fibres are the result of breakdown or dissolution of longer, inhaled chrysotile fibres whereas long amphibole fibres concentrated at "black spots" in the pleura to gain access to effective lymphatic clearance (Boutin *et al.*, 1996).

4.5.2. Cations (Ca^{2+} and Mg^{2+})

The effect of Ca^{2+}, Mg^{2+} and other ions either as constitutional components or contaminants of asbestos fibres cannot be ignored (Gold *et al.*, 1997). Iron accumulation in the lung is induced by redox active, iron-containing amphibole asbestos fibres, and ferruginous 'asbestos' bodies form on these long thin fibres (see Capella *et al.*, 2017, this volume). In lung sections of mice after inhalation of crocidolite asbestos, both iron and calcium have been found on inhaled fibres in a process that occurs concomitantly with inflammatory and hyperplastic changes (Pascolo *et al.*, 2016). Microcalcifications have also been observed after brief exposures to chrysotile asbestos (Brody and Hill, 1982). It was also found (Bergamini *et al.*, 2007) that natural and SYN chrysotile release significant amounts of magnesium which is known to inhibit calcium uptake by exerting a direct effect on the uniporter.

The extent of magnesium leaching from chrysotile fibres is of interest because several researchers have reported that magnesium-depleted chrysotile fibres are less toxic and produce fewer mesothelial cancers in animal studies than normal chrysotile fibres (Monchaux *et al.*, 1981; Jaurand *et al.*, 1987; Langer and Nolan, 1987). Morgan *et al.* (1977) reported that chrysotile fibres 90% depleted in magnesium produced a statistically lower incidence of mesothelial cancers than untreated fibres when injected into the pleura of animals whereas chrysotile depleted of 50% of their magnesium content did not.

Leaching of magnesium from chrysotile fibres under acidic conditions is well known and has been demonstrated using mineral or organic acids, in *in vitro* studies using macrophages, and in rodent lungs where chrysotile fibres dissolve and were converted to smaller fibrils (Hargreaves and Taylor, 1946; Morgan *et al.*, 1977; Monchaux *et al.*, 1981; Spurny *et al.*, 1983; Jaurand *et al.*, 1987; Helsen *et al.*, 1989). Magnesium leaching is probably facilitated by water and carbon dioxide (CO_2) combining to form dilute carbonic acid (Gronow, 1987; Riché *et al.*, 2006). Some authors noted that leaching of magnesium from chrysotile fibres results in a 'porous silica skeleton' (Spurny *et al.*, 1983; Langer and Nolan, 1987) or degraded crystalline lattice (Luys *et al.*, 1982), possibly explaining morphological alterations and loss of its tubular structure.

The leaching of magnesium from chrysotile fibres also occurs under natural weathering conditions or natural acidic conditions created by lichens colonizing an asbestos cement roof (Favero-Longo *et al.*, 2005). Thus, magnesium depletion in leaching of chrysotile is relevant to the detoxification of asbestos according to priorities related to the specific products of the recovery treatment (Valouma *et al.*, 2016).

4.5.3. Iron and iron metabolism

Cytotoxic, genotoxic and proliferative effects of amphibole asbestos have been attributed to their content and the availability of iron which induce both generation of ROS/RNS (reviewed by Mossman and Marsh, 1989; Shukla *et al.*, 2003a; Srivastava *et al.*, 2010) and alter iron metabolism. Links between the importance of iron to these parameters of disease have also been verified after chelation of iron from fibres. As shown above in studies by Bergamini *et al.* (2007), iron ions are associated with mitochondrial damage and impaired cell motility as well as overall toxicity as in the case of iron-rich tremolite fibres from the Italian Western Alps (Rinaudo *et al.*, 2004; Turci *et al.*, 2009). Moreover, iron is linked to the induction of apoptosis (Srivastava *et al.*, 2010) and risk of cancer progression (Luanpitpong *et al.*, 2010). Most convincing are the observations that the iron-containing amphibole asbestos types, crocidolite and amosite, are more carcinogenic in the development of MMs as reviewed above (Toyokuni, 2009). It is clear that these amphibole fibres have greater iron contents and more reactive surface iron than iron-free chrysotile or chrysotile deposits with contamination predominantly by non-reactive iron oxides (Fe_2O_3). Cellular studies have shown that increased iron transport and intracellular iron mobilization occur after exposure to iron-containing asbestos fibres (Carratala *et al.*, 1992).

Asbestos fibres of amphiboles with large iron contents, such as crocidolite and amosite, that are associated with the development of human MMs are potent ROS producers (Mossman *et al.*, 1990; Shukla *et al.*, 2003a). Toxic concentrations of crocidolite asbestos fibres generate ROS *via* iron reactivity on the fibre surface and/or frustrated phagocytosis of long fibres. These ROS/RNS oxidize an intracellular pool of the antioxidant Trx-1, which results in release of thioredoxin interacting protein (TXNIP) and subsequent activation of inflammasomes in human mesothelial cells (Thompson *et al.*, 2014). Thus, oxidation of Trx-1 and disassociation of TXN1P are implicated as mechanisms of asbestos-induced inflammasome activation. Deferoxamine, an iron-chelating agent, inhibits asbestos-induced toxicity in a number of cell models (Chao and Aust, 1994; Thompson *et al.*, 2014) as well as IL-1β secretion (Dostert *et al.*, 2008). Because iron chelation does not inhibit IL-1β secretion following exposure to MSU, this mechanism appears to be specific to reactive iron-containing fibres (Dostert *et al.*, 2008).

4.5.4. Generation of reactive oxygen and nitrogen species (ROS/RNS)

Generation of ROS by asbestos and their role in MM have received much attention, based upon the hypothesis that they can be a 'double-edged sword' in the development of MM (Benedetti *et al.*, 2015), and antioxidants have been proposed as preventive or therapeutic tools in the treatment of this insidious disease. In addition, oxidants may modulate inflammation and epigenetic events that are critical in the development of MMs (IARC, 2012).

In addition to producing oxidants or antioxidants and stimulating inflammasomes after exposures to pathogenic minerals, epithelial cells in normal human lung express iNOS (Donnelly and Barnes, 2002) that can be induced at sites of inflammation by cytokines and lipopolysaccharide (LPS), especially at sites of chronic inflammation. Oxidant-generating enzymes that like NADPH oxidase (Dostert *et al.*, 2008) and myeloperoxidase (Haegens *et al.*, 2005) can also be activated by asbestos in inflammatory cells to produce ROS and RNS which play important roles in the carcinogenic process as they contribute to inflammation and proliferation as well as genomic instability (Murata *et al.*, 2012). *Via* its catalytic action, iNOS can produce and release $NO^{\cdot}$ (Kamp and Weitzman, 1999) that regulates normal airway function as well as airway disease such as asthma, where levels of exhaled $NO^{\cdot}$ are elevated. $NO^{\cdot}$ is a free radical which reacts readily with ROS to form more deleterious species and $NO^{\cdot}$ itself is able to activate transcription factors including NF-κB, STAT and HIF-1α that regulate iNOS expression (Hiraku *et al.*, 2010). *In vitro* studies have reported iNOS expression in human airway epithelial cells, including bronchial BEAS-2B (Felley-Bosco *et al.*, 1994) cells and alveolar A549 cells (Ji *et al.*, 2011). Both crocidolite and chrysotile asbestos fibres up regulate iNOS activity and $NO^{\cdot}$ production in lung epithelial cells and alveolar macrophages, although the significance of these changes in the development of lung cancers and MMs is unclear (Quinlan *et al.*, 1998; Zeidler and Castranova, 2004; Hiraku *et al.*, 2010).

4.6. Thermally transformed asbestos-containing materials

Asbestos cement, the main asbestos-containing material (ACM) manufactured in Italy, is a health hazard the elimination of which is a priority. As shown by Bloise *et al.* (2017, this volume), asbestos fibres can be transformed into potentially non-hazardous silicates by high-temperature treatment *via* complete solid-state transformation (Gualtieri *et al.*, 2008b). The product of the thermal transformation of cement asbestos at 1200°C, named KRY·AS (from Greek: krústallos and ásbeston, crystals from asbestos), is a mixture of crystalline silicate phases the composition of which is very similar to clinker (Gualtieri *et al.*, 2008a). Giacobbe *et al.* (2010) determined that iron in KRY·AS is exclusively in the ferric state in the tetrahedral framework. Ferric iron is probably inside a crystalline phase in four-fold coordination, and this crystalline phase could be Fe-bearing åkermanite-gehlenite (melilite). A recent study of the crystal chemistry of asbestos cement and its high temperature products (Viani and Gualtieri, 2013) revealed that the secondary material produced is composed mainly of SiO_2 and CaO with minor MgO, Al_2O_3 and Fe_2O_3, with chemical and mineralogical features similar to that of a Mg-rich clinker.

SEM images of raw cement asbestos (RCA), chrysotile fibres, and small crystals of cement phases are easily distinguished by their different crystal habits: fibrous asbestiform for asbestos and granular for cement. KRY·AS at a comparable magnification to RCA shows complete pseudomorphic recrystallization of the fibres in the asbestos cement matrix – individual asbestos fibres no longer exist and are not detected in KRY·AS even at higher magnifications.

An *in vitro* experimental design (Giantomassi *et al.*, 2010) was defined to assess the effects of different concentrations of raw cement asbestos (RCA) and KRY·AS. In brief, their effects on A549 cells up to 96 h after exposures were assessed by monitoring cell growth rate and viability, ultrastructural features, ROS production and induction of apoptosis at different time points (Dopp *et al.*, 2005). More severe cytotoxic damage was detected invariably after RCA contact than after KRY·AS treatment at each time point. Dose-dependent cytotoxicity, assayed as relative cell viability by the MTT test at concentrations of 10, 50 and 100 μg/mL resulted in reduced cell viability at all time points – the greatest decrements occurring with RCA. The cytotoxic effects of RCA were evident at all concentrations at 24 h and thereafter. At 72 h, surviving cells still proliferated even though cell numbers had not recovered completely. At 96 h, viability decreased in all treated cultures, the lowest value being found in RCA-exposed cells.

Morphological features from semi-thin sections of control A549 cells showed largely intact monolayers with a typical cuboidal morphology and preservation of most cell junctions (Davoren *et al.*, 2007). Mitotic cells in prophase and metaphase were particularly evident at 48 h and 72 h. In cultures exposed to 50 μg/mL of either mineral, solitary cells were observed often at 48 h and occurred after RCA exposure with necrotic cells and cell debris. Toxicity was not found in KRY·AS-treated cells although mineral internalization was evident. Phagocytized minerals were observed in

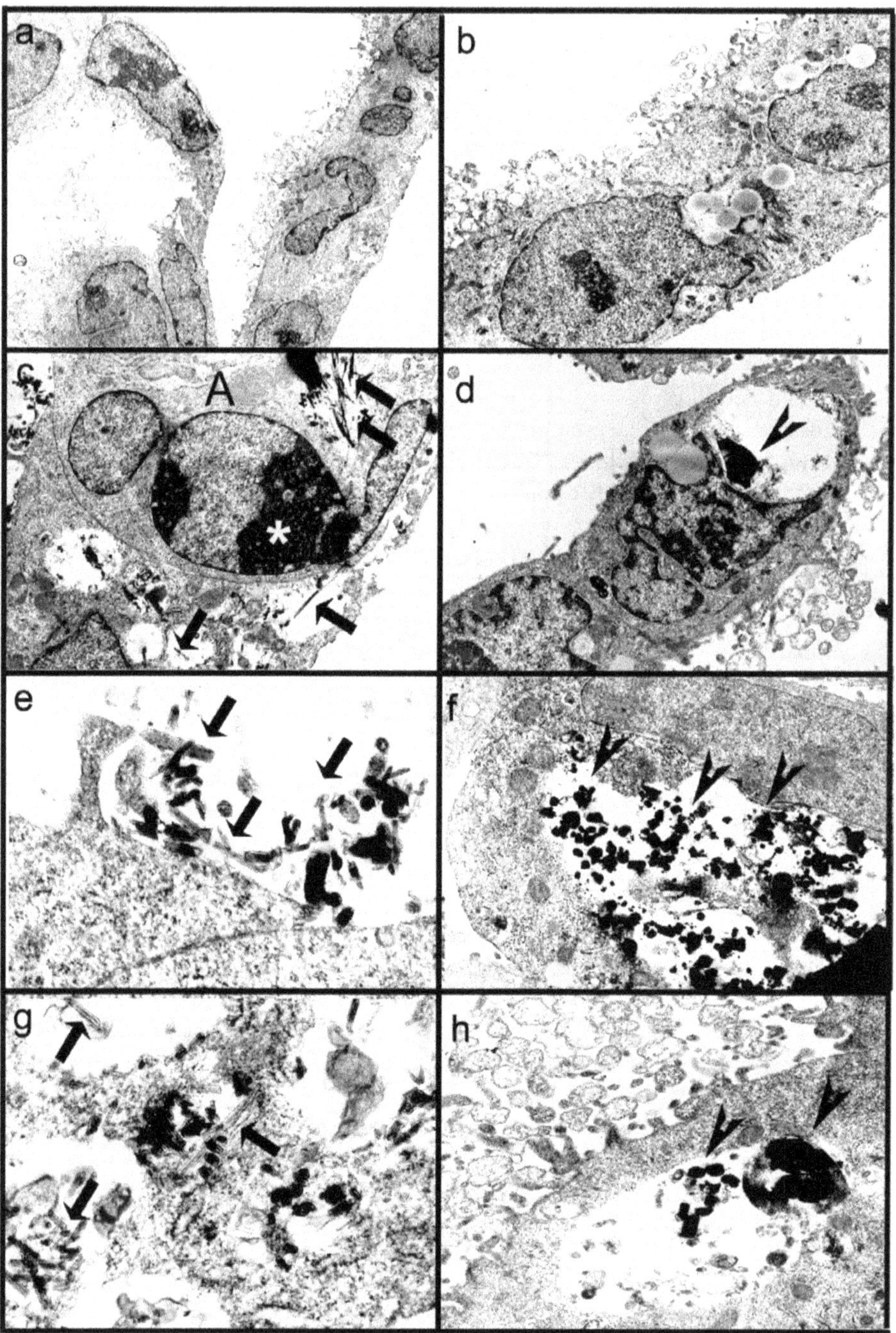
a
b
c
A
*
d
e
f
g
h

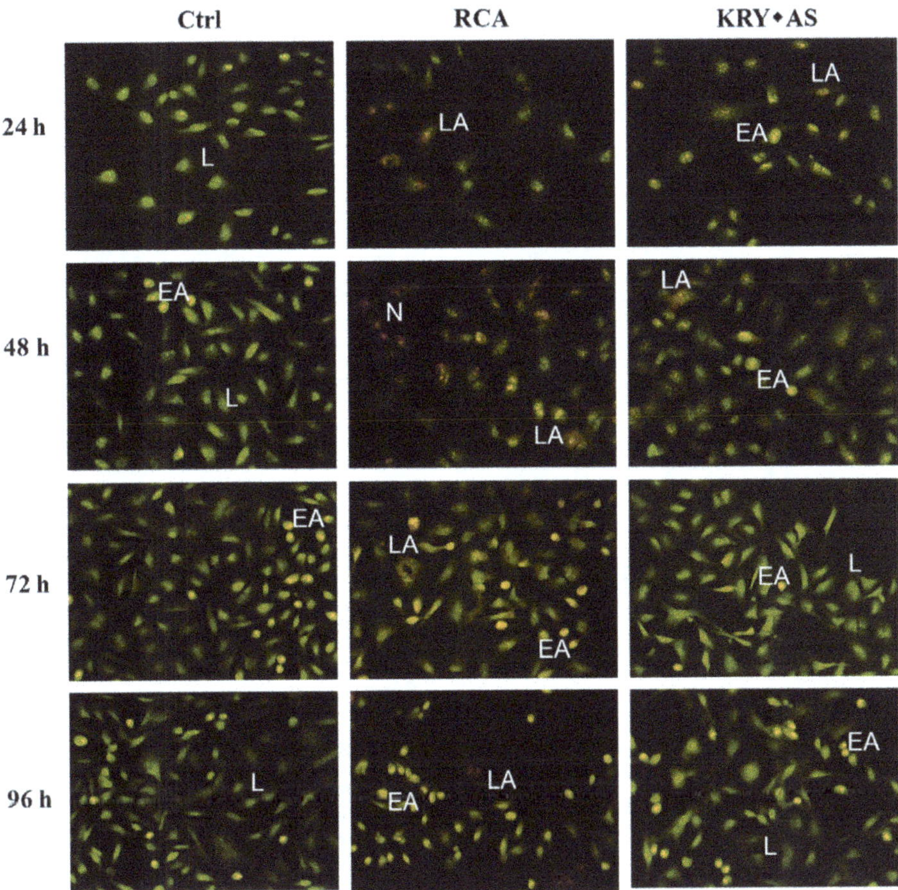

Figure 12. AO/EB double staining of control A549 cells and of cells exposed to 50 μg/mL RCA or KRY·AS for different periods of time. Live (L) cells show a normal green nucleus; early apoptosis (EA) cells display a bright green nucleus with condensed chromatin; in late apoptosis (LA) cells, condensed and fragmented chromatin is orange stained. The nucleus (N) appears orange (20 ×) (from Giantomassi *et al*., 2010; reprinted by permission of *Toxicology In Vitro*).

Figure 11 (facing page). Ultrastructural features of A549 cultures at 72 h. TEM images of control cells (a, b) and cells treated with 50 μg/mL RCA (c, e, g) or KRY·AS (d, f, h). (c) Chrysotile fibres (arrows) are detected in an apoptotic cell (A) showing characteristic marginalization of nuclear chromatin (*). (d) KRY·AS appearing as a compact body (arrowhead in the vacuolated cytoplasmic compartment. (e) RCA (arrows) is clearly visible on the cell surface before internalization and in cytoplasmic vacuoles. (g) RCA in cytoplasmic vacuoles (arrows) in a disrupted cell. (f, h) Intracellular KRY·AS in different stages of degradation (from Giantomassi *et al*., 2010; reprinted with permission from *Toxicology In Vitro*).

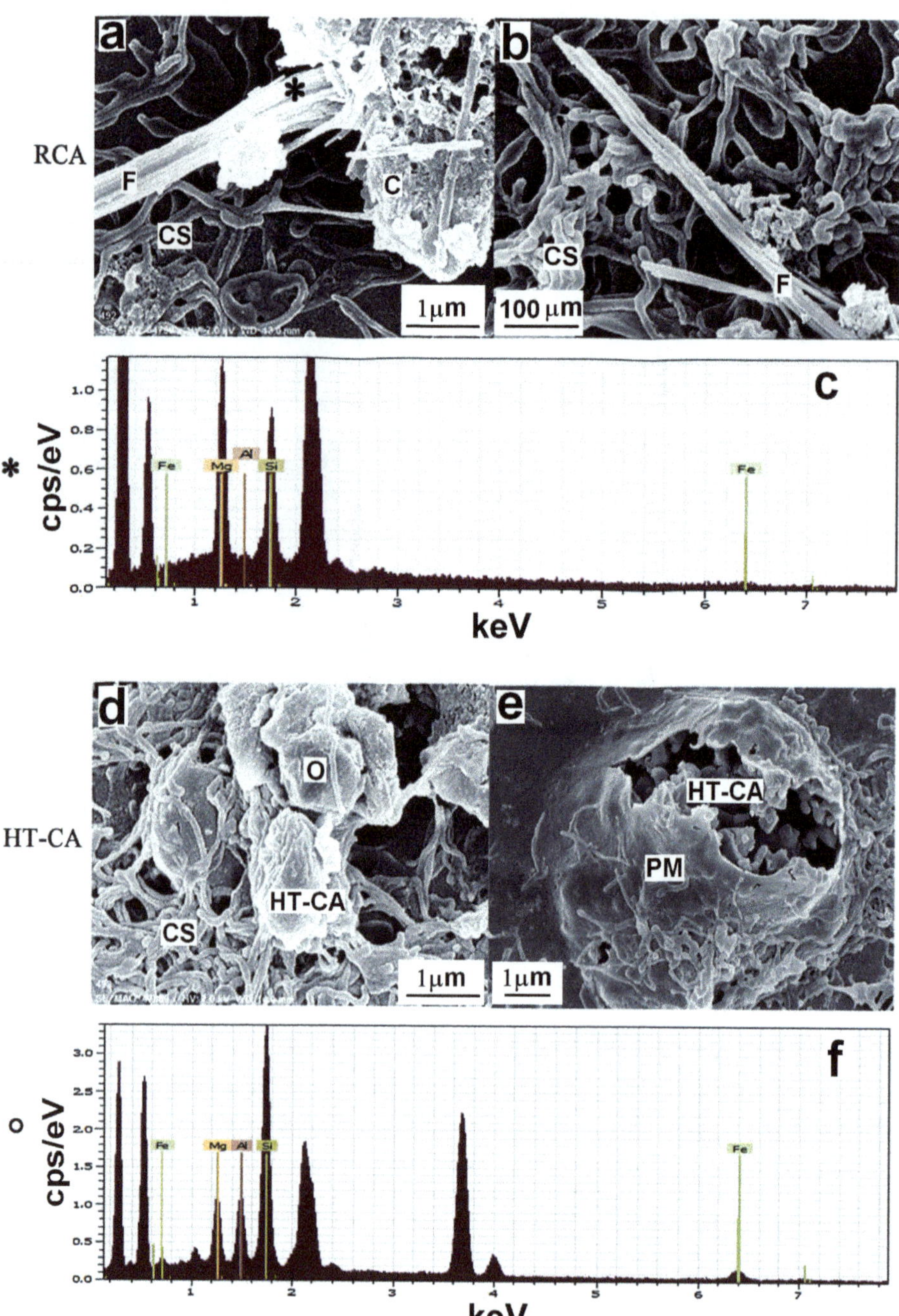
RCA
a
b
F
C
CS
1µm
100µm
c
cps/eV
keV
Fe
Mg
Al
Si
HT-CA
d
e
O
PM
f

Figure 13 (facing page). High-resolution SEM images and EDS investigations. Secondary electron SEM image of A549 cells exposed to RCA (a, b) and HT-CA (d, e) at 48 h. In a and b, cell sprouts (SC) interact closely with chrysotile fibres (F). C: cement fraction. (c) Energy dispersive spectrum EDS-SEM of chrysotile fibres in RCA performed at the * position. (d) Cell interaction with HT-CA. CS: cell sprouts. (e) HT-CA crystals inside the cell surrounded by the plasma membrane (PM). (f) Energy dispersive spectrum (EDS-SEM) of HT-CA performed at the ° location. Chemical components characteristic of RCA and HT-CA were represented by Si, Al, Fe, Mg. Differences were found in element content expressed as wt.% among RCA and HT-CA (see Table 1) (from Pugnaloni *et al.*, 2015; reprinted with permission from *Acta Histochemica*).

cytoplasmic vacuoles at all times (Fig. 11). Asbestos fibres showing the typical ultrastructural features of chrysotile with an asbestiform habit (Falini *et al.*, 2006) were seen in RCA-treated cells undergoing apoptosis and showing chromatin marginalization. However, cells in contact with KRY·AS had no adverse ultrastructural modifications and no signs of apoptosis.

The cell proliferation index (Healy *et al.*, 1995; Stratmann *et al.*, 2014) was assayed by immunohistochemical detection of Ki-67 protein, a non-histone protein expressed in the nuclei of all proliferating cells throughout the cell cycle except during G0 and early G1 phases. Morphometric investigations showed a significant reduction in Ki-67 protein expression at 48 h in RCA-exposed cells due to adverse effects of the raw mineral on cell proliferation. KRY·AS-exposed cells showed no inhibition of normal cell proliferation.

Fluorescent acridine orange/ethidium bromide (AO/EB) staining to assay apoptotic potential indicated morphological features of live cells (L) and cells in early apoptosis (EA), late apoptosis (LA) and necrosis (N) (Fig. 12). Quantitation of the values of these populations showed that RCA caused apoptosis of more severe grade compared with KRY·AS. This finding is of interest as both asbestos-induced apoptosis (Baskic *et al.*, 2006) and necrosis (Jung *et al.*, 2000) contribute to lung toxicity. Although the apoptotic mechanisms in alveolar epithelial cells are not completely clear, the apoptotic cell pathways are known to be activated by ROS production (Kamp, 2009). Supporting these data, intracellular ROS detected with the fluorescent probe, DCFA, were significantly increased in RCA-treated cultures compared to untreated controls, levels did not increase significantly after KRY·AS treatment.

Table 1. Chemical analyses obtained by EDS-SEM of chrysotile fibres in RCA and HT-CA in an *in vitro* culture of A549 cells at 48 h. Mean values ± SD (10 analyses).

Sample (wt.%)	RCA chrysotile fibres	HT-CA
Magnesium	2.41 ± 0.55	1.47 ± 0.15
Aluminium	0.41 ± 0.04	1.21 ± 0.04
Silicon	2.82 ± 0.51	4.06 ± 2.03
Iron	0.34 ± 0.13	1.23 ± 0.27

Further morphological and immunohistochemical investigations were performed on RCA and the inert product, termed high-temperature-cement asbestos (HT-CA) (Pugnaloni *et al.*, 2015) to evaluate the expression of the cell cycle regulatory proteins, p53 and p73, and the TNF-related apoptosis-inducing ligand, TRAIL. p53 mediates asbestos-induced alveolar epithelial cell (AEC) apoptosis as well as apoptosis in a number of other situations (Panduri *et al.*, 2006). Loss of p53 function is caused by mutation, nuclear retention, loss of its upstream activator, p14, and/or amplification of its antagonists (Moll and Slade, 2004). p73-related protein is a novel family member with a similar structure (Moll and Slade, 2004) found over-expressed in many cancers, in particular in non-small cell lung cancer (NSCLC) (Di Vinci *et al.*, 2009).

The tumour necrosis factor (TNF)-related apoptosis-inducing ligand, TRAIL, is a member of the TNF family that induces apoptosis of cancer cells through the activation of a caspase cascade by its two death receptors, DR4 and DR5 (Mellier *et al.*, 2010). The expression of DRs and sensitivity to TRAIL-induced apoptosis found in cancers such as colon carcinomas, bladder cancers, MMs and NSCLC (Spierings *et al.*, 2003) appear strictly related to the activation of p53 that can also act as an antioxidant protein (Zhuang *et al.*, 2013).

Increments of p53/p73 expression, iNOS positive cells and NO concentrations were found with RCA, compared to HT-CA in A549 mainly at 48 h. TRAIL expression showed no differences among groups despite data from literature supporting TRAIL-NO interactions (Cantarella *et al.*, 2007). RCA and HT-CA contact could involve, in particular, TRAIL receptors and for this reason it should therefore be of relevant interest to assay the expression of DR4, DR5 receptors and decoy receptors DcR1, DcR2, that inhibit DR4- and DR5-mediated TRAIL-induced apoptosis (Merino *et al.*, 2006).

Interestingly, ferrous iron was found in RCA as a contaminant. EDS-SEM performed on RCA in contact with cells showed characteristic chemical components of chrysotile fibres represented by Si, Al, Fe and Mg (Fig. 13). EDS spectra of HT-CA in culture showed similar chemical elements of RCA fibres, but in different amounts as testified by the detection of a discrete peak of Fe. Iron due to environmental contamination of RCA, was found at the chrysotile fibre surface and existed mostly in the ferrous phase (Governa *et al.*, 2002). Ferrous ions *via* a Fenton-type reaction catalyse the production of hydroxyl radicals that have many deleterious effects in biological systems.

These data are in line with observations showing that the cytotoxicity of chrysotile asbestos in human cells was decreased by calcination of the mineral fibres (Otero Areán *et al.*, 2001). All these studies support the proviso that the thermal treatment of ACMs may represent a safe approach for storing or recycling hazardous asbestos containing wastes and be an environmentally friendly alternative to their dumping in landfills.

5. Summary, conclusions and future directions

In vitro assays are useful tools for testing the cytotoxic and pre-cancerous effects exerted by asbestos fibres occurring naturally from environmental outcrops or from industrial products. Using proper positive (crocidolite asbestos, erionite) and negative controls (glass beads, fibre glasses or other particulates not associated with human lung cancers or MMs), minerals of interest can be studied using modern tools in cell and molecular biology to compare their intrinsic potential as biohazards. Further *in vitro* studies using human alveolar epithelial cells, bronchial epithelial cells and mesothelial cells as the target cells of lung cancers and MMs could shed more light about the actions of modified minerals as described in detail in this chapter as well as new or historical

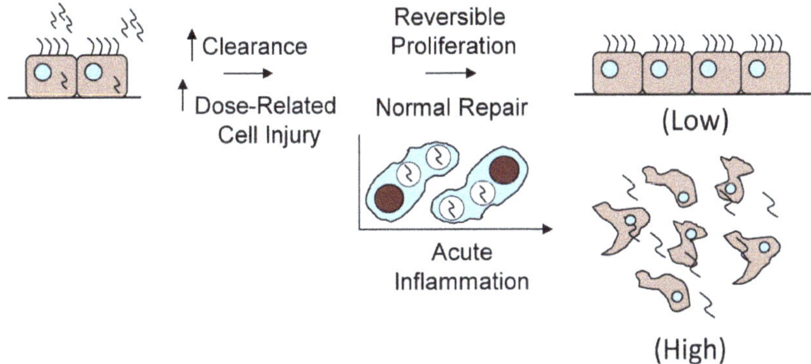

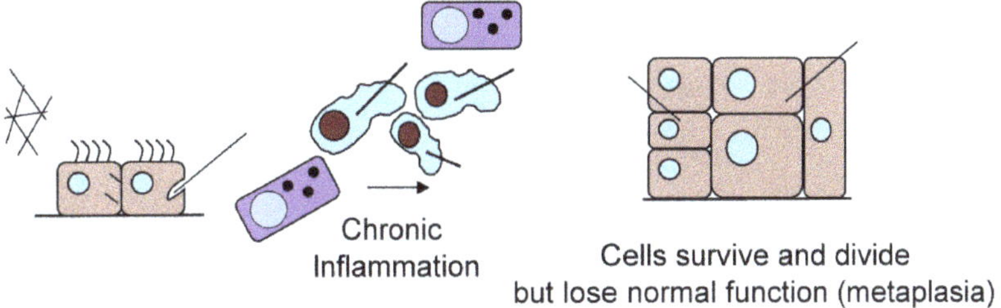

Figure 14. A schematic diagram showing differences in cell responses to naturally occurring long (>5 to 20 μm) chrysotile asbestos fibres (a) (reviewed by Heintz *et al.*, 2010; Mossman *et al.*, 2013; Lippmann, 2014) and the amphibole asbestos fibres, crocidolite and amosite, that are linked to the development of lung cancers and MMs (b). Note that curly chrysotile fibres are cleared more effectively from the lungs, are less apt to penetrate the deep lung, and break down in the lung due to fragmentation into fibrils and leaching of Mg^{2+} due to lysosomal acidity (see text). Dose-related cell injury and reversible proliferation (Warheit *et al.*, 1996) is associated with repair of epithelial and mesothelial cells (Mossman *et al.*, 2011) or acute inflammation (Mossman *et al.*, 2011) at high concentrations of chrysotile asbestos. In contrast, more durable, needle-like crocidolite and amosite asbestos fibres cause more generation of ROS (*via* iron availability and frustrated phagocytosis) and inflammasome activation, leading to chronic inflammation and loss of normal cell function during the long latency periods of lung cancers and MMs.

particulates considered 'inert' from a cytotoxic point of view, such as grey cement that shows no dangerous implications for *in vitro* testing (Pugnaloni *et al.*, 2010). In addition to epidemiological studies, it would be helpful to evaluate mineral fibres comparatively using a number of dose parameters, cell and animal models, and time points of exposure. Most importantly, a comparison of *in vitro* measures with inflammatory potential and disease causation in experimental animal and epidemiologic studies would be helpful in drawing conclusions about the properties of mineral fibres important in health effects.

Another question worthy of attention is the so-called "short-fibre controversy" (Roggli, 2015) which stresses that there is still no convincing evidence for a cancerous or other pathogenic effect for fibres that are 5 μm or less in length. This appears to be in line with data from comparative studies with synthetic tremolite assays (Pacella *et al.*, 2012). Ultrastructual studies with KRY·AS and HT-CA show obvious degradation of the material within the cell as compared to the natural mineral.

Activation of inflammasomes appears to regulate the fibre-induced, dose-related balance between tissue repair and chronic inflammation (Fig. 14). Future studies regarding mechanisms by which more biopersistent mineral fibres produce constitutive and prolonged activation of the inflammasome (in particular the study of toll-like receptors (TLRs) and other PRRs that may sense both external and internal mineral fibres) may enable more precise identification of fibre length and other characteristics that are critical to various asbestos-related diseases including lung cancers, MMs and fibrosis. Clearly, the importance of dimensions and other features of mineral fibres may be different in these diseases.

Lastly, the more recent Italian studies described in this chapter and others are clever approaches to synthesizing asbestos-like fibres in the presence or absence of different metals – these modified fibres are not only valuable tools for allowing an understanding of the association between chemical composition and the toxicity of different asbestos fibres, but may also be used to create 'safer' fibres and modify fibre outcrops and products in the future.

Acknowledgements

The authors appreciate the assistance of Jennifer Díaz, in the technical preparation of this manuscript, and Maximilian MacPherson, both at the UVM College of Medicine, in Burlington, Vermont, USA.

References

Addison, J. and McConnell, E.E. (2008) A review of carcinogenicity studies of asbestos and non-asbestos tremolite and other amphiboles. *Regulatory Toxicology and Pharmacology*, **52**, S187–199.

Altomare, D.A., You, H., Xiao, G.H., Ramos-Nino, M.E., Skele, K.L., De Rienzo, A., Jhanwar, S.C., Mossman, B.T., Kane, A.B. and Testa, J.R. (2005) Human and mouse mesotheliomas exhibit elevated AKT/PKB activity, which can be targeted pharmacologically to inhibit tumor cell growth. *Oncogene*, **24**, 6080–6089.

Alvarez, A.R., Godoy, J.A., Mullendorff, K., Olivares, G.H., Bronfman, M. and Inestrosa, N.C. (2004) Wnt-3a overcomes beta-amyloid toxicity in rat hippocampal neurons. *Experimental Cell Research*, **297**, 186–196.

Ault, J.G., Cole, R.W., Jensen, C.G., Jensen, L.C., Bachert, L.A. and Rieder, C.L. (1995) Behavior of crocidolite asbestos during mitosis in living vertebrate lung epithelial cells. *Cancer Research*, **55**, 792–798.

Ballirano, P., Bloise, A., Gualtieri, A.F., Lezzerini, M., Pacella, A., Perchiazzi, N., Dogan, M. and Dogan, A.U. (2017) Crystal structure of mineral fibres. Pp. 17–64 in: *Mineral Fibres: Crystal Chemistry, Chemical-physical Properties, Biological Interaction and Toxicity* (A.F. Gualtieri, editor). EMU Notes in Mineralogy, **18**. European Mineralogical Union and Mineralogical Society of Great Britain & Ireland, London.

Baskic, D., Popovic, S., Ristic, P. and Arsenijevic, N.N. (2006) Analysis of cycloheximide-induced apoptosis in human leukocytes: fluorescence microscopy using annexin V/propidium iodide versus acridin orange/ ethidium bromide. *Cell Biology International*, **30**, 924–932.

Battaglia, T., Belluso, E., Cappella, S., Fornero, E., Ferraris, G., Bellis, D., Biagini, G., Pugnaloni, A., Panico, A.M. and Cardile, V. (2005) Asbestos Monitoring and Analytical Methods. *Proceedings of the European Conference*, Auditorium Santa Margherita, Venezia, Italy, 5-7 December, 2005.

Bauernfeind, F., Bartok, E., Rieger, A., Franchi, L., Nunez, G. and Hornung, V. (2011) Cutting edge: reactive oxygen species inhibitors block priming, but not activation, of the NLRP3 inflammasome. *Journal of Immunology*, **187**, 613–617.

Belluso, E. and Ferraris, G. (1991) New data on balangeroite and carlosturanite from alpine serpentinites. *European Journal of Mineralogy*, **3**, 559–566.

Belluso, E., Compagnoni, R. and Ferraris, G. (1994) Occurrence of asbestiform minerals in the serpentinites of the Piemonte Zone, Western Alps. *Giornata di studio in ricordo del Prof. Stefano Zucchetti*, 57–66.

Belluso, E., Bellis, D., Biagini, G., Cappella, S., Fato, R. and Ferraris, G. (2004) Fibrous antigorite: multidisciplinary approach to a preliminary evaluation of cytotoxicity meeting. *Proceedings of the from Asbestos to Nanoparticles: Past Experience for Future Challenges and Needs in Particle Toxicology*, University of Torino, 28–30 June 2004, Book of Abstracts, p. 21.

Belluso, E., Cavallo, A. and Halterman, D. (2017) Crystal habit of mineral fibres. Pp. 65–109 in: *Mineral Fibres: Crystal Chemistry, Chemical-physical Properties, Biological Interaction and Toxicity* (A.F. Gualtieri, editor). EMU Notes in Mineralogy, **18**. European Mineralogical Union and Mineralogical Society of Great Britain & Ireland, London.

Ben-Baruch, A. (2006) Inflammation-associated immune suppression in cancer: the roles played by cytokines, chemokines and additional mediators. *Seminars in Cancer Biology*, **16**, 38–52.

Benedetti, S., Nuvoli, B., Catalani, S. and Galati, R. (2015) Reactive oxygen species a double-edged sword for mesothelioma. *Oncotarget*, **6**, 16848–16865.

Bergamini, C., Fato, R., Biagini, G., Pugnaloni, A., Giantomassi, F., Foresti, E., Lesci, G.I., Roveri, N. and Lenaz, G. (2004) Mitochondrial changes induced by natural and synthetic asbestos fibers: studies on isolated mitochondria. *Cellular and Molecular Biology*, **50 Online Pub**, OL691–700.

Bergamini, C., Fato, R., Biagini, G., Pugnaloni, A., Giantomassi, F., Foresti, E., Lesci, G.I., Roveri, N. and Lenaz, G. (2007) Mitochondrial changes induced by natural and synthetic asbestos fibers: studies on isolated mitochondria. *Cellular and Molecular Biology*, **52 Suppl**, OL905–913.

Bergsbaken, T., Fink, S.L. and Cookson, B.T. (2009) Pyroptosis: host cell death and inflammation. *Nature Reviews Microbiology*, **7**, 99–109.

Berman, D.W. and Crump, K.S. (2008) A meta-analysis of asbestos-related cancer risk that addresses fiber size and mineral type. *Critical Reviews in Toxicology*, **38 Suppl 1**, 49–73.

Bernstein, D. and Pavlisko, E. (2017) Differential pathological response and pleural transport of mineral fibres. Pp. 417–434 in: *Mineral Fibres: Crystal Chemistry, Chemical-physical Properties, Biological Interaction and Toxicity* (A.F. Gualtieri, editor). EMU Notes in Mineralogy, **18**. European Mineralogical Union and Mineralogical Society of Great Britain & Ireland, London.

Blanco, D., Vicent, S., Elizegi, E., Pino, I., Fraga, M.F., Esteller, M., Saffiotti, U., Lecanda, F. and Montuenga, L.M. (2004) Altered expression of adhesion molecules and epithelial-mesenchymal transition in silica-induced rat lung carcinogenesis. *Laboratory Investigation*, **84**, 999–1012.

Bloise, A., Fornero, E., Belluso, E., Barrese, E. and Rinaudo, C. (2008) Synthesis and characterization of tremolite asbestos fibres. *European Journal of Mineralogy*, **20**, 1027–1033.

Bloise, A., Kusiorowski, R., Lassinantti Gualtieri, M. and Gualtieri, A.F. (2017) Thermal behaviour of mineral fibres. Pp. 215–260 in: *Mineral Fibres: Crystal Chemistry, Chemical-physical Properties, Biological*

Interaction and Toxicity (A.F. Gualtieri, editor). EMU Notes in Mineralogy, **18**. European Mineralogical Union and Mineralogical Society of Great Britain & Ireland, London.

Boardman, K.C., Aryal, A.M., Miller, W.M. and Waters, C.M. (2004) Actin re-distribution in response to hydrogen peroxide in airway epithelial cells. *Journal of Cellular Physiology*, **199**, 57–66.

Boutin, C., Dumortier, P., Rey, F., Viallat, J.R. and De Vuyst, P. (1996) Black spots concentrate oncogenic asbestos fibers in the parietal pleura. Thoracoscopic and mineralogic study. *American Journal of Respiratory and Critical Care Medicine*, **153**, 444–449.

Boylan, A.M., Sanan, D.A., Sheppard, D. and Broaddus, V.C. (1995) Vitronectin enhances internalization of crocidolite asbestos by rabbit pleural mesothelial cells via the integrin alpha v beta 5. *Journal of Clinical Investment*, **96**, 1987–2001.

Brody, A.R. and Hill, L.H. (1982) Interstitial accumulation of inhaled chrysotile asbestos fibers and consequent formation of microcalcifications. *American Journal of Pathology*, **109**, 107–114.

Bruno, C., Tumino, R., Fazzo, L., Cascone, G., Cernigliaro, A., De Santis, M., Giurdanella, M.C., Nicita, C., Rollo, P.C., Scondotto, S., Spata, E., Zona, A. and Comba, P. (2014) Incidence of pleural mesothelioma in a community exposed to fibres with fluoro-edenitic composition in Biancavilla (Sicily, Italy). *Annali dell'Istituto Superiore Di Sanita*, **50**, 111–118.

Brusselle, G.G., Provoost, S., Bracke, K.R., Kuchmiy, A. and Lamkanfi, M. (2014) Inflammasomes in respiratory disease: from bench to bedside. *Chest*, **145**, 1121–1133.

Cantarella, G., Lempereur, L., D'Alcamo, M.A., Risuglia, N., Cardile, V., Pennisi, G., Scoto, G.M. and Bernardini, R. (2007) Trail interacts redundantly with nitric oxide in rat astrocytes: potential contribution to neurodegenerative processes. *Journal of Neuroimmunology*, **182**, 41–47.

Capella, S., Belluso, E., Bursi, N., Tibaldi, E. and Belpoggi, F. (2017) *In vivo* biological activity of mineral fibres. Pp. 307–345 in: *Mineral Fibres: Crystal Chemistry, Chemical-physical Properties, Biological Interaction and Toxicity* (A.F. Gualtieri, editor). EMU Notes in Mineralogy, **18**. European Mineralogical Union and Mineralogical Society of Great Britain & Ireland, London.

Carbone, M. and Yang, H. (2017) Biological activities of asbestos and other mineral fibres. Pp. 435–445 in: *Mineral Fibres: Crystal Chemistry, Chemical-physical Properties, Biological Interaction and Toxicity* (A.F. Gualtieri, editor). EMU Notes in Mineralogy, **18**. European Mineralogical Union and Mineralogical Society of Great Britain & Ireland, London.

Cardile, V., Lombardo, L., Belluso, E., Panico, A., Renis, M., Gianfagna, A. and Balazy, M. (2007) Fluoro-edenite fibers induce expression of Hsp70 and inflammatory response. *International Journal of Environmental Research Public and Health*, **4**, 195–202.

Cardillo, I., Spugnini, E.P., Verdina, A., Galati, R., Citro, G. and Baldi, A. (2005) Cox and mesothelioma: an overview. *Histology and Histopathology*, **20**, 1267–1274.

Carratala, J., Salazar, A., Mascaro, J. and Santin, M. (1992) Community-acquired pneumonia due to Pseudomonas stutzeri. *Clinical Infectious Diseases*, **14**, 792.

Case, B.W., Abraham, J.L., Meeker, G., Pooley, F.D. and Pinkerton, K.E. (2011) Applying definitions of "asbestos" to environmental and "low-dose" exposure levels and health effects, particularly malignant mesothelioma. *Journal of Toxicology and Environmental Health, Part B: Critical Reviews*, **14**, 3–39.

Chang, J.S., Su, C.Y., Yu, W.H., Lee, W.J., Liu, Y.P., Lai, T.C., Jan, Y.H., Yang, Y.F., Shen, C.N., Shew, J.Y., Lu, J., Yang, C.J., Huang, M.S., Lu, P.J., Lin, Y.F., Kuo, M.L., Hua, K.T. and Hsiao, M. (2015) GIT1 promotes lung cancer cell metastasis through modulating Rac1/Cdc42 activity and is associated with poor prognosis. *Oncotarget*, **6**, 36278–36291.

Chao, C.C. and Aust, A.E. (1994) Effect of long-term removal of iron from asbestos by desferrioxamine B on subsequent mobilization by other chelators and induction of DNA single-strand breaks. *Archives Biochemistry and Biophysics*, **308**, 64–69.

Chen, S., Guttridge, D.C., You, Z., Zhang, Z., Fribley, A., Mayo, M.W., Kitajewski, J. and Wang, C.Y. (2001) Wnt-1 signaling inhibits apoptosis by activating beta-catenin/T cell factor-mediated transcription. *Journal of Cell Biology*, **152**, 87–96.

Cole, R.W., Ault, J.G., Hayden, J.H. and Rieder, C.L. (1991) Crocidolite asbestos fibers undergo size-dependent microtubule-mediated transport after endocytosis in vertebrate lung epithelial cells. *Cancer Research*, **51**, 4942–4947.

Comba, P., Gianfagna, A. and Paoletti, L. (2003) Pleural mesothelioma cases in Biancavilla are related to a new fluoro-edenite fibrous amphibole. *Archives of Environmental Health*, **58**, 229–232.

Coussens, L.M. and Werb, Z. (2002) Inflammation and cancer. *Nature*, **420**, 860–867.

Cui, H., Wu, W., Okuhira, K., Miyazawa, K., Hattori, T., Sai, K., Naito, M., Suzuki, K., Nishimura, T., Sakamoto, Y., Ogata, A., Maeno, T., Inomata, A., Nakae, D., Hirose, A. and Nishimaki-Mogami, T. (2014) High-temperature calcined fullerene nanowhiskers as well as long needle-like multi-wall carbon nanotubes have abilities to induce NLRP3-mediated IL-1beta secretion. *Biochemical and Biophysical Research Communication*, **452**, 593–599.

Cunniff, B., Benson, K., Stumpff, J., Newick, K., Held, P., Taatjes, D., Joseph, J., Kalyanaraman, B. and Heintz, N.H. (2013) Mitochondrial-targeted nitroxides disrupt mitochondrial architecture and inhibit expression of peroxiredoxin 3 and FOXM1 in malignant mesothelioma cells. *Journal of Cellular Physiology*, **228**, 835–845.

Daniel, F.B. (1983) In vitro assessment of asbestos genotoxicity. *Environmental Health Perspectives*, **53**, 163–167.

Davoren, M., Herzog, E., Casey, A., Cottineau, B., Chambers, G., Byrne, H.J. and Lyng, F.M. (2007) In vitro toxicity evaluation of single walled carbon nanotubes on human A549 lung cells. *Toxicology In Vitro*, **21**, 438–448.

Denizeau, F., Marion, M., Chevalier, G. and Cote, M.G. (1985) Inability of chrysotile asbestos fibers to modulate the 2-acetylaminofluorene-induced UDS in primary cultures of rat hepatocytes. *Mutation Research*, **155**, 83–90.

Di Vinci, A., Sessa, F., Casciano, I., Banelli, B., Franzi, F., Brigati, C., Allemanni, G., Russo, P., Dominioni, L. and Romani, M. (2009) Different intracellular compartmentalization of TA and DeltaNp73 in non-small cell lung cancer. *International Journal of Oncology*, **34**, 449–456.

Dodson, R.F., Atkinson, M.A. and Levin, J.L. (2003) Asbestos fiber length as related to potential pathogenicity: a critical review. *American Journal of Industrial Medicine*, **44**, 291–297.

Dogan, A.U., Dogan, M. and Hoskins, J.A. (2008) Erionite series minerals: mineralogical and carcinogenic properties. *Environmental Geochemistry and Health*, **30**, 367–381.

Donaldson, K., Murphy, F.A., Duffin, R. and Poland, C.A. (2010) Asbestos, carbon nanotubes and the pleural mesothelium: a review of the hypothesis regarding the role of long fibre retention in the parietal pleura, inflammation and mesothelioma. *Particle and Fibre Toxicology*, **7**, 5.

Donaldson, K., Poland, C.A., Murphy, F.A., MacFarlane, M., Chernova, T. and Schinwald, A. (2013) Pulmonary toxicity of carbon nanotubes and asbestos – similarities and differences. *Advances in Drug Delivery Reviews*, **65**, 2078–2086.

Donaldson, K. and Tran, C.L. (2002) Inflammation caused by particles and fibers. *Inhalation Toxicology*, **14**, 5–27.

Donnelly, L.E. and Barnes, P.J. (2002) Expression and regulation of inducible nitric oxide synthase from human primary airway epithelial cells. *American Journal of Respiratory Cell and Molecular Biology*, **26**, 144–151.

Dopp, E., Yadav, S., Ansari, F.A., Bhattacharya, K., von Recklinghausen, U., Rauen, U., Rodelsperger, K., Shokouhi, B., Geh, S. and Rahman, Q. (2005) ROS-mediated genotoxicity of asbestos-cement in mammalian lung cells *in vitro*. *Particle and Fibre Toxicology*, **2**, 9.

Dostert, C., Petrilli, V., Van Bruggen, R., Steele, C., Mossman, B.T. and Tschopp, J. (2008) Innate immune activation through Nalp3 inflammasome sensing of asbestos and silica. *Science*, **320**, 674–677.

Douvdevani, A., Einbinder, T., Yulzari, R., Rogachov, B. and Chaimovitz, C. (1996) TNF-receptors on human peritoneal mesothelial cells: regulation of receptor levels and shedding by IL-1 alpha and TNF alpha. *Kidney International*, **50**, 219–228.

Eastman, A., Mossman, B.T. and Bresnick, E. (1983) Influence of asbestos on the uptake of benzo(a)pyrene and DNA alkylation in hamster tracheal epithelial cells. *Cancer Research*, **43**, 1251–1255.

Elkabets, M., Ribeiro, V.S., Dinarello, C.A., Ostrand-Rosenberg, S., Di Santo, J.P., Apte, R.N. and Vosshenrich, C.A. (2010) IL-1beta regulates a novel myeloid-derived suppressor cell subset that impairs NK cell development and function. *European Journal of Immunology*, **40**, 3347–3357.

Enrico Favero-Longo, S., Turci, F., Tomatis, M., Compagnoni, R., Piervittori, R. and Fubini, B. (2009) The effect of weathering on ecopersistence, reactivity, and potential toxicity of naturally occurring asbestos and asbestiform minerals. *Journal of Toxicology and Environmental Health, Part A*, **72**, 305–314.

Etienne-Manneville, S. (2006) In vitro assay of primary astrocyte migration as a tool to study Rho GTPase function in cell polarization. *Methods in Enzymology*, **406**, 565–578.

Falini, G., Foresti, E., Lesci, G. and Roveri, N. (2002) Structural and morphological characterization of synthetic chrysotile single crystals. *Chemical Communications*, 1512–1513.

Falini, G., Foresti, E., Lesci, I.G., Lunelli, B., Sabatino, P. and Roveri, N. (2006) Interaction of bovine serum albumin with chrysotile: spectroscopic and morphological studies. *Chemistry*, **12**, 1968–1974.

Favero-Longo, S.E., Turci, F., Tomatis, M., Castelli, D., Bonfante, P., Hochella, M.F., Piervittori, R. and Fubini, B. (2005) Chrysotile asbestos is progressively converted into a non-fibrous amorphous material by the chelating action of lichen metabolites. *Journal of Environmental Monitoring*, **7**, 764–766.

Felley-Bosco, E., Ambs, S., Lowenstein, C.J., Keefer, L.K. and Harris, C.C. (1994) Constitutive expression of inducible nitric oxide synthase in human bronchial epithelial cells induces c-fos and stimulates the cGMP pathway. *American Journal of Respiratory Cell and Molecular Biology*, **11**, 159–164.

Fenoglio, I., Croce, A., Di Renzo, F., Tiozzo, R. and Fubini, B. (2000) Pure-silica zeolites (Porosils) as model solids for the evaluation of the physicochemical features determining silica toxicity to macrophages. *Chemical Research in Toxicology*, **13**, 489–500.

Ferrara, N. (2000) Vascular endothelial growth factor and the regulation of angiogenesis. *Recent Progress in Hormone Research*, **55**, 15–35; discussion 35–16.

Finley, B.L., Pierce, J.S., Phelka, A.D., Adams, R.E., Paustenbach, D.J., Thuett, K.A. and Barlow, C.A. (2012) Evaluation of tremolite asbestos exposures associated with the use of commercial products. *Critical Reviews in Toxicology*, **42**, 119–146.

Fox, S.A., Richards, A.K., Kusumah, I., Perumal, V., Bolitho, E.M., Mutsaers, S.E. and Dharmarajan, A.M. (2013) Expression profile and function of Wnt signaling mechanisms in malignant mesothelioma cells. *Biochemical and Biophysical Research Communication*, **440**, 82–87.

Franchi, L., Kanneganti, T.D., Dubyak, G.R. and Nunez, G. (2007) Differential requirement of P2X7 receptor and intracellular K+ for caspase-1 activation induced by intracellular and extracellular bacteria. *Journal of Biological Chemistry*, **282**, 18810–18818.

Fubini, B., Ghiazza, M. and Fenoglio, I. (2010) Physico-chemical features of engineered nanoparticles relevant to their toxicity. *Nanotoxicology*, **4**, 347–363.

Fukumaru, K., Yoshii, N., Kanzaki, T. and Kanekura, T. (2007) Immunohistochemical comparison of beta-catenin expression by human normal epidermis and epidermal tumors. *Journal of Dermatology*, **34**, 746–753.

Gaffney, S.H., Grespin, M., Garnick, L., Drechsel, D.A., Hazan, R., Paustenbach, D.J. and Simmons, B.D. (2017) Anthophyllite asbestos: state of the science review. *Journal of Applied Toxicology*, **37**, 38–49.

Giacobbe, C., Gualtieri, A.F., Quartieri, S., Rinaudo, C., Allegrina, M. and Andreozzi, G.B. (2010) Spectroscopic study of the product of thermal transformation of chrysotile-asbestos containing materials (ACM). *European Journal of Mineralogy*, **22**, 535–546.

Gianfagna, A. and Oberti, R. (2001) Fluoro-edenite from Biancavilla (Catania, Sicily, Italy) crystal chemistry of a new amphibole end-member. *American Mineralogist*, **86**, 1489–1493.

Giantomassi, F., Gualtieri, A.F., Santarelli, L., Tomasetti, M., Lusvardi, G., Lucarini, G., Governa, M. and Pugnaloni, A. (2010) Biological effects and comparative cytotoxicity of thermal transformed asbestos-containing materials in a human alveolar epithelial cell line. *Toxicology In Vitro*, **24**, 1521–1531.

Gold, J., Amandusson, H., Krozer, A., Kasemo, B., Ericsson, T., Zanetti, G. and Fubini, B. (1997) Chemical characterization and reactivity of iron chelator-treated amphibole asbestos. *Environmental Health Perspectives*, **105 Suppl 5**, 1021–1030.

Goldberg, J.L., Zanella, C.L., Janssen, Y.M., Timblin, C.R., Jimenez, L.A., Vacek, P., Taatjes, D.J. and Mossman, B.T. (1997) Novel cell imaging techniques show induction of apoptosis and proliferation in mesothelial cells by asbestos. *American Journal of Respiratory Cell and Molecular Biology*, **17**, 265–271.

Goley, E.D. and Welch, M.D. (2006) The ARP2/3 complex: an actin nucleator comes of age. *Nature Reviews Molecular Cell Biology*, **7**, 713–726.

Governa, M., Fenoglio, I., Amati, M., Valentino, M., Bolognini, L., Coloccini, S., Volpe, A.R., Carmignani, M. and Fubini, B. (2002) Cleavage of the fifth component of human complement and release of a split product with C5a-like activity by crystalline silica through free radical generation and kallikrein activation. *Toxicology and Applied Pharmacology*, **179**, 129–136.

Grice, J.D. and Ferraris, G. (2001) New minerals approved in 2000 by the Commission on New Minerals and Mineral Names. IMA (No 2000-a49, p.1001). *European Journal of Mineralogy*, **13**, 995–1002.

Grivennikov, S.I. and Karin, M. (2010) Inflammation and oncogenesis: a vicious connection. *Current Opinions in Genetics and Development*, **20**, 65–71.

Grivennikov, S.I., Greten, F.R. and Karin, M. (2010) Immunity, inflammation, and cancer. *Cell*, **140**, 883–899.

Gronow, J.R. (1987) The dissolution of asbestos fibres in water. *Clay Minerals*, **22**, 21–35.

Gualtieri, A.F. (2017) Introduction. Pp. 1–15 in: *Mineral Fibres: Crystal Chemistry, Chemical-physical Properties, Biological Interaction and Toxicity* (A.F. Gualtieri, editor). EMU Notes in Mineralogy, **18**. European Mineralogical Union and Mineralogical Society of Great Britain & Ireland, London.

Gualtieri, A.F., Cavenati, C., Zanatto, I., Meloni, M., Elmi, G. and Gualtieri, M.L. (2008a) The transformation sequence of cement-asbestos slates up to 1200 degrees C and safe recycling of the reaction product in stoneware tile mixtures. *Journal of Hazardous Materials*, **152**, 563–570.

Gualtieri, A.F., Gualtieri, M.L. and Tonelli, M. (2008b) In situ ESEM study of the thermal decomposition of chrysotile asbestos in view of safe recycling of the transformation product. *Journal of Hazardous Materials*, **156**, 260–266.

Gualtieri, A.F., Mossman, B.T. and Roggli, V.L. (2017) Towards a general model for predicting the toxicity and pathogenicity of mineral fibres. Pp. 501–532 in: *Mineral Fibres: Crystal Chemistry, Chemical-physical Properties, Biological Interaction and Toxicity* (A.F. Gualtieri, editor). EMU Notes in Mineralogy, **18**. European Mineralogical Union and Mineralogical Society of Great Britain & Ireland, London.

Guthrie, G.D. and Mossman, B.T. editors (1993) *Health Effects of Mineral Dusts*. Reviews in Mineralogy and Geochemistry. Mineralogical Society of America and The Geochemical Society, Washington, USA, 584 pp.

Haegens, A., van der Vliet, A., Butnor, K.J., Heintz, N., Taatjes, D., Hemenway, D., Vacek, P., Freeman, B.A., Hazen, S.L., Brennan, M.L. and Mossman, B.T. (2005) Asbestos-induced lung inflammation and epithelial cell proliferation are altered in myeloperoxidase-null mice. *Cancer Research*, **65**, 9670–9677.

Halle, A., Hornung, V., Petzold, G.C., Stewart, C.R., Monks, B.G., Reinheckel, T., Fitzgerald, K.A., Latz, E., Moore, K.J. and Golenbock, D.T. (2008) The NALP3 inflammasome is involved in the innate immune response to amyloid-beta. *Nature Immunology*, **9**, 857–865.

Hanahan, D. and Weinberg, R.A. (2011) Hallmarks of cancer: the next generation. *Cell*, **144**, 646–674.

Hargreaves, A. and Taylor, W.H. (1946) An X-ray examination of decomposition products of chrysotile (asbestos) and serpentine. *Mineralogical Magazine*, **27**, 204–216.

Health Effects Institute-Asbestos Research. (1991) *Asbestos in public and commercial buildings: A literature review and synthesis of current knowledge*. Cambridge, MA, 13 pp.

Healy, E., Angus, B., Lawrence, C.M. and Rees, J.L. (1995) Prognostic value of Ki67 antigen expression in basal cell carcinomas. *British Journal of Dermatology*, **133**, 737–741.

Heintz, N.H., Janssen-Heininger, Y.M. and Mossman, B.T. (2010) Asbestos, lung cancers, and mesotheliomas: from molecular approaches to targeting tumor survival pathways. *American Journal of Respiratory Cell and Molecular Biology*, **42**, 133–139.

Helsen, J.A., van de Velde, P., Kuczumow, A. and Deruyttere, A. (1989) Surface characteristics of asbestos fibers released from asbestos- cement products. *American Industrial Hygiene Association Journal*, **50**, 655–663.

Hillegass, J.M., Miller, J.M., MacPherson, M.B., Westbom, C.M., Sayan, M., Thompson, J.K., Macura, S.L., Perkins, T.N., Beuschel, S.L., Alexeeva, V., Pass, H.I., Steele, C., Mossman, B.T. and Shukla, A. (2013) Asbestos and erionite prime and activate the NLRP3 inflammasome that stimulates autocrine cytokine release in human mesothelial cells. *Particle and Fibre Toxicology*, **10**, 39.

Hiraku, Y., Kawanishi, S., Ichinose, T. and Murata, M. (2010) The role of iNOS-mediated DNA damage in infection- and asbestos-induced carcinogenesis. *Annals of the New York Academy of Sciences*, **1203**, 15–22.

Hla, T. and Neilson, K. (1992) Human cyclooxygenase-2 cDNA. *Proceedings of the National Academy of Sciences of the United States of America*, **89**, 7384–7388.

Hodgson, J.T. and Darnton, A. (2000) The quantitative risks of mesothelioma and lung cancer in relation to asbestos exposure. *Annals of Occupational Hygiene*, **44**, 565–601.

Hommura, F., Furuuchi, K., Yamazaki, K., Ogura, S., Kinoshita, I., Shimizu, M., Moriuchi, T., Katoh, H.,

Nishimura, M. and Dosaka-Akita, H. (2002) Increased expression of beta-catenin predicts better prognosis in nonsmall cell lung carcinomas. *Cancer*, **94**, 752–758.

Hu, B., Elinav, E., Huber, S., Booth, C.J., Strowig, T., Jin, C., Eisenbarth, S.C. and Flavell, R.A. (2010) Inflammation-induced tumorigenesis in the colon is regulated by caspase-1 and NLRC4. *Proceedings of the National Academy of Sciences of the United States of America*, **107**, 21635–21640.

IARC (2012) Arsenic, Metals, Fibres and Dusts: A Review of Human Carcinogens. *Proceedings of the IARC Working Group on the Evaluation of Carcinogenic Risks to Humans*, Lyon, France. International Agency for Research on Cancer, 501 pp.

Ishii, H., Yazawa, T., Sato, H., Suzuki, T., Ikeda, M., Hayashi, Y., Takanashi, Y. and Kitamura, H. (2004) Enhancement of pleural dissemination and lymph node metastasis of intrathoracic lung cancer cells by vascular endothelial growth factors (VEGFs). *Lung Cancer*, **45**, 325–337.

Jablonski, R., Kim, S.J., Cheresh, P. and Kamp, D.W. (2017) Insights into mineral fibre-induced lung epithelial cell toxicity and pulmonary fibrosis. Pp. 447–500 in: *Mineral Fibres: Crystal Chemistry, Chemical-physical Properties, Biological Interaction and Toxicity* (A.F. Gualtieri, editor). EMU Notes in Mineralogy, **18**. European Mineralogical Union and Mineralogical Society of Great Britain & Ireland, London.

Janssen, Y.M., Marsh, J.P., Absher, M., Borm, P.J. and Mossman, B.T. (1990) Increases in endogenous antioxidant enzymes during asbestos inhalation in rats. *Free Radicical Research Communications*, **11**, 53–58.

Janssen, Y.M., Marsh, J.P., Absher, M.P., Hemenway, D., Vacek, P.M., Leslie, K.O., Borm, P.J. and Mossman, B.T. (1992) Expression of antioxidant enzymes in rat lungs after inhalation of asbestos or silica. *Journal of Biological Chemistry*, **267**, 10625–10630.

Janssen, Y.M., Driscoll, K.E., Howard, B., Quinlan, T.R., Treadwell, M., Barchowsky, A. and Mossman, B.T. (1997) Asbestos causes translocation of p65 protein and increases NF-kappa B DNA binding activity in rat lung epithelial and pleural mesothelial cells. *American Journal of Pathology*, **151**, 389–401.

Jaurand, M.C., Fleury, J., Monchaux, G., Nebut, M. and Bignon, J. (1987) Pleural carcinogenic potency of mineral fibers (asbestos, attapulgite) and their cytotoxicity on cultured cells. *Journal of the National Cancer Institute*, **79**, 797–804.

Jensen, C.G. and Watson, M. (1999) Inhibition of cytokinesis by asbestos and synthetic fibres. *Cell Biology International*, **23**, 829–840.

Ji, D.B., Xu, B., Liu, J.T., Ran, F.X. and Cui, J.R. (2011) beta-Escin sodium inhibits inducible nitric oxide synthase expression via downregulation of the JAK/STAT pathway in A549 cells. *Molecular Carcinogenesis*, **50**, 945–960.

Jin, C., Frayssinet, P., Pelker, R., Cwirka, D., Hu, B., Vignery, A., Eisenbarth, S.C. and Flavell, R.A. (2011) NLRP3 inflammasome plays a critical role in the pathogenesis of hydroxyapatite-associated arthropathy. *Proceedings of the National Academy of Sciences of the United States of America*, **108**, 14867–14872.

Jube, S., Rivera, Z.S., Bianchi, M.E., Powers, A., Wang, E., Pagano, I., Pass, H.I., Gaudino, G., Carbone, M. and Yang, H. (2012) Cancer cell secretion of the DAMP protein HMGB1 supports progression in malignant mesothelioma. *Cancer Research*, **72**, 3290–3301.

Jung, M., Davis, W.P., Taatjes, D.J., Churg, A. and Mossman, B.T. (2000) Asbestos and cigarette smoke cause increased DNA strand breaks and necrosis in bronchiolar epithelial cells *in vivo*. *Free Radical Biology and Medicine*, **28**, 1295–1299.

Kamp, D.W. (2009) Asbestos-induced lung diseases: an update. *Translational Research*, **153**, 143–152.

Kamp, D.W. and Weitzman, S.A. (1999) The molecular basis of asbestos induced lung injury. *Thorax*, **54**, 638–652.

Kamp, D.W., Panduri, V., Weitzman, S.A. and Chandel, N. (2002) Asbestos-induced alveolar epithelial cell apoptosis: role of mitochondrial dysfunction caused by iron-derived free radicals. *Molecular and Cellular Biochemistry*, **234-235**, 153–160.

Kanno, S., Hirano, S., Chiba, S., Takeshita, H., Nagai, T., Takada, M., Sakamoto, K. and Mukai, T. (2015) The role of Rho-kinases in IL-1beta release through phagocytosis of fibrous particles in human monocytes. *Archives of Toxicology*, **89**, 73–85.

Langer, A.M. and Nolan, R.P. (1987) The properties of chrysotile asbestos as determinants of biological activity: variations in cohort experience and disease spectra as related to mineral properties. Pp. 30–51 in: *The*

Biological Effects of Chrysotile (J.C. Wagner, editor). J.P. Lippincott Company, Philadelphia, Pennsylvania, USA.

Lee, G.S., Subramanian, N., Kim, A.I., Aksentijevich, I., Goldbach-Mansky, R., Sacks, D.B., Germain, R.N., Kastner, D.L. and Chae, J.J. (2012) The calcium-sensing receptor regulates the NLRP3 inflammasome through Ca2+ and cAMP. *Nature*, **492**, 123–127.

Lemen, R.A. (2006) Epidemiology of asbestos-related diseases and the knowledge that led to what is known today. Pp. 201–308 in: *Asbestos: Risk Assessment, Epidemiology, and Health Effects* (R. F. Dodson and S. P. Hammar, editors). CRC Press, Boca Raton, Florida, USA.

Lenaz, G. (2001) The mitochondrial production of reactive oxygen species: mechanisms and implications in human pathology. *IUBMB Life*, **52**, 159–164.

Li, M., Gunter, M.E. and Fukagawa, N.K. (2012) Differential activation of the inflammasome in THP-1 cells exposed to chrysotile asbestos and Libby "six-mix" amphiboles and subsequent activation of BEAS-2B cells. *Cytokine*, **60**, 718–730.

Lim, C.B., Prele, C.M., Baltic, S., Arthur, P.G., Creaney, J., Watkins, D.N., Thompson, P.J. and Mutsaers, S.E. (2015) Mitochondria-derived reactive oxygen species drive GANT61-induced mesothelioma cell apoptosis. *Oncotarget*, **6**, 1519–1530.

Lin, W.W. and Karin, M. (2007) A cytokine-mediated link between innate immunity, inflammation, and cancer. *Journal of Clinical Investment*, **117**, 1175–1183.

Lippmann, M. (2014) Toxicological and epidemiological studies on effects of airborne fibers: coherence and public [corrected] health implications. *Critical Reviews in Toxicology*, **44**, 643–695.

Luanpitpong, S., Talbott, S.J., Rojanasakul, Y., Nimmannit, U., Pongrakhananon, V., Wang, L. and Chanvorachote, P. (2010) Regulation of lung cancer cell migration and invasion by reactive oxygen species and caveolin-1. *Journal of Biological Chemistry*, **285**, 38832–38840.

Luo, J.L., Maeda, S., Hsu, L.C., Yagita, H. and Karin, M. (2004) Inhibition of NF-kappaB in cancer cells converts inflammation- induced tumor growth mediated by TNFalpha to TRAIL-mediated tumor regression. *Cancer Cell*, **6**, 297–305.

Luys, M.J., de Roy, G., Vansant, E.F. and Adams, F. (1982) Characteristics of asbestos minerals. Structural aspects and infrared spectra. *Journal of the Chemical Society, Faraday Transactions 1: Physical Chemistry in Condensed Phases*, **78**, 3561–3571.

Mariathasan, S., Weiss, D.S., Newton, K., McBride, J., O'Rourke, K., Roose-Girma, M., Lee, W.P., Weinrauch, Y., Monack, D.M. and Dixit, V.M. (2006) Cryopyrin activates the inflammasome in response to toxins and ATP. *Nature*, **440**, 228–232.

Martinon, F., Petrilli, V., Mayor, A., Tardivel, A. and Tschopp, J. (2006) Gout-associated uric acid crystals activate the NALP3 inflammasome. *Nature*, **440**, 237–241.

Mayer, C., Darb-Esfahani, S., Meyer, A.S., Hubner, K., Rom, J., Sohn, C., Braicu, I., Sehouli, J., Hansch, G.M. and Gaida, M.M. (2016) Neutrophil granulocytes in ovarian cancer – induction of epithelial-to-mesenchymal-transition and tumor cell migration. *Journal of Cancer*, **7**, 546–554.

McDonald, J.C. (2010) Epidemiology of malignant mesothelioma–an outline. *Annals of Occupational Hygiene*, **54**, 851–857.

Mellier, G., Huang, S., Shenoy, K. and Pervaiz, S. (2010) TRAILing death in cancer. *Molecular Aspects of Medicine*, **31**, 93–112.

Menges, C.W., Chen, Y., Mossman, B.T., Chernoff, J., Yeung, A.T. and Testa, J.R. (2010) A phosphotyrosine proteomic screen identifies multiple tyrosine kinase signaling pathways aberrantly activated in malignant mesothelioma. *Genes and Cancer*, **1**, 493–505.

Merino, D., Lalaoui, N., Morizot, A., Schneider, P., Solary, E. and Micheau, O. (2006) Differential inhibition of TRAIL-mediated DR5-DISC formation by decoy receptors 1 and 2. *Molecular and Cellular Biology*, **26**, 7046–7055.

Meunier, E., Coste, A., Olagnier, D., Authier, H., Lefevre, L., Dardenne, C., Bernad, J., Beraud, M., Flahaut, E. and Pipy, B. (2012) Double-walled carbon nanotubes trigger IL-1beta release in human monocytes through Nlrp3 inflammasome activation. *Nanomedicine*, **8**, 987–995.

Mirabelli, D. and Cadum, E. (2002) Mortality among patients with pleural and peritoneal tumors in Alta Valle di Susa. *Epidemiology and Prevention*, **26**, 284–286.

Moalli, P.A., MacDonald, J.L., Goodglick, L.A. and Kane, A.B. (1987) Acute injury and regeneration of the mesothelium in response to asbestos fibers. *American Journal of Pathology*, **128**, 426–445.

Moll, U.M. and Slade, N. (2004) p63 and p73: roles in development and tumor formation. *Molecular Cancer Research*, **2**, 371–386.

Monchaux, G., Bignon, J., Jaurand, M.C., Lafuma, J., Sebastien, P., Masse, R., Hirsch, A. and Goni, J. (1981) Mesotheliomas in rats following inoculation with acid-leached chrysotile asbestos and other mineral fibres. *Carcinogenesis*, **2**, 229–236.

Morgan, A., Davies, P., Wagner, J.C., Berry, G. and Holmes, A. (1977) The biological effects of magnesium-leached chrysotile asbestos. *British Journal of Experimental Pathology*, **58**, 465–473.

Mossman, B.T. and Churg, A. (1998) Mechanisms in the pathogenesis of asbestosis and silicosis. *American Journal of Respiratory and Critical Care Medicine*, **157**, 1666–1680.

Mossman, B.T. and Craighead, J.E. (1975) Long-term maintenance of differentiated respiratory epithelium in organ culture I. Medium composition. *Proceedings of the Society of Experimental Biology and Medicine*, **149**, 227–233.

Mossman, B.T. and Craighead, J.E. (1979) Use of hamster tracheal organ cultures for assessing the cocarcinogenic effects of inorganic particulates on the respiratory epithelium. *Progress in Experimental Tumor Research*, **24**, 37–47.

Mossman, B.T. and Craighead, J.E. (1982) Comparative cocarcinogenic effects of crocidolite asbestos, hematite, kaolin and carbon in implanted tracheal organ cultures. *Annals of Occupational Hygiene*, **26**, 553–567.

Mossman, B.T. and Marsh, J.P. (1989) Evidence supporting a role for active oxygen species in asbestos-induced toxicity and lung disease. *Environmental Health Perspectives*, **81**, 91–94.

Mossman, B.T., Ley, B.W. and Craighead, J.E. (1976) Squamous metaplasia of the tracheal epithelium in organ culture. I. Effects of hydrocortisone and beta-retinyl acetate. *Experimental and Molecular Pathology*, **24**, 405–414.

Mossman, B.T., Kessler, J.B., Ley, B.W. and Craighead, J.E. (1977) Interaction of crocidolite asbestos with hamster respiratory mucosa in organ culture. *Laboratory Investigation*, **36**, 131–139.

Mossman, B.T., Adler, K.B. and Craighead, J.E. (1978) Interaction of carbon particles with tracheal epithelium in organ culture. *Environmental Research*, **16**, 110–122.

Mossman, B.T., Craighead, J.E. and MacPherson, B.V. (1980) Asbestos-induced epithelial changes in organ cultures of hamster trachea: inhibition by retinyl methyl ether. *Science*, **207**, 311–313.

Mossman, B.T., Eastman, A., Landesman, J.M. and Bresnick, E. (1983) Effects of crocidolite and chrysotile asbestos on cellular uptake and metabolism of benzo(a)pyrene in hamster tracheal epithelial cells. *Environmental Health Perspectives*, **51**, 331–335.

Mossman, B.T., Eastman, A. and Bresnick, E. (1984) Asbestos and benzo[a]pyrene act synergistically to induce squamous metaplasia and incorporation of [3H]thymidine in hamster tracheal epithelium. *Carcinogenesis*, **5**, 1401–1404.

Mossman, B.T., Bignon, J., Corn, M., Seaton, A. and Gee, J.B. (1990) Asbestos: scientific developments and implications for public policy. *Science*, **247**, 294–301.

Mossman, B.T., Lippmann, M., Hesterberg, T.W., Kelsey, K.T., Barchowsky, A. and Bonner, J.C. (2011) Pulmonary endpoints (lung carcinomas and asbestosis) following inhalation exposure to asbestos. *Journal of Toxicology and Environmental Health, Part B: Critical Reviews*, **14**, 76–121.

Mossman, B.T., Shukla, A., Heintz, N.H., Verschraegen, C.F., Thomas, A. and Hassan, R. (2013) New insights into understanding the mechanisms, pathogenesis, and management of malignant mesotheliomas. *American Journal of Pathology*, **182**, 1065–1077.

Murakami, T., Ockinger, J., Yu, J., Byles, V., McColl, A., Hofer, A.M. and Horng, T. (2012) Critical role for calcium mobilization in activation of the NLRP3 inflammasome. *Proceedings of the National Academy of Sciences of the United States of America*, **109**, 11282–11287.

Murata, M., Thanan, R., Ma, N. and Kawanishi, S. (2012) Role of nitrative and oxidative DNA damage in inflammation-related carcinogenesis. *Journal of Biomedicine and Biotechnology*, **2012**, 623019.

Murphy, F.A., Poland, C.A., Duffin, R., Al-Jamal, K.T., Ali-Boucetta, H., Nunes, A., Byrne, F., Prina-Mello, A., Volkov, Y., Li, S., Mather, S.J., Bianco, A., Prato, M., Macnee, W., Wallace, W.A., Kostarelos, K. and Donaldson, K. (2011) Length-dependent retention of carbon nanotubes in the pleural space of mice initiates sustained inflammation and progressive fibrosis on the parietal pleura. *American Journal of Pathology*, **178**, 2587–2600.

Murphy, F.A., Schinwald, A., Poland, C.A. and Donaldson, K. (2012) The mechanism of pleural inflammation by long carbon nanotubes: interaction of long fibres with macrophages stimulates them to amplify pro-inflammatory responses in mesothelial cells. *Particle and Fibre Toxicology*, **9**, 8.

Musumeci, G., Cardile, V., Fenga, C., Caggia, S. and Loreto, C. (2011) Mineral fibre toxicity: expression of retinoblastoma (Rb) and phospho-retinoblastoma (pRb) protein in alveolar epithelial and mesothelial cell lines exposed to fluoro-edenite fibres. *Cell Biology and Toxicology*, **27**, 217–225.

NatureBeachy, P.A., Karhadkar, S.S. and Berman, D.M. (2004) Tissue repair and stem cell renewal in carcinogenesis. *Nature*, **432**, 324–331.

O'Kane, S.L., Eagle, G.L., Greenman, J., Lind, M.J. and Cawkwell, L. (2010) COX-2 specific inhibitors enhance the cytotoxic effects of pemetrexed in mesothelioma cell lines. *Lung Cancer*, **67**, 160–165.

Okamoto, M., Liu, W., Luo, Y., Tanaka, A., Cai, X., Norris, D.A., Dinarello, C.A. and Fujita, M. (2010) Constitutively active inflammasome in human melanoma cells mediating autoinflammation via caspase-1 processing and secretion of interleukin-1beta. *Journal of Biological Chemistry*, **285**, 6477–6488.

Oppenheimer, S.B. (2006) Cellular basis of cancer metastasis: A review of fundamentals and new advances. *Acta Histochemistry*, **108**, 327–334.

Otero Areán, C., Barcelo, F., Fenoglio, I., Fubini, B., Llabres i Xamena, F.X. and Tomatis, M. (2001) Free radical activity of natural and heat treated amphibole asbestos. *Journal of Inorganic Biochemistry*, **83**, 211–216.

Pacella, A., Andreozzi, G.B., Fournier, J., Stievano, L., Giantomassi, F., Lucarini, G., Rippo, M.R. and Pugnaloni, A. (2012) Iron topochemistry and surface reactivity of amphibole asbestos: relations with *in vitro* toxicity. *Analytical and Bioanalytical Chemistry*, **402**, 871–881.

Pallotti, F. and Lenaz, G. (2001) Isolation and subfractionation of mitochondria from animal cells and tissue culture lines. *Methods in Cell Biology*, **65**, 1–35.

Palomaki, J., Valimaki, E., Sund, J., Vippola, M., Clausen, P.A., Jensen, K.A., Savolainen, K., Matikainen, S. and Alenius, H. (2011) Long, needle-like carbon nanotubes and asbestos activate the NLRP3 inflammasome through a similar mechanism. *ACS Nano*, **5**, 6861–6870.

Panduri, V., Surapureddi, S., Soberanes, S., Weitzman, S.A., Chandel, N. and Kamp, D.W. (2006) P53 mediates amosite asbestos-induced alveolar epithelial cell mitochondria-regulated apoptosis. *American Journal of Respiratory Cell and Molecular Biology*, **34**, 443–452.

Pascolo, L., Zabucchi, G., Gianoncelli, A., Kourousias, G., Trevisan, E., Pascotto, E., Casarsa, C., Ryan, C., Lucattelli, M., Lungarella, G., Cavarra, E., Bartalesi, B., Zweyer, M., Cammisuli, F., Melato, M. and Borelli, V. (2016) Synchrotron X-ray microscopy reveals early calcium and iron interaction with crocidolite fibers in the lung of exposed mice. *Toxicology Letters*, **241**, 111–120.

Petrilli, V., Papin, S., Dostert, C., Mayor, A., Martinon, F. and Tschopp, J. (2007) Activation of the NALP3 inflammasome is triggered by low intracellular potassium concentration. *Cell Death and Differentiation*, **14**, 1583–1589.

Pierce, J.S., Ruestow, P.S. and Finley, B.L. (2016) An updated evaluation of reported no-observed adverse effect levels for chrysotile asbestos for lung cancer and mesothelioma. *Critical Reviews in Toxicology*, **46**, 561–586.

Pugnaloni, A., Giantomassi, F., Lucarini, G., Capella, S., Belmonte, M.M., Orciani, M. and Belluso, E. (2010) Effects of asbestiform antigorite on human alveolar epithelial A549 cells: a morphological and immunohistochemical study. *Acta Histochemistry*, **112**, 133–146.

Pugnaloni, A., Giantomassi, F., Lucarini, G., Capella, S., Bloise, A., Di Primio, R. and Belluso, E. (2013) Cytotoxicity induced by exposure to natural and synthetic tremolite asbestos: an *in vitro* pilot study. *Acta Histochemistry*, **115**, 100–112.

Pugnaloni, A., Lucarini, G., Giantomass, I.F., Lombardo, L., Capella, S., Belluso, E., Zizzi, A., Panico, A.M., Biagini, G. and Cardile, V. (2007) In vitro study of biofunctional indicators after exposure to asbestos-like fluoro-edenite fibres. *Cellular and Molecular Biology*, **53 Suppl**, OL965–980.

Pugnaloni, A., Lucarini, G., Rubini, C., Smorlesi, A., Tomasetti, M., Strafella, E., Armeni, T. and Gualtieri, A.F. (2015) Raw and thermally treated cement asbestos exerts different cytotoxicity effects on A549 cells *in vitro*. *Acta Histochemistry*, **117**, 29–39.

Quinlan, T.R., BeruBe, K.A., Hacker, M.P., Taatjes, D.J., Timblin, C.R., Goldberg, J., Kimberley, P., O'Shaughnessy, P., Hemenway, D., Torino, J., Jimenez, L.A. and Mossman, B.T. (1998) Mechanisms of asbestos-induced nitric oxide production by rat alveolar macrophages in inhalation and *in vitro* models. *Free Radical Biology and Medicine*, **24**, 778–788.

Rapisarda, V., Loreto, C., Ledda, C., Musumeci, G., Bracci, M., Santarelli, L., Renis, M., Ferrante, M. and Cardile, V. (2015) Cytotoxicity, oxidative stress and genotoxicity induced by glass fibers on human alveolar epithelial cell line A549. *Toxicology In Vitro*, **29**, 551–557.

Rastrick, J. and Birrell, M. (2014) The role of the inflammasome in fibrotic respiratory diseases. *Minerva Medica*, **105**, 9–23.

Reid, G. (2015) MicroRNAs in mesothelioma: from tumour suppressors and biomarkers to therapeutic targets. *Journal of Thoracic Disease*, **7**, 1031–1040.

Reinmuth, N., Jauch, A., Xu, E.C., Muley, T., Granzow, M., Hoffmann, H., Dienemann, H., Herpel, E., Schnabel, P.A., Herth, F.J., Gottschling, S., Lahm, H., Steins, M., Thomas, M. and Meister, M. (2008) Correlation of EGFR mutations with chromosomal alterations and expression of EGFR, ErbB3 and VEGF in tumor samples of lung adenocarcinoma patients. *Lung Cancer*, **62**, 193–201.

Reiss, B., Solomon, S., Tong, C., Levenstein, M., Rosenberg, S.H. and Williams, G.M. (1982) Absence of mutagenic activity of three forms of asbestos in liver epithelial cells. *Environmental Research*, **27**, 389–397.

Riché, E., Carrié, A., Andin, N. and Mabic, S. (2006) High-purity water and pH. *American Laboratory*, **38**, 22.

Ridley, A.J. (2001) Rho family proteins: coordinating cell responses. *Trends in Cell Biology*, **11**, 471–477.

Riganti, C., Aldieri, E., Bergandi, L., Fenoglio, I., Costamagna, C., Fubini, B., Bosia, A. and Ghigo, D. (2002) Crocidolite asbestos inhibits pentose phosphate oxidative pathway and glucose 6-phosphate dehydrogenase activity in human lung epithelial cells. *Free Radical Biology and Medicine*, **32**, 938–949.

Rinaudo, C., Belluso, E. and Gastaldi, D. (2004) Assessment of the use of Raman spectroscopy for the determination of amphibole asbestos. *Mineralogical Magazine*, **68**, 455–465.

Robinson, C., Alfonso, H., Woo, S., Olsen, N., Bill Musk, A.W., Robinson, B.W., Nowak, A.K. and Lake, R.A. (2014) Effect of NSAIDS and COX-2 inhibitors on the incidence and severity of asbestos-induced malignant mesothelioma: evidence from an animal model and a human cohort. *Lung Cancer*, **86**, 29–34.

Roggli, V.L. (2015) The so-called short-fiber controversy: Literature review and critical analysis. *Archives of Pathology and Laboratory Medicine*, **139**, 1052–1057.

Rossol, M., Pierer, M., Raulien, N., Quandt, D., Meusch, U., Rothe, K., Schubert, K., Schoneberg, T., Schaefer, M., Krugel, U., Smajilovic, S., Brauner-Osborne, H., Baerwald, C. and Wagner, U. (2012) Extracellular Ca^{2+} is a danger signal activating the NLRP3 inflammasome through G protein-coupled calcium sensing receptors. *Nature Communications*, **3**, 1329.

Rudd, R.M. (2010) Malignant mesothelioma. *British Medical Bulletin*, **93**, 105–123.

Sayan, M. and Mossman, B.T. (2016) The NLRP3 inflammasome in pathogenic particle and fibre-associated lung inflammation and diseases. *Particle and Fibre Toxicology*, **13**, 51.

Schinwald, A., Murphy, F.A., Prina-Mello, A., Poland, C.A., Byrne, F., Movia, D., Glass, J.R., Dickerson, J.C., Schultz, D.A., Jeffree, C.E., Macnee, W. and Donaldson, K. (2012) The threshold length for fiber-induced acute pleural inflammation: shedding light on the early events in asbestos-induced mesothelioma. *Toxicological Sciences*, **128**, 461–470.

Schroder, K. and Tschopp, J. (2010) The inflammasomes. *Cell*, **140**, 821–832.

Sekine, S., Shibata, T., Matsuno, Y., Maeshima, A., Ishii, G., Sakamoto, M. and Hirohashi, S. (2003) Beta-catenin mutations in pulmonary blastomas: association with morule formation. *Journal of Pathology*, **200**, 214–221.

Shukla, A., Gulumian, M., Hei, T.K., Kamp, D., Rahman, Q. and Mossman, B.T. (2003a) Multiple roles of oxidants in the pathogenesis of asbestos-induced diseases. *Free Radical Biology and Medicine*, **34**, 1117–1129.

Shukla, A., Jung, M., Stern, M., Fukagawa, N.K., Taatjes, D.J., Sawyer, D., Van Houten, B. and Mossman, B.T. (2003b) Asbestos induces mitochondrial DNA damage and dysfunction linked to the development of apoptosis. *American Journal of Physiology - Lung Cellular and Molecular Physiology*, **285**, L1018–1025.

Spierings, D.C., de Vries, E.G., Timens, W., Groen, H.J., Boezen, H.M. and de Jong, S. (2003) Expression of

TRAIL and TRAIL death receptors in stage III non-small cell lung cancer tumors. *Clinical Cancer Research*, **9**, 3397–3405.

Spurny, K.R., Pott, F., Stöber, W., Opiela, H., Schörmann, J. and Weiss, G. (1983) On the chemical changes of asbestos fibers and MMMFs in biologic residence and in the environment: part 1. *American Industrial Hygiene Association Journal*, **44**, 833–845.

Srivastava, R.K., Lohani, M., Pant, A.B. and Rahman, Q. (2010) Cyto-genotoxicity of amphibole asbestos fibers in cultured human lung epithelial cell line: role of surface iron. *Toxicology of Industrial Medicine*, **26**, 575–582.

Stegmann, K.A., De Souza, J.B. and Riley, E.M. (2015) IL-18-induced expression of high-affinity IL-2R on murine NK cells is essential for NK-cell IFN-gamma production during murine Plasmodium yoelii infection. *European Journal of Immunology*, **45**, 3431–3440.

Stengel, K. and Zheng, Y. (2011) Cdc42 in oncogenic transformation, invasion, and tumorigenesis. *Cellular Signalling*, **23**, 1415–1423.

Stratmann, A.T., Fecher, D., Wangorsch, G., Gottlich, C., Walles, T., Walles, H., Dandekar, T., Dandekar, G. and Nietzer, S.L. (2014) Establishment of a human 3D lung cancer model based on a biological tissue matrix combined with a Boolean in silico model. *Molecular Oncology*, **8**, 351–365.

Sun, B., Wang, X., Ji, Z., Wang, M., Liao, Y.P., Chang, C.H., Li, R., Zhang, H., Nel, A.E. and Xia, T. (2015) NADPH oxidase-dependent NLRP3 inflammasome activation and its important role in lung fibrosis by multiwalled carbon nanotubes. *Seminars in Cancer Biology*, **11**, 2087–2097.

Suzuki, Y. and Yuen, S.R. (2002) Asbestos fibers contributing to the induction of human malignant mesothelioma. *Annals of the New York Academy of Sciences*, **982**, 160–176.

Suzuki, Y., Yuen, S.R. and Ashley, R. (2005) Short, thin asbestos fibers contribute to the development of human malignant mesothelioma: pathological evidence. *International Journal Hygiene and Environmental Health*, **208**, 201–210.

Tennis, M., Van Scoyk, M. and Winn, R.A. (2007) Role of the wnt signaling pathway and lung cancer. *Journal of Thoracic Oncology*, **2**, 889–892.

Testa, J.R., Cheung, M., Pei, J., Below, J.E., Tan, Y., Sementino, E., Cox, N.J., Dogan, A.U., Pass, H.I., Trusa, S., Hesdorffer, M., Nasu, M., Powers, A., Rivera, Z., Comertpay, S., Tanji, M., Gaudino, G., Yang, H. and Carbone, M. (2011) Germline BAP1 mutations predispose to malignant mesothelioma. *Nature Genetics*, **43**, 1022–1025.

Thompson, J.K., Westbom, C.M., MacPherson, M.B., Mossman, B.T., Heintz, N.H., Spiess, P. and Shukla, A. (2014) Asbestos modulates thioredoxin-thioredoxin interacting protein interaction to regulate inflammasome activation. *Particle and Fibre Toxicology*, **11**, 24.

Timblin, C.R., Janssen, Y.M., Goldberg, J.L. and Mossman, B.T. (1998) GRP78, HSP72/73, and cJun stress protein levels in lung epithelial cells exposed to asbestos, cadmium, or H_2O_2. *Free Radical Biology and Medicine*, **24**, 632–642.

Topping, D.C. and Nettesheim, P. (1980) Two-stage carcinogenesis studies with asbestos in Fischer 344 rats. *Journal of the National Cancer Institute*, **65**, 627–630.

Toyokuni, S. (2009) Mechanisms of asbestos-induced carcinogenesis. *Nagoya Journal of Medical Science*, **71**, 1–10.

Travaglione, S., Bruni, B., Falzano, L., Paoletti, L. and Fiorentini, C. (2003) Effects of the new-identified amphibole fluoro-edenite in lung epithelial cells. *Toxicology in Vitro*, **17**, 547–552.

Tucci, M.G., Lucarini, G., Brancorsini, D., Zizzi, A., Pugnaloni, A., Giacchetti, A., Ricotti, G. and Biagini, G. (2007) Involvement of E-cadherin, beta-catenin, Cdc42 and CXCR4 in the progression and prognosis of cutaneous melanoma. *British Journal of Dermatology*, **157**, 1212–1216.

Turci, F., Tomatis, M., Compagnoni, R. and Fubini, B. (2009) Role of associated mineral fibres in chrysotile asbestos health effects: the case of balangeroite. *Annals of Occupational Hygiene*, **53**, 491–497.

Valouma, A., Verganelaki, A., Maravelaki-Kalaitzaki, P. and Gidarakos, E. (2016) Chrysotile asbestos detoxification with a combined treatment of oxalic acid and silicates producing amorphous silica and biomaterial. *Journal of Hazardous Materials*, **305**, 164–170.

van Dekken, H., Wink, J.C., Vissers, K.J., Franken, P.F., Ruud Schouten, W., WC, J.H., Kuipers, E.J., Fodde, R. and Janneke van der Woude, C. (2007) Wnt pathway-related gene expression during malignant progression in ulcerative colitis. *Acta Histochemistry*, **109**, 266–272.

Viani, A. and Gualtieri, A.F. (2013) Recycling the product of thermal transformation of cement-asbestos for the preparation of calcium sulfoaluminate clinker. *Journal of Hazardous Materials*, **260**, 813–818.

Wagner, J.C., Sleggs, C.A. and Marchand, P. (1960) Diffuse pleural mesothelioma and asbestos exposure in the North Western Cape Province. *British Journal of Industrial Medicine*, **17**, 260–271.

Warheit, D.B., Hartsky, M.A. and Frame, S.R. (1996) Pulmonary effects in rats inhaling size-separated chrysotile asbestos fibres or p-aramid fibrils: differences in cellular proliferative responses. *Toxicology Letters*, **88**, 287–292.

Watanabe, T., Wang, S. and Kaibuchi, K. (2015) IQGAPs as Key Regulators of Actin-cytoskeleton Dynamics. *Cell Structure and Function*, **40**, 69–77.

Wicks, F.J. and O'Hanley, D.S. (1988) Serpentine minerals: structure and petrology. Pp. 91–167 in: *Hydrous Phyllosilicates (Exclusive of Micas)* (S.W. Bailey, editor). Reviews in Mineraology, **19**. Mineralogical Society of America, Washington D.C.

Williams, G.M. (1979) Review of *in vitro* test systems using DNA damage and repair for screening of chemical carcinogens. *Journal of the Association of Official Analytical Chemists*, **62**, 857–863.

Willingham, S.B., Allen, I.C., Bergstralh, D.T., Brickey, W.J., Huang, M.T., Taxman, D.J., Duncan, J.A. and Ting, J.P. (2009) NLRP3 (NALP3, Cryopyrin) facilitates *in vivo* caspase-1 activation, necrosis, and HMGB1 release via inflammasome-dependent and -independent pathways. *Journal of Immunology*, **183**, 2008–2015.

Woodworth, C.D., Mossman, B.T. and Craighead, J.E. (1982) Comparative effects of fibrous and nonfibrous minerals on cells and liposomes. *Environmental Research*, **27**, 190–205.

Woodworth, C.D., Mossman, B.T. and Craighead, J.E. (1983a) Induction of squamous metaplasia in organ cultures of hamster trachea by naturally occurring and synthetic fibers. *Cancer Research*, **43**, 4906–4912.

Woodworth, C.D., Mossman, B.T. and Craighead, J.E. (1983b) Squamous metaplasia of the respiratory tract. Possible pathogenic role in asbestos-associated bronchogenic carcinoma. *Laboratory Investigation*, **48**, 578–584.

Wozniak, H., Wiecek, E. and Stetkiewicz, J. (1988) Fibrogenic and carcinogenic effects of antigorite. *Polish Journal of Occupational Medicine*, **1**, 192–202.

Xiao, H., Tian, Y., Yang, Y., Hu, F., Xie, X., Mei, J. and Ding, F. (2015) USP22 acts as an oncogene by regulating the stability of cyclooxygenase-2 in non-small cell lung cancer. *Biochemical and Biophysical Research Communication*, **460**, 703–708.

Yao, R., Wang, Y., Lubet, R.A. and You, M. (2002) Differentially expressed genes associated with mouse lung tumor progression. *Oncogene*, **21**, 5814–5821.

Zeidler, P.C. and Castranova, V. (2004) Role of nitric oxide in pathological responses of the lung to exposure to environmental/occupational agents. *Redox Report*, **9**, 7–18.

Zhou, R., Yazdi, A.S., Menu, P. and Tschopp, J. (2011) A role for mitochondria in NLRP3 inflammasome activation. *Nature*, **469**, 221–225.

Zhuang, H., Jiang, W., Zhang, X., Qiu, F., Gan, Z., Cheng, W., Zhang, J., Guan, S., Tang, B., Huang, Q., Wu, X., Huang, X., Jiang, W., Hu, Q., Lu, M. and Hua, Z.C. (2013) Suppression of HSP70 expression sensitizes NSCLC cell lines to TRAIL-induced apoptosis by upregulating DR4 and DR5 and downregulating c-FLIP-L expressions. *Journal of Molecular Medicine*, **91**, 219–235.

EMU Notes in Mineralogy, Vol. 18 (2017), Chapter 9, 307–345

In vivo biological activity of mineral fibres

SILVANA CAPELLA[1,*], ELENA BELLUSO[1], NICOLA BURSI GANDOLFI[2], EVA TIBALDI[3], DANIELE MANDRIOLI[3] and FIORELLA BELPOGGI[3]

[1]*Dipartimento di Scienze della Terra, Università di Torino, Via Valperga Caluso, 35, 10125 Torino, Italy, e-mail: silvana.capella@unito.it*
[2]*Dipartimento di Scienze Chimiche e Geologiche, Università di Modena e Reggio Emilia, Via Campi 103, 41125 Modena, Italy*
[3]*Istituto Ramazzini, Centro di Ricerca sul Cancro Cesare Maltoni, Via Saliceto 3, 40010 Bentivoglio Bologna, Italy*
**Corresponding author*

Over time, attention to health effects induced by inorganic fibres has increased, resulting not only from non-occupational exposure to asbestos (*i.e.* environmental, either natural or anthropogenic), but also of exposure to non-asbestos inorganic fibres. The International Agency for Research on Cancer classified all forms of asbestos (chrysotile, crocidolite, amosite, tremolite, actinolite and anthophyllite) as carcinogenic to humans (Group 1). The most effective source of information on the health effects of environmental contaminants in humans would be humans themselves as research subjects. Obviously, for ethical and other reasons this is rarely possible. Carcinogen bioassays, such as the ones performed by the Ramazzini Institute (RI), are currently the most predictive non-human model for studying the carcinogenicity of a substance. An alternative method for exposure assessment consists of using a non-experimental animal model represented by populations of animal sentinel systems (ASS). Even if at present there are relatively few studies in this field, the presence of fibres in samples of lung tissue of these animals shows that it is possible to use ASS as indicators of environmental background exposure. However, caution must be taken in extrapolating results from ASS and considering them in terms of human health. The physical mechanisms rather than chemical reaction between inorganic fibres and cells are still unclear. Joint research conducted by a team of biologists and mineralogists in order to investigate the mechanisms of action of the fibres in biological tissues of rodents *in vivo* have been undertaken by the Department of Chemical and Geological Sciences, University of Modena and Reggio Emilia in collaboration with the Cesare Maltoni Cancer Research Centre of the Ramazzini Institute CMCRC/RI. When planned for precise purposes with attention to the key, general, and specific requisites, and when conducted by standardized methods, experimental studies represent an important tool capable of providing useful information and reducing the uncertainties in terms of risk assessment.

DOI: 10.1180/EMU-notes.18.9

1. Introduction

Asbestos is found worldwide but it represents only a fraction of the many inorganic fibres found in nature to which humans are exposed on a daily basis. The association between occupational exposure of asbestos and asbestos-related diseases has been well documented for decades (IARC, 2012). Since Iceland first introduced a ban on all types of asbestos in 1983, more than 50 countries have implemented similar bans (Collegium Ramazzini, 2016a). The indictment of asbestos as a health hazard focused attention also on exposure to other inorganic fibres (IARC, 2014). Exposure occurs commonly through inhalation, other routes of exposure may be possible and all are related to carcinogenic and other harmful health effects in humans and animals (IARC, 2012). In particular asbestos causes malignant mesothelioma (MM) and cancer of the lung, larynx and ovary.

The fate of fibres depends on both the site of their deposition and their dimensional and chemical characteristics. For example, after inhalation, a portion of the fibres might be removed by clearance. In the medical literature, the term 'clearance' refers to the process by which the human body attempts to rid itself of the dust that we inhale. Briefly, clearance starts first when fibres enter the nose and some of them are trapped there. Other fibres continue on their way through the larynx to the trachea-bronchial airways where the mucociliary escalator moves them back to be swallowed or spat out. The fibres that reach the alveoli might be engulfed by macrophages. Some asbestos fibres can be covered with iron-protein-mucopolysaccharide material and these are named asbestos bodies (ABs). ABs can be used as an indicator of past exposure to asbestos, although failure to detect ABs cannot be used as a criterion for excluding exposure to asbestos (DeVuyst *et al.*, 1998; Collegium Ramazzini, 2016b). ABs have a characteristic microscopic appearance that is readily recognized by the pathologist. In addition, a large number of inorganic fibre types other than asbestos can become coated (*e.g.* sheet silicates, carbon fibres, fibrous aluminium silicate, cosmetic talc, fibrous glass, *etc.*) (Churg and Warnock, 1981); therefore the term 'ferruginous bodies' (FBs) is used generally when the nature of the fibrous core was not known (see Section 5).

In this chapter, we investigate the role of various techniques for the detection of fibres and ABs, then we discuss the possibility of using: (1) populations of animal sentinel system (ASS) as models to assess the environmental exposure to aerodispersed inorganic fibres (asbestos and non-asbestos) and to provide early warning of possible implications for human health (O'Brien *et al.*, 1993; Stahl, 1997); (2) *in vivo* animal experiments as models for reproducing the diseases associated with asbestos exposure; life-span chronic rodent studies are considered to be the standard assay for assessing fibre carcinogenicity (IARC, 2012; Report of an ILSI Risk Science Institute, 2005).

2. Why do we find fibres in living organisms?

The fibres found in lung tissue can be attributed to various kinds of exposure: occupational, environmental natural background and environmental anthropogenic.

Historically, the study of the presence of inorganic fibres in living organisms is related closely to occupational exposure to asbestos, documented primarily by workers in various settings including mining, processing and handling of commercial asbestos. Although the Roman naturalist Pliny the Elder wrote about its harmful biological effects, after observing the 'sickness of the lungs' in the slaves who worked with asbestos, only since the early1900s have asbestos-related diseases been associated with occupational activity. It was during this time that reports of asbestosis began to appear in the literature. The first detailed account of death from asbestosis appeared in 1907 (Murray, 1907), and thereafter, a series of reports described a chronic lung disease in asbestos workers that was neither tuberculosis nor silicosis. In 1917, the radiologic features of this disease were first described. In 1924 the first diagnosis of asbestosis in the UK was made when Cooke (1924) reported the case of a female asbestos worker (Annie Kershaw) and in his paper mentioned the presence of mineral particles of various shapes in her lungs. Subsequently the cause of death was called "asbestosis" (Cooke, 1927). Lynch and Smith (1935) described a lung carcinoma in a patient with asbestosis. The pathologist Gloyne (1933) was the first to describe pleural malignancy in a subject exposed occupationally to asbestos. Reports by Wedler (1943) and Mallory *et al.* (1947) describing further cases of pleural malignancy associated with asbestos exposure followed. By the 1950s, MM became accepted as a distinct clinical-pathological entity. Any remaining doubt concerning the association between pleural MM and asbestos occupational exposure was removed with the studies of Wagner *et al.* (1960) and Selikoff *et al.* (1964).

Nowadays, epidemiological studies of cohorts of workers with occupational exposure to asbestos are very common in the medical literature (see Case and Marinaccio, 2017, this volume). At present, there is a large body of scientific literature on the presence of asbestos and other asbestiform inorganic fibres in human lungs (and other tissues and organs) and the correlation to their pathogenicity both in humans and in animals (Table 1). In animals (mainly rats, but also mice and hamsters) asbestos has

Table 1. Cancers related to asbestos exposure in humans and experimental animals.

Cancer type	In humans	In experimental animals
Lung cancer	+	+
Pleural mesothelioma	+	+
Peritoneal mesothelioma	+	+
Other site mesothelioma and possibly sarcoma	+	+
Pharyngolaryngeal cancer	+	
Gastrointestinal cancer	+	
Kidney cancer	+	

Data from Soffritti *et al.* (2003)

been tested by inhalation, by intraperitoneal, intrapleural and subcutaneous injection, and by ingestion.

Although its rarity makes MM outbreaks the cancer most specifically related to asbestos exposure in terms of public health, lung cancer is the most frequent malignancy in asbestos-exposed workers. In the series of cases described by Selikoff and Seidman (1991) considering a cohort of 17,800 insulation workers in the USA and Canada during the period 1967–1987, while 458 deaths were due to MM, >899 deaths due to lung cancer were registered. Lung cancer and MM in people exposed to asbestos may be preceded by or associated with lung fibrosis and pleural plaques. These changes may represent a marker of asbestos exposure. However, epidemiologic studies on asbestos-exposed workers indicated that lung cancer should not be related necessarily to the presence of pleural plaques (or pleural asbestosis). Many lung cancers are diagnosed without any previous diagnosis of pleural plaques. This is partly because the lung parenchyma and parietal pleura are anatomically and histo-pathologically different. Furthermore, when pleural plaques and lung cancer are diagnosed simultaneously, it is very difficult to establish which pathology arose first (Nurminem and Tossavainen, 1994). Other studies have shown that the relative risk of lung cancer is higher in patients with asbestosis (Hillerdal *et al.*, 1997), albeit lung cancer may be caused by a relatively low exposure to asbestos, and may arise independently of parenchymal fibrosis (Anttila *et al.*, 1993; Karjalainen *et al.*, 1993, Wilkinson *et al.*, 1995). The absence of a sequential relationship between asbestosis and lung cancer obliges us to include smokers (with a limited exposure to these mineral fibres) in our examination of the role of asbestos in the onset of this malignancy (Hillerdal *et al.*, 1997). Outstanding epidemiologic studies have shown that tobacco smoke increases enormously the carcinogenic effect of asbestos on lung cancer (Hammond *et al.*, 1979). The number of occupational groups at risk of asbestos cancer has been growing, and the incidence of asbestos-related cancers has increased in recent years. The Ramazzini Institute (RI) collected a number of cases, observed occasionally among different categories of workers exposed to asbestos. Examples of new risk groups, with an increasing frequency of asbestos-related cancers, include MM among workers exposed to asbestos used in the railways (Maltoni *et al.*, 2002), in sugar refining (Belpoggi *et al.*, 2007), in factories as a heat insulator (Maltoni *et al.*, 1994, 1995, 1998), in the asbestos-cement industry (Magnani *et al.*, 1996; Maltoni *et al.*, 1998), and among shipyard workers (Bianchi *et al.*, 2000; Jemal *et al.*, 2000; Puntoni *et al.*, 2001). Also urinary and biliary duct cancers have been associated with asbestos exposure (Lauriola *et al.*, 2012; Brandi *et al.*, 2008). To date, the last Italian report of ReNaM INAIL, with data related to the number of MMs diagnosed in Italy up to December 2012, shows that the categories of workers most affected are engineering industry workers, with 1,243 cases of MMs, and shipyard workers, in which the number of cases has reached 1,253 units. For railway workers, the number of MMs reached 505 units, while in sugar refineries it is 139 units; the largest number of cases was reported in the Emilia-Romagna, Veneto and Tuscany regions for both categories (INAIL, 5° Rapporto il Registro Nazionale dei Mesoteliomi, 2015).

Over time, attention to health effects resulting from other types of exposure (*i.e.* environmental, either natural or anthropogenic) and from non-asbestos inorganic fibres has also increased. Studies have shown that a number of non-asbestos inorganic fibres can be recovered from human lung tissue samples (Churg, 1983) such as talc, silica, rutile, kaolinite and other silicates, fibrous glass and MMMF.

Asbestos as a natural component of soils or rocks (NOA – Naturally Occurring Asbestos) as opposed to asbestos mining or in commercial products (ACM – Asbestos-Containing Material), can be released by natural weathering processes or routine human activity. Natural and anthropogenic sources of asbestos or other asbestiform inorganic fibres are present worldwide. In some of these areas, a high incidence of cases of MM represented a sentinel event pointing to environmental exposure (IARC, 2014). In the small town of Libby, Montana, a high incidence of asbestos-related mortality and respiratory disease was attributed to amphibole mineral fibres, originally believed to be tremolite intergrown with the vermiculite ore mined and milled from 1920s until 1990 (Meeker *et al.*, 2003). Now we know that several different species of amphiboles (approximately 1/3 with asbestiform habit) occur in that deposit, with tremolite being <10% of the total amphibole population (Meeker *et al.*, 2003), while the majority are winchite and richterite. The significant increase for both asbestosis and MM in workers exposed to vermiculite containing amphiboles was observed (McDonald *et al.*, 2004). One of the main uses of the vermiculite ore mined at Libby was for the insulation of attics and walls of homes in the form of a product called Zonolite. Cases of pleural disease among people exposed only environmentally have also been documented: Whitehouse *et al.* (2008) reported 11 cases of MM among people who were not exposed occupationally.

Erionite and fluoro-edenite are, instead, examples of minerals involved in natural and anthropogenic environmental exposure to non-asbestos fibres. They occur in a fibrous or asbestiform habit and they have been associated with MM or other asbestos-related diseases.

In the literature numerous reports of the incidence of pleural and peritoneal MMs (in both in men and women) amongst the occupants of some villages in the Cappadocia region of Central Anatolia (Baris *et al.*, 1978, 1987a,b; Rohl *et al.*, 1982; Temel and Gündogdu, 1996) revealed fibres of erionite, a fibrous zeolite from the locally occurring volcanic tuff, as being responsible. Lung cancer also appeared to be excessive (Baris *et al.*, 1996).

Epidemiological study has been carried out aimed at investigating the relationship between environmental exposure to erionite fibres and the reported high incidence of MMs. The high incidence of MM and lung cancer has been attributed to the presence of erionite in the soil, road dust and building stones of these villages (Baris *et al.*, 1978; IARC, 1987). It is believed that this material was used first during the Roman Empire to build houses, construct roads, sewage channels and milestones (Mumpton, 1975). Airborne fibre levels are generally low but show a larger proportion of erionite fibres in the villages affected by malignant disease than in a control village (Baris *et al.*, 1987a,b). It is significant that the registered increase of MMs in Sweden is due partly to cases of this neoplasia appearing in Turkish migrant workers, probably exposed to erionite at an early age in their country of origin (Metintas *et al.*, 1999). The hypothesis that erionite is the causative agent of the Turkish MMs, and therefore that it is a human carcinogen, has been supported by experimental evidence. Wagner *et al.* (1985) examined the relationship between the exposure (intrapleural inoculation technique) to asbestiform erionite from this region and MM, using rats. MM was observed in almost 100% of rats exposed to erionite. This result is particularly significant when compared with those obtained following injection of chrysotile asbestos (48% of cases of MM) or inhalation of crocidolite (0% of cases of MM). These outcomes confirm previous (Maltoni *et al.*, 1982; Maltoni and Minardi, 1983) reports of peritoneal and pleural MMs in rats and mice following inhalation exposure or intraperitoneal or intrapleural injection of erionite. Other studies confirmed the carcinogenicity of erionite in rats (Maltoni and Minardi, 1989; Hill *et al.*, 1990; Coffin *et al.*, 1992). Later, lung analysis of sheep exposed environmentally confirmed that the presence of asbestiform erionite is broadly correlated with the geographical distribution of the human disease (Baris *et al.*, 1987a). The demonstration of the carcinogenic effect of erionite is of particular relevance considering the large amount and widespread diffusion of natural fibrous and non-fibrous zeolites, their widespread industrial uses, which are expected to increase, and the production of zeolites for several industrial applications (such as detergents, and as catalysts in the petrochemical and refining industries).

In the monitoring report related to the 1988–1992 period, a significant increase in MM was also reported in the town of Biancavilla (Sicily) located on the lower southwestern flaks of the Etna volcano (Il Calvario and Poggio Rosso areas). Industrial use of asbestos had never occurred in Biancavilla but an environmental survey suggested the stone quarries located in Monte Calvario as a possible source of this asbestiform amphibole. This material was used widely in local building. These cases were characterized by relatively low age upon diagnosis and greater mortality in women than in men. The fibres detected in human lung samples, classified as fluoro-edenite, appeared to be amphibole fibres found in the volcanic products and were classified initially as intermediate phases between tremolite and actinolite. Fluoro-edenite fibres were found in the materials obtained from the local quarries (locally named 'azolo') and were used widely for local building activities for ~60 years. They were associated with their possible adverse health effects due to environmental contamination (Gianfagna *et al.*, 2003). The RI study on this fibre provided by the Istituto Superiore di Sanità (ISS, Italy) confirmed the strong mesotheliomatogenic effect of fluoroedenite (Belpoggi *et al.*, 2011).

Tremolite asbestos, generated by natural weathering or human disturbance of certain rock types, has been identified in the lungs of residents of some areas of two Mediterranean islands, Cyprus (McConnochie *et al.*, 1987, 1989) and northeast Corsica (France) (Magee *et al.*, 1986; Viallat *et al.*, 1991; Rey *et al.*, 1993a,b). Among the population with non-occupational exposure, there is evidence of excess MM (Rey *et al.*, 1993a,b). In both areas, a high concentration of tremolite asbestos has been identified in air specimens (McConnochie *et al.*, 1987, 1989; Rey *et al.*, 1993a,b). An increase in cases of MM by anthropogenic environmental exposure to tremolite asbestos has also been reported in inhabitants of Greece (Constantopoulus *et al.*, 1987; Sakellarious *et al.*, 1996), central Turkey (Yazicioglu *et al.*, 1980; Selçuk *et al.*, 1992) and New Caledonia (Luce *et al.*, 1994, 2004; Browne and Wagner 2001). The people of these villages lived in houses whitewashed and plastered using materials from local friable rock quarried from local outcroppings that contained this kind of amphibole, and where streets were covered with the same materials.

A separate case is that of balangeroite, a mineral present in the serpentinite outcrops occurring in the Italian Western Alps and a contaminant of the chrysotile from mines of Balangero (Piedmont, Italy, the largest disused asbestos mine in western Europe). This asbestiform fibrous silicate appears morphologically similar to amphibole asbestos fibres with demonstrated biopersistence and *in vitro* cytotoxicity, but does not belong to either the serpentine or the amphibole families (Belluso *et al.*, 1997; Groppo *et al.*, 2005). No exposure to pure balangeroite has ever occurred; even when isolated by means of physical methods, traces of chrysotile are present, because of the innate intergrowth between the two minerals. The durability of the balangeroite fibres in simulated body fluids is remarkably high, higher even than that of crocidolite, suggesting that they are biopersistent. In addition, the balangeroite fibres exhibit cytotoxic and oxidative properties similar to those exerted by crocidolite asbestos (Gazzano *et al.*, 2005; Turci *et al.*, 2005). An excess mortality for asbestos-related cancer has been reported in Balangero (Pirastu *et al.*, 2011).

Although the present volume deals with mineral fibres, a small paragraph is devoted to man-made mineral fibres (MMMF), which represent another class of non-asbestos mineral fibres that are identified occasionally in human lung samples. MMMF include fibreglass, glass wool, stone wool, slag wool and refractory ceramic fibres. Data on the carcinogenicity of MMMF is of great public interest because of the various industrial uses (the vast majority as asbestos substitutes). At present more than 5 million t of MMMF are produced annually in more than 100 factories located throughout the world. Glass fibre products comprise >50% of the total. MMMF have properties that make them inhalable and potentially harmful, which have led to studies to assess their pathogenicity. MMMF have some similarities with asbestos fibres, including the same aerodynamic properties. The experimental bioassays on carcinogenicity have been performed on rodents, mostly rats, but also mice and hamsters, in which the materials were administered by inhalation and/or intrapleural and intraperitoneal injection/implantation. Three epidemiological studies on workers employed in glass-fibre manufacturing facilities in Canada (Shannon *et al.*, 1987), the United States (Enterline *et al.*, 1987; Marsh *et al.*, 1990) and Europe (Simonato *et al.*, 1987) came to the conclusion that glass wool fibres play a role in causing the excess of lung cancer risk observed among those employees (Doll, 1987). Several MMMF have been evaluated as possible carcinogens by the IARC (2002). Fibrous glass production and use should therefore be regulated, and prompt measures of prevention should be undertaken (EPA, 2017).

3. The analysis of mineral fibres in biological samples

The identification and quantification of asbestos fibres in airborne and bulk samples are regulated by laws in many countries; on the other hand, however, for biological samples a standard approach to sample preparation and examination and for identification and quantification of inorganic fibres does not exist. This leads to great discrepancy among different laboratories worldwide; it would be useful to standardize a universally accepted method (Collegium Ramazzini, 2016b).

Inorganic fibres in bulk samples are identified using techniques such as X-ray diffractometry, infrared spectroscopy, or polarizing microscopy; these are inadequate for identifying inorganic fibres in biological samples. Instead, bright field light microscopy (LM), phase contrast light microscopy (PCLM), scanning electron microscopy (SEM) and transmission electron microscopy (TEM), these last two coupled with energy dispersive spectroscopy (EDS) were found to be appropriate for this type of investigation.

Conventional LM is a simple and inexpensive technique for the quantification of ABs. They are the histological hallmark of exposure to asbestos (DeVuyst *et al.*, 1998) because they have a characteristic appearance that is readily recognized by the pathologist. PCLM has also been used to quantify the inorganic fibres in lung tissue. Unfortunately, the majority are too fine to be detected with these techniques because they are beyond the resolution limits of the technique (0.2 μm in diameter). So it is impossible to quantify the number/gram of dry lung tissue. Moreover, it is impossible to distinguish the various species chemically.

The use of SEM and TEM with EDS is valid for the identification and quantification of inorganic fibres isolated from human tissues (Gylseth *et al.*, 1981); they offer several advantages over LM and PCLM thanks to both the higher magnification and superior resolution. TEM provides the highest resolution for the identification of the smallest fibrils (Collegium Ramazzini, 2016b), while SEM-EDS is often considered the best method for analysis in terms of cost/benefit (Belluso *et al.*, 2003; Roggli and Sharma, 2014). Furthermore, EDS combined with SEM and TEM allows determination of a qualitative chemical composition for inorganic fibres. This information is useful for classifying a fibre as asbestos or non-asbestos: not infrequently, non-asbestos fibres represent an important proportion of the total fibre burden. TEM has the further advantage that selected area electron diffraction (SAED) provides information regarding the crystalline structure, valuable when the chemical composition of two fibres are similar (*e.g.* chrysotile and antigorite). The disadvantages of TEM are that only a small part of the sample prepared can be observed at a time. Because the amount of material analysed with TEM is very small it is important that the portion of the filter sampled is representative and quantification data is reliable. SEM and TEM analysis of inorganic fibres in lung samples is both time-consuming and expensive and is not performed routinely because of the invasive and potentially risky nature of the lung biopsy, and because the procedure is medically unnecessary (Collegium Ramazzini, 2016b).

An internationally adopted protocol for sampling, preparation, counting technique, identification and quantification of inorganic fibres retained in biological samples does not exist, and each laboratory uses a different procedure (Collegium Ramazzini, 2016b). This is the cause of major difficulties in comparing results from the same sample analysed by different laboratories worldwide (Table 2). In addition, intra-laboratory variation can occur, which may be due either, for example, to different portions of tissue or different sites of the same sample being analysed. Generally, specimens are fixed in formalin, sometimes they are embedded in paraffin, even if using fresh tissue would be the best choice, but this is almost never applicable. Therefore, the analysis of inorganic fibres involves several preparation steps that can influence the analytical result.

Dissolution of organic material (in which inorganic fibres may be embedded) can be effected by chemical digestion of the tissue using sodium or potassium hydroxide, hydrogen peroxide, sodium hypochlorite solution, or by proteolytic enzymes. Most laboratories prefer sodium hypochlorite chemical digestion. Plasma ashing in a muffle furnace at 400–500°C is an alternative approach that is considered unsuitable because it may break long fibres, artificially increasing fibre numbers and decreasing mean fibre lengths. This problem can be avoided by ashing the sample in a low-temperature plasma asher. When digestion is complete, the inorganic residue is collected on an acetate or polycarbonate filter with a pore size of 0.2–0.45 μm. Fibres can be lost during the suspension filtration if the pore-size of the membranes is too large in relation of the size of the inorganic fibres (O'Sullivan *et al.*, 1987).

Tissue embedded in paraffin before chemical digestion has to be deparaffinized and rehydrated, processes that remove some lipids, so that a correlation factor must be applied to equate the values obtained with those from formalin-fixed tissue.

Table 2. Steps employed in the analytical procedure and factors conditioning data.

1. Organic material-removal procedure
   - Wet chemical digestion
   - Low-temperature plasma ashing
2. Recovery procedure
   - Centrifugation
   - Filtration
3. Analytical procedure
   - Microscopic technique (LM, PCLM, SEM, TEM)
   - Magnification used
   - Size of fibres counted
   - Number of fibres or fields counted
4. Results
   - Size of fibres
   - Concentration of asbestos bodies (per gram of wet or dry lung or per cm^3)
   - Concentration of fibres (per gram of wet or dry lung or per cm^3)

To detect inorganic fibres in samples of biological material by SEM-EDS in our laboratory we used the protocol proposed by Belluso *et al.* (2006). The amount of fibres is normalized currently as fibres per gram of dry tissue (ff/gdt), according to the international standard (De Vuyst *et al.*, 1998). The Helsinki Committee attempted to connect the number of fibres in lung tissue to asbestos-related disease (Wolff, 2014), but if finding asbestos fibres in lung tissue is an indicator of past exposure, at the same time failure to find asbestos fibres in lung tissue cannot be used to contradict a reliable occupational history of exposure, in particular in the case of chrysotile that has a short residence time in the lung and only rarely forms ABs (Collegium Ramazzini, 2016b).

4. Fibres in animal lungs

4.1. Animals stand watch over the world's environmental health

The most effective source of information on the health effects of environmental contaminants in humans would be humans themselves as research subjects. Obviously, for ethical and other reasons this is rarely possible.

Therefore, the use of non-human organisms as early warning systems for risk to human health has been adopted. Animals that live in the same environment as we do demonstrate intricate connections between them, us and our surroundings; hence, they can be of help in overcoming these drawbacks. Animals can develop similar environmentally induced diseases and provide information about both humans' exposure levels and potential adverse health effects. The most widely known application of an animal as sentinel system for monitoring is perhaps the canary used in coal mines to warn of potentially lethal carbon monoxide concentrations (Schwabe, 1984). The idea of the "canary in coal mine" has been expanded to monitor environmental health hazards (NRC, 1991; van der Schalie *et al.*, 1999). O'Brien *et al.* (1993) defined as sentinels those "organisms in which changes in known characteristics can be measured to assess the extent of environmental contamination and its implications for human health and to provide early warning of those implications." Stahl (1997) defined a sentinel animal as "any non-human organism that can react to ... an environmental contaminant before the contaminant impacts humans" offering the possibility of expanding our understanding and response to environmental health concerns. The use of the animal sentinel system (ASS) has been slow to gain acceptance in human health risk assessment and public health. One important barrier was the lack of understanding of the underlying mechanisms of toxicity in both the sentinel animal and humans. Another limit is that the anatomy of animal respiratory systems and clearance mechanisms are different from those of humans; there are also anatomical and physiological differences among the various animal species. For example, fibre deposition in the lungs does not occur in the same way (Baker *et al.*, 2001). Jones (1993) examined the available experimental and theoretical results describing deposition and clearance of mineral fibres by animals and humans. This work showed that alveolar deposition efficiency in rat and man is sufficiently similar for particles and fibres with aerodynamic diameters <5 μm. Clearance has been shown to differ between species of animals; it is substantially more rapid in the rat or hamster than in humans, dogs or guinea pigs (Oberdoster, 1988).

Chemical dissolution is likely to occur to a similar extent in the lungs of all species, but disease production is a lifespan phenomenon. Nevertheless, the advantages of the ASS may be extremely useful for the aims of the indirect evaluation of the kind and quantity of airborne fibres of respirable dimensions in the environment. The ASS could be used to map natural environmental risk of asbestos exposure or other kinds of inorganic fibres, which may be classified as non-asbestos but with demonstrable human health effects. There are several potential advantages associated with using sentinel animals as indicators of human hazards:

- the shorter life-time and typically shorter latency periods for the development of some diseases after exposure; this may provide early warning of potential risks before disease develops in the human population (Schilling *et al.*, 1988)
- the absence of typical co-factors present in humans (*e.g.* smoking, occupational exposure) that can bias the interpretation of the aetiology of the diseases (Glickman *et al.*, 1983)
- the greater availability of samples

For these reasons, the use of sentinel animals represents a powerful tool for assessing the exposure to environmental pollution and its effects on health. They are useful in terms of providing evidence for risk assessment, for providing the requisite "early warning"" of a situation requiring further study or for monitoring the course of remedial activities (Maxim, 2013).

4.2. Acrodispersed fibres and sentinel animals

The risk of asbestosis, lung cancer and MM is well documented in humans with occupational exposure to asbestos. Information on high-dose exposure is also obtained after *in vivo* experiments on various kinds of animals, with the aim of understanding the mechanisms that lead to related asbestos diseases. The presence of background airborne inorganic fibres and asbestos at low doses in the natural environment can also determine health risks to the exposed human population, but for a number of reasons, it is more difficult to evaluate this exposure. One reason is that it is difficult to do this kind of investigation directly on humans. Such investigations have been done on animal populations, defined as ASS, which are naturally exposed to various airborne inorganic fibres. Even if at present there are relatively few studies in this field, the presence of fibres in samples of the lung tissue of these animals shows that it is possible to use ASS as indicators of environmental background exposure. However, caution must be taken in extrapolating these results and assessing their relationship with human health.

4.2.1. Examples of ASS involving domestic or companion animals

Pet dogs with MM were used as sentinels to evaluate a possible correlation between environmental risk factors and asbestos-related diseases in their owners (Glickman *et al.*, 1983). Because of the relatively short lifespan of these animals, the latent period for cancer development is decreased. Furthermore they share human spaces, but do not indulge in activities (*e.g.* smoking and working), which complicate interpretation of the results of human epidemiological studies. In this study, 18 dogs with histologically confirmed MMs were diagnosed at the Veterinary Hospital of the University of Pennsylvania, in Philadelphia, from April 1977 to December 1981. Owners were interviewed about their dog's medical history, residence, life style, and exposure to asbestos as well as the occupation and medical history of household members. MM in the dogs were significantly associated with household members' asbestos-related occupation or hobby and the use of flea repellents (powders containing asbestos-contaminated talc). In addition, there was a trend indicating an increased risk of MM with an urban residence. In the same study, lung tissue from three MM patients was analysed using an electron microprobe for fibre content. For comparison, lung tissue was also obtained from three dogs with lung cancer and six with no recognized respiratory or neoplastic disease. The results show that the dogs with MM had, in their lung tissue, the highest levels of chrysotile fibres.

Another study (Harbison and Godleski, 1983) examined retrospectively six dogs with a diagnosis of MM. The clinical and morphologic appearance of canine MM is similar to human MM and may also be associated with exposure to airborne fibres. Lung tissue from these dogs and from control dogs was examined by light microscopy to evaluate the presence of ferruginous bodies. They were found in digested lung tissue from all dogs. The MM cases had significantly more ferruginous bodies than the controls. This quantitative difference suggests that exposure to airborne fibres was not the same in the two groups of dogs.

Chrysotile and tremolite fibres have been detected in the parietal pleura of one dog and two goats living in Corsica. In some areas in the northeast of Corsica there are outcrops of serpentinitic rocks bearing these minerals, and there is an excess of pleural plaques and MMs among local inhabitants (Rey *et al.*, 1993a,b, 1994). Although surface deposits contain both chrysotile and tremolite, airborne asbestos fibres and the asbestos lung burden of exposed inhabitants consisted essentially of tremolite (as assessed by TEM-EDS). In contrast, analysis of the parietal pleura of three animals (one dog and two goats) living in the same area showed a predominance of short chrysotile fibres. Dumortier *et al.* (2002) obtained comparable results – they found evidence of increased concentrations of asbestos fibres in the lung and parietal pleura tissue samples of ten goats from areas with asbestos outcrops in northeast Corsica. Chrysotile and tremolite were the only asbestos fibre types detected both in the lung and in the parietal pleura of all the exposed goats. The TEM-EDS analysis showed a predominance of short chrysotile fibres in the parietal pleura but similar quantities of chrysotile and tremolite asbestos in the lungs of the animals examined. The same type of fibres have been also identified in the lungs of goats from areas with asbestos-contaminated soils (Rey *et al.*, 1993a,b; Dumortier *et al.*, 2002).

4.2.2. ASS from El Dorado Hills (California) and correlation with tremolite asbestos

El Dorado Hills was chosen as the prototype of NOA amphibole exposures. During the development of this area, amphiboles were discovered in the soils near a local school. Lung samples of local pets (two cats and four dogs) that died of natural causes, have been analysed in two different laboratories by SEM-EDS and TEM-EDS, respectively. The fibres retained in the lung samples of these animals were asbestiform tremolite and actinolite (Case and Abraham, 2009).

4.2.3. ASS from Susa Valley and Lanzo Valleys (Italy)

Fornero *et al.* (2009) studied the burden of inorganic fibres in the lungs of cows from two areas in Italy's Western Alps, Susa Valley and Lanzo Valleys. In these valleys there are outcrops of rocks bearing asbestos and other mineral fibres, and there are different grades of anthropization. The Susa Valley is heavily anthropized and shows many outcropping serpentinites. In the Lanzo Valleys there is only a local road and one railway; there is a smaller human impact but it does contain the biggest European chrysotile mine, which has been active since 1990. In cow-lung samples from the Susa Valley, five asbestos species have been found. Tremolite asbestos and actinolite asbestos come from natural sources; amosite (grunerite asbestos) (Fig. 1) and crocidolite come from anthropogenic sources (because they are not present in the local rocks); chrysotile (Fig. 2) may originate from both natural and anthropogenic sources. Chrysotile/antigorite group minerals were present (serpentine group minerals cannot be determinable using the SEM-EDS technique). Numerous non-asbestos species have been found and the most frequent of them correspond to TiO_2. In cow-lung

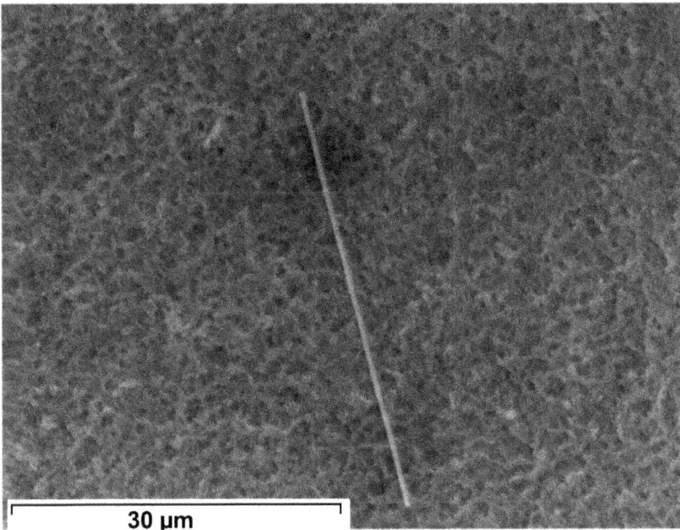

Figure 1. Secondary electron SEM image of amosite (grunerite asbestos) found in the lung of a cow from Susa Valley (TO).

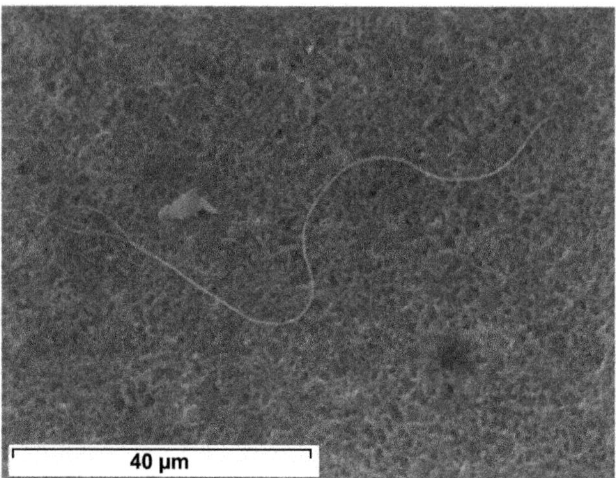

Figure 2. Secondary electron SEM image of a bundle of chrysotile fibres found in the lung of a cow from the Susa Valley (TO).

samples from the Lanzo Valleys, four asbestos species have been found: tremolite asbestos, actinolite asbestos, amosite (grunerite asbestos) and chrysotile. The crysotile/antigorite group is present. Numerous non-asbestos species have been found and the most abundant of these are phyllosilicates. Results demonstrate that >35% of inorganic fibres found by SEM-EDS, both in 20 cows from Susa Valley and 19 cows from Lanzo Valleys, belong to asbestos mineralogical species (tremolite/actinolite asbestos, chrysotile, amosite and crocidolite). Similar concentrations of asbestos from natural sources (tremolite/actinolite asbestos) and asbestos from anthropogenic sources (amosite and crocidolite) were found in the two areas, respectively. Fibres of chrysotile are almost absent from both valleys.

The presence of amosite and crocidolite (commercial asbestos fibres) is compatible with the considerable industrialization of the area and the use of ACM.

4.2.4. ASS from Casale Monferrato (Italy) and correlation with anthropogenic asbestos

In a recent pilot study (Ardizzone *et al.*, 2014), wild rats were used as SSA to identify the areas in Casale Monferrato (Piedmont) with possible background asbestos sources. In Casale Monferrato the oldest and largest Italian factory producing asbestos-cement materials ('Eternit') was active from 1907 to 1986, with serious consequences for the health of workers, their relatives and the general population. Today, ~25 years after the factory closed, there is still asbestos environmental pollution caused by the widespread presence of waste materials in various parts of the town.

Investigation by SEM-EDS allowed the identification of several kinds of inorganic fibres in the rat lungs: asbestos (crocidolite, tremolite/actinolite asbestos, and chrysotile are more frequent and abundant than the other two kinds) (Fig. 3) and

Figure 3. Backscattered electron SEM image of tremolite/actinolite asbestos found in the lung of a rat from Casale Monferrato.

non-asbestos classified (Fig. 4). The proposed method has proved suitable for monitoring the soil contamination level in Casale Monferrato.

4.2.5. ASS from Biancavilla (Italy) and correlation with fluoroedenite

Fluoro-edenite was detected in the lungs of sheep that lived near Biancavilla (Sicily) where this asbestiform amphibole was discovered after the detection of a MM cluster in town residents who had never had any relevant exposure to asbestos during their professional activity. Animal monitoring was also included in an epidemiological environment study: fluoro-edenite was detected in the lungs of sheep. The sheep were chosen as ASS because they are numerous and scattered in this rural district. They are considered suitable because they have a bronchopulmonary anatomy quite similar to that of humans and because two previous studies (McConnichie *et al.*, 1987, 1989; Baris *et al.*, 1987a,b) showed asbestiform fibres from environmental exposure in the lung tissue of sheep. Lung samples of 27 adult animals were prepared for SEM-EDS analysis. Fourteen mineral species were identified, most (feldspars, quartz, clino-pyroxene, ortopyroxene) corresponding to mineral species that occur naturally in the geological environment of Biancavilla; others originate from material dispersed in the environment by human activities (industrial waste, building, *etc.*). Fluoro-edenite fibres were identified in six samples (De Nardo *et al.*, 2004).

Figure 4. Backscattered electron SEM image of TiO_2 fibre found in the lung of a rat from Casale Monferrato.

5. Asbestos bodies and non-asbestos ferruginous bodies

Asbestos bodies (ABs) in human lungs were first described as peculiar pigmented crystals by Marchand (1906), and reported later by Fahr (1914) and Cooke (1927,

1929). Stewart and Haddow (1929) classified them as "asbestosis bodies". Gloyne (1929) showed that such bodies were coating asbestos fibres of various sizes. The term "asbestosis bodies" was substituted by "asbestos bodies" when they were discovered in patients with lung diseases other than asbestosis (Roggli, 1989). In the 1960s, experimental evidence was found to prove that ABs may also grow around Al-silicates, silicon carbide whiskers, talc and glass fibres. Hence, when the nature of the substrate was unknown the term "ferruginous body" was used (Gaensler and Addington, 1969; Gross *et al.*, 1969). Churg and Warnock (1981) finally defined the terms "asbestos body" to indicate bodies containing asbestos fibres, and "ferruginous body" or "pseudoasbestos body" for all non-asbestos-containing structures.

ABs (*e.g.* Fig. 5) display a variety of shapes such as beaded, segmented, dumbbell and others. They are generally 20–50 μm long (Davis, 1970; Langer *et al.*, 1971; Pooley, 1975; Hayashi, 1978; Churg, 1986; Ghio *et al.*, 2004; Roggli, 2014) and 2–5 μm wide (Churg and Warnock, 1981; Ghio *et al.*, 2004). Exceptions occur: they may exceed 200 μm in length (Farley *et al.*, 1977) or display diameter of only 0.5 μm (Roggli, 1989).

5.1. Formation of ABs

Various authors, focusing on coated and uncoated asbestos fibres isolated from human tissues and *in vivo* experiments, postulated early models for the growth mechanism and chemistry of ABs (Davis, 1970; Langer *et al.*, 1971; Pooley, 1975; Hayashi, 1978; Churg, 1986). Morgan and Holmes (1985) showed that ABs could be formed within 2–3 months of exposure, with a time span of formation in animals similar to that in humans (Haque and Kanz, 1988). The formation of an AB was first attributed to hemosiderin organic Fe-storage complex (Richter, 1958) and promoted by the diffusion of dissolved substances from the asbestos fibres (Rath, 1964). Davis (1965) suggested that ABs have a biological origin and their formation is essentially due to an intracellular coating process that begins with deposition of a ferritin layer around the

Figure 5. An AB formed around a chrysotile fibre in rats.

phagocytosed asbestos fibres. According to Suzuki and Churg (1969a,b), formation of ABs deposited in lung parenchyma is due to fibre phagocytosis by alveolar macrophages. Those fibres that are ~20 μm or longer cannot be completely engulfed by a single phagocytic cell (generally a macrophage), prompting 'frustrated phagocytosis' which leads to an inflammatory burst, death of the phagocytic cell and the beginning of the coating process. The coating material could contain fine fibres and clusters of particles, and may be layered. It is evident from the literature that the formation of ABs is a complex mechanism involving a number of factors, such as the nature of the fibre, its morphometry, coating efficiency of the animal host and the instillation process. Concerning the nature of the fibre, Wagner *et al.* (1974) noted differences in accumulation of amphibole *vs.* chrysotile fibres within the lungs of different animals, following long-term inhalational exposure. In this respect, very long chrysotile fibres tend to be curly and are thus more likely to interact with the upper respiratory tract where they are subsequently removed by the mucociliary escalator. Very long amphibole fibres tend to be straight and have a greater chance of penetrating into the gas-exchange regions of the lung (Wagner *et al.*, 1974; Timbrell, 1976; Lee, 1985). Following exposures of 6 weeks or longer, the relative retention of amphibole fibres in the lungs is considerably higher than that of chrysotile (Wagner *et al.*, 1974; Middleton *et al.*, 1979). This is in concert with the much shorter half-life of chrysotile compared to amphiboles that may only become apparent with longer exposures, and with the occurrence of the vast majority of ABs isolated from human lungs with an amphibole asbestos core (Middleton *et al.*, 1979; Churg and Warnock, 1979, 1981; Churg *et al.*, 1979; Morgan and Holmes, 1985; Roggli, 1989). In general, fibres coated by ABs always coexist with uncoated fibres. The latter are invariably more abundant, with the ratio of uncoated to coated fibres being much larger in chrysotile than in amphibole asbestos (Pooley, 1975). The rarity of chrysotile ABs is probably due to its fragmentation into fine and shorter fibrils and to the fact that ABs tend to form only on fibres longer than 20 μm (Morgan and Holmes, 1980; 1985; Dodson *et al.*, 1982). In this regard chrysotile ABs are generally a poor indicator of the burden of pulmonary fibres (Churg and Warnok, 1981; Warnock and Wolery, 1987; Case, 1994). Fibre size and diameter also play a key role. It was observed that ABs generally occur on fibres 10–60 μm long and with a mean diameter greater than 0.5 μm (Gloyne, 1929). They rarely occur on asbestos fibres shorter than 10 μm. Dodson *et al.* (1982) suggested that fibre surface irregularities such as etching and fractures, together with multifibrillar composition may also influence the coating process. *In vivo* studies performed by Vorwald *et al.* (1951) demonstrated that the proliferation of ABs is related to the fibre-coating efficiency of the animal species. Many authors claimed that rats are poor formers of ABs whereas humans, guinea pigs and hamsters are more efficient (Gross and de Treville, 1967; Suzuki and Churg, 1969b; Botham and Holt, 1971, 1972a,b). The choice of fibre injection process (intratracheal or inhalation) has an influence on the development of ABs. For example, Holmes and Morgan (1980) showed that atypical ABs are produced on anthophyllite and crocidolite when inhaled. Holt *et al.* (1964) and Wagner *et al.* (1974) tried to induce the formation of ABs by long-term inhalation of chrysotile dust or other asbestos fibres in rats without success. Harrison and Heath (1986) obtained the best results; they reported that the formation of ABs is promoted by the use of an intratracheal method rather than inhalation. The distribution of dust resulting from instillation is much less homogeneous than that from inhalation, and penetration to the lung periphery is minimal. The resultant inflammatory responses were also quite different (Brain *et al.*, 1976) so caution should be used when extrapolating results based on intratracheal instillation in experimental animals to human inhalation exposures (Roggli and Brody, 1990). In any case, the geometry of the tracheobronchial tree seems to be an important factor ruling the deposition of particulates in the lower respiratory tract (Schlesinger, 1980).

5.2. Chemistry of ABs

Regarding the chemistry of ABs, Harrison *et al.* (1967) showed that the ferritin core is a ferric oxyhydroxide, presumably (FeOOH) or ($FeOOPO_3H_3$), if phosphate is present. Meyer (1970) confirmed the presence of Ca and P in ABs, and recent studies using synchrotron soft X-ray imaging (Pascolo *et al.*, 2011) have demonstrated that magnesium also participates along with Fe in the coating process. Koerten *et al.* (1990) observed the development of ABs after a single injection of crocidolite into the mouse peritoneal cavity and found aggregates, foreign body granulomas and various ABs with a zoned structure composed of Fe-rich dark layers and Ca-rich light layers. According to those authors, AB formation is extracellular and the various configurations found might reflect repeated contact of the same AB with different macrophages. It is important to clarify that not all ABs possess an Fe-rich coating and are composed mostly of Ca and P. μ-Raman spectra taken on ABs of both chrysotile (Fig. 6a) and crocidolite (Fig. 6b) display the typical pattern of apatite (reference pattern from the software library *Omnic 9* of Thermo Scientific). Many authors observed coating of asbestos fibres with spherules of Ca-phosphate (Roggli, 1989; Koerten *et al.*, 1990). De Vuyst *et al.* (1982) and later Ghio *et al.* (2003) described amosite fibres coated with Ca-oxalate crystals. P could be associated with the inorganic core of ferritin (Harrison *et al.*, 1967). The Ca content probably accounts for the structural stability of the body and also plays an important role in the initial stages of formation, acting as an aggregating agent for the ferritin or hemosiderin granules, which accumulate to form the AB. Brody and Hill (1982) observed that the formation of Ca phosphate salts in association with interstitial asbestos fibres is a common reaction to injury in white rats.

5.3. New results on the formation of ABs in chrysotile, crocidolite and erionite

Recently, in order to understand the process of interaction between the cellular environment and mineral fibres which leads to the formation of ABs, a combined systematic FEG-SEM (field emission gun-scanning electron microscopy) and μ-Raman investigation was attempted to study the morphological and chemical characteristics of

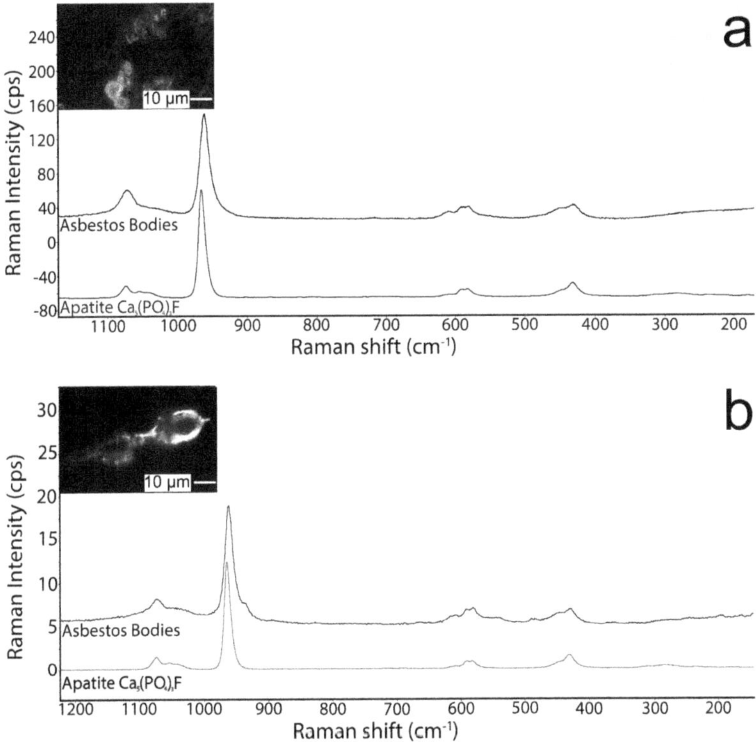

Figure 6. µ-Raman spectra and reference spectra (from the software library *Omnic 9* of Thermo Scientific) shown for comparison of (a) an AB formed from chrysotile after 80 weeks of contact time (ct) with rat organs; (b) an AB formed from crocidolite after 80 weeks of ct with rat organs (modified from Bursi Gandolfi *et al.*, 2016).

both fibres and ABs present in tissues of Sprague-Dawley rats which underwent single intraperitoneal or intrapleural injection of UICC chrysotile, UICC crocidolite and erionite zeolite.

The following discussion is taken mostly from Bursi Gandolfi *et al.* (2016). As far as the fibres found in the rat tissue are concerned chrysotile is characterized by the typical fibrous-asbestiform crystal habit. The mean fibre length ranges from 14.3 to 15.8 µm with diameters in the range 0.45–0.54 µm. The average size of chrysotile fibres that could produce ABs is 29.6 µm long and 0.5 µm in diameter. Leaching of Mg from chrysotile in contact with organic media could occur (Jaurand *et al.*, 1977). Crocidolite fibres display the typical fibrous habit and form stiff bundles. The mean fibre length ranges from 13.7 to 18.6 µm with diameters in the range 0.54–0.71 µm. The average size of crocidolite fibres that could produce ABs is 41.0 µm long and 0.86 µm in diameter. Regarding erionite fibres, they display a typical fibrous habit with stocky short fibres. The calculated mean fibre length ranges from 2.7 to 3.7 mm and the diameters are within the range 0.75–0.98 µm.

Remarkable variations in size and morphology of ABs formed on chrysotile fibres were detected. The length ranges from 1.5 to 20 μm and diameters range from 0.6 to 15 μm. Although the variety of morphologies observed is not great for chrysotile fibres, in Fig. 7 ABs with globular (Fig. 7a,b), granular (Fig. 7c) and curved (Fig. 7d) shapes are shown. All ABs examined have similar chemical compositions with Mg, Si and small Fe contents. Other elements such as P and Ca were associated with the coating. Silicon decreases drastically from the core, around the chrysotile fibre, to the periphery of the AB. Minor elements such as K and S may also be detected inside the ABs. In the case of secondary deposition, the chemistry of the outer layers does not change with respect to the core and no variation of the chemistry and shape of the ABs with time was found (a complete EDS semi-quantitative analysis is presented in Bursi Gandolfi *et al.*, 2016). Uncoated fibres were detected in all the samples examined. The percentage of coated fibres was 3.3%. Such a ratio does not change with time indicating that the number of ABs does not seem to increase with residence time. In UICC crocidolite, the length of ABs varies from 4 to 25 μm, and the diameter from 4 to 8 μm. ABs are formed predominantly on long crocidolite fibres and may occur around a single fibre or clusters of particles. The coating of the fibre may occur in successive steps. The bodies on crocidolite fibres occur in a variety of forms, tend to vary from cylindrical to elliptical,

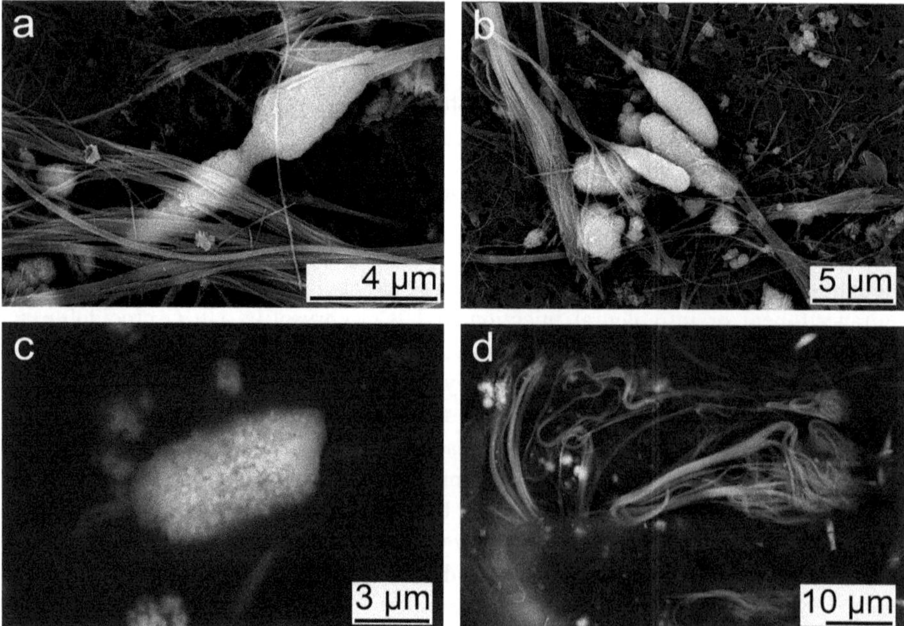

Figure. 7. (a) An AB without a well defined shape formed around chrysotile after 65 weeks of ct with rats' organs; (b) ABs formed at the same time as the termination of chrysotile bundles after 65 weeks of ct with rats' organs; (c) *in situ* picture of granular ABs at the end of the chrysotile fibre after 80 weeks ct with rats' organs; (d) *in situ* picture of chrysotile fibres and ABs after 104 weeks of ct with rats' organs.

and may produce globular clusters at one or both ends of the fibres. Even for crocidolite, most of the fibres observed are uncoated. The percentage of coated fibres was 6.0% and the ratio does not change with time. Thanks to the EDS spectra collected from ABs with different *in vivo* ages, it was observed that Mg, Si and Fe are associated with the crocidolite fibres whereas Fe, P and Ca are associated with the coating. The Fe and Si contents decrease from the core of the fibres to the marginal regions of the ABs. No significant chemical variations of the ABs of different residence times have been recorded (an EDS semi-quantitative analysis is presented in Bursi Gandolfi *et al.*, 2016). Regarding erionite, ABs have not been observed in any sample after any contact time.

We have not seen noteworthy differences in ABs formed in the pleural and peritoneal cavity. ABs appear in chrysotile and crocidolite in <40 weeks. Such short times of formation are in line with literature data (Roggli, 2014). ABs observed around chrysotile and crocidolite fibres are generally smaller (<20 μm and <25 μm for chrysotile and crocidolite, respectively) than those reported in the literature (20–50 μm in length) (Churg and Warnock, 1981; Ghio *et al.*, 2004). On the other hand, diameters are comparable (Churg and Warnock, 1981). These data confirm the finding that most of the fibres are not coated in rats and that ABs are actually a sort of singularity, and as such poorly indicative of the fibre load in the lungs (Collegium Ramazzini, 2016b). The variety of morphologies observed here seems to confirm that a high concentration of fibres in the lungs prompts changes in the shape of ABs and favours the appearance of new forms (Morgan and Holms, 1985).

One of the most striking results is that ABs do not coat erionite fibres. Although the kinetics of formation of ABs in rats and humans are apparently different, this surprising result seems to be in contrast with the literature data. In fact, it is known that fibres of erionite can form bodies morphologically identical to typical ABs even if it has a lower propensity to become coated than tremolite. ABs were observed in the bronchoalveolar lavage fluid of villagers of Tuzköy (Turkey) (Dumortier *et al.*, 2001; Carbone *et al.*, 2007) exposed to erionite. The mean fibre length of those erionite fibres, compared to the erionite sample studied here, was greater (65.7% of the fibres had a diameter greater than 4 μm) and they had a larger Fe content (mean: 1.02 wt.%, standard deviation: 1.21). ABs do not form on the very short fibres (mean length 3.2 μm) used in this study, and phagocytosis does occur (no frustrated phagocytosis as expected in the case of long chrysotile and crocidolite fibres). In this regard ABs act as a secondary defence mechanism aimed at isolating the toxic reactive fibre, and nucleate only on long fibres that cannot be engulfed and hence that cannot be subjected to complete phagocytosis (Bursi Gandolfi *et al.*, 2016). A fibre with a length such that it cannot be completely engulfed will remain in the extracellular space. When such a situation arises, the macrophages or other phagocytes may seal off the fibre and exocytose the content of its lysosome into the space in an attempt to digest the fibre in question extracellularly (Governa *et al.*, 1999). This model is in concert with earlier observations by Koerten *et al.* (1990) showing that the formation of ABs occurs extracellularly similarly to the bone regeneration process provided by osteoclasts. The diminished generation of oxidants and reduced toxicity of coated fibres with respect to uncoated fibres has been reported by many authors (Gross *et al.*, 1969; Roggli, 2014). In this respect, AB is called the 'tombstone' of the asbestos fibre (McLemore *et al.*, 1981). Unfortunately, this second defence mechanism appears to be poorly effective as we observed that only a few fibres are efficiently coated.

In general, the limited number of coated fibres observed should be due to the lack of nutrients to form the asbestos coating, specifically Fe, P and Ca. Because of the chemical stability of the fibres (especially crocidolite and erionite) in contact with the tissues, release of active nutrients such as Fe is poor and cannot promote massive nucleation and growth of ABs (Ghio *et al.*, 2004; Bursi Gandolfi *et al.*, 2016).

5.4. Open issues on ABs

Despite the huge number of papers dealing directly or indirectly with ABs, their nature, origin and control/influence over the cytotoxicity mechanism of asbestos fibres are still a matter of debate. In particular, a number of issues still require clarification:

- Is there a relationship between the type of AB and the nature of the mineral fibre?
- Why are only a few fibres coated? Is there a dependency of fibre coating on fibre geometrical parameters?
- Is AB formation an intracellular or extracellular mechanism?
- Is it only frustrated phagocytosis that triggers the coating process and AB formation?
- Does the chemical nature of the fibre exert a control over the formation of the AB?
- The source of the Fe coating the fibres is unknown: is it derived from haemoglobin/plasma or from the fibres? If it is derived from the fibres, the mineralogical nature of the fibre (Fe-rich crocidolite *vs.* Fe-poor erionite) should play a key role in the formation mechanism.
- Is it reasonable to assume that coated fibres display a minimal potential for cytotoxicity and that Fe coating is bound in such a way that it does not participate efficiently in the generation of reactive oxygen species?

6. *In vivo* animal models studying mineral fibres: The experience of the Ramazzini Institute (RI)

Adequate data on risk assessment and evidence of asbestos carcinogenicity were produced by epidemiological studies and carcinogenicity bioassays on experimental animals. Rodent models have proved to be human equivalent models for reproducing the diseases associated with asbestos exposure; life-span chronic rodent studies are considered the standard assay for assessing fibre carcinogenicity (IARC, 2012; Report of ILSI, 2005).

6.1. Mechanism of carcinogenesis

It has been postulated that physical carcinogens produce cancer by some physical mechanisms rather than by chemical reaction. Such physical mechanisms have been regarded as a mere non-specific irritative effect of hypothetical surface factors on cells, which could cause cellular proliferation, selection of spontaneously occurring transformed clones and, finally, neoplasia. In support of this view, there are several observations and considerations. The ratio between length and diameter of the fibres seems to be crucial in the carcinogenicity of asbestos and man-made mineral fibres (Landrigan, 1991; Infante *et al.*, 1994).

According to IARC, asbestos fibres have been shown to interfere physically with the mitotic apparatus, which may result in aneuploidy or polyploidy, and specific chromosomal alterations characteristic of asbestos-related cancer. In addition to direct clastogenic and aneuploidogenic activities that may be induced following the translocation of asbestos fibres to target cell populations in the lungs, persistent inflammation and macrophage activation can generate secondarily additional reactive oxygen species, and reactive nitrogen species that can indirectly induce genotoxicity (IARC, 2012). Furthermore IARC concluded that "no mutations in oncogenes or cancer-suppressor genes have been linked directly with exposure to asbestos fibre" and "no gene expression signature can be attributed directly to asbestos exposure, and these studies show variable gene expression patterns resulting from limited stability of RNA, contamination of cancer samples with host cells, and use of different microarray platforms" (IARC, 2012).

Inhalation of asbestos or erionite fibres at high doses in humans causes diffuse interstitial fibrosis or asbestosis. This fibrotic reaction develops slowly but progressively, beginning 10 years after the initial exposure. It is hypothesized that oxidants and proteases released from alveolar macrophages and activated by phagocytosis of fibres, damage the alveolar epithelial lining. Fibres may become translocated across the damaged epithelium into the interstitium of the alveolar walls. Upregulation of growth factor expression has been observed at sites of asbestos fibre deposition in rat lungs (Brody *et al.*, 1987; 1997). Pleural plaques and pleural fibrosis are hypothesized to develop after injury to mesothelial cells and destruction of the basement membranes.

6.2. The Ramazzini Institute experimental *in vivo* project on particulate materials

For >40 years, the CMCRC/RI has been engaged in developing a cancer research program which includes the conduct of long-term bioassays on rodents for the identification of potentially carcinogenic agents of industrial and environmental origin.

In the framework of the RI research programme, since the 1980s a large and integrated project of long-term carcinogenicity bioassays on fibrous natural and man-made materials has been running. To date, the compounds studied include: natural and modified asbestos of many types and origins, natural sedimentary (comprising erionite) or hydrothermal and industrial zeolites, glass and ceramic fibres, rock wool, Kevlar, carbon fibres and other organic solid compounds such as silica, alumina, talc, kaolin and bentonite (Maltoni and Minardi, 1983, 1989). In recent years, the CMCRC/RI has studied fluoroedenite, a calcium amphibole with crystals of asbestiform fibre shape, and particulate materials derived from the World Trade Center (WTC) collapse in New York (Belpoggi *et al.*, 2011; Soffritti *et al.*, 2013).

In order to evaluate various fibrous and particulate materials, 17 bioassays were performed by the RI from the 1980s to date, involving >10,000 animals. The compounds studied, number of animals employed and the route of administration are listed in Table 3a,b.

The aims of the systematic long-term experimental project on fibres are:

(1) to identify new potential carcinogenic materials;
(2) to assess the level of carcinogenic risk of a given material in quantitative terms and compare the risks from different materials (relative risk);
(3) to define the role of physical and chemical characteristics of the test materials in the carcinogenic mechanisms;
(4) to predict target organs for the onset of cancer for each material and reconstruct the natural history of the cancers which may be induced by the test compound.

6.2.1. *Test materials*

We describe here the procedures and some relevant results of experiments performed from 1981to 2005 at the CMCRC/RI. The test materials studied had different physical-chemical characteristics and various origins. Among them, we describe those which are the most interesting both for their environmental impact and health consequences for humans; in particular we report results for the following materials:

- UICC samples of crocidolite, amosite, anthophyllite and chrysotile from Canada, Zimbabwe and California; they were provided in 1981 by Professor Irving J. Selikoff of the Mount Sinai School of Medicine, New York, USA;
- Asbestos-cement was provided in 1981 by the 'Association of Cemento-Amianto', Balangero, Turin, Italy;
- Modified asbestos samples of Chrysotile Commercial Grade, 'Paperbestos N. 5', Short Chrysotile and Kevlar fibres were provided in 1982 by Dr Jaques Dunnigan of IRDA (Institute de Recherche et Developpement sur l'Amiante), Université Laval, Québec, Canada;
- Sedimentary erionite samples sourced from Eastgate Churchill County, Nevada, USA, and from Ponza, Italy, were provided in 1981 by Professor G. Gottardi from the Institute of Mineralogy, University of Modena , Italy;
- Rockwool, glass fibres and ceramic refractory fibres (Fibrefrax) were supplied in 1986 by CERAM-SNA Inc., Canada;

Table 3. (a) Long-term carcinogenicity bioassays on natural and man-made mineral materials, performed at the Cesare Maltoni Cancer Research Centre of the European "B. Ramazzini" Institute of Oncology and Environmental Sciences.

Compounds/ agents	Start of the experiment (year)	– Animals – Species	No.	Means of exposure[a]	Publication of results
Asbestos					
Crocidolite	1970, 1975, 1981, 1993	Rat	860	Ing,Int,Ip,Ipl,Sc	Minardi and Maltoni (1988)
Chrysotile (Canada)	1981-1986	Rat, mouse	1,260	Int,Ip,Ipl,Sc	
Chrysotile (Zimbabwe)	1981	Rat	80	Ip	
Chrysotile (California)	1981	Rat	80	Ip	
Amosite	1981	Rat	80	Ip	
Anthophyllite	1981	Rat	80	Ip	
Asbestos-cement	1981	Rat	240	Ip,Ipl,Sc	
Modified asbestos*					
Compound 1	1983	Rat	160	Ip, Ipl	Maltoni and Minardi (1989)
Compound 2	1983	Rat	160	Ip, Ipl	
Compound 3	1983	Rat	160	Ip, Ipl	
Compound 4	1983	Rat	160	Ip, Ipl	
Compound 5	1983, 1986	Rat, mouse	860	Ip, Ipl	
Compound 6	1983, 1986	Rat, mouse	860	Ip, Ipl	
Compound 8	1983, 1986	Rat, mouse	860	Ip, Ipl	
Wollastonite	1981	Rat	240	Ip,Ipl,Sc	–
Rock wool	1983, 1986, 1993	Rat	640	Int,Ip,Ipl	Minardi and Maltoni (1998a)
Kevlar	1983, 1986	Rat	80	Ip,Ipl	Maltoni and Minardi (1989)
Ceramic fibres	1986	Rat	440	Ip,Ipl	Minardi and Maltoni (1998b)
Glass fibres	1993	Rat	200	Int,Ip	–
Crystalline silica	1981	Rat	160	Ip,Sc	–
Amorphous silica	1981	Rat	160	Ip,Sc	–
Alumina	1981	Rat	160	Ip,Sc	–
Talc (pure)	1981	Rat	160	Ip,Sc	Minardi, *et al.* (1990)
Talc (industrial)	1981	Rat	240	Ip,Ipl,Sc	
Kaolin	1981	Rat	160	Ip,Sc	–
Bentonite	1981	Rat	240	Ip,Ipl	–
Erionite	1981, 1993	Rat, mouse	400	Ip,Ipl,Sc	Maltoni *et al.* (1982); Maltoni and Minardi (1983)
Fluoro-edenite	2002	Rat	270	Ip, Ipl	Belpoggi, *et al.* (2011); Soffritti, *et al.* (2004)

Table 3. (b) Long-term carcinogenicity bioassays on natural and man-made mineral materials, performed at the Cesare Maltoni Cancer Research Centre of the European "B.Ramazzini" Institute of Oncology and Environmental Sciences.

Compounds/ agents	Start of the experiment (year)	– Animals – Species	No.	Means of exposure[a]	Publication of results
Other natural zeolites					
Mordenite (sed)	1981	Rat	240	Ip,Ipl,Sc	–
Phillipsite	1981	Rat	240	Ip,Ipl,Sc	–
Clinoptilolite	1981	Rat	160	Ip,Sc	–
Cabasite	1981	Rat	160	Ip,Sc	–
Ferrierite	1981	Rat	160	Ip,Sc	–
Mordenite (crystalline)	1981	Rat	240	Ip,Ipl,Sc	–
Heulandite	1981	Rat	160	Ip,Sc	–
Mesolite	1981	Rat	160	Ip,Sc	–
Natrolite	1981	Rat	160	Ip,Sc	–
Solecite	1981	Rat	160	Ip,Sc	–
Stilbite	1981	Rat	160	Ip,Sc	–
Thomsonite	1981	Rat	160	Ip,Sc	–
Man-made zeolites and precursors					
TE-16460	1982	Rat	360	Ip,Ipl,Sc	–
TE-16461	1982	Rat	360	Ip,Ipl,Sc	–
TE-16462	1982	Rat	360	Ip,Ipl,Sc	–
WGB 1189-2-1 (catalyst)	1982	Rat	360	Ip,Ipl,Sc	–
WGB 1189-2-1 (elutriated)	1982	Rat	360	Ip,Ipl,Sc	–
WGB 1190-2-1 (catalyst)	1982	Rat	360	Ip,Ipl,Sc	–
WGB 1190-2-1 (elutriated)	1982	Rat	360	Ip,Ipl,Sc	–
WGB 1191-2-1 (catalyst)	1982	Rat	360	Ip,Ipl,Sc	–
WGB 1191-2-1 (elutriated)	1982	Rat	360	Ip,Ipl,Sc	–
WGB 1192-1-1	1982	Rat	360	Ip,Ipl,Sc	–
WGB 1193-1-1	1982	Rat	360	Ip,Ipl,Sc	–
WGB 1194-1-1	1982	Rat	360	Ip,Ipl,Sc	–
Paulsboro (catalyst)	1982	Rat	360	Ip,Ipl,Sc	–
Paulsboro (catalyst fines)	1982	Rat	360	Ip,Ipl,Sc	–
Kaiser alumina	1982	Rat	360	Ip,Ipl,Sc	–
Mobil Joliet	1982	Rat	360	Ip,Ipl,Sc	–
Joliet fresh	1982	Rat	360	Ip,Ipl,Sc	–
MS 4A	1981	Rat	160	Ip,Ipl,Sc	–
MS 5A	1981	Rat	160	Ip,Ipl,Sc	–
MS 13X	1981	Rat	160	Ip,Ipl,Sc	–
Carbon fibres (discs)	1981	Rat	140	Implantation	Minardi *et al.* (1991)
Particulate WTC	2005	Rat	480	Int, Ip	Soffritti *et al.* (2013)

* The modified asbestos consists of commercial Chrysotile 'Paperbestos N.5' and Short chrysotile untreated or treated with POCl3 and high temperature

[a] Ing = ingestion; Int = intratracheal instillation; Ip = intraperitoneal injection; Ipl = intrapleural injection; Sc = subcutaneous injection;

- Fluoroedenite samples sourced from the town of Biancavilla, province of Catania in Sicily, Italy, were supplied in 2002 by Professor Antonio Gianfagna of the Institute of Hearth Sciences, University La Sapienza, Rome, Italy, in the framework of a collaboration with the Istituto Superiore di Sanità, Rome, Italy;
- The dust from the collapse of the WTC was supplied in 2005 by Professor Paul Lioy (recently deceased), Director of the Environmental and Occupational Health Sciences Institute (EOHSI) at the State University Rutgers, New Jersey, USA; Professor Lioy was in charge of studying the implications for human health of the disaster for the Municipality of New York City.

6.2.2. Animal model

During about 40 years of experimental research at the CMCRC/RI, male and female Sprague-Dawley (SD) rats from the CMCRC/RI breeding facility were used; Swiss mice were also studied occasionally. The spontaneous cancer incidence in SD rats of the CMCRC/RI colony is well known and the cancer susceptibility is similar to that in humans. In particular, in the CMCRC/RI historical controls, which consist of ~18,000 rats used in more than 500 bioassays, MMs are extremely rare (<1% both for male and female rats). At the beginning of the experiments on all the particulate materials, animals were usually 8 weeks old; the size of the experimental groups was from 40 to 80 animals (50% males, 50% females).

6.2.3. Experimental condition

After weaning at 4–5 weeks of age, the experimental animals were identified by ear punch and randomized in order to have no more than one male and one female from each litter in the same group. The experiments were performed following the principles of Good Laboratory Practices (GLPs). Standard Operating Procedures (SOP) are available from the author for each phase of the experiments. During the experiments, a veterinarian from the National Health Service monitored the health status and behaviour and animal welfare applied in the experimental procedures. The rats were housed at fixed humidity (40–60%) and temperature (mean 21°, range 18–24°C); there were five animals per makrolon cage with stainless steel top and shavings bedding.

The animals were fed the same diet in all experiments, in standard dried pellets, and received *ad libitum* tap water throughout the experiment. Both feed and tap water were examined periodically to exclude possible contaminants which could interfere with the study. The animals were kept under observation until spontaneous death, which generally occurred at ~3 years of age (150 weeks) in the CMCRC colony of SD rats; when requested by the sponsors, some experiments lasted only 2 years (104 weeks). In this case the surviving animals were sacrificed using humanitarian procedures (ethyl ether or CO_2). During the experiments, animals were examined clinically every 2 weeks, to detect all gross changes, and weighed every 4 weeks. A complete necropsy was performed on each animal and histo-pathological examination was carried out on tissues at the site of injection, brain, thymus, lung, heart, liver, spleen, pancreas, kidneys, adrenal glands, stomach, uterus, ovaries or testes, subcutaneous, mediastinal and mesenteric lymph nodes, and any other organs with pathological changes. All organs and tissues were preserved in 70% ethyl alcohol. The specimens were trimmed and processed as paraffin blocks; 3–5 μm sections of every specimen were obtained and routinely stained with haematoxylin-eosin.

6.2.4. Route of administration

The most common route of administration of the test compound (25 mg per dose, *una tantum*, suspended in 1 mL of distilled water) was injection (intrapleural, intraperitoneal or subcutaneous). In some experiments intratracheal instillation or gavage ingestion were also performed. The dose used was identified by analysing the data from *in vivo* experimental studies available in the literature from the 1970s. The test materials were suspended at the selected doses in distilled water. Control groups received only the vehicle (distilled water), with the same means of administration as the corresponding treated group. Injections, intratracheal instillation and gavage were performed following the SOP of the CMCRC.

6.2.5. Results

The overall results of the CMCRC/RI studies are presented in Table 4. The different test compounds studied are described and discussed below.

Asbestos. In the test experimental conditions all the natural asbestiform materials of different type and origin, were proved to be mesotheliomatogenic both for the peritoneum and pleura. MMs were of different patterns and the same patterns have been observed in humans: epithelioid (tubular or papillary), solid and sarcomatous type. In our experimental model the peritoneum appears to be more responsive than the pleura. There were differences in the carcinogenic potency of the various products, both on the basis of the incidence and on the average latency time of peritoneal and pleural MMs. Crocidolite proved to be the most potent among the tested material for the mesotheliomatogenic effect while the less potent carcinogen was found to be asbestos cement (Table 4).

Modified chrysotile asbestos compared to Kevlar and rockwool. The experiment using modified chrysotiles was performed in order to investigate the potential carcinogenicity of these materials, compared to natural ones and other modified non-chrysotile fibrous particles (Rockwool and Kevlar) because the modified type of chrysotile and the other cited fibres are considered to be alternative materials for industrial use. In our experimental conditions all the materials tested, apart the Kevlar fibres, were found to produce MMs in the peritoneum and pleura or in the peritoneum alone. The effects of fibres varied according to the type and origin of the materials and showed that they can be modified by altering the physical-chemical properties of the mineral. Short chrysotile fibres treated with phosphoryl chloride ($POCl_3$) at high temperature proved to be by far the weakest carcinogen among all the tested asbestiform materials (Table 4).

Table 4. Ramazzini Institute Cancer Research program on fibres: comparative incidence and latency time of mesotheliomas induced by single injection of different kinds of fibrous material, in male and female Sprague-Dawley rats[a].

	Mesotheliomas					
	Peritoneal			Pleural		
Material	Incidence (%)	Average latency time (weeks)	Latency time of the first observed tumor (weeks)	Incidence (%)	Average latency time (weeks)	Latency time of the first observed tumor (weeks)
Crocidolite	97.5	59.5	29	45.0	104.8	33
Amosite	90.0	66.7	44	–	–	–
Anthophyllite	87.5	73.3	18[b]	–	–	–
Chrysotile (Rhodesia)	82.5	89.7	48	–	–	–
Fibrous fluoro-edenite	82.5	63.7	36	16.3	79.7	51
Chrysotile (Canada)	80.0	92.2	65	65.0	111.1	80
Chrysotile (California)	72.5	85.3	39	–	–	–
Asbestos-cement	52.5	99.7	60	35.0	113.1	82
Erionite	50.0	106.1	67	87.5	64.2	32
Glass fibres	42.5	90.4	65	–	–	–
Ceramic fibres	32.5	84.9	58	–	–	–
Rock wool	10.0	78.7	69	2.5	105.0	–
Kevlar fibres	0	–	–	–	–	–

[a] from Belpoggi *et al.* (2011); [b] microscopic finding in a live rat for peritoneal mesothelioma.

Erionite. Under our experimental conditions, erionite produced MMs in both the pleura and peritoneum. The pleura was found to be much more responsive and erionite appeared to be the most potent carcinogenic agent for rat pleura ever tested (Maltoni *et al.*, 1982; Maltoni and Minardi, 1983). This result supported the hypothesis that this zeolite is the causative agent for the environmental pleural MMs described in the human population in Turkey (Baris *et al.*, 1987a,b) (Table 4).

Fluoroedenite. The fibrous fluoroedenite proved to have a strong mesotheliomatogenic effect on the peritoneum, which is more responsive than the pleura in our experimental conditions. Peritoneal and pleural MMs increased in incidence and the exposure to fibrous fluoroedenite reduced the latency time. The MMs observed involved the whole abdominal cavity and the visceral and or parietal pleura; microscopically they were polymorphic with a predominance of epithelial-solid and fibrous spindle-cell patterns (the same as those in humans). The prismatic fluoroedenite showed no mesothelioma-togenic effect; this is not surprising if compared to results of other experiments performed in our laboratory on a series of non-fibrous materials. Also, for the fluoroedenite, the experimental study on rats confirmed the epidemiological observation, proving a mesotheliomatogenic effect as observed in the human population in the area of Biancavilla, Sicily (Paoletti *et al.*, 2000; Gianfagna *et al.*, 2003; Belpoggi *et al.*, 2011). On the basis of our experimental study and epidemiological studies, the IARC (2014) classified the fluoroedenite as carcinogenic to humans (Group 1) (Table 4).

Rockwool. Rockwool is carcinogenic and, under the experimental conditions, specifically caused MM of the peritoneum and pleura at the site of injection (Minardi and Maltoni, 1998a). The mesotheliomatogenic potency of rockwool seems to be lower than that of asbestos of various kinds or that of ceramic refractory fibres. Our results are in line with those available in the literature, indicating the mesotheliomatogenic effect of rockwool in experimental animals (Pott *et al.*, 1987; Wagner *et al.*, 1984) (Table 4).

Ceramic fibres ('Fibrefrax'). The data resulting from our experimental bioassay indicate clearly that ceramic fibres are a MM-causing agent with a close relationship between the administered dose (10, 5, 1 mg in 1 mL of water by intraperitoneal injection), latency time and incidence of MM. In the high-dose group an increase of pleural plaques and fibrosis was also observed (Minardi and Maltoni, 1998b) (Table 4).

Glass fibres. In our experimental conditions, glass fibres were administered by intraperitoneal injection; the results showed that they have a mesotheliomatogenic effect similar to that of ceramic fibres for both incidence and latency time (Table 4).

World Trade Center (WTC) dust. In 2005 the Ramazzini Institute started the first long-term bioassay to test the potential carcinogenic effects of a sample of WTC settled dust administered to Sprague-Dawley rats by intratracheal instillation. The first results in terms of the histopathology of the lung showed the onset of some vascular cancers as hemangioma and hemangiosarcoma (Soffritti *et al.*, 2013). These neoplastic lesions are very rare in the historical database of rodent strains and primary hemangiosarcoma is also a very rare cancer in humans. These findings are probably due to the fact that, although 60% of the tested material consisted of particles with a >50 μm diameter, there is also a percentage equal to 1.12% of fine particles (<2.5 μm). The toxic effects of particles of <2.5 μm must be considered because they are known to be able to penetrate deep into the lung and potentially migrate through the blood and lymphatic vessels to other organs and tissues producing effects remote from the point of instillation.

Furthermore, due to the large volume of asbestos employed in the construction of the buildings, we were expecting an increase in lung and pleural lesions. But this was not the case. The high temperature reached during the fire and building collapse probably changed the physicochemical characteristics of asbestos fibres, so neither pleural MM nor lung carcinoma were observed among the treated animals.

7. Conclusions

Experimental animal studies have provided much of our understanding of the mechanisms by which asbestos interacts with biological systems. Different approaches have been used, both *in vitro* and *in vivo*. *In vitro* studies, using various cellular systems, permit the investigation of direct effects of asbestos and other inorganic fibres on cellular function under controlled conditions. *In vivo* models include carcinogen bioassays with different routes of exposure and ASS.

Carcinogen bioassays have proved to be the most predictive model for testing and anticipating fibre carcinogenicity: all forms of asbestos have been shown to be carcinogenic in carcinogen bioassays and afterwards these were confirmed to be known human carcinogens by the IARC (2012). The use of ASS can also contribute to knowledge about human environmental exposure and health. Enough information about variability and biasing factors for the ASS study population must be known. Limited historical data in ASS may make it difficult to separate short-term fluctuations from longer term trends occurring in a given sentinel species or population.

When planned for precise purposes with attention to the key, general and specific requisites, and when conducted using standardized methods, experimental studies represent an important tool capable of providing good information and reducing the uncertainties of risk assessment (Mansevisi *et al.*, 2017).

References

Anttila, S., Karjalainen, A., Taikina-Aho, O., Kyyronen, P. and Vainio, H. (1993) Lung cancer in the lower lobe is associated with pulmonary asbestos fiber count and fiber size. *Environmental Health Perspectives*, **101**, 166–170.

Ardizzone, M., Vizio, C., Bozzetta, E., Pezzolato, M., Meistro, S., Dondo, A., Giorgi, I., Seghesio, A., Mirabelli, D., Capella, S., Vigliaturo, R. and Belluso, E. (2014) The wild rat as sentinel animal in the environmental risk assessment of asbestos pollution: a pilot study. *Science of the Total Environment*, **479-480**, 31–38.

Baker, L.C., Grindem, C.B., Corbett, W.T., Cullins, L. and Unter, J.L. (2001) Pet dogs as sentinels for environmental contamination. *Science of the Total Environment*, **274**, 61–169.

Baris, Y.I., Sahin, A.A., Oezesmi, M., Kerse, E., Ozen, E., Kolacan, B., Altinoers, M. and Goektepeli, A. (1978) An outbreak of pleural mesothelioma and chronic fibrosing pleurisy in the village of Karain/Uerguep in Anatolia. *Thorax*, **33**, 181–192.

Baris, Y., Simonato, L., Artvinli, M., Pooley, F., Saracci, R., Skidmore, J. and Wagner, C. (1987a) Epidemiological and environmental evidence of the health effects of exposure to erionite fibres: a four-year study in the Cappadocian region of Turkey. *International Journal of Cancer*, **39**, 10–17.

Baris, Y., Artvinli, M., Sahim, A.A., Bilir, N., Kalyoncu, F. and Sebastien, P. (1987b) Non-occupational asbestos related chest diseases in a small Anatolian village. *British Journal of Industrial Medicine*, **45**, 841–842.

Baris, B., Demir, A.U., Shehu, V., Karakoca, Y. and Bariş, Y.I. (1996) Environmental fibrous zeolite (erionite) exposure and malignant tumors other than mesothelioma. *Journal of Environmental Pathology, Toxicology and Oncology*, **15**, 183–189.

Belluso, E., Ferraris, G. and Alberico, A. (1997) Amianto, la componente ambientale: dove, quali e come sono gli amianti nelle Alpi? Pp. 167–174 in: *L'amianto: dall'ambiente di lavoro all'ambiente di vita. Nuovi indicatori per futuri effetti.* (C. Minoia, G. Scansetti, G. Piolatto and A. Massola, editors). Fondazione S. Maugeri, IRCCS, I Documenti 12, Pavia, Italia.

Belluso, E., Fornero, E., Capella, S., Bellis, D. and Coverlizza, S. (2003) Preparation of biological samples to identify-quantify inorganic particles by SEM-EDS. Pp. 371–376 in: *Science, Technology and Education of*

Microscopy: an Overview (A. Mendez-Vilas, editors). FORMATEX Microscopy Book Series, **1**. Badajoz, Spain.

Belluso, E., Bellis, D., Fornero, E., Capella, S., Ferraris, G. and Coverlizza, S. (2006) Assessment of inorganic fiber burden in biological samples by scanning electron microscopy-energy dispersive spectroscopy. *Microchimica Acta*, **155**, 95–100.

Belpoggi, F., Soffritti, M., Manservigi, M. and Lauriola, M. (2007) Mesothelioma following asbestos exposure in workers of sugar refinery plants: the Ramazzini Foundation case series. *European Journal of Oncology*, **12**, 101–107.

Belpoggi, F., Tibaldi, E., Lauriola, M., Bua, L., Falcioni, L., Chiozzotto, D., Manservisi, F., Manservigi, M. and Soffritti, M. (2011). The efficacy of long-term bioassays in predicting human risks: mesotheliomas induced by fluoro-edenitic fibres present in lava stone from Etna volcano in Biancavilla, Italy. *European Journal of Oncology*, **16**, 185–196.

Bianchi, C., Brollo, A. and Ramani, L. (2000) Asbestos exposure in a shipyard area, Northeastern Italy. *Industrial Health*, **38**, 301–308.

Botham, S.K. and Holt, P.F. (1971) Development of asbestos bodies on amosite chrysotite and crocidolite fibres in guinea-pig lungs. *Journal Pathology*, **105**, 159–167.

Botham, S.K. and Holt, P.F. (1972a) Asbestos-body formation in the lungs of rats and guinea-pigs after inhalation of anthophyllite. *Journal Pathology*, **107**, 245–252.

Botham, S.K. and Holt, P.F. (1972b) The effects of inhaled crocidolites from Transvaal and North-West Cape mines on the lungs of rats and guinea pigs. *British Journal of Experimental Pathology*, **53**, 612–620.

Brain, J.D., Knudson, D.E., Sorokin, S.P. and Davis, M. A. (1976) Pulmonary distribution of particles given by intratracheal instillation or by aerosol inhalation. *Environmental Research*, **11**, 13–33.

Brandi, G., Di Girolamo, S., Belpoggi, F., Grazi, G., Ercolani, G. and Biasco, G. (2008) Asbestos exposure in patients affected by bile duct tumours. *European Journal of Oncology*, **13**, 171–179.

Brody, A.R. and Hill, L.H. (1982) Interstitial accumulation of inhaled chrysotile asbestos fibres and consequent formation of microcalcifications. *American Journal of Pathology*, **109**, 107–114.

Brody, A.R., Overby, L.H. and Warheit, D.B. (1987) Inhaled asbestos attracts macrophages and induces rapid proliferation of bronchiolar alveolar cells. *Chest*, **91**, 302.

Brody, A.R., Liu, J.Y., Brass, D. and Corti, M. (1997) Analyzing the gene and peptide growth factors in lung cells in vivo and in vitro consequent to asbestos exposure. *Environmental Health Perspectives*, **105**, 1165–1171.

Browne, K. and Wagner, J.C. (2001) Environmental exposure to amphibole-asbestos and mesothelioma. Pp. 21–28 in: *The Health Effects of Chrysotile Asbestos* (R.P. Nolan, A.M. Langer, M. Ross, F.J. Wicks and R.F. Maerin, editors). Contributions of Science to Risk-Management Decisions, *Canadian Mineralogist Special Publication,* **5**.

Bursi Gandolfi, N., Gualtieri, A.F., Pollastri, S., Tibaldi, E. and Belpoggi, F. (2016). Assessment of asbestos body formation by high resolution FEG-SEM after exposure of Sprague-Dawley rats to chrysotile, crocidolite, or erionite. *Journal of Hazardous Materials*, **306**, 95–104.

Carbone, M., Emri, S., Umran Dogan, A., Steele, I., Tuncer, M., Pass, H.I. and Baris, Y.I. (2007) A mesothelioma epidemic in Cappadocia: scientific developments and unex-pected social outcomes. *Nature Reviews*, **7**, 147–154.

Case, B.W. (1994) Biological indicators of chrysotile exposure. *The Annals of Occupational Hygiene*, **38**, 503–518.

Case, B.W. and Abraham, J.L. (2009) Heterogeneity of exposure and attribution of mesothelioma: trends and strategies in two American counties. *Journal of Physics*, Conference series, **151**, 1–16.

Case, B. and Marinaccio, A. (2017) Epidemiological studies on mineral fibres. Pp. 367–416 in: *Mineral Fibres: Crystal Chemistry, Chemical-physical Properties, Biological Interaction and Toxicity* (A.F. Gualtieri, editor). EMU Notes in Mineralogy, **18**. European Mineralogical Union and Mineralogical Society of Great Britain and Ireland, London.

Churg, A.M. (1983) Nonasbestos pulmonary mineral fibers in the general population. *Environmental Research*, **31**, 189–200.

Churg, A.M. (1986) Analysis of asbestos fibers from lung tissue: Research and diagnostic uses. *Seminars in*

Respiratory Medicine, **7**, 281–288.
Churg, A.M. and Warnock, M.L. (1979) Analysis of the cores of ferruginous (asbestos) bodies from the general population. I: Patients with and without lung cancer. *Laboratory Investigation*, **37**, 280–286.
Churg, A.M. and Warnok, M.L. (1981) Asbestos and other ferruginous bodies: their formation and clinical significance. *American Journal Pathology*, **102**, 447–456.
Churg, A.M., Warnock, M.L. and Green, N. (1979) Analysis of the cores of ferruginous (asbestos) bodies from the general population. II. True asbestos bodies and pseudoasbestos bodies. *Laboratory Investigation*, **40**, 31–38.
Coffin, D.L., Cook, P.M. and Creason, J.P. (1992) Relative mesothelioma induction in rats by mineral fibers: Comparison with residual pulmonary mineral fiber number and epidemiology. *Inhalation Toxicology*, **4**, 273–300.
Collegium Ramazzini (2016a) Collegium Ramazzini 18th Statement: "The Global Health Dimensions of Asbestos and Asbestos-Related Diseases". *Journal of Occupational Health*, **58(2)**, 220–223.
Collegium Ramazzini (2016b) Comments on the 2014 Helsinki consensus report on asbestos. *Journal of Occupational Health*, **58**, 224–227.
Constantopoulos, S.H., Saratzis, N.A., Kontogiannis, D., Karantanas, A., Goudevenos J.A. and Katsiotis, P. (1987) Tremolite whitewashing and pleural calcifications. *Chest*, **92**, 709–712.
Cooke, W.E. (1924) Fibrosis of the lungs due to the inhalation of asbestos dust. *British Medical Journal*, **II**, 147.
Cooke, W.E. (1927) Pulmunary asbestosis. *British Medical Journal*, **2(3491)**, 1024–1025.
Cooke, W.E. (1929) Asbestos dust and the curious bodies found in pulmonary asbestosis. *British Medical Journal*, **2**, 578–580.
Davis, J.M.G. (1965) Electron microscope studies of asbestosis in man and animals. *Annals of the New York Academy of Sciences*, **132**, 98–111.
Davis, J.M.G. (1970) Further observations on the ultrastructure and chemistry of the formation of asbestos bodies. *Experimental and Molecular Pathology*, **13**, 346–358.
De Nardo, P., Bruni, B., Paoletti, L., Pasetto, R. and Sirianni, B. (2004) Pulmonary fibre burden in sheep living in the Biancavilla area (Sicily): preliminary results. *Science of the Total Environment*, **325**, 51–58.
De Vuyst, P., Jedwab, J., Robience, Y. and Yernault, J.C., (1982) "Oxalate bodies", another reaction of the human lung to asbestos inhalation? *European Journal of Respiratory Disease*, **63**, 543–549.
De Vuyst, P., Karjalainen, A., Dumortier, P., Pairon, J.C., Mons, E., Brochard, P., Teschler, H., Tossavainen, A. and Gibbs, A. (1998) Guidelines for mineral fibre analyses in biological samples: report of the ERS Working Group. *European Respiratory Journal*, **11**, 1416–1426.
Dodson, R.F., O'Sullivan, M.F., Williams, M.G. and Hurst, G.A. (1982) Analysis of cores of ferruginous bodies from former asbestos workers. *Environmental Reseaarch*, **28**, 171–178.
Doll, R. (1987) Symposium on MMMF, Copenhagen, October 1986: Overview and conclusions. *Annuals of Occupational Hygiene*, **31**, 805.
Dumortier, P., Coplü, L., Broucke, I., Emri, S., Selcuk, T., De Maertelaer, V., De Vuyst, P. and Baris, I. (2001) Erionite bodies and fibres in bronchoalveolar lavage fluid (BALF) of residents from Tuzköy, Cappadocia, Turkey. *Journal of Occupational Environmental Medicine*, **58**, 261–266.
Dumortier, P., Rey, F., Viallat, J.R., Broucke, I., Boutin, C. and De Vuist, P. (2002) Chrysotile and tremolite asbestos fibres in the lungs and parietal pleura of Corsican goats. *Occupational Environmental Medicine*, **59**, 643–646.
Enterline, P.E., Marsh, G.M., Henderson, V. and Callahan, C. (1987) Mortality update of a cohort of US man-made mineral fibre workers. *Annals of Occupational Hygiene*, **31**, 625–656.
EPA – Environmental Protection Agency (2017) Office of environmental health hazard assessment safe drinking water and toxic enforcement act of 1986. Chemicals known to the State to cause cancer or reproductive toxicity. https://oehha.ca.gov/proposition-65/proposition-65-list
Fahr, T. (1914) Demonstrationen: Praparate und Microphotogrammes von einen Falle von Pneumokoniose. *Munchener Medizinische Wochenschrift*, **11**, 625.
Farley, M.L., Greenberg, S.D, Shuford, E.H. Jr, Hurst, G.A., Spivey, C.G. and Christianson, C.S. (1977) Ferruginous bodies in sputa of former asbestos workers. *Acta Cytologica*, **27**, 693–700.
Fornero, E., Belluso, E., Capella, S. and Bellis, D. (2009) Environmental exposure to asbestos and other

inorganic fibres using animal lung model. *Science of the Total Envoronment*, **407**, 1010–1018.

Gaensler, E.A. and Addington, W.W. (1969) Current concepts: Asbestos or ferruginous bodies. *Journal of Occupational Environmental Medicine*, **11**, 701.

Gazzano, E', Riganti, C., Tomatis, M., Turci, F., Bosia, A., Fubini, B. and Ghigo, D. (2005) Potential toxicity of nonregulated asbestiform minerals: balangeroite from the western Alps. Part 3: Depletion of antioxidant defenses. *Journal of Toxicology and Environmental Health*, part A, **68**, 41–49.

Ghio, A.J., Roggli, V.L., Richards, J.H., Crissman, K.M., Stonehuerner, J.D. and Piantadosi, C.A. (2003) Oxalate deposition on asbestos bodies. *Human Pathology*, **34**, 737–742.

Ghio, A.J., Churg, A. and Roggli, V.L. (2004) Ferruginous bodies: implications in the mechanism of fibre and particle toxicity. *Toxicology Pathology*, **32**, 643–649.

Gianfagna, A., Ballirano, P., Bellatreccia, F., Bruni, B., Paoletti, L. and Oberti, R. (2003) Characterization of amphibole fibres linked to mesothelioma in the area of Biancavilla, Eastern Sicily, Italy. *Mineralogical Magazine*, **67**, 1221–1229.

Glickman, L.T., Domanski, C.M., Magmire, T.G., Dubielzig, R.R. and Churg, A. (1983) Mesothelioma in pet dogs associated with exposure of their owners to asbestos. *Environmental Research*, **32**, 305–313.

Gloyne, S.R. (1929) The presence of the asbestos fibre in the lesions of asbestos workers. *Tubercle*, **10(9)**, 404–407.

Gloyne, SR. (1933) The morbid anatomy and histology of asbestosis. *Tubercle*, **14**, 550–558.

Governa, M., Amati, M., Fontana, S., Visonà, I., Botta, G.C., Mollo, F., Bellis, D. and Bo, P. (1999) Role of iron in asbestos-body-induced oxidant radical generation. *Journal of Toxicology and Environmental Health*, **58**, 279–287.

Gross, P. and de Treville, R.T.P. (1967) Experimental asbestosis. Studies on the progressiveness of the pulmonary fibrosis caused by chrysotile dust. *Archives of Environmental Health*, **6**, 638–649.

Gross P., de Treville, R.T.P., Cralley, L.J. and Davis, J.M.G. (1969) Pulmonary ferruginous bodies: Development in response to filamentous dusts and a method of isolation and concentration. *International Journal of Occupational Environmental Medicine*, **11**, 208–209.

Groppo, C., Tomatis, M., Turci, F., Gazzano, E., Ghigo, D., Compagnoni, R. and Fubini, B. (2005) Potential toxicity of nonregulated asbestiform minerals: balengeroite from the Western Alps, Part1: identification and characterization. *Journal of Toxicology and Environmental Health*, **68**, 1–19.

Gylseth, B., Baunan, R.H. and Bruun, R. (1981) Analysis of inorganic fiber concentrations in biological samples by scanning electron microscopy. *Scandinavian Journal of Work, Environment & Health*, **7**, 240–249.

Hayashi, H. (1978) Energy dispersive X-ray analysis of asbestos fibers. *Clay Science*, **5**, 145–154.

Hammond, E.C., Selikoff, I.J. and Saidman, H. (1979) Asbestos exposure, cigarette smoking, and death rates. *Annals of New York Academy of Sciences*; **330**, 473–490.

Harbison, M.L. and Godleski, J.J. (1983) Malignant mesothelioma in urban dogs. *Veterinary Pathology*, **20**, 531–540.

Haque, A.K. and Kanz, M.F. (1988) Asbestos bodies in children's lungs: an association with sudden infant death syndrome and bronchopulmonary dysplasia. *Archives of Pathology, Laboratory Medicine*, **112**, 514–518.

Harrison, P.M., Fischbach, F.A. and Hoy, T.G. (1967) Ferric oxyhydroxide core of ferritin. *Nature*, **216**, 1188–1190.

Harrison, P.T.C. and Heath, J.C. (1986) Asbestos bodies in rat lung following intratracheal instillation of chrysotile. *Laboratory Animals*, **20**, 132–137.

Hayashi, H. (1978) Energy dispersive x-ray analysis of asbestos fibers. *Clay Sciences*, **5**, 145–154.

Hill, R.J., Edwards, R.E. and Carthew, P. (1990) Early changes in the pleural mesothelium following intrapleural inoculation of the mineral fibre erionite and the subsequent development of mesotheliomas. *Journal of Experimental Pathology*, **71**, 105–118.

Hillerdal, G., Henderson, D.W. and Path, M.R.C. (1997) Asbestos, asbestosis, pleural plaques and lung cancer. *Scandinavian Journal of Work, Environment & Health*, **23**, 93–103.

Holmes, A. and Morgan, A. (1980) Clearance of anthophyllite fibres from the rat lung and the formation of asbestos bodies. *Environmental Research*, **22**, 13–21.

Holt, P.F., Mills, J. and Young, D.K. (1964) The early effects of chrysotile asbestos dust on the rat lung. *Journal of Pathology and Bacteriology*, **87**, 15–23.

IARC – International Agency for Research on Cancer (1987) Silica and some silicates. *Monographs on the evaluation of carcinogenic risks to humans.* **42**, Lyon.

IARC – International Agency for Research on Cancer (2002) Man-made Vitreous Fibres. *Monographs on the evaluation of carcinogenic risks to humans.* **81**. Lyon.

IARC – International Agency for Research on Cancer (2012) Arsenic, Metals, Fibres and Dusts. *Monographs on the evaluation of carcinogenic risks to humans.* **100C**. Lyon.

IARC – International Agency for Research on Cancer (2014) Carcinogenicity of fluoro-edenite, silicon carbide fibres and whiskers, and carbon nanotubes. *Monographs on the evaluation of carcinogenic risks to humans.* **111**. Lyon.

INAIL – Istituto Nazionale Assicurazione Infortuni sul Lavoro, 5° Rapporto il Registro Nazionale dei mesoteliomi (2015).

Infante, P.F., Schuman, L.D., Dement, J. and Huff, J. (1994) Fibrous glass and cancer. *American Journal of Industrial Medicine*, **26**, 559–584.

Jaurand, M.C., Bignon, J., Sebastien, P. and Goni, J. (1977) Leaching of chrysotile asbestos in human lungs: Correlation with in vitro studies using rabbit alveolar macrophages. *Environmental Research*, **14**, 245–254.

Jemal, A., Grauman, D. and Deversa, S. (2000) Recent geographic patterns of lung cancer and mesothelioma mortality rates in 49 shipyard counties in the United States, 1970–94. *American Journal of Industrial Medicine*, **37**, 512–521.

Jones, A.D. (1993) Respirable industrial fibres: deposition, clearance and dissolution in animal models. *Annals of Occupational. Hygiene*, **37**, 211–226.

Karjalainen, A., Anttila, S., Heikkila, L., Kyyronen, P. and Vainio, H. (1993) Lobe of origin of lung cancer among asbestos-exposed patients with or without diffuse interstitial fibrosis. *Scandinavian Journal of Work, Environmental & Health*, **19**, 102–107.

Koerten, H.K., Hazekamp, J., Kroon, M. and Daems, W.T. (1990) Asbestos body formation and iron accumulation in mouse peritoneal granulomas after the introduction of crocidolite asbestos fibres. *American Journal of Pathology*, **136**, 141–157.

Landrigan, P.J. (1991) The third wave of asbestos disease: exposure to asbestos in place. Public health control. Introduction. *Annals of The New York Academy of Sciences*, **31**, 643: xv–xvi.

Langer, A.M., Selikoff, I.J. and Sastre, A. (1971) Chrysotile asbestos in the lungs of persons in New York City. *Archives of Environmental Health*, **22**, 348–361.

Lauriola, M., Bua, L., Chiozzotto, D., Manservisi, F., Panetta, A., Martorana, G. and Belpoggi, F. (2012) Urinary apparatus tumours and asbestos: the Ramazzini Institute caseload. *Archivi Italiani di Urologia e Andrologia*, **84**, 189–196.

Lee, K.P. (1985) Lung response to particulates with emphasis on asbestos and other fibrous dusts. *CRC Critical Reviews in Toxicology*, **14**, 33–86.

Lynch, K.M. and Smith, W.A. (1935) Pulmonary asbestosis. III Carcinoma of lung in asbestos-silicosis. *American Journal of Cancer*, **24**, 56.

Luce, D., Brochard, P., Quénel, P., Salomon-Nekiriai, C., Goldberg, P., Billon-Galland and M.A. and Goldberg, M. (1994) Malignant pleural mesothelioma associated with exposure to tremolite. *The Lancet*, **344**, 1777.

Luce, D., Billon-Galland, M.A., Bugel, I., Goldberg, P., Salomon, C., Févotte, J. and Goldberg, M. (2004) Assessment of environmental and domestic exposure to tremolite in New Caledonia. *Archives of Environmental Health*, **59**, 91–100.

Magee, F., Wright, J.L., Chan, N., Lawson, L. and Churg, A. (1986) Malignant mesothelioma caused by childhood exposure to long-fiber low aspect ratio tremolite. *American Journal of Industrial Medicine*, **9**, 529–533.

Magnani, C., Terracini, B., Ivaldi, C., Mancini, A. and Botta, M. (1996) Mortalità per tumori e altre cause tra i lavoratori del cemento-amianto a Casale Monferrato. *Medicina del Lavoro*, **87**, 133.

Mallory, T.B., Castlman, B. and Parris, E.E. (1947) Case records of the Massachusetts General Hospital. *New England Journal of Medicine*, **236**, 407–412.

Maltoni, C. and Minardi, F. (1983) The comparative potency of asbestos and erionite in producing mesothelioma, following intrapleural and intraperitoneal injection into Sprague-Dawley rats. *Acta Oncologica*, **4**, 69.

Maltoni, C. and Minardi, F. (1989) Recent results of carcinogenicity bioassays of fibers and other particulate materials. In: *Non-occupational Exposure to Mineral Fibers*. (E. Bignon, J. Peto and R. Saracci, editors). IARC, Scientific Publications (**90**), Lyon.

Maltoni, C., Minardi, F. and Morisi, L. (1982) Pleural mesotheliomas in Sprague-Dawley rats by erionite: first experimental evidence. *Environmental Research*, **29**, 238–244.

Maltoni, C., Pinto, C., Carnuccio, R., Valenti, D. and Amaducci, E. (1994) Mesothelioma following exposure to asbestos used in railroads: 122 Italian cases In: *The Identification and Control of Environmental and Occupational Diseases: Hazards and Risks of Chemicals in the Oil Refining Industry*. (M.A. Mehlman and A. Upton, editors). Princeton Scientific Publishing, Princeton, New Jersey, USA. 635 pp

Maltoni, C., Pinto, C., Carnuccio, R., Valenti, D., Lodi, P. and Amaduccl, E. (1995) Mesotheliomas following exposure to asbestos used in railroads: 130 Italian cases. *Medicina del Lavoro*, **86**, 461–477.

Maltoni, C., Pinto, C., Valenti, D., Amaducci, E. and Di Bisceglie, M. (1998) Mesotheliomas following exposure to asbestos used in sugar refineries: report of 17 Italian cases in: *25° Anniversario degli Istituti di Chirurgia dell'Università di Chieti diretti dal Professor Vanni Beltrami*. (Vecchio Faggio editor). Scritti degli allievi e degli amici, Chieti.

Maltoni, C., Lambertini, L., Cevolani, D., Minardi, F. and Soffritti, M. (2002) I mesoteliomi da amianto usato nelle ferrovie: resoconto di 199 casi. *European Journal of Oncology*, **7**, 51–55.

Manservisi, F., Marquillas, C.B., Buscaroli, A., Huff, J., Lauriola, M., Mandrioli, D., Manservigi, M., Panzacchi, S., Silbergeld, E.K. and Belpoggi, F. (2017) An integrated experimental design for the assessment of multiple toxicological end points in rat bioassays. *Environmental Health Perspective*, **125**, 289–295.

Marchand, F. (1906) Ueber eigenttimliche Pigmentkristalle in den Lungen. *Verhandlungen der Deutschen Gesellschaft fur Pathologie*, **10**, 223–228.

Marsh, G.M., Enterline, P.E., Stone, R.A. and Henderson, V.L. (1990) Mortality among a cohort of US man-made mineral fiber workers: 1985 follow-up. *Journal of Occupational Medicine*, **32**, 594–604.

Maxim, L. and van der Sluijs, J. (2013) Seed-dressing systemic insecticides and honeybees. *Late lessons from early warnings: Science, precaution, innovation (European Environmental Agency)*, **2**, 369–406.

McConnochie, K., Simonato, L., Mavrides, P., Christofides, P., Pooley, F.D. and Wagner, J.C. (1987) Mesothelioma in Cyprus: The role of tremolite. *Thorax*, **42**, 342–347.

McConnochie, K., Simonato, L., Mavrides, P., Christofides, P., Mitha, R., Griffiths, D.M. and Wagner J.C. (1989) Mesothelioma in Cyprus. Pp. 411–419 in: *Non-Occupational Exposure to Mineral Fibres* (J.Bignon, J. Peto and R. Saracci, editors). International Agency for Research on Cancer.

McDonald, J.C., Harris, J. and Armstrong, B. (2004) Mortality in a cohort of vermiculite miners exposed to fibrous amphibole in Libby, Montana. *Journal of Occupational Environmental Medicine*, **61**, 363–366.

McLemore, T.L., Stevens, P.M., Roggli, V., Greenberg, S.D., Marshall, M.V. and Lawrence, E.C. (1981) Comparison of phagocytosis of uncoated versus coated asbestos fibres by cultured human pulmonary alveolar macrophages. *Chest Journal*, **80**, 39–42.

Meeker, G.P., Bern, A.M., Brownfild, I.K., Lowers, H.A., Sutley, S.J., Hoefen, T.M. and Vance, J.S. (2003) The composition and morphology of amphiboles from the Rainy Creek Complex, near Libby, Montana. *American Mineralogy*, **88**, 1955–1969.

Metintas, M., Hillerdal, G. and Metintas, S. (1999) Malignant mesothelioma due to environmental exposure to erionite: follow-up of a Turkish emigrant cohort. *European Respiratory Journal*, **13**, 523–526.

Meyer, J. (1970) Private Communication. Environmental Research Laboratories, T.N.O., Delft, Holland.

Middleton, A.P., Beckett, S.T. and Davis, J.M. (1979) Further observations on the short-term retention and clearance of asbestos by rats, using UICC reference samples. *Annals of Occupational Hygiene*, **22**, 141–152.

Minardi, F. and Maltoni, C. (1998a) Results of long-term carcinogenicity bioassays of rockwool on Sprague-Dawley rats. *European Journal of Oncology*, **3**, 251–260.

Minardi, F. and Maltoni, C. (1998b) Results of long-term carcinogenicity bioassays of ceramic fibers ("Fiberfrax") on Sprague-Dawley rats. *European Journal of Oncology*, **3**, 241.

Morgan, A. and Holmes, A. (1980) Concentrations and dimensions of coated and uncoated asbestos fibres in the human lung. *British Journal of Industrial Medicine*, **37**, 25–32.

Morgan, A. and Holmes A. (1985) The enigmatic asbestos body: its formation and significance in asbestos-

related disease. *Environmental Research*, **38**, 283–292.
Mumpton, F.A. (1975) Occurrence and utilization of natural zeolites; a review. *International Clay Conference, Mexico*, pp. 215–218.
Murray, H.M. (1907) *Report of the Departmental Committee on Compensation for Industrial Diseases.* H.M.S.O. London.
NCR – National Research Council (1991) Committee on animals as Monitors of environmental health hazards. Animal as sentinels of environmental health hazards. Washington DC: National Academy Press, 160 pp.
Nurminen, M. and Tossavainen, A. (1994) Is there an association between pleural plaques and lung cancer without asbestosis? *Scandinavian Journal of Work, Environment & Health*, **20**, 62–64.
Oberdorster, G. (1988) Lung clearance of inhaled insoluble and soluble particles. *Journal of Aerosol Medicine*, **1**, 289–330.
O'Brien, D.J., Kaneene, J.B. and Poppenga, R.H. (1993) The use of mammals as sentinels for human exposure to toxic contaminants in the environment. *Environmental Health Perspectives*, **99**, 351–368.
O'Sullivan, M.F., Corn, C.J. and Dodson, R.F. (1987) Comparative efficiency of Nucleopore filters of various pore size as used in digestion studies of tissue. *Environmental Research*, **43**, 97–103.
Paoletti, L., Battisti, D., Bruno, C., Di Paola, M., Gianfagna, A., Mastrantoni, M., Nesti, M. and Comba, P. (2000) Unusually high incidence of malignant pleural mesothelioma in a town of eastern Sicily: an epidemiological and environmental study. *Archives of Environmental Health*, **556**, 392–398.
Pascolo, L., Gianocelli, A., Kaulich, B., Rizzardi, C., Schneider, M., Bottin, C., Polentarutti, M., Kiskinova, M., Longoni, A. and Melato, M. (2011) Synchrotron soft X-ray imaging and fluorescence microscopy reveal novel features of asbestos body morphology and composition in human lung tissues. *Particle Fibre Toxicology*, **8**, 7–17.
Pirastu, R., Zona, A., Ancona, C., Bruno, C., Fano, V., Fazzo, L., Iavarone, I., Minichilli, F., Mitis, F., Pasetto, R. and Comba, P. (2011) Mortality results in SENTIERI Project. *Epidemiologia Prevenzione*, **35(5-6 Suppl. 4)**, 29–152.
Pooley, F.D. (1975) The identification of asbestos dust with an electron microscope analyser. *Annals of Occupational Hygiene*, **18**, 181–186.
Pott, F., Ziem, U., Reiffer, F.J., Huth, F., Ernst, H. and Mohr, U. (1987) Carcinogenicity studies on fibres, metal compounds, and some other dusts in rats. *Experimental Pathology*, **32**, 129–152.
Puntoni, R., Merlo, F., Borsa, L., Reggiardo, G., Garrone, E. and Ceppi, M. (2001) A historical cohort mortality study among shipyard workers in Genoa, Italy. *American Journal of Industrial Medicine*, **40**, 363–370.
Rath, R. (1964) Form und Formanderung der Asbestosis. *Korpechen. Beitr. Silikose-Forsch*, **81**, 1–10.
Report of an ILSI Risk Science Institute Working Group (2005) Testing of fibrous particles: short-term assays and strategies. *Inhalation Toxicology*, **17**, 497–537.
Rey, F., Boutin, C., Steinbauer, J., Viallat, J.R., Alessandroni, P., Jutisz, P., Di Giambattista, D., Billon-Galland, M.A., Hereng, P., Dumortier, P. and De Vuyst, P. (1993a) Environmental pleural plaques in an asbestos exposed population of northeast Corsica. *European Respiratory Journal*, **6**, 978–982.
Rey, F., Viallat, J.R., Boutin, C., Farisse, P., Billon-Galland, M.A., Hereng, P., Dumortier, P. and De Vuyst, P. (1993b) Environmental mesotheliomas in northeast Corsica. *Revue Maladies Respiratoires*, **10**, 339–345.
Rey, F., Boutin, C., Viallat, J.R., Steinbauer, J., Alessandroni, P., Jutisz, P., Di Giambattista, D., Billon-Galland, M.A, Hereng, P., Dumortier, P. and De Vuyst, P. (1994) Environmental asbestotic pleural plaques in northeast Corsica: Correlations with airborne and pleural mineralogic analysis. *Environmental Health Perspectives*, **102**, 251–252.
Richter, G.M. (1958) Electron microscopy of hemosiderin: Presence of ferritin and occurrence of crystalline lattices in hemosiderin deposits. *Journal of Biophysical and Biochemical Cytology*, **4**, 55–58.
Roggli, V.L. (1989) Pathology of human asbestosis: A critical review. Pp. 31–60 in: *Advances in Pathology*, **2** (C.M. Fenoglio-Preiser, editor), Chicago, Illinois, USA.
Roggli, V.L. (2014) Asbestos bodies and non-asbestos ferruginous bodies Pp. 25–51 in: *Pathology of Asbestos-associated Diseases* (T.D. Oury, T.A. Sporn and V.L. Roggli, editors). Springer-Verlag, Berlin, Heidelberg.
Roggli, V.L. and Brody, A.R. (1990) The role of electron microscopy in experimental models of pneumoconiosis. *Lung Biology in Health and Disease*, **48**, 315–343.
Roggli, V.L. and Sharma, A. (2014). Analysis of tissue mineral fiber content. Pp. 253–292 in: *Pathology of*

Asbestos-Associated Diseases (T.D. Oury, T.A. Sporn and V.L. Roggli, editors). Springer-Verlag, Berlin, Heidelberg.

Rohl, A.N., Langer, A.M., Moncure, G., Selikoff, I.J. and Fischbein, A. (1982) Endemic pleural disease associated with exposure to mixed fibrous dust in Turkey. *Science*, **216**, 518–520.

Sakellarious, K., Malamou-Mitsi, V., Haritou, A., Koumpaniou, C., Stachouli, C., Dimoliatis, I.D. and Constantopoulus, S.H. (1996) Malignant pleural mesothelioma from non-occupational asbestos exposure in Metsovo (north-west Greece): slow end of an epidemic? *European Respiratory Journal*, **9**, 1206–1210.

Schilling, R.J., Steele, G.H., Harris, A.E., Donahue, J.F. and Ina, R.T. (1988) Canine serum levels of polychlorinated biphenyls (PCBs): a pilot study to evaluate the use of animal sentinels in environmental health. *Archives of Environmental Health*, **43**, 218–221.

Schlesinger, R. (1980) Particle deposition in model systems of human and experimental animal airways. Pp. 553–575 in: *Generation of Aerosols and Facilities for Exposure Experiments* (K. Willeke, editor.). *Annals of Arbor Science*.

Schwabe, C.W. (1984) Animal monitors of the environment. Pp. 562–578 in: *Veterinary Medicine and Human Health*. Williams & Wilkins, Baltimore, Maryland, USA.

Selçuk, Z.T., Cöplü, L., Emri, S., Kalyoncu, A.F., Sahin, A.A. and Bari Y.I. (1992) Malignant pleural mesothelioma due to environmental mineral fiber exposure in Turkey. Analysis of 135 cases. *Chest*, **102**,790–796.

Selikoff, I.J., Churg, J. and Hammond, E.C. (1964) Asbestos exposure and neoplasia. *JAMA*, **188**, 22-6.

Selikoff, I.J. and Seidman, H. (1991) Asbestos-associated deaths among insulation workers in the United States, and Canada, 1967–1987. *Annals of the New York Academy of Sciences*, **643**.

Shannon, H.S., Jamieson, E., Julian, J.A., Muir, D.C.F. and Walsh, C. (1987) Mortality experience of Ontario glass fibers workers – extended follow-up. *Annals of Occupational Hygiene*, **31**, 657–662.

Simonato, L., Fletcher, A.C., Cherrie, J.W., Anderson, A., Bertazzi, P., Charnay, N., Claude, J., Dodgson, J., Esteve, J., Frentzel-Beyme Gardner, M.J., Jensen, O., Olsen, J., Teppo ,L., Winkelmann, R., Westerholm, P. Winter, P.D., Zocchetti, C. and Saracci, R. (1987) The International Agency for Research on Cancer historical cohort study of MMMF production workers in seven European countries: extension of the follow-up. *Annals of Occupational Hygiene*, **31**, 603–623.

Soffritti, M., Minardi, F. and Maltoni, C. (2003) Physical carcinogens. Pp. 313–320 in *Holland-Frei Cancer Medicine* 6th edition, Vol. **1** (D.W. Kufe *et al.*, editors). BC Decker Inc, Toronto, Canada

Soffritti, M., Falcioni, L., Bua, L., Tibaldi, E., Manservigi, M. and Belpoggi, F. (2013) Potential carcinogenic effects of world trade center dust after intratracheal instillation to Sprague-Dawley rats: first observation. *American Journal of Industrial Medicine*, **56**, 155–62.

Stahl, R.G. Jr (1997) Can mammalian and non-mammalian "sentinel species" data be used to evaluate the human health implications of environmental contaminants. *Human and Ecological Risk Assessment*, **3**, 329–335.

Stewart, M.J. and Haddow, A.C. (1929) Demonstration of the peculiar bodies of pulmonary asbestosis ("asbestosis bodies") in material obtained by lung puncture and in the sputum. *Journal of Pathology and Bacteriology*, **32**, no. 172.

Suzuki, Y. and Churg, J. (1969a) Formation of the asbestos body: A comparative study with three types of asbestos. *Environmental Research*, **3**, 107–118.

Suzuki, Y. and Churg, J. (1969b) Structure and development of the asbestos body. *American Journal of Pathology*, **55**, 79–107.

Temel, A. and Gündogdu, M.N. (1996) Zeolite occurrences and the erionite-mesothelioma relationship in Cappadocia, Central Anatolia, Turkey. *Mineralium Deposita*, **31**, 539–547.

Timbrell, V. (1976) Aerodynamic considerations and other aspects of glass fibre. *Occupational Exposure to Fibrous Glass: A Symposium*, 33–50.

Turci, F., Tomatis, M., Gazzano, E., Riganti, C., Martra, G., Bosia, A., Ghigo, D. and Fubini, B. (2005) Potential toxicity of nonregulated asbestiform minerals: balangeroite from the Western Alps. Part 2: oxidant activity of the fibers. *Journal of Toxicology and Environmental Health*, part A, **68**, 21–29.

Van der Schalie, W.H., Gardner, H.S., Bantle, J.A., De Rosa, C.T., Finch, R.A., Reif, J.S. Reuter, R.H. Backer, L.C., Burger, J., Folmar, L.C. and Stokes, W.S. (1999) Animals as sentinels of human health azards of environmental chemicals. *Environmental Health Perspectives*, **107**, 309–315.

Viallat, J.R., Boutin, C., Steinbauer, J., Gaudichet, A. and Dufour, G. (1991) Pleural effects of environmental asbestos pollution in Corsica. *Annals of New York Academy of Sciences*, **643**, 438–443.

Vorwald, A.J., Durkan, T.M. and Pratt, P.C. (1951) Experimental studies of asbestosis. *Archives of Industrial Hygiene Occupational Medicine*, **3**, 1–43.

Wagner, J.C., Sleggs, C.A. and Marchand, P. (1960) Diffuse pleural mesothelioma and asbestos exposure in the North West Cape Province. *British Journal of Industrial Medicine*, **17**, 260–271.

Wagner, J.C., Berry, G., Skidmore and J.W. Timbrell, V. (1974) The effects of the inhalation of asbestos in rats. *British Journal of Cancer*, **29**, 252–269.

Wagner, J.C., Berry, F., Hill, R.J., Munday, D.E. and Skidmore, J.D. (1984) Animal experiments with MMM(V)F – effects of inhalation and intrapleural inoculation in rats. Pp. 209–233 (**2**) in: *Biological Effects of Man-made Mineral Fibres*, **2**. Proceedings of a WHO/IARC Conference, Copenhagen.

Wagner, J.C., Skidmore, J.W., Hill, R.J. and Griffiths, D.M. (1985) Erionite exposure and mesotheliomas in rats. *British Journal of Cancer*, **51**, 727–730.

Warnock, M.L. and Wolery, G. (1987) Asbestos bodies or fibres and the diagnosis of asbestosis. *Environmental Research*, **44**, 29–44.

Wedler, H.W. (1943) Uber den Lungenkrebs bei Asbestose. *Deutsch Archives Clinical Medicine*, **191**, 189–209.

Whitehouse, A.C., Black, B., Heppe, M.S. Ruckdeschel, J. and Levin, S.M. (2008) Environmental exposure to Libby Asbestos and Mesotheliomas. *American Journal of Industrial Medicine*, **51**, 877–880.

Wilkinson, P., Hansell, D.M., Janssens, J., Rubens, M., Rudd, R.M. and Taylor, A.N. (1995) Is lung cancer associated with asbestos exposure when there are no small opacities on the chest radiograph? *Lancet*, **345**, 1074–1078.

Wolff, H., Vehmas, T., Oksa, P., Rantanen, J. and Vainio, H. (2014) Asbestos, asbestosis, and cancer, the Helsinki criteria for diagnosis and attribution 2014: recommendations. *Scandinavian Journal of Work, Environment & Health*, **41**, 5–15.

Yazicioglu, S., Ilçayto, R., Balci, K., Sayli, B.S. and Yorulmaz, B. (1980) Pleural calcification, pleural mesotheliomas, and bronchial cancers caused by tremolite dust. *Thorax*, **35**, 564–569.

EMU Notes in Mineralogy, Vol. 18 (2017), Chapter 10, 347–366

Dissolution and biodurability of mineral fibres

MARISA ROZALÉN[1,*], F. JAVIER HUERTAS[1], ALESSANDRO PACELLA[2] and PAOLO BALLIRANO[2,3]

[1]*Instituto Andaluz de Ciencias de la Tierra (IACT), CSIC-University of Granada, Avda. de las Palmeras 4, 18100 Armilla, Granada, Spain, e-mail: marisarozalen@ugr.es*
[2]*Dipartimento di Scienze della Terra, Sapienza Università di Roma, Piazzale Aldo Moro, 5-I-00185 Rome, Italy*
[3]*Laboratorio Rettorale Fibre e Particolato Inorganico, Sapienza Università di Roma, Piazzale Aldo Moro, 5-I-00185 Rome, Italy*
**Corresponding author*

Dissolution rates of mineral fibres in several environments are obtained as proxies for their biodurability in body fluids. This chapter provides a description of the experimental methods, the parameters and characteristics to be fixed during the design of dissolution experiments in closed (batch reactors) and open systems (flow-through cells), as well as details of the dissolution media. The dissolution of mineral fibres in buffered inorganic solutions is the key to understanding their behaviour during weathering processes because it contributes not only to their chemical transformation, but also to the breakdown of the fibres that may be dispersed in the environment. On the other hand, preparation of fluids representing different interstitial conditions in the lung is described, with particular attention to artificial lysosomal fluid (ALF) employed to mimic the environment that inhaled particles would encounter after phagocytosis by alveolar and interstitial macrophages. Moreover, the use of a neutral fluid such as Gamble's solution (GS) simulates the interstitial lung fluid and airway lining fluid. Finally, the results of studies of mineral-fibre dissolution in inorganic and body fluids, found in the literature, are discussed.

Methodologies for assessing the biodurability of fibres are illustrated, starting from dissolution rate data, and focus on *in vitro* studies. Rate constants are used to assess fibre lifetimes utilizing a fibre-shrinking model equation. Finally, literature studies show differences in biopersistence between serpentine and amphibole asbestos, due to their different crystal structures and dissolution conditions of pH and solution composition.

1. Introduction

The application of geochemical methods to determine mineral-fibre dissolution kinetics represents a powerful tool to provide some insights into how fibres may interact with environmental and biological surroundings. Classic studies of mineral solubility analyse change over time in solution chemistry for a limited number of constituents, such as aqueous silica, magnesium and iron. With this information, coupled with analytical data on the composition of the fibres after leaching, dissolution rates can be inferred based on the total amount of these constituents released over time. The

DOI: 10.1180/EMU-notes.18.10

adjustment of these traditional studies to mimic the physical and chemical conditions encountered in biological fluids allow us to estimate the biodurability of the fibres through calculation of residence times based on the dissolution clearance mechanism.

Although these experiments cannot substitute for *in vivo* experiments due to the complexity of the human body and the multitude of processes that may occur, they do provide a benchmark to evaluate the biological breakdown of these fibres. In order to ensure quality results, it is necessary to prepare dissolution experiments properly; setting up all the parameters that control dissolution rates. Consequently, the calculation of kinetics parameters and the description of the dissolution mechanism should be addressed, taking into account that solubility in biological fluids is influenced strongly by the composition of the fluids, pH, ionic strength and organic molecule content present among cellular and extracellular fluids.

The rate at which inhaled particles dissolve *in vivo* plays an important role in their biodurability, and therefore their ability to trigger fibrosis, cancer and other diseases. Dissolution rates are a complex function of the chemical solubility of particles in body fluids, coupled with factors (such as crystal structure and mineral surface properties) that determine the rate at which particles dissolve.

2. Experimental methodology

In dissolution studies of mineral fibres in the literature, closed (batch reactors) and open systems (flow-through reactors) have been used to obtain dissolution rates, by measuring the amount of cations released to the media over time. Both methods offer a different perspective and provide important insights into different situations. For example Wood *et al.* (2006) suggested the possibility that some portions of the lung environment (*e.g.* inside a macrophage) may behave as an at least partially closed system (batch reactors). Nevertheless, the use of flow-through cells (opened system) is recommendable to avoid reaching equilibrium, which may restrict dissolution and is more representative of dissolution occurring in a biological medium (Eastes and Hadley, 1994). Before describing both methods, it is necessary to consider various factors.

The dissolution process depends on the interaction of the solutions with mineral surfaces. As minerals are anisotropic, there exist different reactive surfaces. The fibrous shape makes sample preparation more difficult in terms of controlling the particle size, which could induce changes in surface area during the dissolution process. Moreover, in the case of amphiboles for example, microscopy studies (Rozalen *et al.*, 2014b) suggest that fibre length decreases by splitting parallel to the main cleavage plane (fibre longitudinal direction). Examples for tremolite or anthophyllite are shown in Fig. 1. For tremolite, two mechanisms of particle breakage coexist during the process of dissolution: fibre length decreases by splitting parallel to the main cleavage plane (fibre longitudinal direction), with developments of kinks along the fibres. Moreover, coalescence of etch pits perpendicular to the *c* axis induces breakage and particle shortening, particularly intense in the presence of ligands. In the case of anthophyllite fibres, splitting parallel to the main cleavage plane unravels the fibre

Figure 1. Morphology of a tremolite and anthophyllite before (a) and after (b) dissolution reaction.

bundles. This leads to variation of surface area during the dissolution process and errors in the calculation of bulk dissolution rates. Therefore, it is more appropriate to normalize dissolution rates to mass (mol $g^{-1}s^{-1}$) than BET surface area (Metz *et al.*, 2005; Rozalen *et al.*, 2008).

Various studies indicate that short fibres are less pathogenic than long fibres, in part because alveolar macrophage clearance is limited (Davis *et al.*, 1991;Van Oss *et al.*, 1999; Holland and Smith, 2001) and possibly in part because of chemical differences between short and long fibres (Graham *et al.*, 1999). Moreover, Crawford (1980) and Hochella (1993) have shown that mineral dissolution can occur preferentially along crystal-structural defects.

Amphibole fibres are straighter and more rigid than typically curved chrysotile fibres or small crystalline silica particles. Some reports conclude that rigid needle-like fibres can penetrate deeper into lung tissue and contribute to interstitial diseases such as mesothelioma (*e.g.* Van Oss *et al.*, 1999). However, other studies show that short chrysotile fibres (<5 μm) may also penetrate and migrate to the pleural and peritoneal spaces and trigger mesothelioma and other diseases in these regions (Suzuki and Yuen, 2001).

Finally, very small particles (<1.5–2 μm) could be dangerous when the clearance capacity *via* macrophage is exceeded by particle deposition rates. Phagocytic cells can transport these particles into the lymph and circulatory systems.

Crystal structure and chemical composition are also key parameters to take into account. Firstly, numerous studies attempt to explain the greater dissolution rate of fibres of chrysotile compared to those of amphiboles due to its unique crystal structure (see Ballirano *et al.*, 2017, this volume): in chrysotile, a silicate mineral belonging to the phyllosilicates subclass, the fibres are constructed by layers coiled around the fibre axis. On the other hand, amphiboles are double-chain inosilicates composed basically of Si, Mg, Fe, Ca and Na in different proportions. Iron usually occurs as Fe^{2+}, which in oxidant media is transformed to Fe^{3+}, leading to a destabilization of the structure and enhancement of dissolution rates. However, much attention has been focused on the potential generation of free radicals for minerals such as crocidolite or erionite (*e.g.* Lund and Aust, 1992; Hardy and Aust, 1995; Werner *et al.*, 1995).

2.1. Batch reactors

A batch reactor is an inert container in which a known mass of fibres is in contact with a fixed volume of fluid. The ions released during mineral dissolution accumulate in the solution, the composition of which evolves and may reach saturation, inhibiting fibre dissolution. Buffer reagents may be added to the solution to keep pH constant as well as inert electrolytes to fix ionic strength. Aliquots of the suspension are periodically withdrawn, filtered and analysed for the desired elements (Si, Al, Mg...) using colorimetric methods, ion chromatography or ICP.

Dissolution rates are calculated as variation of the mineral component j concentration in the reactive solutions as time progresses:

$$\mathrm{Rate_j} = \frac{1}{\nu_j}\frac{V}{SM}\frac{dC_j}{dt} \quad (1)$$

where ν_j is the stoichiometric coefficient of component j in the dissolution reaction, S is the BET surface area ($m^2\ g^{-1}$), M represents the mineral mass (g), V is the volume of solution (L), C_j is the concentration of component j, and t represents time (s).

2.2. Flow-through reactors

Dissolution experiments performed in flow-through cells facilitate the measurement of the dissolution rate under fixed saturation-state conditions by modifying flow rate, initial sample mass and input solution concentration. Temperature may be maintained constant by immersing reactors in a thermostatic water-bath (Fig. 2a). The flow rate is controlled by means of a peristaltic pump that injects the input solution into the cell where the mineral fibres are confined by using two membrane filters (Fig. 2b). The solid/solution ratio is normally in the range 2–10 g L^{-1}. The output solution is sampled at time intervals that should be larger than the solution residence time and stored for analyses of pH and chemical composition (cations, ligands, anions, *etc.*).

In each run, the flow rate and the input pH should be constant until steady-state conditions are achieved. For silicates, the steady state prevails when the Si output concentration remains fairly constant. At steady-state, dissolution is expected to proceed under far-from-equilibrium conditions.

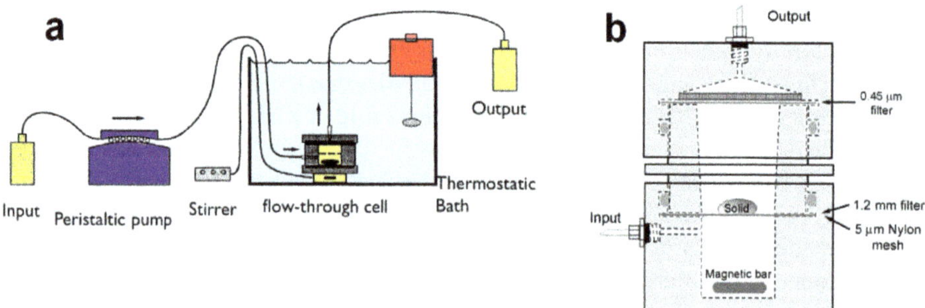

Figure 2. (a) A flow-through dissolution experimental design; (b) detail of a flow-through cell.

Dissolution rates (mol g^{-1} s^{-1}) are calculated based on the mass balance of a given mineral component j. Under steady-state conditions this is given by the following equation (*e.g.* Cama *et al.*, 2002; Rozalen *et al.*, 2008):

$$\text{Rate (mol g}^{-1}\text{s}^{-1}) = \frac{1}{\nu_j}\frac{q}{M}(C_{j,out} - C_{j,in}) \qquad (2)$$

where ν_j is the stoichiometric coefficient of component j in the dissolution reaction, q stands for the volumetric fluid flow through the system, M is the mass of mineral and $C_{j,out}$ and $C_{j,in}$ correspond to the concentrations of component j in the output and input solutions, respectively. The rate is defined as negative for dissolution and positive for precipitation. The dissolution rate in silicate studies is generally calculated from Si concentrations (R_{Si}) in the output solutions as it is considered to be the rate-limiting step in the dissolution reaction (*e.g.* Oelkers and Schott, 2001; Liu *et al.*, 2006; Olsen and Rimstidt, 2008). The use of other cations such as magnesium or calcium can lead to errors due to a variety of reasons such as: non-stoichiometric dissolution, precipitation of secondary phases (*e.g.* hydroxides, carbonates) and the possibility of exchange reactions with protons (Liu *et al.*, 2006; Olsen and Rimstidt, 2008).

A limitation of the flow-through reactors is the use of large volumes of test fluid to maintain flow rate, and the low reproducibility within laboratories due to different experimental conditions being implemented (Utembe *et al.*, 2015).

3. Dissolution media

3.1. Inorganic solutions

Dissolution of mineral fibres in buffered inorganic solutions is of key importance in understanding their behaviour during weathering because it contributes not only to their chemical transformation but also to the breakdown of fibres that may be dispersed in the environment.

Alteration solutions usually contain a combination of inert electrolytes (chlorides or nitrates of alkaline cations) and buffer agents. The use of buffer agents may be complex for some pH ranges. Strong acids, such as hydrochloric or nitric acids, maintain acidic conditions below pH 4–5 in the appropriate concentrations; pH conditions >9 are achieved when carbonates or Na or K hydroxides are used. The buffer agents for intermediate pH conditions may interfere with the dissolution process. Intermediate pH conditions are difficult to buffer as common substances such as organic acids and the corresponding salt, phosphates, *etc.* may enhance or inhibit the dissolution reaction (*e.g.* Golubev *et al.*, 2006; Ramos *et al.*, 2011). Acetate/acetic acid is used for slightly acidic conditions. Bicarbonates have been used for slightly alkaline conditions, with limited success. Different pH could represent different natural or anthropogenic situations. Very acidic media could correspond with acid rain or contaminated waters from industry. The pH of natural waters (~6.5–8.5) is the most difficult to study, because the low solubility of cations such as Al or Mg leads to saturated solutions. Basic media are also very interesting because of the use of serpentines in carbon capture and storage.

3.2. Organic solutions

The role of soil organic acids in enhancing chemical weathering rates has been recognized for decades. Many studies show that polyfunctional acids such as oxalic, citric, malic or salicylic enhance silicate dissolution (Huang and Kiang, 1972; Franklin *et al.*, 1994; Wang *et al.*, 2005; Ramos *et al.*, 2011). Consequently, the dissolution mechanism is now promoted by ligands. For example, the study of Rozalen *et al.* (2013) shows an increase in dissolution rates up to 20 times with respect to ligand-free solutions for tremolite due to the formation of an oxalate complex at pH 4.

3.3. Simulated biological fluids

Human body fluids exhibit a wide range of compositions, as well as pH and Eh, electrolyte content and concentrations of organic species such as amino acids, organic acids and proteins. This compositional variability indicates that any given mineral fibre has the potential to behave quite differently from a geochemical perspective, depending upon the exposure pathway, the resulting body fluid(s) it encounters and how it may modify body-fluid chemistry. Studies of mineral-fibre dissolution can be accomplished using different environmental conditions in the lung. In particular, the artificial lysosomal fluid (ALF) is employed to mimic the environment with which inhaled particles would come into contact after phagocytosis by alveolar and interstitial macrophages (Marques *et al.*, 2011). In addition, the use of a neutral fluid such as Gamble's solution (GS) simulates the interstitial lung fluid and airway lining fluid (Marques *et al.*, 2011). Table 1 shows the composition of ALF and Gamble's solution. More detailed tables including composition of blood plasma, gastric fluids, *etc.* can be found in Plumblee *et al.* (2006).

4. Literature dissolution studies

4.1. Dissolution during weathering

The effect of mineral weathering can be to separate fibres and break them down to fibrils of submicroscopic size that are inhalable easily. Early studies of dissolution of magnesium silicates in aqueous solutions (forsterite, enstatite and serpentine: Luce *et al.*, 1972; bronzite: Grandstaff, 1977; hypersthene: Siever and Woodford, 1979; enstatite, diopside and tremolite: Schott *et al.*, 1981) suggest mostly that the dissolution of magnesium silicates takes place incongruently with the formation of surface phases depleted in magnesium that inhibit the dissolution reaction.

The dissolution of tremolite has been studied poorly and results are incoherent. Schott *et al.* (1981) dissolved it in aqueous solution at 20–60°C and pH 1–6, observing

Table 1. Composition of simulated lung fluids (modified from Marques *et al*., 2011).

Composition	ALF (g/L)	Gamble's solution (g/L)
Magnesium chloride	0.050	0.095
Sodium chloride	3.21	6.019
Potassium chloride	–	0.298
Disodium hydrogen phosphate	0.071	0.126
Sodium sulfate	0.039	0.063
Calcium chloride dihydrate	0.128	0.368
Sodium acetate	–	0.574
Sodium hydrogen carbonate	–	2.604
Sodium citrate dihydrate	0.077	0.097
Sodium hydroxide	6.00	–
Citric acid	20.8	–
Glycine	0.059	–
Sodium tartrate dihydrate	0.090	–
Sodium lactate	0.085	–
Sodium pyruvate	0.086	–
pH	4.5	7.4

a slight depletion of Mg relative to Si and a marked initial Ca depletion followed by congruent behaviour. Using a rate law, they obtained a reaction order of n = 0.11. However, Mast and Drever (1987) reported weak pH dependence (from pH 2 to 5, n = 0). Rozalen *et al*. (2014b) studied the effect of pH on the kinetics of tremolite at 25°C in batch reactors over the pH range 1–13.5, in inorganic buffered solution. From pH 1–6 there is a noticeable rate dependence on pH, obtaining a reaction order of n = 0.185. A minimum is reached around neutral conditions, at pH 6–8. At basic pH, this dependence becomes stronger (n = 0.264) but dissolution takes place with collateral effects of saturation and carbonation. Calcium and Mg are released preferentially in acid media reducing the Mg/Si ratio to the extent that Mg solubility decreases with pH.

Regarding anthophyllite dissolution, Chen and Brantley (1998) obtained a value of n = 0.24 using flow-through reactors. Rozalen *et al*. (2014b) obtained a complete outline of dissolution rate *vs*. pH. They observed a linear dependence from pH 1 to 9.5, obtaining a reaction order of n = 0.20. A minimum is reached around pH 9.5, where Mg^{2+} solubility is at a minimum and there is a change in the reaction stoichiometry from congruent to incongruent. Above pH 9.5 the pH dependence is smoother and the reaction order is halved (n = 0.099).

The results of Jurinski and Rimstidt (2001) for talc dissolution in flow-through cells at 37°C, using inorganic buffers and simulated lung fluids (GS), show that dissolution rates measured from silica release are independent of pH over the pH range 2–8, obtaining a rate constant of 1.4×10^{-11}mol Si/m^{2}g. Their experiments as well as those of Lin and Clemency (1981) showed incongruent dissolution with a preferential release of magnesium. Saldi *et al.* (2007) measured dissolution rates using flow-through experiments between 25 and 150°C and pH 2 to 12. They obtained a reaction order of $n = -0.5$ between pH 2 to 7. Rates at pH above 7 appear to be pH independent. They also used an atomic force microscope to confirm that dissolution proceeds by the removal of T–O–T sheets from talc edges.

Finally, and due to its particular structure, studies for chrysotile are plentiful but they are not focused on dissolution. Since Hargreaves and Taylor (1946) reported that magnesium could be removed completely from the chrysotile structure by treatment with dilute acids, many other studies focused on acid treatments in order to develop environmental remediation processes. Moreover, the mechanism is also of critical importance for engineered carbon capture.

Regarding dissolution, Choi and Smith (1972) studied it at 5–45°C and pH 6. They suggested that the rate-controlling mechanism was diffusion for Mg and OH^- ions from the surface into solution. Later, Bales and Morgan (1985) measured dissolution rates derived from Mg release in short-time batch experiments (5 days) and obtained a reaction order of $n = 0.24$) between pH 7 and 10. However, dissolution rates derived from Si release showed no definite pH dependence. A more recent study from Rozalen *et al.*, (2014a) using batch reactors at 25°C and pH range 1–13.5 showed incongruent dissolution. Dissolution rates have strong pH dependence from pH 1 to 8 (reaction order $n = 0.27$) reaching a minimum around pH 8–8.5, which is consistent with data obtained by Bales and Morgan (1985). Above pH 8.5 the dependence becomes smoother ($n = 0.06$).

Figure 3 shows that chrysotile dissolves faster than amphiboles through the entire pH scale. However, these differences vary depending on the mineral studied. For anthophyllite, the dissolution rate is only half an order of magnitude greater than for amphibole and pyroxene but more than two orders of magnitude greather compared with talc, enstatite and tremolite. The dissolution rates for these series of minerals follow the decreasing sequence:

$$R_{\mathrm{chr}} > R_{\mathrm{ant}} >> R_{\mathrm{tlc}} \approx R_{\mathrm{trm}}$$

Even though the dissolution rates outlined exhibit different patterns for different mineral fibres, the rate pH dependence is similar in acid media ($n_{H^+} = 0.2$) for tremolite, anthophyllite and talc. Thereby, a general mechanism for dissolution in acid media can be proposed (Fig. 4); the relative ease with which Mg–O bonds break compared to that of Si–O bonds and the evidence of Mg–H exchange reactions with the outermost part of the octahedral sheet (Kohler *et al.*, 2005; Saldi *et al.*, 2007) support the suggestion that the breaking of Si–O bonds is the limiting step of the process.

Around neutral and basic pHs, the behaviour varies for different minerals due to saturation and/or precipitation of magnesium carbonates or hydroxides. Recently, Pacella *et al.* (2014, 2015) studied the surface alteration mechanism of both fibrous tremolite and crocidolite in buffered solution at pH 7.4, in the presence of H_2O_2.

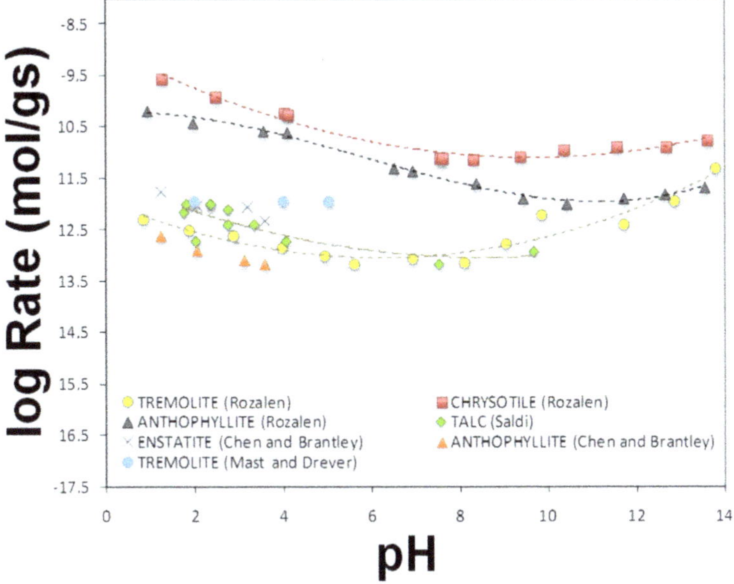

Figure 3. Dissolution rates of different mineral fibres at 25°C and inorganic buffered solutions, normalized to mass, available in literature studies at 25°C for different weathering conditions. Data taken from Mast and Drever (1987), Chen and Brantley (1998), Saldi *et al.* (2007), Rozalen *et al.* (2013, 2014a,b).

Results showed that incongruent dissolution of the amphiboles takes place in the first stage of surface alteration resulting in an amorphous, altered surface layer. Iron oxidation and the formation of FeOOH species may take place together with the congruent dissolution of the altered layer. Iron-rich amorphous nanoparticles on top of the fibres are formed in the last stage. In addition, surface alteration is markedly higher for crocidolite indicating a fibre lifetime which is shorter than that of tremolite.

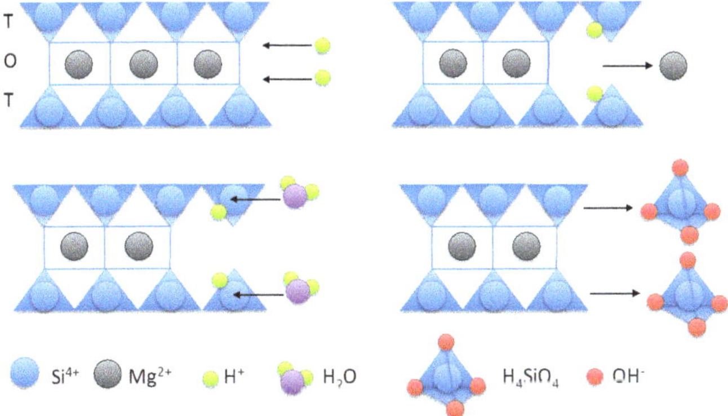

Figure 4. Dissolution mechanism for talc in acid media as proposed by Saldi *et al.* (2007).

4.2. Dissolution in mimicked lung fluids

Although these studies are of paramount importance for obtaining an approach to the biodurability of mineral fibres, just a few of them can be found in the literature. The first authors to use flow-through cells at 37°C with simulated physiological fluids were Hume and Rimstidt (1992) who found that the lifetime of a chrysotile fibre is controlled by the dissolution rate of the silica layer. However, they obtained very high rate values, from 1.07×10^{-9} to 7.17×10^{-11}mol/m^{2}s, probably due to the short duration of the experiments (3 h). In 2001, Jurinski and Rimstidt studied the dissolution of talc at 37°C, using physiological simulated fluids (pH 5.2–7.7). They described the experimental set up for flow-through cells and preparation of physiological fluids, because these are the key for obtaining reliable dissolution data used to calculate biodurability. In 2010, Oze and Solt measured dissolution rates of chrysotile and tremolite in batch reactors using simulated lung (pH 7.4) and gastric fluids (pH 1.2) at 37°C for 30 days. They used calculated rates to model biodurability of both minerals. Finally, Rozalen *et al.* (2013) measured dissolution of tremolite at 37°C and using simulated macrophages (pH 4), interstitial fluids (pH 7.4) and middle check point (pH 5.5). Rozalen *et al.* (2013) also studied the catalyst effect of ligand organics (citrate and oxalate), which are able to enhance dissolution rates up to 20 times for solutions with 15 mmol L^{-1} citrate at pH 4. However, at pH 7.4 the effect is less pronounced and oxalate solutions (15 mmol L^{-1}) enhance dissolution rates only eight times with respect to free organic ligand solutions.

As is shown in Fig. 5, values for the dissolution rates in simulated biological fluids are scarce and incongruent. This could be due to the different experimental conditions applied in the different cases. Consequently, it is necessary to set up ideal conditions that represent closely the interaction between mineral fibres and biological fluids, in order to provide accurate models for predicting biodurability.

5. Biodurability

Biopersistence of mineral particles and fibres is defined as their ability to persist in the human body despite chemical, physical and other physiological clearance mechanisms (Bernstein *et al.*, 2005). Biodurability, defined as the resistance to chemical/biochemical alteration, is a significant contributor to biopersistence (Utembe *et al.*, 2015). Biopersistence and biodurability have the potential to influence the long-term toxicity and pathogenicity of fibres deposited in the body and, therefore, they are considered to be important parameters needed for the risk assessment of fibres. As shown previously, fibre dissolution in body fluids has been used as a measure of their biodurability (Hume, 1991; Guthrie, 1997; Muhle and Bellmann, 1997; Jurinski 1998; Bernstein *et al.*, 2005b; Oze and Solt, 2010). The determination of dissolution rates provides the opportunity to explore further the interaction between particles/fibres and their surrounding biological environment. Dissolution rate of fibres may be determined using *in vivo* (short and long term) and *in vitro* ('acellular' and 'cellular') tests (Morgan *et al.*, 1982; Bernstein *et al.*, 1984, 2005a; Jaurand *et al.*, 1984; Muhle *et al.*, 1986; Hume and Rimstidt, 1992; McClellan and Hesterberg, 1994; Guthrie Jr., 1997; Muhle and Bellmann, 1997; Oze and Solt, 2010; Utembe *et al.*, 2015).

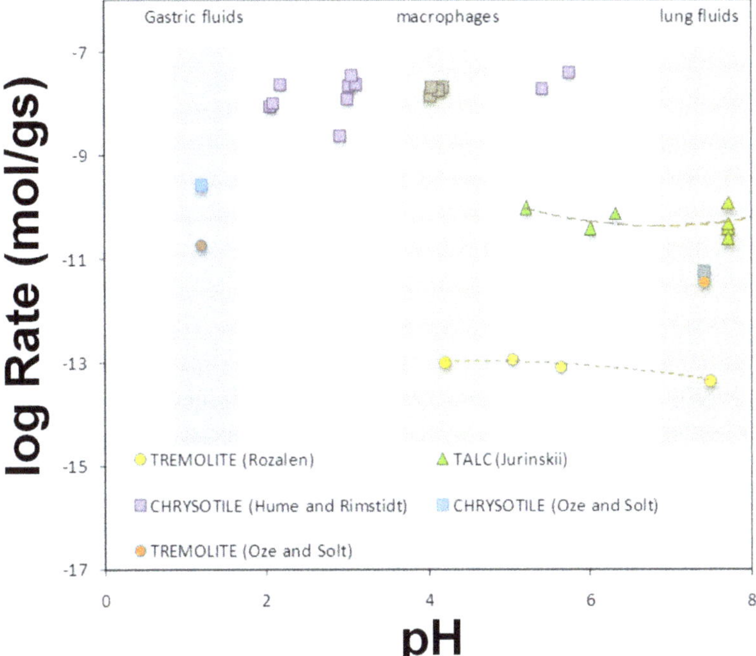

Figure 5. Dissolution rates of different mineral fibres at 37°C in simulated biological fluids. Data taken from Hume and Rimstidt (1992), Oze and Solt (2010), Jurinkski and Rimstidt (2001) and Rozalen *et al.* (2013).

5.1. Methodologies for assessing biodurability

5.1.1. In vivo investigations of biodurability

In vivo investigations of biodurability include inhalation or intratracheal instillation of fibres into rats after which the clearance of fibres from the lung is measured by serial sacrifices of the animals (Nyberg *et al.*, 1992). The biodurability is assessed through the decrease of the fibre diameter. The European Commission (EU, 1997) accepts both inhalation and intratracheal instillation methods. Inhalation is the most appropriate method for delivering fibres into the lungs of rats. However, an advantage of the intratracheal injection is that a precise starting point for kinetics analysis can be defined.

5.1.2. In vitro investigations of biodurability

Cellular *in vitro* investigation for assessing biodurability includes the treatment of cultured cells with fibres, followed by the examination of the intracellular fibres using microscopy to determine the change in their diameter and composition. For cellular dissolution tests, alveolar macrophages are commonly used (Jaurand *et al.*, 1977, 1984). Note that cellular systems have a number of limitations because the cells are not in their normal natural environment and the volumes of media used are small compared to *in vivo* systems (Bernstein *et al.*, 2005b; Nguea *et al.*, 2008).

Acellular *in vitro* investigations include the determination of fibre dissolution rates by their incubation in simulated lung fluids (SLF) as a function of mineral surface. Since dissolution of fibres after inhalation can occur in lungs both intracellularly and extracellularly (Bernstein *et al.*, 2005a), acellular *in vitro* studies involve the use of SLFs representing different interstitial conditions in the lung. In particular, the artificial lysosomal fluid, (ALF, pH ~4.5) is employed to mimic the environment with which inhaled particles would come into contact after phagocytosis by alveolar and interstitial macrophages (Marques *et al.*, 2011). In addition, the use of a neutral fluid such as GS (pH 7.4) simulates the interstitial lung fluid and airway lining fluid (Marques *et al.*, 2011). Table 1 shows the composition of ALF and GS. Although acellular *in vitro* investigations are far from mimicking a real cellular environment, they are easy to carry out and provide a rapid and low-cost choice for promoting the biological breakdown of asbestos that may occur *in vivo* at both acidic and near-neutral pHs (Thelohan and de Meringo, 1994; Searl and Buchanan, 2000). Dissolution experiments are performed at a temperature of 37°C to simulate human body temperature, using either static or dynamic methods. The degree of dissolution is determined by measuring changes in the mass of fibres, and/or the concentration of the ions released into the simulated body fluid (Searl and Buchanan, 2000; Oze and Solt, 2010).

5.2. Lifetime calculation. Approaches

The biodurability of particles and fibres using *in vitro* dissolution tests is based on the determination of the fibre dissolution rate (estimated from the Si concentration released into solution) using non-linear regression. The analysis assumes that the rate of reaction is proportional to the reaction surface area and the reaction order is the power to which the surface area of the fibres is raised (Searl *et al.*, 1999; Oze and Solt, 2010; Utembe *et al.*, 2015): $r = \text{Rate} = -dSi/dt = \text{k}[\text{asbestos}]^n$. The rate law above describes the change in the release of Si into solution over a period of time t (μmol h^{-1}), k is the dissolution rate constant expressed in μmol m^{-2} h^{-1}, and n is the order of reaction. In addition, rate constants may also be used to assess fibre lifetimes by utilizing their fibre shrinking model equation (Hume, 1991; Rozalen *et al.*, 2013; Utembe *et al.*, 2015):

$$t = 3d/(4\text{k}V_\text{m})$$

where t is time (s), d is the diameter (m) of the fibre, V_m is the molar volume (m^3/mol), and k is the rate constant (mol m^{-2} s^{-1}).

5.3. Experimental results

It is well known from studies on the exposure of populations and workers to asbestos that there is a preferential retention in the lung of amphiboles compared to chrysotile (Churg and Warnock, 1980; Langer and Nolan, 1989). Over the last five decades a number of studies have reported experiments on the biodurability of asbestos and many efforts have been devoted to investigating in depth the relationship of fibre clearance from the lungs and fibre type. The earliest experiments were those of Wagner and

Skidmore (1965). These authors indicated that, after 6 weeks of dust exposure of rats over a period of 2 months, the clearance of both chrysotile and amphibole asbestos (amosite and crocidolite) could be described using a single exponential function, but the rate of clearance of chrysotile was greater by a factor of three than that of amphiboles. Subsequent investigations conducted using more prolonged exposure times (6 months) indicated that the lung burden of chrysotile reached a plateau, whereas that of amphibole asbestos continued to increase, further suggesting the greater rate of clearance of chrysotile (Wagner *et al.*, 1974). However, Morgan *et al.* (1975a,b) failed to confirm either the lower initial retention or the fastest clearance of chrysotile. In addition, Middleton *et al.* (1977) indicated that there is no difference in the asbestos clearance rate related to fibre type, even if the observed retention of chrysotile was lower than that of amosite and crocidolite. Later, Middleton *et al.* (1979) confirm that the rate of clearance of asbestos by rats after a 6-week period of inhalation is independent of asbestos type and can be described using an exponential model. However, the depositional rate of chrysotile was about one quarter that of amosite and crocidolite and appeared to be inversely correlated to the asbestos concentration in the airborne dust. It was suggested that the curly fibres of chrysotile tend to aggregate, especially at higher concentrations, and therefore are filtered out more efficiently from the upper airways (Middleton *et al.*, 1979; Timbrell, 1982). However, investigation of the lungs of miners and millers refuted this hypothesis, since they demonstrated the accumulation of both long and short chrysotile fibres in the distal parenchyma under the pleura (Churg *et al.*, 1984; Sebastien *et al.*, 1986). Bolton *et al.* (1983) performed clearance studies on UICC amosite using dust concentrations ranging from 6.0 to ~90 mg/m^3. Rats were exposed for up to 210 h, comprising 7 h per day, 5 days a week, over 6 weeks. The results were consistent with those of Middleton *et al.* (1977) and highlighted that the mechanism of pulmonary fibre clearance depends on the lung burden, which in turns is related to the level of dust exposure. In particular, for low cumulative lung burdens fibre clearance in the first stage is rapid and then slows down, whereas for large cumulative lung burdens, clearance appeared always to be slower. The modification of the chemistry of chrysotile fibres after their ingestion in alveolar macrophages (AM) and pleural mesothelial cultured cells (PMC) was studied by Jaurand *et al.* (1977, 1984) using scanning electron microscopy. Those authors observed a progressive leaching of Mg from the fibres, leading to a decrease of the Mg/Si ratio over time as a consequence of the biodegradation of chrysotile in a biological medium. In particular, AM were more efficient than PMC at leaching incubated fibres, whilst the kinetics of fibre dissolution by AM and PMC were comparable with leaching at pH 4 and 7, respectively. Indeed, Mg leaching from chrysotile fibres has been shown previously to be related to the acidity of the solution by Johan *et al.* (1976) and Papirer *et al.* (1976). According to Parry (1985) the dissolution reaction of chrysotile is: $Mg_3Si_2O_5(OH)_4 + 6H^+ \rightarrow 3Mg^{2+} + 2H_4SiO_4 + H_2O$. From this, the author estimated the rate of dissolution of chrysotile fibres, ingested and phagocytized in cells, by formulating an appropriate rate law, in which the rate of dissolution is dependent on the fibre surface area, mass of solution to which the fibres are exposed within the cells, the rate of transport of H^+ into the cells and the rate of transport of Mg^{2+} and SiO_2 out of cells. The maximum dissolution rate of a single fibre of 10 μm × 1 μm, if cytoplasm is maintained at quartz saturation, was $2.13-10^{-7}$mol h^{-1} kg^{-1}, which corresponds to fibre dissolution of 30% in ~6 months. This estimate is consistent with the results of Hume and Rimstidt (1992) for which a 1 μm diameter of chrysotile fibre would completely dissolve in the human lung in 9.0 months. However, unlike Parry (1985) (who assumed congruent dissolution), Hume and Rimstidt (1992) showed that the dissolution reaction of chrysotile proceeds in two steps: first, the magnesium hydroxide layer dissolves and leaves behind silica that dissolves at a slower rate. Another comparative *in vitro* study on the biodurability of naturally occurring silicate fibres was conducted by Scholze and Conradt (1987) using incubation in simulated extra-cellular fluid, derived from Gamble's solution, under flow conditions. The results showed that erionite is much more biopersistent than both crocidolite and chrysotile. In particular, the fibre dissolution velocity was 0.0002 nm per day for erionite and 0.005 and 0.011 nm per day for the amphibole and the serpentine, respectively. Experiments of intratracheal injection of asbestos in rats were conducted firstly by Muhle *et al.* (1994) to assess the clearance half time of crocidolite fibres that resulted after >300 days. Later, Searl *et al.* (1999) showed, for amosite fibres, that their persistence increased with increasing fibre length, and there was no evidence for any clearance of the longest fibres (>20 μm). This hypothesis was confirmed further by Hesterberg *et al.* (1998), following inhalation experiments in rats through 1 year of post exposure. The results showed that the number of short amosite fibres (<5 μm) in the lung was reduced by 90% after 90 days of recovery, while that of long fibres (>20 μm) was reduced by only 65%. In addition, from 90 days to 1 year, little or no reduction occurred in amosite fibres for any of the length categories. Recently, Bernstein *et al.* (2003, 2004, 2005b) published inhalation biopersistence studies on three different samples of chrysotile asbestos, using the standardized EC biopersistence protocol. These works included Canadian chrysotile (Eastern Township, Quebec, Canada), Brazilian chrysotile (Cana Brava Mine, Gioias State, Brazil) and Calidria chrysotile (Coalinga Mine, New Idria, USA). In all these studies chrysotile was found to be rapidly removed from the lungs with clearance half times of the fibres longer than 20 μm ranging from 0.3 days for Calidria chrysotile, 1.3 days for Brazilian chrysotile and 11.4 days for Canadian chrysotile. In addition, Bernstein *et al.* (2005b) highlighted that chrysotile is considerably less persistent than amphibole asbestos. In fact, no long chrysotile fibres (L > 20 μm) remained in the lung 2 years after exposure; in contrast, with both amosite and tremolite, there was only a small amount of clearance of the long fibres after cessation of exposure, followed by virtually no further clearance (Fig. 6). On this basis, the clearance half time of the long amosite and tremolite fibres resulted in 418 days and infinity, respectively (Table 2). As seen in Table 2, the fibres with length ranging from 5 to 20 μm also clear quickly from the lungs, although more slowly than those with L > 20 μm. The authors reported that this result is influenced strongly by the continuous breaking apart of the longer fibres, which serve therefore as a replenishment source for shorter fibres. Following Bernstein *et al.* (2005b), the observed difference in

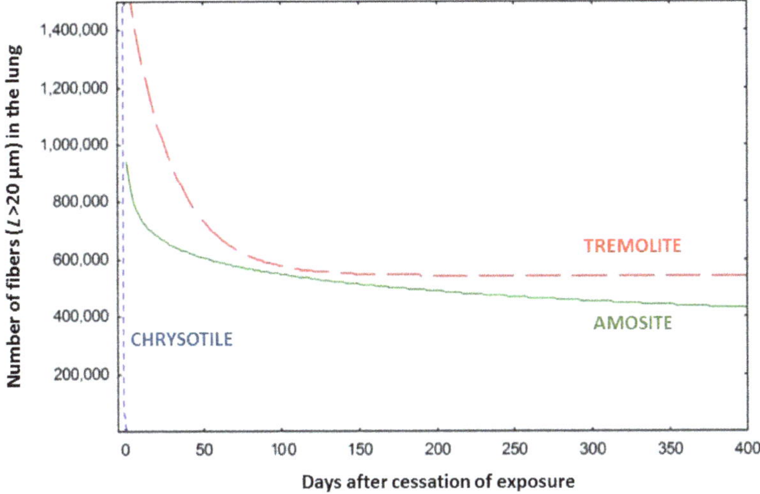

Figure 6. Clearance of tremolite, amosite and chrysotile fibres longer than 20 µm from the lung in inhalation biopersistence studies (after Bernstein and Hoskins, 2006).

biopersistence between serpentine and amphibole asbestos is due to their different crystal structures. Chrysotile fibre is physically a very thin rolled sheet, with individual layers linked to each other by weak van der Waals bonds, much more fragile than the silicate double chains of the amphiboles, in which the bonds between the silica tetrahedra are very strong. Therefore, while amphibole asbestos splits into fibres that are subsequently very stable, the rolled sheets of chrysotile can break easily and the magnesium hydroxide layer dissolve in the lung fluid. In addition, Bernstein *et al.* (2008) showed that the clearance of chrysotile is significantly increased (by a factor of ten) if the fibres are inhaled in combination with fine particles (sanded powder),

Table 2. Comparative clearance half-times of fibres of >20 µm and fibres between 5 and 20 µm for chrysotile and amphibole asbestos (data from Bernstein *et al.*, 2005b).

	—— Clearance half-time (days) ——	
Fibre type	Fibre length >20 µm	Fibre length 5–20 µm
Calidria chrysotile	0.3	7
Brazilian chrysotile	1.3	2.4
Canadian chrysotile	11.4	29.7
Amosite	418	900
Crocidolite	536	262
Tremolite	Infinity	Infinity

because the large number of fine particles deposited in the lung accelerates the recruitment of macrophages. Recently, Oze and Solt (2010) examined the biodurabilities of chrysotile and tremolite asbestos in simulated lung fluid (SLF, Gamble's solution at pH 7.4) and simulated gastric fluid (SGF at pH 1.2) as a function of mineral surface, over 720 h. Provided equivalent surface areas, asbestos tremolite proved to be less bio-durable than chrysotile in lung fluid and *vice versa* in gastric fluid. However, the results confirmed that chrysotile is less bio-durable than tremolite when taking into account the greater surface area of chrysotile, the relative bio-durability being tremolite (SLF) > chrysotile (SLF) > tremolite (SGF) > chrysotile (SGF).

References

Bales, R.C. and Morgan, J.J. (1985) Dissolution kinetics of chrysotile at pH 7 to 10. *Geochimica et Cosmochimica Acta*, **49**, 2281–2288.

Ballirano, P., Bloise, A., Gualtieri, A.F., Lezzerini, M., Pacella, A., Perchiazzi, N., Dogan, M. and Dogan, A.U. (2017) Crystal structure of mineral fibres. Pp. 17–64 in: *Mineral Fibres: Crystal Chemistry, Chemical-physical Properties, Biological Interaction and Toxicity* (A.F. Gualtieri, editor). EMU Notes in Mineralogy, **18**. European Mineralogical Union and Mineralogical Society of Great Britain & Ireland, London.

Bernstein, D.M. and Hoskins, J.A. (2006) The health effect of chrysotile: current perspective based upon recent data. *Regulatory Toxicology and Pharmacology*, **45**, 252–264.

Bernstein, D.M., Drew, R.T., Schidlovsky, G. and Kuschner, M. (1984) Pathogenicity of MMMF and the contrasts with natural fibres. Pp. 169–195 in: *Biological Effects of Man-Made Mineral Fibres*, Vol. 2, World Health Organisation, Copenhagen.

Bernstein, D.M. and Riego Sintes, J.M. (1999) Methods for the determination of the hazardous properties for human health of man made mineral fibres (MMMF). (Report No. EUR 18748), European Commission Joint Research Centre, European Chemicals Bureau, Brussels.

Bernstein, D.M., Mast, R., Anderson, R., Hesterberg, T.W., Musselman, R., Kamstrup, O. and Hadley J. (1994) An experimental approach to the evaluation of the biopersistence of respirable synthetic fibres and minerals. *Environmental and Health Perspectives*, **102 (Suppl. 5)**, 15–18.

Bernstein, D.M., Rogers, R. and Smith, P. (2003) The biopersistence of Canadian chrysotile asbestos following inhalation. *Inhalation Toxicology*, **15**, 1247–1274.

Bernstein, D.M., Rogers, R. and Smith, P. (2004) The biopersistence of Brazilian chrysotile asbestos following inhalation. *Inhalation Toxicology*, **16**, 745–761.

Bernstein, D.M., Castranova, V., Donaldson, K., Fubini, B., Hadley, J., Hesterberg, T., Kaneg, A., Laih, D., McConnell, E.E., Muhle, H., Oberdorsterk, G., Olin, S. and Warheitm, D.B. (2005a) Testing of fibrous particles: short-term assays and strategies. *Inhalation Toxicology*, **17**, 497–537.

Bernstein, D.M., Chevalier, J. and Smith, P. (2005b) Comparison of Calidria chrysotile asbestos to pure tremolite: final results of the inhalation biopersistence and histopathology examination following short-term exposure. *Inhalation Toxicology*, **17**, 427–449.

Bernstein, D.M., Donaldson, K., Decker, U., Gaering, S., Kunzendorf, P., Chevalier, J. and Holm, S.E. (2008) A biopersistence study following exposure to chrysotile asbestos alone or in combination with fine particles. *Inhalation Toxicology*, **20**, 1009–1028.

Bolton, R.E., Vincent, J.H., Jones, A.D., Addison, J. and Beckett, S.T. (1983) An overload hypothesis for pulmonary clearance of UICC amosite fibres inhaled by rats. *British Journal of Industrial Medicine*, **40**, 262–272.

Cama, J., Metz, V. and Ganor, J. (2002) The effect of pH and temperature on kaolinite dissolution rate under acidic conditions. *Geochimica et Cosmochimica Acta*, **66**, 3913–3926.

Chen, Y. and Brantley, S.L., (1998) Diopside and anthophyllite dissolution at 25ºC and 90ºC and acid pH. *Chemical Geology*, **147**, 233–248.

Choi, I. and Smith, R.W. (1972) Kinetic study of dissolution of asbestos fibres in water. *Journal of Colloid and*

Interface Science, **40**, 253–262.
Churg, A. and Warnock, M.L. (1980) Asbestos fibres in the general population. *American Review of Respiratory Disease Journal*, **122**, 669–678.
Churg, A., DePaoli, L., Kempe, B. and Stevens, B. (1984) Lung asbestos content in chrysotile workers with mesothelioma. *American Review of Respiratory Disease Journal*, **130**, 1042–1045.
Crawford, D. (1980) Electron microscopy applied to studies of the biological significance of defects in crocidolite asbestos. *Journal of Microscopy*, **120**, 181–192.
Davis, J.M.G., Addison, J., McIntosh, C., Miller and B.G, Niven, K. (1991) Variations in the carcinogenicity of tremolite dust samples of different morphology. *Annals NY Academy of Science*, 473–490.
Eastes, W. and Hadley, J.G. (1994) Role of fibre dissolution in biological activity in rats. *Regulatory Toxicology and Pharmacology*, **20**, 104–112.
EU (1997) European Commission directive 97/69/EC of 5.XII.97 (23 adaptation) O.J. L 343/1997. European Commission, Brussels.
Franklin, S.P., Hajash A.J., Dewers T.A and Tieh, T.T. (1994) The role of carboxylic acids in albite and quartz dissolution: An experimental study under diagenetic conditions. *Geochimica et Cosmochimica Acta*, **58**, 4259–4279.
Golubev, S.V., Bauer, A. and Pokrovsky, O.S. (2006) Effect of pH and organic ligands on the kinetics of smectite dissolution at 25°C. *Geochimica et Cosmochimica Acta*, **70**, 4436–4451.
Graham, A., Higinbotham, J., Allan, D., Donaldson, K. and Beswick, P.H. (1999) Chemical differences between long and short amosite asbestos: differences in oxidation state and coordination sites of iron, detected by infrared spectroscopy. *Occupational and Environmantal Medicine*, **56**, 606–611.
Grandstaff, D.E. (1977) Some kinetics of bronzite orthopyroxene dissolution. *Geochimica et Cosmochimica Acta*, **41**, 1097–1103.
Guthrie, Jr, G.D. (1997) Mineral properties and their contributions to particle toxicity. *Environmental and Health Perspectives*, **105**, 1003–1011.
Hardy, J.A. and Aust, A.E. (1995) The effect of iron binding on the ability of crocidolite asbestos to catalyze DNA single-strand breaks. *Carcinogenesis*, **16**, 319–325.
Hargreaves, A. and Taylor, W.H. (1946) An X-ray examination of decomposition products of chrysotile (asbestos) and serpentine. *Mineralogical Magazine*, **27**, 204–216.
Hesterberg, T.W., Chase, G., Axten, C., Miller, W.C., Musselman, R.P., Kamstrup, O., Hadley, J., Mordscheidt, J., Bernstein, B.M. and Thevenaz, P. (1998) Biopersistence of synthetic vitreous fibres and amosite asbestos in the rat lung following inhalation. *Toxicology and Applied Pharmacology*, **151**, 262–275.
Hochella, M.F. Jr. (1993) Surface chemistry, structure, and reactivity of hazardous mineral dust. Pp. 275–308 in: *Heath Effects of Mineral Dusts*. (G.D. Guthrie and B.T. Mossman, editors). Reviews in Mineralogy and Geochemistry, **28**, Mineralogical Society of America and The Geochemical Society. Washington, USA.
Holland, J.P. and Smith, D.D. (2001) Asbestos. Pp. 1214–1227 in: *Clinical Environmental Health and Exposures*, 2ndedn. (J.B. Sullivan, Jr. and G. Krieger, editors). Lippincott Williams & Wilkins, Philadelphia, USA.
Hume, L.A. (1991) *The Dissolution Rate of Chrysotile*. Virginia Polytechnic Institute and State University: Blacksburg, Virginia, USA.
Hume L.A. and Rimstidt, J.D. (1992). The biodurability of chrysotile asbestos. *American Mineralogist*, **77**, 1125–1128.
Huang, W.H. and Kiang, W.C. (1972) Laboratory dissolution of plagioclase feldspar in water and organic acids at room temperature. *American Mineralogist*, **57**, 1849–1859.
Jaurand, M.C., Bignon, J., Sebastien, P. and Goni, J. (1977) Leaching of chrysotile asbestos in human lungs: correlation with in vitro studies using rabbit alveolar macrophages. *Environmental Research*, **14**, 245–254.
Jaurand, M.C., Gaudichet, A., Halpern, S. and Bignon, J. (1984) In vitro biodegradation of chrysotile fibres by alveolar macrophages and mesothelial cells in culture: comparison with a pH effect. *British Journal of Industrial Medicine*, **41**, 389–395.
Johan, Z.A, Goni, J., Sarcia, C., Bonnaud, G. and Bignon, J. (1976) Influences de certains acides organiques sur la stabilite du reseau du chrysotile. *6th Congress International Geochemical Organique Technip*. Ed. Paris, pp. 883–903.

Jurinski, J.B. (1998) *Geochemical investigations of respirable particulate matter*. Doctoral thesis. Faculty of the Virginia Polytechnic Institute and State University, Blacksburg, Virginia, USA.

Jurinski, J.B. and Rimstidt, J.D. (2001) Biodurability of talc. *American Mineralogist*, **86**, 392–399.

Kohler, S.J., Bosbach, D. and Oelkers, E.H. (2005) Do clay mineral dissolution rates reach steady state? *Geochimica et Cosmochimica Acta*, **69**, 1997–2006.

Langer, A.M. and Nolan, R.P. (1989) Fibre type and burden in parenchymal tissues of workers occupationally exposed to asbestos in the United States. Pp. 330–335 in: *Non-occupational Exposure to Mineral Fibres* (J. Bignon, J. Peto and R. Saracci, editors). IARC, **90**, International Agency for Research on Cancer, Lyon, France.

Liu, Y., Olsen, A.A. and Rimstidt, J.D. (2006) Mechanism for the dissolution of olivine series minerals in acidic solutions. *American Mineralogist*, **91**, 455–458.

Lin, F. and Clemency, C.V. (1981) The dissolution kinetics of brucite, antigorite, talc, phlogopite at room temperature and pressure. *American Mineralogist*, **66**, 801–806.

Luce, R.W., Bortlett, R.W. and Parks, G.A. (1972) Dissolution kinetics of magnesium silicates. *Geochimica et Cosmochimica Acta*, **36**, 35–50.

Lund, L.G. and Aust, A.E. (1992) Iron mobilization from crocidolite asbestos greatly enhances crocidolite-dependent formation of DNA single-strand breaks in phi X174 RFI DNA. *Carcinogenesis*, **13**, 637–642.

Marques, M.R.C., Loebenberg, R. and Almukainzi, M. (2011) Simulated biological fluids with possible application in dissolution testing. *Dissolution Technologies*, **18**, 15–28.

Mast, M.A. and Drever, J.L. (1987) The effect of oxalate on the dissolution rates of oligoclase and tremolite. *Geochimica et Cosmochimica Acta*, **51**, 2559–2568.

McClellan, R.O. and Hesterberg, T.W. (1994) Role of biopersistence in the pathogenicity of man-made fibres and methods for evaluating biopersistence: a summary of two round-table discussions. *Environmental and Health Perspectives*, **102** (Suppl. 5), 277–283.

Metz, V., Raanan, H., Pieper, H., Bosbach D. and Ganor, J. (2005) Towards the establishment of a reliable proxy for the reactive surface area of smectite. *Geochimica et Cosmochimica Acta*, **69**, 2581–2591.

Middleton, A.P, Beckett, S.T. and Davis, J.M. (1977) A study of the short-term retention and clearance of inhaled asbestos by rats using UICC standard reference samples. Pp. 247–257 in *Inhaled Particles IV* (W.H. Walton, editor). Pergamon Press, Oxford, UK.

Middleton, A.P., Beckett, S.T. and Davis, J.M. (1979) Further observations on the short-term retention and clearance of asbestos by rats, using UICC reference samples. *Annals of Occupational Hygiene*, **22**, 141–152.

Morgan, A., Evans, J.C., Evans, R.J., Hounam, R.F., Holmes, A. and Doyle, S.G. (1975a) Studies on the deposition of inhaled fibrous material in the respiratory tract of the rat and its subsequent clearance using radioactive tracer techniques. *Environmental Research*, **10**, 196–207.

Morgan, A., Evans, J.C. and Holmes, A. (1975b) Deposition and clearance of inhaled fibrous minerals in the rat. Studies using radioactive tracer techniques. *Inhaled Particles*, **4**, 259–274.

Morgan, A., Holmes, A. and Davison, W. (1982) Clearance of sized glass fibres from the rat lung and their solubility *in vivo*. *Annals of Occupational Hygiene*, **25**, 317–331.

Muhle, H. and Bellmann, B. (1997) Significance of the biodurability of man-made vitreous fibres to risk assessment. *Environmental and Health Perspectives*, **105**, 1045–1047.

Muhle, H., Koch, W. and Bellmann, B. (1986) Acute and subchronic effects of intratracheally instilled nickel containing particles in hamsters. Pp. 337–340 in: *Health Hazards and Biological Effects of Welding Fumes and Gases* (R.M. Stein, A. Berlin, A.C. Fletcher and J. Järvisalo, editors). Excerpta Medica, New York.

Muhle, H., Bellmann, B. and Pott, F. (1994) Comparative investigation of the biodurability of mineral fibres in the rat lung. *Environmental and Health Perspectives*, **102**, 163–168.

Nguea, H.D., de Reydellet, A., Le Faou, A., Zaiou, M. and Rihn, B. (2008) Macrophage culture as a suitable paradigm for evaluation of synthetic vitreous fibers. *Critical Reviews in Toxicology*, **38**, 675–95.

Nyberg, K., Johansson, U., Johansson, A. and Camner, P. (1992) Phagolysosomal pH in alveolar macrophages. *Environmental and Health Perspectives*, **97**, 149–152.

Olsen, A.A. and Rimstidt, J.D. (2008) Oxalate-promoted forsterite dissolution at low pH *Geochimica et Cosmochimica Acta*, **72**, 1758–1766.

Oelkers, E.H. and Schott, J. (2001) An experimental study of enstatite dissolution rates as a function of pH, temperature and aqueous Mg and Si concentration, and the mechanism of pyroxene/pyroxenoid dissolution. *Geochimica et Cosmochimica Acta*, **65**, 1219–1231.

Oze, C. and Solt, K. (2010) Biodurability of chrysotile and tremolite asbestos in simulated lung and gastric fluids. *American Mineralogist*, **95**, 825–831.

Pacella, A., Fantauzzi, M., Turci, F., Cremisini, C., Montereali, M.R., Nardi, E., Atzei, D., Rossi, A. and Andreozzi, G.B. (2014) Surface modifications and dissolution reactions of UICC crocidolite in buffered solutions at physiological pH: a combined ICP-OES, XPS and TEM investigation. *Geochimica et Cosmochimica Acta*, **127**, 221–232.

Pacella, A., Fantauzzi, M., Turci, F., Cremisini, C., Montereali, M.R., Nardi, E., Atzei, D., Rossi, A. and Andreozzi, G.B. (2015) Surface alteration mechanism and topochemistry of iron in tremolite asbestos: a step toward understanding the potential hazard of amphibole asbestos. *Chemical Geologist*, **405**, 28–38.

Papirer, E., Dovergne, G. and Leroy, P. (1976) Modifications physico-chimiques du chrysotile par attaque chimique ménagée: I. En milieu aqueux. *Bulletin de la Société Chimique de France*, **5-6**, 651–653.

Parry, W.T. (1985) Calculated solubility of chrysotile asbestos in physiological systems. *Environmental Research*, **37**, 410–418.

Plumblee, G.S. and Ziegler, T.L. (2006) The Medical Geochemistry of dusts, soils and other Earth materials. In: Environmental Geochemistry. *Treatise in Geochemistry*, **9**, 263–310.

Ramos, M.E., Cappelli, C., Rozalen, M., Fiore, S. and Huertas, F.J. (2011) Effect of lactate, glycine, and citrate on the kinetics of montmorillonite dissolution. *American Mineralogist*, **96**, 768–780.

Rozalen, M.L., Huertas, F.J., Brady, P.V., Cama, J., Garcia-Palma, S. and Linares, J. (2008) Experimental study of the effect of pH on the kinetics of montmorillonite dissolution at 25°C. *Geochimica et Cosmochimica Acta*, **72**, 4224–4253.

Rozalen, M., Ramos, M.E., Huertas, F.J., Fiore, S. and Gervilla, F. (2013) Dissolution kinetics and biodurability of tremolite particles in mimicked lung fluids: Effect of citrate and oxalate. *Journal of Asian Earth Sciences*, **77**, 318–326.

Rozalen, M., Ramos, M.E., Fiore, S., Gervilla, F. and Huertas, F.J. (2014a) Effect of oxalate and pH on chrysotile dissolution at 25°C: An experimental study. *American Mineralogist*, **99**, 589–600.

Rozalen, M., Ramos, M.E., Gervilla, F., Kerestedjian, T., Fiore, S. and Huertas, F.J. (2014b) Dissolution study of tremolite and anthophyllite: pH effect on the reaction kinetics. *Applied Geochemistry*, **49**, 46–56.

Saldi, G.D., Kohler, S.J., Marty, N. and Oelkers, E.H. (2007). Dissolution rates of talc as a function of solution composition, pH and temperature. *Geochimica et Cosmochimica Acta*, **71**, 3446–3457.

Scholze, H. and Conradt, R. (1987) An *in vitro* study of the chemical durability of siliceous fibres. *Annals of Occupational Hygiene*, **31**, 683–692.

Schott, J., Berner, R.A. and Sjoberg, E.L. (1981) Mechanism of pyroxene and amphibole weathering. I. Experimental studies of iron-free minerals. *Geochimica et Cosmochimica Acta*, **45**, 2123–2135.

Searl, A. and Buchanan, D. (2000) Measurement of the durability of manmade vitreous fibres. Report HRR, IOM, TM/00/03. Institute of Occupational Medicine. UK.

Searl, A., Buchanan, D., Cullen, R.T., Jones, A.D., Miller, B.G. and Soutar, C.A. (1999) Biopersistence and durability of nine mineral fibre types in rat lung over 12 months. *Annals of Occupational Hygiene*, **43**, 143–153.

Sebastien, P., Begin, R., Case, B.W. and McDonald, J.C. (1986) Inhalation of chrysotile dust. Pp. 19–29 in: *Biological Effects of Chrysotile* (J.C. Wagner, editor). Lippincott, Philadelphia, Pennsylvanis, USA.

Siever, R. and Woodford, N. (1979) Dissolution kinetics and the weathering of mafic minerals. *Geochimica et Cosmochimica Acta*, **43**, 717–724.

Suzuki, Y. and Yuen, S.R. (2001) Asbestos tissue burden study on human malignant mesothelioma. *Industrial Health*, **39**, 150–160.

Thelohan, S. and de Meringo, A. (1994) In vitro dynamic solubility test: influence of various parameters. *Environmental and Health Perspectives*, **102**, 91–96.

Timbrell, V. (1982) Deposition and retention of fibres in the human lung. *Annals of Occupational Hygiene*, **26**, 347–369.

Utembe, W., Potgieter, K., Stefaniak, A.B. and Gulumian, M. (2015) Dissolution and biodurability: Important parameters needed for risk assessment of nanomaterials. *Particle and Fibre Toxicology*, **12**, 11–22.

Van Oss, C.J., Naim, J.O., Costanzo, P.M., Giese, R.F. Jr., Wu, W. and Sorling, A.F. (1999) Impact of different asbestos species and other mineral particles on pulmonary pathogenesis. *Clays and Clay Minerals*, **47**, 679–707.

Wang, X., Li, W., Hu, H., Zhang, T. and Zhou, Y. (2005) Dissolution of kaolinite induced by citric, oxalic and malic acids. *Journal of Colloid and Interface Science*, **290**, 481–488.

Werner, A.J., Hochella, M.F., Guthrie, G.D., Hardy, J.A., Aust, A.E. and Rimstidt, J.D. (1995) Asbestiform riebeckite (crocidolite) dissolution in the presence of Fe chelators: implications for mineral-induced disease. *American Mineralogist*, **80**, 1093–1103.

Wagner, J.C. and Skidmore, J.W. (1965) Asbestos dust deposition and retention in rats. *Annals of the New York Academy of Sciences*, **132**, 77–86.

Wagner, J.C., Berry, G., Skidmore, J.W. and Timbrell, V. (1974) The effect of inhalation of asbestos in rats. *British Journal of Cancer*, **29**, 252–269.

Wood, S.A., Taunton, A.E., Normand, C. and Gunter, M.E. (2006) Mineral–fluid interaction in the lungs: Insights from reaction-path modeling. *Inhalation Toxicology*, **18**, 975–984.

EMU Notes in Mineralogy, Vol. 18 (2017), Chapter 11, 367–416

Epidemiological approaches to health effects of mineral fibres: Development of knowledge and current practice

BRUCE W. CASE[1] and ALESSANDRO MARINACCIO[2]

[1]*Department of Pathology and Combined Department of Epidemiology, Biostatistics and Occupational Health, McGill University, Montreal, Québec, Canada, e-mail: bruce.case@mcgill.ca*
[2]*INAIL (Istituto Nazionale per l'Assicurazione contro gli Infortuni sul Lavoro) Dipartimento di Medicina, Epidemiologia, Igiene del Lavoro e Ambientale, Rome, Italy*

Epidemiology links exposures to health outcomes. For mineral fibres, exposures have been defined inconsistently and characterized poorly in the historical studies upon which most of our epidemiological knowledge is based. This is due in part to the nature and development of epidemiology itself, in part to the shifting understanding of mineralogical definitions and their importance for health outcomes, and very largely to a failure of communication between health scientists and scientists in mineralogy and allied fields. Nevertheless, for a subset of mineral fibres including, but not limited to, commercial asbestos varieties, a great deal is known. This includes on the exposure side physical characteristics of fibre chemistry and fibre dimensions. Exposure concentrations, routes of exposure, internal dose characteristics and effects, and timing of exposures have all been explored with some success. The epidemiology of mineral fibres is overwhelmingly about 'asbestos', and this chapter is largely confined to that set of fibres. There are other fibres of concern (synthetic vitreous fibres, erionite and more exotic amphibole varieties such as 'blueschist' composed of fibrous glaucophane and winchite amphibole, among others). For some of these, there is a very large database of human epidemiological work; for others there is none yet, as potential human exposures have been recognized only recently. All of them, however, are very much related to 'asbestos' in exposure and health-effect characteristics, and this chapter is confined to a discussion of the epidemiology of the asbestos minerals. Summaries of epidemiological study results are available elsewhere, as are comprehensive discussions of individual diseases. Here, the hope is to provide a narrative overview of the historical development and execution of epidemiological studies of asbestos exposure and disease, with an emphasis on exposure measurement problems and techniques. While there are many technical factors involved, occupational status has been and remains a key factor. Exposure measurement has been well recognized as the Achilles Heel of epidemiological study for mineral fibres. Problems in the definition of the diseases themselves have received less attention. Environmental exposures, their measurement, and their relationship to disease status are particularly challenging. For 'naturally occurring asbestos' (NOA) fibres, we are only beginning to understand health effects, although it is reasonable to assume that for the most part they will follow the knowledge developed for those produced in occupational studies. The greatest importance for the future then is not to establish health effects, but to identify possible sources of mineral fibre exposure. Once this is done, past lessons of traditional mineral fibre epidemiology can be applied to prevent future exposure and disease.

DOI: 10.1180/EMU-notes.18.11

1. Introduction

1.1. Basic concepts

1.1.1. Exposure, mechanism and health effects

Epidemiology is a discipline which tests putative relationships between health cause and effect. Proximate in the causal chain of events is exposure, which we can define for mineral fibres operationally as the concentration and character of the material. Characterization for mineral fibres is ideally complete; mineral fibres do not make for simple identification as do many infectious agents. Indeed, simply defining what constitutes a fibre can be a controversial matter. There are mineralogical, industrial and especially regulatory definitions. This is inevitable: no industrial hygiene limit can exist without stating what is being limited. What is limited should in principle be determined by those characteristics which are directly related to health effects. But this requires *a priori* that the health effects and the biological pathways that lead to them be defined and understood as well. Without knowing modes of action and mechanisms that operate in the 'black box' between exposure and disease – and they usually are not known – historically we have had to work backwards from health effects. For mineral fibres, health effects appearing in occupations have been the starting point for analysis. In such circumstances exposures have been reasonably specific, over long periods of time, and concentrated within known populations. Extrapolating knowledge from the study of occupational groups to analogous environmental exposures – or even to other occupations with similar, but not identical, exposures – has proven difficult.

The health effects in question are almost always chronic and are both general – mortality being primary – and specific, with a considerable emphasis on both common and uncommon cancers. When these are identified in occupations working with mineral fibres in any capacity, the characteristics and concentrations of those fibres are uncovered and epidemiology tests the linkages between these varied aspects of exposure and disease.

1.1.2. Exposure, dose and the black boxes of biology

Exposure, which occurs outside the organism, depends on the release of a material in the presence of a human organism which can incorporate the exposure as dose. Dose is delivered by pathways – inhalation, ingestion, surface (skin) exposure, *etc.* For mineral fibres, inhalation is usually the pathway of greatest interest. Exposure is what is usually measured, although dose can in some cases be measured through tissue-destructive techniques. It is ultimately dose which leads to health effects, and dose may differ from exposure in many ways, having been modified in its route to the organ or tissue affected. To fully understand dose-response relationships – as opposed to exposure-response relationships – it is necessary to know what mechanisms operate within the 'black box' between the two. It is also necessary to know the mode of action. These two – mechanism and mode of action – are related, but not identical. 'Mechanisms' are the individual processes that occur in the interactions with biological elements, while a mode of action (MOA) "encompasses a sequence of key events and processes that begin with the interaction between fibres and cells that may result in both cancer and complex non-cancer health effects" (Gwinn, 2011).

We may know more about the importance of an MOA than we do about the mechanistic elements that make it up. We can 'see the biology in action' for some modes of action; some fibres for example elicit inflammatory responses. Even without complete knowledge of what happens in the black boxes of biology it is possible to know that an 'exposure causes a disease' without actually knowing how that happens. Testing such relationships is the principal task of epidemiology. We can, for example, test and accept that 'asbestos causes mesothelioma' without knowing either the mechanisms or the MOA. We can further refine that understanding by better defining 'asbestos' and testing the individual elements (*e.g.* fibre type, dimension, crystallinity, *etc.*) which may be mechanistically involved. Again, it is not necessary to know how (for example) asbestos fibre types differentially produce disease in order to know that they do act in this way. In this way it becomes possible, through epidemiology, to set priorities for exposure limits.

There has been a slow history of prevention of chronic diseases related to mineral fibres, but without further specific knowledge this logic may – and has – had failures. As a convention for example, we accept a 'no-threshold' linear risk model for mineral fibres (and many other exposures) when the MOAs and mechanisms are unknown. The convention however is often misinterpreted as meaning there is no safe level of exposure, which implies that any exposure – even to a single fibre – can cause disease. This is not only untrue; if practiced it could lead to a nihilistic approach to prevention – because at some level (sometimes called an 'ambient' level), prevention of any exposure is precluded by the very nature and ubiquity of mineral fibres in the human environment.

2. Epidemiology of mineral fibres: History and development

2.1. Early stages

Epidemiology is a discipline which grew out of medical science and which has become linked inextricably to public health and regulatory policy. It is more difficult to define than is usually acknowledged, and has evolved since its relatively recent emergence. Historically, medical science linked observations of associating single common sources to disease outbreaks. The story of the Broad Street Pump is a good example (Snow, 1855; McLeod, 2000).

John Snow was a physician and published medical scientist who first studied cholera in relation to an outbreak in Newcastle in 1831–32. In 1849 he distributed a pamphlet 'On the mode of communication of cholera' summarizing the knowledge of and opinions concerning the aetiology and modes of transmission of the disease to that date. Contrary to then-prevailing opinion, Snow concluded 'from the pathology' of the disease that spread from an agent in air which could be inhaled was unlikely. Snow suggested instead that an ingested agent produced this digestive tract disorder. Further, he noted (as had others) that it seemed to spread from person to person and that in many instances this spread was to be found in or close to seaports. In the fall of 1854, there was an outbreak in his own neighbourhood in London. This afforded a chance to test his hypothesis, and by going house-to-house he determined that "Within two hundred and fifty yards of the spot where Cambridge Street joins Broad Street, there were upwards of five hundred fatal attacks of cholera in ten days...on proceeding to the spot I found that nearly all the deaths had taken place within a short distance of the pump.'' (Snow, 1855; Koch, 2008).

Although Snow is often referred to as a founder of Epidemiology as a result of this work, it is a hybrid type of 'epidemiology' which has also been called 'Geographic Pathology'. More recently such studies are referred to as 'ecological studies'; studies of the occurrence of a health effect in a defined space (block, county, city, country, *etc.*) over a defined period of time.

The infectious disease paradigm of epidemiology has a short gap between exposure and disease, and (as in the case of cholera) is often uncomplicated by multiple agents. Such studies can be misleading when applied to chronic diseases or to diseases which are multifactorial. Ecological studies are commonly used to identify possible 'hotspots' of mineral-fibre related disease, especially for mesothelioma, often comparing disease incidence or mortality in geographic or demographic units (*e.g.* counties). Units with highest rates can then be compared as to possible asbestos exposure sources – as Snow did with London water supply districts. Such comparisons can be useful in identifying areas for further study, but chronic diseases such as those caused by exposure to mineral fibres are complicated by the many decades following exposure(s).

2.2. Occupational epidemiology: First steps

In the early 20th century, study of 'dust-related diseases', in particular those affecting the lung or pneumoconioses, came of age in occupational settings. It was in this context that exposure to fibres first came to be linked to lung fibrosis – asbestosis, in the case of asbestos fibres. This was not a discovery of formal epidemiological study, for the tools of chronic disease epidemiology did not yet exist. Instead the development of knowledge of asbestosis depended upon reporting of cases in occupational circumstances of exposure, and the accumulation of these reports. Pathologist William Cooke (Cooke, 1924) described a young woman who worked in the Rochdale textile plant in Lancashire, England at an inquest into her death. He testified that "mineral particles in the lungs originated from asbestos and were, beyond reasonable doubt, the primary cause of the fibrosis of the lungs and therefore of death" (Cooke, 1924; Bartrip, 2004). In 1927 he first used the term "asbestosis" in a publication concerning the same case (Cooke, 1927). In the years immediately following, a number of case reports appeared in the British and South African literature. In 1927, Merewether, the Medical Inspector of Factories in the United Kingdom, was asked to ascertain "whether the occurrence of this disease in an asbestos worker was merely a coincidence, or evidence of a definite health risk in the industry". Merewether concluded that a prolonged duration of exposure to 'asbestos dust' at high concentration produced "definite occupational risk among asbestos workers as a class" (Merewether and Price, 1930). Merewether, others charged with occupational health, and the medical and policy establishments of the time did not have the tools of modern analytical epidemiology. This did not stop them from making simple observations of exposure and disease especially *in workforces*, and this was the true beginning of the epidemiological study of health effects of mineral fibres.

Within three decades, this type of observational epidemiology – observing cases in workers with exposures to asbestos, silica and other materials – made material determinations of hazard which resulted in regulatory regimes and decreased exposures. Unfortunately, despite serial measures reducing such exposures, iterations of these resulted in initial optimism followed by a realization that the measures taken were not enough to reduce and eliminate the diseases in question. These were failures of the existing system, with a lack of precision in identifying hazard – the knowledge that a material can cause disease. There was also failure to quantify risk in terms of exposure-response relationships. As noted later, this was in part a result of imprecise methodologies for assessing exposure itself. This was a problem that continued for decades; as McDonald observed in 2000:

"Least evidence of progress is probably in the quantitative assessment of exposure-response, not because of any lack of concern but because the measurement of exposure in any cumulative sense over long periods of time is the weakest link in the chain of cause and effect, whether in occupational or general epidemiology...With only the most approximate information on past levels of environmental exposure, let alone on effective human dose, all attempts to estimate risk in various circumstances, particularly at low levels, are little better than guesswork" (McDonald, 2000b).

The epidemiology of health effects related to exposures to mineral fibres nonetheless remains the area of occupational and environmental epidemiology for which we have, quantitatively, the most evidence stretching back for the longest time. This adds to the primary problems such as those posed for accurate exposure assessment, since methods have changed over time. This in turn is related to changes in definitions of the exposures to measure, and sometimes even those of the diseases of interest. Inevitably, this has resulted in a field filled with controversy and often consensus has been hard to find.

2.3. Occupational malignancy and mineral fibres: Early observations 1930–1955

During this period no relationship between asbestos exposure and any malignancy was epidemiologically established. Astute observers raised possibilities, for both lung cancer and for mesothelioma. For the former the initial suspicions were lung cancers reported in cases of asbestosis. As noted by Enterline (1978, 1991), the first such published observation was by Wood and Gloyne (1934). Two cases of carcinoma of the lung were reported in autopsies of women who worked at an asbestos plant, among 43 with asbestosis. A link between asbestos, or asbestosis, and the cancer was rejected (Gloyne, 1935). A single case report by Lynch and Smith followed in which the link was suggested as causal (Lynch and Smith, 1935). The authors related the disease in a man with 21 years as a weaver at an asbestos factory to 'asbestosis' rather than asbestos exposure *per se*. The reports were sceptical about a causal relationship because of a lack of epidemiological evidence – which did not yet exist – and of experimental (animal) evidence. These two are the modern cornerstones of the hazard evaluation step in risk assessment (IARC Working Group on the Evaluation of Carcinogenic Risks to Humans *et al.*, 1987; Case and Mattison, 1991).

The first case reports of 'pleural cancer' related to 'asbestosis' seem to have been in Germany by Wedler (1943). Nazi Germany was ahead of the rest of the West in assessment of dust- and smoking-related diseases, and in preventive measures and industrial hygiene (Proctor, 1999). Wedler provided autopsy reports of two asbestos plant workers with both asbestosis and 'pleural cancer'. An accompanying review of 92 autopsies with asbestosis found in the medical literature showed 14 cases of 'lung or pleural cancer'.

A short account by Cartier, a physician at an industrial clinic for chrysotile mines in Quebec, was appended to an article by W.E. Smith, recounting his European travels and visits with Merewether, Gloyne and others (Cartier, 1952; Smith, 1952). Cartier noted eight cases of 'lung cancer' among 4000 asbestos miners and millers over a ten-year period, two of which he called 'pleural mesothelioma', a term not yet in general use. Cartier discounted the small number of lung cancers, but the rarer mesothelioma variant (thought at the time to be a form of lung cancer) led him to speculate that these might have an occupational origin. Other scientists present thought perhaps this rare diagnosis was wrong and that the 'mesotheliomas' were in fact broncho-alveolar carcinomas. In fact, the original autopsy report for one of Cartier's two workers at our hospital indicates that the cancer did appear to arise in a major bronchus. The subject was an office worker who had been a very heavy smoker and whose office had been immediately adjacent to the tremolite-rich Thetford 'original complex' chrysotile mines (McDonald *et al.*, 1997) for almost thirty years (Fig. 1). His son, interviewed decades later in old age, recalled that he had worked in an Ontario mine as a younger man and was a very heavy smoker (Case, 2008).

2.4. Sir Richard Doll and the rise of analytical epidemiology in occupational disease

Isolated observations of cases of lung cancer in asbestos workers and to a lesser extent mesothelioma (which sometimes went by different names) were made between 1930 and 1959. Some of these had some of the trappings of epidemiological study in that crude attempts were made to relate numbers of cases to numbers of workers. At best however, papers simply reviewed numbers of case reports to a given date.

Case reports are useful in delineating questions to be asked, but not in answering them. In the mid-1950s there was a scientific shift from reporting cases and speculation to the solid proof provided by modern analytical epidemiology. Richard Doll, who essentially founded that science, was also the first to apply it to asbestos, as he had only a few years earlier in his proof of tobacco as the principal cause of lung cancer (Doll and Hill, 1950, 1954). In 1955 Doll applied his new epidemiological methods to asbestos workers at Rochdale for lung cancer (Doll, 1955).

Figure 1. Photograph of a girl, ~12 years of age, working on raw chrysotile ore as a cobber at the Johnson Chrysotile Mine, Thetford Mines West, Quebec in the 1930s. Photograph from a publicity brochure from the Johnson Mining Company, Thetford Mines West, found in the New York Public Library by the author. This mine closed in 1960.

He reviewed death certificates for 105 employees of an asbestos plant who had had autopsies from 1935 to 1952. Of 75 who had asbestosis, 15 also had lung cancer, while only three of thirty without asbestosis had lung cancer. He reported on the mortality separately for 113 males who had worked for 20 or more years in areas designated as hazardous under the prevailing Asbestos Industry Regulations of 1931; a legacy of Merewether's work. The latter were found to have lung cancer risk tenfold greater than that expected in the general population.

This study established the relationship between asbestos exposure and lung cancer: "The paper by Doll marked an end to widespread acceptance of purely clinical observations...and opened the way for contributions by epidemiologists and biostatisticians" (Enterline *et al.*, 1978).

2.5. Consolidation of knowledge: The New York conference and beyond

More epidemiological studies followed Doll's, providing further evidence for lung cancer causation. Many suggested a linkage not just between lung cancer and asbestos exposure, but between lung cancer and asbestosis itself. For workers with lung cancer the very high exposures and frequent finding of asbestosis led to a mistaken idea that clinical asbestosis was a "necessary precondition" for lung cancer, a proposition now generally thought due to collinearity of the dose-response relationships in the two diseases (Case, 2006). Asbestos-lung cancer relationships were confirmed for 19 of 1495 asbestos product plant manufacturing workers employed near Lancaster, Pennsylvania during 1938–39: only five were to be expected (Mancuso and Coulter, 1963). Selikoff and colleagues assessed 1,522 New York City Asbestos Workers Union members exposed to 'asbestos dust' for at least 20 years. These insulation workers ('laggers' in Great Britain, 'pipe coverers', 'insulators,' or 'asbestos workers' in the USA) died of lung or 'pleural' cancer in 45 cases *vs.* 6.6 predicted. Further, the authors estimated that "even if *all* our asbestos workers had smoked a pack or more of cigarettes a day...and if exposure to asbestos were of no significance, then their lung cancer death rate would have been about 3.4 times as high as the rate in the general US male population. Clearly, the smoking habits of the asbestos workers cannot account for the fact that their lung cancer death rate was 6.8 times as high as that of white males in the general population" (Selikoff *et al.*, 1964).

The fact that four of the 45 cases were mesothelioma was not lost on the authors either. Four years earlier, Wagner and his colleagues including general physician C.A. Sleggs had published their seminal report on mesothelioma cases in South Africa (Wagner *et al.*, 1960). Sleggs, the Superintendent of the West End Tuberculosis Hospital in Kimberley, began to see an unusual disease among his patients in the area of the Cape Province crocidolite fields around 1956. This was first misdiagnosed as pleural tuberculosis, but his patients did not respond to therapy for that disease. Puzzled, Sleggs turned for help to surgeon Paul Marchand and pathologist Wagner in Johannesburg. Marchand went to Kimberley, and brought back biopsies on the first sixteen cases of what proved to be the first mesothelioma epidemic.

At first the link to crocidolite or 'blue asbestos' seemed tenuous; the first patients were "housewives, shepherds, farmers, lawyers, and insurance agents" (Wagner, 1965). The mines were 200 miles away. Sleggs and Marchand set out to question all the surviving patients and relatives, obtaining detailed life histories. By 1959, they were able to establish an association with the Cape Asbestos fields or the industrial use of asbestos in 32 of 33 patients with histologically proven pleural mesothelioma, and the following year these were published (Wagner *et al.*, 1960). By the time of publication, fourteen additional cases had to be added in a footnote; one year later the number had risen to a total of 87 (two of them peritoneal). Unlike the work of Doll and Selikoff, the cases were not confined to those with occupational exposure but were also found in their households and even in the general surrounding environment.

The work of Wagner and colleagues is not merely a collection of case reports. It was not a prospective cohort study of the nature of Doll's or Mancuso and Coulter's, but it was epidemiological in nature, as we can see from the 1960 abstract (page 260):

"Thirty-three cases (22 males, 11 females, ages 31 to 68) of diffuse pleural mesothelioma are described; all but one have a probable exposure to crocidolite asbestos (Cape blue). In a majority this exposure was in the Asbestos Hills which lie to the west of Kimberley in the north west of Cape Province. The tumour is rarely seen elsewhere in South Africa." This is a classical description of ecological study design.

The paper by Wagner *et al.* (1960) was not analytical epidemiology. Analytical epidemiology has as its principal tool the use of internal or external comparison groups. In *cohort studies*, groups – in this instance usually of workers – are followed over time. Sometimes their exposures were measured, but more frequently in the early studies it was assumed from their occupational status. Disease rates or more often, mortality rates, could then be compared with those in similar occupations without status which would convey exposure. Duration of exposure was often known, and could be used as a partial indicator of extent. Disease status itself – asbestosis or silicosis for example – could often be used as an indicator of higher levels of exposure in studying other diseases such as lung cancer.

For rare diseases such as mesothelioma, *case-control studies* came to be used. These took an opposite approach to cohorts of single occupation: disease was ascertained first, and persons with disease were 'matched' to controls who were similar to the degree possible (for example age, sex, smoking status, *etc.*) to the cases. Cases and controls could then be compared to the disease of interest for their exposures. Usually however, as with cohort studies, the latter was assumed by occupational status alone – exposure was not actually measured but assumed from jobs performed.

By 1964, there were studies of asbestos and cancer underway throughout much of the industrialized world. In June of that year, a landmark conference was held in New York City under the auspices of the New York Academy of Sciences (Selikoff and Churg, 1965). The organizers were Irving Selikoff and John Gilson, the Director of the British Medical Research Council (MRC) Pneumoconiosis Research Unit. Christopher Wagner had done a sabbatical with Gilson at around the time of the paper on mesothelioma by Wagner *et al.* (1960). Wagner joined in the international recruitment of asbestos researchers. One that he approached was the incoming Chair of Epidemiology at McGill University in Montreal, Corbett McDonald. McDonald had also come from the UK and was persuaded to attend the New York meeting.

The meeting itself consolidated knowledge to 1964 and set the agenda for years to come. Greenberg described it half a century later:

"Contributors to the report, with its 705 pages of text, constituted a contemporary International Who's Who of academics, industry experts, and civil servants involved in the fields of research and control of asbestos and its effects. Its contributions varied qualitatively and quantitatively, but overall it constituted an excellent compendium of the state of knowledge of the physical and health aspects of exposure to dusts containing asbestos" (Greenberg, 2003).

By today's standards, the meeting and the publication of the proceedings were remarkably open, with 'discussions' following each section. Selikoff, simultaneously the diplomat and the publicist, is described by Greenberg as having "...no compunction about using the media in an attempt to alert the public and legislators to the hazards of asbestos and the urgent need for its control. He was photogenic and looked the part of a solid conscientious scientist, was good at exposition and a good source of 'sound bites' attractive to journalists". He was a fearless (and feared) advocate, and not without cost to himself and his team; Langer notes that, in this era, "(Selikoff) was something to watch...I was there when we confronted all of the disbelieving forces of industry and the reluctant forces of the federal government. We were in the trenches in gore up to our ankles and sometimes the gore was our own gore" (Stone, 1991). Disagreement on any issue, though, was not tolerated by the tough 'Brooklyn Boy' that Greenberg describes and whose ire Langer later felt; coming to believe that (in the N.Y. Times reporter Joseph Hooper's words) Mount Sinai's 'intransigence on the issue of chrysotile' would ultimately be destructive to health rather than constructive. Langer found himself in a "Luciferian role...cast out of Mt. Sinai...(and doing) work on fibre type and mesothelioma at Brooklyn College" (Hooper, 1990).

But this all came later. In 1964, Selikoff opened the conference with equanimity:

"This journey...has branched into intertwining roads, including those of epidemiology, oncology, physical chemistry, physiology, experimental pathology and many others. Those of us who have been exploring these roads, meet now at a common junction, to exchange experiences and perhaps ask helpful directions" (Selikoff, 1965).

He went on to praise those that had come before like Merewether, Lynch, Gloyne; spoke of Cartier, Vigliani and Vorwald (who were all present) as groundbreakers in the field; and had special words of thanks for the 'resourcefulness and energy' of John Gilson. Gilson followed by giving a nuanced address comparing the ways in which asbestos had saved and continued to save lives through its fire-resistant and friction properties "...set against the adverse effects of its use, with which this conference has been solely concerned...We must all hope that this conference will lead to a much more systematic surveillance of all those exposed to asbestos...(but) Let us not delude ourselves that animal studies are an adequate substitute for the well-planned epidemiological investigations in man...If we are right in concluding, on the evidence presented at this conference, that there are probably important differences in the biological effects of different fibres, we must exploit to the full any opportunities for the investigation of groups of individuals exposed to only one type of fibre" (Gilson, 1965). The role of epidemiology for the study of mineral fibres was solidly ensconced.

It is difficult to describe either the scope or importance of the 1964 meeting and its publication in 1965. Everything that had been known to that point was summarized and virtually everything that would be investigated later came up in the discussions.

Selikoff, Churg and Hammond again reported on the US insulators' experience (Selikoff *et al.*, 1965) pointing out that asbestos insulation work had been largely ignored and the one previous US study in naval workers had been limited to "study of men with relatively short durations of exposure". The investigators pointed out the difficulties in studying such a group as opposed to miners or textile workers; multiple employers; repeated but 'limited and intermittent' asbestos exposure; and the varying nature of insulation materials itself with respect to asbestos content (none, chrysotile, crocidolite and amosite). In addition to the lung cancer and mesothelioma previously reported (Selikoff *et al.*, 1964) Selikoff and Hammond also noted changes in diagnosis site from gastrointestinal cancer to mesothelioma, but none in the opposite direction (Hammond and Selikoff, 1965).

Adding to the US epidemiological evidence for mortality among asbestos-products workers was US public health scientist and biostatistician Philip Enterline, (Enterline, 1965). He examined 2,833 white male workers in asbestos products plants (nine textile plants and one producer of asbestos friction materials) between 1948 and 1951 who were aged 15–64 in 1950. Workers meeting the same criteria in three cotton textile plants ($N = 6{,}242$ total) were used as controls. The groups were also compared using Standardized Mortality Ratios (SMR) for the US white male population. The SMR for all diseases of the respiratory system was 266 for the asbestos workers (25 deaths *vs.* 9.4 expected) and only 69.7 for cotton textile workers (14 *vs.* 20.1 expected). There were 18 asbestosis deaths among the total 286 asbestos worker deaths and none among the 486 cotton textile plant worker deaths. The total respiratory system cancer SMR was 201.7 in asbestos workers and 80.5 in cotton textile workers.

Many other subjects were discussed which presaged later topics of investigation and controversy: fibres and asbestos bodies retained in the lungs of both workers (Nagelschmidt, 1965) and those exposed environmentally (Thomson, 1965); the pathologic diagnosis and histological classification of mesothelioma (Churg *et al.*, 1965; McCaughey, 1965; Webster, 1965); the significance and aetiology of pleural plaques (Cartier, 1965; Kiviluoto, 1965); and proper means of measurement of exposure (*e.g.* the older midget impinger and membrane filter methods: Addingley, 1965; Ayer *et al.*, 1965). The latter was shortly to assume importance in debate over asbestos-exposure assessment (see below). The issue of contamination of inhaled asbestos fibres with carcinogenic 'oils', including benzo-[a]-pyrene, was brought up by many authors, especially two groups financed by the United States and British public health and medical research agencies (Harington and Roe, 1965; Miller *et al.*, 1965).

The issue of relative potency for asbestos fibre type was discussed by many, especially for mesothelioma. Wagner observed that in South Africa, "...although the country produces approximately the same amount of crocidolite, amosite and chrysotile (if Swaziland is excluded), no case of mesothelioma has been reported from either of the other areas in spite of an intensive investigation" (Wagner, 1965). Similarly, in the UK Wagner (1965) observed "...it has not been possible to obtain positive, clear evidence implicating a single type...since the majority of cases have been exposed where several types of asbestos were used. However...no indisputable case of pure chrysotile exposure has yet been discovered". In East Germany, "The asbestos used in the Dresden asbestos industry include(ed) asbestos from Canadian, Soviet and South African sources, in varying amounts at varying times. For this reason we are unable to offer valid observations concerning effects of the different varieties of asbestos." Six pleural mesotheliomas (all in women with asbestosis) and one 'peritoneal cancer' had been observed among 120 deaths between 1952 and 1964 for which cause could be ascertained (Jacob and Anspach, 1965).

A further major accomplishment associated with the New York Conference was goal-setting for future investigations. On October 22nd and 23rd, 1964, a Working

Group was convened in New York to "discuss evidence of an association between exposure to asbestos dust and cancer". Forty delegates from eight countries were involved, including Cartier and McDonald from Canada; Bohlig from East Germany; Peter Elmes, John Gilson, John Harington, J.F. Knox, W.T.E. McCaughey and Wagner from the UK; Sluis-Cremer, Thompson and Ian Webster from South Africa; and Selikoff, Churg, Hammond, Enterline, Mancuso and many others from the US. This "Report and Recommendations of the Working Group on Asbestos and Cancer (of the International Union against Cancer's Geographical Pathology Committee)" established epidemiological investigation as the first item of business under 'Terms of Reference' (page 709 of Proceedings of the New York Conference: Selikoff and Churg (1965)): "1. Epidemiology. A. To investigate the incidence of mesothelial tumours of the pleura and peritoneum in groups and/or regions where exposure to only one type of asbestos fibre has occurred".

Canadians, led by McDonald, were to investigate chrysotile (as were the Russians, Italians and Cypriots); the Finns, anthophyllite; the Americans were to investigate chrysotile and tremolite (which had been discovered at Zonolite Mountain, Montana in 1928, among other US locations, although we find no mention of this in the text); Australians, crocidolite; and South Africa, crocidolite, amosite and chrysotile.

The 'Terms of Reference' list only mesothelioma for this task – perhaps it was thought that this was, as it indeed turned out to be, the clearest point of difference between fibre types in terms of disease aetiology. It was also noted that "Present evidence indicates that the associated carcinoma of the lung is not limited to exposure to any one type of asbestos fibre...In the case of mesothelioma, evidence from several countries suggests that exposure to crocidolite may be of particular importance, but it cannot be concluded that only this type of fibre is concerned with these tumours, and further investigation of this problem is needed" (page 710 of the New York Conference Proceedings).

3. Epidemiological study types and the evolution of asbestos fibre research

3.1. Introduction

Because of the rarity of mesothelioma, cohort studies of worker populations, unless very large and well defined such as those of Quebec chrysotile miners and millers (see below), or the North American insulators, were likely to take considerable time to achieve results. As a result, investigators turned as well to case-control methodology. Between 1965 and 1975, eight such studies were performed in Canada, the UK and Europe, as described by McDonald (2000a):

"All confirmed the association with occupational exposure to asbestos, with relative risks ranging from 2.3 to 7.0; the risk depended inversely on the proportion of controls exposed and employment of men in shipyards was most frequently incriminated".

In the largest study of 165 cases with 165 matched controls in North America, expanded in 1980 to over 600 cases and controls, the greatest relative risk (RR) was for insulation work (RR 46), asbestos product manufacture (6.1), work in heating trades

(4.4), shipyards (2.8), and construction work (2.6) (McDonald *et al.*, 1970; McDonald and McDonald, 1980). Of course, these relative risks do not tell the whole story of burden of disease by occupation or locale. For example, 'trades' notwithstanding, 20% of the cases worked in shipyards.

3.2. Case-control studies

Results of case-control studies in different countries and in different time periods have shifted with changing patterns of fibre type, distribution of employment, and other factors. For example, although 'carpenters' had no apparent excess risk in the 1970/1980 North American case-control studies, carpenters were the occupation most at risk 40 years later in the UK (Rake *et al.*, 2009) with "(Odds Ratio (OR) 36.0, 95% confidence interval (CI) 19.2–67.3), followed by all non-construction high-risk jobs (OR 16.8, 95% CI 9.6–29.3), plumbers, electricians and painters (OR 14.6, 95% CI 8.8–24.4) and other construction workers (OR 7.9, 95% CI 4.7–13.3)". Rake *et al.* believed it likely that a major cause of the "extraordinary risk" to British carpenters was "cutting amosite board with power tools, which was widespread in the UK construction industry through the 1970s and continued into the 1980s".

3.3. Cohort studies

Following the New York Conference several large cohort studies were initiated in workforces with different types of exposure. With large groups of workers in a given industry, it became possible over time to follow patterns of exposure and disease development by updating data over time. Two of many examples are given here in more detail.

In Canada, Corbett and Alison McDonald developed a cohort of Quebec miners, millers and factory workers beginning in 1966 (Case, 2016).

Figure 2. The Jeffrey Chrysotile Mine at Asbestos, Quebec, which expanded from (a) a deep but narrow hole in the ground with dangerous transport to the bottom by derrick and cable before 1900 to (b) the massive pit 2 km wide, a portion photographed here by the author in September, 2000. The mine closed in 2011.

The cohort consisted of ~11,000 men born 1891–1920 who worked at least one month in the chrysotile mines and mills of Quebec, as well as a small subgroup who worked in a products factory near the Jeffrey Mine at Asbestos (Fig. 2), Quebec in which imported commercial amphibole fibre was also used. On occasion, crocidolite was milled as well. Total deaths were 2413 to November 1966 (McDonald *et al.*, 1971); 3216 deaths to 1970 (McDonald *et al.*, 1973); 4463 deaths to 1976 (McDonald *et al.*, 1980), and 7116 deaths to 1989 (McDonald *et al.*, 1993). The final follow-up was published in 1997 (Liddell *et al.*, 1997; McDonald, *et al.*, 1997) and included 8009 deaths; 38 from mesothelioma. All were pleural; five were in the factory workers and 33 among chrysotile miners and millers. The cohort showed progressively greater proportional mortality for mesothelioma over time. No cases were diagnosed before 1950; 0.18%, 0.68% and 1.1% of deaths were from mesothelioma in 1950–74, 1975–84 and 1985–92, respectively.

Overall, one in 200 deaths were from mesothelioma. Mean latency – time from first exposure to diagnosis – was 47 years. Exposure-response showed a relationship to 'net service' (that is, duration of employment) in the largest group of cases, but not to cumulative exposure to mining dust. For the 36 of 38 cases for which cumulative exposure could be estimated to age 55, 14 exceeded 300 million particles per cubic-foot-years (MPCFY) and 27 exceeded 100 MPCFY. For three cases however MPCFY was less than 30. Using the conversion factors for fibres/mL (with caveats as noted), 75% of cases had more than 300 fibres/mL-years cumulative exposure while 8% had less than 100 fibres/mL-years.

Exposure-response calculations were complicated by commercial amphibole exposures in the Johns Manville products factory in Asbestos, Quebec. Miners and millers worked part-time in this factory. In the mills, imported fibre was processed on occasion but this was rare. Individual fibre-type exposures indicated from lung content varied as described in tabular form in the main study (McDonald *et al.*, 1997) and in a companion paper (Case *et al.*, 1997). Mesothelioma death rates in the products factory (five cases) were three times greater than the rate amongst miners and millers, and lung analyses for two of these included four and ten million fibres of crocidolite per gram of dry lung, in keeping with the work practices and exposures there. For miners and millers, lung chrysotile and tremolite were both present – usually in massive amounts. For 19 men with available lung fibre content, 17 had tremolite exceeding chrysotile. The two with chrysotile levels exceeding tremolite were employed for three years or less in the mines or mills. One was a barber for 40 years and the other a salesman for 37 years and a welder for six.

A comparison of cases from the highest tremolite exposure mines and mills of the original complex at Thetford Mines (Bell, Beaver, King and Johnson) to those elsewhere indicated that those with some or all service in the original complex had an odds ratio for mesothelioma of 2.50 for 20 adjusted years net service in that area (90% CI 1.47–4.20). This did not seem to be related to greater cumulative exposure. Nevertheless, there were three mining and milling 'cases' who worked only in 'peripheral' mines or mills believed to have lower potential tremolite exposures.

McDonald and colleagues concluded for mesothelioma that "Of the 38 deaths from mesothelioma in this cohort, 27 were in miners and millers with employment ranging from 20 to 48 years, and we conclude from the evidence presented that these 27 deaths can be attributed with reasonable certainty to occupational exposure in the Quebec chrysotile production industry". The presence of other minerals in or near the ore body – specifically 'fibrous tremolite' – was the best explanation for the regional variation in risk.

Fibrous tremolite (a generic morphological description: the tremolite included both asbestiform and non-asbestiform structures) was also seen to be a factor in lung-cancer mortality. Lung-cancer rates seemed to be dose-related only at the higher exposures. There was synergy with smoking greater than an addition of the asbestos and smoking risks, but less than multiplicative (Liddell and Armstrong, 2002).

Numerous studies in an asbestos textile worker cohort in South Carolina which received chrysotile from the Quebec mines showed greater fibre-for-fibre risk for lung cancer than that seen among the miners and millers. The linear exposure-response was "some 50-fold greater at similar accumulated dust exposures than in Canadian chrysotile mining and milling" (McDonald *et al.*, 1983). Of 863 male deaths in the cohort of 2543 employed for one month or more 1938–1958, only one was from malignant mesothelioma.

Workers forming a differently constituted cohort in the same plant were studied more extensively by Dement and colleagues. This cohort was followed through time, and the greater lung cancer exposure-response rates than those seen in mining continued (Dement *et al.*, 1982, 1983, 1994; Dement, 1991a,b; Hein *et al.*, 2007). Of 1961 deaths in their cohort followed to a later date than in the previous work, three were from mesothelioma. The lung cancer findings and their possible explanations have drawn most attention, with differences in fibre length and width between textile and mining/milling chrysotile exposures of particular importance for lung cancer (McDonald, 1998; Stayner *et al.*, 2008; Loomis *et al.*, 2009, 2010, 2012). Three studies demonstrated unexpected lung content of commercial amphiboles (as well as tremolite) in these chrysotile textile workers (Sébastien *et al.*, 1989; Green *et al.*, 1997; Case *et al.*, 2000), making conclusions more difficult for the rare mesotheliomas observed. On average, textile workers had unusually short employment as well, so some may have had their amphibole exposures elsewhere. Studies of North Carolina textile workers are similarly limited for mesothelioma, although quality data on a small number of archived membrane filter samples from a limited time period extracted from the Carolina mills has greatly advanced our understanding of asbestos-related lung cancer.

3.4. Meta-analysis and the problem of which data sets to use

Over time, analytical epidemiology studies accumulated. These included sequential follow-ups of cohort studies as above, newly constituted cohort studies in various occupations, and a multitude of case-control studies. It becomes difficult to summarize the results of these over time due to their number and to the differences between them in terms of exposure assessment and even case definition. It is not possible to review individually or even catalogue all of these studies in a short chapter, and in any case most simply built on the knowledge accumulated previously.

As in other areas of medical science, one method used increasingly to deal with the hundreds of available studies addressing similar endpoints is to aggregate the results. This can be done informally in review papers, which generally simply list studies and draw editorial conclusions from them. That practice risks conscious or unconscious bias in a field where both ideology and commercial interests on all sides can influence conclusions. However, summarization of studies can also be done formally through the application of systematic review and meta-analysis. The latter methodology was originally adopted for randomized controlled trials (*e.g.* of therapeutic agents), which are a form of 'experiment' and therefore do not have the many problems associated with observational studies (Stroup *et al.*, 2000). Observational studies – whether prospective cohorts, historical cohorts, case-control studies, or even case series – are both numerous and heterogeneous with respect to exposure definitions (if any), case definitions and methodologies. It can be difficult to 'combine' them and it is particularly difficult to decide which studies to include in a meta-analysis.

The major difficulty is one of inclusiveness. One can attempt to have all available studies that address a problem which meet a minimum standard. Alternatively, one can include only 'the best studies' with 'best' being designated for example by the quality of exposure assessment. A hybrid of these approaches grades studies, with weighting assigned to the best and worst studies in the opinions of the authors. Regardless of the approach, inclusions and exclusions must be made, and avoiding intentional or unintentional bias.

It is instructive to look at the set of meta-analyses available for the question of fibre type and risk of lung cancer and mesothelioma for examples of the dilemmas posed. For chrysotile, a particular problem has been how to combine studies of risk for lung cancer in different industries. As evidenced above, risks have been quite different for textile industries and mining industries. While this has been referred to as the 'great textile mystery' (McDonald, 1998), perhaps the only real similarity is that chrysotile is the main form of asbestos used: the type of work itself involves very different fibre characteristics, especially with respect to fibre length. It is quite possible to reconcile the studies when this is taken into account, even with the many other differences between the cohorts (Berman, 2010, 2011; Berman and Case, 2012). When attempting a meta-analysis to determine the risk of chrysotile *vs.* commercial amphiboles *generally* for lung cancer however, the very different results obtained in observational studies in the two industries – reliable and valid within the industries – are so different as to make it difficult to reconcile them to come up with a single estimate of risk for chrysotile. Doing so may be wrong, as "In cases when heterogeneity of outcomes is particularly problematic (in a meta-analysis), a single summary measure may well be inappropriate" (Stroup *et al.*, 2000).

In the real world, however, regulatory bodies are often faced with setting 'one size fits all' recommendations and coming up with general conclusions. In the area of meta-analysis for asbestos fibre types and lung cancer/mesothelioma outcome, a number of studies have been especially influential. Although not a meta-analysis *per se*, an early influential example was the EPA 'IRIS' risk model, sometimes called the Nicholson or Nicholson-Peto model, used by the United States Environmental Protection Agency (Nicholson, 1986; US EPA, 1988). The document recognized "that the risk of mesothelioma is greater with exposure to crocidolite than with amosite or chrysotile exposure alone" and that "differences in fibre size distribution and other process differences may contribute at least as much to the observed variation in risk as does the fibre type itself". However, these differences were excluded from the risk model due to "lack of information on fibre exposure" in the referenced studies, leaving the stated risk as simply for 'asbestos'. More seriously, of mining studies being excluded, IRIS "lacked even a single estimate for (mesothelioma potency) from an environment in which exposure was predominantly to chrysotile" (Berman, 2011). The result has been overprediction of lung cancer (Camus *et al.*, 1998) and mesothelioma (Camus *et al.*, 2002) for chrysotile by at least an order of magnitude. Conversely, "the IRIS protocol may substantially underestimate risk when exposure is primarily to amphibole asbestos" (Berman, 2011).

Later meta-analyses had greater datasets to choose from, and more up-to-date follow-ups from historical cohorts such as those for mining and milling, textile work and insulation workers described above. Particularly influential models include those of Hodgson and Darnton (2000) and Berman and Crump (2008a,b). The latter was more 'inclusive'; the Hodgson-Darnton model for example, in an opposite approach to that of US EPA, included chrysotile mining/milling studies while excluding those of the North American textile workers published to 2000. Berman and Crump took an inclusive approach which accounted for fibre type and dimension, while weighting studies for quality.

Exclusion of studies and weighting of studies are different approaches which can result in quite different results. A good example of an exclusionary approach is that of Lenters *et al.*, which provides five 'quality criteria' for data for exposure assessment in eligible studies and then includes or excludes them in meta-analysis (Lenters *et al.*, 2011). The authors found that incrementally excluding studies with lesser coverage of the exposure history by exposure measurement data and less complete job histories produced "higher meta-estimates of the lung cancer risk per unit of exposure". In practice, this meant taking the opposite approach to that of Hodgson and Darnton in excluding the chrysotile mining studies in Quebec while once again including the South Carolina textile studies. Excluding either of these has a very large effect on the conclusions, with resultant controversy and debate (Berman and Case, 2012, 2013; Lenters *et al.*, 2012; Heederik *et al.*, 2013). The problem cannot be solved by application of 'quality criteria' in any case, due to the inherent subjectivity in choosing such criteria and because, ultimately, "the fundamental issue is whether the exposure estimates used are, in fact, correct, and this could be the case when none of the criteria applies. Furthermore, even when all these criteria do apply, estimates can still be erroneous" (Hodgson, 2013). An alternative approach weights studies by quality, for example by increasing confidence intervals for studies with more uncertainty (Berman and Crump, 2008a; Hodgson, 2013).

A common-sense approach to the dilemma posed above realizes that in fact, study results obtained in miners and millers are most applicable to miners and millers, while study results obtained for asbestos textile workers or asbestos insulators are most applicable to those workers. There are real differences in exposures, in terms of fibre types and dimensions, even when the major fibre source in terms of geographical/ geological location – *e.g.* the Quebec chrysotile mines – is similar. There are also real differences in workforces in terms of other possible exposures, mean duration of work in the industry, smoking, and other lifestyle factors. A problem arises when one attempts to create a 'one size fits all' approach for a particular type of fibre, or even 'all asbestos fibres', in all circumstances. There is good evidence that this is not appropriate. Epidemiological study for chrysotile miners and millers, for example, is convincing that they are at considerably increased risk for mesothelioma, albeit likely in large part from associated amphibole (tremolite) fibres (McDonald *et al.*, 1997), while meta-analysis for professions with low-level short-fibre chrysotile exposure *via* work on vehicular brakes is convincing that these professions lack such risk (Garabrant *et al.*, 2015). A similar dilemma arises in the situation where there is a question as to whether amphibole exposures are to asbestiform fibres or to the more common amphibole species including but not limited to 'cleavage fragments' (see below).

If it is not possible to derive job or industry-associated risk estimates which cross the lines of specific jobs or industries, it is difficult then, for example, to apply such global estimates to new circumstances such as naturally occurring asbestos (NOA) environmental exposures (see below). One can apply the precautionary principle to such situations, taking from the range of available studies those which had the 'worst' results in terms of possible risk. Alternatively, where possible, we can in a sufficiently detailed new exposure situation, match measured exposures to the most applicable past studies. The latter approach would require enough detail in both the newly assessed environment and the historical data to allow this to happen.

3.5. 'Old' data in cohorts: to include or not?

Just as there are difficulties in summarizing epidemiology findings using meta-analysis, problems arise when a cohort of workers has been followed up with multiple updates over a long period. The advantage of following a large occupationally exposed cohort includes more complete follow-up with greater incidence of disease, especially important for diseases of long latency like mesothelioma. This is apparent in the Quebec miners, millers and factory workers where to 1976 there had been ten mesothelioma deaths of 4463 total – less than one in 400 (McDonald *et al.*, 1980). By 1993, 38 had died from mesothelioma among 8009 deaths, more than twice the proportion. In addition to latency (median first exposure to death was 44 years) it is possible that there is better case recognition and death certificate ascertainment with more recent time periods.

Similarly, the most recent update of the North American Insulators' cohort revealed greater effects of smoking on lung cancer risk synergistic with asbestos exposure and a greater effect of the presence of asbestosis on lung cancer risk (controlled for exposure and smoking) than in the original reports (Markowitz *et al.*, 2013). The authors stated that the "added risk of lung cancer conferred by asbestosis was not due to additional exposure to asbestos". Their basis for this can be questioned, because in their opinion "The study group of insulators, unlike most asbestos-exposed groups studied to date, had reasonably homogeneous exposure to asbestos, as measured by the duration of insulation work or based on the nature of their work with asbestos-containing products such as cement, pipe covering, or insulation spray from the 1950s through the 1970s". The latter judgment, spread over a wide variety of industries using a wide variety of fibres in a wide variety of applications, seems speculative: the fact remains that the insulator results are based on no measurements at all.

Unfortunately, even when there are records of measured exposure rather than simple duration of work, earlier time periods are also likely to have less accurate exposure assessment. When exposure characterization improves over time, investigators may have more accurate recent data with lower exposure levels and fewer cases of disease, with less follow-up time. Using only more recent exposure measures will then avoid non-differential misclassification of exposure. A long-held view in epidemiology has been that such misclassification can result in lowered estimates of disease risk, although this is overly simplistic and has been challenged (Moolgavkar *et al.*, 2014). What is certain is that by excluding earlier cases one evaluates fewer cases. This dilemma occurred in the United States EPA's recent risk assessment for 'Libby Amphibole' asbestos (LAA) and the studies considered in establishing the risks presented. Because there was no way at all to measure such exposure other than in Libby vermiculite miners or processing workers, only those could be considered. But even then, as with the chrysotile miners and millers, exposure data before 1969 was limited to dust measurements in limited areas. Actual phase contrast microscopic measurements of fibres were available only for individual job locations from 1968 to 1982. For lung cancer risk, the slope was greater (*i.e.* the risk per unit of exposure was greater) when the fibre measures were used. This led investigators to limit themselves to the more recent data for lung cancer (Benson *et al.*, 2015) – a controversial decision, in that it is well demonstrated that the full data set including earlier cases is of 'high quality' (Lenters *et al.*, 2011). Even greater possible error is introduced by the similar approach taken for mesothelioma cases, which may also be inappropriate because there are so few recent cases in and around Libby. Between 1999 and 2011, only five cases could be identified as mesothelioma – giving SMRs for both sexes combined, males and females, which "were elevated but not statistically significant before controlling for past employment" (Naik *et al.*, 2016). Four were among male workers employed by the vermiculite mining company; only one female case was not so employed. As a result, NIOSH statistics no longer list Lincoln County, Montana among the top 50 for mesothelioma in the United States (NIOSH 2014), at least one case per year being required for inclusion (Case *et al.*, 2011). Only seven mesothelioma cases were actually used in the EPA's IRIS risk assessment, *vs.* the 17 actually available for analysis in the vermiculite miners' cohort. This resulted in an overly simplistic model for mesothelioma which 'effectively precludes' a proper likelihood-based time-to-cancer analysis of mesothelioma (Moolgavkar *et al.*, 2014). The model developed by Peto and Nicholson used in part in earlier EPA models accounts for the exponential effects of time from first exposure rather than simple cumulative exposure. Time from first exposure – missing from the Libby Amphibole model – is well accepted as a potent risk factor as important as amphibole fibre type in mesothelioma risk (Peto *et al.*, 1982; Nicholson, 1986; Berman and Crump, 2008b; Galateau-Salé *et al.*, 2015).

4. Changing definitions of 'Mineral Fibres'

4.1. Fibre counts *vs.* dust measurement

Given the importance of measuring exposure for the purposes of epidemiological study, it would be useful to have hard-and-fast definitions of what to measure. As mineralogists know, definitions for 'fibre' (see Belluso *et al.*, 2017, this volume) and for specific types of fibre have changed over time. It is also not necessary to define what a fibre is to perform epidemiological studies of those exposed. Historically, the most important occupational studies were defined by who was working, not by the fine details of what they worked with. Nevertheless, more precise definitions of exposure, including quantitative and semi-quantitative estimates, have become more and more common over time. This has made standard definitions important. Occupational and environmental exposures are to the whole range of fibres, dusts and chemicals in the occupation or environment in question. Exposures will thus include, but not be limited to, those minerals defined as fibres. In that sense, while industrial hygiene-based studies may include estimates of exposure expressed in terms of fibre concentrations or cumulative exposure (that is, intensity of exposure multiplied by time of exposure), these estimates will include all minerals in the class of interest as well as any associated minerals, and will include those that are not 'fibres' by definition as well as those that are.

For the asbestos minerals this poses a particular problem. Definitions of 'asbestos fibre' are usually related to the means developed to measure them in practice. As noted above these included prior to the mid-to-late 1960s era total dust measurements, because instrumentation to measure (for example) fibres defined as >5 μm long were not used in practice. Dust measurements obtained with equipment such as the Midget Impinger measured particles having diameters of ~1 μm and greater, without regard to length, diameter or aspect ratio. Such measurement was still useful, as having a range of exposures to dusts one could compare groups of workers with greater and lesser exposures. However, this presented problems particularly if converting such dust measurements to true fibre measurements using methods developed later, such as the membrane filter method. Side-by-side comparisons of dust counts using both methods were attempted, but correlations were poor. In a pilot study of 87 paired midget impinger dust counts with membrane filter fibre counts for chrysotile miners and millers in Canada, Gibbs and Lachance found that "no single conversion factor could be applied to all mines or to all work areas within a mine" (Gibbs and LaChance, 1974).

The same was true for plants producing primarily asbestos textiles (Ayer *et al.*, 1965). From fibre preparation, carding, spinning, twisting and weaving operations,

230 sample pairs were compared across plants. Ratios between membrane filter counts and impinger dust counts varied with process and between plants. For weaving, for example, "ratios ranged from 2 to 20 ff/mL for each million particles per cubic foot for four plants". The comparisons improved as fibres in longer size ranges only (*e.g.* >5 μm or >10 μm long) were used. For epidemiological studies using exposure estimates, it seemed that use of a single method throughout would provide more consistent estimates of exposure to 'fibres', even if the measurement method itself was actually for total dust:

"With adequate counts of pairs of samples at selected sites, and with assumptions as to the effects that dust control and changes in ore have had on impinger-filter counts, dose-response relationships based on fibre exposures could be examined...Without this, there is no way of estimating with any accuracy the total fibre exposure of persons in the Quebec asbestos mining and milling industry, or indeed of deciding whether the dose-response relationships would be better or worse than those obtained from particle counts. Until satisfactory conversion factors are derived, it would seem best to base standards, at least for mines and mills, on particle counts, for which there now exists a reasonable body of evidence. A similar argument probably applies equally to all epidemiological studies of asbestos workers in North America in which exposure has been considered" (Gibbs and LaChance, 1974).

4.2. Asbestos or not asbestos? The mineralogical definition problem

Mineral-fibre definition has been a moving target. This has been particularly true for 'asbestos', which was originally more a generic term of commerce than a scientifically defined entity. We have discussed this more fully elsewhere (Case *et al.*, 2011), and other chapters in this book deal with mineralogical classification.

As noted in our 2011 article, distinction between mineralogical groups collectively known as 'asbestos' is important when considering the aetiology of asbestos-related diseases because it demonstrates that the *potential of asbestos to produce disease* is not confined to one crystalline mineral or chemical grouping. The biological potential of asbestos is most likely related directly to the ability of the minerals to form fibrous dust particles, or 'elongated mineral particles' (EMP) in the parlance of the NIOSH Roadmap (NIOSH, 2011). The problems arising are summarized in the NIOSH 2011 Executive Summary:

"Imprecise terminology and mineralogical complexity have affected progress in research. 'Asbestos' and 'asbestiform' are two commonly used terms that lack mineralogical precision. 'Asbestos' is a term used for certain minerals that have crystallized in a particular macroscopic habit with certain commercially useful properties. These properties are less obvious on microscopic scales, and so a different definition of asbestos may be necessary at the scale of the light microscope or electron microscope, involving characteristics such as chemical composition and crystallography. 'Asbestiform' is a term applied to minerals with a macroscopic habit similar to that of asbestos. The lack of precision in these terms and the difficulty in translating macroscopic properties to microscopically identifiable characteristics contribute to miscommunication and uncertainty in identifying toxicity associated with various forms of minerals. Deposits may have more than one mineral habit and transitional minerals may be present, which make it difficult to clearly and simply describe the mineralogy".

Given these problems, it becomes very difficult, especially retrospectively, to know exactly what minerals are being – or have been – studied epidemiologically. Comparably sized (by length, width, aspect ratio, *etc.*) amphibole particles which are considered 'asbestos' or 'not asbestos', including but not limited to 'cleavage fragments', 'transitional fibres', 'acicular structures' and the like have not been tested adequately for toxic potential, and in many real-world situations – tremolite associated with chrysotile mining being one – many health scientists believe that such asbestos analogues need to be treated with similar caution, as long as they meet minimum requirements for fibre length. For regulatory and health-assessment purposes there is no evidence that potentially affected cells can distinguish between 'asbestiform' and 'non-asbestiform' fibres having equivalent dimensions (American Thoracic Society, 1990; Case, 1991a,b; Aust *et al.*, 2011; Case *et al.*, 2011).

This is a particularly important question for naturally occurring asbestos (NOA) and other extant mineral fibre deposits that may produce human exposures in the course of natural or human activity. As pointed out by Wylie and Candela, amphibole minerals in particular are very common and "massive, nonfibrous habits are much more common than fibrous habits" (Wylie and Candela, 2015). The authors further note correctly that (at least insofar as epidemiological proof is concerned) "the hypothesis for similarity in response of cleavage fragments and asbestos fibres of the same dimension put forward by Case *et al.* (2011) is unlikely to be easily evaluated". On the other hand, we do not want to wait for a situation where we can count the numbers of mesothelioma cases exposed to such structures before sounding the alarm. Situations in which epidemiology has identified excesses of mesothelioma in occupations exposed to such fibres have also been contested, two examples being the talc deposits in upstate New York and the taconite mines of Minnesota (Case *et al.*, 2011). A common-sense approach with precautions taken where there are significant populations of cleavage fragment or 'nonasbestiform' amphibole 'fibres' having EMP dimensions which would otherwise be of concern may be the best available option.

5. Emphasis on occupation and the relationship to exposure measures

It bears repeating here before discussion of formal exposure measurements in epidemiology that by far the most-used proxy for exposure has been occupational status itself. Although far from perfect – especially in knowing with any degree of precision the mineralogical parameters including fibre type and dimensions – occupations from the work of Doll linking 'asbestos workers' with lung cancer in 1955 through the development of the Quebec and Carolina cohorts were usually the best available proxy for exposure. Exposure was not measured directly in many cohorts that have given us a great deal of information, such as the North American Insulators. As work has progressed, historical exposure estimates have been made, but often the uncertainty associated with these cannot be assessed with confidence. At times, later follow-up of cohorts will have less disease due to lower levels of exposure, but the exposure itself will be better measured, as noted in the Libby example discussed above.

Case-control studies have also regularly used occupational status as a surrogate for exposure. This works well depending on the accuracy of categorization of work: better job descriptions make for better understanding of exposures within jobs. It is not as useful, for example, to know that workforces were employed in the 'textile industry' as it is to also know what jobs they did in that industry (spinning, weaving, *etc.*), given that there is some understanding of variation of exposure with jobs.

Finally, the ultimate understanding of a rare asbestos-related disease like mesothelioma also comes down to very individual exposure assessment. Workers at the large Jeffrey Mine at Asbestos, Quebec had increased mortality from mesothelioma at very high exposure levels (McDonald *et al.*, 1997). They mined and milled chrysotile, but information at the individual level was also available from job histories and in many instances from lung fibre content (Case *et al.*, 1997). Workers with mesothelioma also worked in a factory that used crocidolite; one miller with mesothelioma milled crocidolite. Men worked in the factory on occasion even when they were employed principally in mining. Fifteen of 23 miners and millers had excesses of crocidolite in their lung tissue constituting >11% of all fibres for the entire group, while in another mining area where there was no factory work, crocidolite was present in the lungs of only one of 18 workers (Case and Sebastien, 1987). Women in another Quebec mining area were exposed not only domestically and environmentally, but in some cases through work cleaning and repairing burlap bags recycled to ship all types of asbestos in a small unventilated shop. Their lungs contained all fibre types, including amosite and crocidolite as well as chrysotile and tremolite at levels as high or higher than those seen in miners and millers (Case *et al.*, 2002).

6. Measurement of exposure: standard approaches

6.1. Approaches based on actual fibre measurements

As noted above, the transition between total dust measurements and actual measures of fibre concentrations in the mid-1960s was an important shift. In asbestos industries which continued up to time periods during which true fibre measurements were taken for regulatory and industrial hygiene purposes, the degree to which we have measurements by occupation varies. Within an industry, job categories also varied for exposure, and within job categories there was variation between industrial processes, including fibre-size distributions.

A good example is the asbestos textile industry. There, membrane filter method fibre-counting began in the 1960s, and there is some direct information from this. For several North Carolina textile plants, Loomis, Dement and colleagues stated that "a stratified random sample of 77 historical dust samples captured on membrane filters was selected from among 333 samples available from industrial hygiene studies the US Public

Health Service conducted in the study plants during 1964–1971" (Loomis *et al.*, 2009). They used this data as well as historically recorded phase contrast (light) microscopy (PCM) industrial hygiene measurements obtained by conversion of particle counts before 1964 or actual PCM measurements between 1964 and 1986 (Dement *et al.*, 2008, 2009). Using TEM for the limited archival filters, 22,776 fibres or fibre bundles were counted and sized. Of these, almost all were chrysotile; the authors found only "16 that were tremolite and actinolite" (Dement, 2012). Results were applied to 181 lung cancer deaths, mesothelioma being too rare in the cohort to allow such analysis. Following this, the investigators broke results down into categories, either by bivariate distributions of fibre length and width or applying predefined biologically based length/width distribution categories suggested from animal models.

In bivariate length/diameter categories, a model for TEM fibres longer than 5 μm, which "correspond(ed) most closely to PCM estimates, fit the data better than models for other TEM exposure indicators". When fibre length and diameter data were categorized "the best model fit and the strongest associations with lung cancer were achieved for cumulative exposure to fibres 20–40 μm in length", fitting well with animal data. 'Biologically-based' distributions predicted well textile workers' lung cancer risk. The best fit to lung cancer risk was obtained with an index suggested by Lippmann (1988, 2014): fibres longer than 10 μm and of diameter 0.3–1.0 μm. For dose-response relationship, an index proposed by Berman *et al.* (1995) for relative lung cancer potency (fibres <0.3 μm in diameter and 5–40 μm in length, >40 μm but <0.3 μm diameter, and >40 μm in length but >3 μm in diameter) also fit the textile workers' results well. Results were similar for a similar exercise in South Carolina asbestos textile workers (Dement *et al.*, 2008; Stayner *et al.*, 2008). Overall, these studies confirm the importance of increasing fibre length and decreasing fibre diameter as human lung cancer potency factors, coupled with exposure concentration (or dose). In the parallel South Carolina textile plant study, strongest associations for lung cancer risk were observed for fibres >10 μm long and <0.25 μm in diameter. The authors concluded that "Assessments of exposure to asbestos should account for fibre sizes, as well as numbers" (Loomis *et al.*, 2009).

6.2. Approaches based on questionnaires

Other than simple occupational status as obtained from employment records, one of the simplest ways to determine what workers and others have been exposed to is to ask them. This is true even when they do not *know* they were exposed, as the investigator may know – and know in some detail – from the responses given.

In its simplest iteration, this means administering a questionnaire, done as early as the work by Wagner *et al.* (1960) and Newhouse and Thompson (1965) described above. At base this will provide simply an indication of occupational, domestic or 'pure' environmental exposures, but adding details provides further information. Place of residence can be related to distance from a mine or factory. Even the number of relatives employed in an asbestos industry can be quantified. In a case-control study of ten women with mesothelioma living around the chrysotile mines of Quebec compared with 15 matched controls to each case, we found that breaking down answers into how many asbestos workers cases lived with could be quantified. Relative risk for mesothelioma was 1.0, 3.4 and 9.0 for no household workers, one or two household workers, or three or more household workers respectively (Case *et al.*, 2002).

Questionnaire-based exposure continues to be used in even the largest studies, and is best when actual cases are alive at the time of questionnaire as opposed to relatives or other 'proxies'. Rake and colleagues were able to study occupational and household exposures well in the largest case-control study of mesothelioma ever published using this method, which minimizes knowledge bias while still subject to faulty recall (Rake *et al.*, 2009).

6.3. Job-exposure matrices and expert-based assessments

Given the importance of occupation (and job within occupation), well described occupational status can be translated into real exposure terms such as exposure intensity, duration and cumulative exposure in terms of fibres/mL-years. This method was developed by Siemiatycki and colleagues at IARC and in Montreal (Siemiatycki, 1990; Siemiatycki *et al.*, 1981). At base it relies on qualitative judgments in given jobs for three parameters: confidence that the exposure actually occurred (possible, probable, definite), frequency of exposure during a normal workweek (<5%, 5–30%, and >30%), and the level of concentration (intensity) of the agent in the work environment (low, medium, high). The method has been applied to occupational lung carcinogens including asbestos (Pintos *et al.*, 2008; Lacourt *et al.*, 2015). It is useful, and to some degree backed up by parallel tissue analyses for subjects where lung tissue was available (Takahashi *et al.*, 1994). Levels of exposure are evidently accurate only relative to one another rather than in absolute terms. Modelling of exposure is improved through the use of expert chemists or industrial hygienists. In one study of mesothelioma in France (Iwatsubo *et al.*, 1998) the five industrial hygiene experts for example applied the 'concentration' parameter as low intensity <1 fibre/mL; medium intensity = 1–2 fibres/mL; high intensity = 2–10 fibres/mL); and very high intensity >10 fibres/mL. Results showed the expected gradient of risk from low to high exposure, but the study received more attention for the 'absolute' values estimated for low exposures when translated into fibres/mL-years ('f/mL-year'). Iwatsubo *et al.* (1998) warned readers that a calculated "Cumulative exposure index was based on subjective assessment, that is, semiquantification of exposure by the experts and selected weighting factors assigned to each category of exposure, with no objective measurement of airborne asbestos levels. Thus, the exposure unit, f/mL-years, is expressed in quotation marks". Despite their warning, the authors' guesses at 'f/mL-year' levels are often referred to as if they were actual measures of exposure.

Job exposure matrix (JEM) and expert-based exposure estimates are best used for assessment of dose-response, or for discovery of new carcinogens. When real measurements are also available over time however, JEM assessment may be obtained with on-the-job real measures to combine the best aspects of both types of exposure assessment. A recent example assessed five lung carcinogens, including asbestos and chromium (Peters *et al.*, 2016). In this study, a database of 27,958 actual personal asbestos fibre measurements where related job codes were available, covering the 1970s until 2009, with sampling duration known to have been between 60 and 600 min. Applying these measurements to JEM type analysis in a complex statistical model, the authors estimated geometric mean asbestos exposure levels for 1980 and 2000 in a variety of jobs. Overall estimates went from 0.061 f/mL in 1980 to 0.004 f/mL in 2000 for 'low exposed jobs' and from 0.074 f/mL to 0.005 f/mL for 'high exposed jobs'. Highest job exposures were seen for "Heating, ventilation, and refrigeration engineering technician" and for "Spinner, thread, and yarn" (within "Manufacture of non-metallic mineral products not elsewhere classified", a category said to include all asbestos-product-making industries). Both were >0.4 fibres/mL in 1980. Heating, ventilation, *etc.* dropped to 0.03 fibres/mL by 2000. No later measures were available for the spinner, thread and yarn jobs.

Of particular interest were calculations in this study determining the overall decrease in exposures over time *vis-à-vis* asbestos bans. Peters *et al.* (2016) showed an annual decrease of 10.7% (95% CI 11.3% to 10.0%) before and no further downward trend after the introduction of such bans. They hypothesize that this "steep decrease is likely the result of the growing awareness of asbestos hazards and the final introduction of bans on asbestos".

6.4. Approaches based on lung-retained fibre analyses

Tissue measures of retained dose can be as helpful as good exposure measurements. In December 1983, Sébastien *et al.* (1988) collected sputum specimens from all but three of 173 then-current workers of the W.R. Grace Libby, Montana vermiculite mine and mill. Cumulative exposure was also determined from job history and available air measurements taken since 1956. Asbestos bodies – the iron- and protein-coated structures that develop on mainly amphibole forms of asbestos *in vivo* – were shown to be present in 128 workers by light microscopy (75.3%). The number of bodies per sample was correlated significantly with age, duration of work and with cumulative exposure. As predictors of disease – indicated by radiological abnormalities – logistic regression corrected for age and smoking showed exposure indices based on body counts were generally higher than those based on cumulative exposure (Sébastien *et al.*, 1988).

This type of analysis, most frequently of lung-retained fibres assessed by SEM or TEM, has proved useful as a means of exposure assessment in epidemiological study of fibres. It is a direct measure of dose, although often taken many years after initial exposure. Fibres which clear rapidly, such as synthetic vitreous fibres (McDonald *et al.*, 1990) or chrysotile asbestos (Case, 1994), will not be represented as accurately as fibres with longer clearance times. However, this property of lack of relative biopersisitence is also related to the relative lack of cancer potency for these fibres. Lung-fibre content has been used as the exposure variable in case-control analyses. McDonald *et al.* (1989) compared 78 matched autopsy mesothelioma cases across Canada to controls without malignant or respiratory disease, matched for autopsy hospital, sex, dates of birth and death, and type of tissue available. Although 25 fibre types were recognized by electron microscopic analysis, almost all fibres longer than 8 μm were of either asbestos or man-made mineral fibre types. Except for asbestos fibres, no significant difference was observed between cases and controls. Both 'long' (>8 μm) and 'short' (<8 μm) fibre distributions showed substantial differences between cases and controls for amosite, crocidolite, and tremolite; much less for anthophyllite/talc and chrysotile. In multivariate analysis, long amphibole fibres accounted for mesothelioma attributable risk of 68%; results suggested that "long amphibole fibres could explain a high proportion of mesotheliomas in both men and women" (McDonald *et al.*, 1989).

More recently, Gilham and colleagues showed that that asbestos fibre burden by TEM in lung samples for 133 UK mesothelioma cases showed a linear increase *vs.* matched lung cancer cases with significant Odds Ratios beginning at 4.81 (95% CI 2.61 to 8.85) for 0.05 to 0.2 million fibres per gram dry lung, up to 30.7 (95% CI 6.38 to 147.7) for more than one million fibres per gram (Gilham *et al.*, 2016). Most asbestos fibres were amosite (75%) or crocidolite (18%). Lung-retained chrysotile and other amphiboles were correlated with total fibre burden and "were too low for their effect on risk to be estimated". Results fit well with previous observations that amosite consumption in the UK was very high (about five times higher than in the US, where mesothelioma incidence and mortality are four to five times less).

As indicators of past exposure and as indicators of ongoing internal dose, lung-retained fibre is close to being a gold standard. However, it is difficult to compare studies against one another due to differing methodologies. International panels continue nevertheless to recommend that we "perform lung fibre content analyses on tissues obtained at autopsy or at time of surgery from residents of (exposed areas) with or without malignant mesothelioma to help in understanding exposure disease relationships" (Carbone *et al.*, 2016).

Analysis of other tissues for fibre content has been more controversial. This is particularly true for (parietal) pleura, which is exceptionally difficult to sample without cross-contamination from the adjacent lung tissue, especially in autopsy specimens. Analysis of true pleura is difficult, due to the sparse and heterogeneous presence of fibres, and such work is best limited to experimental studies (De Vuyst *et al.*, 1998). The problems associated with attempts to study fibre in pleura are best exemplified by often misrepresented studies by Suzuki and colleagues (Suzuki and Yuen, 2002; Suzuki *et al.*, 2005) in which they studied not pleural tissue but tumoural mesothelioma tissues: "Mesotheliomatous tissue was selected from the primary serosal cancer where the cancer was intimately associated with fibrosis and/or hyaline plaque". Further to cross-contamination issues, the studies were biased toward sampling of particles rather than fibres by extending length criteria downwards to the resolution of the TEM. Chrysotile was accepted as 'fibres' at lengths as small as 0.07 μm, and samples were of lung, pleural plaque and tumour, not pleura. Problems with these studies were summarized by Roggli (2015) with a general discussion of erroneous conceptions around 'short fibres'. High fibre levels in control tissues suggest contamination or procedural fragmentation falsely inflating true fibre numbers, and "it is improper to compare cancer tissue with normal pleura in terms of fibre concentration. The only proper comparison for cancers would be analysis of other pleural-based cancers that encase the lung and are not asbestos-related (epithelioid hemangioendothelioma of the pleura or some pseudomesotheliomatous carcinomas of the pleura). This is because blockage of parietal pleural stomata by a diffuse rind of cancer could impede the clearance of short chrysotile fibres and result in their accumulation in any type of cancer with this growth pattern. Indeed, cancer tissues are improper for analysis because they begin as a single cell and grow over time to form billions of cells that comprise the cancer rind. Whatever is present in cancer by necessity must have gotten there after the fact and cannot be considered a contributing factor to the disease...because of issues regarding possible contamination, improper controls, unconventional analytical methodology, and false reasoning concerning causation, the Suzuki papers are not reliable..." (Roggli, 2015).

There are, however, a few good experimental surveys of actual fibres in actual human pleura. These include work showing that areas of the pleura sampled *in vivo* in asbestos-exposed workers containing "concentrated long amphibole fibres, which in some cases had levels exceeding those found in the lung tissue of the same individual. The fibre concentrations were at least an order of magnitude higher in the black spots compared with nearby parietal pleura" (Boutin *et al.*, 1996; Roggli, 2015).

Asbestos fibre-length distributions, whether in air, water, other media or tissue, are always skewed towards greater numbers of structures having shorter lengths. This is true regardless of where one starts a count of fibres in a given distribution: counting fibres in lung tissue beginning at 18 μm results in most of those being counted having length of <22 μm, for example (Case *et al.*, 2000). While fibre lengths within a distribution tend to be interrelated, there is often an erroneous conception that because there are always *more* short structures they must contribute to a disease to a greater degree. This is evident in the Suzuki and Yuen (Suzuki and Yuen, 2002; Suzuki *et al.*, 2005) papers where "Even if one assumes that these short fibres actually derived from the tissue rather than these other sources, the authors seem to be arguing that these fibres must be doing something because there are so many of them. The fallacy of this argument is obvious when one considers that carbon particles are far more numerous in pleural tissues than short chrysotile fibres, so one could equally argue that carbon particulates are the cause of mesothelioma (Roggli, 2015)".

In fact, "there is a strong weight of evidence that asbestos and Synthetic Vitreous Fibres shorter than 5 μm are unlikely to cause cancer in humans" (Eastern Research Group/ATSDR, 2003).

7. Changing definitions of disease

7.1. Asbestosis and cancer

While problems relating to the characterization of mineral fibres, and exposure to those fibres, have been well described as a problem for epidemiology, poor definition of disease is less discussed. Most outside specialized medical fields believe with some reason that disease definitions are more settled.

Nevertheless, there have been problems both early and recent with the definitions of diseases related to exposure to mineral fibres, especially the asbestos fibres. In particular, asbestosis would seem to be a straightforward entity. Early studies largely involved massive exposure and extensive disease, but as exposures were reduced *degrees of disease* became as important as the presence or absence of the disease. This was true first for radiologically defined asbestosis, which was defined by grading chest X-rays on a scale. The most common methods involved reading of a single X-ray, sometimes in series in follow-ups of workers with exposure to dusts and fibres. Classifications of such radiographic appearances "proliferated before and after World War I", and from 1950 International Labour Office (ILO)-published classifications were widely adopted. (Liddell and Morgan, 1978). The ILO U/C (1971) classification of the radiographic appearances of the pneumoconioses became important to grade asbestosis, but was controversial. For epidemiology, the purpose of such readings was "to seek relationships between radiographic abnormality and one or more of dust retention or tissue reaction or symptoms or lung-function impairment or measures of exposure" (Liddell and Morgan, 1978). However, just as exposure had to be defined, radiographic abnormality had to be defined with degrees and qualities of abnormality observed with testable uniformity. A major problem existed in that asbestosis and similar diseases existed on a continuum of radiographic change; this was unlike an all-or-nothing disease like cancer. In addition, technical aspects like training, use of multiple readers, inter- and intra-reader reliability testing, and use of reference X-rays added complexity. Later, use of computed tomography (CT) and high resolution CT (HRCT) of the thorax were seen to identify significantly more abnormalities consistent with asbestosis than were chest X-rays alone, and with better reliability and fewer indeterminate cases (Begin *et al.*, 1993). Nevertheless, plain chest X-ray remains the predominant tool, and findings, especially at the lowest categories of abnormal reading are controversial as to the presence or absence of 'disease' (Gitlin *et al.*, 2004; Oliver *et al.*, 2004). Consensus views as to what is 'abnormal' have also changed over time (ATS, 1986; Guidotti *et al.*, 2004). The most recent American Thoracic Society 'official statement' (Guidotti *et al.*, 2004) reviews this history and puts radiographic findings in context:

"The initial radiographic presentation of asbestosis is typically that of bilateral small primarily irregular parenchymal opacities in the lower lobes bilaterally. Over time, the distribution and density or 'profusion' of opacities may spread through the middle and upper lung zones... A critical distinction is made between films that are suggestive but not presumptively diagnostic (0/1) and those that are presumptively diagnostic but not unequivocal (1/0). This dividing point is generally taken to separate films that are considered to be 'positive' for asbestosis from those that are considered to be 'negative.' However, profusion itself is continuous".

The 1986 ATS statement had specified a higher grade ILO classification of X-ray abnormalities ('opacities' 1/1) as defining asbestosis. However, use of X-ray findings (and HRCT) in diagnosis may be different from use in epidemiological studies.

The 2004 ATS statement by Guidotti *et al.* (2004) required "evidence of structural pathology consistent with asbestos-related disease as documented by imaging or histology", in addition to an identified exposure and exclusion of alternative (and possibly treatable) causes. The statement seemed to suggest that pathologic findings of asbestosis were likely in the absence of radiographic abnormalities, relying heavily upon an older study of American insulation workers with lung cancer (Kipen *et al.*, 1987). In that study, of 138 cases with "histological evidence of parenchymal fibrosis", 25 (18%) had "no radiographic evidence of parenchymal fibrosis". However, parenchymal fibrosis is not necessarily asbestosis, and importantly, the article was applicable to a group of very heavily exposed individuals all of whom had died from lung cancer.

In addition, the 2004 ATS statement depended for its definition and grading of pathologic asbestosis on a 1982 statement from the College of American Pathologists (CAP) (Craighead *et al.*, 1982), which made the minimum criteria necessary for a diagnosis of asbestosis a finding of ''discrete foci of fibrosis in the walls of respiratory bronchioles associated with accumulations of asbestos bodies'' in histologic sections. In 2010, the CAP revised its definition making the minimum histologic criterion more precise and more difficult to meet: "a diagnosis of asbestosis should only be made when there is an acceptable pattern of alveolar septal fibrosis and an average rate of asbestos bodies of at least $2/cm^2$ of lung". It was felt that this eliminated incorrectly diagnosed cases of asbestos airways disease, or bronchiolar wall fibrosis without fibrotic thickening of alveolar septa (Roggli *et al.*, 2010).

To summarize, in the first decade of the 21st century radiological criteria for the amount of abnormality necessary for a diagnosis of asbestosis had been relaxed, while pathologic criteria had become more stringent. Further details here are beyond the scope of this chapter, but both are controversial. Whether revisions are accepted or not in either area, the action leads to clinical and medicolegal controversy for cases having lower degrees of radiological or pathologic change. Which definition to use – as with definitions of mineral fibres themselves – will affect case definitions in epidemiological studies, including retroactive considerations of such studies.

Lung cancer and mesothelioma diagnosis would seem to be less likely to be disputed, especially after 50 years of enhancement of such diagnoses with special techniques such as immunohistochemistry (Case, 2016). Even here, however, both overdiagnosis and underdiagnosis of disease has been reported repeatedly (Labreche *et al.*, 2012). While actual misdiagnosis (in either direction) is possible, for epidemiological studies, use of death certificates as the source for diagnoses has historically posed a larger problem. For lung cancer this does not seem to be as much of a problem currently (Edwards *et al.*, 2014) as for mesothelioma, especially where the disease is rare and poorly reported.

7.2. Pleural plaques – disease, health effect, or marker of exposure

Non-malignant change in the pleura has long been recognized as a result of asbestos exposure. In the most extreme cases, severe fibrosis restricts lung movement. This is called diffuse pleural fibrosis, and is often associated with asbestosis; it is related clearly to cumulative exposure. The most common pleural change, however, is a localized thickening of the parietal pleura – the pleural plaque – which, especially when calcified and bilateral, is a reliable indicator of past asbestos exposure. Clearly such structures on the parietal pleura and diaphragm are abnormal structures, and because they denote exposure they are also associated indirectly with increased risk of malignant disease. Whether plaques are independent risk factors for other diseases, and especially whether they themselves constitute 'disease' is controversial. The answer rests ultimately on the question of whether they interfere with lung function and if so, if such interference is clinically significant. Detection of plaques can be difficult, and they can be confused with fat pads on standard chest X-rays, the traditional means of plaque identification. Computed tomography (CT) scans, especially those of high resolution (HRCT), may improve clarity of identification.

A number of studies using standard lung function tests have shown that small but statistically significant reductions in lung volumes can be associated with the radiological presence of pleural plaques. This type of study has two major problems associated with it. First, it is difficult to remove other asbestos-related disease, particularly subclinical asbestosis, as a contributor to lung-function abnormalities. Second, while it is possible to determine whether decrements in function (*e.g.* lung forced vital capacity (FVC) or forced expiratory volume in one second (FEV1)) differ between those with and without plaques statistically, it is much more difficult to determine whether such differences are *clinically* significant.

According to the British Thoracic Society (2011), "In some studies, subjects with pleural plaques have been shown to have a small but statistically significant reduction in lung volumes of ~5% compared with matched controls (Schwartz *et al.*, 1990; Kouris *et al.*, 1991). Other studies have not confirmed this after controlling for parenchymal changes representing fibrosis (Copley *et al.*, 2001). The fact that plaques are present on the parietal pleura means that they have little effect on lung expansion. The lung function changes (if any) are considered too small, in a legal sense, to attract compensation. Extensive and confluent plaques are uncommon but can result in a restrictive ventilatory defect that results in disability" (Lilis *et al.*, 1991).

Recent meta-analyses have come to opposite conclusions. One grouped 20 previous studies having internal comparison groups and concluded that "The presence of pleural plaques is associated with a small, but statistically significant mean difference in FVC and FEV1 in comparison to asbestos-exposed individuals without plaques or other abnormalities. From a public health perspective, small-group mean decrements in lung function coupled with an increased rate of decline in lung function of the exposed population may be consequential" (Kopylev *et al.*, 2015a). Conversely, Kerper and colleagues concluded that "the weight of evidence indicates that pleural plaques do not impact lung function" after choosing six of 16 HRCT and 16 of 36 chest X-ray studies in the literature based on quality criteria. Although they noted that "Half of the higher quality studies reported small but statistically significant mean lung function decrements associated with plaques" none of the differences was clinically significant (Kerper *et al.*, 2015).

The differences – indeed, opposite conclusions – of these two meta-analyses underline a philosophical difference which comes down to the clinical effect of a tiny decrement in lung function in an individual as opposed to a consistent mean small effect in a large population of exposed individuals. Kopylev and colleagues (2015) in fact found only "statistically significant" decrements in FVC (4.09% of predicted, 95% CI 2.31% to 5.86%) and FEV1 (1.99% of predicted, 95% CI (0.22% to 3.77%). No-one would (or should) argue that such a small difference in an individual has any meaning; indeed, in repeated tests in an individual there might well be a degree of variation of this magnitude. However, Kopylev and colleagues and others argue that the small effect when consistent over a population has wider significance for the exposed population. An analogous situation has been suggested in the small decrease in mean IQ observed for children exposed to lead: it is evident that a 4% decrease of intellect in a population would be of extreme concern, and would further imply that while some individuals within that population might show no effect, others might be severely impaired.

Kopylev and colleagues thus argue: "Small but significant permanent changes in group mean lung function can be indicative of functional impairment at the population level; the focus on population-level versus clinical-level effects is at the heart of our and Kerper *et al.*'s differing interpretations of the observed decrement in lung function" (Kopylev *et al.*, 2015b).

It is hard to resolve this difference scientifically and it has very considerable practical effects on issues of compensation. Issues surrounding pleural plaques have also become important in the area of asbestos risk assessment with US EPA's recent decision to use 'localized pleural thickening' or LPT as the indicator for a "quantitative exposure-response model for the non-cancer effects of Libby Amphibole Asbestos (LAA)" (Benson *et al.*, 2015). This is the *only* non-cancer risk assessment extant in a regulatory regime mathematically based on pleural changes, and its use is controversial. The underlying study, which shows a clear progression of pleural changes in a cohort of amphibole-exposed vermiculite processing workers (Benson, *et al.*, 2015) is not in doubt. There are clearly increased numbers of workers affected by changes in LPT in this cohort over time that are exposure-related.

Extending LAA-associated LPT to all pleural plaques is another matter however, and even within the small universe of LAA-exposed processing workers or miners there is considerable disagreement as to whether there is any clinically significant change in lung function for LPT (Benson *et al.*, 2016). In much of the recent literature, despite EPA's indication that the assessment is "not a complete toxicological review of other amphiboles or of chrysotile asbestos" (US EPA, 2015) LPT and pleural plaques have been used interchangeably. More importantly, there has been a failure to comprehend the very considerable clinical differences between LPT with LAA and other pleural plaques. The disease associated with LAA is "markedly different pleural disease" which is "much more likely to be accompanied by progressive dyspnea, restrictive ventilatory impairment and difficult-to-treat pleuritic pain", all factors which can themselves impair lung function. Unfortunately, there is, as yet, no good pathologic or physiologic characterization of this unique pleural disease, and it should not be confused with pleural plaques of the more common variety.

8. Environmental exposure to asbestos: Is there a role for epidemiology?

8.1. Overview

We know that environmental exposures to amphibole asbestos and similar biopersistent mineral fibres, such as fibrous zeolite, are important in disease genesis, particularly in the case of mesothelioma. Determining the extent of the contribution of such exposures is more difficult than the situation for occupational exposures. The cohort approach is rarely of any use, except in limited situations such as that described for erionite in Turkey where entire populations are so exposed, and even in that situation there is some discussion of a possible genetic component.

We have discussed the breakdown of environmental exposures elsewhere recently, and the reader is referred to that discussion (Case *et al.*, 2011). While 'domestic' or 'take-home' exposures are sometimes described as environmental, those exposures are analogous to those of co-workers or bystanders. This discussion is limited to exposures not related to occupational status either directly or indirectly.

Environmental exposures have been described as causing mesothelioma from the beginnings of the realization that the exposures were related to the disease. Frequently, collections of cases in a confined area in a given time period were ascertained and separated into those which could be shown to have had occupational or domestic exposures and those which did not. The assumption was that the cases were related to environmental exposure without quantifying or measuring it in any way. As described, the first study by Wagner *et al.* (1960) was preceded by years of ascertainment of data in which – at first – many of the cases did not appear to be asbestos-related. Before publication, however, such cases were approached with two questions. In addition to being asked had their work brought them in contact with "asbestos or the blue dust *at any time*", the research team asked whether at any time had they "lived in the vicinity of an asbestos mine or mill" (Case, 2008). Using this additional information, a history of exposure to asbestos dust was established in every case. Some were women who had left the mining villages at a young age with cancer development as much as 50 years later; others had been at the school in the village of Kuruman (see the location in Wagner *et al.*, 1960), "which was opposite the asbestos mills and in the break-time they used to slide down the tailings dump."

At around the same time in the UK, Newhouse and Thompson ascertained mesothelioma cases in a London hospital diagnosed during the previous 50 years. Sources of possible asbestos exposure were obtained through hospital and clinic notes, records of an asbestos factory in the area, and personal interviews with patients or their surviving relatives. Cases were also compared with matched controls from a variety of control groups (Newhouse and Thompson, 1965). Slightly over half of the mesothelioma cases for which information could be obtained (40 of 76) had some past exposure history, compared with 11.8% of one set of controls. Eighteen worked at the asbestos factory, all exposed to crocidolite asbestos, and four others worked at other asbestos factories; eight were asbestos insulators (or 'laggers'), and one made deliveries to the factory, for a total of 31 of 40; nine had a relative who worked with asbestos. The remaining 36 cases had "no history of work or domestic exposure to asbestos", but 11 of these (30.6%) lived within a half mile of the factory while it was operating as compared to only 7.5% of controls, a significant excess.

Newhouse and Thompson's findings have been replicated around asbestos plants and mines ever since, sometimes using the case-control design they applied and at other times simply by listing cases within a given radius of a plant. Crocidolite mines and asbestos cement pipe plants using crocidolite, in particular, have been implicated in causation for 'pure' environmental mesothelioma cases, in Japan (Kurumatani and Kumagai, 2008; Kumagai and Kurumatani, 2009), Italy (Corfiati *et al.*, 2015; Mensi *et al.*, 2015), Australia (Hansen *et al.*, 1993, 1998; Reid *et al.*, 2007, 2008), Louisiana, USA (Case and Abraham, 2009; Case *et al.*, 2011) and many other countries. The situation around the notorious Eternit asbestos cement plant at Casale Monferrato is a much-studied example *e.g.* (Magnani *et al.*, 2001).

Goldberg and Luce reviewed the evidence for 'nonoccupational' mesothelioma, including but not limited to 'pure' environmental exposures (Goldberg and Luce, 2009). The authors estimated that "in industrialized countries about 20% of the mesotheliomas could be caused by environmental exposure to asbestos", while admitting that exposure assessment for such cases is poorly documented and difficult to ascertain, except indirectly. They noted as well that "According to the population under study and the methods of exposure ascertainment, (the) proportion of cases without any identified exposure to asbestos is usually in the range of 10–20% among men, and much higher among women (50–60%)" (Goldberg and Luce, 2009).

8.2. Environmental asbestos exposure using ecological study design

Case-control study design or relationship of cases to point emission sources have often been used to investigate the effects of environmental exposures, but more commonly such studies use an 'ecological' study design. This is not a reference to 'ecology' as environment but simply analysing cases in space and time as to what might be expected by chance. This can be misleading in the absence of knowledge of actual individual exposures, migration patterns, allowing sufficient time to pass after exposures occurred, *etc.* There is a temptation with increasing availability of massive administrative databases and increasing computing power to attempt linkages which are not entirely logical. Study in California attempted to link mesothelioma case status to "Residential proximity to NOA estimated as the distance to the edge of the nearest known body of ultramafic rock" – a crude measure at best (Pan *et al.*, 2005). Critics (Berman *et al.* 2013; Cox *et al.* 2013) suggested in a simulation that such 'associations' are likely to be spurious.

The problem with such ecologic study design is that

> "people tend to choose where they live at least partly intentionally, rather than at random. They live in places they can afford. They live close to work or close to transportation to work... Such differences (which may not even be entirely identifiable) can cause the residential distribution of any two selected groups of

> individuals to differ (*i.e.* can create non-uniform heterogeneity). We hypothesize that such differences can suffice to cause one group or another to appear to live significantly closer to any arbitrary set of sources. If true, such associations would have nothing to do with any putative interaction between source locations and the groups being compared; they would clearly be non-causal" (Berman *et al.*, 2013).

Related problems arise when local (for example, county) data on disease or mortality rates are used to extremes, with subgroups handpicked by sex and age, as is evident in some recent exploratory work in the US (Baumann and Carbone, 2016). A full discussion of the problems surrounding ecological study design is provided elsewhere (Case *et al.*, 2011). This type of study is, however, justified where there is clear *a priori* evidence of a potential human exposure to disease-causing minerals, usually from industrial legacy sources such as the neighbourhood of asbestos cement plants and mines.

Mineralogists and toxicologists can be helpful in determining where this type of study is justified and where it is not, especially where newly suspect minerals are involved. Careful characterization of the actual minerals present and the likelihood that they may be associated with human exposures can be followed, if justified, by animal experimentation to assess toxicity. An example is offered by a recent study finding that palygorskite found in one environmental circumstance in southern Nevada is not a likely contributor to mesotheliomas found in ecological study in age and sex subgroups (Larson *et al.*, 2016).

It is difficult to assess accurately general environmental exposures other than with a considerable potential for error. One example is study of the prevalence of lung cancer and mesothelioma among women living in the chrysotile mining regions of Quebec. Camus and colleagues estimated exposure for women in these areas using a complex model which included recorded area airborne asbestos levels in the asbestos-mining towns from 1972, annual asbestos production volumes for each asbestos mining town from 1900 to 1984, emission controls and stack samples, wind rose patterns, published values of lung content for such women, and other factors. An expert panel modelled exposures, and the authors concluded that the lung cancer rate in the exposed population did not differ from that expected. The standardized mortality ratio for lung cancer (based on 71 deaths) was 0.99 (95 percent confidence interval, 0.78 to 1.25), whereas the standardized proportionate mortality ratio was 1.10 (95 percent confidence interval, 0.88 to 1.38) (Camus *et al.*, 1998). This was tenfold less than the lung cancer rate that would have been expected using the EPA IRIS model given the estimated exposures.

'Pleural cancer', a surrogate measure for mesothelioma, was elevated (SMR ratio 7.63; 95% CI 3.06 to 15.73) based on seven cases in the study interval. While this was again an order of magnitude less than that predicted (Camus *et al.*, 2002), the real questions had to do with the actual individual exposures of cases, both in magnitude and type (occupational *vs.* domestic *vs.* pure environmental and type of fibre). It was known that women in the area who had neither worked in nor lived in the mining areas nonetheless had significantly elevated lung chrysotile content if they lived within 40 km of one of the mines (Case and Sebastien, 1987), and that in another area where tremolite exposure was more common, lung tremolite content in those with 'pure' environmental exposure was related directly to time lived near the mines and inversely to distance lived from the mines (Case, 1991a). Pleural mesothelioma cases occurred exclusively in the women in the higher-tremolite area during the relevant time period. In a case-control study of these, it was demonstrated that all but one of the cases in fact had occupational and/or domestic exposures, including exposures to commercial and non-commercial amphiboles (Case *et al.*, 2002). Five of the ten cases (50%) worked in the asbestos industry (*vs.* 2.7% of controls; relative risk = 53.8). Nine (90%) had lived with one or more asbestos workers (*vs.* 65% of controls); and six (60%) with two or more workers (*vs.* 36% of controls). Mean cumulative exposure (residential, domestic, and occupational combined) was estimated as 226.1 fibres/mL-year (range 84–525) for cases and 84.1 fibres/mL-year (range 0–189) for controls, with a plausible 5-fold error on either side.

8.3. 'Naturally occurring' exposures: a challenge for mineralogy and for epidemiology

Commercial asbestos fibres remain a serious problem both in previous installations in the industrial west and especially in those mostly Asian countries where asbestos is still used heavily. Nevertheless, increased regulation and asbestos bans have made a difference, with measured concentrations reducing significantly between 1980 and 2000 (Peters *et al.*, 2016). In the US, mesothelioma rates have followed suit, with significant year-on-year decreases since 1992. The National Cancer Institute's Surveillance, Epidemiology and End Results (SEER) program shows a decline for male, age-adjusted mesothelioma incidence over age 50 from 8.06 cases per 100,000 population in 1992 to 4.89 cases per 100,000 in 2013, a 39% decline in two decades. This represents an annual per centage change (APC) of −1.5% per year based on rates age-adjusted to the 2000 US standard population ($p<0.05$)[1]. Unfortunately, this is not the case for the UK and some other western European countries, where incidence is much higher (Gilham *et al.*, 2016).

Even though cases due to exposures from our industrial legacy may be decreasing in the US and some other countries, there are newly recognized sources which pose a problem. These are collectively described as 'Naturally Occurring Asbestos' (NOA), an awkward phrase covering natural sources likely to produce risk due to both human

---

[1] As calculated at http://seer.cancer.gov/faststats/selections.php?series=cancer on September 26, 2016 for male mesothelioma incident cases over age 50, age-adjusted. Incidence source: SEER 13 areas (San Francisco, Connecticut, Detroit, Hawaii, Iowa, New Mexico, Seattle, Utah, Atlanta, San Jose-Monterey, Los Angeles, Alaska Native Registry and Rural Georgia). Rates are per 100,000 and are age-adjusted to the 2000 US standard population (19 age groups - Census P25-1130). The modelled rates are the point estimates for the regression lines calculated by the *Joinpoint Regression Program* (Version 4.2.0, April 2015, National Cancer Institute).

activity and natural processes such as erosion. Recognizing the possibility of such exposures is important. Often they may be encountered during large projects such as highway or dam construction. For California's Calaveras Dam (Hetch Hetchy Water System Improvement Program, 2014), which involves the movement of more than seven million cubic yards of rock and earth, a portion of which may contain NOA, an extensive dust control and air monitoring programme was installed to avoid exposing workers or the public to increased levels. Background air quality levels in the area were monitored for 2 years prior to construction and continue around the perimeter and within the project, with publicly available data available for monitoring stations on a website (San Francisco Public Utilities Commission, 2015). At least 27 varied fibrous amphibole species have been detected in air-monitoring samples, and results are reported for both total asbestos and amphiboles. When trigger levels are exceeded, work on the site may be shut down temporarily.

Many communities are potentially liable to be affected by NOA, and awareness is increasing. Epidemiology is of little assistance because, for the most part, only ecological studies (*e.g.* county mesothelioma rate comparisons) can be made, and are sometimes applied without adequate caution and logic (see above). The long latency of such diseases makes finding increased cases unlikely, and their absence in the presence of a clear potential source with a newly exposed population is meaningless. In theory, such methods may work in very narrow areas where a small geographic area with a high proportion of potentially exposed population has existed for a long time. Usually however such areas are recognized only through a combination of factors involving recognition of a hazard by geologists or mineralogists. Even air sampling is of little use unless it is activity-based. Nevertheless, hazard can occasionally be recognized, one example being the asbestiform tremolite in portions of El Dorado County, California (Case and Abraham, 2009). NOA is discussed further in a previous paper (Case *et al.* 2011).

9. Epidemiological surveillance systems for MM incidence and the role of non-occupational exposure to asbestos.

As we have seen in the previous sections of this Chapter, assessment of the extent of non-occupational asbestos exposure and the health effects caused by such exposure continues to be of great public interest. This is because it is related to specific exposure circumstances, *e.g.* living with asbestos workers or living near asbestos mines or manufacturing plants, or sources of naturally occurring asbestos fibres, or in asbestos-insulated buildings, which may be still incompletely regulated (Goldberg and Luce, 2009). The increased risk of cancer for groups of people resident in the neighbourhood of raw asbestos production sites (mines or mills), as well as for people living close to industrial plants manufacturing asbestos products, has been demonstrated (Mirabelli *et al.*, 2008; De Klerk *et al.*, 2013). Throughout the last four decades, excessive incidence of MM and lung cancer has been reported as a consequence of the presence of naturally occurring asbestos (or asbestos-like) material in rural areas of Turkey (Bayram *et al.*,

2013), Greece (Sakellariou *et al.*, 1996), Corsica in France (Boutin *et al.*, 1986), Sicily in Italy (Bruno *et al.*, 2006), New Caledonia (Goldberg *et al.*, 1994) and China (Luo *et al.*, 2003).

In Italy, one of the most involved and sensitive countries in asbestos-related diseases monitoring and control, some 3,748,550 tons of raw asbestos were produced up to the 1992 ban, with a peak between 1976 and 1980 of more than 160,000 tons/year (Marinaccio *et al.*, 2012). Environmental exposure related to residence near asbestos-cement plants has been reported in detail for Casale Monferrato (Magnani *et al.*, 2001), Bari (Musti *et al.*, 2009), Broni (Mensi *et al.*, 2015) and La Spezia (Dodoli *et al.*, 1992). The extent of asbestos exposure in occupational settings is expected to decrease in coming years. Conversely, the contribution of different patterns of non-occupational exposures is probably underestimated, due to their much lower level, which nonetheless are not negligible and possibly sufficient to cause disease.

Since 2002 in Italy, a permanent system of surveillance of mesothelioma incidence has been run by the National Register of Malignant Mesotheliomas (Registro Nazionale dei Mesoteliomi, ReNaM in Italian), identifying cases and assessing asbestos exposure (Marinaccio *et al.*, 2015). Specific surveillance systems with reliable information completeness, exposure assessment and territorial coverage are scarce (Ferrante *et al.*, 2016) and currently ongoing in only three countries: Australia (Leigh *et al.* 2002), France (Goldberg *et al.*, 2006) and South Korea (Jung *et al.*, 2012). The ReNaM has a regional structure: a Regional Operating Centre (COR) has been established gradually in all 20 Italian regions. Each COR works by applying the national standardized methods described in the national Guidelines. They actively search for incidents of MM from health-care institutions that diagnose and treat cases (especially pathology units, lung care and chest surgery wards). Diagnostic coding criteria have been fixed by means of a grid according to three classes of decreasing level of certainty: certain (if histological confirmation is available), probable (if cytological confirmation is available) and possible MM (only radiological and clinical evidence). Occupational history, lifestyle habits and residential history are obtained using a standardized questionnaire, administered by a trained interviewer to the subject or to the next of kin. CORs may consult public local health and safety agencies to obtain supplementary information about occupational and/or residential exposure. In each COR an industrial hygienist, or a panel of industrial hygienists, classifies and codes the exposure and examines the information collected. Moreover, an agreement between the Italian Workers' Compensation Authority (INAIL) and the Italian Social Security Institute (INPS) allows retrieval of pension contributions from personal data. Therefore, in many cases INAIL may provide CORs with information about occupational histories of mesothelioma cases, both as a confirmation of information obtained directly from subjects, and may also be a major source of information when an interview is not available. In this work, classification of occupational exposure is qualitative: it can be assigned as definite, probable or possible. Definite occupational exposure is assigned to the subjects whose work has involved the use of asbestos or materials containing asbestos. Probable occupational exposure is attributed to the subjects who have worked in a firm where asbestos was certainly used, but whose exposure cannot be documented. Possible occupational exposure applies to subjects who have worked in a firm related to an economic sector in which asbestos has been used. Further specific codes are assigned to environmental exposure (residence near a source of asbestos pollution without work-related exposure), familial exposure (when patients have lived with a cohabitant occupationally exposed) and to leisure activities exposures (other non-occupational exposures like those due to leisure-time activities).

In the period 1993 to 2012 a case list of 21,463 MM was collected. A pleural site is reported in 93.3% (19,955 cases) of recorded MM cases; peritoneal and pericardial cases represent 6.5% (1,392 cases) and 0.3% (51 cases) of the whole case-list, respectively. ReNaM provides a reliable estimation of 10% for the proportion of MM incident cases related to non-occupational exposure to asbestos based on more than 15,845 detected cases, of which 12,065 were interviewed individually (Marinaccio *et al.*, 2015). The distribution of modalities of exposure shows that residence near asbestos-cement plants is the predominant factor. However, the analytical description of life conditions (the historical place of residence) and habits (leisure time and hobby activities) involved in the risk of asbestos exposure, remains of great value for primary prevention and public health policies.

Environmental exposure from naturally occurring asbestos contamination of the soil has been documented in Turkey where mesothelioma epidemics due to the presence of tremolite, chrysotile and erionite have been reported extensively (Sébastien *et al.*, 1984; Simonato *et al.*, 1989; Berk *et al.*, 2014). In Greece a high frequency of pleural calcifications was found in patients living in places where no industrial use of asbestos was documented in the past, but tremolite fibres have been found in the soil called 'luto' used intensively in that area (Langer *et al.*, 1987). Natural contamination of soil by tremolite fibres was detected also in Cyprus (McConnochie *et al.*, 1987) and Corsica (Boutin *et al.*, 1986), and was found to be associated with occurrences of asbestos-related diseases without industrial asbestos applications. Similarly, cases of pleural mesothelioma in New Caledonia have been reported as being associated with the use of a material containing primarily tremolite fibres (Luce *et al.*, 2004), as well as in Chinese rural areas where cases were associated with the presence of crocidolite in soil (Luo *et al.*, 2003). Cases of mesothelioma were found in the town of Libby (Montana, USA), close to which the world's largest vermiculite mine operated and which contained Libby Amphibole Asbestos that was used extensively in the surrounding residential zone (Whitehouse *et al.*, 2008). Similar reports come from the township of Wittenoom (Western Australia), where a large crocidolite mine was active from 1943 to 1966, with extensive use of the tailings to pave roads, footpaths, school playgrounds, *etc.* (Reid *et al.*, 2008).

In Italy, three MM cases associated with tremolite pollution in a rural area of Basilicata were reported (Bernardini *et al.*, 2003). In the area of Biancavilla, Etna (Sicily) an excess of mortality for pleural tumours has been observed and subsequent analyses have confirmed a causal role for exposure to a nearby mine extracting a fluoro-edenite-contaminated material used on a large scale for construction (Comba *et al.*, 2014).

Risk of mesothelioma associated with local industrial sources was demonstrated clearly for neighbouring populations (Kurumatani and Kumagai, 2008; Tarres *et al.*, 2013). The spatial variation of mesothelioma risk in an area highly polluted by the asbestos-cement plant of Casale Monferrato, adjusted for occupational and domestic exposure, has been discussed showing that the effect of exposure from industrial sources on the general population decreases with increasing distance from the factory (Maule *et al.*, 2007). Mesothelioma cases due to cohabitation with exposed workers have been attributed to soiled work clothes brought home. An Italian study of a cohort of wives of Casale Monferrato asbestos-cement factory workers showed a large excess of pleural mesothelioma (SMR = 18.00, 21 observed *vs*. 1.2 expected) (Ferrante *et al.*, 2007).

Systematic epidemiological surveillance of the incidence of asbestos-related disease and mortality is a valuable tool for preventing asbestos exposure, improving the efficiency of insurance and welfare systems, and for providing epidemiological evidence. Asbestos exposure in asbestos-containing buildings, such as public offices or schools, is still a real concern even where asbestos has been banned, because subjects involved may not be conscious of direct physical contact with asbestos-containing material (Binazzi *et al.*, 2013).

Disclosures

Dr Case has acted as an expert witness for law firms representing defendants and plaintiffs in asbestos litigation and compensation board proceedings and has been a paid consultant to regulatory agencies and compensation boards in North America.

References

Addingley, C.G. (1965) Dust measurement and monitoring in the asbestos industry. *Annals of the New York Academy of Sciences*, **132**, 298–305.

American Thoracic Society (1990) Health effects of tremolite. This official statement of the American Thoracic Society was adopted by the ATS Board of Directors, June 1990. *The American Review of Respiratory Disease*, **142**, 1453–1458.

ATS (1986) American Thoracic Society. Medical Section of the American Lung Association: The diagnosis of nonmalignant diseases related to asbestos. *American Review of Respiratory Disease*, **134**, 363–368.

Aust, A.E., Cook, P.M. and Dodson, R.F. (2011) Morphological and chemical mechanisms of elongated mineral particle toxicities. *Journal of Toxicology and Environmental Health, Part B: Critical Reviews*, **14**, 40–75.

Ayer, H.E., Lynch, J.R. and Fanney, J.H. (1965) A comparison of impinger and membrane filter techniques for evaluating air samples in asbestos plants. *Annals of the New York Academy of Sciences*, **132**, 274–287.

Bartrip, P.W. (2004) History of asbestos related disease. *Postgraduate Medical Journal*, **80**, 72–76.

Baumann, F. and Carbone, M. (2016) Environmental risk of mesothelioma in the United States: An emerging concern – epidemiological issues. *Journal of Toxicology and Environmental Health, Part B*, **19**, 231–249.

Bayram, M., Dongel, I., Bakan, N.D., Yalçin, H., Cevit, R., Dumortier, P. and Nemery, B. (2013) High risk of malignant mesothelioma and pleural plaques in subjects born close to ophiolites. *Chest*, **143**, 164–171.

Begin, R., Ostiguy, G., Filion, R., Colman, N. and Bertrand, P. (1993) Computed tomography in the early detection of asbestosis. *British Journal of Industrial Medicine*, **50**, 689–698.

Belluso, E., Cavallo, A. and Halterman, D. (2017) Crystal habit of mineral fibres. Pp. 65–109 in: *Mineral Fibres: Crystal Chemistry, Chemical-physical Properties, Biological Interaction and Toxicity* (A.F.

Gualtieri, editor). EMU Notes in Mineralogy, **18**. European Mineralogical Union and Mineralogical Society of Great Britain & Ireland, London.

Benson, R., Berry, D., Lockey, J., Brattin, W., Hilbert, T. and LeMasters, G. (2015) Exposure-response modeling of non-cancer effects in humans exposed to Libby Amphibole Asbestos; update. *Regulatory Toxicology and Pharmacology*, **73**, 780–789.

Benson, R., Berry, D., Lockey, J., Brattin, W., Hilbert, T. and LeMasters, G. (2016) Response to comment on "Exposure-response modeling of non-cancer effects in humans exposed to Libby Amphibole Asbestos; update" by Benson *et al.* (2015) submitted by Goodman *et al.* (2016). *Regulatory Toxicology and Pharmacology*, **80**, 270–271.

Berk, S., Yalcin, H., Dogan, O.T., Epozturk, K., Akkur, I. and Seyfikli, Z. (2014) The assessment of the malignant mesothelioma cases and environmental asbestos exposure in Sivas province, Turkey. *Environmental Geochemistry and Health*, **36**, 55–64.

Berman, D.W. (2010) Comparing milled fiber, Quebec ore, and textile factory dust: has another piece of the asbestos puzzle fallen into place? *Critical Reviews in Toxicology*, **40**, 151–188.

Berman, D.W. (2011) Apples to apples: the origin and magnitude of differences in asbestos cancer risk estimates derived using varying protocols. *Risk Analysis*, **31**, 1308–1326.

Berman, D.W. and Crump, K.S. (2008a) A meta-analysis of asbestos-related cancer risk that addresses fiber size and mineral type. *Critical Reviews in Toxicology*, **38 Suppl. 1**, 49–73.

Berman, D.W. and Crump, K.S. (2008b) Update of potency factors for asbestos-related lung cancer and mesothelioma. *Critical Reviews in Toxicology*, **38 Suppl. 1**, 1–47.

Berman, D.W. and Case, B.W. (2012) Overreliance on a single study: there is no real evidence that applying quality criteria to exposure in asbestos epidemiology affects the estimated risk. *Annals of Occupational Hygiene*, **56**, 869–878.

Berman, D.W. and Case, B.W. (2013) Quality of evidence must guide risk assessment of asbestos, by Lenters, V; Burdorf, A; Vermeulen, R; Stayner, L; Heederik, D. *Annals of Occupational Hygiene*, **57**, 667–669.

Berman, D.W., Crump, K.S., Chatfield, E.J., Davis, J.M. and Jones, A.D. (1995) The sizes, shapes, and mineralogy of asbestos structures that induce lung tumors or mesothelioma in AF/HAN rats following inhalation. *Risk Analysis*, **15** 181–195.

Berman, D.W., Cox, L.A. Jr. and Popken, D.A. (2013) A cautionary tale: the characteristics of two-dimensional distributions and their effects on epidemiological studies employing an ecological design. *Critical Reviews in Toxicology*, **43**, 1–25.

Bernardini, P., Schettino, B., Sperduto, B., Giannandrea, F., Burragato, F. and Castellino, N. (2003) Three cases of pleural mesothelioma and environmental pollution with tremolite outcrops in Lucania. *Giornale Italiano di Medicina del Lavoro e Ergonomia*, **25**, 408–11.

Binazzi, A., Scarselli, A., Corfiati, M., Di Marzio, D., Branchi, C., Verardo, M., Mirabelli, D., Gennaro, V., Mensi, C., Schallenberg, G., Merler, E., De Zotti, R., Romanelli, A., Chellini, E., Pascucci, C., D'Al, D., Forastiere, F., Trafficante, L., Menegozzo, S., Musti, M., Cauzillo, G., Leotta, A., Tumino, R., Melis, M. and Marinaccio, A. (2013) Epidemiologic surveillance of mesothelioma for the prevention of asbestos exposure also in non-traditional settings. *Epidemiologia e Prevenzione*, **37**, 35–42.

Boutin, C., Viallat, J.R., Steinbauer, J., Massey, D.G., Charpin, D. and Mouries, J.C. (1986) Bilateral pleural plaques in Corsica: a non-occupational asbestos exposure marker. *European Journal of Respiratory Diseases*, **69**, 4–9.

Boutin, C., Dumortier, P., Rey, F., Viallat, J.R. and De Vuyst, P. (1996) Black spots concentrate oncogenic asbestos fibers in the parietal pleura. Thoracoscopic and mineralogic study. *American Journal of Respiratory and Critical Care Medicine*, **153**, 444–449.

British Thoracic Society (2011) Pleural Plaques: Information for Health Care Professionals. BTS 299912 2011 Jan 11. https://www.brit-thoracic.org.uk/document-library/clinical-information/mesothelioma/pleural-plaques-information-for-health-care-professionals/ accessed September 23, 2016.

Bruno, C., Comba, P. and Zona, A. (2006) Adverse health effects of fluoro-edenitic fibers: epidemiological evidence and public health priorities. *Annals of New York Academy of Sciences*, **1076**, 778–83.

Camus, M., Siemiatycki, J. and Meek, B. (1998) Nonoccupational exposure to chrysotile asbestos and the risk of lung cancer. *New England Journal of Medicine*, **338**, 1565–1571.

Camus, M., Siemiatycki, J., Case, B.W., Desy, M., Richardson, L. and Campbell, S. (2002) Risk of mesothelioma among women living near chrysotile mines versus US EPA asbestos risk model: Preliminary findings. *Annals of Occupational Hygiene*, **46**, 95–98.

Carbone, M., Kanodia, S., Chao, A., Miller ,A., Wali, A., Weissman, D., Adjei, A., Baumann, F., Boffetta, P., Buck, B., de Perrot, M., Dogan, A.U., Gavett, S., Gualtieri, A., Hassan, R., Hesdorffer, M., Hirsch, F.R., Larson, D., Mao, W., Masten, S., Pass, H.I., Peto, J., Pira, E., Steele, I., Tsao, A., Woodard, G.A., Yang, H. and Malik, S. (2016) Consensus Report of the 2015 Weinman International Conference on Mesothelioma. *Journal of Thoracic Oncology*, **11**, 1246–1262.

Cartier, P. (1952) Transcript of discussion in: Smith, W.E., Survey of some current British and European studies of occupational tumor problems. *AMA Archives of Industrial Hygiene and Occupational Medicine*, **5**, 242–263.

Cartier, P. (1965) Discussion: pleural plaques. *Annals of the New York Academy of Sciences*, **132**, 235–239.

Case, B.W. (1991a) Health effects of tremolite. Now and in the future. *Annals of the New York Academy of Sciences*, **643**, 491–504.

Case, B.W. (1991b) On talc, tremolite, and tergiversation. Ter-gi-ver-sate: 2: to use subterfuges. *British Journal of Industrial Medicine*, **48**, 357–359.

Case, B.W. (1994) Biological indicators of chrysotile exposure. *Annals of Occupational Hygiene*, **38**, 503-518, 410–501.

Case, B.W. (2006) Asbestos, smoking, and lung cancer: interaction and attribution. *Occupational and Environmental Medicine*, **63**, 507–508.

Case, B.W. (2008) From Cotton-Stone to the New York Conference: Asbestos-related diseases 1878–1965. Pp. 3–22 in: *Asbestos and Its Diseases* (J.E. Craighead and A. Gibbs, editors). Oxford University Press, New York.

Case, B.W. (2016) Epidemiology of malignant pleural mesothelioma in the Americas. Pp. 53–72 in: *Malignant Pleural Mesothelioma: Present Status and Future Directions* (T.C. Mineo, editor). Bentham eBooks. http://www.eurekaselect.com/139052/chapter/epidemiology-of-malignant-pleural-mesothelioma-in-the-america. Accessed October 4, 2016.

Case, B.W. and Abraham, J.L. (2009) Heterogeneity of exposure and attribution of mesothelioma: Trends and strategies in two American counties. *Journal of Physics: Conference Series* **151**, 1–15. http://iopscience.iop.org/article/10.1088/1742-6596/151/1/012008/pdf Accessed October 4, 2016.

Case, B.W. and Mattison, D.R. (1991) Risk assessment in pathology education. In: *Environmental and Occupational Disease: A-State-of-the-Art Conference for Pathology Educators, Warren, Vermont* (J.E. Craighead, editor). UAREP monograph, Universities Associated for Research and Education in Pathology, Inc. UAREP, Bethesda, Maryland, USA, 90 pp.

Case, B.W. and Sebastien, P. (1987) Environmental and occupational exposures to chrysotile asbestos: a comparative microanalytic study. *Archives of Environmental Health*, **42**, 185–191.

Case, B.W., Churg, A., Dufresne, A., Sebastien, P., McDonald, A.D. and McDonald, J.C. (1997) Lung fibre content for mesothelioma in the 1891–1920 birth cohort of Quebec chrysotile workers: A descriptive study. *Annals of Occupational Hygiene*, **41**, 231–236.

Case, B.W., Dufresne, A., McDonald, A.D., McDonald, J.C. and Sebastien, P. (2000) Asbestos fiber type and length in lungs of chrysotile textile and production workers: Fibers longer than 18 μm. *Inhalation Toxicology*, **12**, 411–418.

Case, B.W., Camus, M., Richardson, L., Parent, M.-É., Désy, M. and Siemiatycki, J. (2002) Preliminary findings for pleural mesothelioma among women in the Québec chrysotile mining regions. *Annals of Occupational Hygiene*, **46**, 128–131.

Case, B.W., Abraham, J.L., Meeker, G., Pooley, F.D. and Pinkerton, K.E. (2011) Applying definitions of "Asbestos" to environmental and "low-dose" exposure levels and health effects, particularly Malignant Mesothelioma. *Journal of Toxicology and Environmental Health, Part B: Critical Reviews*, **14**, 3–39.

Churg, J., Rosen, S.H. and Moolten, S. (1965) Histological characteristics of mesothelioma associated with asbestos. *Annals of the New York Academy of Sciences*, **132**, 614–622.

Comba, P., Scondotto, S. and Musmeci, L. (2014) The fibres with fluoro-edenitic composition in Biancavilla (Sicily, Italy): health impact and clues for environmental remediation. *Annali dell'Istituto Superiore di Sanità*, **50**, 108–110.

Cooke, W.E. (1924) Fibrosis of the lungs due to the inhalation of asbestos dust. *British Medical Journal*, **2(3317)**, 140–142.

Cooke, W.E. (1927) Pulmonary asbestosis. *British Medical Journal*, **2(3491)**, 1024–1025.

Copley, S.J., Wells, A.U., Rubens, M.B., Chabat, F., Sheehan, R.E., Musk, A.W. and Hansell, D.M. (2001) Functional consequences of pleural disease evaluated with chest radiography and CT. *Radiology*, **220**, 237–243.

Corfiati, M., Scarselli, A., Binazzi, A., Di Marzio, D., Verardo, M., Mirabelli, D., Gennaro, V., Mensi, C., Schallemberg, G., Merler, E., Negro, C., Romanelli, A., Chellini, E., Silvestri, S., Cocchioni, M., Pascucci, C., Stracci, F., Romeo, E., Trafficante, L., Angelillo, I., Menegozzo, S., Musti, M., Cavone, D., Cauzillo, G., Tallarigo, F., Tumino, R., Melis, M., Iavicoli, S., Marinaccio, A. and ReNa, M.W.G. (2015) Epidemiological patterns of asbestos exposure and spatial clusters of incident cases of malignant mesothelioma from the Italian national registry. *BMC Cancer*, **15**, 286.

Cox, L.A. Jr., Popken, D.A. and Berman, D.W. (2013) Causal versus spurious spatial exposure-response associations in health risk analysis. *Critical Reviews in Toxicology*, **43**, 26–38.

Craighead, J.E., Abraham, J.L., Churg, A., Green, F.H., Kleinerman, J., Pratt, P.C., Seemayer, T.A., Vallyathan, V. and Weill, H. (1982) The pathology of asbestos-associated diseases of the lungs and pleural cavities: diagnostic criteria and proposed grading schema. Report of the Pneumoconiosis Committee of the College of American Pathologists and the National Institute for Occupational Safety and Health. *Archives of Pathology and Laboratory Medicine*, **106**, 544–596.

De Klerk, N., Alfonso, H., Olsen, N., Reid, A., Sleith, J., Palmer, L., Berry, G. and Musk, A.B. (2013) Familial aggregation of malignant mesothelioma in former workers and residents of Wittenoom, Western Australia. *International Journal of Cancer*, **132**, 1423–1428.

De Vuyst, P., Karjalainen, A., Dumortier, P., Pairon, J.C., Monso, E., Brochard, P., Teschler, H., Tossavainen, A. and Gibbs, A. (1998) Guidelines for mineral fibre analyses in biological samples: report of the ERS Working Group. European Respiratory Society. *European Respiratory Journal*, **11**, 1416–1426.

Dement, J.M. (1991a) Carcinogenicity of chrysotile asbestos: a case control study of textile workers. *Cell Biology and Toxicology*, **7**, 59–65.

Dement, J.M. (1991b) Carcinogenicity of chrysotile asbestos: evidence from cohort studies. *Annals of the New York Academy of Sciences*, **643**, 15–23.

Dement, J.M. (2012) Videotaped deposition of John M. Dement, United States Bankruptcy Court, Middle District Of North Carolina, Charlotte Division: Case No. 10-Bk-31607 In Re: Garlock Sealing Technologies, LLC, *et al.*, 31–32. https://pl-garlock.s3.amazonaws.com/02-%20Dep%20Designations/Dement,%20John%2012-21-2012.pdf Accessed October 4, 2016.

Dement, J.M., Harris, R.L., Jr., Symons, M.J. and Shy, C. (1982) Estimates of dose-response for respiratory cancer among chrysotile asbestos textile workers. *Annals of Occupational Hygiene*, **26**, 869–887.

Dement, J.M., Harris, R.L., Jr., Symons, M.J. and Shy, C.M. (1983) Exposures and mortality among chrysotile asbestos workers. Part I: exposure estimates. *American Journal of Industrial Medicine*, **4**, 399–419.

Dement, J.M., Brown, D.P. and Okun, A. (1994) Follow-up study of chrysotile asbestos textile workers: cohort mortality and case-control analyses. *American Journal of Industrial Medicine*, **26**, 431–447.

Dement, J.M., Kuempel, E.D., Zumwalde, R.D., Smith, R.J., Stayner, L.T. and Loomis, D. (2008) Development of a fibre size-specific job-exposure matrix for airborne asbestos fibres. *Occupational and Environmental Medicine*, **65**, 605–612.

Dement, J.M., Loomis, D., Richardson, D., Wolf, S. and Myers, D. (2009) Estimates of historical exposures by phase contrast and transmission electron microscopy in North Carolina USA asbestos textile plants. *Occupational and Environmental Medicine*, **66**, 574–583.

Dodoli, D., Del Nevo, M., Fiumalbi, C., Iaia, T.E., Cristaudo, A., Comba, P., Viti, C. and Battista, G. (1992) Environmental household exposures to asbestos and occurrence of pleural mesothelioma. *American Journal of Industrial Medicine*, **21**, 681–687.

Doll, R. (1955) Mortality from lung cancer in asbestos workers. *British Journal of Industrial Medicine*, **12**, 81–86.

Doll, R. and Hill, A.B. (1950) Smoking and carcinoma of the lung; preliminary report. *British Medical Journal*, **2(4682)**, 739–748.

Doll, R. and Hill, A.B. (1954) The mortality of doctors in relation to their smoking habits; a preliminary report. *British Medical Journal*, **1(4877)**, 1451–1455.

Eastern Research Group, Inc. (2003) Agency for Toxic Substances and Disease Registry. Report on the Expert Panel on Health Effects of Asbestos and Synthetic Vitreous Fibres: The Influence of Fibre Length. New York, NY: *Expert Panel on Health Effects of Asbestos and Synthetic Vitreous Fibres*; October 29–30, 2002. 1-65. https://www.atsdr.cdc.gov/hac/asbestospanel/finalpart1.pdf accessed September 26, 2016.

Edwards, J.K., Cole, S.R., Chu, H., Olshan, A.F. and Richardson, D.B. (2014) Accounting for outcome misclassification in estimates of the effect of occupational asbestos exposure on lung cancer death. *American Journal of Epidemiology*, **179**, 641–647.

Enterline, P.E. (1965) Mortality among asbestos products workers in the United States. *Annals of the New York Academy of Sciences*, **132**, 156–165.

Enterline, P.E. (1991) Changing attitudes and opinions regarding asbestos and cancer 1934–1965 [see comments]. *American Journal of Industrial Medicine*, **20**, 685–700.

Enterline, P.E., Sussman, N. and Marsh, G.M. (1978) *Asbestos and Cancer: The First Thirty Years*. Philip E. Enterline, Pittsburgh, USA.

Ferrante, D., Bertolotti, M., Todesco, A., Mirabelli, D., Terracini, B. and Magnani, C. (2007) Cancer mortality and incidence of mesothelioma in a cohort of wives of asbestos workers in Casale Monferrato, Italy. *Environmental Health Perspectives*, **115**, 1401–1405.

Ferrante, P., Binazzi, A., Branchi, C. and Marinaccio, A. (2016) National epidemiological surveillance systems of mesothelioma cases. *Epidemiologia e Prevenzione*, **40**, 215–223

Galateau-Sallé, F., Churg, A., Roggli, V., Chirieac, L.R., Attanoos, R., Borczuk, A., Cagle, P., Dacic, S., Hammar, S., Husain, A.N., Inai, K., Ladanyi, M., Marchevsky, A.M., Naidich, D., Ordonez, N.G., Rice, D.C., Sheaff, M.T., Travis, W.D. and van Meerbeeck, J. (2015) Mesothelial tumours: Diffuse malignant mesothelioma: Epithelioid mesothelioma. Pp. 156–164 in: *WHO Classification of Tumours of the Lung, Pleura, Thymus and Heart* (W.D. Travis, E. Brambilla, A.P. Burke, A. Marx and A.G. Nicholson, editors). Lyon, France, IARC.

Garabrant, D.H., Alexander, D.D., Miller, P.E., Fryzek, J.P., Boffetta, P., Teta, M.J., Hessel, P.A., Craven, V.A., Kelsh, M.A. and Goodman, M. (2015) Mesothelioma among motor vehicle mechanics: An updated review and meta-analysis. *Annals of Occupational Hygiene*, **60**, 8–26.

Gibbs, G.W. and LaChance, M. (1974) Dust-fiber relationships in the Quebec chrysotile industry. *Archives of Environmental Health*, **28**, 69–71.

Gilham, C., Rake, C., Burdett, G., Nicholson, A.G., Davison, L., Franchini, A., Carpenter, J., Hodgson, J., Darnton, A. and Peto, J. (2016) Pleural mesothelioma and lung cancer risks in relation to occupational history and asbestos lung burden. *Occupational and Environmental Medicine*, **73**, 290–299.

Gilson, I.J. (1965) Man and Asbestos. *Annals of the New York Academy of Sciences*, **132**, 9–11.

Gitlin, J.N., Cook, L.L., Linton, O.W. and Garrett-Maye, E. (2004) Comparison of "B" readers' interpretations of chest radiographs for asbestos related changes. *Academic Radiology*, **11**, 843–856.

Gloyne, S.R. (1935) Two cases of squamous carcinoma of the lung occurring in asbestosis. *Tubercle*, **17**, 5.

Goldberg, M. and Luce, D. (2009) The health impact of non-occupational exposure to asbestos: what do we know? *European Journal of Cancer Prevention*, **18**, 489–503.

Goldberg, M., Goldberg, P., Leclerc, A., Chastang, J.F., Marne, M.J. and Dubourdieu, D. (1994) A 10-year incidence survey of respiratory cancer and a case-control study within a cohort of nickel mining and refining workers in New Caledonia. *Cancer Causes and Control*, **5**, 15–25.

Goldberg, M., Imbernon, E., Rolland, P., Gilg Soit Ilg, A., Savès, M., de Quillacq, A., Frenay, C., Chamming's, S., Arveux, P., Boutin, C., Launoy, G., Pairon, J.C., Astoul, P., Galateau-Sallé, F. and Brochard, P. (2006) The French National Mesothelioma Surveillance Program. *Occupational and Environmental Medicine*, **63**, 390–395.

Goodman, J.E., Zu, K., Tao, G. and Kerper, L.E. (2016) Comment on "Exposure-response modeling of non-cancer effects in humans exposed to Libby Amphibole Asbestos; update" by Benson *et al.* (2015). *Regulatory Toxicology and Pharmacology*, **80**, 268–269.

Green, F.H., Harley, R., Vallyathan, V., Althouse, R., Fick, G., Dement, J., Mitha, R. and Pooley, F. (1997) Exposure and mineralogical correlates of pulmonary fibrosis in chrysotile asbestos workers. *Occupational and Environmental Medicine*, **54**, 549–559.

Greenberg, M. (2003) Biological effects of asbestos: New York Academy of Sciences 1964. *American Journal of Industrial Medicine*, **43**, 543–552.

Guidotti, T.L., Miller, A., Christiani, D., Wagner, G., Balmes, J., Harber, P., Brodkin, C.A., Rom, W., Hillerdal, G., Harbut, M. and Green, F.H.Y. (2004) American Thoracic Society. Official Statement. Diagnosis and Initial Management of Nonmalignant Diseases Related to Asbestos. *American Journal of Respiratory and Critical Care Medicine*, **170**, 691–715.

Gwinn, M.R. (2011) Multiple modes of action of asbestos and related mineral fibers. *Journal of Toxicology and Environmental Health. Part B, Critical Reviews*, **14**, 1–2.

Hammond, E.C. and Selikoff, I.J. (1965) Discussion. *Annals of the New York Academy of Sciences*, **132**, 600–601.

Hansen, J., de Klerk, N.H., Eccles, J.L., Musk, A.W. and Hobbs, M.S. (1993) Malignant mesothelioma after environmental exposure to blue asbestos. *International Journal of Cancer*, **54**, 578–581.

Hansen, J., de Klerk, N.H., Musk, A.W. and Hobbs, M.S. (1998) Environmental exposure to crocidolite and mesothelioma: exposure-response relationships. *American Journal of Respiratory and Critical Care Medicine*, **157**, 69–75.

Harington, J.S. and Roe, F.J. (1965) Studies of carcinogenesis of asbestos fibers and their natural oils. *Annals of the New York Academy of Sciences*, **132**, 439–450.

Heederik, D., Lenters, V. and Vermeulen, R. (2013) Reply: response to the letter by Drs Berman and Case. *Annals of Occupational Hygiene*, **57**, 675–677.

Hein, M.J., Stayner, L.T., Lehman, E. and Dement, J.M. (2007) Follow-up study of chrysotile textile workers: cohort mortality and exposure-response. *Occupational and Environmental Medicine*, **64**, 616–625.

Hetch Hetchy Water System Improvement Program. (2014) Calaveras Dam Replacement Project: Naturally Occurring Asbestos Fact Sheet. Pages 1–4. http://sfwater.org/modules/showdocument.aspx?documentid =8055 accessed September 26, 2016.

Hodgson, J.T. (2013) Quality of evidence must guide risk assessment of asbestos, by Lenters, V; Burdorf, A; Vermeulen, R; Stayner, L; Heederik, D. *Annals of Occupational Hygiene*, **57**, 670–674.

Hodgson, J.T. and Darnton, A. (2000) The quantitative risks of mesothelioma and lung cancer in relation to asbestos exposure. *Annals of Occupational Hygiene*, **44**, 565–601.

Hooper, J. (1990) The asbestos mess. In *The New York Times Sunday Magazine*. New York Edition, November 25, 1990. New York. http://www.nytimes.com/1990/11/25/magazine/the-asbestos-mess.html?pagewanted =all Accessed October 4, 2016.

IARC Working Group on the Evaluation of Carcinogenic Risks to Humans, World Health Organization and International Agency for Research on Cancer (1987) Overall evaluations of carcinogenicity : an updating of IARC monographs volumes **1** to **42**, supplement 7. Lyon, International Agency for Research on Cancer.

Iwatsubo, Y., Pairon, J.C., Boutin, C., Menard, O., Massin, N., Caillaud, D., Orlowski, E., Galateau-Salle, F., Bignon, J. and Brochard, P. (1998) Pleural mesothelioma: dose-response relation at low levels of asbestos exposure in a French population-based case-control study. *American Journal of Epidemiology*, **148**, 133–142.

Jacob, G. and Anspach, M. (1965) Pulmonary neoplasia among Dresden asbestos workers. *Annals of the New York Academy of Sciences*, **132**, 536–548.

Jung, S.H., Kim, H.R., Koh, S.B., Yong, S.J., Chung, M.J., Lee, C.H., Han, J., Eom, M.S. and Oh, S.S. (2012) A decade of malignant mesothelioma surveillance in Korea. *American Journal of Industrial Medicine*, **55**, 869–875.

Kerper, L.E., Lynch, H.N., Zu, K., Tao, G., Utell, M.J. and Goodman, J.E. (2015) Systematic review of pleural plaques and lung function. *Inhalation Toxicology*, **27**, 15–44.

Kipen, H.M., Lilis, R., Suzuki, Y., Valciukas, J.A. and Selikoff, I.J. (1987) Pulmonary fibrosis in asbestos insulation workers with lung cancer: a radiological and histopathological evaluation. *British Journal of Industrial Medicine*, **44**, 96–100.

Kiviluoto, R. (1965) Pleural plaques and asbestos: further observations on endemic and other nonoccupational asbestosis. *Annals of the New York Academy of Sciences*, **132**, 235–239.

Koch, T. (2008) John Snow, hero of cholera: RIP. *Canadian Medical Association Journal*, **178**, 1736.

Kopylev, L., Christensen, K.Y., Brown, J.S. and Cooper, G.S. (2015a) A systematic review of the association between pleural plaques and changes in lung function. *Occupational and Environmental Medicine*, **72**, 606–614.

Kopylev, L., Christensen, K.Y., Brown, J.W. and Cooper, G.S. (2015b) Authors' response: A systematic review of the association between pleural plaques and changes in lung function. *Occupational and Environmental Medicine*, **72**, 685–686.

Kouris, S.P., Parker, D.L., Bender, A.P. and Williams, A.N. (1991) Effects of asbestos-related pleural disease on pulmonary function. *Scandinavian Journal of Work, Environment and Health*, **17**, 179–183.

Kumagai, S. and Kurumatani, N. (2009) Asbestos fiber concentration in the area surrounding a former asbestos cement plant and excess mesothelioma deaths in residents. *American Journal of Industrial Medicine*, **52**, 790–798.

Kurumatani, N. and Kumagai, S. (2008) Mapping the Risk of Mesothelioma due to Neighborhood Asbestos Exposure. *American Journal of Respiratory and Critical Care Medicine*, **178**, 624–629.

Labreche, F., Case, B.W., Ostiguy, G., Chalaoui, J., Camus, M. and Siemiatycki, J. (2012) Pleural mesothelioma surveillance: validity of cases from a tumour registry. *Canadian Respiratory Journal*, **19**, 103–107.

Lacourt, A., Pintos, J., Lavoue, J., Richardson, L. and Siemiatycki, J. (2015) Lung cancer risk among workers in the construction industry: results from two case-control studies in Montreal. *BMC Public Health*, **15**, 941.

Langer, A.M., Nolan, R.P., Constantopoulos, S.H. and Moutsopoulos, H.M. (1987) Association of Metsovo lung and pleural mesothelioma with exposure to tremolite containing whitewash. *The Lancet*, **1**, 965–967.

Larson, D., Powers, A., Ambrosi, J.-P., Tanji, M., Napolitano, A., Flores, E.G., Baumann, F., Pellegrini, L., Jennings, C.J., Buck, B.J., McLaurin, B.T., Merkler, D., Robinson, C., Morris, P., Dogan, M., Umran, A.D., Pass, H.I., Pastorino, S., Carbone, M., and Yang H. (2016) Investigating palygorskite's role in the development of mesothelioma in southern Nevada: Insights into fiber-induced carcinogenicity. *Journal of Toxicology and Environmental Health Part B*, **19**, 213–230.

Leigh, J., Davidson, P., Hendrie, L. and Berry, D. (2002) Malignant mesothelioma in Australia, 1945–2000. *American Journal of Industrial Medicine*, **41**, 188–201.

Lenters, V., Vermeulen, R., Dogger, S., Stayner, L., Portengen, L., Burdorf, A. and Heederik, D. (2011) A meta-analysis of asbestos and lung cancer: Is better quality exposure assessment associated with steeper slopes of the exposure-response relationships? *Environmental Health Perspectives*, **119**, 1547–1555.

Lenters, V., Burdorf, A., Vermeulen, R., Stayner, L. and Heederik, D. (2012) Quality of evidence must guide risk assessment of asbestos. *Annals of Occupational Hygiene*, **56**, 879–887.

Liddell, F.D. and Morgan, W.K. (1978) Methods of assessing serial films of the pneumoconioses: a review. *Journal of the Society of Occupational Medicine*, **28**, 6–15.

Liddell, D. and Armstrong, B. (2002) The combination of effects on lung cancer of cigarette smoking and exposure in Quebec chrysotile miners and millers. *Annals of Occupational Hygiene*, **46**, 5–13.

Liddell, F.D., McDonald, A.D. and McDonald, J.C. (1997) The 1891–1920 birth cohort of Quebec chrysotile miners and millers: development from 1904 and mortality to 1992. *Annals of Occupational Hygiene*, **41**, 13–36.

Lilis, R., Miller, A., Godbold J., Chan, E. and Selikoff, I.J. (1991) Pulmonary function and pleural fibrosis: quantitative relationships with an integrative index of pleural abnormalities. *American Journal of Industrial Medicine*, **20**, 145–161.

Lippmann, M. (1988) Asbestos exposure indices. *Environmental Research*, **46**, 86–106.

Lippmann, M. (2014) Toxicological and epidemiological studies on effects of airborne fibers: coherence and pubic health implications. *Critical Reviews in Toxicology*, **44**, 643–695.

Loomis, D., Dement, J.M., Wolf, S.H. and Richardson, D.B. (2009) Lung cancer mortality and fibre exposures among North Carolina asbestos textile workers. *Occupational and Environmental Medicine*, **66**, 535–542.

Loomis, D., Dement, J., Richardson, D. and Wolf, S. (2010) Asbestos fiber dimensions and lung cancer mortality among workers exposed to chrysotile. *Occupational and Environmental Medicine*, **67**, 580–584.

Loomis, D., Dement, J.M., Elliott, L., Richardson, D., Kuempel, E.D. and Stayner, L. (2012) Increased lung cancer mortality among chrysotile asbestos textile workers is more strongly associated with exposure to long thin fibres. *Occupational and Environmental Medicine*, **69**, 564–568.

Luce, D., Billon-Galland, M.A., Bugel, I., Goldberg, P., Salomon, C., Févotte, J. and Goldberg, M. (2004)

Assessment of environmental and domestic exposure to tremolite in New Caledonia. *Archives of Environmental Health*, **59**, 91–100.

Luo, S., Liu, X., Mu, S., Tsai, S.P. and Wen, C.P. (2003) Asbestos related diseases from environmental exposure to crocidolite in Da-yao, China. I. Review of exposure and epidemiological data. *Occupational and Environmental Medicine*, **60**, 35–41.

Lynch, K.M. and Smith, W.A. (1935) Pulmonary asbestosis III: Carcinoma of the lung in asbesto-silicosis. *American Journal of Cancer Research*, **24**, 56–64.

Magnani, C., Dalmasso, P., Biggeri, A., Ivaldi, C., Mirabelli, D. and Terracini, B. (2001) Increased risk of malignant mesothelioma of the pleura after residential or domestic exposure to asbestos: a case-control study in Casale Monferrato, Italy. *Environmental Health Perspectives*, **109**, 915–919.

Mancuso, T.F. and Coulter, E.J. (1963) Methodology in industrial health studies: The cohort approach with special reference to an asbestos company. *Archives of Environmental Health*, **6**, 210–226.

Marinaccio, A., Binazzi, A., Di Marzio, D., Scarselli, A., Verardo, M., Mirabelli, D., Gennaro, V., Mensi, C., Riboldi, L., Merler, E., Zotti, R.D., Romanelli, A., Chellini, E., Silvestri, S., Pascucci, C., Romeo, E., Menegozzo, S., Musti, M., Cavone, D., Cauzillo, G., Tumino, R., Nicita, C., Melis, M. and Iavicoli, S. (2012) Pleural malignant mesothelioma epidemic. Incidence, modalities of asbestos exposure and occupations involved from the Italian national register. *International Journal of Cancer*, **130**, 2146–2154.

Marinaccio, A., Binazzi, A., Bonafede, M., Corfiati, M., Di Marzio, D., Scarselli, A., Verardo, M., Mirabelli, D., Gennaro, V., Mensi, C., Schallemberg, G., Merler, E., Negro, C., Romanelli, A., Chellini, E., Silvestri, S., Cocchioni, M., Pascucci, C., Stracci, F., Ascoli, V., Trafficante, L., Angelillo, I., Musti, M., Cavone, D., Cauzillo, G., Tallarigo, F., Tumino, R. and Melis, M. (2015) Malignant mesothelioma due to non-occupational asbestos exposure from Italian national surveillance system (ReNaM): epidemiology and public health issues. *Occupational and Environmental Medicine*, **72**, 648–655.

Markowitz, S.B., Levin, S.M., Miller, A. and Morabia, A. (2013) Asbestos, asbestosis, smoking, and lung cancer. New findings from the North American insulator cohort. *American Journal of Respiratory and Critical Care Medicine*, **188**, 90–96.

Maule, M.M., Magnani, C., Dalmasso, P., Mirabelli, D., Merletti, F. and Biggeri, A. (2007) Modeling mesothelioma risk associated with environmental asbestos exposure. *Environmental Health Perspectives*, **115**, 1066–1071.

McCaughey, W.T. (1965) Asbestos and neoplasia: diffuse mesothelial tumors. Criteria for diagnosis of diffuse mesothelial tumors. *Annals of the New York Academy of Sciences*, **132**, 603–613.

McConnochie, K., Simonato, L., Mavrides, P., Christofides, P., Pooley, F.D. and Wagner, J.C. (1987) Mesothelioma in Cyprus: The role of tremolite. *Thorax*, **42**, 342–347.

McDonald, J.C. (1998) Unfinished business: the asbestos textiles mystery [editorial]. *Annals of Occupational Hygiene*, **42**, 3–5.

McDonald, C. (2000a) Asbestos. Pp. 85–108 in *Epidemiology of Work-related Diseases* (C. McDonald, editor). BMJ Press, London.

McDonald, C. (2000b) Introduction: Occupational epidemiology. Pp. 1–6 in: *Epidemiology of Work-related Diseases* (C. McDonald, editor). BMJ Press, London.

McDonald, A.D. and McDonald, J.C. (1980) Malignant mesothelioma in North America. *Cancer*, **46**, 1650–1656.

McDonald, A.D., Harper, A., McDonald, J.C. and el-Attar, O.A. (1970) Epidemiology of primary malignant mesothelial tumors in Canada. *Cancer*, **26**, 914–919.

McDonald, J.C., McDonald, A.D., Gibbs, G.W., Siemiatycki, J. and Rossiter, C.E. (1971) Mortality in the chrysotile asbestos mines and mills of Quebec. *Archives of Environmental Health*, **22**, 677–686.

McDonald, J.C., Rossiter, C.E., Eyssen, G. and McDonald, A.D. (1973) Mortality in the chrysotile producing industry of Quebec: a progress report. Pp. 234–237 in *Proceedings of the IVth International Pneumoconiosis Conference*. Apimondia, International Labour Office, Ministerul Muncii, Bucharest.

McDonald, J.C., Liddell, F.D., Gibbs, G.W., Eyssen, G.E. and McDonald, A.D. (1980) Dust exposure and mortality in chrysotile mining, 1910-75. *British Journal of Industrial Medicine*, **37**, 11–24.

McDonald, A.D., Fry, J.S., Woolley, A.J. and McDonald, J. (1983) Dust exposure and mortality in an American chrysotile textile plant. *British Journal of Industrial Medicine*, **40**, 361–367.

McDonald, J.C., Armstrong, B., Case, B., Doell, D., McCaughey, W.T.E., McDonald, A.D. and Sebastien, P. (1989) Mesothelioma and asbestos fiber type. Evidence from lung tissue analyses. *Cancer*, **63**, 1544–1547.

McDonald, J.C., Case, B.W., Enterline, P.E., Henderson, V., McDonald, A.D., Plourde, M. and Sebastien, P. (1990) Lung dust analysis in the assessment of past exposure of man-made mineral fibre workers. *Annals of Occupational Hygiene*, **34**, 427–441.

McDonald, J.C., Liddell, F.D., Dufresne, A. and McDonald, A.D. (1993) The 1891–1920 birth cohort of Quebec chrysotile miners and millers: mortality 1976–88. *British Journal of Industrial Medicine*, **50**, 1073–1081.

McDonald, A.D., Case, B.W., Churg, A., Dufresne, A., Gibbs, G.W., Sébastien, P. and McDonald, J.C. (1997) Mesothelioma in Quebec chrysotile miners and millers: epidemiology and aetiology. *Annals of Occupational Hygiene*, **41**, 707–719.

McLeod, K.S. (2000) Our sense of Snow: the myth of John Snow in medical geography. *Social Science and Medicine*, **50**, 923–935.

Mensi, C., Riboldi, L., De Matteis, S., Bertazzi, P.A. and Consonni, D. (2015) Impact of an asbestos cement factory on mesothelioma incidence: global assessment of effects of occupational, familial, and environmental exposure. *Environment International*, **74**, 191–199.

Merewether, E. and Price, C. (1930) Report on effects of asbestos dust on the lungs and dust suppression in the asbestos industry. London, H.M.S.O., 31 pp.

Miller, L., Smith, W.E. and Berliner, S.W. (1965) Tests for effect of asbestos on benzo[a]pyrene carcinogenesis in the respiratory tract. *Annals of the New York Academy of Sciences*, **132**, 489–500.

Mirabelli, D., Calisti, R., Barone-Adesi, F., Fornero, E., Merletti, F. and Magnani, C. (2008) Excess of mesotheliomas after exposure to chrysotile in Balangero, Italy. *Occupational and Environmental Medicine*, **65**, 815–819.

Moolgavkar, S.H., Anderson, E.L., Chang, E.T., Lau, E.C., Turnham, P. and Hoel, D.G. (2014) A review and critique of U.S. EPA's risk assessments for asbestos. *Critical Reviews in Toxicology*, **44**, 499–522.

Musti, M., Pollice, A., Cavone, D., Dragonieri, S. and Bilancia, M. (2009) The relationship between malignant mesothelioma and an asbestos cement plant environmental risk: a spatial case-control study in the city of Bari (Italy). *International Archives of Occupational and Environmental Health*, **82**, 489–497.

Nagelschmidt, G. (1965) Some observations of the dust content and composition in lungs with asbestosis, made during work on coal miners pneumoconiosis. *Annals of the New York Academy of Sciences*, **132**, 64–76.

Naik, S.L., Lewin, M., Young, R., Dearwent, S.M. and Lee, R. (2016) Mortality from asbestos-associated disease in Libby, Montana 1979–2011. *Journal of Exposure Science and Environmental Epidemiology*, 1–7.

NIOSH (2011) Asbestos fibres and other elongate mineral particles: State of the science and roadmap for research. Revised Edition. NIOSH Current Intelligence Bulletin 62, DHHS (NIOSH) Publication No. 2011–159. https://www.cdc.gov/niosh/docs/2011-159/pdfs/2011-159.pdf, accessed September 24, 2016.

NIOSH (2014) Work-related lung disease surveillance system (eWoRLD). 2014-806 U.S. Department of Health and Human Services, Centers for Disease Control and Prevention, National Institute for Occupational Safety and Health, Respiratory Health Division, Morgantown, WV. Available at: <https://wwwn.cdc.gov/eworld/Data/806> accessed September 21, 2016.

Newhouse, M.L. and Thompson, H. (1965) Mesothelioma of pleura and peritoneum following exposure to asbestos in the London area. *British Journal of Industrial Medicine*, **22**, 261–269.

Nicholson, W.J. (1986) Airborne asbestos health assessment update, report EPA-600/8-84-003F. Washington, D.C., Office of Health and Environmental Assessment, US Environmental Protection Agency, pp. 256, Research Triangle Park, North Carolina, USA.

Oliver, L.C., Welch, L.S. and Harbut, M.R. (2004) Comparison of B readers' interpretations of chest radiographs for asbestos related changes. *Academic Radiology*, **11**, 1397–1399; author reply 1402–1394.

Pan, X.L., Day, H.W., Wang, W., Beckett, L.A. and Schenker, M.B. (2005) Residential proximity to naturally occurring asbestos and mesothelioma risk in California. *American Journal of Respiratory and Critical Care Medicine*, **172**, 1019–1025.

Peters, S., Vermeulen, R., Portengen, L., Olsson, A., Kendzia, B., Vincent, R., Savary, B., Lavoue, J., Cavallo, D., Cattaneo, A., Mirabelli, D., Plato, N., Fevotte, J., Pesch, B., Bruning, T., Straif, K. and Kromhout, H. (2016) SYN-JEM: A Quantitative job-exposure matrix for five lung carcinogens. *Annals of Occupational Hygiene*, **60**, 795–811.

Peto, J., Seidman, H. and Selikoff, I.J. (1982) Mesothelioma mortality in asbestos workers: implications for models of carcinogenesis and risk assessment. *British Journal of Cancer*, **45**, 124–135.

Pintos, J., Parent, M.E., Rousseau, M.C., Case, B.W. and Siemiatycki, J. (2008) Occupational exposure to asbestos and man-made vitreous fibers, and risk of lung cancer: evidence from two case-control studies in Montreal, Canada. *Journal of Occupational and Environmental Medicine*, **50**, 1273–1281.

Proctor, R. (1999) *The Nazi War on Cancer*. Princeton University Press, Princeton, New Jersey, USA, 392 pp.

Rake, C., Gilham, C., Hatch, J., Darnton, A., Hodgson, J. and Peto, J. (2009) Occupational, domestic and environmental mesothelioma risks in the British population: a case-control study. *British Journal of Cancer*, **100**, 1175–1183.

Reid, A., Berry, G., de Klerk, N., Hansen, J., Heyworth, J., Ambrosini, G., Fritschi, L., Olsen, N., Merler, E. and Musk, A.W. (2007) Age and sex differences in malignant mesothelioma after residential exposure to blue asbestos (crocidolite). *Chest*, **131**, 376–382.

Reid, A., Heyworth, J., de Klerk, N.H. and Musk, B. (2008) Cancer incidence among women and girls environmentally and occupationally exposed to blue asbestos at Wittenoom, Western Australia. *International Journal of Cancer*, **122**, 2337–2344.

Roggli, V.L. (2015) The so-called short-fiber controversy: Literature review and critical analysis. *Archives of Pathology and Laboratory Medicine*, **139**, 1052–1057.

Roggli, V.L., Gibbs, A.R., Attanoos, R., Churg, A., Popper, H., Cagle, P., Corrin, B., Franks, T.J., Galateau-Salle, F., Galvin, J., Hasleton, P.S., Henderson, D.W. and Honma, K. (2010) Pathology of asbestosis – An update of the diagnostic criteria: Report of the asbestosis committee of the college of american pathologists and pulmonary pathology society. *Archives of Pathology and Laboratory Medicine*, **134**, 462–480.

Sakellariou, K., Malamou-Mitsi, V., Haritou, A., Koumpaniou, C., Stachouli, C., Dimoliatis, I.D. and Constantopoulos, S.H. (1996) Malignant pleural mesothelioma from non occupational asbestos exposure in Mestovo (north-west Greece): slow end of an epidemic? *European Respiratory Journal*, **9**, 1206–1210.

San Francisco Public Utilities Commission. (2015) Calaveras Dam Replacement Project Air Monitoring Results. Calaveras Dam Replacement Project: Naturally Occurring Asbestos. https://sfwater.org/index.aspx?page=530 accessed September 26, 2016.

Schwartz, D.A., Fuortes, L.J., Galvin, J.R., Burmeister, L.F., Schmidt, L.E., Leistikow, B.N., LaMarte, F.P. and Merchant, J.A. (1990) Asbestos-induced pleural fibrosis and impaired lung function. *The American Review of Respiratory Disease*, **141**, 321–326.

Sébastien, P., Awad, L., Bignon, J., Petit, G. and Barris, Y.I. (1984) Ferruginous bodies in sputum as an indication of exposure to airborne mineral fibers in the mesothelioma villages of Cappadocia. *Archives of Environmental Health: An International Journal*, **39**, 18–23.

Sébastien, P., Armstrong, B., Case, B., Barwick, H., Keskula, H. and McDonald, J.C. (1988) Estimation of amphibole exposure from asbestos body and macrophage counts in sputum: a survey in vermiculite miners. *Annals of Occupational Hygiene*, **32**, 195–201.

Sébastien, P., McDonald, J.C., McDonald, A.D., Case, B. and Harley, R. (1989) Respiratory cancer in chrysotile textile and mining industries: exposure inferences from lung analysis. *British Journal of Industrial Medicine*, **46**, 180–187.

Selikoff, I.J. (1965) Opening remarks. *Annals of the New York Academy of Sciences*, **132**, 7–8.

Selikoff, I.J. and Churg, J. (editors) (1965) Biological effects of asbestos. *Annals of the New York Academy of Sciences*, **132**, 705 pp.

Selikoff, I.J., Churg, J. and Hammond, E.C. (1964) Asbestos exposure and neoplasia. *Journal of the American Medical Association*, **188**, 22–26.

Selikoff, I.J., Churg, J. and Hammond, E.C. (1965) The occurrence of asbestosis among insulation workers in the United States. *Annals of the New York Academy of Sciences*, **132**, 139–155.

Siemiatycki, J. (1990) Discovering occupational carcinogens in population-based case-control studies: review of findings from an exposure-based approach and a methodologic comparison of alternative data collection strategies. *Recent Results in Cancer Research*, **120**, 25–38.

Siemiatycki, J., Day, N.E., Fabry, J. and Cooper, J.A. (1981) Discovering carcinogens in the occupational environment: a novel epidemiologic approach. *Journal of the National Cancer Institute*, **66**, 217–225.

Simonato, L., Baris, R., Saracci, R., Skidmore, J. and Winkelmann, R. (1989) Relation of environmental exposure to erionite fibres to risk of respiratory cancer. *IARC Scientific Publications*, **90**, 398–405.

Smith, W.E. (1952) Survey of some current British and European studies of occupational tumor problems. *AMA Archives of Industrial Hygiene and Occupational Medicine*, **5**, 242–63.

Snow, J. (1855) *On the Mode of Communication of Cholera*. John Churchill, London, 162 pp.

Stayner, L., Kuempel, E., Gilbert, S., Hein, M. and Dement, J. (2008) An epidemiological study of the role of chrysotile asbestos fibre dimensions in determining respiratory disease risk in exposed workers. *Occupational and Environmental Medicine*, **65**, 613–619.

Stone, R. (1991) No meeting of the minds on asbestos. *Science*, **254**, 928–931.

Stroup, D.F., Berlin, J.A., Morton, S.C., Olkin, I., Williamson, G.D., Rennie, D., Moher, D., Becker, B.J., Sipe, T.A. and Thacker, S.B. (2000) Meta-analysis of observational studies in epidemiology: a proposal for reporting. Meta-analysis of Observational Studies in Epidemiology (MOOSE) group. *JAMA*, **283**, 2008–2012.

Suzuki, Y. and Yuen, S.R. (2002) Asbestos fibers contributing to the induction of human malignant mesothelioma. *Annals of the New York Academy of Sciences*, **982**, 160–176.

Suzuki, Y., Yuen, S.R. and Ashley, R. (2005) Short, thin asbestos fibers contribute to the development of human malignant mesothelioma: pathological evidence. *International Journal of Hygiene and Environmental Health*, **208**, 201–210.

Takahashi, K., Case, B.W., Dufresne, A., Fraser, R., Higashi, T. and Siemiatyck, J. (1994) Relation between lung asbestos fibre burden and exposure indices based on job history. *Occupational and Environmental Medicine*, **51**, 461–469.

Tarrés, J., Albertí, C., Martínez-Artés, X., Abós-Herràndiz, R., Rosell-Murphy, M., García-Allas, I., Krier, I., Cantarell, G., Gallego, M., Canela-Soler, J. and Orriols, R. (2013) Pleural mesothelioma in relation to meteorological conditions and residential distance from an industrial source of asbestos. *Occupational and Environmental Medicine*, **70**, 588–790.

Thomson, J.G. (1965) Asbestos and the urban dweller. *Annals of the New York Academy of Sciences*, **132**, 196–214.

US EPA (1988) National Center for Environmental Assessment Integrated Risk Information System (IRIS): Asbestos; CASRN 1332–21-4.

US EPA (2015) Fact sheet and frequent questions for the Libby Amphibole asbestos assessment: Libby amphibole asbestos frequently asked questions. Pp. 1–3. https://cfpub.epa.gov/ncea/iris/iris_documents/documents/supdocs/LAA_Frequent_Questions.pdf Last updated on September 30, 2015. Accessed September 23, 2016.

Wagner, J.C. (1965) Epidemiology of diffuse mesothelial tumors: evidence of an association from studies in South Africa and the United Kingdom. *Annals of the New York Academy of Sciences*, **132**, 575–578.

Wagner, J.C., Sleggs, C.A. and Marchand, P. (1960) Diffuse pleural mesothelioma and asbestos exposure in the North Western Cape Province. *British Journal of Industrial Medicine*, **17**, 260–271.

Webster, I. (1965) Mesotheliomatous tumors in South Africa: pathology and experimental pathology. *Annals of the New York Academy of Sciences*, **132**, 623–646.

Wedler, H.W. (1943) Lung cancer in asbestosis patients [German Uber den Lungenkrebs bei Asbestos]. *Deutsch Archive Klinik Medizin*, **191**, 189–209.

Whitehouse, A.C., Black, C.B., Heppe, M.S., Ruckdeschel, J. and Levin, S.M. (2008) Environmental exposure to Libby Asbestos and mesotheliomas. *American Journal of Industrial Medicine*, **51**, 877–880.

Wood, W.B. and Gloyne, S.R. (1934) Pulmonary asbestos: A review of one hundred cases. *Lancet*, **2**, 1383.

Wylie, A.G. and Candela, P.A. (2015) Methodologies for determining the sources, characteristics, distribution, and abundance of asbestiform and nonasbestiform amphibole and serpentine in ambient air and water. *Journal of Toxicology and Environmental Health. Part B, Critical Reviews*, **18**, 1–42.

EMU Notes in Mineralogy, Vol. 18 (2017), Chapter 12, 417–434

Differential pathological response and pleural transport of mineral fibres

DAVID M. BERNSTEIN[1] and ELIZABETH N. PAVLISKO[2]

[1]*Consultant in Toxicology, 40 chemin de la Petite-Boissière, 1208 Geneva, Switzerland, e-mail: davidb@itox.ch*
[2]*Duke University Medical Center, 40 Duke Medicine Circle, Durham, North Carolina 27710, USA*

The evaluation of mineral fibres within the pleural cavity and the possible subsequent pathological response in this region have the potential to provide important kinetic and pathological information. The pleural cavity, however, is a very dynamic environment with highly regulated pressures and fluid flows between the lung, the pleural space, and the chest wall. The difficulty of sampling the pleural cavity for fibre content stems from the dynamic conditions which maintain the physiological balance in the pleural cavity which changes dramatically *post mortem*.

1. Pleural anatomy

The visceral pleura in humans is ~20 to 80 μm thick and consists of a mesothelial layer and a sub-pleural connective tissue layer. The alveoli and the pulmonary circulation lie beneath the visceral pleural membrane. Both the bronchial microvessels and lymphatics are farther from the pleural space *vs.* the parietal pleura (20 to 50 μm *vs.* 10 to 12 μm, respectively) (Albertine *et al.*, 1982; Wang, 1998). The visceral pleura is very rich in lymphatic vessels, with an intercommunicating "network" arranged over the lung surface and penetrating into the lung parenchyma *via* the secondary interlobular septa to join the bronchial lymph vessels with drainage to the various hilar nodes. The larger lymphatic vessels in the visceral pleura have one-way valves which also direct flow towards the hilar regions of the lung (Light and Gary Lee, 2008).

As described by Lai-Fook (2004) and Zocchi (2002) and as illustrated in Fig. 1 (reproduced from Zocchi, 2002) pleural fluid is infiltrated mainly from capillaries in the parietal pleura lining the chest wall. Fluid filters into the pleural space driven by Starling forces through the parietal pleura. The pressure differential across the visceral pleura is in favour of fluid absorption (Agostoni, 1972; Agostoni and Zocchi, 1991, 1998; Negrini, 1995; Miserocchi and Negrini, 1997).

For several reasons, the systemic blood supply of the parietal pleura is thought to be the major source of normal pleural fluid. First, despite the great differences in visceral pleural anatomy and blood supply, measured rates of pleural fluid production are similar among different species, suggesting parietal pleura as the constant source

DOI: 10.1180/EMU-notes.18.12

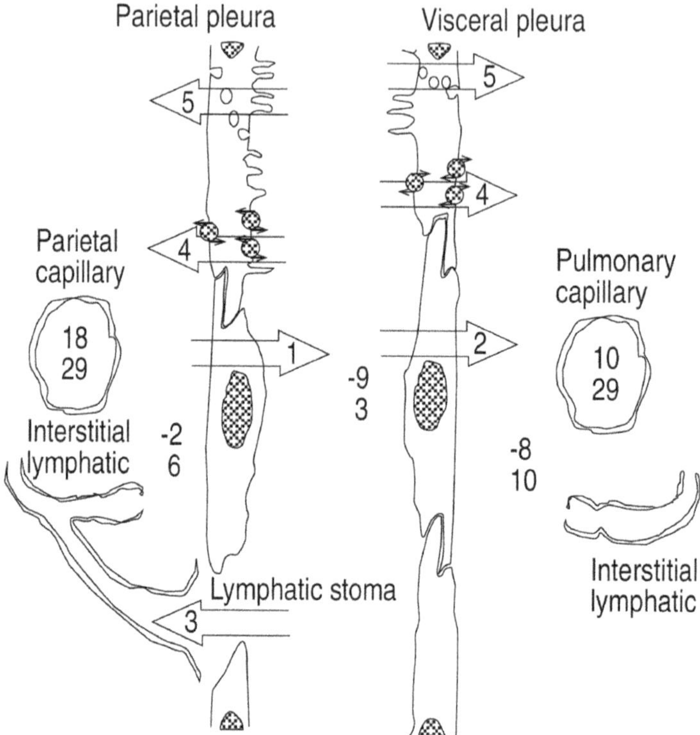

Figure 1. Schematic representation of transpleural liquid flows (large arrows) under physiological conditions in rabbit. Mesothelial cells and adjoining interstitia, with embedded capillaries and lymphatics, are shown on each side; the mesothelial cell at the top displays cellular mechanisms involved in pleural liquid turnover (microvilli, vesicles, electrolyte transporters (grey circles)). Arrow size is not proportional to corresponding flow magnitude. (1) Starling filtration at the parietal pleura (tentative estimate:~0.15–0.20 mL h^{-1} kg^{-1}); (2) starling absorption at the visceral pleura (tentative estimate: ~0.1 mL h^{-1} kg^{-1}); (3) direct drainage through lymphatic stomas (not known; tentative estimate of total pleural lymphatic flow (through stoma interstitial): ~0.07 mL h^{-1} kg^{-1}); (4) electrolyte-coupled liquid outflow (measured (Agostoni *et al.*, 1993): 0.07 mL h^{-1} kg^{-1}); (5) vesicular flow of liquid accompanying protein transcytosis (not known; recent estimates have shown it to be 0.02 mL h^{-1} kg^{-1}). Pairs of large numbers are hydraulic (above) and colloidosmotic (below) pressures (cmH_2O) involved in Starling balance, pertaining to (from left to right): parietal capillary, parietal interstitium, pleural liquid, visceral interstitium, visceral capillary. Hydraulic pressures refer to heart level; they are functional mean values for capillaries, and averaged value over a breath for pleural liquid, while interstitial values are indirect and uncertain. Interstitial colloidosmotic pressures are uncertain. See text for details, reasons for uncertainty and cautionary comments (Figure and caption reproduced from Zocchi, 2002 with the permission of the European Respiratory Society).

(Wiener-Kronish *et al.*, 1984; Broaddus and Araya, 1992). Second, the parietal pleural microvasculature is closer to the pleural space (10–15 mm) than the microvasculature of the visceral pleura (20–50 mm) (Albertine *et al.*, 1982, 1984). Finally, the parietal pleural vessels are likely to have a higher microvascular pressure due to their drainage into systemic venules while the visceral bronchial vessels drain into lower-resistance pulmonary venules (Staub *et al.*, 1985).

Normal pleural fluid is a microvascular filtrate; its volume and composition are controlled tightly. Fluid enters the pleural space through the parietal pleura down a net filtering pressure gradient, and is removed by an absorptive pressure gradient through the visceral pleura, by lymphatic drainage through parietal pleura stomas and by cellular mechanisms (active transport of solutes by mesothelial cells) (Staub *et al.*, 1985; Broaddus and Light, 2000). The transpleural pressures and one-way lymphatic valves which direct flow towards the hilar regions of the lung are such that short, thin fibres which may be present in the lymphatic system in the lung would not drain into the pleural space.

Assessing fibre translocation into the pleura accurately is challenging as the fluid dynamics of the respiratory system change dramatically *post mortem*. As summarized by Zocchi (2002), direct *in vitro* measurements of the biophysical properties of the pleura appear unreliable because the mesothelium is an active tissue that reacts to injury (Fentie *et al.*, 1986; Peng *et al.*, 1994; Agostoni, 1998; Agostoni and Zocchi, 1998), and the procedures required to obtain the specimens are likely to damage it (Agostoni, 1998). Studies performed on pleura specimens stripped from the underlying tissues (Kim *et al.*, 1979; Payne *et al.*, 1988) provided extremely high values of permeability to water and solutes, thought to be secondary to the experimental conditions (Zocchi *et al.*, 1991, 1992, 1998; Agostoni and Zocchi, 1993; Negrini *et al.*, 1994; Negrini, 1995; Miserocchi and Negrini, 1997; Agostoni *et al.*, 1999; Bodega *et al.*, 2000). Additionally, the pressure differentials described above cease to exist *post mortem*. The visceral pleura is permeable and, as illustrated in Fig. 2, underlying the mesothelium is a sub-pleural layer containing blood and lymph vasculature (Bernaudin and Fleury, 1985). Although the larger lymphatic vessels of the visceral pleura are equipped with one-way valves, the lymphatic flow in smaller vessels is governed by pressure gradients which are no longer present *post mortem*. In addition, the visceral pleura in humans is ~20 to 80 μm thick, and can be breached easily during intervention at autopsy.

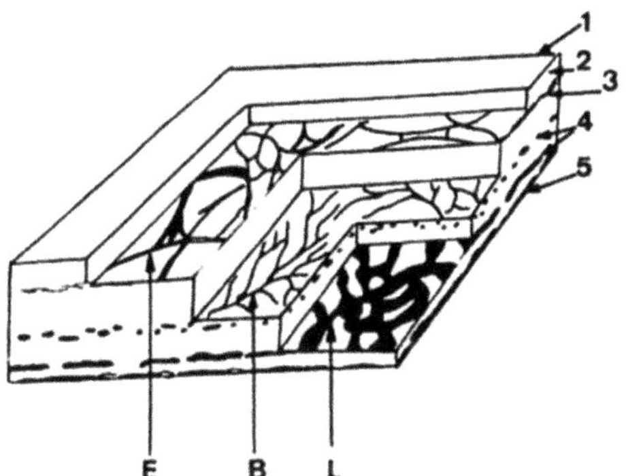

Figure 2. Schematic drawing of the human visceral pleura according to Naigaishi (1972). The different constitutive layers are: (1) mesothelium; (2) sub mesothelial tissue; (3) external elastic layer with elastic fibres (E); (4) a subpleural layer containing the blood (B) and lymph (L) vessels; and (5) internal elastic layer. The visceral pleural is permeable and, as illustrated, underlying the mesothelium is a sub-pleural layer containing the blood and lymph vessels (figure and caption reproduced with the permission of the Taylor and Francis Group from Bernaudin and Fleury, 1985).

2. The difficulty in sampling the pleural space

The delicate balance that is maintained within the pleural cavity *in vivo* can be perturbed easily through either surgical manipulation or *post mortem* examination.

The visceral pleural surface of the lung consists of a collagen layer populated on the pleural side with mesothelial cells. Directly below the visceral pleura on the side of the lung is a dense network of lymphatic channels termed 'subpleural lymphatic plexus' (Agur and Dalley, 2013). These lymphatic channels have one-way valves that ensure that the lymph flow is away from the visceral pleural surface. The lymphatic channels are one of the routes by which shorter fibres are removed from the lung surface (Bernstein *et al.*, 2004). Because of their size, the longer fibres are not able to enter the lymphatic channels. Thus, with this very dense network of lymphatic channels below the pleural visceral surface which would contain only shorter fibres, if, in manipulation or sampling or *post mortem*, the visceral surface is breached, then there would be the possibility for extensive infiltration of the shorter fibres into the pleural space thereby providing an incorrect image of what is present *in vivo*.

Lymphatic vessels of the lung originate in the subpleural (superficial) and deep lymphatic plexuses.

- The subpleural lymphatic plexus is superficial, lying deep to the visceral pleura, and drains lymph from the surface of the lung to the bronchopulmonary (hilar) nodes;
- The deep lymphatic plexus is in the lung and follows the bronchi and pulmonary vessels to the pulmonary, and subsequently bronchopulmonary, lymph nodes located at the hilum of the lung;
- All lymph from the lungs enters the inferior (carinal) and superior tracheobronchial nodes and then continues to the right and left bronchomediastinal trunks which drain into the venous system *via* the right lymphatic and thoracic ducts; lymph from the left inferior lobe passes largely to the right side (Agur and Dalley, 2013).

3. Historical studies

Moalli *et al.* (1987) performed studies in mice where asbestos fibres and other mineral particles were injected into the peritoneum. Observations of the mesothelial-lined diaphragms took place following sacrifice of the mice. While this study has some flaws, the results correlate quite well with more recent studies designed to avoid artefacts. Moalli *et al.* (1987) observed that short fibres were cleared by lymphatics whereas long fibres remained trapped and an intense inflammatory response, dominated by macrophages, resulted. This inflammatory response was not seen in their short-fibre only model. They also observed reactive change in the mesothelial cells themselves.

Choe *et al.* (1997) investigated macrophage recruitment and activation in the setting of asbestos-induced injury to the pleura. In their model, they subjected rats to intermittent inhalation of crocidolite or chrysotile and compared them to a control group exposed to filtered room air. They observed increased pleural macrophages in the groups exposed to asbestos as well as greater amounts of nitric oxide and TNα. This study was one of the first to confirm translocation of inhaled asbestos fibres to the pleural cavity.

Prior studies in animals and humans have reported the translocation of short fibres to the pleural cavity. Gelzleichter *et al.* (1996) assessed lung and pleural fibre burdens in rats following short-term inhalation exposure (6 h/day for 5 days) to ceramic fibres (RCF1) at an exposure concentration of 6206 total fibres/cm^3, 2645 WHO fibres/cm^3. Following sacrifice, the abdominal cavity was exposed and agarose/SDS (sodium dodecyl sulfate) was injected into the pleural cavity with a 16-gauge needle inserted through the diaphragm adjacent to the sternum. After removal from the thoracic cavity, the lung and pleural cast were separated carefully and frozen to −20°C for subsequent recovery of fibres. The authors reported that "there appeared to be a strict size limitation, whereby only those fibres <5 μm long (and <0.35 μm in diameter) were capable of migration to the pleural compartment within the time frame of these experiments". Later, Bermudez *et al.* (2003) reported on a similar study evaluating MMVF10a glass fibres in which agarose/SDS (sodium dodecyl sulfate) was injected into the pleural cavity to determine pleural fibre burdens. Animals were exposed for 4 h on 5 consecutive days per week for a maximum of 12 consecutive weeks. Air-control and fibre-exposed groups of animals were euthanized immediately following 4 or 12 weeks of exposure or following a recovery period of 12 weeks. The authors reported that "the total number of fibres in the pleural compartment was at least a thousand times lower than in the lung. A shift in the distribution of fibre length toward longer fibres was noted in the pleural compartment, relative to the lung, such that the majority of fibres were of lengths greater than 5 μm." Suzuki and Yuen (2001, 2002) and Suzuki *et al.* (2005) reported on the analysis of lung and mesothelial tissues from human mesothelioma patients. They sent autopsy or biopsy tissue samples to their principal author for pathologic review over a period of 15 years. Tissue was then archived at the Mount Sinai School of Medicine. They concluded that as the majority of fibres found in their analyses were short and thin, that "short, thin asbestos fibres appear to contribute to the causation of human malignant mesothelioma". They also stated that "chrysotile was the most common asbestos type to be categorized as short, thin asbestos fibres." The difficulty with such analyses, however, is that in any autopsy or surgical intervention, the physiological balance has long since been disrupted. The autopsy will likely mix the fluids of the lung and lymphatics into the pleural space, and in biopsy samples there is a high likelihood of sampling at one of the bronchial bifurcations or at the alveolar duct area. Both of these locations are sites for bronchial-associated lymphoid tissue which can accumulate those fibres absorbed into the lymphatics. In addition, as fibre lengths follow log-normal distributions, the majority of fibres present in the lung will be short. Additional points of concern with these studies, as pointed out by Roggli (2015), include: (1) lack of proper controls; (2) possible contamination; (3) use of unconventional methodology; (4) the improper logic that pathology is related to thin, short fibres because they are so abundant. As discussed above, such thin and short fibres are readily cleared through the lymphatic system.

The visceral pleural thickness in the rat is estimated to be ~5–10 μm (Payne *et al.*, 1988; Lai-Fook, 2004). If the injected agarose/SDS cast adhered to even a few μm of the visceral pleura, there is a high probability that upon separation of the pleural cast from the lung there would be release of lymphatic fluid in the subpleural lymphatic plexus. In such an event, fibres small enough to be cleared by the lymphatics (<5 μm in length and thinner than 0.35 μm in diameter) would be released. It is still uncertain whether the findings reported above by Gelzleichter *et al.* (1996) and Bermudez *et al.* (2003) reflect the fibres in the subpleural lymphatic plexus or the pleural cavity.

Like lymphatics in other organ systems, the pulmonary lymphatic vessels provide a unidirectional drainage system that is designed to maintain homeostasis of the interstitial areas by transporting excess tissue fluids (lymph) back into the bloodstream. This continuous drainage prevents accumulation of fluids and proteins that normally leak across the alveolar capillary membrane (Mayerson, 1963). Thus, the continuous outward flow of fluids from alveolar capillaries is toward the interstitium and into the pulmonary lymphatics (Guyton *et al.*, 1971, 1976; Staub, 1974a,b; Leak and Jamuar, 1983).

4. Non-invasive evaluation of pleura fibre translocation and response

Recent work has sought to design procedures for evaluation of the pleural space while avoiding as much as possible the types of artefacts mentioned above. This was especially important as rat visceral pleura is even thinner than that of humans. Bernstein *et al.* (2010, 2011, 2014, 2015) developed two independent methods for examining the translocation of fibres to the pleural cavity and any associated inflammatory response following exposure to either chrysotile and sanded material or to amosite asbestos. These methods included the examination of the diaphragm as a parietal pleural tissue and the *in situ* examination of the lungs and pleural space in deep frozen rats.

The parietal pleura covers the chest wall, diaphragm and mediastinum. The most interesting and unusual features of the parietal pleura are the lymphatic stomata, holes of 2 to 6 μm in diameter that open onto the pleural space. Each stoma is formed by a gap in the otherwise continuous mesothelial cell layer, where the mesothelial cells join with the endothelial cells of the lymphatics. Each lymphatic joins others, forming a lake or lacuna; from the lacunae, collecting lymphatics join intercostal trunk lymphatics, which travel to the parasternal and periaortic lymph nodes (Negrini and Moriondo, 2013). The diaphragm was chosen as a representative parietal pleural tissue because at necropsy, it could be removed carefully within minutes of sacrifice without disturbing or altering the visceral lung surface. The area of the diaphragm chosen for examination included an important lymphatic drainage site (stomata) on the diaphragmatic surface. The use of both confocal microscopy and scanning electron microscopy enables the identification of fibres as well as the examination for possible inflammatory response. Examination *in situ* of the pleural space including the lung, visceral pleura, parietal pleura, and intercostal muscles in rats deep frozen immediately after termination provides a non-invasive method for determining fibre location and inflammatory response. The confocal imaging is especially useful in that it permits the quantification of not only the number of fibres present but also the amount of collagen in the same region (Rogers *et al.*, 1999). As fibres inhaled into the lung have to traverse the visceral pleural wall to enter into the pleural cavity, the number of fibres present near the visceral pleural wall provides an indication of the fibres that can potentially enter the pleural cavity. In addition, the cellular and collagen response in the same region provides a direct measure of the pathogenic response to these fibres.

4.1. Fibre transport to the parietal pleura

In a recent study, Bernstein *et al.* (2014, 2015) reported on the effects following short-term exposure (5 day, 6 h/day) of brake dust from automobile brakes manufactured with chrysotile as one component both alone and mixed with added chrysotile to increase the chrysotile fibre concentration. This was compared to crocidolite asbestos exposed at a comparable long-fibre (>20 μm) exposure concentration. The crocidolite used in this study was unique in that it was a commercial sample obtained from the mine and was not ground in advance as were most other samples used in toxicology studies (*e.g.* Union for International Cancer Control (UICC) or US National Institute of Environmental Health Sciences (NIEHS) crocidolite: Bernstein *et al.*, 2014). The results indicate the very rapid translocation of longer amphibole fibres to the pleural cavity with crocidolite fibres observed on the diaphragm within hours following cessation of the 5-day exposure. As illustrated in Fig. 3 (reproduced from Negrini and Moriondo, 2013) and Fig. 4 (reproduced from Bernstein *et al.*, 2015), these longer fibres are seen embedded in the stomata on the diaphragm. By 90 days after cessation of exposure a more intense inflammatory response was observed. In contrast, only a very occasional short chrysotile fibre was observed on the diaphragm and in no case was the fibre associated with a cellular response.

The morphology of the pleural lymphatic network of the diaphragm has been described by Negrini and Moriondo (2013). As shown in plates d and e of Fig. 3, the pleural stomata feed into the complex network of subpleural lymphatic channels. While a fibre that has traversed into the pleural cavity from the lung may be thin enough to enter a stoma, depending on its length it may become blocked because of its interception with the lymphatic bifurcations below the surface of the diaphragm. The crocidolite fibre shown in plate 4 of Fig. 4, appears to be partially encased with a cellular network, suggesting that it is embedded in the stomata. This may be a result of impacting the bifurcations of the lymphatic channels below the surface. In contrast, the diaphragms of chrysotile-exposed animals were similar to those of the air-control animals.

Boutin *et al.* (1996) and Mitchev *et al.* (2002) have described the existence of 'black spots' that appear to concentrate asbestos fibres in the parietal pleura in individuals who have been exposed to asbestos. Mitchev *et al.* (2002) described them as "Black spots of the parietal pleura are deposits of opaque particles located under an intact mesothelial layer. They were usually associated with concentrations of lymphocytes, plasmocytes,

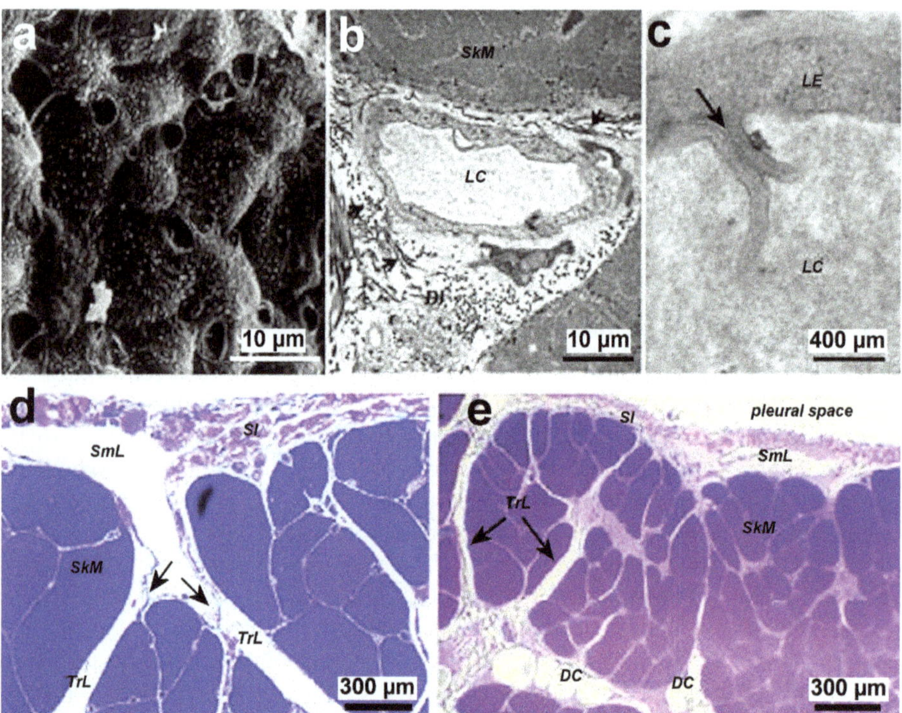

Figure 3. Morphology of the pleural lymphatic network of the diaphragm. (a) Scanning electron microscopy image of the muscular surface of the pleural diaphragm. Lymphatic stomata of diameter ranging from ~1 to ~10 µm appear as bordered holes at the junctions between mesothelial cells, covered densely by microvilli. (b) Transmission electron microscopy image of a lymphatic capillary (LC) in the diaphragmatic interstitium (DI) among skeletal muscle (SkM) fibres. The external surface of the LC lies in close proximity to the interstitial collagen fibrils (arrows). (c) High-magnification transmission electron microscopy image of the wall of a LC. Leaflets of lymphatic endothelium (LE) protrude into the vessel lumen to form a primary unidirectional valve (arrow). (d) Semi-thin cross section of the pleural side of the diaphragm. A large submesothelial lymphatic (SmL) collects the pleural fluid which entered through the stomata (not visible in this image) and from the pleural submesothelial interstitium (SI). Unidirectional secondary valves (arrows), formed by leaflets of LE, convey the newly formed lymph centripetally along transverse lymphatics (TrL) running through the SkM cells. (e) Semi-thin cross section of the pleural side of the diaphragm showing the entire diaphragmatic lymphatic network, including the deep collectors (DC), probably receiving lymph from both the pleural and peritoneal cavities (from Negrini and Moriondo, 2013, fig. 1, reproduced with the permission of J. Wiley & Sons).

and macrophages, indicating a chronic inflammatory reaction. A minority of cases had no inflammation." While both chrysotile and amphibole fibres were observed, Boutin *et al.* (1996) stated that all fibres $\geqslant 5$ µm long were amphiboles (crocidolite, amosite and tremolite). These areas of accumulation may correspond to the lacuna below the stomata where fluid can accumulate. It may be possible that this is aided by subsequent one-way valves which may inhibit some of the fibres from passing through the lymphatic channels (Light and Gary Lee, 2008).

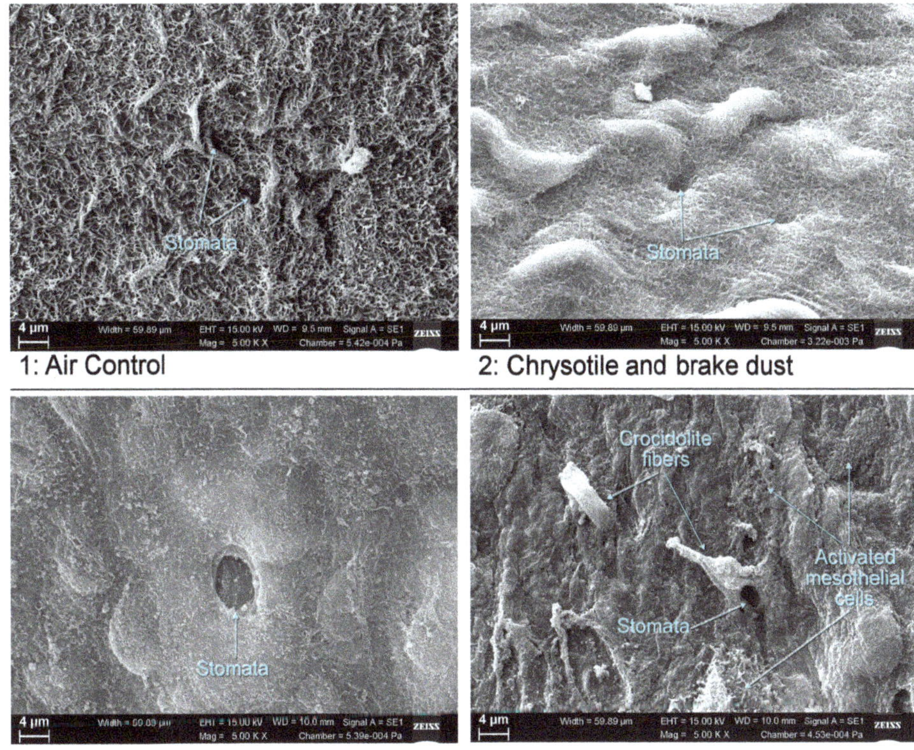

Figure 4. Scanning electron micrographs of the diaphragm at 0 days following cessation of exposure. Exposure groups 1–4 are shown in each of the corresponding plates. Plates 1, 2 and 3 show a normal diaphragm structure with a stoma. Plate 4 shows crocidolite fibres penetrating the stomata of the diaphragm with adjacent activated mesothelial cells (from Bernstein *et al.*, 2015, fig. 11).

The second method developed by Bernstein *et al.* (2011, 2014) involves the use of confocal microscopy to image the lung, pleural cavity and chest wall. Two confocal methods were developed, one for the lung and the other for the pleural cavity.

4.2. Transport to and pathological response in the visceral pleura

Inhaled thin fibres with aerodynamics suitable for deposition in alveoli adjacent to the visceral pleura would have the greatest potential of interacting with the visceral pleura. Fibres deposited in less peripheral alveoli could potentially migrate with time towards the visceral pleura. Both fibre length and bio-solubility can mediate the potential.

Very thin and shorter fibres can be cleared by the fluid flow of the lung directly into the lymphatic system (Bernstein *et al.*, 2004). The interstitial fluid which is formed from the extravasation of solute and fluid from the capillaries, enters blind-ended sacs composed only of an endothelial layer that is tethered to the interstitial matrix. These initial lymphatics possess overlapping endothelial cells that behave collectively like a valve, only permitting unidirectional entry of fluid, solute and cells into the lumen of these vessels. From there, the lymph flows through microlymphatics and then onto the larger collecting lymphatic vessels. Valves present in the collecting lymphatics ensure that the lymph flow is unidirectional. These valves differ from the initial lymphatic 'valve' in that they consist of two modified endothelial cell leaflets that meet in the vessel lumen (Scallan *et al.*, 2010). The subpleural lymphatics network provides a clearance route for these shorter fibres which reach the adjacent alveoli. The unidirectional valves of the collecting lymphatics provide a continuous intrinsic pumping mechanism which moves the fluid from the peripheral lung *via* larger lymphatic trunks and lymph nodes towards the neck where the fluid is eventually returned to the central circulation by emptying into the subclavian veins (Schmid-Schonbein, 1990; Breslin, 2014).

Fibres which are too large to enter into the lymphatic system and pass through the unidirectional valves will remain in the alveoli if they are biopersistent. The shorter fibres can be phagocytized by macrophages and cleared. The longer durable fibres, which lack the ability to be effectively phagocytosed and cleared, will either remain in the alveoli or can have the potential to penetrate the interstitial wall and translocate towards the visceral pleura. Because of the difficulty in actually sampling the pleural cavity area without bias *post mortem*, a method was developed which involved quick freezing of the intact rat chest wall in liquid nitrogen within minutes of termination (Bernstein *et al.*, 2010, 2011, 2014, 2015). For each chest wall a series of frozen cross sectional slabs 4–5 mm thick were subsequently cut. Following freeze substitution, cross sections of chest wall were transferred to anhydrous methanol at –20°C and brought to room temperature. Chest wall slabs and lung pieces were then stained with Lucifer yellow-CH (0.0001%) (Rogers *et al.*, 1999), infiltrated in Spurr epoxy resin then heat cured. The sections were then polished to be glass smooth. Using confocal microscopy, the cellular constituents and fibres (and particles) were imaged simultaneously with this arrangement with each 'exposure' producing two digital images in perfect register with one another (Bernstein *et al.*, 2010, 2011, 2014, 2015).

The confocal methodology enables the imaging of the fibres and cellular response *in situ*, the quantification of the number and dimensions of the fibres and in addition, the quantification of the amount of collagen present (fibrotic response). In an earlier study in which rats were exposed by inhalation to amosite asbestos, as shown in Fig. 5 an amphibole fibre was seen penetrating the visceral pleural surface and extending into the pleural cavity (Bernstein *et al.*, 2011).

Lung tissue is in constant movement as it inflates and deflates during each respiration. The alveoli expand non-uniformly and increase in size by a mean of 14% during respiration (Perlman and Bhattacharya, 2007). With non-uniform expansion, weak points can be formed during respiration. As shown in Fig. 6, amphibole fibres (crocidolite) have sharp edges. Because they are insoluble in the lung, these fibres remain sharp and can facilitate the fibres penetrating through the interstitial barrier possibly through such weak points. When this occurs in alveoli adjacent to the visceral pleural surface these long biopersistent fibres could traverse into the pleural cavity.

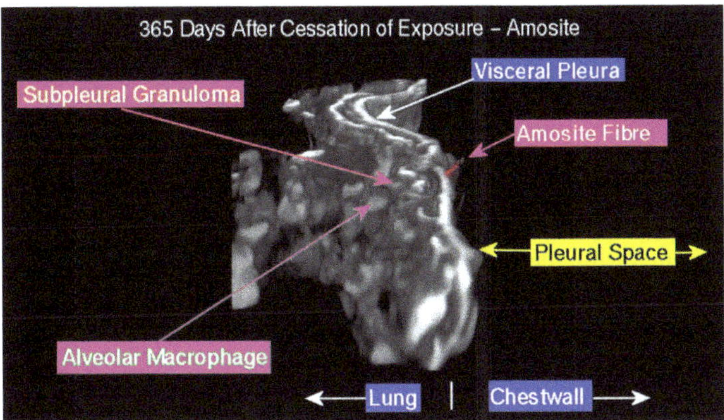

Figure 5. View of the pleural space from an animal exposed to amosite asbestos at 365 days post exposure. An amosite fibre penetrating the visceral pleural wall into the pleural space is seen. On the side of the lung, a well developed subpleural granuloma is seen with alveolar macrophages on the surface (from Bernstein *et al.*, 2011, fig. 19).

Chrysotile, which is mineralogically quite different, is formed as a rolled sheet and this morphology appears to inhibit the fibres' ability to penetrate though the tissues. In a subsequent study, Bernstein *et al.* (2014, 2015) examined in greater detail the visceral pleura to obtain a qualitative evaluation of the number of fibres, the amount of collagen and the associated pathological response at the visceral pleura again using confocal microscopy. Figures 7 and 8 (reproduced from Bernstein *et al.*, 2015) illustrate the differential response between the air control, brake dust (from brakes manufactured with chrysotile), the same brake dust with added chrysotile, and crocidolite asbestos at 14 and 91 days following a 5-day inhalation exposure. The confocal images show the

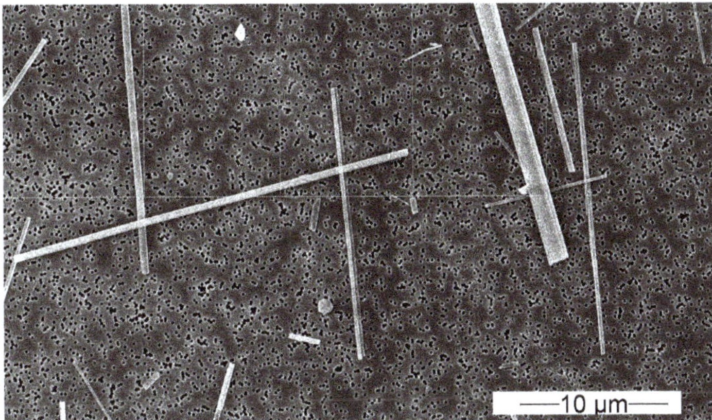

Figure 6. SEM image of crocidolite asbestos on Nuclepore filter (SN 110.609, Whatman Ltd, pore size 0.8 μm).

lung, visceral pleura, pleural cavity and parietal pleura. Both brake-dust groups are similar to the air control showing no effect. In contrast, the crocidolite asbestos group shows, at 14 days post exposure (Fig. 8), extensive cellular infiltrate into the sub-visceral pleura alveoli surrounding a long crocidolite fibre. The lymphatic channels below the parietal pleura are expanded, suggesting an important pleural inflammatory response to the crocidolite. At 91 days (Fig. 9) post exposure, a crocidolite fibre is seen within the pleural space. In addition, numerous pleural macrophages and neutrophils are also present surrounded by greyish wisps, which is probably coalesced pleural protein that was a result of the freezing process. Activated mesothelium is seen on the visceral pleura, which now has a dense collagen matrix (bright white area). The parietal pleura is not seen in this image. In this study, the percent fibrosis (% connective tissue) in the visceral pleural wall was measured by confocal microscopy as shown in Fig. 7. The brake-dust groups are nearly identical to the air-control group. In the crocidolite asbestos group, the mean percent connective tissue per field of view increased from a mean of 6.5 ± 2.4% on day 14 to 16.1 ± 9.4% on day 365 which was statistically different from the air-control and the brake-dust groups (analysis of variance, $p < 0.01$). The authors reported that no fibres were observed at any time point in the visceral pleura region of air control group and one short fibre of 3.3 μm was observed in the brake-dust group at 365 days. In the chrysotile and brake-dust group, fibres up to 17.9 μm long were observed at day 14, however, only two short fibres of 3 and 4.7 μm were observed at day 91 and at day 365 one 4.3 μm long fibre was observed which is consistent with the disintegration of the longer chrysotile fibres and their subsequent clearance. In the

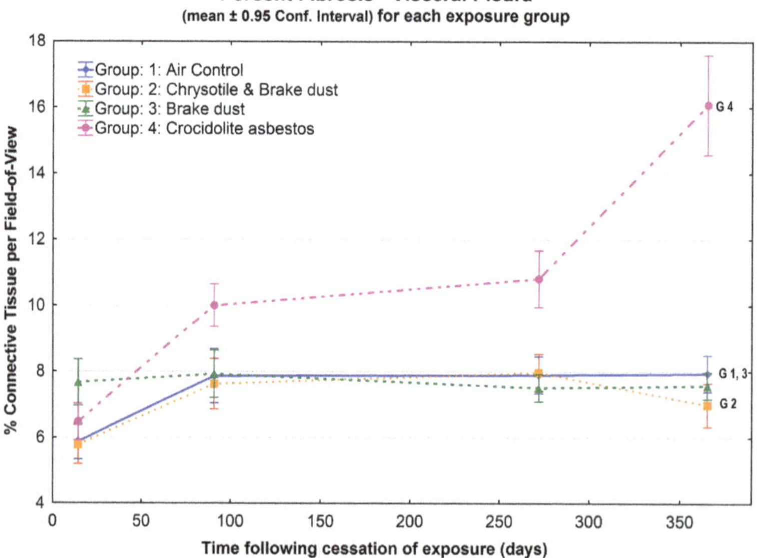

Figure. 7. Percent fibrosis (% connective tissue) in the visceral pleural wall measured by confocal microscopy (from Bernstein *et al.*, 2015, fig. 5).

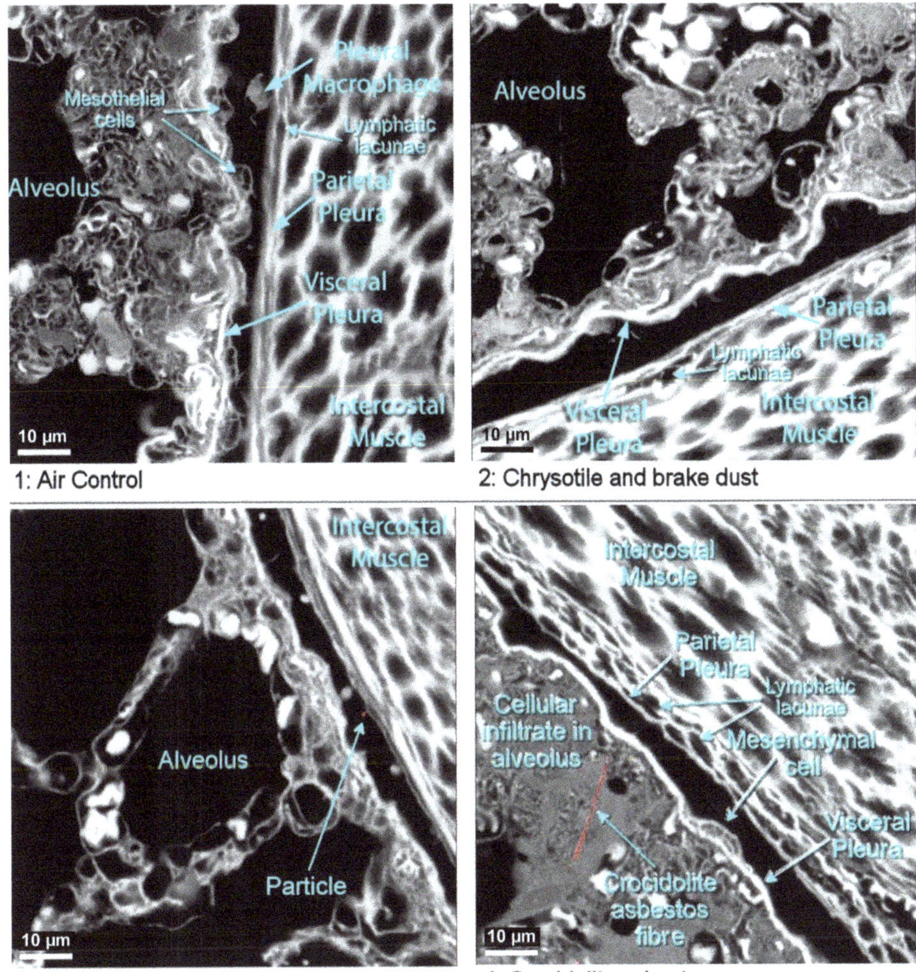

Figure 8. Confocal images of the pleural cavity 14 days after cessation of exposure. The images were obtained from snap frozen chestwall sections which preserved the tissue and cellular and spatial orientation of any particles or fibres present. The intercostal muscle which runs between the ribs and is mainly involved in the mechanical aspect of breathing is seen on the right, adjacent to the parietal pleura (when present). Opposite is the visceral pleura wall and the alveolar region of the lung. Details of each image are provided in the text (from Bernstein *et al.*, 2015, fig. 8).

crocidolite-asbestos exposed rats, fibres up to 26 µm long were observed at 14 days post exposure with no systematic clearance of the fibres through 365 days post exposure. At 365 days post exposure, 15 fibres were observed in the visceral pleura region surveyed ranging up to 22.2 µm in length.

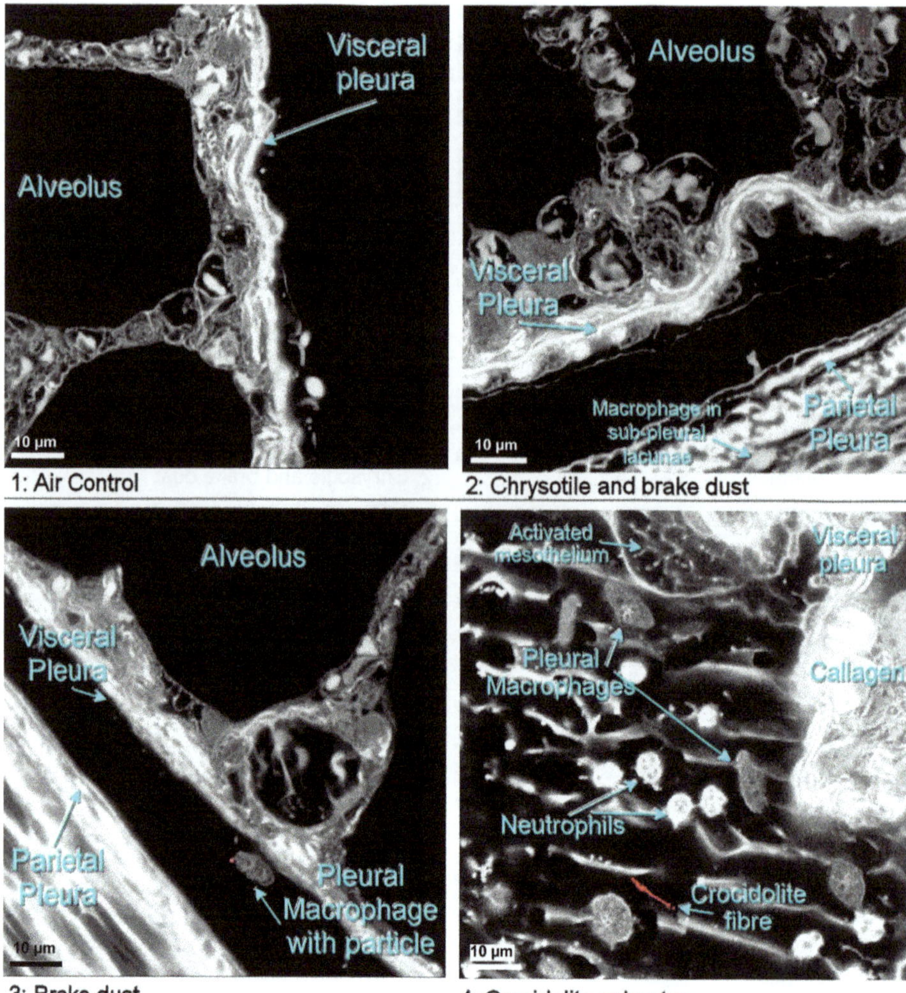

Figure 9. Confocal images of the pleural cavity 91 days after cessation of exposure. The images were obtained from snap frozen chest wall sections which preserved the tissue and cellular and spatial orientation of any particles or fibres present. The intercostal muscle which runs between the ribs and is mainly involved in the mechanical aspect of breathing is seen on the right, adjacent to the parietal pleura (when present). Opposite is the visceral pleura wall and the alveolar region of the lung. Details of each image are provided in the text (from Bernstein *et al.*, 2015, fig. 9).

5. Summary

While the dynamics and complexity of the pleural cavity physiology have been known for some time, it has only been more recently that this information has been integrated into our understanding of fibre pleural translocation and clearance. The triad of fibre pathogenicity includes diameter, length and biopsersistance. Pathologic responses to fibres include fluid accumulation (pleura or peritoneal effusions), pleural fibrosis and neoplasia of the serosal (pleural or peritoneal) cavity.

Following inhalation, fibre diameter plays a key role in aerodynamics and if thin enough, the fibres can pass through the ciliated tracheobronchial tree into the non-ciliated alveolar region. When deposited in the ciliated airways the fibres can be cleared by ciliated mucus transport. Fibres which deposit in the non-ciliated alveolar region can be cleared by macrophage clearance if they are short enough to be fully phagocytized. Very short thin fibres, which are small enough to pass through the pulmonary lymphatic valves, can also be cleared through the pulmonary lymphatic system. Depending upon their solubility these shorter fibres may accumulate within the bronchus-associated lymphoid tissue (BALT) and the lymph nodes. For longer fibres which cannot be fully phagocytosed by the macrophage, bio-solubility is the important determinant for potential toxicity. Biopsersistance, or resistance to degeneration by the host, allows for persistence and then accumulation (Donaldson *et al.*, 2010). In contrast, if these longer fibres are biosoluble in the lung fluid (pH 7.4) or the macrophage (~pH 4), they can either dissolve or break apart into shorter fibres and be cleared. Longer fibres which are not biosoluble will persist in the lung and initiate an inflammatory response (through frustrated phagocytosis followed by neutrophil response). Those fibres which deposit in the alveoli adjacent to the visceral pleural wall will be most available with the cyclic expansion of the lung with respiration to penetrate through the interstitium and visceral pleural wall into the pleural cavity

In the vicinity of the visceral pleural wall, these persistent longer fibres quickly initiate a fibrotic response and an increase in thickness of the visceral wall. Persistent long fibres have been seen following inhalation to penetrate partially through the visceral pleural wall with the remainder of the fibre embedded in a dense cellular matrix on the side of the lung of the visceral wall. With each respiration, these fibres would effectively scrape the opposing parietal pleural surface with scarring and inflammatory response. When penetrating into the pleural cavity, these fibres have been seen to initiate a sustained macrophage and neutrophil response with what appears to be a thickening of the pleura.

Pleural fluid clears from the pleural cavity through an array of stomata on the parietal pleural surface. These stomata feed into a network of branching lymphatic channels immediately below the parietal surface which drain largely to the mediastinal lymph nodes. The pleural lymphatics are equipped with two types of unidirectional valves. The first set of valves ensures that the flow is unidirectional from the interstitial space into the initial lymphatics, while preventing fluid backflow. The second set of valves orient lymph progression unidirectionally within the network towards larger collecting lymphatics. Particles and or fibres which enter into the pleural cavity would follow the fluid flow towards these stomata. If fibres are too large to enter the stomata, then they would remain on the surface, accumulate and potentially elicit an inflammatory response. Short and thin fibres can enter into the stomata can be cleared through the lymphatic system. However, if the fibres are too long to follow the branching of the lymphatic channels, they could become blocked at the lymphatic bifurcations below the surface. Longer fibres can also become blocked at the lymphatic valves. In these situations, if they are not biosoluble, they will probably initiate an inflammatory response. With persistent longer fibres, this inflammatory response can lead to the development of pleural fibrosis and over time, neoplastic processes.

References

Agostoni, E. (1972) Mechanics of the pleural space. *Physiology Review*, **52**, 57–128.

Agostoni, E. and Zocchi, L. (1991) Starling forces and lymphatic drainage in pleural liquid and protein exchanges. *Respiration Physiology*, **86**, 271–281.

Agostoni, E. and Zocchi, L. (1993) Active Naz transport and coupled liquid outflow from hydrothoraces of various size. *Respiration Physiology*, **92**, 101–113.

Agostoni, E. and Zocchi, L. (1998) Mechanical coupling and liquid exchanges in the pleural space. *Clinics in Chest Medicine*, **19**, 241–60.

Agostoni, E., Bodega, F. and Zocchi, L. (1999) Equivalent radius of paracellular "pores" of the mesothelium. *Journal of Applied Physiology*, **87**, 538–544.

Agur, A.M.R. and Dalley, A.F. (2013) *Grant's Atlas of Anatomy*, 13th edition, Lippincott Williams & Wilkins.

Albertine, K.H., Wiener-Kronish, J.P., Roos, P.J. and Staub, N.C. (1982) Structure, blood supply, and lymphatic vessels of the sheep's visceral pleura. *American Journal of Anatomy*, **165**, 277–294.

Albertine, K.H., Wiener-Kronish, J.P. and Staub, N.C. (1984) The structure of the parietal pleura and its relationship to pleural liquid dynamics in sheep. *The Anatomical Record*, **208**, 401–409.

Bermudez, E., Mangum, J.B., Moss, O.R., Wong, B.A. and Everitt, J.I. (2003) Pleural dosimetry and pathobiological responses in rats and hamsters exposed subchronically to MMVF 10a fiberglass. *Toxicology Science*, **74**, 165–173.

Bernaudin, J.F. and Fleury, J. (1985) Anatomy of the blood and lymphatic circulation of the pleural serosa. Pp. 101–124 in: *The Pleura in Health and Disease* (J. Chretien, J. Bignon and A. Hirsch, editors). Marcel Dekker Inc. New York, USA.

Bernstein, D.M., Rogers, R. and Smith, P. (2004) The biopersistence of brazilian chrysotile asbestos following inhalation. *Inhalation Toxicology*, **16**, 745–761.

Bernstein, D.M., Rogers, R.A., Sepulveda, R., Donaldson, K., Schuler, D., Gaering, S., Kunzendorf, P., Chevalier, J. and Holm, S.E. (2010) The pathological response and fate in the lung and pleura of chrysotile in combination with fine particles compared to amosite asbestos following short-term inhalation exposure: interim results. *Inhalation Toxicology*, **22**, 937–962.

Bernstein, D.M., Rogers, R.A., Sepulveda, R., Donaldson, K., Schuler, D., Gaering, S., Kunzendorf, P., Chevalier, J. and Holm, S.E. (2011) Quantification of the pathological response and fate in the lung and pleura of chrysotile in combination with fine particles compared to amosite-asbestos following short-term inhalation exposure. *Inhalation Toxicology*, **23**, 372–391.

Bernstein, D.M., Rogers, R., Sepulveda, R., Kunzendorf, P., Bellmann, B., Ernst, H. and Phillips, J.I. (2014) Evaluation of the deposition, translocation and pathological response of brake dust with and without added chrysotile in comparison to crocidolite asbestos following short-term inhalation: interim results. *Toxicology and Applied Pharmacology*, **276**, 28–46.

Bernstein, D.M., Rogers, R.A., Sepulveda, R., Kunzendorf, P., Bellmann, B., Ernst, H., Creutzenberg, O. and Phillips, J.I. (2015) Evaluation of the fate and pathological response in the lung and pleura of brake dust alone and in combination with added chrysotile compared to crocidolite asbestos following short-term inhalation exposure. *Toxicology and Applied Pharmacology*, **283**, 20–34.

Bodega, F., Zocchi, L. and Agostoni, E. (2000) Macromolecule transfer through mesothelium and connective tissue. *Journal of Applied Physiology*, **89**, 2165–2173.

Boutin, C., Dumortier, P., Rey, F., Viallat, J.R. and De Vuyst, P. (1996) Black spots concentrate oncogenic asbestos fibers in the parietal pleura. Thoracoscopic and mineralogic study. *American Journal of*

Respiratory and Critical Care Medicine, **153**, 444–449.
Breslin, J.W. (2014) Mechanical forces and lymphatic transport. *Microvascular Research*, **96**, 46–54.
Broaddus, V.C. and Araya, M. (1992) Liquid and protein dynamics using a new, minimally invasive pleural catheter in rabbits. *Journal of Applied Physiology*, **72**, 851–857.
Broaddus, V.C. and Light, R.W. (2000) Pleural effusion. Pp. 1913–1960 in: *Murray Nadel's Textbook of Respiratory Medicine* (R.J. Mason, J.T. Murray, V.C. Broaddus and J.A. Nadel, editors). 4th edition, Saunders, Philadelphia, Pennsylvania, USA.
Choe, N., Tanaka, S., Zia, W., Hemenaway, D.R., Roggli, V.L. and Kagan, E. (1997) Pleural macrophage recruitment and activation in asbestos-induced pleural injury. *Environmental Health Perspectives*, **105**, 1257–1260.
Donaldson, K., Murphy, F.A., Duffin, R. and Poland, C.A. (2010) Asbestos, carbon nanotubes and the pleural mesothelium: a review of the hypothesis regarding the role of long fibre retention in the parietal pleura, inflammation and mesothelioma. *Particle and Fibre Toxicology*, **7**, 5.
Fentie, I.H., Allen, D.J., Schenck, M.H. and Didio, L.J. (1986) Comparative electron microscopic study of bovine, porcine and human parietal pericardium, as materials for cardiac valve bioprostheses. *Journal of Submicroscopic Cytology*, **18**, 53–65.
Gelzleichter, T.R., Bermudez, E., Mangum, J.B., Wong, B.A., Moss, O.R. and Everitt, J.I. (1996) Pulmonary and pleural responses in Fischer 344 rats following short-term inhalation of a synthetic vitreous fiber. II. Pathobiologic responses. *Fundamental and Applied Toxicology*, **30**, 39–46.
Guyton, A.C., Granger, H.J. and Taylor, A.E. (1971) Interstitial fluid pressure. *Physiological Reviews*, **51**, 527–563.
Guyton, A.C., Taylor, A.E. and Brace, R.A. (1976) A synthesis of interstitial fluid regulation and lymph formation. *Federation Proceedings*, **35**, 1881–1885.
Kim, K.J., Critz, A.M. and Crandall, E.D. (1979) Transport of water and solutes across sheep visceral pleura. *American Review of Respiratory Disease*, **120**, 883–892.
Lai-Fook, S.J. (2004) Pleural mechanics and fluid exchange. *Physiological Reviews*, **84**, 385–410.
Leak, L.V. and Jamuar, M.P. (1983) Ultrastructure of pulmonary lymphatic vessels. *American Review of Respiratory Disease*, **128(2P2)**, 559–565.
Light, R.W. and Gary Lee, Y.C. (2008) *Textbook of Pleural Diseases*. 2nd edition. CRC Press, Boca Raton, Florida, USA.
Mayerson, H.S. (1963) The physiological importance of lymph. Pp. 1035–1073 in: *Handbook of Physiology. Section 2* (J. Field, editor). Williams & Wilkins, Baltimore, Maryland, USA.
Miserocchi, G. and Negrini, D. (1997) Pleural space: pressure and fluid dynamics. Pp. 1217–1225 in: *The Lung, Scientific Foundations* (R.G. Crystal, J.B. West, E.R. Weibel and P.J. Barnes, editors). Lippincott-Raven Publisher, New York.
Mitchev, K., Dumortier, P. and De Vuyst, P. (2002) 'Black Spots' and hyaline pleural plaques on the parietal pleura of 150 urban necropsy cases. *The American Journal of Surgery Pathology*, **26**, 1198–1206.
Moalli, P.A., Macdonald, J.L., Goodglick, L.A. and Kane, A.B. (1987) Acute injury and regeneration of the mesothelium in response to asbestos fibers. *American Journal of Pathology*, **128**, 426–455.
Negrini, D. (1995) Pulmonary microvascular pressure profile during development of hydrostatic edema. *Microcirculation*, **2**, 173–180.
Negrini, D. and Moriondo, A. (2013) Pleural function and lymphatics. *Acta Physiologica (Oxford)*, **207**, 244–259.
Negrini, D., Venturoli, D., Townsley, M.I. and Reed, R.K. (1994) Permeability of parietal pleura to liquid and proteins. *Journal of Applied Physiology*, **76**, 627–633.
Payne, D.K., Kinasewitz, G.T. and Gonzalez, E. (1988) Comparative permeability of canine visceral and parietal pleura. *Journal of Applied Physiology*, **65**, 2558–2564.
Peng, M.J., Wang, N.S., Vargas, F.S. and Light, R.W. (1994) Subclinical surface alterations of human pleura. A scanning electron microscopic study. *Chest*, **106**, 351–353.
Perlman, C.S. and Bhattacharya, J. (2007) Alveolar expansion imaged by optical sectioning microscopy. *Journal of Applied Physiology*, **103**, 1037–1044.
Rogers, R.A., Antonini, J.M., Brismar, H., Lai, J., Hesterberg, T.W., Oldmixon, E.H., Thevenaz, P. and Brain,

J.D. (1999) In situ microscopic analysis of asbestos and synthetic vitreous fibers retained in hamster lungs following inhalation. *Environmental Health Perspectives*, **107**, 367–375.

Roggli, V.L. (2015) The so-called short-fiber controversy: Literature review and critical analysis. *Archives of Pathology and Laboratory Medicine*, **139**, 1052–1057.

Scallan, J., Huxley, V.H. and Korthuis, R.J. (2010) *Capillary Fluid Exchange Regulation, Functions, and Pathology*. Morgan & Claypool Life Sciences. University of Missouri-Columbia, San Rafael, California, USA.

Schmid-Schonbein, G.W. (1990) Microlymphatics and lymph flow. *Physiological Reviews*, **70**, 987–1028.

Staub, N.C. (1974a) Pathogenesis of pulmonary edema (abstract). *American Review of Respiratory Disease*, **109**, 358–372.

Staub, N.C. (1974b) Pulmonary edema. *Physiological Reviews*, **54**, 678–811.

Staub, N.C., Wiener-Kronish, J.P. and Albertine, K.H. (1985) Transport through the pleura: physiology of normal liquid and solute exchange in the pleural space. Pp. 174–175 in: *The Pleura in Health and Disease* (J. Chretien, J. Bignon and A. Hirsch, editors). Marcel Dekker, Inc., New York.

Suzuki, Y. and Yuen, S.R. (2001) Asbestos tissue burden study on human malignant mesothelioma. *Industrial Health*, **39**, 150–160.

Suzuki, Y. and Yuen, S.R. (2002) Asbestos fibers contributing to the induction of human malignant mesothelioma. *Annals of New York Academy of Science*, **982**, 160–176.

Suzuki, Y., Yuen, S.R. and Ashley, R. (2005) Short, thin asbestos fibres contribute to the development of human malignant mesothelioma: pathological evidence. *International Journal of Hygiene and Environmental Health*, **208**, 201–210.

Wang, N.S. (1998) Anatomy of the pleura. *Clinics in Chest Medicine*, **19**, 229–240.

Wiener-Kronish, J.P., Albertine, K.H., Licko, V. and Staub, N.C. (1984) Protein egress and entry rates in pleural fluid and plasma in sheep. *Journal of Applied Physiology*, **56**, 459–463.

Zocchi, L. (2002) Physiology and pathophysiology of pleural fluid turnover. *European Respiratory Journal*, **20**, 1545–1558.

Zocchi, L., Raffaini, A., Agostoni, E. and Cremaschi, D. (1998) Diffusional permeability of rabbit mesothelium. *Journal of Applied Physiology*, **85**, 471–477.

Zocchi, L., Agostoni, E. and Cremaschi, D. (1991) Electrolyte transport across the pleura of rabbits. *Respiration Physiology*, **86**, 125–138.

EMU Notes in Mineralogy, Vol. 18 (2017), Chapter 13, 435–445

Biological activities of asbestos and other mineral fibres

MICHELE CARBONE and HAINING YANG

Thoracic Oncology Program, University of Hawaii Cancer Center, Honolulu, Hawaii USA, e-mail: mcarbone@cc.hawaii.edu

The mechanisms of asbestos carcinogenesis are related largely to the chronic inflammatory response to these fibres when they are lodged in tissues. Mesothelial cells are much more susceptible than other cell types to asbestos and thus individuals exposed to asbestos characteristically develop malignant mesothelioma (MM). The chronic inflammatory response is driven by the release of high mobility group box 1 protein (HMGB1) in the extracellular space by mesothelial cells and by inflammatory cells undergoing programmed cell necrosis caused by asbestos. Genetics influences the individual response to asbestos, and thus individual susceptibility to asbestos carcinogenesis varies. HMGB1 represents a promising biomarker of asbestos exposure to MM and strategies aimed at interfering with HMGB1 and chronic inflammation may help prevent/delay the onset of MM in individuals exposed to asbestos or other carcinogenic fibres. Individuals carrying germline BAP1 mutations can develop any type of cancer, including MM, regardless of asbestos exposure. However, animal experiments suggest that these individuals are also more susceptible to asbestos carcinogenesis, even at low exposure levels, compared to the population at large.

1. Introduction

There are >400 mineral fibres in nature. Six of them were used commercially, and millions of people were exposed to them. These six mineral fibres include crocidolite, amosite, anthophyllite, actinolite, tremolite and chrysotile (see Gualtieri, 2017 and Ballirano *et al.*, 2017, this volume, for details). In the late 1970s in the US and soon after in Western Europe, the use of these six mineral fibres was first regulated and later almost totally banned. These commercial mineral fibres were collectively called "asbestos" (Baumann *et al.*, 2013). Besides this family of six regulated asbestos minerals, there are about 400 additional non-regulated fibres that possess physical and chemical structures that are similar to industrial asbestos, and may also be carcinogenic (Baumann *et al.*, 2013; Wylie and Candela, 2015) (see Gualtieri, 2017 and Ballirano *et al.*, 2017, this volume, for details). Since then, some of these fibres, such as erionite and antigorite, have been shown to be as or more potent than asbestos fibres in causing human MM, while others, such as palygorskite were found not to be oncogenic (Carbone *et al.*, 2016; Larson *et al.*, 2016). Some of the non-regulated fibrous minerals have been used commercially, *e.g.* erionite was used to pave hundreds of miles of dirt roads in North Dakota (Carbone *et al.*, 2011), antigorite was used to pave roads in North Caledonia: exposure to both types of fibres has been linked to the development of human MM (Larson *et al.*, 2016). In addition, in areas

DOI: 10.1180/EMU-notes.18.13

where deposits of fibrous minerals are present, processes of erosion, often enhanced by human activities, disturb soils, releasing mineral fibres into the air (Carbone *et al.*, 2011; Baumann *et al.*, 2015; Wylie and Candela, 2015). Once fibres are dispersed in the environment and become inhalable, individuals may be exposed, and over time develop MM and other malignancies. In the United States, and particularly in the more arid portions of Nevada and California, both natural dust emissions and human activities such as off-road-vehicle use, or road construction and development of rural areas have significantly increased emissions of mineral fibre-containing dust (Goossens and Buck, 2009, 2011; Goossens *et al.*, 2012; Baumann *et al.*, 2015), and increased the risk of mineral fibres-related diseases. Therefore, it is important to determine the carcinogenic potential of all types of asbestiform fibres present in the environment. Chronic inflammation contributes to the initiation and progression of MM and of other cancers (Hillegass *et al.*, 2010; Carbone and Yang, 2012; Noy and Pollard, 2014; Cyphert *et al.*, 2015). Within days of exposure to carcinogenic mineral fibres, a vigorous pro-inflammatory immune response is elicited. An influx of macrophages occurs as these cells try to engulf and clear asbestos fibres, but cannot and therefore, undergo a process of frustrated phagocytosis (Miller and Shukla, 2012; Hillegass *et al.*, 2013; Kodavanti *et al.*, 2014). This leads to the release of pro-inflammatory cytokines and chemotactic factors resulting in the recruitment of more macrophages and neutrophils (Hillegass *et al.*, 2010, 2013; Miller and Shukla, 2012). Recruited macrophages and neutrophils release reactive oxygen species (ROS) and reactive nitrogen species (RNS), which are mutagenic (Hillegass *et al.*, 2010; Carbone and Yang, 2012; Kagan, 2013). Other factors released during this inflammatory response, such as high mobility group box 1 (HMGB1) and TNF-α, promote the survival of damaged mesothelial cells, increasing the odds that a malignant cell may develop (Fig. 1) (Yang *et al.*, 2006, 2010; Carbone and Yang, 2012). Fibres that do not induce the release of HMGB1 and TNF-α, such as palygorskite, are not carcinogenic (Larson *et al.*, 2016).

2. Asbestos carcinogenicity in humans

Although injection and inhalation studies in animals suggest that all forms of asbestos are equally carcinogenic and that there is a clear dose-response relationship between asbestos exposure and MM risk, epidemiologic studies in humans do not support these observations. Most human studies have indicated that amosite and crocidolite exposure carries a much greater risk than does exposure to chrysotile (Mossman *et al.*, 1990), and that chrysotile is frequently combined with tremolite, an amphibole, which may be most responsible for MM development (Smith and Wright, 1996). It is possible that the discrepancy between chrysotile carcinogenicity in animals and humans is related to the types of exposure in each group. In animal studies, exposure is usually intense and over a short period, whereas humans tend to be exposed to asbestos in smaller concentrations over longer periods.

Results of lung-content analysis studies, which are the most reliable indicators of exposure to date, may underestimate the amount of chrysotile exposure, because this

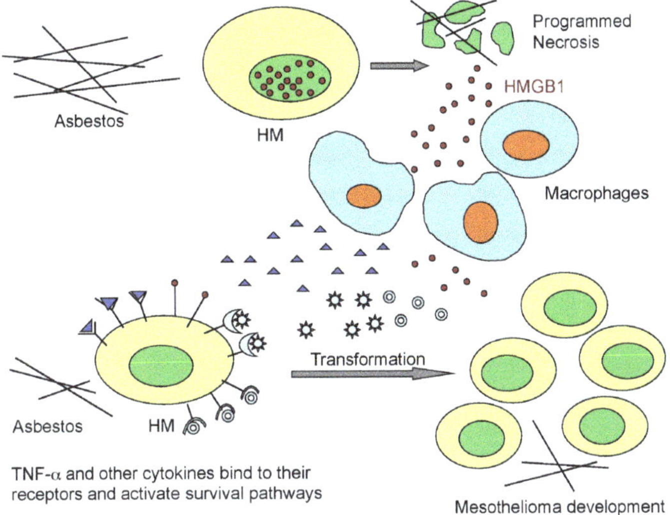

Figure 1. Mechanisms during asbestos-induced inflammatory activity leading to the formation of malignant mesothelioma. Legend: HM = human mesothelial cells.

mineral may dissolve over time (see Case and Marinaccio, 2017, this volume). We tested the carcinogenicity of chrysotile *vs.* crocidolite by exposing human mesothelial cells *in vitro* to these fibres and by injecting asbestos intraperitoneally in mice, and testing the genetic alterations induced by these fibres and chronic inflammation. We found that following a single exposure to these fibres, they induce similar genetic alterations and that both fibres induce HMGB1 release *in vitro* and *in vivo*. However, the effects of crocidolite are long lasting, while those of chrysotile are limited in time as gene alterations and HMGB1 serum levels reverted to normal within a few weeks of crysotile exposure. However, when exposure was continuous, *e.g.* by multiple intraperitoneal injection of low doses of fibres, then both crocidolite and chrysotile induced similar genetic changes and HMGB1 release over time (Qi *et al.*, 2013).

In humans, there is no clear dose-response curve linking asbestos exposure to MM. However, it is generally accepted in the scientific community that background levels of exposure, such as those found in the lungs of almost all individuals, do not increase the risk of MM, probably because these background asbestos levels are insufficient to sustain the chronic inflammation that leads over the course of decades to MM. It is estimated that ~5% of asbestos miners exposed to high levels of asbestos developed MM. However, the wives of some of these workers have developed MM after they were presumably exposed to lower levels of asbestos compared to their husbands, while handling and washing their clothes (Vianna and Polan, 1978). Similarly, MM development occurs commonly in workers in occupations in which asbestos exposure is higher than in the general population but much lower than in asbestos miners or insulators. The carpenters, electricians and construction workers who have developed MM are a reflection of this (Yates *et al.*, 1997). These data suggest that high levels of exposure are not correlated directly with increased risk of malignancy compared to moderate asbestos exposure, arguing against a classic dose-response relationship and suggesting the existence of co-factors that modulate asbestos carcinogenesis and MM.

3. Mechanism of mineral-fibres carcinogenesis: asbestos-induced chronic inflammation

Epidemiological and laboratory studies during the past decades concentrated almost exclusively on studies of "asbestos" fibres. These studies have demonstrated a clear connection between exposure to asbestos fibres and the subsequent development of MM (Wagner *et al.*, 1960; Spirtas *et al.*, 1994; Boffetta, 2007). The risk of developing MM following exposure to asbestos fibres is underscored by the observation that only ~5% of workers with 10 or more years of heavy asbestos exposure such as those working in crocidolite asbestos mines in South Africa, insulation or shipbuilding industries developed MM (Sluis-Cremer, 1991; Carbone *et al.*, 2015a). At the same time, these findings show that, fortunately, most individuals exposed for prolonged time to high levels of asbestos and those exposed only for very short periods of time do not develop MM (Sluis-Cremer, 1991). It has been estimated that ~60–70% of pleural MM are associated with a history of asbestos exposure (Spirtas *et al.*, 1994).

A key mechanism by which asbestos causes the transformation of mesothelial cells has recently been elucidated. Working with primary human mesothelial (HM) cells, Yang *et al.* (2006; 2010) discovered that asbestos induces necrotic cell death with the resultant release of HMGB-1 in the extracellular space. HMGB1 is a nuclear protein that acts as a damage-associated molecular pattern (DAMP) molecule upon its release into the extracellular milieu from necrotic, damaged, or severely stressed cells and by some cancer cell types (Carbone and Yang, 2012). HMGB-1 release in the extracellular space causes a chronic inflammatory response, macrophage accumulation and the secretion of the cytokine TNF-α which in turn activates NF-kβ, leading to the survival of HM cells that have accumulated genetic damage because of asbestos exposure (Fig. 1). We found that six of seven different MM cell lines tested actively secreted HMGB1 in tissue culture, in contrast to HM (Jube *et al.*, 2012). We demonstrated that the cancer phenotype (*i.e.* cancer cell invasion, growth in soft agar, *etc.*) of HMGB1-secreting human MM cells requires HMGB1, and therefore, abrogation of HMGB1 function may have therapeutic efficacy (Jube *et al.*, 2012). The role of HMGB1 and of chronic inflammation (Hillegass *et al.*, 2010) in asbestos carcinogenesis and MM progression (Carbone and Yang, 2012) has been established. These studies provided a mechanistic rationale to the observation by Liu *et al.* (1998) that asbestos is not pathogenic in transgenic mice that do not express the TNF-α receptor, and of Altomare *et al.* (2009) linking TNF-α to MM. In other words, asbestos carcinogenesis is linked to the chronic inflammatory process caused by the deposition of a sufficient number of asbestos fibres and the consequent release of cytokines – especially HMGB-1 and TNF-α by macrophages and mesothelial cells. On one hand, TNF-α protects mesothelial cells from asbestos-induced cell death and on the other hand promotes the growth of mesothelial cells that have accumulated DNA damage, following asbestos exposure. Asbestos causes DNA damage directly, possibly by mechanically interfering with the segregation of chromosomes during mitosis (Olofsson and Mark, 1989), and indirectly by causing the accumulation of reactive macrophages that release reactive mutagenic oxygen species –ROS (Choe *et al.*, 1997; Xu *et al.*, 1999, 2002). Amphibole asbestos fibres are often coated with iron-rich proteins, such as ferritin and hemosiderin, that appear to favour the production of mutagenic ROS (Heintz *et al.*, 2010).

4. The link with HMGB1 provides opportunities to interfere with mineral-fibre carcinogenesis

We discovered recently that total serum HMGB1 is a sensitive and specific marker to distinguish asbestos workers from non-asbestos exposed healthy individuals (Napolitano *et al.*, 2016b). Furthermore, when HMGB1 is released passively by cells undergoing programmed cell necrosis it is not acetylated, while active secretion of HMGB1 in the extracellular environment by MM cells requires HMGB1 acetylation. We tested the hypothesis that these two isoforms of HMGB1 may identify, among asbestos exposed cohorts, those who had MM. We found that hyper-acetylated HMGB1 accurately identifies MM patients from asbestos-exposed individuals with 100% specificity and sensitivity (Napolitano *et al.*, 2016b). A multi-centre study to validate these exciting findings should start in the coming year. Proteomic-based screening has provided encouraging results and may be integrated with HMGB1 screening for early detection of mesothelioma (Ostroff *et al.*, 2012).

We hypothesized that agents that possess anti-inflammatory properties may decrease serum HMGB1 levels and suppress or delay MM growth. Aspirin (acetyl salicylic acid, ASA) is used frequently as an anti-inflammatory and cardio-protective drug and has been shown to decrease and or delay cancer growth (Ulrich *et al.*, 2006). Aspirin was developed by the French chemist Gerhardt who acetylated salicylic acid (SA) to improve its adverse gastrointestinal (GI) effects. Bayer commercialized ASA as Aspirin at the beginning of the twentieth century. Aspirin remains, to date, the most widely used drug worldwide (Grossner *et al.*, 2011). ASA is absorbed by the stomach and upper intestine and in about 15–30 min is deacetylated to form SA. SA has a half-life of several hours in human plasma; thus, much of the aspirin bioactivity is attributed to SA (Grossner *et al.*, 2011). The primary mechanism of action of aspirin in mammals has been attributed to the disruption of eicosanoid biosynthesis through the irreversible inhibition *via* acetylation of cyclooxygenases (COX) 1 and 2, thereby altering the levels of prostaglandins (Ekinci *et al.*, 2011). Aspirin's protective effect against cancer has therefore been associated with the inhibition of COX-1/2. Studies have indeed shown that: (1) other non-steroidal anti-inflammatory drugs (NSAIDs), which like aspirin inhibit COX-1/2's cyclooxygenase activity, also reduce cancer risk; (2) cox-2 is over expressed in many cancers; and (3) cox-2 null mice have much reduced numbers of pre-cancerous colon polyps (reviewed by Williams *et al.*, 1999; Brown and DuBois,

2005). In contrast, other studies challenge the hypothesis that COX inhibition is responsible for the anti-cancer effects of aspirin and other NSAIDs (reviewed by Gurpinar *et al.*, 2013). For example: (1) inhibition of cancer cell growth by NSAIDs cannot be reversed by addition of prostaglandins; (2) genetic suppression of cox-1 and/or cox-2 expression often did not alter sensitivity to NSAIDs; and (3) some derivatives, metabolites, or enantiomers of several NSAIDs retain anti-cancer activity despite having lost their ability to inhibit COX-1/2. Together these studies indicate that mechanism(s) independent of inhibition of the cyclooxygenase activity of COX-1/2 are, at least partially, responsible for the anti-cancer activity of aspirin and other NSAIDs, a conclusion also reached by others (*e.g.* Phillips *et al.*, 2013).

We discovered that SA, the pharmacologically active component of aspirin, and ASA, interacts with HMGB1 (Yang *et al.*, 2015). SA and ASA inhibit HMGB1-induced cell migration at physiologically relevant concentrations, and reduce HMGB1-induced transcription of inflammatory cytokines and chemokines. We also found that HMGB1 induces COX-2 transcription, and that SA reduces this effect of HMGB1. Finally, we found that aspirin and SA reduce effectively the growth of human MM cells, which were previously shown to be autocrinally dependent on extracellular HMGB1, both *in vitro* and in a mouse model of MM. These findings are encouraging. Although, to the best of our knowledge, the possible therapeutic effects of aspirin in MM have not been studied, we found some evidence supporting the role of aspirin in preventing or possibly delaying the growth, and thus the diagnosis, of MM. Specifically, the Physician's Health Study suggests a possible association between ASA use and reduced MM incidence (Anonymous, 1989). In this study, 22,067 physicians were followed for 24 years with 17 reported cases of MM. An intention to treat analysis comparing ASA (325 mg on alternate days) and placebo treatment revealed a relative risk of 0.7 (0.3–1.9 95% confidence interval), indicating a 30% reduction in MM risk. Although the small number of MM cases resulted in insufficient power for a statistically significant analysis, these research findings together with our preliminary data suggest that ASA may have potential as an effective drug for MM by targeting the inflammatory response elicited by asbestos and other mineral fibres. The beneficial effects of aspirin detected in our study were observed treating animals during the early stages of cancer cell growth. Therefore, similarly to what has been proposed for colon cancer (Rothwell *et al.*, 2012), we propose that in human MM, the beneficial effects of aspirin may be observed better in a preventive/early-stage setting. Thus, aspirin administration to individuals at high risk of developing MM, such as those with a history of asbestos and or erionite exposure (Carbone *et al.*, 2011; Baumann *et al.*, 2015) or BAP1 mutation carriers (Carbone *et al.*, 2015b; Napolitano *et al.*, 2016b), may delay the growth of MM, thus improving their quality of life and possibly increasing life expectancy.

5. Co-carcinogenesis

Genetics and viral infection may increase the individual susceptibility to mineral fibre carcinogenesis. Recent studies demonstrated that germline mutations of the BAP1 gene are associated with a very high incidence of MM (Bononi *et al.*, 2015; Carbone *et al.*, 2015b; 2016). The critical role of BAP1 in MM is underscored by the finding that over 60% of sporadic – *i.e.* not genetically related – MM contain BAP1 somatic mutations (Guo *et al.*, 2015; Nasu *et al.*, 2015). In addition to MM, carriers of germline BAP1 mutations develop a large incidence of eye and skin melanomas and skin carcinomas, as well as other malignancies (Carbone *et al.*, 2013). We tested our hypothesis that carriers of germline BAP1 mutations were more susceptible to low doses of asbestos in mice carrying only one wild-type BAP1 allele (mice $BAP1^{+/-}$). We found that over 30% of $BAP1^{+/-}$ mice developed MM when exposed to low doses of crocidolite asbestos (0.5 mg by intraperitoneal injection) compared to an incidence of 7% in wild mice (Napolitano *et al.*, 2016a). These experimental findings in mice suggest that individuals carrying germline BAP1 mutations may also be more susceptible to asbestos compared to the population at large. In addition these findings suggest the possibility that in addition to BAP1 there may be other genes that influence individual susceptibility to asbestos carcinogenesis.

Simian virus 40 (SV40) DNA sequences have been detected prevalently in human mesothelioma, lymphoma, bone and brain cancers; these same cancer types are induced when SV40 is injected into hamsters (Carbone, 1999; Carbone *et al.*, 2000; Gazdar and Carbone, 2003). These findings raised concerns about the possible role of SV40 in the pathogenesis of these same cancers in humans. We discovered that primary human mesothelial cells (HM) and primary astrocytes (Ast) are unusually susceptible to SV40-mediated malignant transformation (Carbone *et al.*, 2008; Zhang *et al.*, 2010). The complex of SV40 large T antigen (Tag) with p53 promotes human cell transformation and malignant cell growth through its ability to bind and activate the IGF-1 promoter (Bocchetta *et al.*, 2008). SV40 has been shown to be a co-carcinogen with asbestos in causing MM in animals and malignant transformation of mesothelial cells in tissue culture (Kroczynska *et al.*, 2006). In addition, the data showed that in the presence of SV40 smaller amounts of asbestos were sufficient to cause MM (Kroczynska *et al.*, 2006). Co-carcinogenesis was mediated through the activation of the ERKs kinases and AP-1 activity that led to cell proliferation and stromal invasion (Kroczynska *et al.*, 2006).

References

Anonymous (1989) Final report on the aspirin component of the ongoing Physicians' Health Study. Steering Committee of the Physicians' Health Study Research Group. *The New England Journal of Medicine*, **321**, 129–135.

Altomare, D.A., Menges, C.W., Pei, J., Zhang, L., Skele-Stump, K.L., Carbone, M., Kane, A.B. and Testa, J.R. (2009) Activated TNFa/NFkB signaling via down regulation of Fas-associated Factor 1 in asbestos-induced mesotheliomas from Arf knockout mice. *Proceedings of the National Academy of Sciences of the USA*, **106**, 3420–3025.

Ballirano, P., Bloise, A., Gualtieri, A.F., Lezzerini, M., Pacella, A., Perchiazzi, N., Dogan, M. and Dogan, A.U. (2017) The crystal structure of mineral fibres. Pp. 17–64 in: *Mineral Fibres: Crystal Chemistry, Chemical-physical Properties, Biological Interaction and Toxicity* (A.F. Gualtieri, editor). EMU Notes in Mineralogy, **18**. European Mineralogical Union and Mineralogical Society of Great Britain & Ireland, London.

Baumann, F., Ambrosi, J.P. and Carbone, M. (2013) Asbestos is not just asbestos: an unrecognized health hazard. *Lancet Oncology*, **14**, 576–578.

Baumann, F., Buck, B.J., Metcalf, R.V., McLaurin, B.T., Merkler, D. and Carbone, M. (2015) The presence of asbestos in the natural environment is likely related to mesothelioma in young individuals and women from Southern Nevada. *Journal of Thoracic Oncology*, **10**, 731–737.

Bocchetta, M., Eliasz, S., Arakelian De Marco, M., Rudzinski, J., Zhang, L. and Carbone, M. (2008) The SV40 large T Antigen-p53 complexes bind and activate the insulin-like growth factor-I promoter stimulating cell growth. *Cancer Research*, **68**, 1022–1029.

Boffetta, P. (2007) Epidemiology of peritoneal mesothelioma: a review. *Annals of Oncology*, **18**, 985–990.

Bononi, A., Napolitano, A., Pass, H.I., Yang, H. and Carbone, M. (2015) Latest developments in our understanding of the pathogenesis of mesothelioma and the design of targeted therapies. *Expert Review of Respiratory Medicine*, **9**, 633–654.

Brown, J.R. and DuBois, R.N. (2005) COX-2: a molecular target for colorectal cancer prevention. *Journal of Clinical Oncology*, **23**, 2840–2855.

Carbone, M. (1999) Simian virus 40 and human tumors: It is time to study mechanisms. *Journal of Cellular Biochemistry*, **76**, 189–193.

Carbone, M., Rizzo, P. and Pass, H. (2000) Simian virus 40: The link with human malignant mesothelioma is well established. *Anticancer Research*, **20**, 875–878.

Carbone, M., Pannuti, A., Zhang, L., Testa, J.R. and Bocchetta, M. (2008) A novel mechanism of late gene silencing drives SV40 transformation of human mesothelial cells. *Cancer Research*, **68**, 9488–9496.

Carbone, M., Baris, Y.I., Bertino, P., Brass, B., Comertpay, S., Dogan, A.U., Gaudino, G., Jube, S., Kanodia, S., Partridge, C.R., Pass, H.I., Rivera, Z.S., Steele, I., Tuncer, M., Way, S., Yang, H. and Miller, A. (2011) Erionite exposure in North Dakota and Turkish villages with mesothelioma. *Proceedings of the National Academy of Sciences of the USA*, **108**, 13618–13623.

Carbone, M. and Yang, H. (2012) Molecular pathways: targeting mechanisms of asbestos and erionite carcinogenesis in mesothelioma. *Clinical Cancer Research*, **18**, 598–604.

Carbone, M., Yang, H., Pass, H.I., Krausz, T., Testa, J.R. and Gaudino, G. (2013) Bap1 and cancer. *Nature Reviews Cancer*, **13**, 153–159.

Carbone, M., Gaudino, G. and Yang, H. (2015a) Recent insights emerging from malignant mesothelioma genome sequencing. *Journal of Thoracic Oncology*, **10**, 409–411.

Carbone, M., Flores, E.G., Emi, M., Johnson, T.A., Tsunoda, T., Behner, D., Hoffman, H., Hesdorffer, M., Nasu, M., Napolitano, A., Power, A., Minaai, M., Baumann, F., Bryant-Greenwood, P., Lauk, O., Kirschner, M.B., Weder, W., Opitz, I., Pass, H.I., Gaudino, G., Pastorino, S. and Yang, H. (2015b) Combined genetic and genealogic studies uncover a large BAP1 cancer syndrome kindred, tracing back nine generations to a common ancestor from the 1700s. *PLOS Genetics*, **11**, e1005633.

Carbone, M., Kanodia, S., Chao, A., Miller, A., Wali, A., Weissman, D., Adjei, A., Baumann, F., Boffetta, P., Buck, B., de Perrot, M., Dogan, A.U., Gavett, S., Gualtieri, A., Hassan, R., Hesdorffer, M., Hirsch, F.R., Larson, D., Mao, W., Masten, S., Pass, H.I., Peto, J., Pira, E., Steele, I., Tsao, A., Woodard, G.A., Yang, H. and Malik, S. (2016) Consensus Report of the 2015 Weinman International Conference on Mesothelioma. *Journal of Thoracic Oncology*, **11**, 1246–1262.

Case, B.W. and Marinaccio, A. (2017) Epidemiological approaches to health effects of mineral fibres: Development of knowledge and current practice. Pp. 367–416 in: *Mineral Fibres: Crystal Chemistry, Chemical-physical Properties, Biological Interaction and Toxicity* (A.F. Gualtieri, editor). EMU Notes in Mineralogy, **18**. European Mineralogical Union and Mineralogical Society of Great Britain & Ireland, London.

Choe, N., Tanaka, S., Xia, W., Hemenway, D.R., Roggli, V.L. and Kagan, E. (1997) Pleural macrophage recruitment and activation in asbestos-induced pleural injury. *Environmental Health Perspectives*, **105** , 1257–60.

Cyphert, J.M., Carlin, D.J., Nyska, A., Schladweiler, M.C., Ledbetter, A.D., Shannahan, J.H., Kodavanti, U.P. and Gavett, S.H. (2015) Comparative long-term toxicity of Libby amphibole and amosite asbestos in rats after single or multiple intratracheal exposures. *Journal of Toxicology and Environmental Health A*, **78**, 151–165.

Ekinci, D., Sentürk, M. and Küfrevioğlu, Ö.İ. (2011) Salicylic acid derivatives: synthesis, features and usage as therapeutic tools. *Expert Opinion on Therapeutic Patents*, **21**, 1831–1841.

Gazdar, A.F. and Carbone, M. (2003) Molecular pathogenesis of malignant mesothelioma and its relationship to simian virus 40. *Clinical Lung Cancer*, **5**, 177–181.

Goossens, D. and Buck, B. (2009) Dust dynamics in off-road vehicle trails: Measurements on 16 arid soil types, Nevada, USA. *Journal of Environmental Management*, **90**, 3458–3469.

Goossens, D. and Buck, B. (2011) Effects of wind erosion, off-road vehicular activity, atmospheric conditions and the proximity of a metropolitan area on PM10 characteristics in a recreational site. *Atmospheric Environment*, **45**, 94–107.

Goossens, D., Buck, B. and McLaurin, B. (2012) Contributions to atmospheric dust production of natural and anthropogenic emissions in a recreational area designated for off-road vehicular activity (Nellis Dunes, Nevada, USA). *Journal of Arid Environments*, **78**, 80–99.

Grossner, T., Smyth, E.M. and Fitzgerald, G.A. (2011) Anti-inflammatory, antipyretic, and analgesic agents: pharmacotherapy of gout. Pp. 959–1004 in: *Goodman and Gilman's The Pharmacological Basis of Therapeutics*, 12th edition (L.L. Brunton, B.A. Chabner and B.C. Knollman, editors). McGraw-Hill, New York.

Gualtieri, A.F. (2017) Introduction. Pp. 1–15 in: *Mineral Fibres: Crystal Chemistry, Chemical-physical Properties, Biological Interaction and Toxicity* (A.F. Gualtieri, editor). EMU Notes in Mineralogy, **18**. European Mineralogical Union and Mineralogical Society of Great Britain & Ireland, London.

Guo, G., Chmielecki, J., Goparaju, C., Heguy, A., Dolgalev, I., Carbone, M., Seepo, S., Meyerson, M. and Pass, H.I. (2015) Whole exome sequencing reveals frequent genetic alterations in BAP1, NF2, CDKN2A and CUL1 in malignant pleural mesothelioma. *Cancer Research*, **75**, 264–269.

Gurpinar, E., Grizzle, W.E. and Piazza, G.A. (2013) COX-independent mechanisms of cancer chemoprevention by anti-inflammatory drugs. *Frontiers in Oncology*, **3**, 181.

Heintz, N.H., Janssen-Heininger, Y.M. and Mossman, B.T. (2010) Asbestos, lung cancers, and mesotheliomas: from molecular approaches to targeting tumor survival pathways. *American Journal of Respiratory Cell and Molecular Biology*, **42**, 133–139.

Hillegass, J.M., Shukla, A., Lathrop, S.A., MacPherson, M.B., Beuschel, S.L., Butnor, K.J., Testa, J.R., Pass, H.I., Carbone, M., Steele, C. and Mossman, B.T. (2010) Inflammation precedes the development of human malignant mesotheliomas in a SCID mouse xenograft model. *Annals of the New York Academy of Sciences*, **1203**, 7–14.

Hillegass, J.M., Miller, J.M., MacPherson, M.B., Westbom, C.M., Sayan, M., Thompson, J.K., Macura, S.L., Perkins, T.N., Beuschel, S.L., Alexeeva, V., Pass, H.I., Steele, C., Mossman, B.T. and Shukla, A. (2013) Asbestos and erionite prime and activate the NLRP3 inflammasome that stimulates autocrine cytokine release in human mesothelial cells. *Particle and Fibre Toxicology*, **10**, 39.

Jube, S., Rivera, Z., Bianchi, M.E., Powers, A., Wang, E., Pagano, I., Pass, H.I., Gaudino, G., Carbone, M. and Yang, H. (2012) Cancer cell secretion of the DAMP protein HMGB1 supports progression in malignant mesothelioma. *Cancer Research*, **72**, 3290–3301.

Kagan, E. (2013) Asbestos-induced mesothelioma: is fiber biopersistence really a critical factor? *American Journal of Pathology*, **183**, 1378–81.

Kodavanti, U.P., Andrews, D., Schladweiler, M.C., Gavett, S.H., Dodd, D.E. and Cyphert, J.M. (2014) Early and delayed effects of naturally occurring asbestos on serum biomarkers of inflammation and metabolism. *Journal of Toxicology and Environmental Health A*, **77**, 1024–1039.

Kroczynska, B., Cutrone, R., Bocchetta, M., Yang, H., Elmishad, A.G., Vacek, P., Ramos-Nino, M., Mossman, B.T., Pass, H.I. and Carbone, M. (2006) Crocidolite asbestos and SV40 are co-carcinogens in human mesothelial cells and in causing mesothelioma in hamsters. *Proceedings of the National Academy of Sciences of the USA*, **103**, 14128–14133.

Larson, D., Powers, A., Ambrosi, J.P., Tanji, M., Napolitano, A., Flores, E.G., Baumann, F., Pellegrini, L., Jennings, C.J., Buck, B.J., McLaurin, B.T., Merkler, D., Robinson, C., Morris, P., Dogan, M., Dogan, A.U., Pass, H.I., Pastorino, S., Carbone, M. and Yang, H. (2016) Investigating palygorskite's role in the development of mesothelioma in southern Nevada: Insights into fiber-induced carcinogenicity. *Journal of Toxicology and Environmental Health, Part B*, 19, 2016 Oct; Epub.

Liu, J.Y., Brass, D.M., Hoyle, G.W. and Brody, A.R. (1998) TNF-alpha receptor knockout mice are protected from the fibroproliferative effects of inhaled asbestos fibres. *American Journal of Pathology*, **153**, 1839–1847.

Miller, J. and Shukla, A. (2012) The role of inflammation in development and therapy of malignant mesothelioma. *American Medical Journal*, **3**, 240–248.

Mossman, B.T., Bignon, J., Corn, M., Seaton, A. and Gee, J.B. (1990) Asbestos: scientific developments and implications for public policy. *Science*, **247(4940)**, 294–301.

Napolitano, A., Pellegrini, L., Dey, A., Larson, D., Tanji, M., Flores, E.G., Kendrick, B., Lapid, D., Powers, A., Kanodia, S., Pastorino, S., Pass, H.I., Dixit, V., Yang, H. and Carbone, M. (2016a) Minimal asbestos exposure in germline BAP1 heterozygous mice is associated with deregulated inflammatory response and increased risk of mesothelioma. *Oncogene*, **35**, 1996–2002.

Napolitano, A., Antoine, D.J., Pellegrini, L., Baumann, F., Pagano, I.S., Pastorino, S., Goparaju, C.M., Prokrym, K., Canino, C., Pass, H.I., Carbone, M. and Yang, H. (2016b) HMGB1 and its hyper-acetylated isoform are sensitive and specific serum biomarkers to detect asbestos exposure and to identify mesothelioma patients. *Clinical Cancer Research*, **22**, 3087–3096.

Nasu, M., Emi, M., Pastorino, S., Tanji, M., Powers, A., Baumann, F., Zhang, Y.A., Gazdar, A., Kanodia, S., Tiirikainen, M., Flores, E., Gaudino, G., Becich, G.J,, Pass, H.I., Yang, H. and Carbone, M. (2015) High incidence of somatic BAP1 alterations in sporadic malignant mesothelioma. *Journal of Thoracic Oncology*, **10**, 565–576.

Noy, R. and Pollard, J.W. (2014) Tumor-associated macrophages: from mechanisms to therapy. *Immunity*, **41**, 49–61.

Olofsson, K. and Mark, J. (1989) Specificity of asbestos-induced chromosomal aberrations in short-term cultured human mesothelial cells. *Cancer Genetics and Cytogenetics*, **41**, 33–39.

Ostroff, R.M., Mehan, M.R., Stewart, A., Ayers, D., Brody, E.N., Williams, S.A., Levin, S., Black, B., Harbut, M., Carbone, M., Goparaju, C. and Pass, H.I. (2012) Early detection of malignant pleural mesothelioma in asbestos-exposed individuals with a noninvasive proteomics-based surveillance tool. *PLoS One*, **7**, e46091.

Qi, F., Okimoto, G., Jube, S., Napolitano, A., Pass, H.I., Laczko, R., DeMay, R.M., Khan, G., Tiirikainen, M., Rinaudo, C., Croce, A., Yang, H., Gaudino, G. and Carbone, M. (2013) Continuous exposure to chrysotile asbestos can cause transformation of human mesothelial cells via HMGB1 and TNF-α signaling. *American Journal of Pathology*, **183**, 1654–1666.

Phillips, I., Langley, R., Gilbert, D. and Ring, A. (2013) Aspirin as a treatment for cancer. *Clinical Oncology Journal (Royal College of Radiologists)*, **25**, 333–335.

Rothwell, P.M., Wilson, M., Price, J.F., Belch, J.F., Meade, T.W. and Mehta, Z. (2012) Effect of daily aspirin on risk of cancer metastasis: a study of incident cancers during randomised controlled trials. *Lancet*, **379(9826)**, 1591–1601.

Sluis-Cremer, G.K. (1991) Asbestos disease at low exposures after long residence times. *Annals of the New York Academy of Sciences*, **643**, 182–193.

Smith, A.H. and Wright, C.C. (1996) Chrysotile asbestos is the main cause of pleural mesothelioma. *American Journal of Industrial Medicine*, **30**, 252–266.

Spirtas, R., Heineman, E.F., Bernstein, L., Beebe, G.W., Keehn, R.J., Stark, A., Harlow, B.L. and Benichou, J. (1994) Malignant mesothelioma: attributable risk of asbestos exposure. *Occupational and Environmental Medicine*, **51**, 804–811.

Ulrich, C.M., Bigler, J., and Potter, J.D. (2006) Non-steroidal anti-inflammatory drugs for cancer prevention: promise, perils and pharmacogenetics. *Nature Reviews Cancer*, **6**, 130–140.

Vianna, N.J. and Polan, A.K. (1978) Non-occupational exposure to asbestos and malignant mesothelioma in females. *Lancet*, **1(8073)**, 1061–1063.

Wagner, J.C., Sleggs, C.A. and Marchand, P. (1960) Diffuse pleural mesothelioma and asbestos exposure in the North Western Cape Province. *British Journal of Industrial Medicine*, **17**, 260–271.

Williams, C.S., Mann, M. and DuBois, R.N. (1999) The role of cyclooxygenases in inflammation, cancer, and development. *Oncogene*, **18**, 7908–7916. Review.

Wylie, A.G., and Candela, P.A. (2015) Methodologies for determining the sources, characteristics, distribution, and abundance of asbestiform and nonasbestiform amphibole and serpentine in ambient air and water.

Journal of Toxicology and Environmental Health, B Critical Reviews, **18**, 1–42.

Xu, A., Wu, L.J., Santella, R.M. and Hei, T.K. (1999) Role of oxyradicals in mutagenicity and DNA damage induced by crocidolite asbestos in mammalian cells. *Cancer Research*, **59**, 5922–5926.

Xu, A., Zhou, H., Yu, D.Z. and Hei, T.K. (2002) Mechanisms of the genotoxicity of crocidolite asbestos in mammalian cells: implication from mutation patterns induced by reactive oxygen species. *Environmental Health Perspectives*, **110**, 1003–1008.

Yang, H., Bocchetta, M., Kroczynska, B., Elmishad, A.G., Chen, Y., Liu, Z., Bubici, C., Mossman, B.T., Pass, H.I., Testa, J.R., Franzoso, G. and Carbone, M. (2006) TNF-alpha inhibits asbestos-induced cytotoxicity via a NF-kappaB-dependent pathway, a possible mechanism for asbestos-induced oncogenesis. *Proceedings of the National Academy of Sciences of the USA*, **103**, 10397–10402.

Yang, H., Rivera, Z., Jube, S., Nasu, M., Bertino, P., Goparaju, C., Franzoso, G., Lotze, M.T., Krausz, T., Pass. H.I., Bianchi, M.E. and Carbone, M. (2010) Programmed necrosis induced by asbestos in human mesothelial cells causes high-mobility group box 1 protein release and resultant inflammation. *Proceedings of the National Academy of Sciences of the USA*, **107**, 12611–12616.

Yang, H., Pellegrini, L., Napolitano, A., Giorgi, C., Jube, S., Preti, A., Jennings, C.J., De Marchis, F., Flores, E.G., Larson, D., Pagano, I., Tanji, M., Powers, A., Kanodia, S., Gaudino, G., Pastorino, S., Pass, H.I., Pinton, P., Bianchi, M.E. and Carbone, M. (2015) Aspirin delays mesothelioma growth by inhibiting HMGB1-mediated tumor progression. *Cell Death and Disease*, **6**, e1786.

Yates, D.H., Corrin, B., Stidolph, P.N. and Browne, K. (1997) Malignant mesothelioma in south east England: clinicopathological experience of 272 cases. *Thorax*, **52**, 507–512.

Zhang, L., Strianese, O., Gaudino, G., Morris, P., Pass, H.I., Yang, H., Nerurkar, V.R., Bocchetta, M. and Carbone, M. (2010) Tissue tropism of SV40 transformation of human cells: role of the viral regulatory region and of cellular oncogenes. *Genes and Cancer*, **1**, 1008–1020.

EMU Notes in Mineralogy, Vol. 18 (2017), Chapter 14, 447–500

Insights into mineral fibre-induced lung epithelial cell toxicity and pulmonary fibrosis

RENEA P. JABLONSKI[1,2], SEOK-JO KIM[1,2], PAUL CHERESH[1,2], GANG LIU[3] and DAVID W. KAMP[1,2]

[1]*Department of Medicine, Division of Pulmonary & Critical Care Medicine, Jesse Brown VA Medical Center, Chicago, Illinois, USA,*
e-mail: renea.jablonski@northwestern.edu
[2]*Department of Medicine, Northwestern University Feinberg School of Medicine, Chicago, Illinois 60611, USA*
[3]*Clinical Research Center, Affiliated Hospital of Guangdong Medical College, Zhanjiang, China*

Asbestos mineral-fibre exposure is a well established cause of pulmonary fibrosis (asbestosis) and malignancies (bronchogenic carcinoma and mesothelioma). In this chapter, we review the work of numerous groups employing asbestos fibres to investigate the early molecular events promoting alveolar epithelial cell (AEC) injury, a key event for inciting and advancing pulmonary fibrosis. First, we summarize the accumulating evidence implicating the crucial role of the AEC in the pathobiology of pulmonary fibrosis, including asbestosis. We summarize briefly the mechanisms by which AEC toxicity from asbestos exposure results from the generation of reactive oxygen species (ROS) from at least three sources: (1) by redox reactions occurring on the fibre surface; (2) by surrounding inflammatory cells, especially macrophages; and (3) by AECs, especially from the mitochondria. Second, we review the emerging evidence showing that asbestos fibre-derived ROS induce AEC mtDNA damage important for promoting mitochondrial dysfunction and mitochondria (intrinsic)- and p53-regulated apoptosis, events important in facilitating lung fibrosis and malignant transformation. We focus on a novel role for AEC mtDNA damage repair by 8-oxoguanine DNA glycosylase (OGG1), a base excision repair enzyme, and mitochondrial aconitase (ACO-2) in preserving mtDNA integrity important in preventing asbestos-induced AEC apoptosis and lung fibrosis. Third, we examine studies implicating autophagy (lysosomal degradation of cytosolic materials), autophagy of mitochondria (mitophagy), and kinase signalling pathways in modulating lung epithelial cell toxicity following asbestos exposure and in patients with idiopathic pulmonary fibrosis (IPF). Finally, we review the evidence implying that an "exaggerated" aging lung is important in the pathogenesis of IPF and asbestosis. We focus on the role of the sirtuins (SIRT), especially SIRT3, in maintaining mitochondrial integrity important for preventing mtDNA damage and fibrosis. Collectively, the asbestos paradigm is informing our understanding of the cellular and molecular mechanisms underlying AEC mitochondrial dysfunction, autophagy/mitophagy and apoptosis that can promote lung fibrosis. Importantly, these studies are providing the scientific basis for novel therapeutic targets that may prove useful for the management of asbestos pulmonary toxicity with broader implications for other age-related diseases, including IPF and lung cancer, for which more effective treatments are urgently required.

DOI: 10.1180/EMU-notes.18.14

Acronyms: Alveolar epithelial cell (AEC); alveolar epithelial type I cell (AT1); alveolar epithelial type II cell (AT2); Alveolar macrophage (AM); autophagy protein 5 (ATG5); base excision repair (BER); bronchoalveolar duct (BAD); cigarette smoke (CS); c-Jun N-terminal kinase (JNK); endoplasmic reticulum (ER); epidermal growth factor receptor (EGFR); extracellular matrix (ECM); extracellular superoxide dismutase (EC-SOD); electron transport chain (ETC); extracellular signal-related kinase (ERK); genome-wide association studies (GWAS); idiopathic pulmonary fibrosis (IPF); inositol requiring kinase 1 alpha (IRE1α); mammalian target of rapamycin (mTOR); mitochondria-associated ER membrane (MAM); mitochondrial aconitase (ACO-2); mitochondrial DNA (mtDNA); mitochondrial membrane potential ($\Delta\psi_m$); mitogen-activated protein kinase (MAPK); multi-walled carbon nanotubes (MWCN); nicotinamide adenine dinucleotide phosphate (NADPH) oxidase 4 (Nox-4); non-small cell lung cancer (NSCLC); 8-oxoguanine-DNA glycosylase 1 (OGG1); particulate matter (PM); protein kinase-like ER kinase (PERK); PTEN-induced putative kinase 1 (PINK1); reactive nitrogen species (RNS); reactive oxygen species (ROS); sirtuin (SIRT); sirtuin 3 (SIRT3); surfactant protein C (SPC); transforming growth factor beta (TGF-β); thioredoxin interacting protein (TXNIP); tricarboxylic acid (TCA) cycle; unfolded protein response (UPR).

1. Introduction

Epidemiological studies have established clearly that asbestos-fibre exposure is associated with an increased risk of pulmonary toxicity that includes pulmonary fibrosis (asbestosis), pleural abnormalities (effusions and plaques) and malignancies (bronchogenic carcinoma and mesothelioma) (see for reviews: Broaddus *et al.*, 2011; Bunderson-Schelvan *et al.*, 2011; Huang *et al.*, 2011; Mossman *et al.*, 2011; Gulati and Redlich, 2015; Roe and Stella, 2015). The World Health Organization (WHO) estimates asbestos exposure during asbestos mining, milling, or industrial use approaches 125 million workers world-wide and 1.3 million workers in the USA, resulting in 13,885 deaths due to asbestosis world-wide between 1994 and 2010 (Diandini *et al.*, 2013; Stayner *et al.*, 2013). Although the use of asbestos has been banned in many industrial countries, some, including the United States and Canada, still permit trace amounts of the mineral in consumer products. Owing to the long latency period between exposure and disease and the continued presence of asbestos in buildings pre-dating the decline in asbestos use in the 1970s, asbestos-induced pulmonary diseases are likely to remain a challenge for clinicians. In this chapter, we focus on the pathobiology of asbestosis and, in particular, we review the key insights emerging regarding the role of asbestos-induced lung epithelial cell injury important for promoting pulmonary fibrosis.

Pulmonary fibrosis, including that due to asbestos exposure, is characterized by an excess accumulation of extracellular matrix (ECM) collagen in the distal lung interstitial tissue in association with an injured overlying epithelium and activated macrophages and myofibroblasts. Extensive investigations over the past several decades have identified many of the important cellular and molecular pathogenic mechanisms by which asbestos fibres cause asbestosis, yet the precise mechanisms involved as well as the cross-talk between implicated pathways are not fully understood. There are currently no effective treatments for asbestosis while two drugs have been shown to slow the progression of idiopathic pulmonary fibrosis (IPF). Clearly, more effective management approaches are needed urgently.

Numerous investigators, including the present group, have been investigating asbestos-induced AEC injury and lung fibrosis to better inform our understanding of the pathobiology of pulmonary fibrosis. The purpose of this review is to highlight our current understanding of how asbestos fibres injure the pulmonary epithelium, especially alveolar epithelial cells (AEC), which subsequently can trigger a fibrotic response in the lungs. We focus on AEC toxicity from asbestos but, where appropriate, we mention scientific insights garnered from studies exploring lung epithelial cell toxicity following exposure from other toxins (*i.e.* silica, airborne particulate matter [PM], carbon nanotubes, *etc.*). Although asbestos fibres and other airborne pulmonary toxins activate remarkably similar pathobiological pathways in the lung epithelium, we highlight relevant distinctions between toxins. We begin by summarizing briefly the evidence implying that the extent of AEC injury and inadequate repair are critical determinants of the fibrogenic potential of toxins, such as asbestos. We discuss the studies establishing that lung epithelial cell asbestos fibre uptake and subsequent epithelial cell injury are among the earliest abnormalities noted following asbestos exposure. We argue that a better understanding of how asbestos fibres promote AEC injury that results in lung fibrosis will probably have important broader implications for informing malignant transformation pathways, including lung cancer. In this regard, we summarize briefly the robust epidemiological studies that have established a direct relationship between pulmonary fibrosis (*i.e.* asbestosis, IPF, silicosis, and others) and risk of bronchogenic lung cancer. We also review evidence afforded from lung epithelial gene profiling studies following asbestos and other airborne toxin exposure.

We next review the mechanisms by which asbestos-induced AEC toxicity arises from the generation of ROS, especially from the mitochondria of the lung epithelium or surrounding inflammatory cells (*i.e.* macrophages). We examine the scientific basis supporting AEC mtDNA damage as a key target of asbestos fibre-derived ROS that subsequently promotes mitochondrial dysfunction and mitochondria (intrinsic)- and p53-regulated apoptosis, events that are critical for augmenting lung fibrosis and may also promote malignant epithelial cell transformation. We focus on an innovative role for AEC mtDNA damage repair by 8-oxoguanine DNA glycosylase (OGG1), a base excision repair (BER) enzyme, and mitochondrial aconitase (ACO-2) in preserving mtDNA integrity important in preventing asbestos-induced AEC apoptosis and lung fibrosis. We explore the emerging evidence on the important cross-talk between mitochondrial ROS production, mtDNA damage, p53 activation, OGG1, and ACO-2 acting as a mitochondrial redox-sensor involved in mtDNA maintenance in animal models of lung fibrosis. As compared to Mendelian nuclear genetic principles, mutations in maternally inherited mtDNA in alveolar type II (AT2) cells, which are considered the stem cells of the distal lung epithelium, may better account for the complex clinical-pathological features of degenerative fibrotic lung diseases, such as IPF and asbestosis (see for reviews: Cheresh *et al.*, 2013; Liu *et al.*, 2013; Wallace, 2013; Schumacker *et al.*, 2014). We reason that AT2 cells exposed to oxidative stress, as occurs following asbestos exposure, can result from defective mtDNA encoding for key genes regulating mitochondrial energy-generating oxidative phosphorylation. Although we concentrate on the effects of asbestos on AEC mtDNA given its prominent role of AEC in the pathophysiology of lung fibrosis, we recognize that oxidative mtDNA damage in other cell types (*i.e.* vascular endothelial cells, macrophages, fibroblasts, *etc.*) are probably important.

We summarize the emerging evidence suggesting a key role for altered autophagy, mitophagy and kinase signalling pathways in regulating lung epithelial cell toxicity following asbestos exposure as well as in patients with IPF. Each of these pathways appears crucial for modulating the AEC's appropriate response to oxidative stress. Finally, we review the evidence suggesting that an "exaggerated" aging lung and, in particular, AEC mitochondrial dysfunction is important in the pathogenesis of IPF and asbestosis (see for review: Selman and Pardo, 2014). In AECs, at least five of the nine aging pathways are implicated in humans with IPF and asbestosis including: DNA damage, activation of epigenetic signalling, shortened alveolar type II cell (AT2 – the distal lung epithelial stem cell) telomeres, mitochondria-mediated (intrinsic) apoptosis, and activated endoplasmic reticulum (ER) stress response in apoptotic AECs (Weiss *et al.*, 2010; Mossman *et al.*, 2011; Noble *et al.*, 2012; Tanjore *et al.*, 2012; Thannickal, 2013; Liu *et al.*, 2013; Uhal and Nguyen, 2013; Selman and Pardo, 2014). Interestingly, genome-wide association studies (GWAS) have shown a prominent role for aberrant DNA repair pathways in patients with IPF and asbestosis (Fingerlin *et al.*, 2013; Uhal and Nguyen, 2013; Selman and Pardo, 2014; Liu *et al.*, 2015a). We explore emerging evidence implicating sirtuins (SIRT), especially SIRT3, in maintaining mitochondrial integrity important for preventing mtDNA damage and fibrosis (see for reviews: Bohr *et al.*, 2002; Kroemer *et al.*, 2007; Kincaid and Bossy-Wetzel, 2013; Chen *et al.*, 2014b). SIRT3 is considered the 'gate-keeper' of mitochondrial integrity because it is the major mitochondrial deactylase optimizing protein function important in mitochondrial metabolism crucial for maintaining mtDNA integrity and the prevention of aging. A simplified conceptual model of the interconnected pathways reviewed herein in modulating lung epithelial fibrogenic signalling is shown in Fig. 1.

Collectively, the asbestos paradigm is informing our understanding of the fundamental mechanisms underlying AEC mitochondrial dysfunction, autophagy, mitophagy, altered kinase signalling pathways, and apoptosis crucial for promoting lung fibrosis and, perhaps, malignant transformation. Importantly, these studies are providing the scientific basis for the identification of novel therapeutic targets that may prove useful for the management of asbestosis as well as other age-related diseases, including IPF and lung cancer, for which more effective treatments are urgently required.

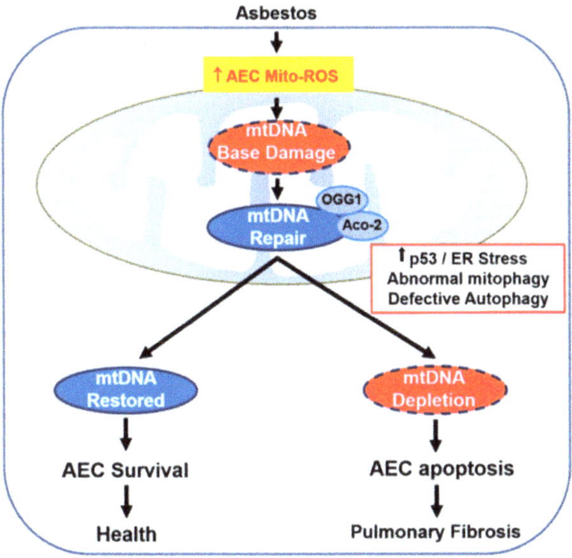

Figure 1. A hypothetical model depicting some of the key pathways reviewed.

2. Lung pathology: the role of the epithelium

2.1. Asbestos fibre types

As we have seen in the first two chapters of this book, asbestos refers to a naturally occurring group of silicate minerals that have in common the eponymous 'asbestiform' crystal habit comprising long, thin fibrous aggregates with high tensile strength and an ability to withstand heat and many chemicals. Asbestos fibre types, which have been reviewed in detail elsewhere in this publication (see Gualtieri, 2017 and Ballirano *et al.*, 2017, this volume), have all been shown to cause lung epithelial cell toxicity and will be mentioned only briefly here. Currently, the World Health Organization (WHO), the National Institute for Occupational Safety (NIOSH) and the Occupational Safety and Health Administration (OSHA) define the regulated form of asbestos as a fibre whose length is greater than 5 μm and whose length:width ratio is 3:1. Six mineral types are defined by the United States Environmental Protection Agency (EPA) as asbestos and all six are group I (definite) human carcinogens as defined by the International Agency for Research on Cancer (IARC, 1987).

Chrysotile is the sole member of the serpentine family to be included in the asbestos fibres. It makes up 95% of the asbestos used previously in industrial applications in the United States owing to the fact that its flexible nature lends itself for use in woven fabrics such as pipe insulation and brake pads.

The amphibole class of asbestos fibres is characterized by a rigid, needle-like shape; its members include amosite, crocidolite, fibrous tremolite, fibrous anthophyllite and fibrous actinolite. Notably, amphibole asbestos fibres have a longer biopersistence than chrysotile and as such amosite and crocidolite are considered to be the most hazardous types of asbestos. However, all are known to cause human disease (Straif *et al.*, 2009; Huang *et al.*, 2011). Crocidolite, or Cape Blue asbestos so named for its blue colour, is an amphibole asbestos the chemical formula of which is $Na_2(Fe^{3+})_2(Fe^{2+})_3Si_8O_{22}(OH)_2$. Due to its physical properties, crocidolite is considered to be the most hazardous form of asbestos. Amosite, or brown asbestos, derives its name from the Asbestos Mines of South Africa Company and is a vitreous fibre with a unit structure $(Fe^{2+},Mg)_7Si_8O_{22}(OH)_2$. Other amphibole asbestos fibre types include tremolite $Ca_2Mg_5Si_8O_{22}(OH)_2$, anthophyllite $(Mg,Fe^{2+})_7Si_8O_{22}(OH)_2$ and actinolite $Ca_2(Mg,Fe)_5Si_8O_{22}(OH)_2$. Both amphibole asbestos species and chrysotile may contain redox-active iron (ranging from 27 to 30% and 1 to 6%, respectively) which, as will be addressed later, plays a key role in mineral fibre toxicity.

2.2. Role of lung epithelial cell fibre uptake

All inhaled particles invoke an inflammatory response in the lung that limits lung epithelial cell fibre uptake and subsequent injury. Respirable fibres are those that can be inhaled into and deposited on the lower respiratory tract; whether a fibre is respirable depends on its diameter, density and length. The WHO defines a fibre, for the measurement of exposure in industrialized workers, to be a structure greater than 5 μm in length with an aspect ratio (length:diameter) of at least 3:1. Fibres with a diameter less than 3 μm are respirable and may be deposited in the alveolar space with an increasing efficiency of deposition with decreasing diameter to 1 μm. Determinants of fibre toxicity include the dose, dimension, durability and bioavailability of the inhaled agent as well as the genetic background and underlying susceptibility of the exposed host. While a detailed analysis of many of these factors is beyond the scope of this chapter, recent reviews are available elsewhere (Yao *et al.*, 2010; Aust *et al.*, 2011; Case *et al.*, 2011; Mossman *et al.*, 2011).

Particle uptake depends primarily on the ability of a particle to reach a particular location or cell type of interest. Clearly, a fibre must be respirable in order to reach the distal airways and lung epithelium and cause injury. Studies in rodents following brief exposure to inhaled particles have given insight into the location of respirable fibre deposition within the lung. Brody *et al.* (1981) described the primary site of chrysotile asbestos deposition on the alveolar epithelium at bifurcations of the alveolar duct in rats following a single inhalational exposure. Fibres were deposited in both central and peripheral locations with the longer, more fibrinogenic/carcinogenic fibres ($\geqslant$16 μm in length) showing significant peripheral deposition (Coin *et al.*, 1992). Importantly, the viscous layer of mucus that lines the central airways is significantly thicker than that in the distal airways and the ciliary beat frequency and mucus transport rates are faster, all of which potentially account for differential fibre deposition (van As, 1980). In experimental animals, clearance from the bronchoalveolar duct (BAD) junction is slower than from tubular airway segments, probably playing a role in increased fibre uptake at alveolar duct bifurcations due to prolonged contact time between the particle and the underlying cells. Comparable human studies to evaluate the site of inhaled particle deposition are lacking due to difficulties in design; unfortunately the existing studies that have attempted to answer this question through use of autoradiography detecting inhaled radioactive particles have given inconsistent results (Little *et al.*, 1965). Though less is known about the location of asbestos fibre deposition in the human lung, the initial fibrotic lesion in asbestosis arises at a similar location at the BAD junction and autopsy studies demonstrate preferential deposition at airway bifurcations suggesting that some aspects of particle deposition between the human and rodent lungs may be similar.

Particle uptake, both in cultured cells and *in vivo*, is a rapid process and occurs generally within hours following exposure. Uptake of inhaled particulate matter, including asbestos fibres, is influenced by particle size with the available data suggesting greater retention and slower clearance of smaller particles. However, the relevance of these findings to asbestos uptake in the lung epithelium is unclear as all studies evaluating the differential uptake of large *vs.* small particles dose particles by weight, so naturally a larger number of smaller (or lighter) particles will be administered raising the consideration of normal uptake *vs.* particle overload. Upon deposition, fibres can be taken up by alveolar macrophages and cleared *via* the mucociliary escalator or alveolar epithelial type I (AT1) pneumocytes through which they are transported to the basement membrane (Brody *et al.*, 1981). From the basement membrane, asbestos fibres may be moved into interstitial cells (such as fibroblasts or macrophages), capillary endothelial cells or the lymphatic system.

Experimental analysis of asbestos fibres, in particular, demonstrates rapid uptake of short chrysotile and crocidolite fibres by type I AECs. Interestingly, the overall fibre length increased with time in rat model of inhalational exposure, suggesting that the smaller and less pathogenic fibres are effectively cleared while larger fibres persist and contribute to ongoing lung cell injury (Roggli and Brody, 1984; Roggli *et al.*, 1987). In addition, long chrysotile fibres undergo longitudinal splitting over time which may further enhance their negative biological effects due to an effective increase in the overall fibre number (Coin *et al.*, 1994). For asbestos in particular, serum promotes rapid asbestos uptake into cultured lung epithelial cells resulting in toxicity (Kamp *et al.*, 1990). Furthermore, $\alpha v\beta 5$ integrin-mediated endocytosis appears to play a key role in asbestos uptake in lung epithelial cells and its subsequent toxic effects (Pande *et al.*, 2006).

The biopersistence of asbestos fibres in the lungs depends on multiple factors, including the length and solubility of the fibre. Asbestos fibres are deposited in the BAD junction of the sub-pleural alveoli and at the level of the visceral pleura following inhalation (Bernstein *et al.*, 2015). Bernstein and colleagues showed in rodents that while soluble chrysotile fibres (combined with brake dust or alone) were cleared readily with an estimated clearance time of 42 and 29 days, respectively, insoluble amosite fibres persist with an estimated clearance time of >1000 days (Bernstein *et al.*, 2015). Hence, the long biopersistence of amphibole fibres in the lung, as compared to chrysotile, may account for the observation that amphibole asbestos exposure is more likely to cause lung epithelial cell toxicity and fibrosis.

Cigarette smoke (CS) exposure can augment lung epithelial cell uptake of asbestos which may result subsequently in adverse effects on the lung. Patients exposed to asbestos often have a significant history of cigarette smoking and it can be difficult clinically to unravel which exposure may be the primary driver of the subsequent pulmonary disease (Bledsoe *et al.*, 2015). A CS-induced increase in particle uptake has been similarly demonstrated for some, but not all, particulate matter studied including titanium dioxide, talc and fibrous silicon carbide. In isolated tracheal epithelial cells, exposure to CS was shown to increase epithelial cell asbestos fibre uptake in a dose-dependent manner even in the setting of a significant time delay between smoke and asbestos exposure (Churg *et al.*, 1990). The uptake of particulate matter can be enhanced by the presence of ROS (including that from CS or another common oxidant, ozone) and further reducing ROS *via* iron chelation with deferoxamine or the addition of the H_2O_2 scavenger catalase decreases uptake of amosite asbestos *in vitro* (Chang *et al.*, 1988; Hardy and Aust, 1995). Co-exposure to CS similarly increased retention of particles and asbestos-induced pulmonary fibrosis in a guinea pig model (McFadden *et al.*, 1986; Tron *et al.*, 1987) which mirrors the increased lung asbestos fibre burden and incidence of asbestosis and bronchogenic carcinoma in asbestos workers who also smoke (Churg and Stevens, 1995).

2.3. Asbestos-induced lung epithelial cell toxicity

Although the diseases caused by asbestos inhalation have been well described and the mechanisms underlying the toxic effects of asbestos on the lung epithelium have been studied intensely, the precise mechanisms involved remain incompletely understood. Injury to the lung epithelium is prominently implicated in mediating asbestosis and lung cancer (Huang *et al.*, 2011; Mossman *et al.*, 2011; Liu *et al.*, 2013). Interestingly, epidemiological studies have established that patient cohorts with the most robust asbestosis are at the greatest risk of lung cancer, suggesting an important association between ongoing asbestos-induced epithelial cell toxicity and subsequent lung fibrosis as well as risk for lung cancer (Weiss, 1999). For example, Wu and colleagues reported an overall excess of malignancy in a cohort of ship workers, especially that of the oesophagus, trachea, bronchus and lung, which increased in a dose-dependent fashion with increasing asbestos exposure (Wu *et al.*, 2015). Although no single biological mechanism is responsible fully for the toxic effects of asbestos inhalation on the lung epithelium, we focus our attention on asbestos-induced ROS generation (see Section 3) leading to subsequent DNA damage and apoptosis (see Section 4) which are believed to be crucial in the pathogenesis of asbestos-induced pulmonary epithelial cell toxicity.

AEC injury is an important early event in the pathogenesis of chronic interstitial lung fibrosis including IPF and asbestosis (Brody and Overby, 1989; McGavran and Brody, 1989; Gardner *et al.*, 1997; Korfei *et al.*, 2008; Kamp, 2009; Huang *et al.*, 2011; Mossman *et al.*, 2011; Cheresh *et al.*, 2013; Liu *et al.*, 2013; Uhal and Nguyen, 2013; Selman and Pardo, 2014). The alveolar epithelium comprises alveolar epithelial type I (AT1) cells, which cover ~90% of the alveolar surface area and are subject to oxidative stress, and AT2 cells, which produce surfactant and can proliferate and differentiate into AT1 cells. The early stage of asbestosis is characterized by discrete foci of fibrosis within the respiratory bronchiole walls and alveolar duct bifurcations.

Apoptosis, or programmed cell death, eliminates cells with extensive DNA damage. Accumulating evidence has shown convincingly that AEC apoptosis is an important pathophysiological event leading to pulmonary fibrosis and asbestosis (Kamp, 2009; Huang *et al.*, 2011; Mossman *et al.*, 2011; Cheresh *et al.*, 2013; Uhal and Nguyen, 2013; Selman and Pardo, 2014). First, the alveolar epithelium of patients with IPF and rodents with asbestos-induced lung fibrosis demonstrate both DNA damage and apoptosis (Kuwano *et al.*, 1996; Korfei *et al.*, 2008; Liu *et al.*, 2013; Uhal and Nguyen, 2013; Bueno *et al.*, 2015; Patel *et al.*, 2015). Second, AEC apoptosis is both necessary and sufficient for inducing pulmonary fibrosis after exposure to bleomycin (Li *et al.*, 2003) or transforming growth factor-β (TGF-β) (Lee *et al.*, 2004) and blocking apoptosis can prevent the development of lung fibrosis (Budinger *et al.*, 2006). Third, preventing activation of latent TGF-β with an antibody directed against epithelial cell-wall integrin αvβ6 prevents pulmonary fibrosis (Horan *et al.*, 2008). Fourth, as recently reviewed (Uhal and Nguyen, 2013), considerable evidence shows convincingly a primary role for AEC death and inadequate epithelial cell repair in causing pulmonary fibrosis as first pointed out 35 years ago by Haschek and Witschi (1979). This hypothesis has since been confirmed elegantly by various transgenic murine models of pulmonary fibrosis and genetic mutations in ten different surfactant protein C (SPC) BRICHOS domain mutations that are only evident in the AT2 cells of humans with interstitial pulmonary fibrosis. Taken together, these studies indicate firmly that AEC apoptosis and the ensuing inflammatory response are important factors in the pathobiological response of the lung epithelium in IPF as well as lung fibrosis resulting from asbestos exposure. Below we review in detail the intensive investigations over the last decade exploring the molecular mechanisms that underlie asbestos-induced AEC apoptosis (see Sections 4 and 5) and how this is modulated by autophagy/kinase signalling (see Section 6) and aging (see Section 7).

2.4. Epithelial cell gene profiling studies

The lung epithelium is not solely a physical barrier, as previously thought, but instead plays an active role in particle uptake, recruitment of inflammatory cells and secretion of pro-fibrotic growth factors. Gene-profiling studies have provided additional insight into the complex molecular events that underlie the pathological response to mineral fibre inhalation.

In vitro studies using gene expression profiling have led to greater understanding of the potential transcriptional pathways that are altered following exposure to inhaled particles. Two groups used gene expression microarray techniques in cultured lung epithelial and mesothelial cells with hierarchical clustering analyses and a systems biology approach to study asbestos induced whole genome expression profiles (>50,000 genes) (Nymark *et al.*, 2007; Hevel *et al.*, 2008). These studies suggest a prominent role for p53 activation, which is involved in the DNA damage response of cells as reviewed below (see Section 4) along with nearly 2500 additional genes involved in diverse functions including cancer suppression, cell cycle arrest, apoptosis and cell survival. Nymark and colleagues exposed the A549, BEAS-2B and Met5A human lung cell lines to crocidolite asbestos and subjected them to gene expression microarray and gene ontogeny analysis to show decreased expression across all cell lines of BNIP3L, a potential cancer suppressor gene, and TXNDC, which plays a role in cell redox homeostasis and the response to ER stress (Nymark *et al.*, 2007). Another group assessed gene profiles in human bronchial epithelial cells following exposure to asbestos in the presence and absence of the carcinogen BPDE and showed that both exposures had in common regulation of genes involved in proliferation, DNA damage recognition and nucleotide excision repair, cytokines and apoptosis (Belitskaya-Levy *et al.*, 2007).

Similarly, comparative analysis of signalling pathways has been used to elucidate potential mechanisms underlying the hazardous effects of inhaled particles. Perkins *et al.* (2015) treated primary human bronchial epithelial and BEAS-2B cells with asbestos and silica, another common cause of pneumoconiosis, and used hierarchical cluster analysis to analyse differences in gene expression. Interestingly, while both asbestos and silica induced genes governing cell adhesion and migration, inflammation and the cellular stress response, high doses of silica were associated with a stronger inflammatory and fibrotic response while asbestos exposure favoured genes characteristic of aberrant proliferation and carcinogenesis. Differences in gene expression following exposure to fibrous dust (crocidolite and chrysotile asbestos) or granular dust (titanium dioxide, zirconium dioxide, hematite) included induction of genes associated with proliferation and carcinogenesis *vs.* inflammation, respectively, suggesting that *in vitro* analysis of gene expression may be a tool useful in the prediction of hazards associated with new materials (Helmig *et al.*, 2014). Comparison of gene expression following asbestos *vs.* multi-walled carbon nanotube (MWCN) exposure, which may also lead to fibrosis and neoplasia following inhalational exposure (Donaldson *et al.*, 2013), showed that both induced changes in genes associated with mesothelioma and lung cancer when compared to inert controls (Kim *et al.*, 2012). These studies have been performed in cultured single cell lines and thus are unable to inform us of the more complex interaction of the pulmonary epithelium with the surrounding interstitial and inflammatory cells which play roles in the complex pathogenesis of lung fibrosis and malignancy. Although these studies have not clearly identified the pathway responsible for the fibrotic and carcinogenic effects of mineral fibres, they have provided potential targets for further investigation. Additionally, *in vitro* studies of gene expression may help to inform of the potential toxigenic effects of exposure to new fibres and particles prior to their widespread adoption and use.

Gene profiling studies in animals and humans have also provided insight into the varied mechanisms by which exposures such as asbestos or CS may lead to carcinogenesis. For example, Sabo-Attwood and associates used C57Bl/6 mice in an inhalation model of asbestos exposure to show a marked induction of *mclca3* expression with an increase in mucus production as well as genes involved in inflammation, matrix remodelling and mitogenesis (Sabo-Attwood *et al.*, 2005). Manning and colleagues described the use of laser capture microdissection and quantitative RT-PCR to study particular cell types of interest in tissue from asbestos-exposed animals (Manning *et al.*, 2006). In part using these techniques, Manning and colleagues showed an important role for mitogen-activated protein kinase (MEK) which is upstream of the oxidant signal-regulated kinase 1 and 2 (ERK1/2) in regulating epithelial cell proliferation. Laser capture techniques were also employed by Yin and co-workers to show a dose-dependent increase in TNF-α, TGF-β, PDGFA, PDGFB and α-1 pro-collagen at the site of asbestos fibre deposition at the BAD junction compared to the same regions in untreated mice (Yin *et al.*, 2007). In an analysis of cancer samples from patients exposed to asbestos, an increase in p53 mutations was seen whereas both loss of heterozygosity and p53 mutations were seen in cancers isolated from patients with both asbestos and CS exposures, suggesting that asbestos acts as a cancer promoter in the lung (Inamura *et al.*, 2014).

3. Asbestos-induced ROS production – effect on the lung epithelium

As described above, ROS and RNS generated by asbestos fibres have been implicated as crucial second messengers in asbestos-induced epithelial cell toxicity. ROS can be generated in multiple ways including: (1) through reactions occurring on the surface of the mineral fibre; (2) by activation of macrophages or neutrophils following fibre phagocytosis; or (3) through mitochondrial dysfunction in target cells such as lung epithelial cells. As the mechanisms underlying asbestos-induced ROS production have been reviewed extensively elsewhere (Kamp *et al.*, 1992; Shukla *et al.*, 2003a; Aust *et al.*, 2011; Cheresh *et al.*, 2013; Liu *et al.*, 2013), we will briefly review the key concepts and highlight the role of the lung epithelium both in terms of ROS generation and subsequent modulation of inflammatory signalling and cell death which are integral in promoting lung fibrinogenesis and carcinogenesis.

3.1. Asbestos-fibre surface reactivity

Experiments using electron spin resonance trapping methods revealed that asbestos-derived iron is capable of generating free radicals in a cell-free system that contain a reducing agent (Weitzman and Graceffa, 1984). Following inhalation, asbestos fibres and other inhaled particles can become coated in redox active iron and mucopolysaccharide, creating asbestos or ferruginous bodies. Even trace amounts of iron (<2% total weight) complexed with asbestos fibres can augment the generation of ROS. The redox active iron is able to cycle back and forth between the reduced and oxidized forms leading to generation of the highly reactive HO· *via* the Fenton-catalysed Haber-Weiss reaction (Weitzman and Graceffa, 1984; Kamp *et al.*, 1992; Turci *et al.*, 2011):

Step 1: $Fe^{3+} + \cdot O_2^- \rightarrow Fe^{2+} + O_2$

Step 2: $Fe^{2+} + H_2O_2 \rightarrow Fe^{3+} + OH^- + \cdot OH$

Net reaction: $\cdot O_2^- + H_2O_2 \rightarrow \cdot OH + OH^- + O_2$

Multiple lines of evidence argue for a role of abnormal iron homeostasis in the lungs of asbestos-exposed patients and animals. Patients exposed to asbestos have increased levels of iron, ferritin, lactoferrin, transferrin and transferrin receptors in bronchoalveolar lavage fluid (Ghio *et al.*, 2008). Interestingly, iron concentrations are increased in tissues with age and cigarette smoking (Ghio *et al.*, 1994) which may explain in part the increased susceptibility to asbestos-induced lung disease with age and tobacco use. Although unclear in lung epithelial cells, asbestos fibres can induce the expression of a subunit of ferritin, the ferritin heavy chain (FHC), which acts as an anti-apoptotic protein capable of restricting redox active iron by oxidizing Fe^{2+} to Fe^{3+} and which may play a role in the development of oncogenesis following asbestos exposure (Aung *et al.*, 2007).

Iron-derived ROS leading to oxidant stress and DNA damage following asbestos exposure has been demonstrated both *in vitro* and *in vivo* in a variety of cells, including in lung epithelial cells (for review, see: Kamp *et al.*, 1992; Shukla *et al.*, 2003a; Liu *et al.*, 2010, 2013; Aust *et al.*, 2011; Huang *et al.*, 2011; Mossman *et al.*, 2011; Cheresh *et al.*, 2013). These injurious effects occur in an asbestos dose-dependent manner and can be attenuated by treatment with the iron chelators deferoxamine (Goodglick and Kane, 1990) and phytic acid (Kamp *et al.*, 1995) or augmentation of antioxidant enzyme (*e.g.* catalase) expression (Mossman *et al.*, 1990; Murthy *et al.*, 2009). In a similar fashion, MWCN can also be pathogenic on the lung in part due to contamination by surface transition metals capable of reducing ferric to redox active iron as shown above (Muller *et al.*, 2008).

3.2. Inflammatory cell-derived ROS and TGF-β – effects on the lung epithelium

Another important source of ROS production is inflammatory cells which are recruited to the site of asbestos deposition. As described previously, some fibres that deposit on the respiratory epithelium following inhalation undergo phagocytosis by alveolar macrophages (AM), which engulf the fingers and migrate to the airways where they can be cleared *via* the mucociliary escalator. 'Frustrated phagocytosis' occurs when particles cannot be cleared by the usual mechanism, either due to particle overload or excess particle length (>15–20 μm). Such frustrated phagocytosis leads to augmented generation of ROS through continued activation of the respiratory burst and may augment oxidant-induced injury. Similarly to asbestos fibres, MWCN can induce frustrated phagocytosis resulting in potent pro-inflammatory and pro-fibrotic responses in a MWCN length-dependent manner which is as potent as or worse than a similar dose of asbestos fibres (Boyles *et al.*, 2015). However, asbestos fibres and other particles that can be completely engulfed by macrophages are similarly able to stimulate H_2O_2 secretion, suggesting other mechanisms of oxidant stress. Using cultured lung epithelial cells *in vitro*, our group showed that neutrophils augment cell toxicity while AM are protective, demonstrating potentially important differences between these inflammatory cells in host defence against asbestos (Kamp *et al.*, 1994).

Asbestos-induced ROS generation *via* the respiratory burst has been linked to inflammatory and fibrotic signalling *via* multiple lines of evidence. First, all forms of asbestos induce ROS production from AMs and neutrophils (Kamp *et al.*, 1992). In particular, considerable evidence suggests that mitochondria-generated ROS from inflammatory cells, especially macrophages, play a prominent role in the pathogenesis of asbestos-induced lung toxicity in part *via* deleterious effects on the surrounding lung epithelium. AMs exposed to asbestos produce H_2O_2 that can be blocked *via* exogenous administration of catalase or mitigation of Ras-related C3 botulinum toxin substrate (Rac1) expression (Murthy *et al.*, 2009). Rac1 is localized in the mitochondria of AMs isolated from patients with asbestos (Osborn-Heaford *et al.*, 2012) where it augments mitochondrial H_2O_2 production and suppresses MMP-9 transcriptional activity (Murthy *et al.*, 2010). In addition to its role in generation of ROS, Rac1 is required for regulation of actin polymerization and the activity of lammellipodia, which may be an additional mechanism by which *Rac1*$^{-/-}$ mice are protected from developing lung fibrosis as macrophages are unable to be recruited to areas of asbestos deposition (Wittmann *et al.*, 2003). Second, extracellular superoxide dismutase (EC-SOD) inhibits oxidant-induced shedding of syndecan-1, which promotes aberrant wound healing and neutrophil chemotaxis and, as such, attenuates inflammation and fibrosis (Kliment *et al.*, 2009). Indeed, mice deficient in EC-SOD have enhanced asbestos-induced lung fibrosis (Fattman *et al.*, 2006). Third, although unclear with lung epithelial cells, H_2O_2 release by asbestos-exposed mesothelial cells leads to high-mobility group box 1 (HMGB1) translocation to the extracellular space and secretion of TNF-α by macrophages, resulting in a chronic inflammatory response (Yang *et al.*, 2010).

Finally, and perhaps most importantly, asbestos and silica can each induce activation of the nucleotide oligomerization domain (NOD)-like receptor protein 3 (NLRP3) inflammasome *via* ROS-dependent mechanisms resulting in pro-inflammatory and pro-fibrotic signalling through mediators such as IL-1β, TNF-α, bFGF and HMGB1 *via* effects on the lung epithelium (Dostert *et al.*, 2008). The inflammasome, an innate immune system sensor that regulates activation of caspase-1 and leads to pyroptosis (a form of programmed cell death), is activated following particle phagocytosis. Inflammasome sensing in the setting of fibre exposure appears to require both particle uptake and generation of ROS as an upstream sensor for activation (Naik and Dixit, 2011; Zhou *et al.*, 2011). Release of thioredoxin interacting protein (TXNIP) following asbestos-induced oxidation of Trx1 may be one mechanism by which asbestos exposure activates the inflammasome (Thompson *et al.*, 2014) though the specific mechanism by which particles are recognized and lead to inflammasome activation remains incompletely understood. Notably, chrysotile asbestos and Libby six-mix (a mixture of several forms of amphibole asbestos) had different effects on inflammasome activation where chrysotile exposure lead to aggressive lysosomal rupture and classical inflammasome activation whereas the smaller Libby six-mix fibres generated significantly more ROS and an alternative mode of inflammasome triggering (Li *et al.*, 2012). Similar to chrysotile asbestos, crystalline silica activated the NLRP3 inflammasome in human bronchial epithelial cells and this was associated with release of the pro-fibrotic molecules bFGF and HMGB1 (Peeters *et al.*, 2013). Fascinatingly, in a mouse model of bleomycin- and silica-induced lung fibrosis, treatment with an IL-1b receptor antagonist reduced lung collagen deposition and was able to reverse lung fibrosis when given up to 25 days following the fibrotic stimulus (Piguet *et al.*, 1993).

Dos Santos and colleagues (dos Santos *et al.*, 2015) demonstrated recently a key role for vimentin, the most abundant intermediate filament protein, in regulating NLRP3-induced lung inflammation and fibrosis following asbestos and bleomycin exposure *via* a direct protein–protein interaction between NLRP3 and vimentin. Although mitochondrial ROS production has been implicated in NLRP3 priming and activation (Bauernfeind *et al.*, 2011; Zhou *et al.*, 2011; Lawlor and Vince, 2014) and vimentin-deficient mice have increased mitochondrial ROS production (Mor-Vaknin *et al.*, 2003), a prominent role for macrophages was suggested based upon the protective effects of bone marrow chimeric mice lacking vimentin against bleomycin-induced lung fibrosis. Importantly, the downstream pro-inflammatory mediators IL-1β and TNF-α generated by inflammasome sensing can promote tissue fibrosis as well as malignant transformation (Wang *et al.*, 2004). Taken together, accumulating evidence suggests that selective targeting of downstream NLRP3 signalling pathways activated by asbestos and inhaled particulate matter-induced oxidative stress may provide novel therapeutic targets for fibrotic lung disease as well as asbestos-induced malignancies. However, further studies are warranted to better understand the precise molecular pathways through which these various asbestos-induced NLRP3 effects occur as well as the cross-talk between lung epithelial cells and macrophages.

Asbestos-induced ROS in pulmonary epithelial cells and AMs can lead to the activation of TGF-β signalling, a mediator of central importance in the generation of extracellular matrix and epithelial-mesenchymal transition that are hallmarks of lung fibrosis. Latent TGF-β1 is constitutively produced in a complex where it is covalently bound to latent-associated peptide (LAP). TGF-β can be activated through various mechanisms in response to particle inhalation or cell injury. Exposure to asbestos leads to ROS generation and oxidant damage to LAP which is no longer able to form a complex with TGF-β1 thus releasing it in its active form (Pociask *et al.*, 2004). Epithelial cell-restricted αvβ6 integrin, the expression of which is increased in response to cell injury or fibrotic stimuli, binds to latent TGF-β and is also able to convert it to its active form (Munger *et al.*, 1999). Partial inhibition of TGF-β activation using antibodies directed against αvβ6 integrin or mice deficient in αvβ6 integrin was able to block bleomycin-induced pulmonary inflammation in a murine model of lung fibrosis (Horan *et al.*, 2008).

Interestingly, exposure to MWCN leads to epithelial cell TGF-β release and increased fibroblast TGF-β1 receptor expression, leading to fibroblast activation and promoting epithelial-mesenchymal transition (Chen *et al.*, 2014a; Mishra *et al.*, 2015). In contrast, asbestos-induced lung epithelial cell TGF-β production promotes epithelial cell plasticity and EMT through activation of the MAPK/Erk signalling pathway (Tamminen *et al.*, 2012). Enhanced TFG-β expression may lead to other detrimental effects as enhanced TGF-β expression in lung A549 cells resulted in decreased expression of the master regulator of mitochondrial biogenesis transcription factor PGC-1a as well as expression of antioxidant enzymes such as SOD2 and MCAD (Sohn *et al.*, 2012). In addition to driving pro-fibrotic signalling pathways, enhanced epithelial TGF-β production following particle inhalation induces Treg cells and suppresses anti-cancer immune function leading to enhanced cancer growth in a murine model of malignant mesothelioma (Maeda *et al.*, 2014). Further study of the mechanisms by which the lung epithelium modulates TGF-β release and inflammatory signalling following asbestos exposure will be of considerable interest.

3.3. Lung epithelial cell mitochondrial ROS production

The vital role of the lung epithelium in the pathogenesis of lung fibrosis, which was reviewed above (see Sections 2 and 3), represents the initial step in a pathway which culminates in fibrotic lung disease, including asbestosis. Using hamster tracheal epithelial explants and a [^{75}Se]-selenomethionine epithelial cell labelling technique in a model of asbestos exposure, Mossman and colleagues demonstrated that epithelial cell toxicity was greatest with long chrysotile asbestos fibres (>10 μm compared to <2 μm) and partially attenuated by the addition of antioxidants including SOD and dimethylthiourea (Mossman and Landesman, 1983; Mossman *et al.*, 1986). Since these early studies, numerous groups have confirmed that all forms of asbestos fibres can injure the lung epithelium (bronchial epithelium and AEC) as assessed by various techniques including inducing a proliferation blockade, LDH release, DNA damage and apoptosis (see for review: Huang *et al.*, 2011; Cheresh *et al.*, 2013). A role for mitochondrial ROS production by AECs is suggested by several lines of evidence including: (1) in ρ^0–A549 cells incapable of mitochondrial ROS production, asbestos-induced DNA damage, mitochondrial dysfunction and intrinsic apoptosis are all attenuated as compared to wide-type cells (Panduri *et al.*, 2004); (2) using highly sensitive ratiometric sensors (RoGFP) to detect ROS production in the mitochondria and cytosol, asbestos fibres preferentially induce mitochondrial ROS production (Panduri *et al.*, 2009; Soberanes *et al.*, 2009); and (3) Euk-134, a cell-permeable combined SOD and catalase mimetic, prevents mitochondrial ROS production as well as toxin-induced ER stress response and intrinsic apoptosis in a variety of cell types including primary isolated rat AT2 cells (Kamp *et al.*, 2013). Although the precise site of mitochondrial free radical generation in asbestos-exposed lung epithelial cells is unknown, further studies are warranted in the hopes of identifying a potentially modifiable target.

Given the multiple sources of endogenous ROS following exposure to asbestos or other inhaled particulate matter, augmenting antioxidant defences would appear to be a rational therapeutic strategy for the prevention or repair of fibre-induced lung injury and fibrosis. However, this has met with limited success to date in animal models and supportive human data are lacking. One endogenous defence is secretion of EC-SOD, which is expressed by multiple cell types in the lung and depleted in murine models of asbestos-induced lung fibrosis (Tan *et al.*, 2004; Fattman *et al.*, 2006). Manni and colleagues used bone marrow chimeric mice incapable of pulmonary EC-SOD expression to demonstrate that leukocyte-derived EC-SOD was ineffective in preventing early fibrotic injury following asbestos exposure (Manni *et al.*, 2012). Lending further support to the idea that asbestos-derived ROS play a pathogenic role in

AEC injury and ongoing lung injury, multiple *in vitro* and *in vivo* studies have demonstrated a protective effect of exogenous antioxidants (see for review: Kamp *et al.*, 1992; Shukla *et al.*, 2003a; Aust *et al.*, 2011; Huang *et al.*, 2011; Cheresh *et al.*, 2013; Liu *et al.*, 2013). Despite the accepted role that increased mitochondrial ROS and downstream signalling plays in the pathogenesis of asbestos-induced lung fibrosis, there are no studies demonstrating a protective effect of augmenting antioxidant defences in asbestos-exposed workers. Supplementation with the antioxidant N-acetylcysteine did not preserve lung function or reduce indicators of inflammation or oxidative stress in patients with IPF (Izumi *et al.*, 2012; Martinez *et al.*, 2014) or affect serum thiol/antioxidant levels in a small study of asbestos-exposed individuals ($n = 66$) in a double-blind, randomized, placebo-controlled clinical trial of patients followed for four months (Alfonso *et al.*, 2015). In addition to the inadequate ability of these agents to act as an antioxidant as suggested from *in vitro* and murine studies, another explanation for these discordant findings is that mitochondrial ROS can have both pivotal pro-survival and death-promoting roles depending upon the levels generated and the cellular context in which this occurs (Diebold and Chandel, 2016). Nymark and colleagues argue that when studying the toxicity of asbestos and newly developed synthetic fibres, specific free radical formation and scavenging properties as well as the influence of culture medium and other additives should be taken into account (Nymark *et al.*, 2014) to better help predict *in vivo* toxicity based on *in vitro* modelling.

3.4. Nuclear and mitochondrial DNA damage

ROS derived from either the inflammatory cells or the lung epithelium following asbestos exposure leads to DNA damage in the lung epithelium that can impact lung epithelial cell function (*e.g.* proliferation blockade, apoptosis, mutagenesis) (Huang *et al.*, 2011). Moreover, there is evidence that asbestos-induced lung epithelial cell DNA damage is most prominent in cells containing fibres *in vitro* (Nygren *et al.*, 2004) or in murine models of asbestos-induced lung fibrosis at the BAD junction where fibres are initially deposited (Brody and Overby, 1989; Aljandali *et al.*, 2001; Kido *et al.*, 2008; Cheresh *et al.*, 2013). An alveolar epithelial-capillary *in vitro* co-culture study showed enhanced sensitivity as compared to monocultures for detecting silica-induced DNA damage and apoptosis (Kasper *et al.*, 2011). Accumulating evidence implicates asbestos-induced oxidative stress and DNA damage in apoptotic AEC death (see for review: Huang *et al.*, 2011; Cheresh *et al.*, 2013; Kim *et al.*, 2015). Mitochondrial DNA damage depletes electron transport constituents important for ATP production resulting in mitochondrial dysfunction and intrinsic apoptosis. Using a PCR-based assay, we have shown that ROS generated from asbestos exposure leads to a greater degree of mitochondrial, compared to nuclear, DNA damage which subsequently leads to intrinsic apoptosis (Kim *et al.*, 2014). Furthermore, defects in mitochondrial DNA repair can lead to cell death whereas augmenting mitochondrial DNA repair protects cells from apoptosis. Work by others as well as the present group studying the mechanisms of mitochondrial DNA repair, especially the role of mitochondrial OGG1 and ACO-2, is addressed in further detail in Section 5 of this chapter.

Asbestos, a recognized carcinogen, can also cause damage to the nuclear DNA which may lead to apoptosis or mutagenesis. There are multiple mechanisms by which exposure to asbestos or other inhaled fibres or particulate matter can lead to nuclear DNA alterations and damage which are beyond the scope of this chapter (see for review: Huang *et al.*, 2011; Kim *et al.*, 2015). Exposure of bronchial epithelial cells to asbestos or urban PM augments cellular ROS levels, oxidative DNA damage and cell cycle arrest with damage to the spindle apparatus occurring within hours of exposure; all of these deleterious effects were inhibited by the addition of antioxidants (Longhin *et al.*, 2013). Glass fibres also induced a dose-dependent genotoxic effect on human AECs, which correlated significantly with the generation of both ROS and RNS implicating oxidant stress as the driver of DNA damage in this system (Rapisarda *et al.*, 2015). Nuclear DNA damage from asbestos exposure can also occur due to the direct effect of the fibre on the genome. Analysis of lung cancers from asbestos-exposed patients using fluorescence *in situ* hybridization revealed a substantial increase in loss of the 19p region, suggesting that asbestos may be able to induce specific chromosomal alterations leading to carcinogenesis (Ruosaari *et al.*, 2008). Another analysis of lung cancers similarly identified an asbestos fibre-dependent copy number alteration and allelic imbalance in multiple regions, including 19p13, which may aid in the identification of asbestos-induced pulmonary malignancy (Nymark *et al.*, 2013).

The mutagenic effect of asbestos fibres may, in part, be cell type-specific; a single study comparing human bronchial epithelial and mesothelial cells exposed to equivalent doses of crocidolite asbestos demonstrated a doubling of DNA damage in the mesothelial cells which, in contrast to bronchial epithelial cells, did not decrease over time (Nygren *et al.*, 2004). Human AECs had a 3.5-fold increase in glutathione expression following asbestos exposure when compared to a pleural mesothelial cell line, suggesting that the differing sensitivity to asbestos-induced genotoxicity may, in part, be mediated by differences in endogenous antioxidant enzyme expression (Puhakka *et al.*, 2002). However, additional studies are warranted to better understand key differences between lung epithelial and mesothelial cell responses to asbestos that might inform the pathobiology of asbestos-induced pulmonary toxicity.

4. Asbestos-induced mitochondrial and p53-regulated lung epithelial cell apoptosis

4.1. Mitochondria-regulated apoptosis

Mitochondria are maternally inherited, double membrane-bound organelles found in all eukaryotic cells; they serve to generate ATP through cellular respiration and are therefore considered to be the "powerhouse of the cell". The number of mitochondria in a cell is variable, depending on the organism, cell and tissue type (Miller *et al.*, 2003; Held and Houtkooper, 2015). In addition, mitochondria play critical roles in other cellular functions including cell cycle control, proliferation and the complex signalling pathways apoptosis (Lim *et al.*, 2015).

Cells have two mechanisms that dictate programmed cell death or apoptosis: the extrinsic (death receptor) and the intrinsic (mitochondria-regulated) pathways. The mitochondria-regulated death pathway is activated by diverse stimuli causing mitochondrial injury which leads to increased permeability of the outer mitochondrial membrane, a reduction in the mitochondrial membrane potential ($\Delta\psi_m$) and release of cytochrome c (Galluzzi *et al.*, 2012). In particular, mitochondrial DNA damage has been implicated in triggering apoptotic cell death (Santos *et al.*, 2003). A hypothetical model by which asbestos-induced mitochondrial ROS induces mtDNA damage resulting in mitochondrial dysfunction and intrinsic apoptosis is depicted in Fig. 2. The B cell lymphoma (Bcl-2) family of proteins governs outer mitochondrial membrane permeability and has members with both pro-apoptotic (Bax, BAD, Bak, Bok) and anti-apoptotic (Bcl-2, Bcl-xL, Bcl-w) effects. Bax- and Bak-induced permeabilization of the outer mitochondrial membrane is considered to be the point of no return in the mitochondria-regulated death pathway in the lung epithelium, leading to release of apoptogenic molecules and activation of the caspase cascade that culminates in cell death (Kroemer *et al.*, 2007). Substantial evidence shows that diverse stimuli including ROS, DNA damage and exposure to asbestos and other fibres activate the intrinsic death pathway in the lung epithelium and as such that will be the focus of this section (see for review: Noble *et al.*, 2012; Cheresh *et al.*, 2013; Liu *et al.*, 2013).

Work by the present group and others using the asbestos paradigm has informed our understanding of the mechanisms by which fibre exposure and oxidative stress promote the AEC injury and mitochondria-regulated apoptosis important for developing pulmonary fibrosis. Evidence for the crucial role of AEC mitochondrial ROS production due to asbestos exposure was obtained through the use of a highly sensitive Rho-GFP (green fluorescent protein) probe targeted to the mitochondria (Panduri *et al.*, 2009). Asbestos exposure results in a dose- and time-dependent reduction in $\Delta\psi_m$ associated with release of cytochrome c and caspase-9 activation in both human A549 and rat isolated primary AT2 cells (Panduri *et al.*, 2003; Kido *et al.*, 2008). Further, both the iron chelator phytic acid and the ROS scavenger sodium benzoate blocked

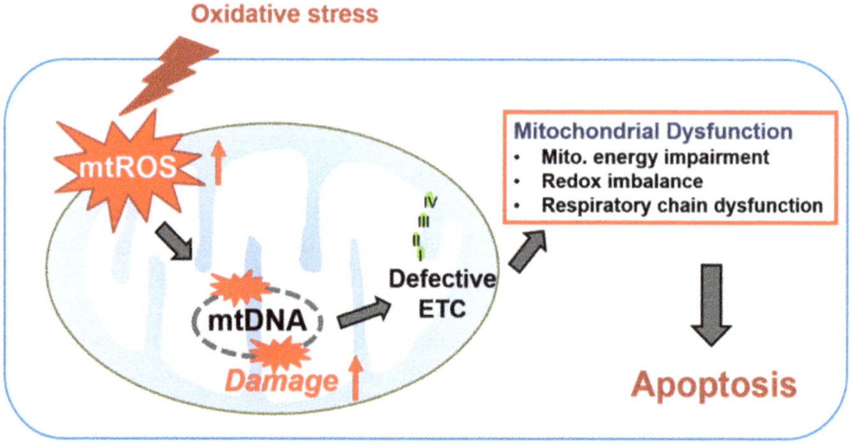

Figure 2. Proposed model mtDNA damage in mediating intrinsic apoptosis.

asbestos-induced $\Delta\psi_m$ and caspase-9 activation in this same study, demonstrating the importance of iron-derived ROS and the intrinsic death pathway. Asbestos-mediated apoptosis was similarly attenuated in A549 cells engineered to over-express the anti-apoptotic Bcl-xL (Panduri *et al.*, 2003) and this protective effect was associated with preservation of outer mitochondrial membrane integrity and a reduction in DNA fragmentation. Finally, studies in ρ^0-A549 cells, which lack a functional electron transport chain (*etc.*) and thus cannot generate mitochondrial ROS, are protected against reductions in asbestos-induced caspase-9 activation and intrinsic apoptosis (Panduri *et al.*, 2004). Mossman and colleagues showed that mitochondrial localization of activated protein kinase delta (PKCδ) occurred in lung epithelial cells following *in vitro* and *in vivo* asbestos exposure (Lounsbury *et al.*, 2002). Moreover, they showed that PKCδ-dependent protein kinase D phosphorylation by asbestos modulates asbestos-induced distal bronchiolar epithelial cell apoptosis *via* downstream effects on the ERK1/2 and JNK1/2 signalling pathways (Zhang *et al.*, 2007; Manning *et al.*, 2008).

Using the bleomycin model of lung fibrosis, the central role of AEC apoptosis in the development of lung fibrosis was suggested by Budinger and colleagues where mice lacking the pro-apoptotic Bid were protected from development of AEC apoptosis as well as bleomycin-induced fibrosis (Budinger *et al.*, 2006). Recently two groups have established that PTEN-induced putative kinase 1 (PINK1) deficiency impairs AEC mitochondrial function in patients with IPF and shown that mice deficient in PINK1 have increased AEC intrinsic apoptosis and lung fibrosis following viral- or bleomycin-induced injury (Bueno *et al.*, 2015; Patel *et al.*, 2015). However, the role of PINK1 in modulating asbestos-induced AEC apoptosis and lung fibrosis is unknown. Together these findings implicate mitochondrial ROS and the intrinsic cell death pathway in the development of lung fibrosis, including asbestosis. Many questions remain, however, regarding the precise role of the mitochondria and the pathobiological consequence of AEC apoptosis in the context of both asbestosis and malignant transformation following asbestos exposure. In this regard, development of lung epithelial cell targeted transgenic mice will be of considerable interest in better understanding the cross-talk between AEC and macrophages.

4.2. Mitochondria-endoplasmic reticulum cross-talk

Emerging data suggest that there is important cross-talk between the mitochondria and the ER in mediating AEC apoptosis and lung fibrosis, including that following asbestos exposure. The ER is an extension of the nuclear envelope which forms a system of interconnected sac-like structures called *cisternae* within the cytoplasm. Its role encompasses the synthesis, folding and export of protein and membrane lipids, though it also acts as a regulator of calcium homeostasis and, importantly, intrinsic apoptosis through physical and functional interconnection with the mitochondria at the mitochondrion-associated membrane (MAM) (see for review: Heath-Engel *et al.*, 2008; Kim *et al.*, 2008; Rong and Distelhorst, 2008; de Brito and Scorrano, 2010). Cellular perturbations including both chemical and structural stressors can trigger a specific ER response, the unfolded protein response (UPR), characterized by accumulation of unfolded proteins in the ER. ER stress can be reduced by the UPR through three mechanisms: (1) induction of transcription of ER chaperones to aid in protein folding; (2) attenuation of protein translation thus blocking synthesis of new proteins; and (3) degradation of unfolded or misfolded proteins *via* the ubiquitin-proteasome pathway. The UPR is vitally important for coordinating numerous adaptive cellular responses that enhance cell survival in the setting of oxidative stress or other noxious stimuli, though in the setting of severe or prolonged ER stress, apoptosis or autophagy may ensue (Malhotra and Kaufman, 2011). Pro-survival signalling is controlled by transient release of Ca^{2+}; in contrast, apoptotic signalling is characterized by prolonged Ca^{2+} release and mitochondrial Bax/Bak binding (Kroemer *et al.*, 2007; de Brito and Scorrano, 2010; Malhotra and Kaufman, 2011).

The importance of calcium homeostasis in intrinsic apoptosis is demonstrated by studies showing that over-expression of sarcoplasmic ER Ca^{2+} ATPase (SERCA) in Bax/Bak double-knockdown embryonic fibroblasts, which are resistant to apoptosis, leads to restoration of ER Ca^{2+} levels and oxidant stress-induced cellular apoptosis (Husgafvel-Pursiainen *et al.*, 1997; Lin *et al.*, 2000; Nakano and Vousden, 2001; Matsuoka *et al.*, 2003; Kopnin *et al.*, 2004). Bcl-2 acts to regulate ER Ca^{2+} mobilization *via* two mechanisms: (1) by blocking the homodimerization of Bax at the mitochondrial membrane, which leads to displacement of SERCA; and (2) by regulating phosphorylation of the inositol-1,4,5-triphosphate receptors, which are the channels that lead to Ca^{2+} release. However, ER Ca^{2+} release to the mitochondria is necessary but not sufficient for the induction of intrinsic apoptosis and may lead to intrinsic apoptosis *via* other mechanisms such as *via* the inositol-requiring transmembrane kinase/endonuclease-TNF receptor-associated factor-c-Jun N-terminal kinase (IRE-Iα-TRAF-JNK) stress signalling pathway (Klee *et al.*, 2009). Our group has shown that A549 and primary isolated rat AT2 cells treated with asbestos release ER Ca^{2+} within minutes, leading to activation of the UPR signalling proteins IRE-1α and XBP1 (Kamp *et al.*, 2013). Interestingly, pre-treatment with the ER chaperone 4-phenyl butyric acid prevents the increase in IRE-1α and XBP1 but does not alter calcium mobilization or subsequent apoptosis. However, Bcl-xL over-expression and treatment with EUK-134, a superoxide dismutase and catalase mimetic, each blocked asbestos-induced ER Ca^{2+} release, induction of the UPR and intrinsic apoptosis (Kamp *et al.*, 2013). Alveolar macrophages treated with the intracellular calcium chelator BAPTA, which depletes ER Ca^{2+} stores, had increased asbestos-induced XBP1 splicing and BiP expression suggesting that alveolar macrophages similarly undergo ER stress following asbestos exposure (Ryan *et al.*, 2014).

Activation of apoptosis *via* the UPR is mediated by of one of three ER stress-activated transcription factors: inositol requiring kinase 1 alpha (IRE-1a), protein-like kinase endoplasmic reticulum kinase (PERK), or activating transcription factor 6 (ATF6). Regardless of the signalling pathway involved, ER stress-initiated apoptosis is modulated by pro- and anti-apoptotic Bcl-2 family members. The sigma-1 receptor (Sig-1R), which is located at the MAM, promotes cell survival by regulating the ROS-activated transcription factor nuclear factor kappa light chain enhancer of activated B cells (NF-κB) (Meunier and Hayashi, 2010). There is also evidence that PM-induced AEC cell apoptosis is modulated by oxidative and ER stress (Liu *et al.*, 2015c). Together, these studies suggest that intrinsic apoptosis, including asbestos-induced AEC apoptosis, is mediated through the close interaction of the mitochondria and ER where the presence of ER stress promotes Ca^{2+} transfer to the mitochondria rendering a cell sensitive to oxidative stress and intrinsic apoptosis.

Accumulating data supports an important role for the AEC ER stress response in patients with IPF which may also provide insight into the mechanisms underlying asbestos-induced pulmonary toxicity. Several studies have demonstrated co-localization of markers of both ER stress and intrinsic apoptosis in AECs of patients with IPF (Korfei *et al.*, 2008; Lawson *et al.*, 2008; Tanjore *et al.*, 2012; Mulugeta *et al.*, 2015). Notably, aging, a recognized risk factor for the development of lung diseases including fibrosis (Faner *et al.*, 2012; Thannickal *et al.*, 2015), renders the lung more susceptible to γ herpesvirus 68-induced ER stress, AT2 cell apoptosis and development of lung fibrosis in a mouse model (Torres-Gonzalez *et al.*, 2012). ER stress promotes the epithelial-mesenchymal transition (EMT) which generates myofibroblasts, an important effector cell in IPF (Erster *et al.*, 2004; Liu *et al.*, 2005; Zhong *et al.*, 2011; Torres-Gonzalez *et al.*, 2012). ER stress and EMT are associated with the misfolded surfactant protein C mutant $SP\text{-}C^{\Delta exon4}$, which, interestingly, is also associated with a familial form of pulmonary fibrosis (Cregan *et al.*, 2004; Yamaguchi *et al.*, 2004; Liu *et al.*, 2005; Naik *et al.*, 2007; Mulugeta *et al.*, 2015). The crucial role of AT2 cells in particular has been highlighted by Lawson and coworkers who used transgenic mice with AT2 cell-restricted conditional expression of $SP\text{-}C^{L118Q}$ to show both tunicamycin-induced ER stress and bleomycin-induced injury was needed for development of pulmonary fibrosis (Lawson *et al.*, 2011). In a similar fashion, ROS generated by asbestos-exposed fibroblasts led to expression of the pro-fibrotic protein TGF-β and increased expression of the ER stress signals XBP-1 and ATF6α which was associated with increased α-SMA and collagen type 1 expression (Baek *et al.*, 2012). Alveolar macrophages isolated from the bronchoalveolar lavage fluid of patients with asbestosis express several markers of ER stress (Ryan *et al.*, 2014), suggesting a pathobiological link between AM ER stress and lung fibrosis. In our model of asbestos-induced lung fibrosis, we detect extensive evidence of ER stress in the lungs in association with AEC mtDNA damage and apoptosis (Cheresh *et al.*, 2015). Taken together, these data show that various inhaled toxins, including asbestos, promote ER stress and lead to fibrotic remodelling in the lung. Further study is required to elucidate the underlying mechanisms by which ER stress leads to intrinsic AEC apoptosis and whether modification of this pathway could lead to new therapeutic targets for lung fibrosis.

4.3. p53-regulated gene expression and apoptosis

A major mechanism by which asbestos fibres cause epithelial cell toxicity, pulmonary fibrosis and malignancy involves activation of p53 (see for review: Huang *et al.*, 2011; Cheresh *et al.*, 2013; Liu *et al.*, 2013). p53 is considered to be the 'gatekeeper' of genome owing to its role in the integration of various signals controlling cell cycle arrest, differentiation, senescence and apoptosis (see for review: Vousden and Prives, 2009; Liu *et al.*, 2010; Huang *et al.*, 2011). Upon exposure to asbestos or other injurious stimuli, a normal p53 response can prevent the accumulation of gene mutations by arresting the cell cycle to allow time for DNA repair (Vousden and Prives, 2009). If extensive DNA damage ensues, activation of p53 promotes transcription of pro-apoptotic molecules and inhibit the transcription of anti-apoptotic molecules (Chipuk and Green, 2006) leading to mitochondrial dysfunction and apoptosis.

The mechanisms by which p53 regulates apoptosis are complex and incompletely understood (see for review: Janicke *et al.*, 2008; Vousden and Prives, 2009) though they appear to involve both p53-dependent and independent transcriptional regulation. p53 can induce intrinsic apoptosis by increasing gene expression of the pro-apoptotic molecules BAX, NOXA and PUMA and inhibiting the expression of the anti-apoptotic Bcl-2 family (Oda *et al.*, 2000; Nakano and Vousden, 2001; Wallace, 2013). The p53 protein can also interact physically with the anti-apoptotic protein Bcl-xL to increase Bax/Bak oligomerization and permeabilization of the outer mitochondrial membrane to promote cytochrome c release and intrinsic apoptosis in a transcriptionally independent fashion (Erster *et al.*, 2004; Macip *et al.*, 2003; Marchenko *et al.*, 2000).

Exposure of pulmonary AEC to asbestos leads to rapid activation of p53 promoter activity and protein expression (Panduri *et al.*, 2006; Nymark *et al.*, 2007; Hevel *et al.*, 2008). Our group has shown previously, using confocal microscopy and immuno-blotting of AEC mitochondrial protein, that asbestos induces mitochondrial translocation of p53 (Panduri *et al.*, 2006). Pulmonary A549 cells rendered incapable of mitochondrial ROS production are protected from asbestos-induced mitochondrial p53 translocation and intrinsic apoptosis, highlighting the crucial interaction between mitochondrial ROS production and p53 activation following asbestos exposure (Panduri *et al.*, 2006). Amosite asbestos-triggered mitochondrial p53 and pro-apoptotic Bax translocation were similarly blocked *in vitro* using an iron chelator, phytic acid (Panduri *et al.*, 2006; Soberanes *et al.*, 2006). Asbestos-induced mitochondrial p53 expression noted at the BAD junction, the site of initial fibre deposition, was similarly attenuated by phytic acid exposure (Kamp *et al.*, 1995; Mishra *et al.*, 1997). Collectively, these data demonstrate that asbestos triggers p53-regulated intrinsic apoptosis though the *in vivo* relevance of these findings as well as the precise mechanism by which asbestos-induced ROS modulate p53 signalling await future studies.

4.4. p53 and asbestos-induced lung fibrosis and cancer formation

Abnormal p53 expression has been implicated in the pathogenesis of fibrotic lung diseases including IPF and asbestosis, as well as asbestos-associated malignancies such as bronchogenic lung cancer (Burmeister *et al.*, 2004; Plataki *et al.*, 2005; Huang *et al.*, 2011). Transcriptional activation of p53 and p21, which functions as a regulator of cell cycle progression, following asbestos exposure leads to cell cycle arrest in lung epithelial and mesothelial cells (Mishra *et al.*, 1997; Nelson *et al.*, 2001; Burmeister *et al.*, 2004; Plataki *et al.*, 2005). Increased p53 protein expression can be detected in the bronchiolar and alveolar epithelium of rats exposed to asbestos and humans with IPF (Kuwano *et al.*, 1996; Mishra *et al.*, 1997; Nelson *et al.*, 2001; Burmeister *et al.*, 2004; Plataki *et al.*, 2005; Panduri *et al.*, 2006). Further, increased p53 levels are evident in the lungs of patients with asbestosis (Nuorva *et al.*, 1994) and point mutations in p53 are noted in the respiratory epithelium of patients exposed to CS and asbestos (Husgafvel-Pursiainen *et al.*, 1997). Crocidolite asbestos promotes cellular transformation and p53 gene mutations predominantly in exons 9-11 in BALB/c-3T3 cells (Lin *et al.*, 2000). Pulmonary cells from lung-specific dominant-negative p53 mice exposed to chrysotile asbestos showed increased expression of TGF-β and other pro-fibrotic factors as well as augmented formation of lung adenocarcinomas (Yee *et al.*, 2008). As noted earlier, gene expression microarray studies have illustrated the crucial role of p53 in activation of nearly 2500 genes involved in cell cycle arrest, cancer suppression, cell survival and apoptosis (Nymark *et al.*, 2007; Hevel *et al.*, 2008). Together, these studies highlight the prominent role for p53 in mediating asbestos-induced AEC mitochondrial dysfunction and apoptosis that are probably important in the pathogenesis of asbestos pulmonary toxicity.

5. Role of mitochondrial DNA damage and repair in the lung epithelium exposed to asbestos

5.1. Mitochondrial DNA damage and repair mechanisms

Because ~1–5% of the total molecular oxygen used for cellular respiration in eukaryotic cells results in generation of ROS, the mitochondria and its DNA are one of the key targets of oxidative damage (see for reviews: Bohr *et al.*, 2002; Gredilla *et al.*, 2010; Wallace, 2013; Schumacker *et al.*, 2014; Kim *et al.*, 2015). Efficient repair of oxidant-induced mitochondrial DNA (mtDNA) damage from exposures such as asbestos is essential as unrepaired mtDNA mutations may lead to mitochondrial dysfunction and apoptosis, development of a range of diseases, malignant transformation and aging (see for reviews: Bohr *et al.*, 2002; Gredilla *et al.*, 2010; Wallace, 2013; Schumacker *et al.*, 2014; Kim *et al.*, 2015). It is important to note that, compared to nuclear DNA, mtDNA is ~50-fold more sensitive to oxidative damage, probably due to its proximity to the source of ROS generation, the ETC, and owing to the absence of protective histones. Oxidant stress-induced mtDNA damage, which carries a mutation rate that is 10-fold greater than that of nuclear DNA, has been implicated prominently in various degenerative diseases including pulmonary fibrosis and malignant transformation (Bohr *et al.*, 2002; Enns, 2003; Brandon *et al.*, 2006; Kroemer *et al.*, 2007; Tuppen *et al.*, 2010; Cline, 2012; Kim *et al.*, 2015). Indeed, enhanced asbestos-induced mtDNA damage, as compared to nuclear DNA damage, has been established in both lung epithelial cells (Kim *et al.*, 2014) and mesothelial cells (Shukla *et al.*, 2003b). Thus, a more detailed understanding of how asbestos fibres induce mtDNA damage and alter mtDNA repair mechanisms is crucial not only for providing insights into the pathogenesis of asbestos-induced lung toxicity but also for the development of management strategies.

Though oxidative damage can cause multiple DNA lesions, the most common is the DNA adduct 8-hydroxyguanine (8-oxo-G), which can lead to mispairing of the 8-oxo-G base with adenine rather than cytosine, resulting in a transversion mutation (Bohr *et al.*, 2002). Notably, these mtDNA oxidant-induced adducts have been implicated in human aging and the development of cancer (Ralph *et al.*, 2010; Schumacker *et al.*, 2014; Kim *et al.*, 2015). Further, polymorphisms in DNA repair genes have been proposed as a means to identify asbestos workers at risk of malignancy (Dianzani *et al.*, 2006; Dusinska *et al.*, 2006; Zhao *et al.*, 2006). Multiple lines of evidence implicate oxidative damage to the mtDNA as an important trigger of the epithelial cell apoptosis that promotes fibrogenesis and inflammation-associated malignancies including: (1) cell death is more highly correlated with the extent of mtDNA oxidative lesions than with nuclear DNA damage, (2) mtDNA damage leads to ATP depletion and mitochondrial dysfunction that, as discussed above, are important determinants of apoptotic cell death, (3) defective mtDNA repair predisposes to cell death and (4) enhancing mtDNA repair, particularly by augmenting AEC OGG1 expression, can prevent asbestos-induced apoptosis (see for reviews: Bohr *et al.*, 2002; Wallace, 2013; Schumacker *et al.*, 2014; Kim *et al.*, 2015).

Of note, a major factor underlying the susceptibility of mtDNA to oxidative stress-induced damage and subsequent mutations is its relatively limited mechanisms for repairing mtDNA damage. Interestingly, the mitochondrial proteins involved in mtDNA repair, which are beyond the scope of this review, are all nuclear encoded and heavily dependent on the nuclear DNA repair machinery for effective mitochondrial translocation (see for reviews: Bohr *et al.*, 2002; Wallace, 2013; Schumacker *et al.*, 2014; Kim *et al.*, 2015). Herein we focus primarily on the most well characterized mtDNA repair enzyme, OGG1, which repairs the 8-hydroxyguanine lesion through base excision repair (BER). The BER pathway, which is the primary mechanism of mtDNA repair, involves four sequential steps: recognition and excision of the lesion by a DNA glycosylase, removal of the residual abasic site, gap filling by a DNA polymerase and, finally, DNA ligation to seal the gaps and complete the repair process (see for reviews: Bohr *et al.*, 2002; Wallace, 2013; Schumacker *et al.*, 2014; Kim *et al.*, 2015; Prakash and Doublie, 2015).

OGG1, which recognizes and removes 8-oxo-G, is a bifunctional DNA glycosylase that also has an AP lyase activity, which cleaves DNA at abasic sites through β-elimination (Aburatani *et al.*, 1997; Bjoras *et al.*, 1997). *Ogg1* has two isoforms (α and β), created by alternative splicing of the *Ogg1* transcript. Although levels of the β isoform are nearly 20-fold greater in the mitochondria, it lacks DNA repair activity (Hashiguchi *et al.*, 2004). The important role of OGG1 in mtDNA repair has been highlighted by both *in vitro* and *in vivo* studies. Mice deficient in OGG1 ($Ogg1^{-/-}$) have accumulation of 8-oxo-G in both nuclear and mtDNA with a 20-fold increase in hepatic mitochondrial 8-oxo-G levels (Bohr *et al.*, 2002). In experiments using fluorometric techniques, the site of oxidant-induced OGG1 repair activity was localized to the mitochondria much more robustly than the nucleus (Mirbahai *et al.*, 2010). The presence of *Ogg1* gene polymorphisms or mutations is associated with an increased risk of multiple malignancies in humans including head and neck, lung, gastric and colorectal cancers as well as leukemia (Chevillard *et al.*, 1998). Numerous groups, including our own, have shown that over-expression of a mitochondria-targeted OGG1 (mt-OGG1) prevents oxidant-induced mitochondria-regulated apoptosis, including AEC apoptosis following asbestos exposure (Shukla *et al.*, 2003b; Ruchko *et al.*, 2005; Youn *et al.*, 2007; Panduri *et al.*, 2009; Chouteau *et al.*, 2011; Kim *et al.*, 2014). Collectively, these studies firmly support a pivotal role for OGG1 in protecting the lung epithelium exposed to oxidative stress, including that from asbestos fibres, critical for mitigating lung fibrosis and malignant transformation. Below, we focus on emerging studies showing the importance of OGG1-regulated mtDNA repair in limiting asbestos-induced AEC apoptosis and lung fibrosis.

5.2. Role of Ogg1 and ACO-2 in preventing asbestos-induced apoptosis

Using OGG1 over- and under-expression *in vitro* studies, our group established that OGG1 is necessary for maintaining AEC mtDNA integrity crucial for preventing intrinsic apoptosis following oxidative stress from either asbestos or H_2O_2 (Kim *et al.*, 2014). The curious finding of the abundant β-OGG1 isoform lacking DNA glycosylase activity in the mitochondria led our group to suggest that OGG1 may have a cellular role independent of DNA repair (Panduri *et al.*, 2009). We reported that over-expression of an α-OGG1 mutant, which lacks 8-oxo-G DNA repair activity, was as effective as wild-type OGG1 at preventing cleaved caspase-9 activation and intrinsic apoptosis following either intrinsic (H_2O_2) or extrinsic (asbestos) oxidant stress (Panduri *et al.*, 2009). In this study, we identified a new function where mt-OGG1 did not alter mitochondrial ROS production but interestingly protected ACO-2 from oxidative degradation. Notably, both WT and mutant OGG1 co-precipitate with ACO-2, suggesting an important ACO-2 chaperone function of OGG1 that is akin to how frataxin protects ACO-2 from oxidative damage as discussed below (Bulteau *et al.*, 2004).

ACO-2 is a tricarboxylic acid (TCA) cycle enzyme that is responsible for the conversion of citrate to isocitrate. It is an important mediator in cellular energy biosynthesis given that loss of ACO-2 activity decreases cell survival (Singh *et al.*, 2006). ACO-2, an iron-sulfur protein susceptible to oxidative inactivation that results in the release of redox-active iron, acts as a mitochondrial redox sensor (Bulteau *et al.*, 2003; Chen *et al.*, 2005). Intact ACO-2 appears important for survival. For example, inactivation of ACO-2 is associated with decreased life span in the fruit fly *Drosophila* (Yan *et al.*, 1997) and is associated with multiple neurodegenerative diseases including Huntington disease (Tabrizi *et al.*, 1999), progressive supranuclear palsy (Park *et al.*, 2001) and Friedreich's ataxia (Bradley *et al.*, 2000).

Accumulating evidence suggests that ACO-2 is important for maintaining mtDNA integrity, including in AEC exposed to asbestos fibres both *in vitro* and *in vivo* (Chen *et al.*, 2005; Panduri *et al.*, 2009; Cheresh *et al.*, 2015; Kim *et al.*, 2015). Studies in yeast demonstrate that ACO-2 maintains mtDNA integrity independent of its catalytic activity, suggesting that ACO-2 has a dual role in the TCA cycle and preservation of mtDNA (Chen *et al.*, 2005). ACO-2 co-precipitates with frataxin, an iron chaperone protein, thereby reducing the degree of oxidant-induced inactivation of ACO-2 and promoting ACO-2 enzyme reactivation by converting the inactive $[3Fe\text{-}4S]^{1+}$ enzyme to the active $[4Fe\text{-}4S]^{2+}$ form (Bulteau *et al.*, 2004). As noted above, our group has shown that both wild-type OGG1 and a mutant mt-OGG1 incapable of DNA repair co-precipitate with ACO-2 and, remarkably, completely block oxidant-induced decreases in ACO-2 protein expression and activity in lung epithelial cells (Panduri *et al.*, 2009). Using ACO-2 over- and under-expression *in vitro* studies, our group recently established that ACO-2 is important for maintaining mtDNA integrity which prevents subsequent oxidant-induced intrinsic apoptosis following either asbestos or H_2O_2 exposure (Kim *et al.*, 2014). Notably, our finding that AEC mtDNA is maintained in the setting of ACO-2 silencing and over-expression of the WT mt-OGG1, but not mt-OGG1 mutant, suggests that ACO-2 is not involved with mtDNA repair but rather simply acts as a chaperone of mtDNA from oxidative damage akin to histone protection of nuclear DNA (Kim *et al.*, 2014). We also found that Ogg1 silencing using short hairpin RNA techniques decreases basal levels of ACO-2 and augments asbestos- and H_2O_2-induced AEC apoptosis which is in accordance with studies in fibroblasts demonstrating increased ROS generation and intrinsic apoptosis following loss of OGG1 (Bacsi *et al.*, 2007; Youn *et al.*, 2007).

The *in vivo* relevance of these findings has been the subject of a recent investigation by our group, which showed that $Ogg1^{-/-}$ mice are more susceptible to asbestos-induced pulmonary fibrosis when compared to their WT counterparts (Ruchko *et al.*, 2005). Interestingly, primary AT2 cells isolated from $Ogg1^{-/-}$ mice had increased mtDNA damage and p53 expression and decreased ACO-2 expression under baseline conditions which was further augmented following a single intratracheal instillation of crocidolite asbestos. Collectively, our findings suggest that OGG1 and ACO-2 have key roles in maintaining AEC mtDNA integrity in the setting of oxidative stress from asbestos or H_2O_2 that are important for maintaining AEC mitochondrial function and limiting AEC intrinsic apoptosis and lung fibrosis. These data are in agreement with studies from other groups showing that mtDNA damage plays a crucial pathophysiologic role in numerous degenerative diseases including atherosclerosis, cardiac fibrosis and heart failure, mechanical ventilator-induced diaphragmatic dysfunction and malignancies (Elahi *et al.*, 2002; Tsutsui *et al.*, 2011; Wang *et al.*, 2011; Duan *et al.*, 2012; Picard *et al.*, 2012; Ding *et al.*, 2013; Przybylowska *et al.*, 2013). Figure 3 shows a simplified model by which OGG1 and ACO-2 function in concert to protect AEC mtDNA from asbestos-induced mitochondrial ROS and thereby limit apoptosis.

The precise molecular mechanisms underlying the interaction between OGG1 and ACO-2, the crucial binding sites and the manner by which they act in concert to maintain mtDNA integrity remain poorly understood. Mt-OGG1 may inhibit the oxidant-induced modification of ACO-2 that leads to degradation by mitochondrial Lon protease (Bota and Davies, 2002; Bota *et al.*, 2005). Interestingly, Lon protease degrades ACO-2 which has undergone oxidative inactivation at a significantly higher rate than unexposed ACO-2, suggesting that oxidative modification of ACO-2 may be

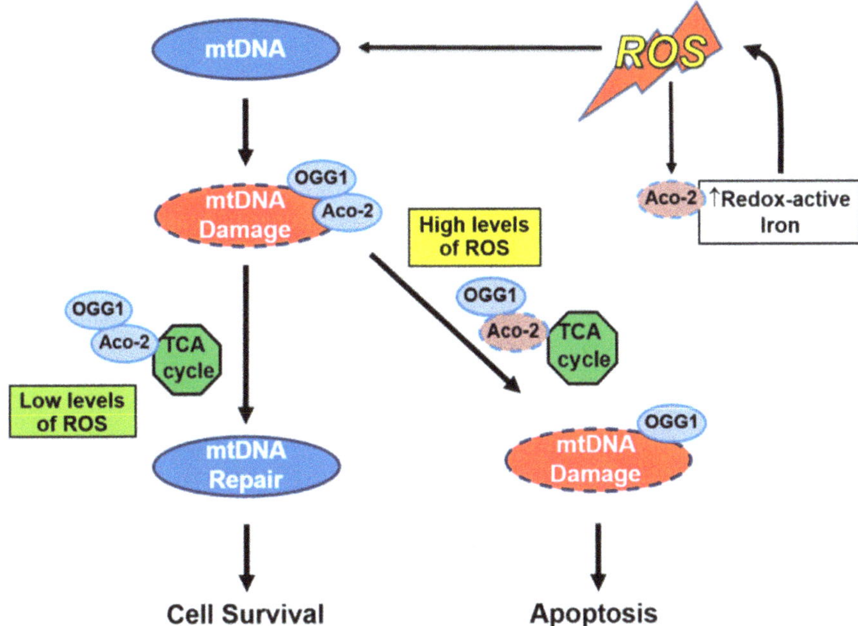

Figure 3. Hypothetical model depicting the role of OGG1 and ACO-2 in mantaining AEC mtDNA integrity necessary for preventing intrinsic apopotosis. Legend: BER = base excision repair; UNG1 = uracil-DNA glycosylase 1: MYH = mutY homologue; OGG1 = 8-oxoguanine-DNA glycosylase 1; NTH1 = endonuclease III-like protein 1; NEIL1 = endonuclease VIII-like 1; APE1 = apurinic/apyrimidicin endonuclease 1; LP-BER = long-patch bese excision repair; FEN1 = flap endonuclease 1; EXOG = exonuclease G; LIG3 = DNA ligase 3.

important in mitochondrial dysfunction and be a signal to ensure the cell undergoes intrinsic apoptosis. An inhibitor of Lon, MG132, attenuates asbestos-induced reductions in ACO-2 activity, lending further support to this hypothesis (Panduri *et al.*, 2006). There is also evidence supporting an important association between p53, OGG1 and ACO-2 that may protect AEC exposed to asbestos. p53 regulates *Ogg1* gene transcription in colon and renal epithelial cells (Youn *et al.*, 2007). p53-deficient cells have reduced expression and activity of both OGG1 and ACO-2 (Youn *et al.*, 2007; Tsui *et al.*, 2011) and p53 activation is required to induce apoptosis in OGG1-deficient fibroblasts (Chouteau *et al.*, 2011). Finally, a novel p53 construct localized to the mitochondrial matrix showed that increased mitochondrial p53, such as occurs following oxidative stress, enhanced the sensitivity of HepG2 cells to oxidant-induced injury (Koczor *et al.*, 2013). However, the relevance of these findings to AEC mtDNA damage and apoptosis following asbestos exposure, as well as their significance *in vivo*, remain unclear. Future studies are clearly necessary to better understand the detailed molecular mechanisms involved as well as the translational significance of these findings in murine models of lung fibrosis as well as in humans with IPF and asbestosis.

6. Role of autophagy and kinase signalling

6.1. Autophagy – role in lung fibrosis and cancer

Work by others, as well as by our group, is reviewed in greater detail below and implicates a prominent role for deficient AEC autophagy and mitophagy in mediating the mitochondrial dysfunction and apoptosis important in the pathobiology of IPF as well as asbestos-exposed AEC. Autophagy, an evolutionarily conserved process first described more than 50 years ago, is also a fundamental cellular homeostatic process that enables cells to clean up portions of the cytoplasm and degrade the constituents in a regulated fashion (Klionsky and Emr, 2000). During this process, cytoplasmic cargo is sequestered within autophagosomes and degraded after fusion of the double-membraned autophagosome with the lysosome (Ashford and Porter, 1962). Autophagy is now widely recognized as a key regulator of innate and adaptive immunity including regulation of inflammation, antigen presentation and bacterial clearance (Levine *et al.*, 2011). Due to this wide range of functions that autophagy can affect, it is intricately linked to a variety of health and disease states (Shintani and Klionsky, 2004). Much of the current knowledge regarding the role of autophagy in cellular homeostasis has been obtained in studies in yeast and mice, whereas its role in human diseases has been investigated only recently.

Under normal conditions, mammalian target of rapamycin (mTOR) inhibits the formation of a complex between autophagy-related gene 5 (Atg5) and UNC 51-like kinase (ULK), the Atg/ULK complex, in effect blocking autophagy. Various stimuli, including hypoxia and starvation, inhibit mTOR and allow the Atg/ULK kinase complex to become activated and initiate formation of an isolation membrane called a phagophore. The completion of this critical step is driven by vascular sorting protein 34, a class III PI3K that is bound to beclin-1 and other ATG proteins (*e.g.* ATG14), which generates PI3K to catalyse the production of phosphatidylinositol-3-phosphage (Chan, 2009). The mTOR pathway is the most widely used pathway in the therapeutic approach to enhance autophagy through drugs such as rapamycin. Autophagy can also be induced by reducing the intracellular levels of IP3 in an mTOR-dependent manner (Williams *et al.*, 2002).

The role of autophagy in human fibrotic lung disease is an emerging area of investigative interest (see for review: Haspel and Choi, 2011; Araya *et al.*, 2013; Patel *et al.*, 2013). Epithelial cell senescence and myofibroblast differentiation, recognized to be key steps in the pathobiology underlying lung fibrosis, are both consequences of insufficient autophagy. In the lung, insufficient autophagy appears to be one mechanism underlying the accelerated epithelial cell senescence and EMT that lead to lung fibrosis (Araya *et al.*, 2013). Recently, Mi and coworkers showed that IL-17A inhibited autophagy in mouse lung epithelial cells and promoted collagen synthesis and EMT in a TGF-β-dependent manner. Interestingly, blockade of IL-17A protected mice against the development of silica- and bleomycin-induced lung fibrosis, suggesting that functional epithelial cell autophagy may play a protective role in the development of fibrosis (Mi *et al.*, 2011). Similarly, TGF-β-mediated autophagy may play a protective role in lung fibroblasts. Patel and colleagues showed increased TGF-β1-induced expression of fibronectin and α-smooth muscle actin in fibroblasts following silencing of the key autophagy protein beclin-1 (Patel *et al.*, 2012). Additionally, suppression of the transcription factor FoxO3a leads to transcriptional activation of LC3B, a marker of autophagosomal activity, leading to abnormal fibroblast proliferation (Im *et al.*, 2015). Autophagy and apoptosis are linked by the autophagy-related gene 5 (Atg5) which, following cleavage, is translocated from the cytosol to the mitochondria where it binds Bcl-xL and triggers cytochrome c release and caspase activation (Yousefi *et al.*, 2006). Regulation of both apoptosis and autophagy share common upstream signals (Eisenberg-Lerner *et al.*, 2009; Eisenberg-Lerner and Kimchi, 2009), though the cross-talk between these two important cellular pathways remains incompletely understood. However, it is unclear whether asbestos activates similar pathways in the lung epithelium. Furthermore, a more detailed understanding of the mechanisms underlying the protective effect of autophagy in lung fibrosis as well as the use agents that induce autophagy, such as rapamycin (Korfhagen *et al.*, 2009), to target therapeutically this pathway, will be of considerable interest.

Mitophagy, the selective engulfment of dysfunctional mitochondria by autophagosomes, has recently been implicated in the pathobiology of IPF (Bueno *et al.*, 2015; Patel *et al.*, 2015). The underlying molecular mechanisms responsible for identifying damaged mitochondria and initiating the mitophagy pathway in IPF has been identified recently as dependent on PINK1 (Valente *et al.*, 2004). In healthy mitochondria, the mitochondrial membrane potential regulates import of PINK1 to the inner mitochondrial membrane, where it is cleaved by presenilins-associated rhomboid-like protein (PARL). Damage to the mitochondria and loss of the mitochondrial membrane potential leads to accumulation of PINK1 on the outer mitochondrial membrane which recruits parkin and leads to mitophagy (Chu, 2010). Transmission electron microscopy (TEM) of lung tissue from IPF patients demonstrates accumulation of enlarged and dysmorphic mitochondria in AT2 cells which co-localized with expression of the autophagy marker LC3 *via* immunofluorescence (Bueno *et al.*, 2015). Interestingly, mice genetically deficient in PINK1 ($PINK1^{-/-}$) developed similarly dysmorphic mitochondria in their AT2 cells and were more susceptible than WT mice to herpesvirus- and bleomycin-induced lung fibrosis. In human lung fibroblasts, PINK1 expression was downregulated by TGF-b1 signalling resulting in decreased mitophagy and an increase in cellular ROS suggesting that TGF-b1-driven abnormal mitophagy may have injurious effects which promote fibrosis in multiple cell types (Sosulski *et al.*, 2015). The precise mechanisms by which loss of PINK1 leads to mitophagy and fibrotic lung disease, as well as whether asbestos fibres induce similar changes in AEC, remain unclear. Further studies in this area are necessary to better inform our understanding of the significance deficiencies in AEC autophagy/mitophagy, mitochondrial dysfunction and susceptibility to fibrotic lung disease, including that following asbestos exposure. Moreover, investigative efforts in this area should provide novel therapeutic targets for lung fibrosis.

The regulation of autophagy overlaps closely with signalling pathways dictating carcinogenesis and, as such, may lead to significant insights into asbestos-induced pulmonary toxicity. Interestingly, the autophagy pathway may also aid in cell survival in nutrient-poor conditions, such as those found in rapidly growing cancers, through lysosomal recycling of intracellular nutrients to allow time for the development of adoptive changes in gene expression and cellular metabolism. Similarly, it may also render malignant cells invulnerable to chemotherapy. Autophagy probably plays a dual role in cancer, acting as a stop-gap to prevent accumulation of damaged proteins and organelles as well as a cell survival mechanism that can promote the growth of established cancer cells (Yang *et al.*, 2011). In human non-small cell lung cancer (NSCLC), expression of the autophagy marker beclin-1 was inversely correlated with cancer size and primary cancer stage (Won *et al.*, 2012). Several cancer suppressor genes involved in the upstream regulation of TOR signalling, including PTEN and the tuberous sclerosis genes 1 and 2 (TSC1 and 2), stimulate autophagy. Another cancer suppressor gene recognized to be integral to development of asbestos-induced fibrosis and cancer formation, p53, positively regulates autophagy through AMPK activation of the TSC1/TSC2 complex (Feng *et al.*, 2005; Crighton *et al.*, 2006).

Modulation of the autophagy pathway may play a strategic role in cancer therapeutics as some cancers, such as those driven by the oncogene Ras, are autophagy-dependent in order to maintain the substrates necessary for mitochondrial acetyl-CoA synthesis (for review see: White, 2012). Acquired resistance to certain classes of chemotherapy and radiation may be acquired in NSCLC through loss of PTEN; *in vitro* loss of PTEN in a NSCLC cell line was accompanied by repression of radiation-induced cytotoxic autophagy (Kim *et al.*, 2013). Treatment of the cells with the mTOR inhibitor rapamycin released the repression of cytotoxic autophagy, suggesting that modulation of the autophagy pathway may be an effective strategy to radiosensitize NSCLC harbouring certain mutations. In a mouse model of KRas-driven lung cancer, tissue-specific inactivation of Atg5, which is required for the formation of autophagosomes, resulted in a reduction in cancer mass and increased survival (Rao *et al.*, 2014). At the same time, disabled autophagy led to a marked acceleration in the onset of lung cancers which was associated with other deleterious effects including impaired mitochondrial energy homeostasis, oxidative stress and a constitutively active DNA damage response. Further study to clarify the multifaceted effects of the autophagy pathway in lung fibrosis and lung cancer following asbestos exposure will be important for facilitating the development of targeted therapies for these diseases.

6.2. Role of autophagy in mediating asbestos-induced lung epithelial cell apoptosis

As discussed above, autophagy is a context-specific process which can either be adaptive and pro-survival or conversely lead to malignancy, aging and various disease processes. Recently, the present research group has demonstrated that chrysotile asbestos causes autophagy in human-lung epithelial cells and murine embryonic fibroblasts (MEF) as demonstrated by the formation of autophagosomes, imaged using TEM, and increased mRNA and protein expression of the autophagy marker LC3-II (Lin *et al.*, 2014). The increase in LC3-II expression was associated with decreased phosphorylation of AKT and mTOR and not seen in double knock-out AKT1/AKT2

MEF cells, suggesting that the effect of chrysotile asbestos on autophagy is mediated at least in part by inhibition of the AKT/mTOR signalling pathway. Further, inhibition of the autophagy cascade using an autophagy inhibitor or si-RNA targeted to autophagy-related gene 5 did not block asbestos-induced apoptosis, suggesting that asbestos-mediated autophagy may be an adaptive rather than pro-survival response.

ROS, including that following exposure to asbestos, are important regulators of autophagy (Li *et al.*, 2011; Filomeni *et al.*, 2015) and can induce directly dephosphorylation of mTOR in glioma cells (Gibson, 2010). Interestingly, use of the antioxidant N-acetylcysteine reduced chrysotile asbestos-induced dephosphorylation of AKT in A549 cells and blocked activation of phospho-JNK and subsequent autophagy, suggesting that autophagy following asbestos exposure is mediated in part by ROS (Lin *et al.*, 2014). ROS-induced activation of the autophagy pathway may also underlie the detrimental effects of exposure to other inhaled particles on the pulmonary epithelium. Similar to asbestos, exposure to PM2.5 particulate matter led to an increase in lung epithelial cell (A549) ROS which was associated with increased expression of the autophagy markers Atg5 and beclin-1, suggesting that the injurious effects of PM2.5 may be in part mediated by oxidant stress-induced autophagy (Deng *et al.*, 2013). Wang and colleagues demonstrated in human lung epithelial cells (A549) that induction of autophagy following PM2.5 exposure was dependent on AMP kinase (AMPK) activation (Wang *et al.*, 2015). Inhibition of autophagy increased p53-mediated apoptotic cell death whereas induction of autophagy with rapamycin protected cells from apoptosis, suggesting that PM2.5-induced autophagy was a pro-survival mechanism similar to that following asbestos. Figure 4 illustrates a hypothetical model by which asbestos-induced alterations may prevent AEC apoptosis and lung fibrosis. Although autophagy is important in clearing oxidatively damaged cells, it remains unclear how ROS alter autophagy as well as how autophagy signalling senses DNA damage (Filomeni *et al.*, 2015). Furthermore, the detailed molecular mechanisms by which asbestos and other particulates induce autophagy as well as the *in vivo* relevance of these findings await further study. We reason that such studies will be important for informing our understanding of the pathobiology of asbestos pulmonary toxicity and may lead to advanced therapies for fibrotic lung disease as well as lung cancer.

6.3. Asbestos-induced lung epithelial kinase signalling mechanisms

Asbestos induces AEC DNA damage and intrinsic apoptosis which is in part mediated by oxidative stress. Studies over the last 10 to 20 years have elucidated a crucial role for asbestos-induced signalling pathways in modulating lung epithelial cell function, including autophagy and apoptosis. Two groups working with lung epithelial and mesothelial cells have demonstrated that asbestos induced apoptosis is due in part to activation of epidermal growth factor receptor (EGFR) signalling, which results in activation of the MAPK pathway (Zanella *et al.*, 1996) including phosphorylation of c-Jun N-terminal kinase (JNK), p38 and ERK1/2 (Swain *et al.*, 2004). Asbestos-induced phosphorylation of EGFR was decreased following treatment with the iron chelator

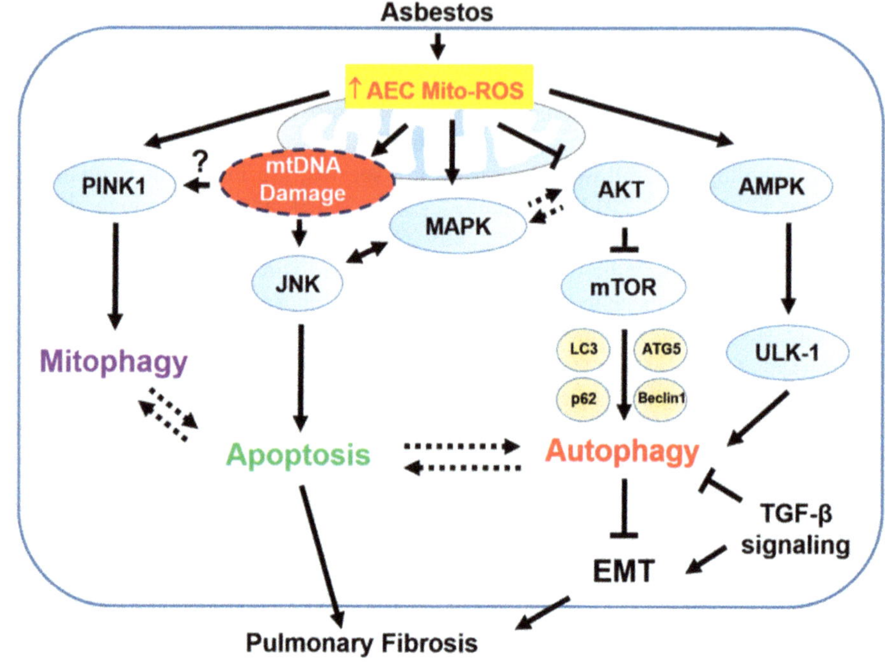

Figure 4. Role of autophagy in mediating asbestos-induced pulmonary fibrosis.

desferrioxamine B and phytic acid, suggesting that EGFR phosphorylation is in part mediated by iron and possibly iron-induced ROS (Baldys and Aust, 2005). Fibroblast growth factor-10, an AT2 cell mitogen, decreased asbestos-induced AEC DNA damage, loss of mitochondrial membrane potential, cleaved caspase-9 activation and apoptosis by modulating MAPK/ERK signalling (Upadhyay *et al.*, 2005). Chrysotile asbestos-induced AEC intrinsic apoptosis was blocked by a specific JNK inhibitor, which further supports an important role for the JNK signalling cascade in mediating apoptosis (Li *et al.*, 2015). A JNK inhibitor also blocks chrysotile asbestos-induced lung epithelial cell autophagy (Lin *et al.*, 2014). An additional mechanism of asbestos-induced JNK activation that drives AEC intrinsic apoptosis occurs *via* protein kinase delta (PKCδ)-dependent mechanisms that promote phosphorylation of JNK leading to expression of pro-apoptotic Bim (Buder-Hoffmann *et al.*, 2009).

Notably, while JNK activation appears to lead autophagy and apoptosis in lung epithelial cells, activation of the MAPK/ERK pathway favours cell homeostasis or pro-survival signalling. The mechanisms governing such variable responses are not entirely clear, though Buder-Hoffman and co-workers showed that PKCδ-mediated phosphorylation of ERK1/2 and JNK1/2 by induction of mitochondrial ROS favours apoptosis in the setting of high-dose asbestos exposure and cell proliferation following low-dose asbestos exposure (Buder-Hoffmann *et al.*, 2009). In addition to mediating mitochondria-regulated apoptosis, MAPK/ERK signalling in human lung epithelial cells (A549) plays a role in pro-fibrotic epithelial cell plasticity and EMT (Tamminen *et al.*, 2012). Taken together, these data demonstrate the complexity of asbestos-induced kinase signalling pathways in mediating autophagy/apoptosis or cell survival suggesting that they are asbestos dose-dependent and cell type-specific. Future studies will be important in better understanding how targeting MAPK signalling pathways can be used for the prevention and treatment of asbestos-induced lung disease.

7. The role of aging in mediating asbestos-induced lung epithelial cell toxicity

7.1. Aging and its effect on lung epithelial cell toxicity, lung fibrosis and lung cancer

Accumulating evidence implicates the role of "exaggerated" lung aging in the pathogenesis of multiple pulmonary processes including COPD and lung fibrosis (see for review: Thannickal, 2013; Selman and Pardo, 2014), though the detailed molecular mechanisms underlying this susceptibility remain poorly understood. There are nine proposed hallmarks of the aging phenotype, including genomic instability, telomere shortening, epigenetic alterations, abnormal proteostasis, dysregulated nutrient sensing, mitochondrial dysfunction, abnormal intracellular communication, stem cell depletion and cellular senescence/apoptosis (Lopez-Otin *et al.*, 2013). In fact, as previously discussed throughout this chapter, many of these abnormalities have been noted in IPF including AEC DNA damage, mitochondria-regulated apoptosis and ER stress. The most common among the many gene mutations found to underlie a familial predisposition to IPF includes the telomerase mutations TERT and TERC (Armanios, 2012) and one study evaluating AT2 cells (the distal lung epithelial stem cell) from IPF patients revealed that 97% had shortened telomeres (Alder *et al.*, 2008). However, the fact that telomere shortening alone did not trigger spontaneous lung fibrosis, or worsen bleomycin-induced lung fibrosis in mice (Degryse *et al.*, 2012), suggests that additional genetic and/or environmental factors are critical in the susceptibility to lung fibrosis.

Swift repair of mtDNA damage is an important protective factor given the wealth of evidence showing that mtDNA damage and mutations are associated with various disease conditions, including lung fibrosis. Mutations in maternally-inherited mtDNA, which encode for genes involved in oxidative phosphorylation, account better for the complex clinical-pathological features of many common degenerative diseases as compared to genetic inheritance through classic Mendelian principles (Wallace and Chalkia, 2013). MtDNA mutations accumulate in tissues with aging, in part due to dysfunctional mtDNA repair, and are thought to modulate aging and longevity (Kujoth *et al.*, 2005). The accumulation of mtDNA mutations due to aging, environmental exposure and chronic oxidative stress support this "mitochondrial theory of aging" that may lead to depletion of key stem cell populations, such as the pulmonary AT2 cell, and promote the development of degenerative diseases and malignancy (Wallace, 2013; Schumacker *et al.*, 2014; Kim *et al.*, 2015).

Given the growing body of evidence implicating IPF as a disease of aging, Sueblinvong and colleagues evaluated the susceptibility of young (3 months) *vs.* old

(24 months) mice to bleomycin-induced lung fibrosis, showing a significant increase in fibrosis at 14 days following injury (Sueblinvong *et al.*, 2012). At baseline, the lungs of older mice exhibited a pro-fibrotic phenotype characterized by higher levels of TGF-β1 receptor and increased MMP-2 and MMP-9 mRNA expression suggesting possible mechanisms underlying the well documented age-related susceptibility to fibrotic lung disease. In a similar manner, Hecker and coworkers demonstrated that while older mice developed a similar degree of fibrotic injury at 3 weeks following bleomycin treatment as their young counterparts, the aged mice were unable to resolve fibrosis in the lungs at 3 months (Hecker *et al.*, 2014). Previous work by Dr Hecker's group implicated the pro-fibrotic mediator, NADPH oxidase 4 (Nox4), as the source of TGF-β1-induced H_2O_2 leading to myofibroblast differentiation, extracellular matrix production and contractility in pulmonary cells and further showed upregulation of NOX4 in lungs obtained from IPF patients (Hecker *et al.*, 2009). Both genetic and pharmacological inhibition of NOX4 in older mice with established bleomycin-induced lung fibrosis attenuated the senescent, anti-apoptotic myofibroblast phenotype and led to reversal of the existing fibrosis (Hecker *et al.*, 2014). A deficiency in nuclear factor erythroid 2-related factor 2 (Nrf2) expression, leading to changes in the cellular redox state, was also observed in aged mice and senescent fibroblasts as has been demonstrated in lung tissue homogenate from IPF patients (Artaud-Macari *et al.*, 2013). Although the role of the NOX4/Nrf2 pathway in the alveolar epithelium and mtDNA damage response is unknown, these studies suggest that targeting the NOX4/Nrf2 imbalance to restore the cellular redox state in lung myofibroblasts may be a novel therapeutic approach for fibrotic lung disease.

7.2. SIRT3

The highly conserved yeast silent information regulator 2 (Sir2) protein has been linked to longevity in a variety of lower organisms including *C. elegans* and *D. melanogaster* (Vaquero, 2009; Verdin *et al.*, 2010; Bosch-Presegue and Vaquero, 2014). The mammalian homologue are the sirtuins, or SIRTs, a group of seven proteins with NAD^+-dependent lysine deacetylase activity localized to various intracellular compartments including the nucleus (SIRT1, SIRT6, SIRT7), cytoplasm (SIRT2) and the mitochondria (SIRT3-5). The sirtuin family, considered by many to be the "guardians of the genome" (Bosch-Presegue and Vaquero, 2014), play a role in maintenance of genomic stability by adapting cellular metabolic flow and energy demand (Verdin *et al.*, 2010) and participating in DNA repair and maintenance of chromatin structure and function through histone deacetylation (Vaquero, 2009).

One member of the sirtuin family, the mitochondrial SIRT3, has been shown to play a crucial role in regulating the mitochondrial redox balance, cellular energy homeostasis and maintenance of mtDNA integrity critical to the development of age-related diseases. As such, SIRT3 has been referred to as the "gatekeeper" of mitochondrial integrity (Kincaid and Bossy-Wetzel, 2013). In light of the relationship between chronic oxidative stress and aging, it is worth noting that SIRT3 is the only sirtuin that has been associated with longevity in humans based on the presence of a variable nucleotide tandem repeat (VNTR) enhancer within the SIRT3 gene associated with increased survival in the elderly (Bellizzi *et al.*, 2005).

As discussed above, mtDNA repair is an important mechanism for protecting against oxidant-induced mitochondrial dysfunction and mutagenesis, including that due to inhaled particle exposure. Emerging evidence suggests that SIRT3 functions as an integral part of the mtDNA repair pathway as its many deacetylase targets include all the members of the TCA cycle, members of the ETC, enzymes involved in cellular antioxidant defences and OGG1 (Verdin *et al.*, 2010; Cheng *et al.*, 2013; Kincaid and Bossy-Wetzel, 2013). Studies in non-lung cell models suggest that SIRT3 deficiency promotes intrinsic apoptosis through several mechanisms including: (1) augmenting mitochondrial ROS production due to acetylation and inactivation of MnSOD and IDH2 (Kincaid and Bossy-Wetzel, 2013; Chen *et al.*, 2014b); (2) increasing cyclophilin D activity leading to loss of the $\Delta\psi_m$ and formation of the mitochondrial permeability transition pore (MPTP) (Sundaresan *et al.*, 2008; Shulga and Pastorino, 2016); (3) promoting Bax-mediated apoptosis through increased acetylation of Ku70 (Sundaresan *et al.*, 2008); and (4) attenuating p53-mediated growth arrest (Kawamura *et al.*, 2010; Li *et al.*, 2010). Notably, a recent study using renal epithelial cancer and glioma cells showed that OGG1 is a SIRT3 deacetylase target and loss of SIRT3 was associated with reduced mtDNA BER resulting in mtDNA damage and intrinsic apoptosis (Cheng *et al.*, 2013). Although the role of SIRT3 in modulating asbestos-induced lung epithelial cell toxicity is unclear, these findings suggest that SIRT3 may be important in maintaining mtDNA integrity and cell survival in the setting of oxidative stress, including that from asbestos.

Chronic oxidative stress leading to SIRT3 depletion and abnormalities in mitochondrial function and biogenesis have been shown to underlie the pathobiology of many age-related diseases. Although young *Sirt3*-deficient mice (*Sirt3*$^{-/-}$) have no discernible phenotype, they are more susceptible to many diseases of aging including the metabolic syndrome, hearing loss, neurodegeneration, cardiac hypertrophy and fibrosis, radiation-induced fibrosis and malignancy (Someya *et al.*, 2009, 2010; Sundaresan *et al.*, 2009; Kincaid and Bossy-Wetzel, 2013; Coleman *et al.*, 2014; Chen *et al.*, 2015; Morigi *et al.*, 2015). More recently, Gupta and colleagues showed that *Sirt3*$^{-/-}$ mice that were old (*i.e.* 18 months) as compared to young (*i.e.* 2 months) developed greater levels of lung fibrosis and that young *Sirt3*$^{-/-}$ mice had enhanced bleomycin-induced lung fibrosis as compared to their WT counterparts (Sundaresan *et al.*, 2016). *Sirt3*$^{-/-}$ mice display striking increases in global protein lysine acetylation not seen in mice deficient in the other mitochondrial sirtuins SIRT4 and SIRT5, suggesting that SIRT3 is indeed the key mitochondria deacetylase (Lombard *et al.*, 2007). In a mouse model of cardiac fibrosis, *Sirt3*$^{-/-}$ mice developed severe cardiac interstitial fibrosis and hypertrophy following application of a hypertrophic stimulus, whereas mice genetically engineered to over-express SIRT3 were protected (Sundaresan *et al.*, 2009). Interestingly, SIRT3 appeared to exert its cardio-protective effect through a FOXO3a-dependent increase in transcription of the antioxidants catalase and MnSOD. Recently, another group demonstrated that the age-related cardiac dysfunction in *Sirt3*$^{-/-}$ mice was associated with increased lysine acetylation of multiple energy metabolic proteins and subsequent depletion of cardiac mitochondrial respiratory capacity and ATP synthesis (Koentges *et al.*, 2015). Several groups have investigated small molecules that can induce SIRT3 expression (*i.e.* resveratrol, viniferin, honokiol) and shown protection against oxidant-induced degenerative changes in the heart and other non-pulmonary cells (Sener *et al.*, 2007; Fu *et al.*, 2012; Desquiret-Dumas *et al.*, 2013; Haohao *et al.*, 2015; Liu *et al.*, 2015b; Pillai *et al.*, 2015). Given the importance of aging in lung fibrosis and asbestosis, we reason that SIRT3 deficiency may play an important role in promoting AEC mtDNA damage and apoptosis following exposure to asbestos. Recent work by the present research group has demonstrated that Sirt3$^{-/-}$ mice are more susceptible to asbestos-induced lung fibrosis, AEC protein lysine acetylation and AT2 cell mtDNA damage (Jablonski *et al.*, 2017). Notably, AT2 cells of IPF lungs have increased acetylation of the known SIRT3 target, MnSOD, suggesting that functional SIRT3 deficiency may play a role in the pathogenesis of IPF (Jablonski *et al.*, 2017). Further studies are needed to determine whether maintenance of SIRT3 function impacts AEC response to oxidative stress (*i.e.* asbestos and other particulate matter) thereby preventing mtDNA damage and ensuing pulmonary fibrosis and whether novel, small-molecule agents may provide exciting new therapies for IPF and asbestosis.

8. Conclusions

As reviewed herein, studies over the last two decades have shown convincingly that asbestos fibres induce lung epithelial cell injury and apoptosis which is important in the pathobiology of pulmonary fibrosis (Fig. 1). Further, all forms of asbestos fibres can injure the pulmonary epithelium, especially alveolar epithelial cells (AEC), which subsequently can trigger a fibrotic response in the lungs.

It is now widely accepted that the extent of AEC injury and inadequate repair are critical determinants of the fibrogenic potential of toxins, such as asbestos. We reviewed briefly how asbestos fibres induce ROS production *via* reactions occurring on the fibre surface, release from recruited inflammatory cells, and from the mitochondria of target cells, such as the lung epithelium and macrophages. A particularly important role for mitochondrial ROS is suggested by the protective effects noted in genetically manipulated lung epithelial cells incapable of mitochondrial ROS production (ρ0-cells), through over-expressing anti-apoptotic Bcl-xL or by enhancing mtDNA repair (mtOGG1). Multiple lines of evidence firmly show that asbestos fibres cause a wide variety of DNA damage to both nuclear and mitochondrial DNA that impacts mitochondrial function. We reviewed the accumulating recent evidence that implicates the levels of asbestos-induced mtDNA damage, as opposed to nuclear DNA damage, in driving AEC p53 activation and intrinsic apoptosis that is supported by some *in vivo* data in *Ogg1*-deficient mice. Although we focused on asbestos-induced AEC toxicity it is evident that lung epithelial cell toxicity following exposure from other toxins (*i.e.* silica, airborne PM, MWCN, *etc.*) occurs in a remarkably similar manner, though some distinct molecular pathways are activated that are probably due to differences in the chemical structure and lung deposition patterns.

There are numerous areas identified in this review for which further information is needed urgently to better inform our understanding of the pathobiology of asbestos-induced lung epithelial cell toxicity. First among the key information gaps for which additional studies are warranted is a better understanding of how asbestos fibres augment AEC mitochondrial ROS production which results in AEC apoptosis and lung fibrosis as this will probably have important broader implications for informing malignant transformation pathways, including lung cancer. Next, more studies on the molecular pathways promoting and mitigating asbestos-induced AEC mtDNA damage, in particular, and whether strategies aimed at augmenting mtOGG1 or ACO-2 in preserving mtDNA integrity in murine lung fibrosis models are crucial in preventing asbestos-induced AEC apoptosis and lung fibrosis. An important caveat to this line of study is the recognition that the complexity of mtDNA repair following asbestos exposure is poorly understood and that future studies will probably identify multiple potentially more important therapeutic strategies. Third, additional work is necessary in order to understand the detailed molecular mechanisms by which asbestos causes abnormalities in lung epithelial cell autophagy and mitophagy important in modulating apoptosis, an area that should also be amenable to novel therapeutic approaches. Finally, studies exploring the role of 'exaggerated' lung aging following asbestos exposure and, in particular, greater understanding of the roles of AEC mtDNA damage, the ER stress response, intrinsic apoptosis, abnormalities in SIRT3 activity and alterations in telomeres should afford insights into the pathogenesis of IPF/asbestosis and lung cancer.

In summary, the asbestos paradigm continues to inform our understanding of the fundamental mechanisms underlying AEC dysfunction, important for augmenting lung fibrosis and, perhaps, malignant transformation. Importantly, past and future investigative efforts noted above will be important for continuing to enlighten our understanding of the scientific basis of asbestos-induced lung epithelial cell toxicity. This will be especially significant for the rational development of novel therapeutic targets that may prove useful for the management of asbestosis as well as other age-related diseases, including IPF and lung cancer, for which more effective treatments are clearly needed.

Acknowledgements

This work was supported by VA Merit (2 I01 BX000786-05A2) and NIH grant RO1 ES020357 to David W. Kamp, and NIH/NHLBI training grant 2T32HL076139-11A1 to Renea P. Jablonski.

References

Aburatani, H., Hippo, Y., Ishida, T., Takashima, R., Matsuba, C., Kodama, T., Takao, M., Yasui, A., Yamamoto, K. and Asano, M. (1997) Cloning and characterization of mammalian 8-hydroxyguanine-specific DNA glycosylase/apurinic, apyrimidinic lyase, a functional mutM homologue. *Cancer Research*, **57**, 2151–2156.

Alder, J.K., Chen, J.J., Lancaster, L., Danoff, S., Su, S.C., Cogan, J.D., Vulto, I., Xie, M., Qi, X., Tuder, R.M., Phillips III, J.A., Lansdorp, P.M., Loyd, J.E. and Armanios, M.Y. (2008) Short telomeres are a risk factor for idiopathic pulmonary fibrosis. *Proceedings of the National Academy of Sciences of the United States of America*, **105**, 13051–13056.

Alfonso, H., Franklin, P., Ching, S., Croft, K., Burcham, P., Olsen, N., Reid, A., Joyce, D., de Klerk, N. and Musk, A.B. (2015) Effect of N-acetylcysteine supplementation on oxidative stress status and alveolar inflammation in people exposed to asbestos: a double-blind, randomized clinical trial. *Respirology (Carlton, Vic)*, **20**, 1102–1107.

Aljandali, A., Pollack, H., Yeldandi, A., Li, Y., Weitzman, S.A. and Kamp, D.W. (2001) Asbestos causes apoptosis in alveolar epithelial cells: role of iron-induced free radicals. *The Journal of Laboratory and Clinical Medicine*, **137**, 330–339.

Araya, J., Kojima, J., Takasaka, N., Ito, S., Fujii, S., Hara, H., Yanagisawa, H., Kobayashi, K., Tsurushige, C., Kawaishi, M., Kamiya, N., Hirano, J., Odaka, M., Morikawa, T., Nishimura, S.L., Kawabata, Y., Hano, H., Nakayama, K. and Kuwano, K. (2013) Insufficient autophagy in idiopathic pulmonary fibrosis. *American Journal of Physiology Heart and Circulatory Physiology*, **304**, L56–69.

Armanios, M. (2012) Telomerase and idiopathic pulmonary fibrosis. *Mutation Research*, **730**, 52–58.

Artaud-Macari, E., Goven, D., Brayer, S., Hamimi, A., Besnard, V., Marchal-Somme, J., Ali, Z.E., Crestani, B., Kerdine-Romer, S., Boutten, A. and Bonay, M. (2013) Nuclear factor erythroid 2-related factor 2 nuclear translocation induces myofibroblastic dedifferentiation in idiopathic pulmonary fibrosis. *Antioxidants & Redox Signaling*, **18**, 66–79.

Ashford, T.P. and Porter, K.R. (1962) Cytoplasmic components in hepatic cell lysosomes. *The Journal of Cell Biology*, **12**, 198–202.

Aung, W., Hasegawa, S., Furukawa, T. and Saga, T. (2007) Potential role of ferritin heavy chain in oxidative stress and apoptosis in human mesothelial and mesothelioma cells: implications for asbestos-induced oncogenesis. *Carcinogenesis*, **28**, 2047–2052.

Aust, A.E., Cook, P.M. and Dodson, R.F. (2011) Morphological and chemical mechanisms of elongated mineral particle toxicities. *Journal of Toxicology and Environmental Health Part B, Critical Reviews*, **14**, 40–75.

Bacsi, A., Chodaczek, G., Hazra, T.K., Konkel, D. and Boldogh, I. (2007) Increased ROS generation in subsets of OGG1 knockout fibroblast cells. *Mechanisms of Ageing and Development*, **128**, 637–649.

Baek, H.A., Kim, D.S., Park, H.S., Jang, K.Y., Kang, M.J., Lee, D.G., Moon, W.S., Chae, H.J., and Chung, M.J. (2012) Involvement of endoplasmic reticulum stress in myofibroblastic differentiation of lung fibroblasts. *American Journal of Respiratory Cell and Molecular Biology*, **46**, 731–739.

Baldys, A., and Aust, A.E. (2005) Role of iron in inactivation of epidermal growth factor receptor after asbestos treatment of human lung and pleural target cells. *American Journal of Respiratory Cell and Molecular Biology*, **32**, 436–442.

Ballirano, P., Bloise, A., Gualtieri, A.F., Lezzerini, M., Pacella, A., Perchiazzi, N., Dogan, M. and Dogan, A.U. (2017) The crystal structure of mineral fibres. Pp. 17–64 in: *Mineral Fibres: Crystal Chemistry, Chemical-physical Properties, Biological Interaction and Toxicity* (A.F. Gualtieri, editor). EMU Notes in Mineralogy, **18**. European Mineralogical Union and Mineralogical Society of Great Britain & Ireland, London.

Bauernfeind, F., Bartok, E., Rieger, A., Franchi, L., Nunez, G. and Hornung, V. (2011) Cutting edge: reactive oxygen species inhibitors block priming, but not activation, of the NLRP3 inflammasome. *Journal of Immunology (Baltimore, Md : 1950)*, **187**, 613–617.

Belitskaya-Levy, I., Hajjou, M., Su, W.C., Yie, T.A., Tchou-Wong, K.M., Tang, M.S., Goldberg, J.D. and Rom, W.N. (2007) Gene profiling of normal human bronchial epithelial cells in response to asbestos and benzo(a)pyrene diol epoxide (BPDE) *Journal of Environmental Pathology, Toxicology and Oncology: Official Organ of the International Society for Environmental Toxicology and Cancer*, **26**, 281–294.

Bellizzi, D., Rose, G., Cavalcante, P., Covello, G., Dato, S., De Rango, F., Greco, V., Maggiolini, M., Feraco, E., Mari, V., Franceschi, C., Passarino, G. and De Benedictis, G. (2005) A novel VNTR enhancer within the SIRT3 gene, a human homologue of SIR2, is associated with survival at oldest ages. *Genomics*, **85**, 258–263.

Bernstein, D.M., Rogers, R.A., Sepulveda, R., Kunzendorf, P., Bellmann, B., Ernst, H., Creutzenberg, O. and Phillips, J.I. (2015) Evaluation of the fate and pathological response in the lung and pleura of brake dust alone and in combination with added chrysotile compared to crocidolite asbestos following short-term inhalation exposure. *Toxicology and Applied Pharmacology*, **283**, 20–34.

Bjoras, M., Luna, L., Johnsen, B., Hoff, E., Haug, T., Rognes, T. and Seeberg, E. (1997) Opposite base-dependent reactions of a human base excision repair enzyme on DNA containing 7,8-dihydro-8-oxoguanine and abasic sites. *The EMBO Journal*, **16**, 6314–6322.

Bledsoe, J.R., Christiani, D.C. and Kradin, R.L. (2015) Smoking-associated fibrosis and pulmonary asbestosis. *International Journal of Chronic Obstructive Pulmonary Disease*, **10**, 31–37.

Bohr, V.A., Stevnsner, T. and de Souza-Pinto, N.C. (2002) Mitochondrial DNA Repair of oxidative damage in mammalian cells. *Gene*, **286**, 127–134.

Bosch-Presegue, L. and Vaquero, A. (2014) Sirtuins in stress response: guardians of the genome. *Oncogene*, **33**, 3764–3775.

Bota, D.A. and Davies, K.J. (2002) Lon protease preferentially degrades oxidized mitochondrial aconitase by an ATP-stimulated mechanism. *Nature Cell Biology*, **4**, 674–680.

Bota, D.A., Ngo, J.K. and Davies, K.J. (2005) Downregulation of the human Lon protease impairs mitochondrial structure and function and causes cell death. *Free Radical Biology & Medicine*, **38**, 665–677.

Boyles, M.S., Young, L., Brown, D.M., MacCalman, L., Cowie, H., Moisala, A., Smail, F., Smith, P.J., Proudfoot, L., Windle, A.H. and Stone, V. (2015) Multi-walled carbon nanotube induced frustrated phagocytosis, cytotoxicity and pro-inflammatory conditions in macrophages are length dependent and greater than that of asbestos. *Toxicology in vitro: an International Journal published in association with BIBRA*, **29**, 1513–1528.

Bradley, J.L., Blake, J.C., Chamberlain, S., Thomas, P.K., Cooper, J.M. and Schapira, A.H. (2000) Clinical, biochemical and molecular genetic correlations in Friedreich's ataxia. *Human Molecular Genetics*, **9**, 275–282.

Brandon, M., Baldi, P. and Wallace, D.C. (2006) Mitochondrial mutations in cancer. *Oncogene*, **25**, 4647–4662.

Broaddus, V.C., Everitt, J.I., Black, B. and Kane, A.B. (2011) Non-neoplastic and neoplastic pleural endpoints following fiber exposure. *Journal of Toxicology and Environmental Health Part B, Critical Reviews*, **14**, 153–178.

Brody, A.R. and Overby, L.H. (1989) Incorporation of tritiated thymidine by epithelial and interstitial cells in bronchiolar-alveolar regions of asbestos-exposed rats. *The American Journal of Pathology*, **134**, 133–140.

Brody, A.R., Hill, L.H., Adkins, B., Jr. and O'Connor, R.W. (1981) Chrysotile asbestos inhalation in rats: deposition pattern and reaction of alveolar epithelium and pulmonary macrophages. *The American Review of Respiratory Disease*, **123**, 670–679.

Buder-Hoffmann, S.A., Shukla, A., Barrett, T.F., MacPherson, M.B., Lounsbury, K.M. and Mossman, B.T. (2009) A protein kinase Cdelta-dependent protein kinase D pathway modulates ERK1/2 and JNK1/2 phosphorylation and Bim-associated apoptosis by asbestos. *The American Journal of Pathology*, **174**, 449–459.

Budinger, G.R., Mutlu, G.M., Eisenbart, J., Fuller, A.C., Bellmeyer, A.A., Baker, C.M., Wilson, M., Ridge, K., Barrett, T.A., Lee, V.Y. and Chandel, N.S. (2006) Proapoptotic bid is required for pulmonary fibrosis. *Proceedings of the National Academy of Sciences of the United States of America*, **103**, 4604–4609.

Bueno, M., Lai, Y.C., Romero, Y., Brands, J., St Croix, C.M., Kamga, C., Corey, C., Herazo-Maya, J.D., Sembrat, J., Lee, J.S., Duncan, S.R., Rojas, M., Shiva, S., Chu, C.T. and Mora, A.L. (2015) PINK1 deficiency impairs mitochondrial homeostasis and promotes lung fibrosis. *The Journal of Clinical Investigation*, **125**, 521–538.

Bulteau, A.L., Ikeda-Saito, M. and Szweda, L.I. (2003) Redox-dependent modulation of aconitase activity in intact mitochondria. *Biochemistry*, **42**, 14846–14855.

Bulteau, A.L., O'Neill, H.A., Kennedy, M.C., Ikeda-Saito, M., Isaya, G. and Szweda, L.I. (2004) Frataxin acts as an iron chaperone protein to modulate mitochondrial aconitase activity. *Science (New York, NY)*, **305**, 242–245.

Bunderson-Schelvan, M., Pfau, J.C., Crouch, R. and Holian, A. (2011) Nonpulmonary outcomes of asbestos exposure. *Journal of Toxicology and Environmental Health Part B, Critical Reviews*, **14**, 122–152.

Burmeister, B., Schwerdtle, T., Poser, I., Hoffmann, E., Hartwig, A., Muller, W.U., Rettenmeier, A.W., Seemayer, N.H. and Dopp, E. (2004) Effects of asbestos on initiation of DNA damage, induction of DNA-strand breaks, P53-expression and apoptosis in primary, SV40-transformed and malignant human mesothelial cells. *Mutation Research*, **558**, 81–92.

Case, B.W., Abraham, J.L., Meeker, G., Pooley, F.D. and Pinkerton, K.E. (2011) Applying definitions of "asbestos" to environmental and "low-dose" exposure levels and health effects, particularly malignant mesothelioma. *Journal of Toxicology and Environmental Health Part B, Critical Reviews*, **14**, 3–39.

Chan, E.Y. (2009) mTORC1 phosphorylates the ULK1-mAtg13-FIP200 autophagy regulatory complex. *Science Signaling*, **2**, pe51.

Chang, L.Y., Overby, L.H., Brody, A.R. and Crapo, J.D. (1988) Progressive lung cell reactions and extracellular matrix production after a brief exposure to asbestos. *The American Journal of Pathology*, **131**, 156–170.

Chen, X.J., Wang, X., Kaufman, B.A. and Butow, R.A. (2005) Aconitase couples metabolic regulation to mitochondrial DNA maintenance. *Science (New York, NY)*, **307**, 714–717.

Chen, T., Nie, H., Gao, X., Yang, J., Pu, J., Chen, Z., Cui, X., Wang, Y., Wang, H. and Jia, G. (2014a) Epithelial-mesenchymal transition involved in pulmonary fibrosis induced by multi-walled carbon nanotubes *via* TGF-beta/Smad signaling pathway. *Toxicology Letters*, **226**, 150–162.

Chen, Y., Fu, L.L., Wen, X., Wang, X.Y., Liu, J., Cheng, Y. and Huang, J. (2014b) Sirtuin-3 (SIRT3), a therapeutic target with oncogenic and tumor-suppressive function in cancer. *Cell Death & Disease*, **5**, e1047.

Chen, T., Li, J., Liu, J., Li, N., Wang, S., Liu, H., Zeng, M., Zhang, Y. and Bu, P. (2015) Activation of SIRT3 by resveratrol ameliorates cardiac fibrosis and improves cardiac function *via* the TGF-beta/Smad3 pathway. *American Journal of Physiology Heart and Circulatory Physiology*, **308**, H424–434.

Cheng, Y., Ren, X., Gowda, A.S., Shan, Y., Zhang, L., Yuan, Y.S., Patel, R., Wu, H., Huber-Keener, K., Yang, J.W., Liu, D., Spratt, T.E. and Yang, J.M. (2013) Interaction of Sirt3 with OGG1 contributes to repair of mitochondrial DNA and protects from apoptotic cell death under oxidative stress. *Cell Death & Disease*, **4**, e731.

Cheresh, P., Kim, S.J., Tulasiram, S. and Kamp, D.W. (2013) Oxidative stress and pulmonary fibrosis. *Biochimica et Biophysica Acta*, **1832**, 1028–1040.

Cheresh, P., Morales-Nebreda, L., Kim, S.J., Yeldandi, A., Williams, D.B., Cheng, Y., Mutlu, G.M., Budinger, G.R., Ridge, K., Schumacker, P.T., Bohr, V.A. and Kamp, D.W. (2015) Asbestos-induced pulmonary fibrosis is augmented in 8-oxoguanine DNA glycosylase knockout mice. *American Journal of Respiratory Cell and Molecular Biology*, **52**, 25–36.

Chevillard, S., Radicella, J.P., Levalois, C., Lebeau, J., Poupon, M.F., Oudard, S., Dutrillaux, B. and Boiteux, S. (1998) Mutations in OGG1, a gene involved in the repair of oxidative DNA damage, are found in human lung and kidney tumours. *Oncogene*, **16**, 3083–3086.

Chipuk, J.E. and Green, D.R. (2006) Dissecting p53-dependent apoptosis. *Cell Death and Differentiation*, **13**, 994–1002.

Chouteau, J.M., Obiako, B., Gorodnya, O.M., Pastukh, V.M., Ruchko, M.V., Wright, A.J., Wilson, G.L. and Gillespie, M.N. (2011) Mitochondrial DNA integrity may be a determinant of endothelial barrier properties in oxidant-challenged rat lungs. *American Journal of Physiology Heart and Circulatory Physiology*, **301**, L892–898.

Chu, C.T. (2010) A pivotal role for PINK1 and autophagy in mitochondrial quality control: implications for Parkinson disease. *Human Molecular Genetics*, **19**, R28–37.

Churg, A. and Stevens, B. (1995) Enhanced retention of asbestos fibers in the airways of human smokers. *American Journal of Respiratory and Critical Care Medicine*, **151**, 1409–1413.

Churg, A., Hobson, J. and Wright, J. (1990) Effects of cigarette smoke dose and time after smoke exposure on uptake of asbestos fibers by rat tracheal epithelial cells. *American Journal of Respiratory Cell and Molecular Biology*, **3**, 265–269.

Cline, S.D. (2012) Mitochondrial DNA damage and its consequences for mitochondrial gene expression. *Biochimica et Biophysica Acta*, **1819**, 979–991.

Coin, P.G., Roggli, V.L. and Brody, A.R. (1992) Deposition, clearance, and translocation of chrysotile asbestos from peripheral and central regions of the rat lung. *Environmental Research*, **58**, 97–116.

Coin, P.G., Roggli, V.L. and Brody, A.R. (1994) Persistence of long, thin chrysotile asbestos fibers in the lungs of rats. *Environmental Health Perspectives*, **102(Suppl. 5)**, 197–199.

Coleman, M.C., Olivier, A.K., Jacobus, J.A., Mapuskar, K.A., Mao, G., Martin, S.M., Riley, D.P., Gius, D. and Spitz, D.R. (2014) Superoxide mediates acute liver injury in irradiated mice lacking sirtuin 3. *Antioxidants & Redox Signaling*, **20**, 1423–1435.

Cregan, S.P., Arbour, N.A., Maclaurin, J.G., Callaghan, S.M., Fortin, A., Cheung, E.C., Guberman, D.S., Park, D.S. and Slack, R.S. (2004) p53 activation domain 1 is essential for PUMA upregulation and p53-mediated neuronal cell death. *The Journal of Neuroscience: The Official Journal of the Society for Neuroscience*, **24**, 10003–10012.

Crighton, D., Wilkinson, S., O'Prey, J., Syed, N., Smith, P., Harrison, P.R., Gasco, M., Garrone, O., Crook, T. and Ryan, K.M. (2006) DRAM, a p53-induced modulator of autophagy, is critical for apoptosis. *Cell*, **126**, 121–134.

de Brito, O.M. and Scorrano, L. (2010) An intimate liaison: spatial organization of the endoplasmic reticulum-mitochondria relationship. *The EMBO Journal*, **29**, 2715–2723.

Degryse, A.L., Xu, X.C., Newman, J.L., Mitchell, D.B., Tanjore, H., Polosukhin, V.V., Jones, B.R., McMahon, F.B., Gleaves, L.A., Phillips III, J.A., Cogan, J.D., Blackwell, T.S. and Lawson, W.E. (2012) Telomerase deficiency does not alter bleomycin-induced fibrosis in mice. *Experimental Lung Research*, **38**, 124–134.

Deng, X., Zhang, F., Rui, W., Long, F., Wang, L., Feng, Z., Chen, D. and Ding, W. (2013) PM2.5-induced oxidative stress triggers autophagy in human lung epithelial A549 cells. *Toxicology in vitro: an International Journal published in association with BIBRA*, **27**, 1762–1770.

Desquiret-Dumas, V., Gueguen, N., Leman, G., Baron, S., Nivet-Antoine, V., Chupin, S., Chevrollier, A., Vessieres, E., Ayer, A., Ferre, M., Bonneau, D., Henrion, D., Reynier, P. and Procaccio, V. (2013) Resveratrol induces a mitochondrial complex I-dependent increase in NADH oxidation responsible for sirtuin activation in liver cells. *The Journal of Biological Chemistry*, **288**, 36662–36675.

Diandini, R., Takahashi, K., Park, E.K., Jiang, Y., Movahed, M., Le, G.V., Lee, L.J., Delgermaa, V. and Kim, R. (2013) Potential years of life lost (PYLL) caused by asbestos-related diseases in the world. *American Journal of Industrial Medicine*, **56**, 993–1000.

Dianzani, I., Gibello, L., Biava, A., Giordano, M., Bertolotti, M., Betti, M., Ferrante, D., Guarrera, S., Betta, G.P., Mirabelli, D., Matullo, G. and Magnani, C. (2006) Polymorphisms in DNA Repair genes as risk factors for asbestos-related malignant mesothelioma in a general population study. *Mutation Research*, **599**, 124–134.

Diebold, L. and Chandel, N.S. (2016) Mitochondrial ROS regulation of proliferating cells. *Free Radical Biology & Medicine*, **100**, 86–93.

Ding, Z., Liu, S., Wang, X., Khaidakov, M., Dai, Y. and Mehta, J.L. (2013) Oxidant stress in mitochondrial DNA damage, autophagy and inflammation in atherosclerosis. *Scientific Reports*, **3**, 1077.

Donaldson, K., Poland, C.A., Murphy, F.A., MacFarlane, M., Chernova, T. and Schinwald, A. (2013) Pulmonary toxicity of carbon nanotubes and asbestos - similarities and differences. *Advanced Drug Delivery Reviews*, **65**, 2078–2086.

dos Santos, G., Rogel, M.R., Baker, M.A., Troken, J.R., Urich, D., Morales-Nebreda, L., Sennello, J.A., Kutuzov, M.A., Sitikov, A., Davis, J.M., Lam, A.P., Cheresh, P., Kamp, D.W., Shumaker, D.K., Budiger, G.R.S. and Ridge, K.M. (2015) Vimentin regulates activation of the NLRP3 inflammasome. *Nature Communications*, **6**, 6574.

Dostert, C., Petrilli, V., Van Bruggen, R., Steele, C., Mossman, B.T. and Tschopp, J. (2008) Innate immune activation through Nalp3 inflammasome sensing of asbestos and silica. *Science (New York, NY)*, **320**, 674–677.

Duan, W.X., Hua, R.X., Yi, W., Shen, L.J., Jin, Z.X., Zhao, Y.H., Yi, D.H., Chen, W.S. and Yu, S.Q. (2012) The association between OGG1 Ser326Cys polymorphism and lung cancer susceptibility: a meta-analysis of 27 studies. *PloS One*, **7**, e35970.

Dusinska, M., Dzupinkova, Z., Wsolova, L., Harrington, V. and Collins, A.R. (2006) Possible involvement of XPA in repair of oxidative DNA damage deduced from analysis of damage, repair and genotype in a human population study. *Mutagenesis*, **21**, 205–211.

Eisenberg-Lerner, A. and Kimchi, A. (2009) The paradox of autophagy and its implication in cancer etiology and therapy. *Apoptosis: an International Journal on Programmed Cell Death*, **14**, 376–391.

Eisenberg-Lerner, A., Bialik, S., Simon, H.U. and Kimchi, A. (2009) Life and death partners: apoptosis, autophagy and the cross-talk between them. *Cell Death and Differentiation*, **16**, 966–975.

Elahi, A., Zheng, Z., Park, J., Eyring, K., McCaffrey, T. and Lazarus, P. (2002) The human OGG1 DNA repair enzyme and its association with orolaryngeal cancer risk. *Carcinogenesis*, **23**, 1229–1234.

Enns, G.M. (2003) The contribution of mitochondria to common disorders. *Molecular Genetics and Metabolism*, **80**, 11–26.

Erster, S., Mihara, M., Kim, R.H., Petrenko, O. and Moll, U.M. (2004) In vivo mitochondrial p53 translocation triggers a rapid first wave of cell death in response to DNA damage that can precede p53 target gene activation. *Molecular and Cellular Biology*, **24**, 6728–6741.

Faner, R., Rojas, M., Macnee, W. and Agusti, A. (2012) Abnormal lung aging in chronic obstructive pulmonary disease and idiopathic pulmonary fibrosis. *American Journal of Respiratory and Critical Care Medicine*, **186**, 306–313.

Fattman, C.L., Tan, R.J., Tobolewski, J.M. and Oury, T.D. (2006) Increased sensitivity to asbestos-induced lung injury in mice lacking extracellular superoxide dismutase. *Free Radical Biology & Medicine*, **40**, 601–607.

Feng, Z., Zhang, H., Levine, A.J. and Jin, S. (2005) The coordinate regulation of the p53 and mTOR pathways in cells. *Proceedings of the National Academy of Sciences of the United States of America*, **102**, 8204–8209.

Filomeni, G., De Zio, D. and Cecconi, F. (2015) Oxidative stress and autophagy: the clash between damage and metabolic needs. *Cell Death and Differentiation*, **22**, 377–388.

Fingerlin, T.E., Murphy, E., Zhang, W., Peljto, A.L., Brown, K.K., Steele, M.P., Loyd, J.E., Cosgrove, G.P., Lynch, D., Groshong, S., Collard, H.R., Wolters, P.J., Bradford, W.Z., Kossen, K., Seiwert, S.D. du Bois, R.M., Garcia, K.M., Devine, M.S., Gudmundsson, G., Isaksson, H.I., Kaminski, N., Zhang, Y., Gibson, K.F., Lancaster, L.H., Cogan, J.D., Mason, W.R., Maher, T.M., Molyneaux, P.L., Wells, A.U., Moffatt, M.F., Selman, M., Pardo, A., Kim, D.S., Crapo, J.D., Make, B.J., Regan, E.A., Walek, D.S., Daniel, J.J., Kamatani, Y., Zelenika, D., Smith, K., McKean, D., Pedersen, B.S., Talbert, J., Kidd, R.N., Markin, C.R., Beckman, K.B., Lathrop, M., Schwarz, M.I. and Schwartz, D.A. (2013) Genome-wide association study identifies multiple susceptibility loci for pulmonary fibrosis. *Nature Genetics*, **45**, 613–620.

Fu, J., Jin, J., Cichewicz, R.H., Hageman, S.A., Ellis, T.K., Xiang, L., Peng, Q., Jiang, M., Arbez, N., Hotaling, K., Ross, C.A. and Duan, W. (2012) trans-(-)-epsilon-Viniferin increases mitochondrial sirtuin 3 (SIRT3), activates AMP-activated protein kinase (AMPK), and protects cells in models of Huntington Disease. *The Journal of Biological Chemistry*, **287**, 24460–24472.

Galluzzi, L., Vitale, I., Abrams, J.M., Alnemri, E.S., Baehrecke, E.H., Blagosklonny, M.V., Dawson, T.M., Dawson, V.L., El-Deiry, W.S., Fulda, S., Gottlieb, S.E., Green, D.R., Hengartner, M.O., Kepp, O., Knight, R.A., Kumar, S., Lipton, S.A., Lu, X., Madeo, F., Malorni, W., Mehlen, P., Nuez, G., Peter, M.E., Piacentini, M., Rubinsztein, D.C., Shi, Y., Simon, H.U., Vandenabeele, P., White, E., Yuan, J., Zhivotovsky, B., Melino, G. and Kroemer, G. (2012) Molecular definitions of cell death subroutines: recommendations of the Nomenclature Committee on Cell Death 2012. *Cell Death and Differentiation*, **19**, 107–120.

Gardner, S.Y., Brody, A.R., Mangum, J.B. and Everitt, J.I. (1997) Chrysotile asbestos and H2O2 increase permeability of alveolar epithelium. *Experimental Lung Research*, **23**, 1–16.

Ghio, A.J., Stonehuerner, J. and Quigley, D.R. (1994) Humic-like substances in cigarette smoke condensate and lung tissue of smokers. *The American Journal of Physiology*, **266**, L382–388.

Ghio, A.J., Stonehuerner, J., Richards, J. and Devlin, R.B. (2008) Iron homeostasis in the lung following asbestos exposure. *Antioxidants & Redox Signaling*, **10**, 371–377.

Gibson, S.B. (2010) A matter of balance between life and death: targeting reactive oxygen species (ROS)-induced autophagy for cancer therapy. *Autophagy*, **6**, 835–837.

Goodglick, L.A. and Kane, A.B. (1990) Cytotoxicity of long and short crocidolite asbestos fibers in vitro and in vivo. *Cancer Research*, **50**, 5153–5163.

Gredilla, R., Bohr, V.A. and Stevnsner, T. (2010) Mitochondrial DNA repair and association with aging – an update. *Experimental Gerontology*, **45**, 478–488.

Gualtieri, A.F. (2017) Introduction. Pp. 1–15 in: *Mineral Fibres: Crystal Chemistry, Chemical-physical Properties, Biological Interaction and Toxicity* (A.F. Gualtieri, editor). EMU Notes in Mineralogy, **18**. European Mineralogical Union and Mineralogical Society of Great Britain & Ireland, London.

Gulati, M. and Redlich, C.A. (2015) Asbestosis and environmental causes of usual interstitial pneumonia. *Current Opinion in Pulmonary Medicine*, **21**, 193–200.

Haohao, Z., Guijun, Q., Juan, Z., Wen, K. and Lulu, C. (2015) Resveratrol improves high-fat diet induced insulin resistance by rebalancing subsarcolemmal mitochondrial oxidation and antioxidantion. *Journal of Physiology and Biochemistry*, **71**, 121–131.

Hardy, J.A. and Aust, A.E. (1995) Iron in asbestos chemistry and carcinogenicity. *Chemical Reviews*, **95**, 97–118.

Haschek, W.M. and Witschi, H.P. (1979) Pulmonary fibrosis: a possible mechanism. *Toxicology and Applied Pharmacology*, **51**, 475–487

Hashiguchi, K., Stuart, J.A., de Souza-Pinto, N.C. and Bohr, V.A. (2004) The C-terminal alphaO helix of human Oggl is essential for 8-oxoguanine DNA glycosylase activity: the mitochondrial beta-Oggl lacks this domain and does not have glycosylase activity. *Nucleic Acids Research*, **32**, 5596–5608.

Haspel, J.A. and Choi, A.M. (2011) Autophagy: a core cellular process with emerging links to pulmonary disease. *American Journal of Respiratory and Critical Care Medicine*, **184**, 1237–1246.

Heath-Engel, H.M., Chang, N.C. and Shore, G.C. (2008) The endoplasmic reticulum in apoptosis and autophagy: role of the BCL-2 protein family. *Oncogene*, **27**, 6419–6433.

Hecker, L., Vittal, R., Jones, T., Jagirdar, R., Luckhardt, T.R., Horowitz, J.C., Pennathur, S., Martinez, F.J. and Thannickal, V.J. (2009) NADPH oxidase-4 mediates myofibroblast activation and fibrogenic responses to lung injury. *Nature Medicine*, **15**, 1077–1081.

Hecker, L., Logsdon, N.J., Kurundkar, D., Kurundkar, A., Bernard, K., Hock, T., Meldrum, E., Sanders, Y.Y. and Thannickal, V.J. (2014) Reversal of persistent fibrosis in aging by targeting Nox4-Nrf2 redox imbalance. *Science Translational Medicine*, **6**, 231–247.

Held, N.M. and Houtkooper, R.H. (2015) Mitochondrial quality control pathways as determinants of metabolic health. *BioEssays: News and Reviews in Molecular, Cellular and Developmental Biology*, **37**, 867–876.

Helmig, S., Dopp, E., Wenzel, S., Walter, D. and Schneider, J. (2014) Induction of altered mRNA expression profiles caused by fibrous and granular dust. *Molecular Medicine Reports*, **9**, 217–228.

Hevel, J.M., Olson-Buelow, L.C., Ganesan, B., Stevens, J.R., Hardman, J.P. and Aust, A.E. (2008) Novel functional view of the crocidolite asbestos-treated A549 human lung epithelial transcriptome reveals an intricate network of pathways with opposing functions. *BMC Genomics*, **9**, 376.

Horan, G.S., Wood, S., Ona, V., Li, D.J., Lukashev, M.E., Weinreb, P.H., Simon, K.J., Hahm, K., Allaire, N.E., Rinaldi, N.J., Goyal, J., Feghali-Bostwick, C.A., Matteson, E.L., O'Hara, C., Lafyatis, R., Davis, G.S., Huang, X., Sheppard, D. and Violette, S.M. (2008) Partial inhibition of integrin alpha(v)beta6 prevents pulmonary fibrosis without exacerbating inflammation. *American Journal of Respiratory and Critical Care Medicine*, **177**, 56–65.

Huang, S.X., Jaurand, M.C., Kamp, D.W., Whysner, J. and Hei, T.K. (2011) Role of mutagenicity in asbestos fiber-induced carcinogenicity and other diseases. *Journal of Toxicology and Environmental Health Part B, Critical Reviews*, **14**, 179–245.

Husgafvel-Pursiainen, K., Kannio, A., Oksa, P., Suitiala, T., Koskinen, H., Partanen, R., Hemminki, K., Smith, S., Rosenstock-Leibu, R. and Brandt-Rauf, P.W. (1997) Mutations, tissue accumulations, and serum levels of p53 in patients with occupational cancers from asbestos and silica exposure. *Environmental and Molecular Mutagenesis*, **30**, 224–230.

IARC (1987) Overall evaluations of carcinogenicity: an updating of IARC Monographs. In IARC monographs on the evaluation of carcinogenic risks to humans Supplement / World Health Organization, International Agency for Research on Cancer, **1-42**, 1–440.

Im, J., Hergert, P. and Nho, R.S. (2015) Reduced FoxO3a expression causes low autophagy in idiopathic pulmonary fibrosis fibroblasts on collagen matrices. *American Journal of Physiology Heart and Circulatory Physiology*, **309**, L552–561.

Inamura, K., Ninomiya, H., Nomura, K., Tsuchiya, E., Satoh, Y., Okumura, S., Nakagawa, K., Takata, A., Kohyama, N. and Ishikawa, Y. (2014) Combined effects of asbestos and cigarette smoke on the development of lung adenocarcinoma: different carcinogens may cause different genomic changes. *Oncology Reports*, **32**, 475–482.

Izumi, S., Iikura, M. and Hirano, S. (2012) Prednisone, azathioprine, and N-acetylcysteine for pulmonary fibrosis. *The New England Journal of Medicine*, **367**, 870; author reply 870–871.

Jablonski, R.P., Kim, S.J., Cheresh, P., Williams, D.B., Morales-Nebreda, L., Cheng, Y., Yeldandi, A., Bhorade,

S., Pardo, A., Selman, M., Ridge, K., Gius, D., Budinger, G.R. and Kamp, D.W. (2017) SIRT3 deficiency promotes lung fibrosis by augmenting alveolar epithelial cell mitochondrial DNA damage and apoptosis. *FASEB Journal*, doi: 10.1096/fj.201601077R.

Janicke, R.U., Sohn, D. and Schulze-Osthoff, K. (2008) The dark side of a tumor suppressor: anti-apoptotic p53. *Cell Death and Differentiation*, **15**, 959–976.

Kamp, D.W. (2009) Asbestos-induced lung diseases: an update. *Translational research: The Journal of Laboratory and Clinical Medicine*, **153**, 143–152.

Kamp, D.W., Dunne, M., Anderson, J.A., Weitzman, S.A. and Dunn, M.M. (1990) Serum promotes asbestos-induced injury to human pulmonary epithelial cells. *The Journal of Laboratory and Clinical Medicine*, **116**, 289–297.

Kamp, D.W., Graceffa, P., Pryor, W.A. and Weitzman, S.A. (1992) The role of free radicals in asbestos-induced diseases. *Free Radical Biology & Medicine*, **12**, 293–315.

Kamp, D.W., Dunn, M.M., Sbalchiero, J.S., Knap, A.M. and Weitzman, S.A. (1994) Contrasting effects of alveolar macrophages and neutrophils on asbestos-induced pulmonary epithelial cell injury. *The American Journal of Physiology*, **266**, L84–91.

Kamp, D.W., Israbian, V.A., Yeldandi, A.V., Panos, R.J., Graceffa, P. and Weitzman, S.A. (1995) Phytic acid, an iron chelator, attenuates pulmonary inflammation and fibrosis in rats after intratracheal instillation of asbestos. *Toxicologic Pathology*, **23**, 689–695.

Kamp, D.W., Liu, G., Cheresh, P., Kim, S.J., Mueller, A., Lam, A.P., Trejo, H., Williams, D., Tulasiram, S., Baker, M., Ridge, K., Chandel, N.S. and Beri, R. (2013) Asbestos-induced alveolar epithelial cell apoptosis. The role of endoplasmic reticulum stress response. *American Journal of Respiratory Cell and Molecular Biology*, **49**, 892–901.

Kasper, J., Hermanns, M.I., Bantz, C., Maskos, M., Stauber, R., Pohl, C., Unger, R.E. and Kirkpatrick, J.C. (2011) Inflammatory and cytotoxic responses of an alveolar-capillary coculture model' to silica nanoparticles: comparison with conventional monocultures. *Particle and Fibre Toxicology*, **8**, 6.

Kawamura, Y., Uchijima, Y., Horike, N., Tonami, K., Nishiyama, K., Amano, T., Asano, T., Kurihara, Y. and Kurihara, H. (2010) Sirt3 protects in vitro-fertilized mouse preimplantation embryos against oxidative stress-induced p53-mediated developmental arrest. *The Journal of Clinical Investigation*, **120**, 2817–2828.

Kido, T., Morimoto, Y., Asonuma, E., Yatera, K., Ogami, A., Oyabu, T., Tanaka, I. and Kido, M. (2008) Chrysotile asbestos causes AEC apoptosis *via* the caspase activation in vitro and in vivo. *Inhalation Toxicology*, **20**, 339–347.

Kim, I., Xu, W. and Reed, J.C. (2008) Cell death and endoplasmic reticulum stress: disease relevance and therapeutic opportunities. *Nature Reviews Drug Discovery*, **7**, 1013–1030.

Kim, J.S., Song, K.S., Lee, J.K., Choi, Y.C., Bang, I.S., Kang, C.S. and Yu, I.J. (2012) Toxicogenomic comparison of multi-wall carbon nanotubes (MWCNTs) and asbestos. *Archives of Toxicology*, **86**, 553–562.

Kim, E.J., Jeong, J.H., Bae, S., Kang, S., Kim, C.H. and Lim, Y.B. (2013) mTOR inhibitors radiosensitize PTEN-deficient non-small-cell lung cancer cells harboring an EGFR activating mutation by inducing autophagy. *Journal of Cellular Biochemistry*, **114**, 1248–1256.

Kim, S.J., Cheresh, P., Williams, D., Cheng, Y., Ridge, K., Schumacker, P.T., Weitzman, S., Bohr, V.A. and Kamp, D.W. (2014) Mitochondria-targeted Ogg1 and aconitase-2 prevent oxidant-induced mitochondrial DNA damage in alveolar epithelial cells. *The Journal of Biological Chemistry*, **289**, 6165–6176.

Kim, S.J., Cheresh, P., Jablonski, R.P., Williams, D.B. and Kamp, D.W. (2015) The role of mitochondrial DNA in mediating alveolar epithelial cell apoptosis and pulmonary fibrosis. *International Journal of Molecular Sciences*, **16**, 21486–21519.

Kincaid, B. and Bossy-Wetzel, E. (2013) Forever young: SIRT3 a shield against mitochondrial meltdown, aging, and neurodegeneration. *Frontiers in Aging Neuroscience*, **5**, 48.

Klee, M., Pallauf, K., Alcala, S., Fleischer, A. and Pimentel-Muinos, F.X. (2009) Mitochondrial apoptosis induced by BH3-only molecules in the exclusive presence of endoplasmic reticular Bak. *The EMBO Journal*, **28**, 1757–1768.

Kliment, C.R., Englert, J.M., Gochuico, B.R., Yu, G., Kaminski, N., Rosas, I. and Oury, T.D. (2009) Oxidative stress alters syndecan-1 distribution in lungs with pulmonary fibrosis. *The Journal of Biological Chemistry*, **284**, 3537–3545.

Klionsky, D.J. and Emr, S.D. (2000) Autophagy as a regulated pathway of cellular degradation. *Science (New York, NY)*, **290**, 1717–1721.

Koczor, C.A., Torres, R.A., Fields, E.J., Boyd, A. and Lewis, W. (2013) Mitochondrial matrix P53 sensitizes cells to oxidative stress. *Mitochondrion*, **13**, 277–281.

Koentges, C., Pfeil, K., Schnick, T., Wiese, S., Dahlbock, R., Cimolai, M.C., Meyer-Steenbuck, M., Cenkerova, K., Hoffmann, M.M., Jaeger, C., Odening, K.E., Kammerer, B., Hein, L., Bode, C. and Bugger, H. (2015) SIRT3 deficiency impairs mitochondrial and contractile function in the heart. *Basic Research in Cardiology*, **110**, 36.

Kopnin, P.B., Kravchenko, I.V., Furalyov, V.A., Pylev, L.N. and Kopnin, B.P. (2004) Cell type-specific effects of asbestos on intracellular ROS levels, DNA oxidation and G1 cell cycle checkpoint. *Oncogene*, **23**, 8834–8840.

Korfei, M., Ruppert, C., Mahavadi, P., Henneke, I., Markart, P., Koch, M., Lang, G., Fink, L., Bohle, R.M., Seeger, W., Weaver, T. and Guenther, A. (2008) Epithelial endoplasmic reticulum stress and apoptosis in sporadic idiopathic pulmonary fibrosis. *American Journal of Respiratory and Critical Care Medicine*, **178**, 838–846.

Korfhagen, T.R., Le Cras, T.D., Davidson, C.R., Schmidt, S.M., Ikegami, M., Whitsett, J.A. and Hardie, W.D. (2009) Rapamycin prevents transforming growth factor-alpha-induced pulmonary fibrosis. *American Journal of Respiratory Cell and Molecular Biology*, **41**, 562–572.

Kroemer, G., Galluzzi, L. and Brenner, C. (2007) Mitochondrial membrane permeabilization in cell death. *Physiological Reviews*, **87**, 99–163.

Kujoth, G.C., Hiona, A., Pugh, T.D., Someya, S., Panzer, K., Wohlgemuth, S.E., Hofer, T., Seo, A.Y., Sullivan, R., Jobling, W.A., Morrow, J.D., Van Remmen, H., Sedivy, J.M., Yamasoba, T., Tanokura, M., Weindruch, R., Leeuwenburgh, C. and Prolla, T.A. (2005) Mitochondrial DNA mutations, oxidative stress, and apoptosis in mammalian aging. *Science (New York, NY)*, **309**, 481–484.

Kuwano, K., Kunitake, R., Kawasaki, M., Nomoto, Y., Hagimoto, N., Nakanishi, Y. and Hara, N. (1996) P21Waf1/Cip1/Sdi1 and p53 expression in association with DNA strand breaks in idiopathic pulmonary fibrosis. *American Journal of Respiratory and Critical Care Medicine*, **154**, 477–483.

Lawlor, K.E. and Vince, J.E. (2014) Ambiguities in NLRP3 inflammasome regulation: is there a role for mitochondria? *Biochimica et Biophysica Acta*, **1840**, 1433–1440.

Lawson, W.E., Crossno, P.F., Polosukhin, V.V., Roldan, J., Cheng, D.S., Lane, K.B., Blackwell, T.R., Xu, C., Markin, C., Ware, L.B., Miller, G.G., Loyd, J.E. and Blackwell, T.S. (2008) Endoplasmic reticulum stress in alveolar epithelial cells is prominent in IPF: association with altered surfactant protein processing and herpesvirus infection. *American Journal of Physiology Heart and Circulatory Physiology*, **294**, L1119–1126.

Lawson, W.E., Cheng, D.S., Degryse, A.L., Tanjore, H., Polosukhin, V.V., Xu, X.C., Newcomb, D.C., Jones, B.R., Roldan, J., Lane, K.B., Morrisey, E.E., Beers, M.F., Yull, F.E. and T.S. Blackwell (2011) Endoplasmic reticulum stress enhances fibrotic remodeling in the lungs. *Proceedings of the National Academy of Sciences of the United States of America*, **108**, 10562–10567.

Lee, C.G., Cho, S.J., Kang, M.J., Chapoval, S.P., Lee, P.J., Noble, P.W., Yehualaeshet, T., Lu, B., Flavell, R.A., Milbrandt, J., Homer, R.J. and Elias, J.A. (2004) Early growth response gene 1-mediated apoptosis is essential for transforming growth factor beta1-induced pulmonary fibrosis. *The Journal of Experimental Medicine*, **200**, 377–389.

Levine, B., Mizushima, N. and Virgin, H.W. (2011) Autophagy in immunity and inflammation. *Nature*, **469**, 323–335.

Li, X., Rayford, H. and Uhal, B.D. (2003) Essential roles for angiotensin receptor AT1a in bleomycin-induced apoptosis and lung fibrosis in mice. *The American Journal of Pathology*, **163**, 2523–2530.

Li, S., Banck, M., Mujtaba, S., Zhou, M.M., Sugrue, M.M. and Walsh, M.J. (2010) p53-induced growth arrest is regulated by the mitochondrial SirT3 deacetylase. *PloS One*, **5**, e10486.

Li, Z.Y., Yang, Y., Ming, M. and Liu, B. (2011) Mitochondrial ROS generation for regulation of autophagic pathways in cancer. *Biochemical and Biophysical Research Communications*, **414**, 5-8.

Li, M., Gunter, M.E. and Fukagawa, N.K. (2012) Differential activation of the inflammasome in THP-1 cells exposed to chrysotile asbestos and Libby "six-mix" amphiboles and subsequent activation of BEAS-2B cells. *Cytokine*, **60**, 718–730.

Li, P., Liu, T., Kamp, D.W., Lin, Z., Wang, Y., Li, D., Yang, L., He, H. and Liu, G. (2015) The c-Jun N-terminal kinase signaling pathway mediates chrysotile asbestos-induced alveolar epithelial cell apoptosis. *Molecular Medicine Reports*, **11**, 3626–3634.

Lim, C.B., Prele, C.M., Baltic, S., Arthur, P.G., Creaney, J., Watkins, D.N., Thompson, P.J. and Mutsaers, S.E. (2015) Mitochondria-derived reactive oxygen species drive GANT61-induced mesothelioma cell apoptosis. *Oncotarget*, **6**, 1519–1530.

Lin, F., Liu, Y., Liu, Y., Keshava, N. and Li, S. (2000) Crocidolite induces cell transformation and p53 gene mutation in BALB/c-3T3 cells. *Teratogenesis, Carcinogenesis, and Mutagenesis*, **20**, 273–281.

Lin, Z., Liu, T., Kamp, D.W., Wang, Y., He, H., Zhou, X., Li, D., Yang, L., Zhao, B. and Liu, G. (2014) AKT/mTOR and c-Jun N-terminal kinase signaling pathways are required for chrysotile asbestos-induced autophagy. *Free Radical Biology & Medicine*, **72**, 296–307.

Little, J.B., Radford, E.P., Jr., McCombs, H.L. and Hunt, V.R. (1965) Distribution of polonium-210 in pulmonary tissues of cigarette smokers. *The New England Journal of Medicine*, **273**, 1343–1351.

Liu, Z., Lu, H., Shi, H., Du, Y., Yu, J., Gu, S., Chen, X., Liu, K.J. and Hu, C.A. (2005) PUMA overexpression induces reactive oxygen species generation and proteasome-mediated stathmin degradation in colorectal cancer cells. *Cancer Research*, **65**, 1647–1654.

Liu, G., Beri, R., Mueller, A. and Kamp, D.W. (2010) Molecular mechanisms of asbestos-induced lung epithelial cell apoptosis. *Chemico-Biological Interactions*, **188**, 309–318.

Liu, G., Cheresh, P. and Kamp, D.W. (2013) Molecular basis of asbestos-induced lung disease. *Annual Review of Pathology*, **8**, 161–187.

Liu, C.Y., Stucker, I., Chen, C., Goodman, G., McHugh, M.K., D'Amelio, A.M., Jr., Etzel, C.J., Li, S., Lin, X. and Christiani, D.C. (2015a) Genome-wide gene-asbestos exposure interaction association study identifies a common susceptibility variant on 22q13.31 associated with lung cancer risk. *Cancer Epidemiology, Biomarkers & Prevention: a publication of the American Association for Cancer Research, cosponsored by the American Society of Preventive Oncology*, **24**, 1564–1573.

Liu, L., Peritore, C., Ginsberg, J., Kayhan, M. and Donmez, G. (2015b) SIRT3 attenuates MPTP-induced nigrostriatal degeneration *via* enhancing mitochondrial antioxidant capacity. *Neurochemical Research*, **40**, 600–608.

Liu, Y., Chen, Y.Y., Cao, J.Y., Tao, F.B., Zhu, X.X., Yao, C.J., Chen, D.J., Che, Z., Zhao, Q.H. and Wen, L.P. (2015c) Oxidative stress, apoptosis, and cell cycle arrest are induced in primary fetal alveolar type II epithelial cells exposed to fine particulate matter from cooking oil fumes. *Environmental Science and Pollution Research International*, **22**, 9728–9741.

Lombard, D.B., Alt, F.W., Cheng, H.L., Bunkenborg, J., Streeper, R.S., Mostoslavsky, R., Kim, J., Yancopoulos, G., Valenzuela, D., Murphy, A., Yang, Y., Chen, Y., Hirschey, M.D., Bronson, R.T., Haigis, M., Guarente, L.P., Farese Jr., R.V., Weissman, S., Verdin, E. and Schwer, B. (2007) Mammalian Sir2 homolog SIRT3 regulates global mitochondrial lysine acetylation. *Molecular and Cellular Biology*, **27**, 8807–8814.

Longhin, E., Holme, J.A., Gutzkow, K.B., Arlt, V.M., Kucab, J.E., Camatini, M. and Gualtieri, M. (2013) Cell cycle alterations induced by urban PM2.5 in bronchial epithelial cells: characterization of the process and possible mechanisms involved. *Particle and Fibre Toxicology*, **10**, 63.

Lopez-Otin, C., Blasco, M.A., Partridge, L., Serrano, M. and Kroemer, G. (2013) The hallmarks of aging. *Cell*, **153**, 1194–1217.

Lounsbury, K.M., Stern, M., Taatjes, D., Jaken, S. and Mossman, B.T. (2002) Increased localization and substrate activation of protein kinase C delta in lung epithelial cells following exposure to asbestos. *The American Journal of Pathology*, **160**, 1991–2000.

Macip, S., Igarashi, M., Berggren, P., Yu, J., Lee, S.W. and Aaronson, S.A. (2003) Influence of induced reactive oxygen species in p53-mediated cell fate decisions. *Molecular and Cellular Biology*, **23**, 8576–8585.

Maeda, M., Chen, Y., Hayashi, H., Kumagai-Takei, N., Matsuzaki, H., Lee, S., Nishimura, Y. and Otsuki, T. (2014) Chronic exposure to asbestos enhances TGF-beta1 production in the human adult T cell leukemia virus-immortalized T cell line MT-2. *International Journal of Oncology*, **45**, 2522–2532.

Malhotra, J.D. and Kaufman, R.J. (2011) ER stress and its functional link to mitochondria: role in cell survival and death. *Cold Spring Harbor Perspectives in Biology*, **3**, a004424.

Manni, M.L., Epperly, M.W., Han, W., Blackwell, T.S., Duncan, S.R., Piganelli, J.D. and Oury, T.D. (2012) Leukocyte-derived extracellular superoxide dismutase does not contribute to airspace EC-SOD after interstitial pulmonary injury. *American Journal of Physiology Heart and Circulatory Physiology*, **302**, L160–166.

Manning, C.B., Mossman, B.T. and Taatjes, D.J. (2006) Analysis of asbestos-induced gene expression changes in bronchiolar epithelial cells using laser capture microdissection and quantitative reverse transcriptase-polymerase chain reaction. *Methods in Molecular Biology (Clifton, NJ)*, **319**, 231–236.

Manning, C.B., Sabo-Attwood, T., Robledo, R.F., Macpherson, M.B., Rincon, M., Vacek, P., Hemenway, D., Taatjes, D.J., Lee, P.J. and Mossman, B.T. (2008) Targeting the MEK1 cascade in lung epithelium inhibits proliferation and fibrogenesis by asbestos. *American Journal of Respiratory Cell and Molecular Biology*, **38**, 618–626.

Marchenko, N.D., Zaika, A. and Moll, U.M. (2000) Death signal-induced localization of p53 protein to mitochondria. A potential role in apoptotic signaling. *The Journal of Biological Chemistry*, **275**, 16202–16212.

Martinez, F.J., de Andrade, J.A., Anstrom, K.J., King, T.E., Jr. and Raghu, G. (2014) Randomised trial of acetylcysteine in idiopathic pulmonary fibrosis. *The New England Journal of Medicine*, **370**, 2093–2101.

Matsuoka, M., Igisu, H. and Morimoto, Y. (2003) Phosphorylation of p53 protein in A549 human pulmonary epithelial cells exposed to asbestos fibers. *Environmental Health Perspectives*, **111**, 509–512.

McFadden, D., Wright, J.L., Wiggs, B. and Churg, A. (1986) Smoking inhibits asbestos clearance. *The American Review of Respiratory Disease*, **133**, 372–374.

McGavran, P.D. and Brody, A.R. (1989) Chrysotile asbestos inhalation induces tritiated thymidine incorporation by epithelial cells of distal bronchioles. *American Journal of Respiratory Cell and Molecular Biology*, **1**, 231–235.

Meunier, J. and Hayashi, T. (2010) Sigma-1 receptors regulate Bcl-2 expression by reactive oxygen species-dependent transcriptional regulation of nuclear factor kappaB. *The Journal of Pharmacology and Experimental Therapeutics*, **332**, 388–397.

Mi, S., Li, Z., Yang, H.Z., Liu, H., Wang, J.P., Ma, Y.G., Wang, X.X., Liu, H.Z., Sun, W. and Hu, Z.W. (2011) Blocking IL-17A promotes the resolution of pulmonary inflammation and fibrosis *via* TGF-beta1-dependent and -independent mechanisms. *Journal of Immunology (Baltimore, MD: 1950)*, **187**, 3003–3014.

Miller, F.J., Rosenfeldt, F.L., Zhang, C., Linnane, A.W. and Nagley, P. (2003) Precise determination of mitochondrial DNA copy number in human skeletal and cardiac muscle by a PCR-based assay: lack of change of copy number with age. *Nucleic Acids Research*, **31**, e61.

Mirbahai, L., Kershaw, R.M., Green, R.M., Hayden, R.E., Meldrum, R.A. and Hodges, N.J. (2010) Use of a molecular beacon to track the activity of base excision repair protein OGG1 in live cells. *DNA Repair*, **9**, 144–152.

Mishra, A., Liu, J.Y., Brody, A.R. and Morris, G.F. (1997) Inhaled asbestos fibers induce p53 expression in the rat lung. *American Journal of Respiratory Cell and Molecular Biology*, **16**, 479–485.

Mishra, A., Stueckle, T.A., Mercer, R.R., Derk, R., Rojanasakul, Y., Castranova, V. and Wang, L. (2015) Identification of TGF-beta receptor-1 as a key regulator of carbon nanotube-induced fibrogenesis. *American Journal of Physiology Heart and Circulatory Physiology*, **309**, L821–833.

Mor-Vaknin, N., Punturieri, A., Sitwala, K. and Markovitz, D.M. (2003) Vimentin is secreted by activated macrophages. *Nature Cell Biology*, **5**, 59–63.

Morigi, M., Perico, L., Rota, C., Longaretti, L., Conti, S., Rottoli, D., Novelli, R., Remuzzi, G. and Benigni, A. (2015) Sirtuin 3-dependent mitochondrial dynamic improvements protect against acute kidney injury. *The Journal of Clinical Investigation*, **125**, 715–726.

Mossman, B.T. and Landesman, J.M. (1983) Importance of oxygen free radicals in asbestos-induced injury to airway epithelial cells. *Chest*, **83**, 50s–51s.

Mossman, B.T., Marsh, J.P. and Shatos, M.A. (1986) Alteration of superoxide dismutase activity in tracheal epithelial cells by asbestos and inhibition of cytotoxicity by antioxidants. *Laboratory Investigation; a Journal of Technical Methods and Pathology*, **54**, 204–212.

Mossman, B.T., Marsh, J.P., Sesko, A., Hill, S., Shatos, M.A., Doherty, J., Petruska, J., Adler, K.B., Hemenway, D., Mickey, R., Vacek, P. and Kagan, E. (1990) Inhibition of lung injury, inflammation, and interstitial pulmonary fibrosis by polyethylene glycol-conjugated catalase in a rapid inhalation model of asbestosis. *The American Review of Respiratory Disease*, **141**, 1266–1271.

Mossman, B.T., Lippmann, M., Hesterberg, T.W., Kelsey, K.T., Barchowsky, A. and Bonner, J.C. (2011) Pulmonary endpoints (lung carcinomas and asbestosis) following inhalation exposure to asbestos. *Journal of Toxicology and Environmental Health Part B, Critical Reviews*, **14**, 76–121.

Muller, J., Huaux, F., Fonseca, A., Nagy, J.B., Moreau, N., Delos, M., Raymundo-Pinero, E., Beguin, F., Kirsch-Volders, M., Fenoglio, I., Fubini, B. and Lison, D. (2008) Structural defects play a major role in the acute lung toxicity of multiwall carbon nanotubes: toxicological aspects. *Chemical Research in Toxicology*, **21**, 1698–1705.

Mulugeta, S., Nureki, S. and Beers, M.F. (2015) Lost after translation: insights from pulmonary surfactant for understanding the role of alveolar epithelial dysfunction and cellular quality control in fibrotic lung disease. *American Journal of Physiology Heart and Circulatory Physiology*, **309**, L507–525.

Munger, J.S., Huang, X., Kawakatsu, H., Griffiths, M.J., Dalton, S.L., Wu, J., Pittet, J.F., Kaminski, N., Garat, C., Matthay, M.A., Rifkin, D.B. and Sheppard, D. (1999) The integrin alpha v beta 6 binds and activates latent TGF beta 1: a mechanism for regulating pulmonary inflammation and fibrosis. *Cell*, **96**, 319–328.

Murthy, S., Adamcakova-Dodd, A., Perry, S.S., Tephly, L.A., Keller, R.M., Metwali, N., Meyerholz, D.K., Wang, Y., Glogauer, M., Thorne, P.S. and Carter, A.B. (2009) Modulation of reactive oxygen species by Rac1 or catalase prevents asbestos-induced pulmonary fibrosis. *American Journal of Physiology Heart and Circulatory Physiology*, **297**, L846–855.

Murthy, S., Ryan, A., He, C., Mallampalli, R.K. and Carter, A.B. (2010) Rac1-mediated mitochondrial H2O2 generation regulates MMP-9 gene expression in macrophages *via* inhibition of SP-1 and AP-1. *The Journal of Biological Chemistry*, **285**, 25062–25073.

Naik, E. and Dixit, V.M. (2011) Mitochondrial reactive oxygen species drive proinflammatory cytokine production. *The Journal of Experimental Medicine*, **208**, 417–420.

Naik, E., Michalak, E.M., Villunger, A., Adams, J.M. and Strasser, A. (2007) Ultraviolet radiation triggers apoptosis of fibroblasts and skin keratinocytes mainly *via* the BH3-only protein Noxa. *The Journal of Cell Biology*, **176**, 415–424.

Nakano, K. and Vousden, K.H. (2001) PUMA, a novel proapoptotic gene, is induced by p53. *Molecular Cell*, **7**, 683–694.

Nelson, A., Mendoza, T., Hoyle, G.W., Brody, A.R., Fermin, C. and Morris, G.F. (2001) Enhancement of fibrogenesis by the p53 tumor suppressor protein in asbestos-exposed rodents. *Chest*, **120**, 33s–34s.

Noble, P.W., Barkauskas, C.E. and Jiang, D. (2012) Pulmonary fibrosis: patterns and perpetrators. *The Journal of Clinical Investigation*, **122**, 2756–2762.

Nuorva, K., Makitaro, R., Huhti, E., Kamel, D., Vahakangas, K., Bloigu, R., Soini, Y. and Paakko, P. (1994) p53 protein accumulation in lung carcinomas of patients exposed to asbestos and tobacco smoke. *American Journal of Respiratory and Critical Care Medicine*, **150**, 528–533.

Nygren, J., Suhonen, S., Norppa, H. and Linnainmaa, K. (2004) DNA damage in bronchial epithelial and mesothelial cells with and without associated crocidolite asbestos fibers. *Environmental and Molecular Mutagenesis*, **44**, 477–482.

Nymark, P., Lindholm, P.M., Korpela, M.V., Lahti, L., Ruosaari, S., Kaski, S., Hollmen, J., Anttila, S., Kinnula, V.L. and Knuutila, S. (2007) Gene expression profiles in asbestos-exposed epithelial and mesothelial lung cell lines. *BMC Genomics*, **8**, 62.

Nymark, P., Aavikko, M., Makila, J., Ruosaari, S., Hienonen-Kempas, T., Wikman, H., Salmenkivi, K., Pirinen, R., Karjalainen, A., Vanhala, E., Kuosma, E., Anttila, S. and Kuttunen, E. (2013) Accumulation of genomic alterations in 2p16, 9q33.1 and 19p13 in lung tumours of asbestos-exposed patients. *Molecular Oncology*, **7**, 29–40.

Nymark, P., Jensen, K.A., Suhonen, S., Kembouche, Y., Vippola, M., Kleinjans, J., Catalan, J., Norppa, H., van Delft, J. and Briede, J.J. (2014) Free radical scavenging and formation by multi-walled carbon nanotubes in cell free conditions and in human bronchial epithelial cells. *Particle and Fibre Toxicology*, **11**, 4.

Oda, E., Ohki, R., Murasawa, H., Nemoto, J., Shibue, T., Yamashita, T., Tokino, T., Taniguchi, T. and Tanaka, N. (2000) Noxa, a BH3-only member of the Bcl-2 family and candidate mediator of p53-induced apoptosis. *Science (New York, NY)*, **288**, 1053–1058.

Osborn-Heaford, H.L., Ryan, A.J., Murthy, S., Racila, A.M., He, C., Sieren, J.C., Spitz, D.R. and Carter, A.B. (2012) Mitochondrial Rac1 GTPase import and electron transfer from cytochrome c are required for pulmonary fibrosis. *The Journal of Biological Chemistry*, **287**, 3301–3312.

Pande, P., Mosleh, T.A. and Aust, A.E. (2006) Role of alphavbeta5 integrin receptor in endocytosis of crocidolite and its effect on intracellular glutathione levels in human lung epithelial (A549) cells. *Toxicology and Applied Pharmacology*, **210**, 70–77.

Panduri, V., Weitzman, S.A., Chandel, N. and Kamp, D.W. (2003) The mitochondria-regulated death pathway mediates asbestos-induced alveolar epithelial cell apoptosis. *American Journal of Respiratory Cell and Molecular Biology*, **28**, 241–248.

Panduri, V., Weitzman, S.A., Chandel, N.S. and Kamp, D.W. (2004) Mitochondrial-derived free radicals mediate asbestos-induced alveolar epithelial cell apoptosis. *American Journal of Physiology Heart and Circulatory Physiology*, **286**, L1220–1227.

Panduri, V., Surapureddi, S., Soberanes, S., Weitzman, S.A., Chandel, N. and Kamp, D.W. (2006) P53 mediates amosite asbestos-induced alveolar epithelial cell mitochondria-regulated apoptosis. *American Journal of Respiratory Cell and Molecular Biology*, **34**, 443–452.

Panduri, V., Liu, G., Surapureddi, S., Kondapalli, J., Soberanes, S., de Souza-Pinto, N.C., Bohr, V.A., Budinger, G.R., Schumacker, P.T., Weitzman, S.A. and Kamp, D.W. (2009) Role of mitochondrial hOGG1 and aconitase in oxidant-induced lung epithelial cell apoptosis. *Free Radical Biology & Medicine*, **47**, 750–759.

Park, L.C., Albers, D.S., Xu, H., Lindsay, J.G., Beal, M.F. and Gibson, G.E. (2001) Mitochondrial impairment in the cerebellum of the patients with progressive supranuclear palsy. *Journal of Neuroscience Research*, **66**, 1028–1034.

Patel, A.S., Lin, L., Geyer, A., Haspel, J.A., An, C.H., Cao, J., Rosas, I.O. and Morse, D. (2012) Autophagy in idiopathic pulmonary fibrosis. *PloS One*, **7**, e41394.

Patel, A.S., Morse, D. and Choi, A.M. (2013) Regulation and functional significance of autophagy in respiratory cell biology and disease. *American Journal of Respiratory Cell and Molecular Biology*, **48**, 1-9.

Patel, A.S., Song, J.W., Chu, S.G., Mizumura, K., Osorio, J.C., Shi, Y., El-Chemaly, S., Lee, C.G., Rosas, I.O., Elias, J.A., Choi, A.M.K. and Morse, D. (2015) Epithelial cell mitochondrial dysfunction and PINK1 are induced by transforming growth factor-beta1 in pulmonary fibrosis. *PloS One*, **10**, e0121246.

Peeters, P.M., Perkins, T.N., Wouters, E.F., Mossman, B.T. and Reynaert, N.L. (2013) Silica induces NLRP3 inflammasome activation in human lung epithelial cells. *Particle and Fibre Toxicology*, **10**, 3.

Perkins, T.N., Peeters, P.M., Shukla, A., Arijs, I., Dragon, J., Wouters, E.F., Reynaert, N.L. and Mossman, B.T. (2015) Indications for distinct pathogenic mechanisms of asbestos and silica through gene expression profiling of the response of lung epithelial cells. *Human Molecular Genetics*, **24**, 1374–1389.

Picard, M., Jung, B., Liang, F., Azuelos, I., Hussain, S., Goldberg, P., Godin, R., Danialou, G., Chaturvedi, R., Rygiel, K., Matecki, S., Jaber, S., Des Rosiers, C., Karpati, G., Ferri, L., Burelle, Y., Turnbull, D.M., Taivassalo, T. and Petrof, B.J. (2012) Mitochondrial dysfunction and lipid accumulation in the human diaphragm during mechanical ventilation. *American Journal of Respiratory and Critical Care Medicine*, **186**, 1140–1149.

Piguet, P.F., Vesin, C., Grau, G.E. and Thompson, R.C. (1993) Interleukin 1 receptor antagonist (IL-1ra) prevents or cures pulmonary fibrosis elicited in mice by bleomycin or silica. *Cytokine*, **5**, 57–61.

Pillai, V.B., Samant, S., Sundaresan, N.R., Raghuraman, H., Kim, G., Bonner, M.Y., Arbiser, J.L., Walker, D.I., Jones, D.P., Gius, D. and Gupta, M.P. (2015) Honokiol blocks and reverses cardiac hypertrophy in mice by activating mitochondrial Sirt3. *Nature Communications*, **6**, 6656.

Plataki, M., Koutsopoulos, A.V., Darivianaki, K., Delides, G., Siafakas, N.M. and Bouros, D. (2005) Expression of apoptotic and antiapoptotic markers in epithelial cells in idiopathic pulmonary fibrosis. *Chest*, **127**, 266–274.

Pociask, D.A., Sime, P.J. and Brody, A.R. (2004) Asbestos-derived reactive oxygen species activate TGF-beta1. *Laboratory Investigation; a Journal of Technical Methods and Pathology*, **84**, 1013–1023.

Prakash, A. and Doublie, S. (2015) Base excision repair in the mitochondria. *Journal of Cellular Biochemistry*, **116**, 1490–1499.

Przybylowska, K., Kabzinski, J., Sygut, A., Dziki, L., Dziki, A. and Majsterek, I. (2013) An association selected polymorphisms of XRCC1, OGG1 and MUTYH gene and the level of efficiency oxidative DNA damage

repair with a risk of colorectal cancer. *Mutation Research*, **745-746**, 6–15.
Puhakka, A., Ollikainen, T., Soini, Y., Kahlos, K., Saily, M., Koistinen, P., Paakko, P., Linnainmaa, K. and Kinnula, V.L. (2002) Modulation of DNA single-strand breaks by intracellular glutathione in human lung cells exposed to asbestos fibers. *Mutation Research* **514**, 7–17.
Ralph, S.J., Rodriguez-Enriquez, S., Neuzil, J., Saavedra, E. and Moreno-Sanchez, R. (2010) The causes of cancer revisited: "mitochondrial malignancy" and ROS-induced oncogenic transformation – why mitochondria are targets for cancer therapy. *Molecular Aspects of Medicine*, **31**, 145–170.
Rao, S., Tortola, L., Perlot, T., Wirnsberger, G., Novatchkova, M., Nitsch, R., Sykacek, P., Frank, L., Schramek, D., Komnenovic, V., Sigl, V., Aumayr, K., Schmauss, G., Fellner, N., Handschuh, S., Glosmann, M., Pasierbek, P., Schlederer, M., Resch, G.P., Ma, Y., Yang, H., Popper, H., Kenner, L., Kroemer, G. and Penninger, J.M. (2014) A dual role for autophagy in a murine model of lung cancer. *Nature Communications*, **5**, 3056.
Rapisarda, V., Loreto, C., Ledda, C., Musumeci, G., Bracci, M., Santarelli, L., Renis, M., Ferrante, M. and Cardile, V. (2015) Cytotoxicity, oxidative stress and genotoxicity induced by glass fibers on human alveolar epithelial cell line A549. *Toxicology in vitro: an International Journal published in association with BIBRA*, **29**, 551–557.
Roe, O.D. and Stella, G.M. (2015) Malignant pleural mesothelioma: history, controversy and future of a manmade epidemic. *European Respiratory Review: an Official Journal of the European Respiratory Society*, **24**, 115–131.
Roggli, V.L. and Brody, A.R. (1984) Changes in numbers and dimensions of chrysotile asbestos fibers in lungs of rats following short-term exposure. *Experimental Lung Research*, **7**, 133–147.
Roggli, V.L., George, M.H. and Brody, A.R. (1987) Clearance and dimensional changes of crocidolite asbestos fibers isolated from lungs of rats following short-term exposure. *Environmental Research*, **42**, 94–105.
Rong, Y. and Distelhorst, C.W. (2008) Bcl-2 protein family members: versatile regulators of calcium signaling in cell survival and apoptosis. *Annual Review of Physiology*, **70**, 73–91.
Ruchko, M., Gorodnya, O., LeDoux, S.P., Alexeyev, M.F., Al-Mehdi, A.B. and Gillespie, M.N. (2005) Mitochondrial DNA damage triggers mitochondrial dysfunction and apoptosis in oxidant-challenged lung endothelial cells. *American Journal of Physiology Heart and Circulatory Physiology*, **288**, L530–535.
Ruosaari, S.T., Nymark, P.E., Aavikko, M.M., Kettunen, E., Knuutila, S., Hollmen, J., Norppa, H. and Anttila, S.L. (2008) Aberrations of chromosome 19 in asbestos-associated lung cancer and in asbestos-induced micronuclei of bronchial epithelial cells in vitro. *Carcinogenesis*, **29**, 913–917.
Ryan, A.J., Larson-Casey, J.L., He, C., Murthy, S. and Carter, A.B. (2014) Asbestos-induced disruption of calcium homeostasis induces endoplasmic reticulum stress in macrophages. *The Journal of Biological Chemistry*, **289**, 33391–33403.
Sabo-Attwood, T., Ramos-Nino, M., Bond, J., Butnor, K.J., Heintz, N., Gruber, A.D., Steele, C., Taatjes, D.J., Vacek, P. and Mossman, B.T. (2005) Gene expression profiles reveal increased mClca3 (Gob5) expression and mucin production in a murine model of asbestos-induced fibrogenesis. *The American Journal of Pathology*, **167**, 1243–1256.
Santos, J.H., Hunakova, L., Chen, Y., Bortner, C. and Van Houten, B. (2003) Cell sorting experiments link persistent mitochondrial DNA damage with loss of mitochondrial membrane potential and apoptotic cell death. *The Journal of Biological Chemistry*, **278**, 1728–1734.
Schumacker, P.T., Gillespie, M.N., Nakahira, K., Choi, A.M., Crouser, E.D., Piantadosi, C.A. and Bhattacharya, J. (2014) Mitochondria in lung biology and pathology: more than just a powerhouse. *American Journal of Physiology Heart and Circulatory Physiology*, **306**, L962–974.
Selman, M. and Pardo, A. (2014) Revealing the pathogenic and aging-related mechanisms of the enigmatic idiopathic pulmonary fibrosis. an integral model. *American Journal of Respiratory and Critical Care Medicine*, **189**, 1161–1172.
Sener, G., Topaloglu, N., Sehirli, A.O., Ercan, F. and Gedik, N. (2007) Resveratrol alleviates bleomycin-induced lung injury in rats. *Pulmonary Pharmacology & Therapeutics*, **20**, 642–649.
Shintani, T. and Klionsky, D.J. (2004) Autophagy in health and disease: a double-edged sword. *Science (New York, NY)*, **306**, 990–995.
Shukla, A., Gulumian, M., Hei, T.K., Kamp, D., Rahman, Q. and Mossman, B.T. (2003a) Multiple roles of

oxidants in the pathogenesis of asbestos-induced diseases. *Free Radical Biology and Medicine*, **34**, 1117–1129.

Shukla, A., Jung, M., Stern, M., Fukagawa, N.K., Taatjes, D.J., Sawyer, D., Van Houten, B. and Mossman, B.T. (2003b) Asbestos induces mitochondrial DNA damage and dysfunction linked to the development of apoptosis. *American Journal of Physiology Heart and Circulatory Physiology*, **285**, L1018–1025.

Shulga, N. and Pastorino, J.G. (2016) Ethanol sensitizes mitochondria to the permeability transition by inhibiting deacetylation of cyclophilin-D mediated by sirtuin-3. *Journal of Cell Science*, **129**, 2685.

Singh, K.K., Desouki, M.M., Franklin, R.B. and Costello, L.C. (2006) Mitochondrial aconitase and citrate metabolism in malignant and nonmalignant human prostate tissues. *Molecular Cancer*, **5**, 14.

Soberanes, S., Panduri, V., Mutlu, G.M., Ghio, A., Bundinger, G.R. and Kamp, D.W. (2006) p53 mediates particulate matter-induced alveolar epithelial cell mitochondria-regulated apoptosis. *American Journal of Respiratory and Critical Care Medicine*, **174**, 1229–1238.

Soberanes, S., Urich, D., Baker, C.M., Burgess, Z., Chiarella, S.E., Bell, E.L., Ghio, A.J., De Vizcaya-Ruiz, A., Liu, J., Ridge, K.M., Kamp, D.W., Chandel, N.S., Shumacker, P.T., Mutlu, G.M. and Budinger, G.R.S. (2009) Mitochondrial complex III-generated oxidants activate ASK1 and JNK to induce alveolar epithelial cell death following exposure to particulate matter air pollution. *The Journal of Biological Chemistry*, **284**, 2176–2186.

Sohn, E.J., Kim, J., Hwang, Y., Im, S., Moon, Y. and Kang, D.M. (2012) TGF-beta suppresses the expression of genes related to mitochondrial function in lung A549 cells. *Cellular and Molecular Biology (Noisy-le-Grand, France)*, **Suppl. 58**, OI 1763–1767.

Someya, S., Xu, J., Kondo, K., Ding, D., Salvi, R.J., Yamasoba, T., Rabinovitch, P.S., Weindruch, R., Leeuwenburgh, C., Tanokura, M. and Prolla, T.A. (2009) Age-related hearing loss in C57BL/6J mice is mediated by Bak-dependent mitochondrial apoptosis. *Proceedings of the National Academy of Sciences of the United States of America*, **106**, 19432–19437.

Someya, S., Yu, W., Hallows, W.C., Xu, J., Vann, J.M., Leeuwenburgh, C., Tanokura, M., Denu, J.M. and Prolla, T.A. (2010) Sirt3 mediates reduction of oxidative damage and prevention of age-related hearing loss under caloric restriction. *Cell*, **143**, 802–812.

Sosulski, M.L., Gongora, R., Danchuk, S., Dong, C., Luo, F. and Sanchez, C.G. (2015) Deregulation of selective autophagy during aging and pulmonary fibrosis: the role of TGFbeta1. *Aging Cell*, **14**, 774–783.

Stayner, L., Welch, L.S. and Lemen, R. (2013) The worldwide pandemic of asbestos-related diseases. *Annual Review of Public Health*, **34**, 205–216.

Straif, K., Benbrahim-Tallaa, L., Baan, R., Grosse, Y., Secretan, B., El Ghissassi, F., Bouvard, V., Guha, N., Freeman, C., Galichet, L. and Cogliano, V. (2009) A review of human carcinogens – Part C: metals, arsenic, dusts, and fibres. *The Lancet Oncology*, **10**, 453–454.

Sueblinvong, V., Neujahr, D.C., Mills, S.T., Roser-Page, S., Ritzenthaler, J.D., Guidot, D., Rojas, M. and Roman, J. (2012) Predisposition for disrepair in the aged lung. *The American Journal of the Medical Sciences*, **344**, 41–51.

Sundaresan, N.R., Samant, S.A., Pillai, V.B., Rajamohan, S.B. and Gupta, M.P. (2008) SIRT3 is a stress-responsive deacetylase in cardiomyocytes that protects cells from stress-mediated cell death by deacetylation of Ku70. *Molecular and Cellular Biology*, **28**, 6384–6401.

Sundaresan, N.R., Gupta, M., Kim, G., Rajamohan, S.B., Isbatan, A. and Gupta, M.P. (2009) Sirt3 blocks the cardiac hypertrophic response by augmenting Foxo3a-dependent antioxidant defense mechanisms in mice. *The Journal of Clinical Investigation*, **119**, 2758–2771.

Sundaresan, N.R., Bindu, S., Pillai, V.B., Samant, S., Pan, Y., Huang, J.Y., Gupta, M., Nagalingam, R.S., Wolfgeher, D., Verdin, E. and Gupta, M.P. (2016) SIRT3 blocks aging-associated tissue fibrosis in mice by deacetylating and activating glycogen synthase kinase 3beta. *Molecular and Cellular Biology*, **36**, 678–692.

Swain, W.A., O'Byrne, K.J. and Faux, S.P. (2004) Activation of p38 MAP kinase by asbestos in rat mesothelial cells is mediated by oxidative stress. *American Journal of Physiology Heart and Circulatory Physiology*, **286**, L859–865.

Tabrizi, S.J., Cleeter, M.W., Xuereb, J., Taanman, J.W., Cooper, J.M. and Schapira, A.H. (1999) Biochemical abnormalities and excitotoxicity in Huntington's disease brain. *Annals of Neurology*, **45**, 25–32.

Tamminen, J.A., Myllarniemi, M., Hyytiainen, M., Keski-Oja, J. and Koli, K. (2012) Asbestos exposure induces alveolar epithelial cell plasticity through MAPK/Erk signaling. *Journal of Cellular Biochemistry*, **113**, 2234–2247.

Tan, R.J., Fattman, C.L., Watkins, S.C. and Oury, T.D. (2004) Redistribution of pulmonary EC-SOD after exposure to asbestos. *Journal of Applied Physiology (Bethesda, MD: 1985)*, **97**, 2006–2013.

Tanjore, H., Blackwell, T.S. and Lawson, W.E. (2012) Emerging evidence for endoplasmic reticulum stress in the pathogenesis of idiopathic pulmonary fibrosis. *American Journal of Physiology Heart and Circulatory Physiology*, **302**, L721–729.

Thannickal, V.J. (2013) Mechanistic links between aging and lung fibrosis. *Biogerontology*, **14**, 609–615.

Thannickal, V.J., Murthy, M., Balch, W.E., Chandel, N.S., Meiners, S., Eickelberg, O., Selman, M., Pardo, A., White, E.S., Levy, B.D., Busse, P.J., Tuder, R.M., Antony, V.B., Sznajder, J.I. and Budiger, G.R.S. (2015) Blue journal conference. Aging and susceptibility to lung disease. *American Journal of Respiratory and Critical Care Medicine*, **191**, 261–269.

Thompson, J.K., Westbom, C.M., MacPherson, M.B., Mossman, B.T., Heintz, N.H., Spiess, P. and Shukla, A. (2014) Asbestos modulates thioredoxin-thioredoxin interacting protein interaction to regulate inflammasome activation. *Particle and Fibre Toxicology*, **11**, 24.

Torres-Gonzalez, E., Bueno, M., Tanaka, A., Krug, L.T., Cheng, D.S., Polosukhin, V.V., Sorescu, D., Lawson, W.E., Blackwell, T.S., Rojas, M. and Mora, A.L. (2012) Role of endoplasmic reticulum stress in age-related susceptibility to lung fibrosis. *American Journal of Respiratory Cell and Molecular Biology*, **46**, 748–756.

Tron, V., Wright, J.L., Harrison, N., Wiggs, B. and Churg, A. (1987) Cigarette smoke makes airway and early parenchymal asbestos-induced lung disease worse in the guinea pig. *The American Review of Respiratory Disease*, **136**, 271–275.

Tsui, K.H., Feng, T.H., Lin, Y.F., Chang, P.L. and Juang, H.H. (2011) p53 downregulates the gene expression of mitochondrial aconitase in human prostate carcinoma cells. *The Prostate*, **71**, 62–70.

Tsutsui, H., Kinugawa, S. and Matsushima, S. (2011) Oxidative stress and heart failure. *American Journal of Physiology Heart and Circulatory Physiology*, **301**, H2181–2190.

Tuppen, H.A., Blakely, E.L., Turnbull, D.M. and Taylor, R.W. (2010) Mitochondrial DNA mutations and human disease. *Biochimica et Biophysica Acta*, **1797**, 113–128.

Turci, F., Tomatis, M., Lesci, I.G., Roveri, N. and Fubini, B. (2011) The iron-related molecular toxicity mechanism of synthetic asbestos nanofibres: a model study for high-aspect-ratio nanoparticles. *Chemistry (Weinheim an der Bergstrasse, Germany)*, **17**, 350–358.

Uhal, B.D. and Nguyen, H. (2013) The Witschi Hypothesis revisited after 35 years: genetic proof from SP-C BRICHOS domain mutations. *American Journal of Physiology Heart and Circulatory Physiology*, **305**, L906–911.

Upadhyay, D., Panduri, V. and Kamp, D.W. (2005) Fibroblast growth factor-10 prevents asbestos-induced alveolar epithelial cell apoptosis by a mitogen-activated protein kinase-dependent mechanism. *American Journal of Respiratory Cell and Molecular Biology*, **32**, 232–238.

Valente, E.M., Abou-Sleiman, P.M., Caputo, V., Muqit, M.M., Harvey, K., Gispert, S., Ali, Z., Del Turco, D., Bentivoglio, A.R., Healy, D.G., Albanese, A., Nussbaum, R., González-Maldonado, R., Deller, T., Salvi, S., Cortelli, P., Gilks, W.P., Latchman, D.S., Harvey, R.J., Dallapiccola, B., Auburger, G. and Wood, N.W. (2004) Hereditary early-onset Parkinson's disease caused by mutations in PINK1. *Science (New York, NY)*, **304**, 1158–1160.

van As, A. (1980) Pulmonary airway defence mechanisms: an appreciation of integrated mucociliary activity. *European Journal of Respiratory Diseases Supplement*, **111**, 21–24.

Vaquero, A. (2009) The conserved role of sirtuins in chromatin regulation. *The International Journal of Developmental Biology*, **53**, 303–322.

Verdin, E., Hirschey, M.D., Finley, L.W. and Haigis, M.C. (2010) Sirtuin regulation of mitochondria: energy production, apoptosis, and signaling. *Trends in Biochemical Sciences*, **35**, 669–675.

Vousden, K.H. and Prives, C. (2009) Blinded by the light: The growing complexity of p53. *Cell* **137**, 413–431.

Wallace, D.C. (2013) A mitochondrial bioenergetic etiology of disease. *The Journal of Clinical Investigation*, **123**, 1405–1412.

Wallace, D.C. and Chalkia, D. (2013) Mitochondrial DNA genetics and the heteroplasmy conundrum in evolution and disease. *Cold Spring Harbor Perspectives in Biology*, **5**, a021220.

Wang, Y., Faux, S.P., Hallden, G., Kirn, D.H., Houghton, C.E., Lemoine, N.R. and Patrick, G. (2004) Interleukin-1beta and tumour necrosis factor-alpha promote the transformation of human immortalised mesothelial cells by erionite. *International Journal of Oncology*, **25**, 173–178.

Wang, J., Wang, Q., Watson, L.J., Jones, S.P. and Epstein, P.N. (2011) Cardiac overexpression of 8-oxoguanine DNA glycosylase 1 protects mitochondrial DNA and reduces cardiac fibrosis following transaortic constriction. *American Journal of Physiology Heart and Circulatory Physiology*, **301**, H2073–2080.

Wang, Y., Lin, Z., Huang, H., He, H., Yang, L., Chen, T., Yang, T., Ren, N., Jiang, Y., Xu, W., Kamp, D.W., Liu, T. and Liu, G. (2015) AMPK is required for PM2.5-induced autophagy in human lung epithelial A549 cells. *International Journal of Clinical and Experimental Medicine*, **8**, 58–72.

Weiss, W. (1999) Asbestosis: a marker for the increased risk of lung cancer among workers exposed to asbestos. *Chest*, **115**, 536–549.

Weiss, C.H., Budinger, G.R., Mutlu, G.M. and Jain, M. (2010) Proteasomal regulation of pulmonary fibrosis. *Proceedings of the American Thoracic Society*, **7**, 77–83.

Weitzman, S.A. and Graceffa, P. (1984) Asbestos catalyzes hydroxyl and superoxide radical generation from hydrogen peroxide. *Archives of Biochemistry and Biophysics*, **228**, 373–376.

White, E. (2012) Deconvoluting the context-dependent role for autophagy in cancer. *Nature Reviews Cancer*, **12**, 401–410.

Williams, R.S., Cheng, L., Mudge, A.W. and Harwood, A.J. (2002) A common mechanism of action for three mood-stabilizing drugs. *Nature*, **417**, 292–295.

Wittmann, T., Bokoch, G.M. and Waterman-Storer, C.M. (2003) Regulation of leading edge microtubule and actin dynamics downstream of Rac1. *The Journal of Cell Biology*, **161**, 845–851.

Won, K.Y., Kim, G.Y., Lim, S.J. and Kim, Y.W. (2012) Decreased Beclin-1 expression is correlated with the growth of the primary tumor in patients with squamous cell carcinoma and adenocarcinoma of the lung. *Human Pathology*, **43**, 62–68.

Wu, W.T., Lin, Y.J., Li, C.Y., Tsai, P.J., Yang, C.Y., Liou, S.H. and Wu, T.N. (2015) Cancer attributable to asbestos exposure in shipbreaking workers: A matched-cohort study. *PloS one*, **10**, e0133128.

Yamaguchi, H., Chen, J., Bhalla, K. and Wang, H.G. (2004) Regulation of Bax activation and apoptotic response to microtubule-damaging agents by p53 transcription-dependent and -independent pathways. *The Journal of Biological Chemistry*, **279**, 39431–39437.

Yan, L.J., Levine, R.L. and Sohal, R.S. (1997) Oxidative damage during aging targets mitochondrial aconitase. *Proceedings of the National Academy of Sciences of the United States of America*, **94**, 11168–11172.

Yang, H., Rivera, Z., Jube, S., Nasu, M., Bertino, P., Goparaju, C., Franzoso, G., Lotze, M.T., Krausz, T., Pass, H.I., Bianchi, M.E. and Carbone, M. (2010) Programmed necrosis induced by asbestos in human mesothelial cells causes high-mobility group box 1 protein release and resultant inflammation. *Proceedings of the National Academy of Sciences of the United States of America*, **107**, 12611–12616.

Yang, Z.J., Chee, C.E., Huang, S. and Sinicrope, F.A. (2011) The role of autophagy in cancer: therapeutic implications. *Molecular Cancer Therapeutics* **10**, 1533–1541.

Yao, S., DellaVentura, G. and Petibois, C. (2010) Analytical characterization of cell-asbestos fiber interactions in lung pathogenesis. *Analytical and Bioanalytical Chemistry*, **397**, 2079–2089.

Yee, H., Yie, T.A., Goldberg, J., Wong, K.M. and Rom, W.N. (2008) Immunohistochemical study of fibrosis and adenocarcinoma in dominant-negative p53 transgenic mice exposed to chrysotile asbestos and benzo(a)pyrene. *Journal of Environmental Pathology, Toxicology and Oncology: Official Organ of the International Society for Environmental Toxicology and Cancer*, **27**, 267–276.

Yin, Q., Brody, A.R. and Sullivan, D.E. (2007) Laser capture microdissection reveals dose-response of gene expression in situ consequent to asbestos exposure. *International Journal of Experimental Pathology*, **88**, 415–425.

Youn, C.K., Song, P.I., Kim, M.H., Kim, J.S., Hyun, J.W., Choi, S.J., Yoon, S.P., Chung, M.H., Chang, I.Y. and You, H.J. (2007) Human 8 oxoguanine DNA glycosylase suppresses the oxidative stress induced apoptosis through a p53-mediated signaling pathway in human fibroblasts. *Molecular Cancer Research: MCR*, **5**, 1083–1098.

Yousefi, S., Perozzo, R., Schmid, I., Ziemiecki, A., Schaffner, T., Scapozza, L., Brunner, T. and Simon, H.U.

(2006) Calpain-mediated cleavage of Atg5 switches autophagy to apoptosis. *Nature Cell Biology*, **8**, 1124–1132.

Zanella, C.L., Posada, J., Tritton, T.R. and Mossman, B.T. (1996) Asbestos causes stimulation of the extracellular signal-regulated kinase 1 mitogen-activated protein kinase cascade after phosphorylation of the epidermal growth factor receptor. *Cancer Research*, **56**, 5334–5338.

Zhang, J., Ghio, A.J., Chang, W., Kamdar, O., Rosen, G.D. and Upadhyay, D. (2007) Bim mediates mitochondria-regulated particulate matter-induced apoptosis in alveolar epithelial cells. *FEBS letters*, **581**, 4148–4152.

Zhao, X.H., Jia, G., Liu, Y.Q., Liu, S.W., Yan, L., Jin, Y. and Liu, N. (2006) Association between polymorphisms of DNA Repair gene XRCC1 and DNA damage in asbestos-exposed workers. *Biomedical and Environmental Sciences : BES*, **19**, 232–238.

Zhong, Q., Zhou, B., Ann, D.K., Minoo, P., Liu, Y., Banfalvi, A., Krishnaveni, M.S., Dubourd, M., Demaio, L., Willis, B.C., Kim, K.J., duBois, R.M., Crandall, E.D., Beers, M.F. and Borok, Z. (2011) Role of endoplasmic reticulum stress in epithelial-mesenchymal transition of alveolar epithelial cells: effects of misfolded surfactant protein. *American Journal of Respiratory Cell and Molecular Biology*, **45**, 498–509.

Zhou, R., Yazdi, A.S., Menu, P. and Tschopp, J. (2011) A role for mitochondria in NLRP3 inflammasome activation. *Nature*, **469**, 221–225.

EMU Notes in Mineralogy, Vol. 18 (2017), Chapter 15, 501–532

Towards a general model for predicting the toxicity and pathogenicity of mineral fibres

ALESSANDRO F. GUALTIERI[1,*], BROOKE T. MOSSMAN[2] and VICTOR L. ROGGLI[3]

[1]*Università di Modena e Reggio Emilia, Dipartimento di Scienze Chimiche e Geologiche, Via Campi 103, I-41125, Modena, Italy,*
e-mail: alessandro.gualtieri@unimore.it
[2]*University of Vermont College of Medicine, Department of Pathology, 89 Beaumont Avenue, Courtyard S265 Burlington, Vermont 05405, USA*
[3]*Duke University School of Medicine DUMC 3712 Durham, North Carolina 27710, USA*
**Corresponding author*

This chapter provides a comprehensive description of the physical, chemical, biological and mineralogical parameters that play a role in determining the toxicity and pathogenicity of mineral fibres. The first steps towards a general toxicity/pathogenicity model of mineral fibres are described here. Eventually the model can be generalized and may be applied to biodurable man-made mineral fibres and other natural and synthetic fibres in addition to silicates. Because of the complexity of the topic, a truly multidisciplinary approach is essential. A concept that will be stressed in the final notes of the chapter is that a full understanding of the toxicity/pathogenicity of mineral fibres aimed at finding effective solutions for the prevention and treatment of asbestos-related diseases can only be the outcome if an holistic approach is applied which takes advantage of synergistic research activity and communication between biochemists, mineralogists/crystallographers, pathologists, physicians, physicists and toxicologists, all sharing their distinct but interrelated perspectives. This is a great challenge for all such scientific individuals to work together to resolve and develop predictive models that incorporate their research findings and conclusions.

1. Introduction

For the last forty years, asbestos fibres have been subjected to extensive research activity which embraces different scientific fields, from geology to medicine. The outcome of this massive endeavour (consider that after 1995, the start of the "golden age" of nano-science, for every four papers containing the word "nanoparticle" in the title, one paper containing the word "asbestos" in the title has been published: source Google scholar) is the discovery of different parameters and mechanisms that individually or in synergy prompt cyto-, geno-toxicity and pathogenicity of mineral fibres. Nevertheless, the precise cause-and-effect relationships between fibre exposure

DOI: 10.1180/EMU-notes.18.15

and the initiation or onset of malignant mesothelioma (MM) and other lung diseases remain unclear, and a comprehensive, universally shared model of toxicity and pathogenicity of mineral fibres has not yet been devised because of the different chemico-physical properties and biological effects of various types of asbestos fibres.

One of the major intricacies arises from the fact that mineral fibres are associated naturally with other contaminants such as polycyclic hydrocarbons (Harington and Roe, 1965) and can occur as mixtures of various types of asbestos and other minerals. Furthermore, naturally occurring mineral fibres can display great variability in terms of their chemistry, crystal structure, fibre dimensions and surface activity (Pollastri *et al.*, 2014). Their ability to generate reactive oxygen and nitrogen species (ROS/RNS), differential clearance or dissolution, inflammatory potential and biopersistence at sites of cancer development (Donaldson *et al.*, 2010; Mossman *et al.*, 2011; Pollastri *et al.*, 2014; Benedetti *et al.*, 2015 and references therein) are other complicating factors, so that devising correct, universally accepted models of toxicity and pathogenicity are no more than pipe-dreams at present. The lack of simple models and the many different scientific views on the application of these models have slowed the delivery of global directives and prudent use of asbestos fibres.

Scientific research aimed at discovering the mechanisms by which asbestos fibres interact with the biosphere and which will aid in prevention and treatment of asbestos-caused diseases is the legacy of occupational exposures during the 20th century. For example, a long-term Italian project which started in 2011 (PRIN 2010-2011 *Interaction between minerals and biosphere: consequences for environment and human health*) has been funded to conduct a systematic investigation of the crystal-chemistry, bio-interaction and potential toxicity of mineral fibres. The project allowed scientists to identify and, individually or cooperatively, investigate factors that play a major role in the toxicity and carcinogenicity of mineral fibres such as iron content, morphometric parameters including length and aspect ratio, crystallinity, surface chemistry and biodurability. The final outcome of the project is a preliminary attempt to devise a general toxicity/pathogenicity model and will be introduced briefly in this chapter together with a systematic description of all the parameters influencing the toxicity or pathogenicity of mineral fibres. The model can be generalized and is thus also applicable to man-made mineral fibres which can be durable or soluble and other natural and/or synthetic fibres.

Although this is the concluding chapter in this volume we should remind ourselves that certain terms need definition as they can be confusing to scientists in different disciplines. 'Toxicity' is a very broad term that is defined medically as a state of being poisonous and encompassing injury and death of cells, termed 'cytotoxicity'. 'Genotoxicity' is a term covering many alterations in DNA, chromosomes and other nuclear alterations that can be associated with 'toxicity', cell death, or the initiation, promotion, and progression of cancers. 'Carcinogenicity' is the ability of an agent(s) to induce these latter changes, thus favouring the neoplastic process and development of cancers. Both lung cancers and MMs, although they may exhibit cancer-specific mutations which can be heritable now appear to have more important 'epigenetic' changes, *i.e.* that are not related directly to nucleotide sequences (DNA) of the genome and can be reversible (reviewed by Toyooka *et al.*, 2008; Christensen *et al.*, 2009; Brzezianska *et al.*, 2013; Reid, 2015). In this regard, both lung cancers (Pastuszak-Lewandoska *et al.*, 2015) and MMs have several mutations in cancer suppressor genes which may also be epigenetically regulated (Kobayashi *et al.*, 2008), but no common mutations in 'driver' oncogenes have been found that are integral to the initiation or development of MMs (Sugarbaker *et al.*, 2008; Christensen and Marsit, 2011). It is important to realize that all of these biological parameters of tumorigenesis are probably dose-dependent and interrelated with the type, duration of exposure, dimensions and durability of asbestos fibres. Lastly, 'pathogenicity' is a broad term that encompasses tissue damage and asbestos-related diseases in general, including asbestosis, pleural fibrosis, lung cancers and MMs.

2. Parameters influencing the toxicity and pathogenicity of a fibre

In this paragraph we provide a systematic description of all the parameters that should be considered in the formulation of toxicity/pathogenicity models of mineral fibres, including their role and cross-correlations. Parameters are grouped into four macro-sections: (1) fibre size and morphometric parameters; (2) iron and surface elements; (3) biodurability and crystallinity; and (4) surface charge.

2.1. Fibre size and morphometric parameters

1a *Length and diameter*

The geometry of a fibrous particle is defined by the length (L) and the diameter (d_p) or by the aspect ratio ($\beta = L/d_p$). Since the early 1980s, these physical parameters have been recognized as playing a key role in the toxicity, inflammatory potential, carcinogenicity and fibrogenic potential of mineral fibres. For example, d_p is important in defining the equivalent aerodynamic diameter, D_{ae}, (see the next parameter **1b**) and its influence on pulmonary and pleural deposition (Kreyling *et al.*, 2007; Donaldson *et al.*, 2010; Murphy *et al.*, 2012, 2013; Schinwald *et al.*, 2012). In this context, the influence of L on D_{ae} is negligible for thin fibres (Walton, 1991). On the other hand, evidence demonstrating that L is a key factor in toxicity, inflammation and pathogenicity of fibres, especially in MM development, comes from a number of *in vitro* and *in vivo* toxicological studies (Donaldson *et al.*, 2010). The famous 'Stanton hypothesis' (Stanton *et al.*, 1981) is based on observations of experimental animals following injection and implantation of fibres, indicating that the optimum morphology for inducing intrapleural tumours by these routes of administration is $d_p \leqslant 0.25$ μm and $L > 8$ μm. The 'Stanton hypothesis' is based upon the fact that needle-shaped particles with $L > 8$ μm (so-called 'Stanton fibres') cannot be eliminated by phagocytic cells such as macrophages (Churg, 1993), leading to a process known as "frustrated phagocytosis". This is confirmed by the studies of Pott (1978). Some aspects of the dosimetry of the carcinogenic potency of asbestos and other fibrous dusts are found in Pott (1978) and Spurny *et al.* (1979) who found that asbestos fibres with the most striking effects on MM development in rodents had $L \geqslant 10$ μm and $d_p \leqslant 0.25$ μm or

$L \geqslant 5$ μm and $d_p \leqslant 0.25$ μm, respectively. Davies *et al.* (1986), who exposed rats to airborne clouds of equal mass concentration of either long or short amosite fibres, observed tumour development parallel to fibrotic responses – both were found only in rats exposed to long fibres ($L \geqslant 5$ μm). The 'Stanton hypothesis' is supported by results from a number of laboratories showing that cytotoxicity, inflammatory potential and the development of asbestos-related MMs, lung cancers, and fibrosis are observed with long fibres (⩾5 μm long) that are selectively retained in the lung or pleura (reviewed by Roggli, 2014, 2015). A recent review (Lippmann, 2014) amalgamating results from toxicological and epidemiological studies concluded that, for biopersistent airborne fibres, length dictated their ability to cause MM and pulmonary fibrosis or lung cancers. For example, MMs are caused by fibres thinner than 0.1 μm and longer than 5 μm whereas excess lung cancers and pulmonary fibrosis occur with fibres longer than 20 μm.

A physical parameter that should be considered is the geometrical relationship between fibre (particle) diameter and curvature of the surface, or better, the radius of curvature, the reciprocal of the curvature. For definition, the radius of curvature for surfaces is the radius of a circle that best fits a normal section thereof. For cylindrical lattices like that of chrysotile, the radius of curvature is large whereas for the other lattices, the radius of curvature tends to zero. It has been observed for nanoparticles that small changes in diameter could affect protein binding to form both "soft" and "hard" corona layers (Deng *et al.*, 2012) and hence biological responses. It was also observed that protein adsorption (see **1e**) on curved surfaces like that of chrysotile or very small nanoparticles (such as 7 nm diameter gold nanoparticles: Deng *et al.*, 2012) is suppressed up to the point when it no longer occurs, an effect likely to be confined to larger proteins, offering a route to differential control of protein adsorption (Lynch and Dawson, 2008).

1b *Equivalent aerodynamic diameter (D_{ae})*

There are many definitions of equivalent aerodynamic diameter (D_{ae}). The simplest one refers to the diameter of a sphere of unit density (1 g/cm^3) that has the same terminal settling velocity as the subject particle:

$$D_{ae} = D_p \left(\frac{\rho_p}{\rho_0 \chi} \right)^{\frac{1}{2}}$$

where D_p = the projected area diameter, the diameter of a sphere that has the same projected area as the particle in question; ρ_p = particle density; ρ_0 = unit density (1 g/cm^3); χ = dynamic shape factor (being ~1 for a fibre) (Scheuch and Heyder, 1990). Instead of the projected area diameter, the equivalent geometrical diameter can be used (Walton, 1991). It is defined as the diameter of a sphere that has the same volume as the particle (an ellipsoid in the case of a fibre) in question $D_p = \sqrt[3]{2Ld_p}$.

A more robust definition was given by Gonda and Abd El Khalik (1985):

$$D_{ae} = D_p \sqrt{\left(\frac{1}{\frac{2}{9}\left(\frac{1}{(\ln 2\beta - 0.5)}\right) + \frac{8}{9}\left(\frac{1}{(\ln 2\beta + 0.5)}\right)} \right) \left(\frac{\rho_p}{\rho_0} \right)}$$

The initial deposition of a mineral fibre is related directly to its D_{ae} which determines in which region of the respiratory tract (extrathoracic nasal and laryngeal, bronchial, alveolar) it will be deposited (Heyder *et al.*, 1986). Deposition in the bronchial region occurs mostly *via* impaction whereas gravitational sedimentation and diffusion dominates as deposition mechanisms in the alveolar region.

Considering the figures reported by Yeh *et al.* (1976) and Heyder *et al.* (1986) on the total respiratory tract deposition with both mouth and nasal breathing, a minimum deposition occurs for particles with $0.4 \leqslant D_{ae} \leqslant 1$ μm. Particles with $D_{ae} > 5$ μm are deposited in the nasal respiratory tract, particles with $3 \leqslant D_{ae} \leqslant 15$ μm are deposited in the laryngeal and bronchial respiratory tract, and maximum alveolar deposition is found for particles with $D_{ae} \approx 2-3$ μm and <0.2 μm. If the 'Stanton hypothesis' can be considered a length-based factor, the aerodynamics of the fibre are governed mostly by the diameter. Applying the above equation (Gonda and Abd El Khalik, 1985) to chrysotile ($\rho_p = 2.53$ g/cm^3) with $d_p = 0.25$ μm, minimum deposition is observed for $1 \leqslant L \leqslant 100$ μm (*i.e.* $0.4 \leqslant D_{ae} \leqslant 1$ μm). Instead, fibres with $d_p = 1$ μm and $L > 20$ μm are deposited in the laryngeal and bronchial respiratory tract (*i.e.* $D_{ae} > 3$ μm) whereas fibres with $d_p = 2$ μm and $L > 15$ μm are intercepted in the nose (*i.e.* $D_{ae} > 5$ μm). The 3D plot in Fig. 1 is built using the same equation and demonstrates the insensitivity of D_{ae} to fibre length. Fibres with $d_p > 1.5$ μm are deposited invariably in the nasal respiratory tract and consequently are less toxic and pathogenic than fibres with $d_p < 1.0$ μm which are deposited in the alveolar space.

1c *Crystal habit*

There are two attributes of crystal habit that influence indirectly the toxicity of a fibre: the asbestiform character of fibre and, in the case of chrysotile, its fibril morphology. Fibres may be curled or needle-like (asbestiform *vs.* acicular), leading to different depositional behaviours once inhaled. Compared to needle-like fibres, curled ones such as chrysotile tend to be deposited in the upper airways, particularly at bronchial or bronchiolar bifurcations (Harris and Timbrell, 1977). There, they come in contact with epithelial cells and may be taken up and translocated to the interstitium or cleared through the mucociliary elevator (Evans *et al.*, 1973). The curled crystal habit thus appears to be a positive factor because fibres do not reach alveolar regions. From this perspective, the potential toxicity and pathogenicity of chrysotile is lower than that of other needle-like fibres such as amphiboles in the deep lung and pleura.

Concerning chrysotile, the fibril morphology in terms of rolling mode is also important. A cylindrical concentric fibre (Whittaker, 1955), such as that shown in Fig. 2, may dissolve more slowly in the acidic environment generated by the phagolysosomes during macrophage phagocytosis, compared to a fibre with a cylindrical spiral lattice (Whittaker, 1955) (see Fig. 2b). The spiral structure possesses a permanent ledge produced by the emergence of a screw dislocation at the crystal surface whereas ledges are absent from the circular structure. As discussed by Veblen and Wylie (1993), surface ledges may affect dissolution kinetics and it is likely that their presence on spiral fibrils enhances their dissolution rates compared to those of circular fibrils (Morgan, 1997). Hence, chrysotile spiral fibrils are potentially less toxic

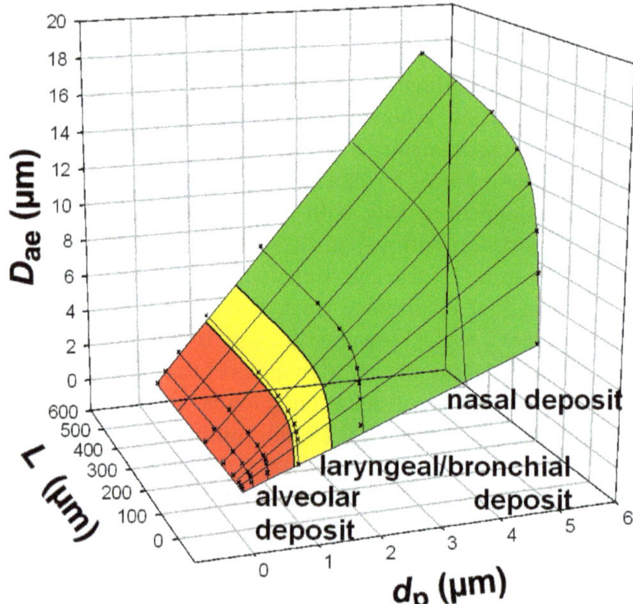

Figure 1. 3-D plot of the equivalent aerodynamic diameter D_{ae} calculated using the equation by Gonda and Abd El Khalik (1985), fibre length L and diameter d_p = highlighting the strong dependence of D_{ae} on d_p and the relatively insensitivity with respect to L.

than chrysotile concentric fibrils as they dissolve more quickly. High-resolution imaging of the cross-section perpendicular to the fibre axis is required to evaluate the rolling mode. Examples of cylindrical circular (concentric) and cylindrical spiral lattice rolls can be found in the literature. Transmission electron microscopy (TEM) images of thin sections of concentric fibril structures were published by Mellini (1986) for the Italian chrysotile from Balangero, whereas cylindrical spiral lattice rolls of Canadian chrysotile were reported by Yada (1967). Considering the dissolution model described by Morgan (1997) and discussed above, Canadian chrysotile should dissolve more quickly than chrysotile from Balangero.

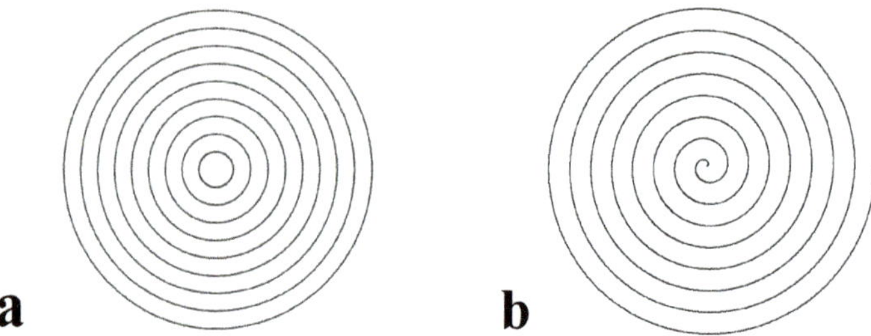

Figure 2. The concept of fibril morphology in terms of rolling mode. (a) cylindrical circular (concentric) lattice; (b) cylindrical spiral lattice.

1d *Density*

The ideal density (ρ_p) of a fibre is included in the calculation of D_{ae} (see the definition in Section **1b**). Hence, it influences the depth of deposition of inhaled particles in the airways (Yeh *et al.*, 1976). For example, an erionite fibre with $\rho_p = 2.11$ g/cm^3, $L = 20$ μm, $d_p >$ 0.25 μm has a $D_{ae} = 0.80$ μm whereas an amosite fibre with the same dimensions but with $\rho_p = 3.45$ g/cm^3 has a $D_{ae} = 1.02$ μm. The values of ρ_p (g/cm^3) of some other relevant fibres are: chrysotile 2.53, crocidolite 3.19, tremolite 3.05, wollastonite 2.84. Although this physical factor has a smaller influence on D_{ae}, compared to d_p (see equation in Section **1b**) it still causes a different depositional depth in the respiratory tract, being deeper for denser fibres (*i.e.* larger D_{ae}). Note that the ideal density (ρ_p) is of little physical and applicative significance as mass transport in suspension is dominated in most cases by agglomerates consisting of multiple primary particles as well as trapped suspension fluid and associated proteins (DeLoid *et al.*, 2014). For such agglomerates, only the effective density of the agglomerate should be considered, *i.e.* the density of the agglomerate unit, which includes both particles and media components, as opposed to the density of the primary particle, which is simply the density of the raw material (DeLoid *et al.*, 2014). Because the entrapped media and proteins usually have considerably lower density than the primary particles, the effective density of the agglomerate unit can be significantly lower than that of the raw material (Cohen *et al.*, 2012). The Sterling equation, based on a fractal model of agglomerate structure is used to obtain rough estimates of agglomerate effective density (Zook *et al.*, 2011).

1e *Surface chemistry*

Hydrophilicity was defined in terms of interfacial free energy of interaction between particles immersed in water (van Oss and Giese, 1995). When the energy is positive, the particles are hydrophilic and repel each other in water. On the contrary, negative interfacial free energy results in hydrophobic particles that attract each other in water. This chemical character, which depends on the crystal structure and microstructure of the fibres as well as on solvent chemistry (pH), influences the surface interaction with organic molecules such as proteins. This is a key factor as it is now universally accepted that a particle is invariably covered by proteins immediately upon contact with a physiological environment (Lynch and Dawson, 2008). Recent studies have demonstrated that the interface between proteins and particles is actually a "protein corona" instead of a fixed layer, with a composition that is determined by the concentration of a huge number of proteins (Lynch and Dawson, 2008). Adsorbed proteins are not always bound indefinitely to the surface, and the surface assemblage of proteins may change over time (this is the so-called 'Vroman effect').

Van Oss *et al.* (1999) reported a variable hydrophobicity for selected chrysotile samples and a clear hydrophobicity for crocidolite, amosite and erionite. Exceptionally, a chrysotile sample from Globe (Arizona, USA) was found to be hydrophilic.

There are two factors related to the hydrophilicity of the particle surface that have the potential to influence toxicity and probably pathogenicity: a capacity to adsorb biopolymers and interaction with human phagocytic cells. Hydrophobic surfaces adsorb biopolymers more strongly than hydrophilic surfaces (van Oss *et al.*, 1999). The hydrophobic surface allows interaction with hydrophobic domains and residues in the protein and the process occurs with a gain of entropy, during the subsequent release of unfavourably organized water at the surface. Highly hydrophobic particles, such as small particles of talc, are more readily and successfully engulfed by human phagocytic cells than hydrophilic particles like silica (van Oss *et al.*, 1999). Hence, as far as the chemical character of the surface is concerned, particles with stronger hydrophobic character are more prone to cell uptake.

1f *Surface area*

The surface area is usually defined as specific or reactive (Fischer *et al.*, 2014). The former is defined as the total surface area of a sample divided by its mass and is generally determined by gas adsorption and the Brunauer-Emmett-Teller (BET) equation (Brunauer *et al.*, 1938). It is important to note that the specific surface area (SSA) of the same crystal or sample powder can vary depending on the crystallographic form and the grain size of the sample material. The term reactive surface area is elusive and very difficult to measure. It is commonly used to distinguish the portions of the surface that contribute dominantly to a measured reaction-related parameter (*e.g.* dissolution) from portions that do not (Rufe and Hochella, 1999).

Whatever the definition of surface area is, it is certainly one factor that affects fibre biodurability/biopersistence (see the definition of the parameter in group **3**). In particular, the dissolution rate is related directly to the surface area and is included as a factor in equations describing dissolution kinetics (see for example Fisher *et al.*, 2014). At a first approximation, other things being equal, fibres with a greater surface area are less biodurable and should be considered less toxic. However, as will be discussed later for the parameters of group **3**, other factors are affected by the surface area and dissolution rate such as: iron or magnesium release induced by dissolution, production of silica relics after rapid dissolution of chrysotile (and with a much lower rate, amphiboles and/or erionite), and release of metals (both toxic and non-toxic).

2.2. Iron and toxic elements

The role of iron in asbestos toxicity is very complex and not yet fully understood. It has long been known that iron (basically Fe^{2+}) at the surface of asbestos minerals and the iron content of amphibole types of asbestos promote the formation of hydroxyl radicals ($HO^{\bullet}$) that can damage DNA and induce protein oxidation and lipid peroxidation. Moreover, iron may not just behave as a catalytic site but can also influence the interaction between mineral fibres and cells in a more complex way (Gazzano *et al.*, 2007). For instance, it can inhibit G6PD formation and the subsequent impairment of the antioxidant defences of the cells (Gazzano *et al.*, 2005) at high or overload concentrations of asbestos fibres, leading to further generation of free radicals and more severe cell damage. However, at lower concentrations of asbestos fibres, antioxidants can scavenge ROS/RNS as an important defence and repair mechanism (reviewed by Shukla *et al.*, 2003; Heintz *et al.*, 2010; Benedetti *et al.*, 2015).

$HO^{\bullet}$ are a predominant damaging oxidant species *in vivo* (Pryor, 1998). Content and speciation (oxidation state and coordination environment) of iron as well as intracellular iron mobilization and transport were shown to be important factors of amphibole asbestos-induced toxicity (see for example, Hardy and Aust, 1995; Werner *et al.*, 1995; Bergamini *et al.*, 2007; Turci *et al.*, 2011).

The surface activity of iron as a source of $HO^{\bullet}$ relies on the Haber-Weiss cycle:

$$Fe^{2+} + O_2 \rightarrow Fe^{3+} + O_2^{\bullet-}$$
$$O_2^{\bullet-} + 2H^+ + e^- + H^+ \rightarrow H_2O_2$$
$$Fe^{2+} + H_2O_2 \rightarrow Fe^{3+} + OH^- + {}^{\bullet}OH \text{ (the Fenton reaction)}$$

The above chain reaction is thought to be a major process responsible for the formation of $HO^{\bullet}$. H_2O_2, the radical species superoxide (O_2^{-}) and free oxygen are released *in vivo* by the macrophages during the inflammatory burst which accompanies frustrated phagocytosis of long amphibole fibres and inflammasome activation (Dostert *et al.*, 2008; Kamp, 2009).

Note that, in principle, during the Haber-Weiss cycle, Fe^{3+} can also be reduced by H_2O_2 (or radical species O_2^{-} or free oxygen):

$$Fe^{3+} + H_2O_2 \rightarrow Fe^{2+} + HO_2^{\bullet}/O_2^{\bullet} + H^+$$

or

$$Fe^{3+} + HO_2^{\bullet}/O_2^{\bullet} \rightarrow Fe^{2+} + O_2 + H^+$$

and the chain propagates through the reactions:

$$H_2O_2 + HO^{\bullet} \rightarrow HO_2^{\bullet}/O_2^{\bullet} + H_2O$$

Although it is correlated strongly with other factors, the structural iron content plays a major role in defining the toxicity of a mineral fibre. The proof that iron content (not just its presence) is a key factor was found by Gazzano *et al.* (2007) who reported that synthetic stoichiometric chrysotile nanofibres, devoid of iron or any other contaminant, did not exert genotoxic and cytotoxic effects nor did they elicit oxidative stress in a murine alveolar macrophage cell line. On the contrary, the same nanofibres, loaded with 0.57% and 0.94% (w/w) iron, induced DNA strand breaks, lipoperoxidation, inhibition of redox metabolism and alterations of cell integrity.

A rational interpretation of the literature data highlights that the iron-related parameters playing a role in the toxicity of mineral fibres are: iron content and nuclearity of the fibre (**2a**); surface *vs.* bulk iron (**2b**); Fe^{2+}/Fe^{3+} ratio (**2c**).

2a *Iron content and nuclearity*

Even if the total amount of iron in a mineral fibre is much less important than the other two iron-related toxicity factors (*i.e.* surface *vs.* bulk iron Fe, Fe^{2+}/Fe^{3+} ratio), a direct correlation exists with the number of potential active catalytic sites for the formation of $HO^{\bullet}$. Zalma *et al.* (1987) calculated that a chrysotile fibre with $d_p = 0.1$ μm and $L = 5$ μm may generate a minimum of 10^4 $HO^{\bullet}$. It is important to make a distinction between the iron content of a fibre and the iron content of a particle. Toxicity due to release of $HO^{\bullet}$ is related to the iron content of a fibre as iron on long fibres may prompt the activation and release of H_2O_2 (or radical species O_2^{-} or free oxygen) during macrophage-frustrated phagocytosis. On the other hand, the iron in a short-fibre, sub-spherical particle or nanoparticle such as magnetite (α-$Fe^{2+}OFe_2^{3+}O_3$) will not be active (and hence toxic) as it is phagocytized successfully and cleared by macrophages, with no H_2O_2-mediated release of $HO^{\bullet}$. For this reason, the iron content of a fibre should exclude iron contained in impurities such as magnetite. As seen below, another reason for the inactivity of iron oxides is their surface iron speciation (**2b** and **2c**). With these premises, it is intuitive that iron-rich fibrous amphiboles such as amosite or crocidolite are considered more toxic and pathogenic than chrysotile because they will display many more potentially active sites.

An anomalous case is represented by fibrous erionite which is considered to be more potent than crocidolite, amosite or chrysotile asbestos in causing mesothelioma (Baris *et al.*, 1978; Wagner *et al.*, 1985; and reviewed by Carbone *et al.*, 2007). Although its toxicity can be explained by the high biodurability/biopersistence (see group **3**), production of iron-catalysed hydroxyl radicals should be negligible because the iron content is small and is mostly Fe^{3+}. On the other hand, iron seems to be concentrated at the surface of the erionite fibres in the form of Fe^{3+}-rich oxide nanoparticles (see for example Ballirano *et al.*, 2009; Pollastri *et al.*, 2015; Gualtieri et al., 2016) with possible minor Fe^{2+} ions at the window of the zeolite micropores (Fig. 3) that acted as a centre of nucleation for the nanoparticles. The surface nuclei may be active later as low nuclearity Fenton $Fe(OH)^{2+}$ groups if the thin iron nano-layer is dissolved in the acidic environment created during phagocytosis. Moreover, release of $HO^{\bullet}$ from the active Fe^{2+} extraframework sites should occur as a single burst because such sites are accessible by H_2O_2 and they are assimilated to surface sites.

The activity of iron sites is also related to their nuclearity, *i.e.* the number of iron atoms joined in a single coordination entity by bridging ligands indicated by monomeric (single iron atom, no other iron atoms in the second shell coordination), dinuclear or dimeric (a cluster of two iron atoms, connected by a bridging oxygen atom), trinuclear of trimeric (a cluster of three iron atoms, connected by bridging

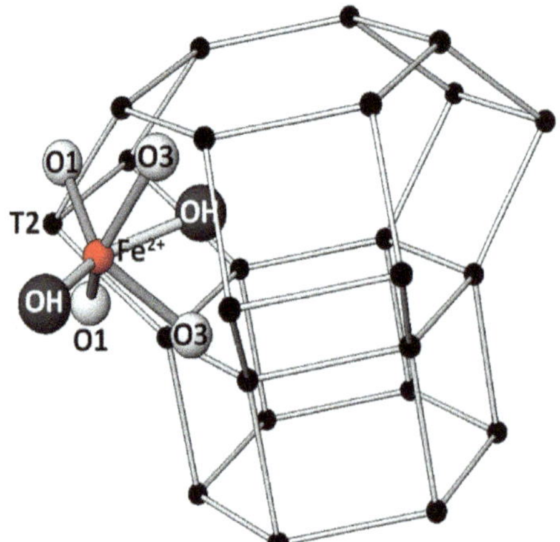

Figure 3. A possible active site for Fe^{2+} ions at the window of the 6-membered ring in the zeolite erionite.

oxygen atoms), and so on. According to Zecchina *et al.* (2007), the sites that have been described for erionite are isolated $(FeO)^{2+}$ structures and are the preferred candidate active site $[(H_2O)_5FeO]^{2+}$ as they have a low iron nuclearity. Minor fractions of paired Fe^{2+}-Fe^{2+} active species or Fe_xO_y clusters entrapped in the zeolite framework cavities are considered less active. Clustered iron with high nuclearity should be much less active or indeed inactive. The limiting case is that of iron oxides with a structure composed of many neighbouring iron atoms.

2b *Surface vs. bulk iron*

To be active, iron sites must be situated at the fibre interface with the cell or cell medium. Hence, factor **2a** is latent and may be considered negligible if iron is not made available at the surface.

However, creation of active sites is possible, *e.g.* due to dissolution phenomena *in vitro* and *in vivo*. In this case, the number of active iron sites is strictly correlated with the specific surface area (**1f**) as well as with other dissolution-related parameters (**3**). The activity of iron as a catalytic site for the production of $HO^{\bullet}$ is possible only when the surface of the mineral fibre (or particle) is fresh. Aging, oxidation and protein coating neutralize surface activity. This proviso was first postulated by Zalma *et al.* (1987): the emergence of fresh surfaces (not oxidized) is a source of active sites. An active material may be an Fe^{2+}-containing material (see below), but this condition is not sufficient if there is easy reconstruction of the solid surface to an oxidized sheet, impermeable to air even in aqueous medium (in the case of magnetite).

2c *Fe^{2+}/Fe^{3+} ratio*

Fubini and Mollo (1995) stated that both a defined coordination and redox state appear to be involved in the toxicity of mineral dusts, keeping in mind that the surface activity of iron as source of $HO^{\bullet}$ relies on the Haber-Weiss cycle and also on the availability of surface Fe^{2+}. In the past, some controversy has arisen as to which oxidation state is potentially more dangerous but it is now well established that Fe^{2+} is the most active, dangerous and hence toxic species.

As a matter of fact, both hematite and magnetite showed very little free radical release activity despite their large iron contents (Fubini *et al.*, 1997). Hematite was always at the level of the blank while magnetite exhibited a little activity only when ground (Fubini *et al.*, 1997). These results are explained by the fact that hematite contains only Fe^{3+} and although magnetite contains both Fe^{2+} and Fe^{3+}, the particles are invariably coated with a nanophasic layer of either maghemite (γ-Fe_2O_3) or amorphous matter with iron in an oxidized state (see for example Frison *et al.*, 2013).

Aging, oxidation or selective extraction destroy the active site (Fubini *et al.*, 1997) as iron is invariably oxidized. In the same vein, a coating of Fe^{3+} was reported to inactivate the release of free radicals from crocidolite (Gulumian *et al.*, 1989). Hence, the toxicity potential of a mineral fibre in terms of release of $HO^{\bullet}$ is related directly to the existence of a large Fe^{2+}/Fe^{3+} ratio at the particle surface whereas mineral fibres with a low or very low Fe^{2+}/Fe^{3+} ratio are poorly active or even inactive. The role of Fenton chemistry in the production of $HO^{\bullet}$ can also explain why soluble, man-made fibres caused only small amounts of DNA strand breakage in an isolated plasmid DNA assay (Donaldson *et al.*, 1996) and are apparently inactive in pathogenicity as compared to PM10 and amphibole asbestos. As man-made fibres such as rock wool or glass wool are produced by melting a volcanic rock at 1500°C, iron is already fully oxidized during the process.

2d *Toxic elements other than iron*

A factor that could play a role in fibre toxicity and therefore should be evaluated quantitatively is the amount of trace elements or mineral oils capable of inducing lung cancer (see for example Harington and Roe, 1965; Bargagli *et al.*, 2008; Wei *et al.*, 2014). There are both epidemiological and experimental indications that trace metals are carcinogenic (Nemery, 1990), and some researchers have claimed that asbestos fibres may play a passive role in producing diseases as carriers of trace elements or hydrocarbons (Dixon *et al.*, 1970). Harington and Roe (1965) suggested that asbestos carcinogenesis effects may be due to the presence of natural mineral oils and trace metals (*e.g.* Cr and Ni). In a later work, Cralley *et al.* (1968) showed that trace metals (Ni, Cr and Mn) could be responsible for the development of asbestos cancers in textile workers and bovine species. Gross *et al.* (1969) lent additional support to the hypothesis, showing that rats exposed to mixed dusts of asbestos-containing large concentrations of Cr, Co and Ni developed lung cancer.

Rare earth elements (*REE*) may also be toxic as they have been associated with redox reactivity (Pagano *et al.*, 2015a,b), involving the formation of reactive oxygen species (ROS). On the basis of clinical observations of workers exposed to *REE* (see for example Rim *et al.*, 2013), it was confirmed that these elements are potentially hazardous and cause lung diseases.

Recently, Bloise *et al.* (2016) investigated the concentration of trace elements (Li, Be, Sc, V, Cr, Mn, Co, Ni, Cu, Zn, As, Rb, Sr, Y, Sb, Cs, Ba, La, Pb, Ce, Pr, Nd, Sm, Eu, Gd, Tb, Dy, Ho, Er, Tm, Yb, Lu, Th, U) in mineral fibres. Their concentrations are highly variable among the different species, due to the different geochemical processes involved in their formation. However, chrysotile, amphiboles and erionite all have a large capacity for hosting trace elements. In amphiboles and chrysotile, trace metals occur as substituting cations for Mg and Fe in the octahedrally coordinated sites. Chrysotile from Valmalenco contains the largest amounts of Co, Ni and Cu; chrysotile from Balangero has the largest Cr content; amosite the largest Mn content and anthophyllite asbestos the largest Zn content. On the whole, anthophyllite hosts the largest amounts of trace metals, followed by Balangero and Valmalenco chrysotile (Bloise *et al.*, 2016).

Because chrysotile is less bio-persistent (see group **3**) in the lungs, trace metals may be released quickly in the extracellular medium. If this proviso is correct, the potential toxicity of metal-rich chrysotile asbestos should be reconsidered. The large amount of Be, Pb and Ce present in crocidolite should be taken into account. Tremolite contains more *REE* than amosite, crocidolite or anthophyllite. Erionite, tremolite and chrysotile from Valmalenco have the greatest concentrations of La, Ce and Gd; erionite also has appreciable amounts of As, Be and Pb. Because there is convincing evidence of a relationship between lung-cancer mortality and cumulative As, Be and Pb exposure, the toxicity of fibrous erionite could, in principle, be related to the synergistic effects of these contaminants (Bloise *et al.*, 2016).

2.3. Biodurability

As is well described by Rozalen *et al.* (2017, this volume), biopersistence is defined as the ability of a particle of fibre to persist in the human body regardless of chemical, physical or other physiological clearance mechanisms (Bernstein *et al.*, 2005) or as the persistence of a fibre in human lung fluids over a lifetime (Oberdörster, 2000). Biodurability is defined as the resistance to chemical/biochemical alteration and is a significant contributor to biopersistence (Utembe *et al.*, 2015). Both biopersistence and biodurability are parameters invoked to explain the toxicity of mineral fibres. The relationship of biopersistence to both toxicity and cancer responses in rats after chronic inhalation or intraperitoneal injection of synthetic fibres (Bernstein *et al.*, 2001) is a model that may also be applied to mineral fibres. If the fibre dissolves or breaks down rapidly in lung fluids, it has a low biopersistence and is assumed to have a low toxicity and pathogenic potential. For example, soluble synthetic fibres and chrysotile were shown to be removed rapidly from the lungs following inhalation in experimental animals (Bernstein *et al.*, 2001, 2005). Albin *et al.* (1994) analysed mineral fibres found in the lungs of workers exposed to chrysotile contaminated with amphibole, and the results showed that most of those fibres were amphiboles despite greater exposure to chrysotile. In fact, amphiboles invariably show a higher biodurability than chrysotile (Bernstein *et al.*, 2005). Erionite also has a very high biodurability (Dogan *et al.*, 2008). The determination of biodurability/biopersistence of a fibre can be done according to the following different procedures:

- *in vitro* acellular studies, generally using simulated lung fluids (SLF) such as Gamble's solution (see for example Marques *et al.*, 2011) that mimic the environment in lung fluids. A solution buffered at pH 7–7.4 is used to simulate the extracellular environment whereas a pH of 4–4.5 is used to simulate the intracellular environment (phagolysosome vacuoles engulfing the fibre during phagocytosis) (biodurability);
- *in vitro* cellular studies exposing fibres to cultured phagocytic cell cultures (see for example Luoto *et al.*, 1997) (biodurability);
- *in vivo* studies, where dissolution of asbestos fibres is monitored in the lungs of laboratory animals (*e.g.* rats) (biopersistence);
- *in vivo ex post* lung analysis of workers exposed to mineral fibres: the 'Gold Standard' for human studies (biopersistence).

Unfortunately the design of these studies and the subsequent interpretation of the data obtained from them are often confounded by the different microscopic and other methods used to determine fibre-size distribution, the ratio of longer fibres to shorter fibres, the presence of non-fibrous particles, the lack of dose-response studies, and the use of fibre overload concentrations in animal or *in vitro* studies (Bernstein and Hoskins, 2006).

3a *Biodurability as a function of decreased fibre dissolution rate*
The high biodurability of a mineral fibre such as crocidolite explains its high toxicity and disease potential because it induces chronic inflammation in the lungs whereas the low biodurability of a fibre such as chrysotile explains its lower toxicity.

An original *in vitro* acellular study (Gualtieri *et al.*, 2017) showed (Fig. 4) that representative chrysotile samples (UICC and Italian samples from Valmalenco and Balangero) were dissolved after ~6 months in contact with Gamble's solution at pH 4 and 37°C. An exception was the UICC chrysotile, which was found to dissolve more rapidly than the other samples, being completely dissolved after 2 months. The half lives (50% of dissolution) were very short, being 13–15 h for the Italian chrysotiles and only 4 h for the UICC chrysotile. The results are comparable with those published by Morgan (1997) but show differences from the results of Oze and Solt (2010). Those authors calculated a total dissolution time for chrysotile from Lancaster, Pennsylvania (USA) with specific surface area of 6.5 m^2/g of ~24 h in acid aqueous solution (pH 1.2).

Table 1, modified after Gualtieri *et al.* (2012) shows the average *in vitro* acellular dissolution time (h) of several samples of chrysotile and fibrous amphiboles. It is clear that the biodurability of chrysotile is significantly less than that of the amphiboles. This difference is invoked to explain the different toxic and pathogenic potential of these fibre species (Ilgren and Chatfield, 1998; Bernstein *et al.*, 2008; Ilgren, 2008).

The difference in dissolution rate between the UICC chrysotile and the chrysotiles from Balangero and Valmalenco cannot be explained by their chemistry or by their

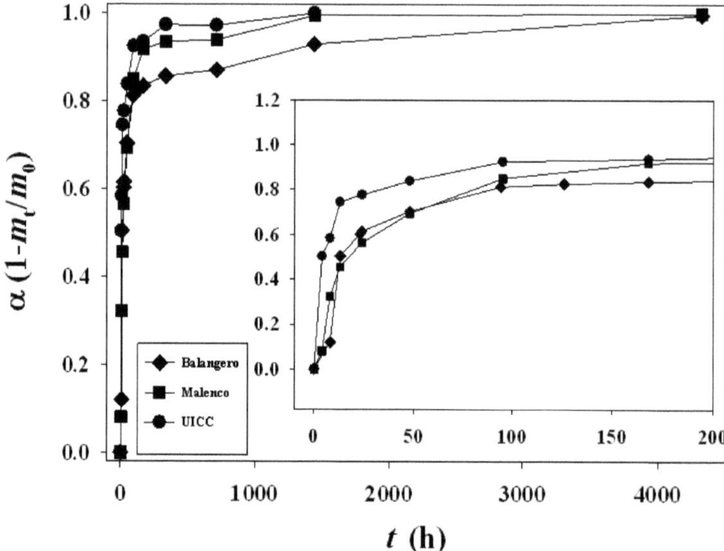

Figure 4. The kinetics of *in vitro* acellular dissolution (Gamble's solution, pH = 4, T = 37°C) for representative chrysotile samples (UICC, Valmalenco and Balangero). Data were renormalized with respect to the different specific surface. The inner graph is an enlargement of the time interval 0–200 h.

Table 1. Selected literature data on the acellular dissolution *in vitro* of mineral fibres.

Fibre species	Morphometric parameters	pH of the solution	Experimental mode	Mean dissolution time (h)	Reference
Amosite	$d^* = 0.5$ µm	4.5	dynamic	60000**	Guldberg *et al.* (1998)
Crocidolite	$d = 0.5$ µm	4.5	dynamic	120000**	Guldberg *et al.* (1998)
Tremolite	$L^{***} < 10$ µm	1.2	static	6576	Oze and Solt (2010)
Chrysotile UICC B		0, 2N HCl	static	287[a]	Morgan and Talbot (1997)
Chrysotile Jeffrey 4T-30 Canada		0, 2N HCl	static	320[a]	Morgan and Talbot (1997)
Chrysotile Coalinga, California (USA)		0, 2N HCl	static	302[a]	Morgan and Talbot (1997)
Chrysotile NIST	$d = 1.0$ µm; $L = 10$ µm	4.5	static	7152(432)	Gualtieri *et al.* (2012)
Chrysotile Lancaster, Pennsylvania (USA)	$L < 10$ µm	1.2	static	24	Oze and Solt (2010)
Chrysotile Bell Mine, Thetford, Quebec (Canada)	$L < 325$ µm SSA = 33.5 m^2/g	2.0–6.0, 37°C	static	6480	Hume and Rimstidt (1992)
Chrysotile UICC	$L = 5(2)$ µm; SSA = 35.5 m^2/g	4.5	static	1440	Gualtieri *et al.* (2012)
Chrysotile Balangero (Italy)	$L = 6(1)$ µm; SSA = 60.0 m^2/g	4.5	static	~4320	Gualtieri *et al.* (2012)
Chrysotile Valmalenco (Italy)	$L = 10(5)$ µm; SSA = 56.5 m^2/g	4.5	static	~4320	Gualtieri *et al.* (2012)

* d = mean diameter of the fibre; **calculated from literature data; ***L = fibre length; [a] calculated on the basis of the Mg dissolution rate only.

specific surface area. A fascinating model that explains the different *in vitro* reactivity concerns the fibril morphology (Morgan, 1997). According to this model, it is possible that the structure of a chrysotile sample composed of cylindrical concentric fibrils causes it to dissolve more slowly than a similar sample composed of fibrils with a cylindrical spiral roll. In fact "Fibrils having a spiral structure possess a permanent 'ledge' produced by the emergence of the screw dislocation at the surface of the crystal. Such ledges, absent from fibrils with cylindrical layers, probably affect dissolution kinetics, and it is probable that their presence on spiral fibrils enhances their dissolution rates compared to those of cylindrical fibrils" (Veblen and Wylie, 1993).

3b *Biodurability and rate of iron release*

Dissolution of iron-rich fibres would result in bulk iron becoming available at their reacting surface. As we have seen above (Section 2.2.), the surface activity of Fe^{2+} can be a source of $HO^{\bullet}$ during the Haber-Weiss cycle. Ideally, fast dissolution of an iron-rich fibre may have a negative side effect, *i.e.* enhancing the production of $HO^{\bullet}$.

Despite the huge difference in iron content between chrysotile (very minor) and fibrous iron-rich amphiboles (major) such as crocidolite and amosite, the much faster dissolution rate of chrysotile compared to amphiboles prompts a much greater release of available active surface iron in the same dissolution time.

3c *Production of silica relics during dissolution*

Pundsack (1955) was the first to find that a more acidic environment promotes dissociation of the surface of chrysotile because of the interaction of surface hydroxyl groups with hydrogen ions, eventually leading to rapid disintegration. According to this author, the process leads to shorter fibres and a large number of exposed amorphous silica particles, both of which contribute to potential toxicity. Later, Wypych *et al.* (2005) showed that acid-leaching of chrysotile fibres removes the brucite-like sheets leaving amorphous silica particles. Recently, it was observed that the first step of chrysotile dissolution is a 'pseudomorphic' amorphization (Fig. 5) process followed by dissolution *sensu stricto* (physical-chemical dissolution of the fibres). The formation of a silica-rich fibre skeleton after amorphization of chrysotile, characterized by silanol groups (Si–OH) and ionized silanol groups ($Si{-}O^{-}$), may prompt production of $HO^{\bullet}$ in synergy with surface iron species (Pollastri *et al.*, 2016). If this surmise is correct, chrysotile may be much more reactive and cytotoxic in short-term assays than crocidolite and erionite (see Mossman and Pugnaloni, 2017, this volume). The fact that chrysotile at small concentrations induces injury that can be repaired, as opposed to genotoxicity and cell death at large concentrations, points to a difference between toxic and carcinogenic mechanisms with this asbestos type in MMs and the importance of performing dose-response studies (reviewed by Mossman *et al.*, 2011). One of the parameters that should be considered when rating the toxicity potential of a mineral fibre is whether or not it forms a metastable silica-rich fibre skeleton during the dissolution process and has a positive surface charge (see below).

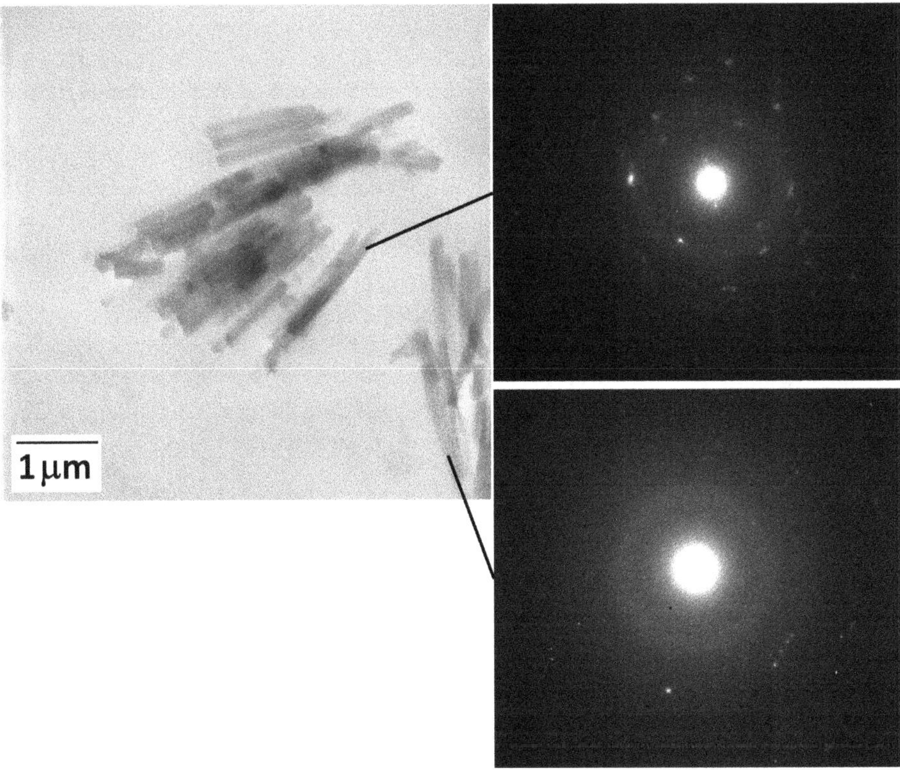

Figure 5. TEM imaging and selected area diffraction pattern of UICC chrysotile that underwent partial dissolution in an acidic environment. The overall crystal habit is preserved but the diffraction pattern evidences partial or complete amorphization ('pseudomorphosis').

3d *Biodurability and rate of release of toxic elements*

There are other metals (**2d**) besides iron that can be released from mineral fibres during their dissolution. Many of these elements may be like iron and display a catalytic surface activity, with production of $HO^•$ and other reactive species. Copper is a redox-active metal that acts as a catalyst for the formation of ROS and catalyses the peroxidation of lipids (Kelly and Mudway, 2007). Manganese acts as catalyst for the dismutation of superoxide to H_2O_2, but does not promote the formation of $HO^•$ in the presence of H_2O_2 to any great extent (Halliwell *et al.*, 1992). Vanadium has several oxidation states, and vanadate production of $HO^•$ has been demonstrated (Shi and Dalal, 1992). It also promotes decomposition of lipid peroxides and ROS production in general (Kelly and Mudway, 2007). Particulate Ni in excess is a carcinogen in humans. It causes oxidation of lipids, proteins, nucleic acids, intracellular radical production and glutathione depletion, but there is little evidence that it undergoes redox-cycling reactions like iron (Kelly and Mudway, 2007). Chromium is a carcinogen which reacts with H_2O_2 to yield $HO^•$ (Galaris and Evangelou, 2002). As far as the non-redox metals are concerned, Zn can catalyse Fe- and Cu-induced oxidation; Al aggravates lipid peroxidation by ferrous iron and promotes iron ROS generation (Zatta *et al.*, 2002); Pb induces oxidative stress, catalyses peroxidative reactions and ROS formation (Kelly and Mudway, 2007).

It is important to note that, albeit stable in both intracellular and extracellular environments, fibrous zeolites like erionite may release or exchange cations in an organic medium, interfering with cell cycles. Ballirano and Cametti (2015) observed that leaching erionite fibres with Gamble's solution induces a complex ionic-exchange process resulting in a temporary partial replacement of Na^+ by Ca^{2+}. This phenomenon is confined to a relatively short leaching time (<2 weeks) and produces a significant ionic exchange between fibres and the interstitial fluid deep within the lung, simulated by Gamble's solution. However, for a longer leaching time, the structure releases Ca^{2+} ions back to the SLF, reacquiring Na^+ ions. Owing to this final exchange process, fibres return approximately to their original composition and structure.

2.4. Surface charge

4a ξ *potential*

A particle dispersed in an electrolyte solution is charged due to ion adsorption or ionization of surface groups, leading to the formation of the so-called 'electrical double layer' (Stern, 1924). The first corona, called the 'Stern layer' with a 'Stern potential', is closely fixed to the particle surface. The second corona, called the 'diffuse layer', penetrates to a certain extent into the liquid phase.

Within this second layer, there exists a theoretical boundary where ions and particles form a stable entity. The potential at this boundary is known as the ξ potential. A large absolute value of ξ potential (*i.e.* >+30 mV or <−30 mV) results in colloidal stability, because electrostatic repulsions, which prevent aggregation of dispersed particles, occur. When the absolute value of the potential is low, coagulation and flocculation are favoured. The ξ potential can be determined from the electrosmotic velocity (von Smoluchowski, 1903):

$$\xi = -\frac{\nu_{eo}\eta}{E\varepsilon_0\varepsilon}$$

where ν_{eo} = electro-osmotic velocity; η = viscosity of the medium; E = applied electric field strength; ε_0 = permittivity of vacuum; ε = dielectric constant of the medium, assuming that ε and η have the same values in the double layer as in the bulk solution.

In double-distilled water, chrysotile shows positive ξ potential values, while crocidolite and erionite show negative values. In contact with Gamble's solution, all fibres display negative ξ potential, showing clearly that this parameter cannot be considered a discriminating factor (Pollastri *et al.*, 2014).

High-profile studies by Light and Wei (1977) have shown that the surface charge on fibres, as measured by the ξ potential, correlates directly with the fibres' haemolytic activity. When chrysotile asbestos, normally displaying a strong positive surface charge at pH 7.4, is leached with 0.1 M HCl, both the ξ potential and hemolysis are decreased due to leaching of Mg. Alternatively, crocidolite asbestos which is normally neutral in charge, becomes more negatively charged in 0.1 M HCl, correlating with increased haemolytic activity. The positive surface charge of chrysotile has been linked to cell death and hypersecretion of mucin, which can be prevented by the use of different lectins, *i.e.* proteins which bind to selected carbohydrate residues on the plasma membrane of airway epithelial cells (Mossman *et al.*, 1982).

Leaching of Mg from chrysotile fibres which occurs in acidic phagolysosomes, *in vitro* and in the lung, has been linked to their rapid dissolution (Jaurand *et al.*, 1977, 1984).

There are other parameters of toxicity and pathogenicity influenced by the ξ potential of minerals that include haemolysis, release of free radicals, fibre encapsulation and apoptosis. According to Gabor and Anca (1975), the mechanism for haemolysis induced by the fibres relies on the formation of lipid-bilayer clusters that normally extend homogeneously throughout the entire lipid bilayer.

Clustering occurs because of the electrostatic repulsion that occurs between negatively charged sialic acid groups and negatively charged fibres. Although ion permeability through lipid bilayers is very limited, ions can penetrate readily those regions of the erythrocyte membrane in which the clusters of glycoproteins are formed, thereby causing haemolytic activity.

Regarding the formation of free radicals, a negatively charged surface (negative ξ potential) may prompt their formation through the reaction with peroxide: $H_2O_2 + e^- \rightarrow HO^{\bullet} + OH^-$. The mechanism is identical to that described for freshly cut quartz surfaces in biological media (Donaldson and Borm, 1998) where silanol groups (Si–OH) and ionized silanol groups ($Si{-}O^-$) on the surface play a major role in interaction with membranes (Fubini *et al.*, 1985).

Regarding fibre encapsulation, negatively charged reactive surfaces, such as those displayed by amphibole fibres, favour the binding of collagen and redox-activated Fe-rich proteins. Hence the nature of the fibre surface certainly plays an active role during collagen deposition and asbestos encapsulation by mucopolysaccharides and Fe-rich proteins (ferritin, hemosiderin) that can be redox activated (reviewed by Liu *et al.*, 2013).

Finally, the ξ potential can also affect cross-talk phenomena and apoptosis. According to the model of mitochondria-endoplasmic reticulum cross-talk in asbestos-induced apoptosis (reviewed in Liu *et al.*, 2013), endoplasmic reticulum stress due to the fibre interaction causes activation of an unfolded protein response and Ca^{2+} release which, in turn, lead to activation of mitochondria-regulated apoptosis. Because the Ca^{2+} release to the mitochondria is necessary for the induction of intrinsic apoptosis, the whole cellular response cycle may be broken if the negatively charged corona of the mineral fibres attract and fix the Ca^{2+} ions (Pollastri *et al.*, 2014).

4b *Fibre aggregation*

The ξ potential of mineral fibres at pH 4.5 and 7.0 displays values in the range −10 to −26 mV, with no remarkable difference between the various fibre species (Pollastri *et al.*, 2014). At such low absolute values, particle agglomeration is favoured as discussed above. This is a critical point as culture conditions having the most agglomeration induce the greatest biological responses (Sharma *et al.*, 2014). Hence, fibres with low absolute values of ξ potential (*i.e.* agglomerated) are virtually more toxic than fibres with high absolute values of ξ potential (*i.e.* stable).

The matrix of qualitative correlations of all the parameters that play a role in the toxicity of mineral fibres is reported in Table 2. The 16×16 matrix is symmetrical. To build a quantitative model of the toxicity of mineral fibres, this matrix should be converted into an expression that takes into account both variance-covariance terms and the weight of each parameter. Because we live in a world with many fibres (one example in thousands is the silicate pumpellyite-Al portrayed in the SEM image of fig. 5b of Gualtieri, 2017, this volume) and have millions of asbestos and other fibres/particles in our lungs due to environmental exposure, a general quantitative model to predict *a priori* the toxicity potential of each mineral fibre is a powerful tool of risk assessment and prevention. A predictive universal model is of paramount importance in reducing exposure to naturally occurring asbestos (NOA), preventing asbestos-related diseases like those observed in certain regions of Turkey caused by erionite and in Italy by fluoro-edenite; and classifying fibre species that, due to their crystal chemistry, crystal habit and dimensions, and occurrence may be potentially toxic. However, it should be cautioned that a different matrix may apply to different fibre types, dimensions and diseases, and that a correlation between toxicity (cell injury or death) and pathogenicity (the development of tissue changes or diseases) may not exist for

Table 2. A sample correlation matrix of all the parameters that play a role in the toxicity of mineral fibres (the explanation of each code is in the text).

	1a	**1b**	**1c**	**1d**	**1e**	**1f**	**2a**	**2b**	**2c**	**2d**	**3a**	**3b**	**3c**	**3d**	**4a**	**4b**
1a		X				X		X			X	X		X	X	X
1b				X							X	X		X	X	X
1c					X											X
1d																
1e											X	X		X	X	X
1f								X			X	X	X	X		
2a								X	X							
2b									X		X	X				
2c											X	X				
2d																
3a												X	X	X		
3b																
3c																
3d																
4a																X
4b																

certain fibres. Moreover, dose-related responses may dictate different endpoints. For example, the toxicity of most chrysotile preparations is greater than that of amphibole asbestos types in hemolysis or cell assays at equal fibre numbers or surface area concentrations, but their biological or pathogenic effects in man, especially in the causation of MMs, are less (IARC 1989, 2012). Conversely, the amphibole types of asbestos, crocidolite and amosite, cause cell proliferation and metaplasia, two initial steps in cancer development, at low concentrations, and cell death at high concentrations in lung and mesothelial cells (Craighead *et al.*, 1980; Mossman *et al.*, 1980, 1996; Woodworth *et al.*, 1983). It is challenging but essential that mineralogists and clinical scientists work together on basic research to resolve and develop predictive models that incorporate their research findings and conclusions.

3. Asbestos bodies: state of the art

3.1. The nature of asbestos bodies

Although asbestos bodies (ABs) have been described by Capella *et al.*, 2017, this volume, it should be remembered that they consist of asbestos fibres that were coated with an iron-protein-mucopolysaccharide matrix, resulting in a beaded dumbbell or lancet-shaped structure that is recognized readily with the light microscope. Asbestos bodies are nearly always 20 μm or greater in length, suggesting that this dimension somehow triggers the coating process, perhaps through a mechanism of incomplete phagocytosis. Although the major proportion of asbestos used commercially in the United States was chrysotile, the vast majority of asbestos bodies occur on amphibole cores (mostly amosite or crocidolite). Asbestos bodies show a strong correlation with uncoated amphibole fibres 5 μm or more long, which in turn correlate well with the occurrence of asbestos-related diseases (Roggli, 2014).

3.2. Non-asbestos ferruginous bodies

A variety of other mineral fibres of sufficient length and biodurability may also become coated with iron, leading some to suggest the more non-committal term of ferruginous body (Table 3). These include various layer silicates (talc, mica) which tend to have broad yellow cores, and various metal oxides (iron, rutile, aluminium, stainless steel) which tend to have dark central cores. Carbon-based ferruginous bodies, commonly seen in coal miners, also have dark cores. Some ferruginous bodies, however, have thin translucent cores (erionite, refractory ceramic fibres) and are thus indistinguishable from true asbestos bodies at the light microscopic level.

3.3. Mechanistic considerations

The purpose of this coating process has long been questioned. As noted above, iron associated with particles and fibres is available for redox reactions which can lead to tissue damage through oxidation of proteins, lipids and nucleic acids. However, early

Table 3. Composition of non-asbestos ferruginous bodies.

Core composition	Percent
Talc	33
Iron	18
Silica	11
FeKAlSi	9
Aluminium silicate	6
Iron chromium (stainless steel)	5
Potassium aluminium silicate	5
Rutile	3
Fibrous glass	3
Magnesium aluminium silicate	3
Aluminium	2
FeKCaMgAlSi	1

Ferruginous body cores examined by scanning electron microscopy (SEM) and energy dispersive spectrometry (EDS). Modified from Roggli (2014).

studies of asbestos bodies *in vitro* indicated that they were less toxic than asbestos fibres (McLemore *et al.*, 1981). Subsequent studies by Ghio *et al.* (1997) indicated that some of the iron on the surface of asbestos bodies was in the form of ferritin and hence unavailable for surface redox reactions. Thus the coating process seems to serve as a protective mechanism against redox-related tissue injury (Ghio *et al.*, 2004).

3.4. Proteins in asbestos bodies

Recently, Perkins *et al.* (2017) used state-of-the-art mass spectrometry and microdissection techniques to analyse the proteins present in the coating material of asbestos bodies. Eleven candidate proteins were found to be specific to or up-regulated in the asbestos body samples compared to non-AB control samples. Of these 11 proteins, six were determined to be specific to AB. These proteins are involved in calcium-dependent protein cross-linking (Calmodulin-like protein 5, Transglutaminase-3), tight junction formation (Desmoglein-1), serine protease inhibition (Serpin-B12), cell proliferation (Histone H2A type-B/E), and inflammation (Lipocalin-1). Proteins that were not specific to but increased in AB samples included more proteins associated with tight junction formation (Desmoplakin, Junction Plakoglobin), cathespin inhibition (Cystatin-A), and squamous differentiation (keratinocyte proline-rich protein).

4. Malignant mesothelioma: aetiological considerations

4.1. Malignant mesothelioma and asbestos

Malignant mesothelioma (MM) is strongly related to previous exposure to asbestos. Roggli and Sharma (2014) reported that in a study of >500 histologically confirmed mesotheliomas, an elevated lung fibre content compared with a well characterized control population was present in 84% of cases. This observation correlates well with epidemiological studies of MM (Pavlisko and Sporn, 2014). The percentage is highest for pleural MMs in men (86%) and lowest for peritoneal MMs in women (16%). These observations indicate that there are probably other causes of MM (Table 4).

Table 4. Aetiological factors in malignant mesothelioma other than asbestos.

Mineral fibres
Erionite
Fluoro-edenite
Balangeroite
Libby amphibole
Therapeutic irradiation
Chronic inflammation
Genetic factors

* Includes winchite, richterite and tremolite

4.2. Non-asbestos related mesothelioma: other mineral fibres

Reports of mineral fibres other than asbestos causing mesothelioma emphasize the importance of the properties of mineral fibres in the pathogenesis of this disease (Comba *et al.*, 2003; Oczypok *et al.*, 2016). These include erionite (a fibrous zeolite), fluoro-edenite, balangeroite and the Libby amphibole. Studies have shown that these fibres have many of the physical-chemical properties of amphibole asbestos with mechanistic actions similar to those described above for asbestos minerals (Cardile *et al.*, 2004; Duncan *et al.*, 2010; Hillegass *et al.*, 2013). These other potential exposures should be kept in mind in cases of MM with no clearly identifiable asbestos causation.

4.3. Non-asbestos related mesothelioma: other factors

It is recognized that there is a percentage of mesotheliomas that cannot be related to exposure to asbestos or other mineral fibres. In this regard, a likely causative factor that has been identified is prior therapeutic irradiation. Patients at risk appear to be those with a fairly common malignancy which is often treated with radiation therapy and with long term survival in many patients. These include lymphomas, breast cancer and testicular malignancies (Neugut *et al.*, 1997; Travis *et al.*, 2005; Tward *et al.*, 2006; Teta *et al.*, 2007; Li *et al.*, 2015). Based upon the consultation files of one of the authors (V.L. Roggli, unpublished observations), radiation-related MMs account for just under 1% of cases (32/4000; 0.8%). An even less common association is with chronic inflammation (Hillerdal and Berg, 1985; Butnor *et al.*, 2016). It is likely that the mechanism of MM induction in chronic inflammation involves activation of the inflammasome in both phagocytic cells (Dostert *et al.*, 2008) and mesothelial cells (Hillegas *et al.*, 2013). Analysis of cases in which a role for mineral fibres has been excluded indicates that the patients tend to be younger, female, with MM occurring as an epithelial variant at peritoneal sites (Kraynie *et al.*, 2016).

4.4. MM and genetic factors

The prevalence of MM of less than 10% in the most heavily exposed cohorts as well as the occurrence of familial clustering of MM indicate that there may be an element of genetic susceptibility in the development of MM. Considerable attention has been given in this regard to BAP-1 (BRCA1-associated protein-1), which is located at 3p21.1 and regulates the BRCA1 growth pathway as a cancer-cell growth suppressor through deubiquitination of DNA. Individuals with germline mutations of BAP-1 are at risk for MM (Carbone and Yang, 2017, this volume), lung cancer, renal cancer, meningiomas, multiple myeloma and ocular melanoma (Carbone and Yang, 2012). Familial clustering of MM has been described in some cohorts with germline BAP-1 mutations (Testa *et al.*, 2011). It is probable that other genes predisposed to the development of MM will be discovered in the not-too-distant future.

5. Therapies/prevention

Standard treatment for MM offers little survival benefit, although some centres have experienced success with cyto-reductive surgery followed by multimodality chemo-radiotherapy (Pavlisko and Sporn, 2014). Therefore, considerable attention has been focused on molecular properties of mesothelioma through whole genome sequencing of cancer specimens, in the hope of finding specific driver mutations which may be targeted by chemotherapy (De Rienzo *et al.*, 2016). Some studies have implied that iron homeostasis mechanisms may be disrupted in both asbestosis and in mesothelial cells following talc pleurodesis (Ghio *et al.*, 2012, 2015). Thus a better understanding of mechanisms by which fibres disrupt iron homeostasis may suggest therapeutic options not currently available for those exposed to mineral fibres in the distant past.

6. Conclusions and future research trends

In this final chapter, the parameters that may be used to construct a general toxicity/pathogenicity model of mineral fibres have been described. A complete and satisfactory understanding of the toxicity/pathogenicity of mineral fibres aimed at finding solutions effective in the prevention and treatment of asbestos-related diseases can be achieved only by an holistic approach which takes advantage of collaboration amongst biochemists, pathologists, physicians, physicists, toxicologists and mineralogists/crystallographers all sharing their distinct but interrelated perspectives. Bringing together those engaged in these scientific fields to collaborate in terms of finding global solutions to the asbestos problem presents a significant challenge but an essential one if we are to solve it.

It is anticipated that there will be further investigation and characterization of the properties of mineral fibres that contribute to tissue injury and disease. Additional research into mechanistic pathways such as the inflammasome, reactive oxygen species generation, cytokine and growth factor pathways, and iron homeostasis will, hopefully, provide investigators with an understanding that will suggest improved methodologies for diagnosis and treatment of fibre-induced disease. Finally, for those unfortunate individuals who develop MM, it is expected that whole genome sequencing of resected cancers may provide clues regarding the progression of cancers and suggest mechanisms for novel patient-specific treatments.

References

Albin, M., Pooley, F.D., Stromberg, U., Attewell, R., Mitha, R., Johansson, L. and Welinder, H. (1994) Retention patterns of asbestos fibres in lung tissue among asbestos cement workers. *Occupational and Environmental Medicine*, **51**, 205–211.

Ballirano, P., Andreozzi, G.B., Dogan, M. and Dogan, A.U. (2009) Crystal structure and iron topochemistry of erionite-K from Rome, Oregon, USA. *American Mineralogist*, **94**, 1262–1270.

Ballirano, P. and Cametti, G. (2015) Minerals in the human body. Crystal chemical and structural modifications of erionite fibers leached with simulated lung fluids. *American Mineralogist*, **100**, 1003–1012.

Bargagli, E., Monaci, F., Bianchi, N., Bucci, C. and Rottoli, P. (2008) Analysis of trace elements in bronchoalveolar lavage of patients with diffuse lung diseases. *Biological Trace Element Research*, **124**, 225–235.

Baris, Y.I., Sahin, A.A., Ozesmi, M., Kerse, I., Ozen, E., Kolacan, B., Altinors, M. and Goktepeli, A. (1978) An outbreak of pleural mesothelioma and chronic fibrosing pleurisy in the village of Karain/Urgup in Anatolia. *Thorax*, **33**, 181–192.

Benedetti, S., Nuvoli, B., Catalani, S. and Galati, R. (2015) Reactive oxygen species a double-edged sword for mesothelioma. *Oncotarget*, **6**, 16848–16865.

Bergamini, C., Fato, R., Biagini, G., Pugnaloni, A., Giantomassi, F., Foresti, E., Lesci, G.I., Roveri, N. and Lenaz, G. (2007) Mitochondrial changes induced by natural and synthetic asbestos fibers: studies on isolated mitochondria. *Cellular and Molecular Biology*, **52 Suppl**, OL905–913.

Bernstein, D.M., Chevalier, J. and Smith, P. (2005) Comparison of Calidria chrysotile asbestos to pure tremolite: final results of the inhalation biopersistence and histopathology examination following short-term exposure. *Inhalation Toxicology*, **17**, 427–449.

Bernstein, D.M. and Hoskins, J.A. (2006) The health effects of chrysotile: current perspective based upon recent data. *Regulatory Toxicology and Pharmacology*, **45**, 252–264.

Bernstein, D.M., Riego Sintes, J.M., Ersboell, B.K. and Kunert, J. (2001) Biopersistence of synthetic mineral fibers as a predictor of chronic intraperitoneal injection tumor response in rats. *Inhalation Toxicology*, **13**, 851–875.

Bernstein, D.M., Donaldson, K., Decker, U., Gaering, S., Kunzendorf, P., Chevalier, J. and Holm, S.E. (2008) A biopersistence study following exposure to chrysotile asbestos alone or in combination with fine particles. *Inhalation Toxicology*, **20**, 1009–1028.

Bloise, A., Barca, D., Gualtieri, A.F., Pollastri, S. and Belluso, E. (2016) Trace elements in hazardous mineral fibres. *Environmental Pollution*, **216**, 314–323.

Brunauer, S., Emmett, P.H. and Teller, E. (1938) Adsorption of gases in multimolecular layers. *Journal of the American Chemical Society*, **60**, 309–319.

Brzezianska, E., Dutkowska, A. and Antczak, A. (2013) The significance of epigenetic alterations in lung carcinogenesis. *Molecular Biology Reports*, **40**, 309–325.

Butnor, K.J., Pavlisko, E.N., Sporn, T.A. and Roggli, V.L. (2016) Malignant peritoneal mesothelioma and Crohn disease. *Journal of Clinical Pathology*, **70**, 228–232.

Capella, S., Belluso, E., Bursi, N., Tibaldi, E. and Belpoggi, F. (2017) *In vivo* biological activity of mineral fibres. Pp. 307–345 in: *Mineral Fibres: Crystal Chemistry, Chemical-physical Properties, Biological Interaction and Toxicity* (A.F. Gualtieri, editor). EMU Notes in Mineralogy, **18**. European Mineralogical Union and Mineralogical Society of Great Britain & Ireland, London.

Carbone, M. and Yang., H. (2012) Molecular pathways: targeting mechanisms of asbestos and erionite carcinogenesis in mesothelioma. *Clinical Cancer Research*, **18**, 598–604.

Carbone, M., Emri, S., Dogan, A.U., Steele, I., Tuncer, M., Pass, H.I. and Baris, Y.I. (2007) A mesothelioma epidemic in Cappadocia: scientific developments and unexpected social outcomes. *Nature Reviews Cancer*, **7**, 147–154.

Cardile, V., Renis, M., Scifo, C., Lombardo, L., Gulino, R., Mancari, B. and Panico, A. (2004) Behaviour of the new asbestos amphibole fluoro-edenite in different lung cell systems. *International Journal of Biochemistry and Cell Biology*, **36**, 849–860.

Christensen, B.C., Houseman, E.A., Godleski, J.J., Marsit, C.J., Longacker, J.L., Roelofs, C.R., Karagas, M.R.,

Wrensch, M.R., Yeh, R.F., Nelson, H.H., Wiemels, J.L., Zheng, S., Wiencke, J.K., Bueno, R., Sugarbaker, D.J. and Kelsey, K.T. (2009) Epigenetic profiles distinguish pleural mesothelioma from normal pleura and predict lung asbestos burden and clinical outcome. *Cancer Research*, **69**, 227–234.

Christensen, B.C. and Marsit, C.J. (2011) Epigenomics in environmental health. *Frontiers in Genetics*, **2**, 84.

Churg, A. (1993) Asbestos lung burden and disease patterns in man. Pp. 409–426 in: *Health Effects of Mineral Dust* (G.D. Guthrie and B.T. Mossmann, editors). Reviews in Mineralogy and Geochemistry, **28**. Mineralogical Society of America, Chantilly, Virginia, USA.

Cohen, J., DeLoid, G., Pyrgiotakis, G. and Demokritou, P. (2012) Interactions of engineered nanomaterials in physiological media and implications for in vitro dosimetry. *Nanotoxicology*, **7**, 417–431.

Comba, P., Gianfagna, A. and Paoletti, L. (2003) Pleural mesothelioma cases in Biancavilla are related to a new fluoro-edenite fibrous amphibole. *Archives of Environmental Health*, **58**, 229–232.

Craighead, J.E., Mossman, B.T. and Bradley, B.J. (1980) Comparative studies on the cytotoxicity of amphibole and serpentine asbestos. *Environmental Health Perspectives*, **34**, 37–46.

Cralley, L.J., Keenan, R.G., Kupel, R.E., Kinser, R.E. and Lynch, J.R. (1968) Characterization and solubility of metals associated with asbestos fibers. *American Industrial Hygiene Association Journal*, **29**, 569–573.

Davis, J.M., Addison, J., Bolton, R.E., Donaldson, K., Jones, A.D. and Smith, T. (1986) The pathogenicity of long versus short fibre samples of amosite asbestos administered to rats by inhalation and intraperitoneal injection. *British Journal of Experimental Pathology*, **67**, 415–430.

DeLoid, G., Cohen, J.M., Darrah, T., Derk, R., Rojanasakul, L., Pyrgiotakis, G., Wohlleben, W. and Demokritou, P. (2014) Estimating the effective density of engineered nanomaterials for *in vitro* dosimetry. *Nature Communications*, **5**, 3514.

Deng, Z.J., Liang, M.L., Toth, I., Monteiro, M.J. and Michin, R.F. (2012) Molecular interaction of Poly(acrylic acid) gold nanoparticles with human fibrinogen. *ACS Nano*, **6**, 8962–8969.

De Rienzo, A., Archer, M.A., Yeap, B.Y., Dao, N., Sciaranghella, D., Sideris, A.C., Zheng, Y., Holman, A.G., Wang, Y.E., Dal Cin, P.S., Fletcher, J.A., Rubio, R., Croft, L., Quackenbush, J., Sugarbaker, P.E., Munir, K.J., Battilana, J.R., Gustafson, C.E., Chirieac, L.R., Ching, S.M., Wong, J., Tay, L.C., Rudd, S., Hercus, R., Sugarbaker, D.J., Richards, W.G. and Bueno, R. (2016) Gender-specific molecular and clinical features underlie malignant pleural mesothelioma. *Cancer Research*, **76**, 319–328.

Dixon, J.R., Lowe, D.B., Richards, D.E., Cralley, L.J. and Stokinger, H.E. (1970) The role of trace metals in chemical carcinogenesis: asbestos cancers. *Cancer Research*, **30**, 1068–1074.

Dogan, A.U., Dogan, M. and Hoskins, J.A. (2008) Erionite series minerals: mineralogical and carcinogenic properties. *Environmental Geochemistry and Health*, **30**, 367–381.

Donaldson, K., Beswick, P.H. and Gilmour, P.S. (1996) Free radical activity associated with the surface of particles: a unifying factor in determining biological activity? *Toxicology Letters*, **88**, 293–298.

Donaldson, K. and Borm, P.J. (1998) The quartz hazard: a variable entity. *Annals of Occupational Hygiene*, **42**, 287–294.

Donaldson, K., Murphy, F.A., Duffin, R. and Poland, C.A. (2010) Asbestos, carbon nanotubes and the pleural mesothelium: a review of the hypothesis regarding the role of long fibre retention in the parietal pleura, inflammation and mesothelioma. *Particle and Fibre Toxicology*, **7**, 5.

Dostert, C., Petrilli, V., Van Bruggen, R., Steele, C., Mossman, B.T. and Tschopp, J. (2008) Innate immune activation through Nalp3 inflammasome sensing of asbestos and silica. *Science*, **320**, 674–677.

Duncan, K.E., Ghio, A.J., Dailey, L.A., Bern, A.M., Gibbs-Flournoy, E.A., Padilla-Carlin, D.J., Roggli, V.L. and Devlin, R.B. (2010) Effect of size fractionation on the toxicity of amosite and Libby amphibole asbestos. *Toxicology Science*, **118**, 420–434.

Evans, J.C., Evans, R.J., Holmes, A., Hounam, R.F., Jones, D.M., Morgan, A. and Walsh, M. (1973) Studies on the deposition of inhaled fibrous material in the respiratory tract of the rat and its subsequent clearance using radioactive tracer techniques: 1. UICC crocidolite asbestos. *Environmental Research*, **6**, 180–201.

Fischer, C., Kurganskaya, I., Schäfer, T. and Lüttge, A. (2014) Variability of crystal surface reactivity: What do we know? *Applied Geochemistry*, **43**, 132–157.

Frison, R., Cernuto, G., Cervellino, A., Zaharko, O., Colonna, G.M., Guagliardi, A. and Masciocchi, N. (2013) Magnetite-maghemite nanoparticles in the 5–15 nm range: correlating the core-shell composition and the surface structure to the magnetic properties. A total scattering study. *Chemistry of Materials*, **25**, 4820–4827.

Fubini, B., Barcelo, F. and Otero Arean, C. (1997) Ferritin adsorption on amosite fibers: possible implications in the formation and toxicity of asbestos bodies. *Journal of Toxicology and Environmental Health*, **52**, 343–352.

Fubini, B., Bolis, V. and Giamello, E. (1985) Description of surface structures by adsorption microcalorimetry. *Thermochimica Acta*, **85**, 23–26.

Fubini, B. and Mollo, L. (1995) Role of iron in the reactivity of mineral fibers. *Toxicology Letters*, **82-83**, 951–960.

Gabor, S. and Anca, Z. (1975) Effect of asbestos on lipid peroxidation in the red cells. *British Journal of Industrial Medicine*, **32**, 39–41.

Galaris, D. and Evangelou, A. (2002) The role of oxidative stress in mechanisms of metal-induced carcinogenesis. *Critical Reviews of Oncology/Hematology*, **42**, 93–103.

Gazzano, E., Foresti, E., Lesci, I.G., Tomatis, M., Riganti, C., Fubini, B., Roveri, N. and Ghigo, D. (2005) Different cellular responses evoked by natural and stoichiometric synthetic chrysotile asbestos. *Toxicology and Applied Pharmacology*, **206**, 356–364.

Gazzano, E., Turci, F., Foresti, E., Putzu, M.G., Aldieri, E., Silvagno, F., Lesci, I.G., Tomatis, M., Riganti, C., Romano, C., Fubini, B., Roveri, N. and Ghigo, D. (2007) Iron-loaded synthetic chrysotile: a new model solid for studying the role of iron in asbestos toxicity. *Chemical Research in Toxicology*, **20**, 380–387.

Ghio, A.J., LeFurgey, A. and Roggli, V.L. (1997) *In vivo* accumulation of iron on crocidolite is associated with decrements in oxygen generation by the fiber. *Journal of Toxicology Environmental Health*, **50**, 125–142.

Ghio, A.J., Churg, A. and Roggli, V.L. (2004) Ferruginous bodies: implications in the mechanism of fiber and particle toxicity. *Toxicologic Pathology*, **32**, 643–649.

Ghio, A.J., Soukup, J.M., Daily, L.A., Richards, J.H., Turi, J.L., Pavlisko, E.N. and Roggli, V.L. (2012) Disruption of iron homeostasis in mesothelial cells following talc pleurodesis. *American Journal of Respiratory Cell and Molecular Biology*, **46**, 80–86.

Ghio, A.J., Pavlisko, E.N. and Roggli, V.L. (2015) Iron and iron-related proteins in asbestosis. *Journal of Environmental Pathology, Toxicology, and Oncology*, **34**, 277–285.

Gonda, I. and Abd El Khalik, A.F. (1985) On the calculation of aerodynamic diameters of fibers. *Aerosol Science and Technology*, **4**, 233–238.

Gross, P., De Treville, R.T.P., Cralley, L.J. and Shapiro, H.A. (1969) Studies on the carcinogenic effects of asbestos dust. *Proceedings of the International Pneumoconiosis Conference*, Johannesburg. Oxford University Press, London, pp. 220–224.

Gualtieri, A.F. (2017) Introduction. Pp. 1–15 in: *Mineral Fibres: Crystal Chemistry, Chemical-physical Properties, Biological Interaction and Toxicity* (A.F. Gualtieri, editor). EMU Notes in Mineralogy, **18**. European Mineralogical Union and Mineralogical Society of Great Britain & Ireland, London.

Gualtieri, A.F., Viani, A., Sgarbi, G. and Lusvardi, G. (2012) In vitro biodurability of the product of thermal transformation of cement-asbestos. *Journal of Hazardous Materials*, **205–206**, 63–71.

Gualtieri, A.F., Gandolfi, N.B., Pollastri, S., Pollok, K. and Langenhorst, F. (2016) Where is iron in erionite? A multidisciplinary study on fibrous erionite-Na from Jersey (Nevada, USA). *Scientific Reports*, **6**.

Gualtieri, A.F., Pollastri, S. and Bursi Gandolfi, N. (2017) Dissolution of mineral fibres. A comparative acellular in vitro study. *Environmental Health Perspectives* (submitted).

Guldberg, M., Christensen, V.R., Perander, M., Zoitos, B., Koenig, A.R. and Sebastian, K. (1998) Measurement of in-vitro fibre dissolution rate at acidic pH. *Annals of Occupational Hygiene*, **42**, 233–243.

Gulumian, M., van Wyk, J.A. and Kolk, B. (1989) Detoxified crocidolite exhibits reduced radical generation which could explain its lower toxicity: ESR and Mossbauer studies. Pp. 197–204 in: *Effects of Mineral Dusts on Cells* (B.T. Mossman and R.O. Begin, editors). NATO ASI Series H: Cell Biology. **30**. Springer-Verlag, Berlin.

Halliwell, B., Gutteridge, J.M. and Cross, C.E. (1992) Free radicals, antioxidants, and human disease: where are we now? *Journal of Laboratory and Clinical Medicine*, **119**, 598–620.

Hardy, J.A. and Aust, E. (1995) Iron in asbestos chemistry and carcinogenicity. *Chemical Reviews*, **95**, 97–118.

Harington, J.S. and Roe, F.J. (1965) Studies of carcinogenesis of asbestos fibers and their natural oils. *Annals of the New York Academy of Sciences*, **132**, 439–450.

Harris, R.L. and Timbrell, V. (1977) Relation of alveolar deposition to the diameter and length of glass fibres.

Proceedings of the Inhaled Particles IV: Proceedings of an International Symposium Organized by the British Occupational Hygiene Society, Edinburgh, 1977, Pergamon Press Oxford, 411 pp.

Heintz, N.H., Janssen-Heininger, Y.M. and Mossman, B.T. (2010) Asbestos, lung cancers, and mesotheliomas: from molecular approaches to targeting tumor survival pathways. *American Journal of Respiratory Cell and Molecular Biology*, **42**, 133–139.

Heyder, J., Gebhart, J., Rudolf, G., Schiller, C.F. and Stahlhofen, W. (1986) Deposition of particles in the human respiratory tract in the size range 0.005–15 μm. *Journal of Aerosol Science*, **17**, 811–825.

Hillegass, J.M., Miller, J.M., MacPherson, M.B., Westborn, C.M., Sayan, M., Thompson, J.K., Macura, S.L., Perkins, T.N., Beuschel, S.L., Alexeeva, V., Pass, H.I., Steele, C., Mossman, B.T. and Shukla, A. (2013) Asbestos and erionite prime and activate the NLRP3 inflammasome that stimulates autocrine cytokine release in human mesothelial cells. *Particle and Fibre Toxicology*, **10**, 39–52.

Hillerdal, G. and Berg, J. (1985) Malignant mesothelioma secondary to chronic inflammation and old scars: 2 cases and a review of the literature. *Cancer*, **55**, 1968–1972.

Hume, L.A. and Rimstidt, J.D. (1992) The biodurability of chrysotile asbestos. *American Mineralogist*, **77**, 1125–1128.

IARC (1989) *Non-Occupational Exposure to Mineral Fibres, No. 90.* International Agency for Research on Cancer, Lyon, France, 540 pp.

IARC (2012) Arsenic, Metals, Fibres and Dusts: A Review of Human Carcinogens. *Proceedings of the IARC Working Group on the Evaluation of Carcinogenic Risks to Humans*, Lyon, France, International Agency for Research on Cancer, 501 pp.

Ilgren, E. (2008) Coalinga chrysotile: Dissolution, concentration, regulation and general relevance. *Indoor and Built Environment*, **17**, 42–57.

Ilgren, E. and Chatfield, E. (1998) Coalinga fibre – a short, amphibole-free chrysotile. Part 2: evidence for lack of tumourigenic activity. *Indoor and Built Environment*, **7**, 18–31.

Jaurand, M.C., Bignon, J., Sebastien, P. and Goni, J. (1977) Leaching of chrysotile asbestos in human lungs. Correlation with in vitro studies using rabbit alveolar macrophages. *Environmental Research*, **14**, 245–254.

Jaurand, M.C., Gaudichet, A., Halpern, S. and Bignon, J. (1984) In vitro biodegradation of chrysotile fibres by alveolar macrophages and mesothelial cells in culture: comparison with a pH effect. *British Journal of Industrial Medicine*, **41**, 389–395.

Kamp, D.W. (2009) Asbestos-induced lung diseases: an update. *Translational Research*, **153**, 143–152.

Kelly, F.J. and Mudway, I.S. (2007) Particle-mediated extracellular oxidative stress in the lung. Pp. 89–118 in: *Particle Toxicology* (K. Donaldson and P.J.A. Borm, editors). CRC Press, Taylor and Francis Group, Boca Raton, Florida, USA.

Kobayashi, N., Toyooka, S., Yanai, H., Soh, J., Fujimoto, N., Yamamoto, H., Ichihara, S., Kimura, K., Ichimura, K., Sano, Y., Kishimoto, T. and Date, H. (2008) Frequent p16 inactivation by homozygous deletion or methylation is associated with a poor prognosis in Japanese patients with pleural mesothelioma. *Lung Cancer*, **62**, 120–125.

Kraynie, A., de Ridder, G., Sporn, T., Pavlisko, E. and Roggli, V.L. (2016) Malignant mesothelioma not related to asbestos exposure: Analytical scanning electron microscopic analysis of 83 cases and comparison with 442 asbestos-related cases. *Ultrastructural Pathology*, **40**, 142–146.

Kreyling, W.G., Möller, W., Semmler-Behnke, M. and Oberdörster, G. (2007) Particle dosimetry: deposition and clearance from the respiratory tract and translocation towards extrapulmonary sites. Pp. 48–69 in: *Particle Toxicology* (K. Donaldson and P.J.A. Borm, editors). CRC Press, Taylor and Francis Group, Boca Raton, Florida, USA.

Li, X., Brownlee, N.A., Sporn, T.A., Mahar, A. and Roggli, V.L. (2015) Malignant diffuse mesothelioma in patients with haematological malignancies: A clinicopathologic study of 45 cases. *Archives of Pathology and Laboratory Medicine*, **139**, 1129–1136.

Light, W.G. and Wei, E.T. (1977) Surface charge and asbestos toxicity. *Nature*, **265**, 537–539.

Lippmann, M. (2014) Toxicological and epidemiological studies on effects of airborne fibers: coherence and public [corrected] health implications. *Critical Reviews in Toxicology*, **44**, 643–695.

Liu, G., Cheresh, P. and Kamp, D.W. (2013) Molecular basis of asbestos-induced lung disease. *Annual Review of Pathology*, **8**, 161–187.

Luoto, K., Holopainen, M., Sarataho, M. and Savolainen, K. (1997) Comparison of cytotoxicity of man-made vitreous fibres. *Annals of Occupational Hygiene*, **41**, 37–50.

Lynch, I. and Dawson, K.A. (2008) Protein-nanoparticle interactions. *Nano Today*, **3**, 40–47.

Marques, M.R.C., Loebenberg, R. and Almukainzi, M. (2011) Simulated biological fluids with possible application in dissolution testing. *Dissolution Technologies*, **18**, 15–28.

McLemore, T.L., Roggli, V.L., Marshall, M.V., Lawrence, E.C., Greenberg, S.D. and Stevens, P.M. (1981) Comparison of phagocytosis of uncoated vs. coated asbestos fibers by cultured human pulmonary alveolar macrophages. *Chest*, **80**, 39S–42S.

Mellini, M. (1986) Chrysotile and polygonal serpentine from the Balangero serpentinite. *Mineralogical Magazine*, **50**, 301–306.

Morgan, A. (1997) Acid leaching studies of chrysotile asbestos form mines in the coalinga region of California and from Quebec and British Columbia. *Annals of Occupational Hygiene*, **41**, 249–268.

Morgan, A. and Talbot, R.J. (1997) Acid leaching studies of neutron-irradiated chyrsotile asbestos. *Annals of Occupational Hygiene*, **41**, 269–279.

Mossman, B.T. and Pugnaloni, A. (2017) *In vitro* biological activity and mechanisms of lung and pleural cancers induced by mineral fibres. Pp. 261–306 in: *Mineral Fibres: Crystal Chemistry, Chemical-physical Properties, Biological Interaction and Toxicity* (A.F. Gualtieri, editor). EMU Notes in Mineralogy, **18**. European Mineralogical Union and Mineralogical Society of Great Britain & Ireland, London.

Mossman, B.T., Adler, K.B., Jean, L. and Craighead, J.E. (1982) Mechanisms of hypersecretion in rodent tracheal explants after exposure to chrysotile asbestos: studies using lectins. *Chest*, **81**, 23S–24S.

Mossman, B.T., Craighead, J.E. and MacPherson, B.V. (1980) Asbestos-induced epithelial changes in organ cultures of hamster trachea: inhibition by retinyl methyl ether. *Science*, **207**, 311–313.

Mossman, B.T., Kamp, D.W. and Weitzman, S.A. (1996) Mechanisms of carcinogenesis and clinical features of asbestos-associated cancers. *Cancer Investigation*, **14**, 466–480.

Mossman, B.T., Lippmann, M., Hesterberg, T.W., Kelsey, K.T., Barchowsky, A. and Bonner, J.C. (2011) Pulmonary endpoints (lung carcinomas and asbestosis) following inhalation exposure to asbestos. *Journal of Toxicology and Environmental Health, Part B: Critical Reviews*, **14**, 76–121.

Murphy, F.A., Schinwald, A., Poland, C.A. and Donaldson, K. (2012) The mechanism of pleural inflammation by long carbon nanotubes: interaction of long fibres with macrophages stimulates them to amplify pro-inflammatory responses in mesothelial cells. *Particle and Fibre Toxicology*, **9**, 8.

Murphy, F.A., Poland, C.A., Duffin, R. and Donaldson, K. (2013) Length-dependent pleural inflammation and parietal pleural responses after deposition of carbon nanotubes in the pulmonary airspaces of mice. *Nanotoxicology*, **7**, 1157–1167.

Nemery, B. (1990) Metal toxicity and the respiratory tract. *European Respiratory Journal*, **3**, 202–219.

Neugut, A.I., Ahsan, H. and Antman, K. (1997) Incidence of malignant pleural mesothelioma after thoracic radiation therapy. *Cancer*, **80**, 948–950.

Oberdörster, G. (2000) Determinants of the pathogenicity of man-made vitreous fibers (MMVF). *International Archives of Occupational and Environmental Health*, **73 Suppl**, S60–68.

Oczypok, E.A., Sanchez, M.S., van Orden, D.R., Berry, G.J., Pourtabib, K., Gunter, M.E., Roggli, V.L., Kraynic, A.M. and Oury, T.D. (2016) Erionite-associated malignant pleural mesothelioma in North America. *International Journal of Clinical and Experimental Pathology*, **9**, 5722–5732.

Oze, C. and Solt, K.L. (2010) Biodurability of chrysotile and tremolite asbestos in simulated lung and gastric fluids. *American Mineralogist*, **95**, 825–831.

Pagano, G., Aliberti, F., Guida, M., Oral, R., Siciliano, A., Trifuoggi, M. and Tommasi, F. (2015a) Rare earth elements in human and animal health: State of art and research priorities. *Environmental Research*, **142**, 215–220.

Pagano, G., Guida, M., Tommasi, F. and Oral, R. (2015b) Health effects and toxicity mechanisms of rare earth elements-Knowledge gaps and research prospects. *Ecotoxicology and Environmental Safety*, **115**, 40–48.

Pastuszak-Lewandoska, D., Kordiak, J., Migdalska-Sek, M., Czarnecka, K.H., Antczak, A., Gorski, P., Nawrot, E., Kiszalkiewicz, J.M., Domanska, D. and Brzezianska-Lasota, E. (2015) Quantitative analysis of mRNA expression levels and DNA methylation profiles of three neighboring genes: FUS1, NPRL2/G21 and RASSF1A in non-small cell lung cancer patients. *Respiratory Research*, **16**, 76.

Pavlisko, E.N. and Sporn, T.A. (2014) Mesothelioma. Pp. 81–140 in: *Pathology of Asbestos-Associated Diseases, 3rd Edition* (T.D. Oury, T.A. Sporn and V.L. Roggli, editors). Springer, New York.

Perkins, T.N., Toftgaard Poulsen, E., Enghild, J.J., Ghio, A., Roggli, V. and Oury, T.D. (2017) Determination of asbestos body-associated proteins by laser scanning micro-dissection and mass spectrometry, in prep.

Pollastri, S., D'Acapito, F., Trapananti, A., Colantoni, I., Andreozzi, G.B. and Gualtieri, A.F. (2015) The chemical environment of iron in mineral fibres. A combined X-ray absorption and Mossbauer spectroscopic study. *Journal of Hazardous Materials*, **298**, 282–293.

Pollastri, S., Gualtieri, A.F., Gualtieri, M.L., Hanuskova, M., Cavallo, A. and Gaudino, G. (2014) The zeta potential of mineral fibres. *Journal of Hazardous Materials*, **276**, 469–479.

Pollastri, S., Gualtieri, A.F., Ignatyev, K., Strafella, E., Pugnaloni, A. and Croce, A. (2016) Stability of mineral fibres in contact with human cell cultures. An in situ μXANES, μXRD and XRF iron mapping study. *Chemosphere*, **164**, 547–557.

Pott, F. (1978) Some aspects on the dosimetry of the carcinogenic potency of asbestos and other fibrous dusts. *Staub Reinhaltung der Luft*, **38**, 486–490.

Pryor, W.A. (1988) Why is the hydroxyl radical the only radical that commonly adds to DNA? Hypothesis: it has a rare combination of high electrophilicity, high thermochemical reactivity, and a mode of production that can occur near DNA. *Free Radical Biology and Medicine*, **4**, 219–223.

Pundsack, F.L. (1955) The properties of asbestos. I. The colloidal and surface chemistry of chrysotile. *Journal of Physical Chemistry*, **59**, 892–895.

Reid, G. (2015) MicroRNAs in mesothelioma: from tumour suppressors and biomarkers to therapeutic targets. *Journal of Thoracic Disease*, **7**, 1031–1040.

Rim, K.T., Koo, K.H. and Park, J.S. (2013) Toxicological evaluations of rare earths and their health impacts to workers: a literature review. *Safety and Health at Work*, **4**, 12–26.

Roggli, V.L. (2014) Asbestos bodies and non-asbestos ferruginous bodies. Pp. 25–51 in: *Pathology of Asbestos-Associated Diseases*, 3rd Edition (T.D. Oury, T.A. Sporn and V.L. Roggli, editors). Springer, New York.

Roggli, V.L. (2015) The so-called short-fiber controversy: Literature review and critical analysis. *Archives of Pathology and Laboratory Medicine*, **139**, 1052–1057.

Roggli, V.L. and Sharma, A. (2014) Analysis of tissue mineral fiber content. Pp. 253–291 in: *Pathology of Asbestos-Associated Diseases*, 3rd Edition (T.D. Oury, T.A. Sporn and V.L. Roggli, editors). Springer, New York.

Rozalén, M., Huertas, F., Pacella, A. and Ballirano, P. (2017) Dissolution and biodurability of mineral fibres. Pp. 347–366 in: *Mineral Fibres: Crystal Chemistry, Chemical-physical Properties, Biological Interaction and Toxicity* (A.F. Gualtieri, editor). EMU Notes in Mineralogy, **18**. European Mineralogical Union and Mineralogical Society of Great Britain & Ireland, London.

Rufe, E. and Hochella, M.F., Jr. (1999) Quantitative assessment of reactive surface area of phlogopite during acid dissolution. *Science*, **285**, 874–876.

Scheuch, G. and Heyder, J. (1990) Dynamic shape factor of nonspherical aerosol particles in the diffusion regime. *Aerosol Science and Technology*, **12**, 270–277.

Schinwald, A., Murphy, F.A., Prina-Mello, A., Poland, C.A., Byrne, F., Movia, D., Glass, J.R., Dickerson, J.C., Schultz, D.A., Jeffree, C.E., Macnee, W. and Donaldson, K. (2012) The threshold length for fiber-induced acute pleural inflammation: shedding light on the early events in asbestos-induced mesothelioma. *Toxicological Sciences*, **128**, 461–470.

Sharma, G., Kodali, V., Gaffrey, M., Wang, W., Minard, K.R., Karin, N.J., Teeguarden, J.G. and Thrall, B.D. (2014) Iron oxide nanoparticle agglomeration influences dose rates and modulates oxidative stress-mediated dose-response profiles in vitro. *Nanotoxicology*, **8**, 663–675.

Shi, X.L. and Dalal, N.S. (1992) The role of superoxide radical in chromium (VI)-generated hydroxyl radical: the Cr(VI) Haber-Weiss cycle. *Archives of Biochemistry and Biophysics*, **292**, 323–327.

Shukla, A., Gulumian, M., Hei, T.K., Kamp, D., Rahman, Q. and Mossman, B.T. (2003) Multiple roles of oxidants in the pathogenesis of asbestos-induced diseases. *Free Radical Biology and Medicine*, **34**, 1117–1129.

Spurny, K.R., Stober, W., Opiela, H. and Weiss, G. (1979) Size-selective preparation of inorganic fibers for biological experiments. *American Industrial Hygiene Association Journal*, **40**, 20–38.

Stanton, M.F., Layard, M., Tegeris, A., Miller, E., May, M., Morgan, E. and Smith, A. (1981) Relation of particle dimension to carcinogenicity in amphibole asbestoses and other fibrous minerals. *Journal of the National Cancer Instisute*, **67**, 965–975.

Stern, O. (1924) The theory of the electrolytic double-layer. *Zeitschrift für Elektrochemie und Angerwande*, **30**, 1014–1020.

Sugarbaker, D.J., Richards, W.G., Gordon, G.J., Dong, L., De Rienzo, A., Maulik, G., Glickman, J.N., Chirieac, L.R., Hartman, M.L., Taillon, B.E., Du, L., Bouffard, P., Kingsmore, S.F., Miller, N.A., Farmer, A.D., Jensen, R.V., Gullans, S.R. and Bueno, R. (2008) Transcriptome sequencing of malignant pleural mesothelioma tumors. *Proceedings of the National Academy of Sciences of the United States of America*, **105**, 3521–3526.

Teta, M.J., Lau, E., Sceurman, B.K. and Wagner, M.E. (2007) Therapeutic radiation for lymphoma. Risk of malignant mesothelioma. *Cancer*, **109**, 1432–1438.

Testa, R.J., Cheung, M., Pei, J., Below, J.E., Tan, Y., Sementino, E., Cox, N.J., Dogan, U., Pass, H.I., Trusa, S., Hesdorffer, M., Nasu, M., Powers, A., Rivera, Z., Comertpay, S., Tanji, M. Gaudino, G., Yang, H. and Carbone, M. (2011) Germline BAP1 mutations predispose to malignant mesothelioma. *Nature Genetics*, **43**, 1022–1025.

Toyooka, S., Kishimoto, T. and Date, H. (2008) Advances in the molecular biology of malignant mesothelioma. *Acta Medica Okayama*, **62**, 1–7.

Travis, L.B., Fosså, S.D., Schonfeld, S.J., McMaster, M.L., Lynch, C.F., Storm, H., Hall, P., Holowaty, E., Andersen, A., Pukkala, E., Andersson, M., Kaijser, M., Gospodarowicz, M., Joensuu, T., Cohen, R.J., Boice, J.D. Jr., Dores, G.M., Gilbert, E.S. (2005) Second cancers among 40,576 testicular cancer patients: focus on long-term survivors. *Journal of National Cancer Institute*, **97**, 1354–1365.

Turci, F., Tomatis, M., Lesci, I.G., Roveri, N. and Fubini, B. (2011) The iron-related molecular toxicity mechanism of synthetic asbestos nanofibres: A model study for high aspect ratio nanoparticles. *Chemistry – A European Journal*, **17**, 350–358.

Tward, J.D., Wenland, M.M., Shrieve, D.C., Sazabo, A. and Gaffney, D.K. (2006) The risk of secondary malignancies over 30 years after the treatment of non-Hodgkin lymphoma. *Cancer* **107**, 108–115.

Utembe, W., Potgieter, K., Stefaniak, A.B. and Gulumian, M. (2015) Dissolution and biodurability: Important parameters needed for risk assessment of nanomaterials. *Particle and Fibre Toxicology*, **12**, 11.

van Oss, C.J. and Giese, R.F. (1995) The hydrophilicity and hydrophobicity of clay minerals. *Clays and Clay Minerals*, **43**, 474–477.

van Oss, C.J., Naim, J.O., Costanzo, P.M., Giese, R.F., Wu, W. and Sorling, A.F. (1999) Impact of different asbestos species and other mineral particles on pulmonary pathogenesis. *Clays and Clay Minerals*, **47**, 697–707.

Veblen, D.R. and Wylie, A.G. (1993) Mineralogy of amphiboles and 1: 1 layer silicates. Pp. 61–131 in: *Health Effects of Mineral Dusts* (G.D. Guthrie Jr. and B.T. Mossman, editors). Reviews in Mineralogy and Geochemistry, **28**, Mineralogical Society of America, Washington D.C.

von Smoluchowski, M. (1903) Contribution à la théorie de l'endosmose électrique et de quelques phénomènes corrélatifs. *Bulletin de l'Academie des Sciences de Cracovie*, 182–199 (in French).

Wagner, J.C., Skidmore, J.W., Hill, R.J. and Griffiths, D.M. (1985) Erionite exposure and mesotheliomas in rats. *British Journal of Cancer*, **51**, 727–730.

Walton, W.H. (1991) Airborne dust. Pp. 55–78 in: *Mineral Fibers and Health* (D. Liddell and K. Miller, editors). CRC Press, Boca Raton, Florida, USA.

Wei, B., Yang, L., Zhu, O., Yu, J. and Jia, X. (2014) Multivariate analysis of trace elements distribution in hair of pleural plaques patients and health group in a rural area from China. *Hair: Therapy and Transplantation*, **4**, 2167–3118.

Werner, A.J., Hochella, M.F., Jr., Guthrie, G.D., Hardy, J.A. and Aust, A.E. (1995) Asbestiform riebeckite (crocidolite) dissolution in the presence of Fe chelators: implications for mineral-induced disease. *American Mineralogist*, **80**, 1093–1103.

Whittaker, E.J.W. (1955) A classification of cylindrical lattices. *Acta Crystallographica*, **8**, 571–574.

Woodworth, C.D., Mossman, B.T. and Craighead, J.E. (1983) Induction of squamous metaplasia in organ cultures of hamster trachea by naturally occurring and synthetic fibers. *Cancer Research*, **43**, 4906–4912.

Wypych, F., Adad, L.B., Mattoso, N., Marangon, A.A. and Schreiner, W.H. (2005) Synthesis and characterization of disordered layered silica obtained by selective leaching of octahedral sheets from chrysotile and phlogopite structures. *Journal of Colloid and Interface Science*, **283**, 107–112.

Yada, K. (1967) Study of chrysotile asbestos by a high resolution electron microscope. *Acta Crystallographica*, **23**, 704–707.

Yeh, H.C., Phalen, R.F. and Raabe, O.G. (1976) Factors influencing the deposition of inhaled particles. *Environmental Health Perspectives*, **15**, 147–156.

Zalma, R., Bonneau, L., Guignard, J., Pezerat, H. and Jaurand, M.-C. (1987) Formation of oxy radicals by oxygen reduction arising from the surface activity of asbestos. *Canadian Journal of Chemistry*, **65**, 2338–2341.

Zatta, P., Kiss, T., Suwalsky, M. and Berthon, G. (2002) Aluminium (III) as a promoter of cellular oxidation. *Coordination Chemistry Reviews*, **228**, 271–284.

Zecchina, A., Rivallan, M., Berlier, G., Lamberti, C. and Ricchiardi, G. (2007) Structure and nuclearity of active sites in Fe-zeolites: comparison with iron sites in enzymes and homogeneous catalysts. *Physical Chemistry Chemical Physics*, **9**, 3483–3499.

Zook, J.M., Rastogi, V., Maccuspie, R.I., Keene, A.M. and Fagan, J. (2011) Measuring agglomerate size distribution and dependence of localized surface plasmon resonance absorbance on gold nanoparticle agglomerate size using analytical ultracentrifugation. *ACS Nano*, **5**, 8070–8079.

EMU Notes in Mineralogy, Vol. 18 (2017), Index, 533–536

Subject and Author index

Numbers refer to the page where a definition or explanation of, and/or a *figure* or a **table** for, a given subject is found. Author names are given with the number of the first page of the paper in which they are involved.

DOI: 10.1180/EMU-notes.18.index

R

S

T

V

X

Y

www.ingramcontent.com/pod-product-compliance
Ingram Content Group UK Ltd.
Pitfield, Milton Keynes, MK11 3LW, UK
UKHW050923290726
14058UKWH00011B/684